Expert Views on HPV Infection

Special Issue Editors

Alison A. McBride
Karl Munger

MDPI • Basel • Beijing • Wuhan • Barcelona • Belgrade

MDPI

Special Issue Editors

Alison A. McBride
National Institute of Allergy and Infectious Diseases
National Institutes of Health
USA

Karl Munger
Tufts University School of Medicine
USA

Editorial Office
MDPI AG
St. Alban-Anlage 66
Basel, Switzerland

This edition is a reprint of the Special Issue published online in the open access journal *Viruses* (ISSN 1999-4915) from 2017–2018 (available at: http://www.mdpi.com/journal/viruses/special_issues/hpv).

For citation purposes, cite each article independently as indicated on the article page online and as indicated below:

Lastname, F.M.; Lastname, F.M. Article title. *Journal Name* **Year**, *Article number*, page range.

First Edition 2018

ISBN 978-3-03842-837-4 (Pbk)
ISBN 978-3-03842-838-1 (PDF)

Cover photo courtesy of Karl Munger.

Table of Contents

About the Special Issue Editors

Alison McBride, Ph.D. received a B.Sc. Hons in Molecular Biology from the University of Glasgow, Scotland, and a Ph.D. in Biochemistry from the Imperial Cancer Research Fund/Imperial College, UK. She began studying papillomaviruses as a postdoctoral fellow in the National Cancer Institute and joined the National Institutes of Allergy and Infectious Diseases (NIAID) as an Investigator in 1994. She is currently Chief of the DNA Tumor Virus Section in NIAID. Dr. McBride's group studies the mechanisms of HPV replication during various stages of the viral life cycle, and the consequences of viral genome integration in HPV-associated cancers. She is a fellow of the American Academy of Microbiology and an Editor for PLOS Pathogens, Virology and Current Protocols in Microbiology.

Karl Munger obtained his Ph.D. degree in Biochemistry from the University of Zurich, Switzerland, followed by a postdoctoral fellowship at the National Cancer Institute in Bethesda, Maryland. He was a faculty member at Harvard Medical School from 1993 to 2014 and he is currently a Professor in the Department of Developmental, Molecular and Chemical Biology at Tufts University School of Medicine in Boston, Massachusetts. Research in his group is focused on delineating the molecular mechanisms of human papillomavirus-associated cancer development. His group not only discovered and validated multiple cellular targets of the HPV E6 and E7 oncoproteins but has also uncovered specific cellular vulnerabilities that arise as a consequence of HPV oncogene expression. He has mentored more than 70 students, fellows and visiting faculty in his group and has published more than 200 articles. He is a fellow of the American Academy of Microbiology.

Preface to "Expert Views on HPV Infection"

There are over four-hundred different papillomavirus (PV) types, which replicate in mucosal and cutaneous stratified epithelial surfaces giving rise to a wide range of lesions. Papillomaviruses have a remarkable lifestyle that relies on the differentiation state of the host epithelium; they infect the basal cells of the epithelium and establish a quiescent infection in the proliferative cells. As the infected cells differentiate, the productive life cycle is activated, and viral-laden squames are eventually released from the surface of the epithelium. To support this lifestyle, PVs interact with, and manipulate, many key cellular pathways.

Individual chapters of this book are written by noted experts and thought leaders and provide viewpoints on unresolved, controversial or emerging topics related to the natural history, evolution, biology, and disease association of papillomavirus infections. Puustusmaa and colleagues present an intriguing study in which they searched the biosphere for distant homologs of PV protein domains in a quest to discover the origin of papillomaviruses [1]. Suarez and Travé describe insights obtained from a review of PV E6 and E7 structural data [2] and Campos reviews the remarkable abilities of the minor capsid protein L2 to deliver the viral genome to the nucleus upon infection [3]. Moody describes how PVs interface with signaling pathways to provide the virus with a replication-competent environment in differentiating cells [4], and Graham describes how PV late gene expression is regulated by keratinocyte differentiation [5]. MmuPV1, a virus capable of infecting laboratory strains of mice, was first described in 2011, and Hu et al. review the remarkable progress made using this valuable model [6].

Over three-hundred human papillomavirus (HPV) types have been described and HPV infection is ubiquitous. However, many questions remain about infection, progression and resolution of HPV-associated disease. Gravitt and Weiner present a natural history model across the lifespan of an infected individual, with a particular focus on the role of viral latency [7]. Alizon and colleagues review our current knowledge about acute/transient infections to provide insight as to why some infections are efficiently cleared while others become persistent [8]. The article by Herfs et al. explains why mucosal junction cells in epithelial transition zones are particularly susceptible to HPV infection and carcinogenic progression [9], while Spurgeon and Lambert describe the role of the stroma and microenvironment in these processes [10]. Continuing in this theme, Strati reviews the role of stem cell dynamics in HPV infection [11].

A subset of alpha-HPVs are oncogenic and are the causative agent of approximately 5% human cancers. Viral manipulation of host pathways can inadvertently promote oncogenesis and several articles in the Special Issue address this. Katzenellenbogen describes the role of telomerase activation in HPV infection and oncogenesis [12], while Warren and colleagues discuss the role of APOBEC3 induction in these processes [13]. Guenat et al. review recent studies showing that HPV regulates the content of exosomes and discuss how this might promote carcinogenesis [14]. Khoury and colleagues explain why the study of HPV infection in individuals prone to cancer due to mutations in DNA repair pathways provides an opportunity to uncover viral and host susceptibility factors [15]. Mirabello et al. report on a meeting of HPV experts that convened to discuss the intersection of HPV epidemiology, genomics and mechanistic studies of HPV-mediated cervical carcinogenesis [16]. Only HPVs from the alpha genus have been officially declared carcinogenic, but there is much discussion about the potential role of beta-HPVs in the initiation of non-melanoma skin cancer. Hufbauer and Akgül describe beta-HPV oncogenic mechanisms that may be relevant for the development of skin cancer [17].

HPV-associated cancers acquire profound changes and phenotypes that are important for carcinogenesis and could impact prognosis and treatment. Morgan and colleagues reevaluate the status of integrated and extrachromosomal HPV genomes in head and neck cancer [18] and Litwin et al. review somatic cell mutations that frequently occur in HPV-driven cancers [19]. Soto and colleagues review epigenetic alterations in HPV-associated cancers and explain why these reversible modifications might be amenable to epigenetic therapy [20]. Hoppe-Seyler et al. describe how many HPV-associated cancers have regions of hypoxia containing dormant cancer cells with no viral oncogene expression and explain why this has important consequences for treatment [21]. Finally, two articles review how HPVs modulate factors and pathways important for viral persistence and discuss therapies that could target these key processes.

Shanmugasundaram and You describe the mechanisms required for viral genome persistence and discuss how small molecule therapeutics could disrupt this process [22]. Smola reviews the complex interplay between HPV-infected cells and the local immune microenvironment and discusses the potential of related diagnostics and immunotherapies [23].

We hope that this book will serve as inspiration for future investigation into some of these fascinating, and often understudied research areas of papillomvirus research.

References

1. Puustusmaa, M.; Kirsip, H.; Gaston, K.; Abroi, A. The enigmatic origin of papillomavirus protein domains. Viruses 2017, 9, 240.
2. Suarez, I.; Trave, G. Structural insights in multifunctional papillomavirus oncoproteins. Viruses 2018, 10, 37.
3. Campos, S.K. Subcellular trafficking of the papillomavirus genome during initial infection: The remarkable abilities of minor capsid protein L2. Viruses 2017, 9, 370.
4. Moody, C. Mechanisms by which HPV induces a replication competent environment in differentiating keratinocytes. Viruses 2017, 9, 261.
5. Graham, S.V. Keratinocyte differentiation-dependent human papillomavirus gene regulation. Viruses 2017, 9, 245.
6. Hu, J.; Cladel, N.M.; Budgeon, L.R.; Balogh, K.K.; Christensen, N.D. The mouse papillomavirus infection model. Viruses 2017, 9, 246.
7. Gravitt, P.E.; Winer, R.L. Natural history of HPV infection across the lifespan: Role of viral latency. Viruses 2017, 9, 267.
8. Alizon, S.; Murall, C.L.; Bravo, I.G. Why human papillomavirus acute infections matter. Viruses 2017, 9, 293.
9. Herfs, M.; Soong, T.R.; Delvenne, P.; Crum, C.P. Deciphering the multifactorial susceptibility of mucosal junction cells to HPV infection and related carcinogenesis. Viruses 2017, 9, 85.
10. Spurgeon, M.E.; Lambert, P.F. Human papillomavirus and the stroma: Bidirectional crosstalk during the virus life cycle and carcinogenesis. Viruses 2017, 9, 219.
11. Strati, K. Changing stem cell dynamics during papillomavirus infection: Potential roles for cellular plasticity in the viral lifecycle and disease. Viruses 2017, 9, 221.
12. Katzenellenbogen, R. Telomerase induction in hpv infection and oncogenesis. Viruses 2017, 9, 180.
13. Warren, C.J.; Westrich, J.A.; Doorslaer, K.V.; Pyeon, D. Roles of APOBEC3A and APOBEC3B in human papillomavirus infection and disease progression. Viruses 2017, 9, 233.

14.   Guenat, D.; Hermetet, F.; Pretet, J.L.; Mougin, C. Exosomes and other extracellular vesicles in HPV transmission and carcinogenesis. Viruses 2017, 9, 211.

15.   Khoury, R.; Sauter, S.; Butsch Kovacic, M.; Nelson, A.; Myers, K.; Mehta, P.; Davies, S.; Wells, S. Risk of human papillomavirus infection in cancer-prone individuals: What we know. Viruses 2018, 10, 47.

16.   Mirabello, L.; Clarke, M.A.; Nelson, C.W.; Dean, M.; Wentzensen, N.; Yaeger, M.; Cullen, M.; Boland, J.F.; NCI HPV Wrokshop, Schiffman, M.; Burk, R.D. The intersection of HPV epidemiology, genomics and mechanistic studies of HPV-mediated carcinogenesis. Viruses 2018, 10, 80.

17.   Hufbauer, M.; Akgul, B. Molecular mechanisms of human papillomavirus induced skin carcinogenesis. Viruses 2017, 9, 187.

18.   Morgan, I.M.; DiNardo, L.J.; Windle, B. Integration of human papillomavirus genomes in head and neck cancer: Is it time to consider a paradigm shift? Viruses 2017, 9, 208.

19.   Litwin, T.R.; Clarke, M.A.; Dean, M.; Wentzensen, N. Somatic host cell alterations in HPV carcinogenesis. Viruses 2017, 9, 206.

20.   Soto, D.; Song, C.; McLaughlin-Drubin, M.E. Epigenetic alterations in human papillomavirus-associated cancers. Viruses 2017, 9, 248.

21.   Hoppe-Seyler, K.; Mandl, J.; Adrian, S.; Kuhn, B.J.; Hoppe-Seyler, F. Virus/host cell crosstalk in hypoxic HPV-positive cancer cells. Viruses 2017, 9, 174.

22.   Shanmugasundaram, S.; You, J. Targeting persistent human papillomavirus infection. Viruses 2017, 9, 229.

23.   Smola, S. Immunopathogenesis of hpv-associated cancers and prospects for immunotherapy. Viruses 2017, 9, 254.

Alison A. McBride, Karl Munger

Special Issue Editors

viruses

MDPI

Article

The Enigmatic Origin of Papillomavirus Protein Domains

Mikk Puustusmaa [1,†] (ID), Heleri Kirsip [1,†], Kevin Gaston [2] and Aare Abroi [3,4,*] (ID)

1   Department of Bioinformatics, University of Tartu, Riia 23a, Tartu 51010, Estonia;
    mikk.puustusmaa@ut.ee (M.P.); heleri16@ut.ee (H.K.)
2   School of Biochemistry, University of Bristol, Bristol BS8 1TD, UK; kevin.gaston@bristol.ac.uk
3   Estonian Biocentre, Riia 23b, Tartu 51010, Estonia
4   Institute of Technology, University of Tartu, Nooruse 1, Tartu 50411, Estonia
*   Correspondence: aabroi@ebc.ee; Tel.: +372-737-5045
†   Both authors contributed equally to this work.

Academic Editors: Alison A. McBride and Karl Munger
Received: 18 July 2017; Accepted: 19 August 2017; Published: 23 August 2017

Abstract: Almost a century has passed since the discovery of papillomaviruses. A few decades of research have given a wealth of information on the molecular biology of papillomaviruses. Several excellent studies have been performed looking at the long- and short-term evolution of these viruses. However, when and how papillomaviruses originate is still a mystery. In this study, we systematically searched the (sequenced) biosphere to find distant homologs of papillomaviral protein domains. Our data show that, even including structural information, which allows us to find deeper evolutionary relationships compared to sequence-only based methods, only half of the protein domains in papillomaviruses have relatives in the rest of the biosphere. We show that the major capsid protein L1 and the replication protein E1 have relatives in several viral families, sharing three protein domains with *Polyomaviridae* and *Parvoviridae*. However, only the E1 replication protein has connections with cellular organisms. Most likely, the papillomavirus ancestor is of marine origin, a biotope that is not very well sequenced at the present time. Nevertheless, there is no evidence as to how papillomaviruses originated and how they became vertebrate and epithelium specific.

Keywords: papillomaviruses; protein domains; structural domains; origin

---

1. Introduction

Members of the *Papillomaviridae* taxonomic family have a small circular double-stranded DNA genome of around 8kb in length that is packaged in a non-enveloped icosahedral capsid. Papillomaviruses (PVs) have been particularly well-studied in humans due to their association with multiple disease states including cervical cancer and other malignancies. Well over 200 human papillomavirus (HPV) types have been identified to date. Historically, the first discovered PV was Cottontail rabbit PV (current name SfPV1), which was also the first DNA tumour virus described [1]. PVs infect most mammal species (both terrestrial and marine), several birds, reptiles, and fish [2,3]. In well-studied host species, some PV type infections are asymptomatic; therefore, in-depth study of vertebrates' epithelial viromes may significantly increase the number of known PVs. After the first fully sequenced PV genomes were published [4,5], the first sequence analyses of PVs were also performed [6–10]. Subsequently, there have been several studies of the ancestral and more recent evolution of PVs [11–15]. However, the evolutionary origin of PVs is not well understood, although it is assumed to be ancient.

PV sequences can be found in different nucleotide databases: in ENA (European Nucleotide Archive) there are ~25,000 sequences, and in NCBI (National Center for Biotechnology Information)

there are 25,189 sequences with the taxonomic restriction *Papillomaviridae*. In NCBI 1686 entries are found with length 6300 to 9500 nucleotides and with taxonomic restriction *Papillomaviridae*, mostly corresponding to PV complete genomes (this redundant set includes isolates, etc.). "NCBI refseq", which is a subset of the NCBI nucleotide collection containing only reference genomes (a non-redundant database), contains 135 reference PV genomes. In the UniProtKB (UniProt Knowledgebase) database, there are 556 entries in the manually annotated SwissProt and 12,302 in the computer-annotated TrEMBL (TrEMBL contains the translations of all coding sequences present in the EMBL Nucleotide Sequence Database not yet integrated in Swiss-Prot). In UniProt "complete proteomes" ("complete proteome"—all proteins annotated for species or isolate), 97 PV proteomes can be found, including 37 "reference proteomes" ("reference proteomes" are a representative cross-section of the taxonomic diversity to be found within UniProtKB "complete proteome", they include the proteomes of well-studied model organisms and other proteomes of interest for biomedical and biotechnological research; for more details, see [16,17]. In the PAVE (Papillomavirus Episteme [2]) database, which was curated by experts in the field, 340 PV types with 3150 protein sequences are found (as of 8 June 2017) [3]. However, whether sequence information alone is enough to tell us something about the deep evolutionary history of PVs and their origin is open to debate.

Viruses are fast evolving units. PV coding sequences have been estimated to evolve ~5 times faster on average compared to their mammalian host nuclear coding sequences [14]. The evolutionary rate of the PV E1 protein is estimated to be 1.76×10^{-8} substitutions/nt/year for Lambdapapillomaviruses infecting Felidae; 7.1×10^{-9} substitutions/nt/year for mammalian PVs; and 1.1×10^{-8} substitutions/nt/year for nonmammalian amniote PVs [11,18–20] compared to 2.2×10^{-9} for mammalian nuclear coding sequences [21]. In general, the short-term evolutionary rates of viruses (and other genomes) are much faster than long-term evolutionary rates due in part at least to the loss of deleterious mutations from the population [22]. Thus, the sequence space sampled by viruses is even larger than that expected from long-term evolutionary rates. It is estimated that PVs have existed at least ~315 million years [23]. Considering this, PV proteins may still have homologs in the biosphere (outside of PVs), but without significant sequence similarity.

It has been known for more than three decades that structure is more conserved than sequence [24,25]. Challis and Schmidler have shown that including structural information enables better phylogenetic inference for distant relationships [26]. Additionally, Herman et al. have shown that including structural information reduces significantly the uncertainty of alignments and topologies of phylogenetic trees, indicating that structure contains more information than can be obtained from sequences alone [27]. This is especially important in the case of viruses, which are able to sample a huge amount of sequence space and loose sequence similarity within a relatively short time (compared to organisms). Thus, it is essential to include structural information in order to study deep evolutionary relationships.

A common view of proteins is that they are composed of domains—independent functional, evolutionary and structural units often linked by unstructured polypeptide chain. A protein (polypeptide chain) can be virtually chopped into domains on multiple criteria and domain borders depend on the domain assignment method. Domains are more monophyletic compared to proteins as one protein may consist of many domains with very different phylogenetic histories. Thus, protein domains, and especially structural domains, can be used to study the evolutionary history (origin) of viral proteins.

In this study, the structural information of protein domains was used to find distant homologs to PV proteins and to shed more light on the evolutionary history of PVs. Our results show that only half of the PV protein domains have a relative in the rest of the sequenced biosphere. E1 replication protein shows the most connections with cellular organisms and viruses alike. Capsid protein L1 has evolutionary relationship with rest of the virosphere. However, for a number of PV protein domains, distant homologs could not be detected.

2. Materials and Methods

2.1. PfamA_28

In this study (if not mentioned otherwise), locally downloaded version of PfamA_28 (based on Swiss 2014_07 + SP-TrEMBL 2014_07) was used instead of the newest version of PfamA for reasons described in the Supplementary Data [28].

Protein domain models in PfamA, and also in SUPERFAMILY [29], are based on profile Hidden Markov models (profile-HMMs), which are widely used for modelling protein or nucleotide consensus sequence. A profile-HMM is constructed from a multiple sequence alignment, which is called the seed alignment, containing a set of representative members of the protein domain family. A query sequence that has a significant score against the profile-HMM is considered homologous to the (seed) sequences that were used to build the profile-HMM. In PfamA, the whole protein domain (PfamA entry) is described by a single HMM. PfamA_28 contains a diverse collection of protein domain families mapped to all available UniProt sequences.

By default, the non-redundant "complete proteomes" subset of UniProt is used here because of the quality of the data and because the coverage of the data can be confidently interpreted. Full UniProt is highly redundant and biased, which makes interpretation of coverage of the data questionable. However, to broaden the scope of our analyses and to evaluate the occurrence of PV_PfamA protein domains in non-complete proteomes, the full Uniprot was used.

2.2. HMMER "Hmmsearch"

PfamA_28 is based on the sequence data from summer 2014 (UniProt version 2014_07). To look for the occurrence of PV protein domains in recently added sequences in databases, we used HMMER web tool [30] to perform "hmmsearch" (searching protein alignment/profile-HMM from protein sequence database) against UniProt "complete proteome", full UniProt and Ensemble databases with PfamA and SUPERFAMILY profile-HMM models listed in Supplementary Table S2 as queries [31]. "Hmmsearch" was performed in March/April 2017 with default settings.

2.3. Criteria for Considering PfamA_28 Database Hits and "Hmmsearch" Hits as True Positives

In PfamA, which is based on high throughput data, every specific case needs to be analysed in detail to avoid including false positives and making premature conclusions. We applied the following additional criteria to PfamA_28 database hits and to HMMER hits (with PfamA models) before considering them as true positives (and to exclude them as false positives if not satisfied):

1. Sequence annotation is valid (not showing evidence for viral contamination);
2. The size and protein coding potential of the cellular contig/scaffold should exclude the possibility of viral contamination by small viruses (applied to complete genome/proteomes);
3. "hmmscan" (protein sequence vs. profile-HMM database with HMMER) gives reciprocal best hit to query PfamA model; and
4. 3D structure prediction by threading meta server LOMETS gives best modelling templates from PV structures at least with one algorithm [32].

Protein 3D structure prediction has been used before to validate sequence based hits of non-vertebrate polyomaviruses [33]. LOMETS meta server is based on multiple primary algorithms predicting 3D structure (algorithms listed in Supplementary Data) [32]. A criterion for true positives was applied when a number of hits in superkingdom or in viruses did not exceed 50 species.

2.4. Galaxy of Folds

Location of PV structural domains in global structure space was visualised in the "galaxy of folds", which is based on the sequence similarity of a non-redundant set of SCOP domains [34]. SCOP database (the Structural Classification of Proteins) is a classification of protein structural domains

(SCOP domains) based on similarities of their structures and amino acid sequences [35]. Alva et al. conducted an all-against-all comparison of SCOP domains with <20% pairwise identity. Domains were clustered using a force-directed procedure, and the statistical significance of pairwise comparisons was used to assign attractive and repulsive forces to each profile pair in a two-dimensional map [34]. Because of the force directed clustering procedure, domains find their equilibrium position on the map not only by attraction to similar domains but also by repulsion of different ones. "Galaxy of folds" visualisation tool was used to map PV domains to the structural space [34,36].

2.5. SUPERFAMILY Database

SUPERFAMILY database was locally downloaded (October 2014) and based on SCOP 1.75 [29,37]. Protein domain (and domain pair) existence in PVs and their distribution in Archaea, Bacteria, and Eukaryota were obtained from the "len_supra" table. Option include = "y" was used in queries against cellular complete genomes, to remove isolates, strains, etc. Information about PV_SF distribution in viruses and plasmids was obtained from "sublen_supra" table, option genome = "vl" or genome = "pla" was used respectively. To extend the queries to non-complete genomes, "sublen_supra" table was used with option genome = "up" and with respective taxonomic restriction.

2.6. Criteria for Considering Hits from SUPERFAMILY Database and from "Hmmsearch" as True Positives

As in PfamA data, we applied criteria to avoid false positives. Similar criteria to PfamA data were used:

(1) Sequence annotation is correct (for UniProt data);
(2) The size and protein coding potential of the cellular contig/scaffold exclude viral contamination by small viruses (applied to complete genomes);
(3) "hmmscan" gives reciprocal best hit to query SF model; and
(4) 3D structure prediction by threading meta server gives best modelling templates from respective SF at least with one algorithm.

For true positive eukaryotic hits in UniProt sequences, annotations of corresponding nucleic acid sequences (as provided by UniProt homepage) were examined to find more information about the origin of the sequences (coded by eukaryotic mitochondria, eukaryotic plasmids, etc.). Criteria for true positives were applied when a number of hits in superkingdom or in viruses did not exceed 50 species.

3. Results

3.1. PfamA Protein Domains Found in PV

PfamA is one widely used protein domain database. As a first approximation, PfamA is sequence and function based. According to the PfamA_28 database, 12 PfamA domains are found in PVs (collectively named PV_PfamA). On average, about 90% of proteins in PVs are covered by at least one PfamA domain (Table 1). In addition, about 84% of amino acids in PVs are covered by PfamA domains (Table 1). Compared to cellular superkingdoms and double-stranded DNA (dsDNA) viruses, PVs are very well covered with PfamA domains (Table 1).

Excluding short N- and C-terminal regions, only two regions internal to the PV proteins are not assigned to PfamA domains. The short region between "PPV_E1_N" and "PPV_E1_C", and the E2 "hinge region" (Figure 1). However, the E2 "hinge region" also encodes the E4 part of the E1^E4 protein, although in another reading frame with respect to E2. In UniProt "complete proteomes" the E4 open reading frame (ORF) (and E1^E4 protein) is not annotated at all in many PV genomes. Additionally, in several UniProt "complete proteomes" (and in UniProt), many non-canonical PV ORFs are annotated, but not yet experimentally characterised, hence they might be misannotations. The potentially misannotated proteins reduce the "the percentage of coverage". However, moving from the redundant set of tens of thousands of sequences to protein domains, we end up with less than 20 evolutionary units (protein domains).

Table 1. Domain coverage comparison in UniProt "complete proteomes" in PfamA and SUPERFAMILY database.

	PfamA_28 *			SUPERFAMILY		
	Sequence Coverage [1]	Residue Coverage [2]	No. of Genomes	Sequence Coverage [1]	Residue Coverage [2]	No. of Genomes
Archaea	73.8	58.0	182	64.4	61.1	122
Bacteria	82.0	63.3	3513	67.6	62.6	1153
Eukaryota	67.9	38.6	422	56.9	38.8	440
Viruses	84.4	65.7	1198	34.3	28.1	4041
dsDNA viruses	62.5	52.9	270	24.8	25.4	1758
Papillomaviridae	90.8	83.8	76	69.5	57.5	125
Polyomaviridae	92.5	70.3	10	60.2	65.3	50
Parvoviridae	74.7	56.3	23	69.5	55.0	81
Geminiviridae	97.0	79.9	34	18.5	15.1	332
Herpesviridae	74.2	53.6	28	27.6	20.7	57

[1] Sequence coverage shows the percentage of proteins in a genome which are covered by at least one domain. [2] Residue coverage shows the percentage of amino acids from all proteins of a genome which are within domain models. * PfamA_28 data from "complete genomes" subset.

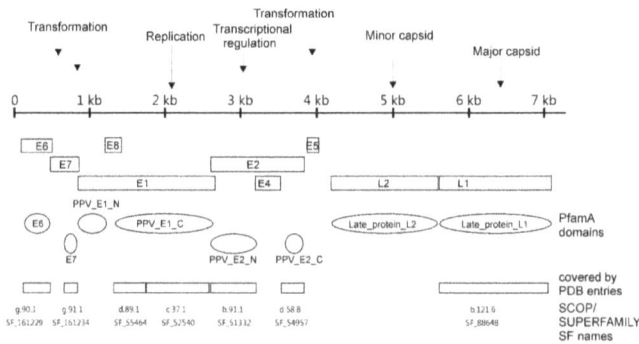

Figure 1. Location of Papillomavirus (PV) proteins and protein domains using Bovine PV type 1 as an example. Bovine PV type 1 encodes 9 proteins including the oncoproteins E6, E7 and E5, the viral helicase E1, the helicase loading factor and transcription factor E2, and the L1 and L2 coat proteins. E8ˆE2 and E1ˆE4 proteins are not shown on the figure. Location of open reading frames (ORFs) does not correspond to reading frames.

3.2. Relationships of PVs to the Sequenced Biosphere According to PfamA Domains

Domains are more monophyletic compared to proteins, as one protein may consist of many domains with very different phylogenetic histories. Thus, the protein domains can be used to study the evolutionary history of viral proteins. In the PfamA database, only a single true positive hit (see Materials and Methods) for the PPV_E1_C domain is found (in PfamA "complete genomes") (Table 2). PPV_E1_N gives only two true positive hits in PfamA "complete genomes" (Table 2). PPV_E1_N was found in two *Nosema* species among Fungi proteins: C4V8V5_NOSCE and R0MJR2_NOSB1 (Supplementary Materials, File S1). However, as no structure for PPV_E1_N is available, we cannot confirm homology via predicted structure similarity (see Material and Methods). It should be noted that this region in PV E1 sequences is not very well conserved and has low complexity. PPV_E1_C gives one hit to Bacteria (Table 2, Supplementary Materials, File S1), namely *Dickeya dadantii* protein E0SH87_DICD3. None of the PV_PfamA domains are found in any viruses outside *Papillomaviridae* in the database used (UniProt "complete proteomes").

Table 2. PV_PfamA domain occurrence in biosphere.

	Papillomaviridae [1,5]	PDB PfamA_28 [2]	PfamA Domain Length [3]	PDB PfamA_31 [2]	Best Coverage of PfamA by PDB [3] (% aa)	Eukaryota (Proteomes) [1]	Bacteria (Proteomes) [1]	Archaea (Proteomes) [1]	Viruses [1,4]	Eukaryota (Full up) [1]	Bacteria (Full up) [1]	Archaea (Full up) [1]	Viruses (Full up) [1,4,6]	HMMER E [1]	HMMER B [1]	HMMER A [1]	HMMER V [1,6]
PF00500 Late_protein_L1	76	10	498	18	0.96	–	–	–	–	–	–	–	–	–	–	–	–
PF00508 PPV_E2_N	76	8	200	8	0.98	–	–	–	–	–	–	–	–	–	–	–	–
PF00511 PPV_E2_C	76	16	80	16	0.96	–	–	–	–	–	–	–	–	–	–	–	–
PF00513 Late_protein_L2	76	0	525	0	–	–	–	–	–	–	–	–	–	–	–	–	–
PF00518 E6	71	7	110	8	0.99	–	–	–	–	–	20	–	–	–	–	–	–
PF00519 PPV_E1_C	74	0	432	8	0.96	–	1	–	–	–	–	–	1	–	1	–	1
PF00524 PPV_E1_N	72	0	121	4	–	2	–	–	–	4	–	–	–	–	–	–	–
PF00527 E7	71	3	93	0	0.50	–	–	–	–	–	–	–	–	–	–	–	–
PF02271 Pap_E4	25	0	95	0	–	–	–	–	–	–	–	–	–	–	–	–	–
PF03025 Papilloma_E5	9	0	72	0	–	–	–	–	–	–	–	–	–	–	–	–	–
PF06576 Papilloma_E5A	5	0	91	0	–	–	–	–	–	–	–	–	–	–	–	–	–
PF08135 EPV_E5	3	0	43	0	–	–	–	–	–	–	–	–	–	–	–	–	–

"–" No true positive hits were found. [1] Number of distinct proteomes/species in database with given taxonomic restrictions coding respective domain. [2] Number of Protein Data Bank (PDB) entries for respective PfamA domain. [3] Model length. [4] Excluding papillomaviruses. [5] 76 PV proteomes in this database. [6] Excluding *Polyomaviridae* and *Parvoviridae*.

The PfamA_28 "complete proteomes" contains 76 PV proteomes. However, the "PPV_E1_C" domain was not found in two PV complete proteomes (HPV53 and HPV56; E1 protein is not annotated for these PV types) and "PPV_E1_N" was not found in 4 complete proteomes (*Fringilla coelebs* papillomavirus (isolate Chaffinch/Netherlands/Dutch), *Psittacus erithacus timneh* papillomavirus (isolate African grey parrot), HPV53, and HPV56). Detailed examination of DNA sequences for HPV53 and HPV56 clearly shows that the absence of "PPV_E1_C" domain is caused by misannotations. For HPV53 and HPV56 the reference genome/proteome is based on the first published sequence and in both types the first sequence has missing nucleotides in the E1 coding region. Most if not all isolates of HPV53 and HPV56 have annotated full-length E1 protein. Thus, misannotations are one reason why PAVE-like activities are important.

To extend the search to non-complete genomes of organisms and viruses, we looked at the presence of PV_PfamA domains in the full UniProt (excluding PVs) in PfamA database. It is expected that UniProt contains more misannotations and partial sequences compared to "complete proteomes". Therefore, more false positives should be expected. The amount of all false positive hits can be seen in Table S1. We performed the analysis and tested for false positives as described in Materials and Methods. In general, the results were similar to "complete genomes" set—only PV_PfamA domains from the E1 protein gave significant hits. PPV_E1_C gave 20 hits to Bacteria, mostly from *Enterobacteriaceae* (Table 2, for more information of positive hits, including species name and full taxonomy see Supplementary Materials, File S1). PPV_E1_N gives hits to four eukaryotes: 3 *Nosema* species (Fungi) (including two species/protein from "complete genome" and protein T0L8A9_9MICR) and one Spermatophyta (Viridiplantae) (protein V7BKU5_PHAVU).

PfamA_28 is based on UniProt release 2014_07, therefore to acquire more up to date data, "hmmsearch" was used with PfamA HMMs listed in Table 2 as queries (HMM version numbers listed in Table S2). After thorough analysis of all hits, only one positive bacterial hit (Planctomycetaceae bacterium SCGC AG-212-D15, protein A0A177Q2P3_9PLAN) and one viral hit to Planaria asexual element, protein Q91S73_9VIRU for PPV_E1_C remained (Table 2, Supplementary Materials, File S1). No other PV_PfamA model gave a true positive hit to cellular sequences.

In "full UniProt" viruses, PPV_E1_C gives highly significant matches to *Polyomaviridae* Large-T and *Parvoviridae* NS1 proteins. This similarity has been observed previously, mostly based on shared common helicase motifs [38]. The reasons why sequences of *Polyomaviridae* and *Parvoviridae* have the best score for PPV_E1_C HMM are described in Supplementary Materials. However, a phylogenetic tree clearly separates *Polyomaviridae*, *Parvoviridae* and *Papillomaviridae* replication protein sequences into three distinct protein families (data not shown). With the exception of the E1 domains described above, other PV_PfamA HMM models did not give any true positive hits to proteins in viruses outside PV sequences.

3.3. Location of PV Domains in the "Galaxy of Folds"

Occurrence of PV_PfamA domains in sequenced biosphere showed only weak connections with cellular organisms and other viruses. Therefore, structural information was included in our analysis. All PfamA domains found in PVs having a structural representative in Protein Data Bank (PDB) [39], are almost completely covered by longest PDB chain sequence, except E7, which is covered by about 50% (Table 2 and Figure 1). Overall, PVs are structurally very well characterised, especially among dsDNA viruses (see Supplementary in Reference [40]). Protein sequences in UniProt (or in other) databases can be chopped into domains on multiple criteria. Protein chains in PDB entries can be divided into domains according to criteria obtained from their 3D structure. As an example, this is done by hierarchical classification of protein domains in Structural Classification of Proteins (SCOP) and CATH databases [35,41]. In this work, SCOP database was used because it is more suitable for evolutionary studies. In addition to sequence similarity, SCOP protein domains are grouped together according to their structural similarity, according to the packaging of the core of the protein domain. SCOP has different hierarchical levels and one of them is Superfamily (SF) level. According to the

SCOP authors, SF level is the highest level with confident homologous relationships. In PV protein structures, the SCOP domains cover most of the PDB chain (Figure 1, Table 2). However, this is not always the case. As noted earlier, proteins can be virtually chopped into domains on multiple criteria and domain borders depend on the assignment method. In PVs, there is good agreement (accordance) between the PfamA and SCOP domains (Figure 1). Only PfamA "PPV_E1_C" is separated into two domains in SCOP where E1 DNA-binding domain (DBD) forms a separate domain from E1 helicase domain (the latter includes a hexamerisation subdomain). In *Polyomaviridae*, Large-T protein and *Parvoviridae* NS1 protein the DBD and helicase domains are classified as separate domains in both PfamA and SCOP. In PVs, seven SCOP domains are identified altogether (Figures 1 and 2).

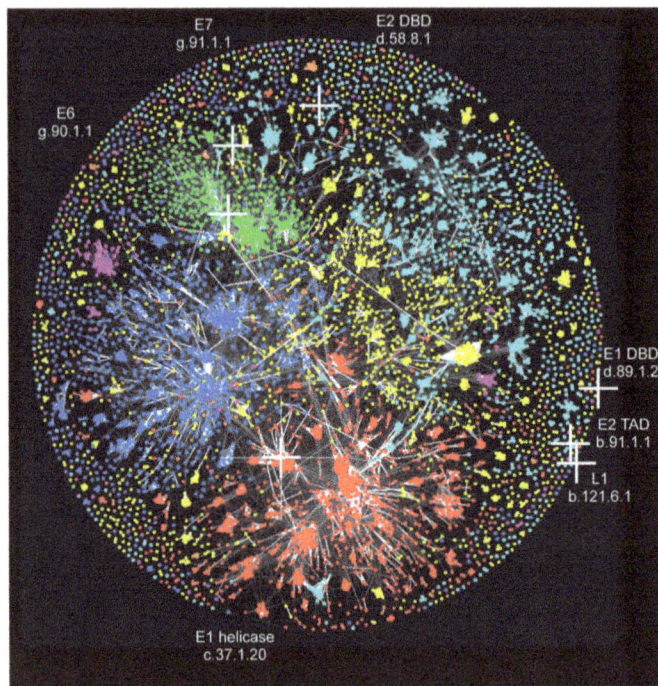

Figure 2. Location of PV domains in the "Galaxy of folds". PV structural domains are marked by white crosses and visualised on protein domain space. Domains in Structural Classification of Proteins (SCOP) were clustered using the software CLANS based on their all-against-all pairwise similarities, as measured by HHsearch *p*-values [34]. Domains are coloured according to their SCOP class: all-a (blue); all-b (cyan); a/b (red); a + b (yellow), small proteins (green); multi-domain proteins (orange); and membrane proteins (magenta). PV protein domain name and SCOP identifier are indicated.

To visualise the global relationship of PV protein structural domains to all other structural domains, "Galaxy of folds" toolkit was used (see Materials and Methods). This "structure space" was generated by Alva et al. to study the homologous origin of SCOP SF and FOLDs (FOLD is another SCOP level; SFs are assigned to FOLDs) [34]. Three PV domains, L1, E2 transactivation domain (TAD) and E1 DBD are located on the sparse periphery of "Galaxy of folds" space where the repulsive forces (i.e., dissimilarity) is dominant over the attractive force (i.e., similarity). E2 DBD and E7 are located at intermediate positions (still a sparse region) and E6 is located in a dense region. E1 helicase domain is located in a very dense region containing many different P-loop ATPases (including other hexameric helicases). Therefore, only the E1 helicase domain has a significant evolutionary relationship to other known structural domains.

3.4. Structural Domains Found in PV Proteins According to SUPERFAMILY Analysis

"Galaxy of Folds" is based on solved structures. Thus, the apparent loss of connections with other structures might be because the relatives are not yet structurally characterised or not yet in a database. To overcome (at least partially) this problem, we used data from SUPERFAMILY resource [29,37]. The SUPERFAMILY resource incorporates SCOP structural domain assignments (based on HMM models) at SF level to all annotated proteins in fully sequenced genomes [37,42]. If a protein with a similar structure to already solved structure is found in another fully sequenced organism, the SUPERFAMILY approach should recognise and classify it accordingly. Additionally, assignments to SFs are also applied to "NCBI viral genomes" and UniProt sequences. Hence, it is possible to evaluate the phylogenomic distribution of structural protein domains without the need of solved structures for each individual organism. The only drawback is that at least one representative structure for a protein domain must be solved. In SUPERFAMILY resource, 7 SCOP domains are found in PV sequences (Table 3 and Figure 1) (collectively named PV_SFs, i.e., SCOP superfamilies found in PVs). SCOP database has a hierarchical tree-structure—protein domains are classified into families and families are assigned to superfamilies, which, in turn, are classified to FOLDs (henceforth capitalised FOLD means SCOP hierarchical level) and then to classes. E2 TAD, E2 DBD, L1, E6, and E7 domains are classified into SFs that have only one family, thereby being the only representatives of the superfamily (Table 3). In addition, E7, E6, and E2 TAD have their own FOLD (i.e., 1 family per SF and 1 SF per FOLD) and thus, they do not have close structural relatives according to SCOP in the current database. L1 protein domain is a member of SF_88648, which together with 4 other viral capsid protein SFs, "Nucleoplasmin-like core domain" (SF_69203) and "PHM/PNGaseF" (SF_49742) form the FOLD called "Nucleoplasmin-like/VP (viral coat and capsid proteins)". E2 DBD is a member of SF_54957 which has one family per SF and the respective SF belongs to the "Ferredoxin-like" FOLD together with 58 other SFs. However, we note that according to SCOP authors the SF level in SCOP is the highest level of confident homologous relationship (so, the SF belonging to the same FOLD might be or might not be evolutionarily related). E1 helicase domain belongs to the highly populated family "Extended AAA-ATPase domain", which, together with 23 other families, forms SF_52540 ("P-loop containing nucleoside triphosphate hydrolases"). E1 DBD is a member of SF_55464 ("Origin of replication-binding domain, RBD-like") and forms its own family. This SF also consists of four other families. Three of them are clearly virus related: polyomavirus Large-T DBD, geminiviral Rep protein DBD and parvoviral Rep protein nuclease domain. The fourth family is "Relaxase domain", a domain with DNA nicking activity responsible for the conjugation of bacterial plasmids and bacterial DNA. Respective domain in parvoviral and geminiviral Rep proteins and Relaxase domain belongs to Rolling Circle Replication (RCR) proteins with endonuclease activity [43].

Table 3. PV_SF domain occurrence in biosphere.

SCOP/SF ID	Classification	SF/FOLD	Families/SF	Description	PV	Viruses[1]	Plasmids[2]	Archaea	Bacteria	Eukaryota	HMMER A	HMMER B	HMMER E	HMMER V[1]
55464	d.89.1	1	5	Origin of replication-binding domain, RBD-like (E1 DBD)	123	424/15*	420	-	134	8	-	4038	32	1563/169*
52540	c.37.1	1	24	P-loop containing nucleoside triphosphate hydrolases (E1 helicase)	123	2346	19971	122	1153	440	ND	ND	ND	ND
51332	b.91.1	1	1	E2 regulatory, transactivation domain (E2 TAD)	123	-	-	-	-	-	-	-	-	-
54957	d.58.8	59	1	Viral DNA-binding domain (E2 DBD)	123	4	-	-	-	-	-	-	-	6
88648	b.121.6	7	1	Group I dsDNA viruses (L1)	123	50/-*	-	-	-	-	-	-	-	170/-*
161229	g.90.1	1	1	E6 C-terminal domain-like	115	-	-	-	-	-	-	1?	-	-
161234	g.91.1	1	1	E7 C-terminal domain-like	108	-	-	-	-	-	-	-	-	-
55464-52540				DBD + helicase	123	7	356	-	119	5	-	ND	10	-

"-" No true positive hits were found. "ND" Not determined. "Underlined" Number of primary hits. "Bold" Number of true positive hits. * Number of true positive hits without *Polyomaviridae*, *Parvoviridae* and *Geminiviridae*. "?" Questionable result. [1] Excluding papillomaviruses. [2] Number of proteins. Non-redundant set of genomes contain 122 Archaeal, 1153 Bacterial, and 440 Eukaryotic species) (i.e., redundant strains and isolates removed). DBD: DNA-binding domain; TAD: transactivation domain. For more detailed information, see Table S3.

3.5. Phylogenetic Distribution of PV_SF Domains

To evaluate the evolutionary history and potential origin of PV structural domains we analysed the phylogenomic distribution of PV (structural) domains using the SUPERFAMILY resource. As shown in Table 3 (see Table S3) five domains (E2_TAD, E2_DBD, L1, E6, and E7) are not found in cellular "complete genomes". Thus, these domains do not have confident homologs in completely sequenced cellular organisms, even including structure-based homology. From the seven SCOP domains, only domains from E1 protein are found in cellular genomes. SF_52540 representatives (E1 helicase domain and relatives, including all P-loop NTPases) are found in every cellular genome in the database and SF_55464 (E1 DBD domain and relatives) is found in 13 sequences of 8 eukaryotes (distinct NCBI taxonomy IDs) and in 261 sequences of 134 bacterial genomes (out of 1153 bacterial genomes in database) (Table 3, Table S3, Supplementary Materials, File S1). E1 DBD distant relatives are present in 5 fungi, 1 Alveolata, 1 Ameobozoa and 1 Viridiplantae and they are most likely relatives of Geminiviral Rep (Table 3, Supplementary Materials, File S1). The phylogenomic distribution of these hits is very sparse. Three out of five fungal hits are among Basidiomycota, but other 45 sequenced Basidiomycota in SUPERFAMILY database do not contain E1 DBD relatives (Figure S1).

To extend the search to non-complete genomes, UniProt sequences were used within SUPERFAMILY database. Additionally, HMM models of PV_SFs were run against all available databases using "hmmsearch". This increased the number of hits of SF_55464 within the bacterial and eukaryotic sequences. For example, in eukaryotes, additional 23 species were found that coded potential SF_55464 homologs, increasing the number of Fungi species by 7 (including two close relatives of previously identified species), the number of Viridiplanate species by 11 (including 9 closely related *Dioscorea* species), two from Stramenopiles and two from Rhodophyta, and one in Rhizaria. Detailed analyses of the annotations of respective coding sequences show that in two Stramenopiles this domain is coded in mitochondrion and in *Rhodophyta* these sequences belong to algal plasmids (Supplementary Materials, File S1). The only domain, excluding E1, that seems to have a true positive hit in cellular organisms is E6 (SF_161229), which gives a hit to bacterium *Achromobacter xylosoxidans* AXX-A "Uncharacterised protein" F7T9H3_ALCXX (Table 3, Supplementary Materials, File S1). This sequence fits equally well into E6 structure and into ferredoxin structures according to LOMETS [32], a protein threading meta server which was used to verify sequence based homology predictions. Thus, only domains from the E1 protein show confident deeper evolutionary connection to cellular proteins.

Both domains found in cellular organisms (SF_52540 and SF_55464) are also found in other viruses (including all members of *Polyomaviridae*) and plasmids. In addition, representatives of SF_88648 (L1 protein) and SF_54957 (E2 DBD and relatives) are found only in viruses. Homologs of L1 protein (SF_88648) are found only in *Polyomaviridae*. E2 DBD relatives are found in a subset of gammaherpesviruses. In "NCBI viral genomes" the SF_55464 is also found in several *Parvoviridae*, *Geminiviridae*, in two *Betaherpesvirinae*, in one *Circoviridae* and *Siphoviridae* (relaxase domain); and in nine viruses recently classified as *Genomoviridae* (Ge—for geminivirus-like, nomo—for no movement protein) [44]. Among *Genomoviridae*, three sequences are classified into Gemycocircularvirus genus (Gemini-like myco-infecting circular virus) [44]. The members of SF_55464 (more precisely, mostly the relaxase domain) are also found in more than 400 bacterial plasmid sequences and notably, only in a single bacterial virus. In plasmid subset of SUPERFAMILY sequences, the SF_55464 is also found in one eukaryotic plasmid pPT4-NU with red algal host *Pyropia tenera* (this sequence was found also in SUPERFAMILY UniProt sequences and in HMMER search). In addition to different viral families and very few eukaryotes, E1 DBD connects PVs confidently with bacteria and bacterial plasmids.

3.6. Occurence of PV Protein Domains in Three Superkingdoms

As shown above, PVs do have a connection with other superkingdoms on some levels. To visualise the occurrence of PV domains (and other similar small viruses) in cellular superkingdoms, we generated Figure 3. This is based on raw data because performing controls similar to PV_SF subset to all of the viruses in the figure would be extremely time-consuming. Figure 3 shows how many protein domains in corresponding viral family are found in the genomes of cellular superkingdoms (shown in percentage). In general, bimodal distribution can be observed (more viral families are covered in the Figure S2), which means that the shared protein domains between viruses and superkingdoms can usually be found in a small percentage of the cellular genomes or in most of them. For example, PVs have one domain (SF_52540) which is found in almost all organisms (value 1 on *x*-axis on Figure 3, panel *Papillomaviridae*) and one domain (SF_55464) is found in more than 0% and less than 10% of Eukaryotic genomes in the database. The rest of the domains (SF_51332, SF_54957, SF_88648, SF_161229 and SF_161234) are not found in any Eukaryotes. Similarly, two domains (SF_161229 and SF_161234) are found in more than 0% and less than 10% of bacteria and one domain (SF_55464) is found in more than 10% and less than 20% of bacterial genomes (Figure 3 and Table S3). Potential PV relatives (*Polyomaviridae*, *Geminiviridae*, and *Parvoviridae*) have a similar bimodal distribution, with a high fraction of domains found only in viruses and very few in cellular genomes. Additionally, *Polyomaviridae* encodes chaperone DnaJ domain (SF_46565) which is also found in half of Archaea genomes.

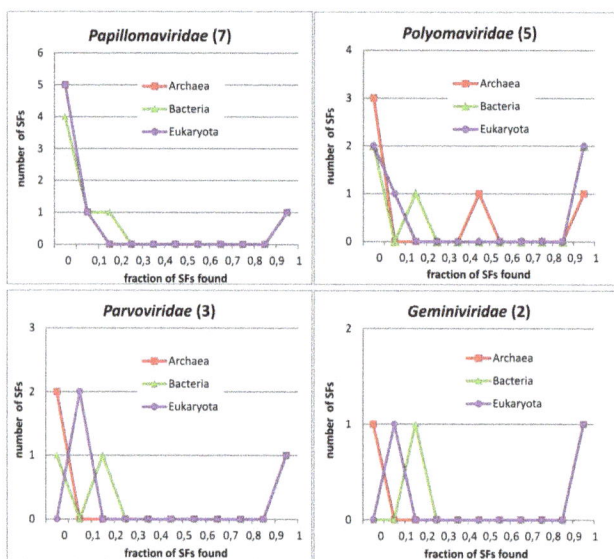

Figure 3. Distribution of protein domains in viral families by superkingdoms. Each figure shows data for the corresponding viral family. The number in the parentheses on titles corresponds to the number of distinct domains (SF) found in the respective viral family. The *y*-axis shows the number of domains (SF) from the viral family, covered by any of the three superkingdoms. The *x*-axis shows the decile of the genomes where the viral protein domains are found by superkingdoms. In panel *Papillomaviridae*, the lines for Archaea and Eukaryota overlap.

Bimodal distribution is not specific for small DNA viruses. Large DNA viruses, like members of *Herpesviridae* (and *Gammaherpesvirinae*, sharing E2 DBD with PVs) also have a bimodal distribution. However, they have a much higher fraction of proteins found in almost every cellular organism (data shown in Figure S2). Collectively, RNA viruses have a higher fraction of domains (either SF domains or PfamA domains) found only in viruses [40]. However, several dsDNA viruses like PVs (point $x = 0.7$, $y = 0.5$ in Abroi 2015 Figure 3), *Polyomaviridae* (0.6; 0.4), *Herpesviridae* (0.8; 0.05) and *Adenoviridae* (0.65; 0.1), have a fraction of virosphere-specific domains (i.e., domains found exclusively in the virosphere) as high as RNA viruses have (Figure S3).

3.7. Phylogenomic Distribution of the E1 SF_55464:SF_52540 Domain Pair

The P-loop NTPase (SF_52540) domain is very abundant in nature and therefore not very informative without much deeper analyses. However, PV E1 protein contains SF_55464 and SF_52540 domain, forming a domain pair. Thus, we decided to examine whether this domain pair is found elsewhere in the biosphere. In the SUPERFAMILY version used, SF_55464 and SF_52540, if on the same protein, are always in the same order, SF_55464 N-terminal and SF_52540 C-terminal, agreeing with previous studies showing that convergent evolution of protein architectures is rare [42]. As expected, this combination is found in all PVs and in all polyomaviruses (when we exclude database misannotations) (Tables 3 and 4). In *Parvoviridae* species, which have an annotated SF_55464 domain, SF_52540 is also present. This domain pair is also found in bacterial plasmids and in more than 100 bacterial species, but not in any Archaea and only in some eukaryotes (Table 3, Supplementary Materials, File S1). We note that databases often do not discriminate between bacterial chromosome and plasmid (sometimes there is no clear border between them either). Most (but not all) of the plasmid (and bacterial) sequences having SF_55464 also have SF_52540 (Table 4). Among the 20 plasmid sequences with a single SF_52540 (i.e., domain architecture similar to PV E1 protein), nine belong to phytoplasma (obligate bacterial parasites of plant phloem tissue) plasmids (Supplementary Materials, File S1). These nine phytoplasma plasmid sequences have domain organisation most similar to PVs. E1_DBD relatives (SF_55464) together with P-loop NTPase (SF_52540) are found in very few eukaryotes with very sparse phylogenomic distribution. In the SUPERFAMILY database, this combination is found in three Fungi (all in Basidiomycota), one Alveolata and one Amoebozoa. Excluding PVs, *Polyomaviridae* and *Parvoviridae*, this combination is found in 21 viruses (21 distinct NCBI taxonomy IDs) including 14 members of *Geminiviridae*. From seven remaining sequences in the "NCBI viral genomes" dataset with both SF_55464 and SF_52540, three belong to *Genomoviridae*, two to *Herpesviridae* and one each to *Circoviridae* and *Siphoviridae* (Supplementary Materials, File S1). Thus, according to SUPERFAMILY data the PV replicative helicase has evolutionary connections with *Polyomaviridae* and *Parvoviridae*, as well as deeper connections with *Geminiviridae*, bacterial conjugative plasmids, including phytoplasma plasmids, and with bacteria.

Table 4. Number of sequences containing SF_55464 with different domain architectures.

No. of 52540 Domains	PV [1] 123 *	*Polyomaviridae* [2] 50 *	*Parvoviridae* 81 *	*Geminiviridae* [3] 332 *	Other Viruses	Plasmids	Bacteria	Eukaryota
0	1	0	0	350	10	64	35	4
1	122	49	33	14	6	20	20	5
2	0	0	0	0	1	334	183	0
3	0	0	0	0	0	2	1	0

[1] In HPV53, only DBD part of E1 is annotated. [2] *Polyomaviridae* Merkel cell polyomavirus does not have annotated full-length Large-T protein in this version of the database used (in current version of NCBI viral genomes it already has). [3] Geminiviruses have often more than one replication protein isoform annotated. * Number of genomes in the respective viral family.

4. Discussion

Several aspects of the molecular biology of PVs are quite well known, however, the origin and the evolutionary relationship to other organisms is still enigmatic. In this work, the occurrence of PV protein domains was used to study the relations of PV domains with other domains characterised so far and to study the origin and/or evolution of PV proteins and PVs.

PVs, similar to several other viral families, encode proteins without detectable structural homologs in cellular organisms [45]. This trend can be quantitatively evaluated in different ways [40]. As shown in Figure 3 (see also Figure S3) and in the analysis of protein domain occurrence at higher taxonomic levels in citation [40], PVs have a high fraction of protein domains not found in cellular superkingdoms or are found in a small fraction of cellular genomes. In this aspect (location of PV in Figure S3 and shape of the PV lines in Figure 3 and Figure S2 compared to Figure 4 in Reference [40]), PVs and *Polyomaviridae* are more similar to RNA viruses and ssDNA viruses than dsDNA viruses. That kind of bimodal or U-shape distribution is confirmed also independently at the structural level. Relationship of PV protein structural domains to other structural domains was assessed and visualised with "Galaxy of folds" toolkit. Only E1 helicase domain locates at a densely populated region (close relationship) and at least four domains locate at very sparse regions (Figure 2).

4.1. SUPERFAMILY Limitations

The SUPERFAMILY resource is a useful tool for deep evolutionary studies; unfortunately, it has its own limitations. Different HMM models of SCOP families from the same SF may not recognise easily the sequences from (structural) sibling family, especially in the case of viruses. For example, when using HMM model of PV E1 DBD domain and searching it against all the known sequences, it does not recognise Large-T antigen DBD sequences from *Polyomaviridae*. However, PV E1 DBD and *Polyomaviridae* Large-T antigen DBD are classified to the same SF in SCOP. In SUPERFAMILY, the SF hits are collected as a union of all of the respective SF HMM results [37]. In addition, SUPERFAMILY is limited to protein structural domains classified in SCOP. Unfortunately, not all protein structures of interest are in the SCOP database (not in SCOP 1.75 [35], SCOP2 [46] or SCOPe [47]).

Because of the gap between structural classification and current data in PDB database, biologically/virologically suspicious results were re-evaluated using most recent data. For example, SF_52540 was found in most *Parvoviridae* species but SF_55464 only in a subset. The structure of respective domain in SCOP is solved for Adeno-associated virus (*Dependoparvovirus*, *Parvovirinae*) (SCOP and PDB representative "1m55"). The HMM model based on "1m55" recognises protoparvoviruses (*Parvovirinae*) but not bocaparvoviruses (*Parvovirinae*) on HMMER "hmmsearch". However, based on published structures the structural and functional similarity of bocaparvovirus "4kw3", dependoparvovirus "1m55" and protoparvovirus "4pp4" gives evidence that bocaparvoviruses and probably *Densovirinae* (another subfamily of *Parvoviridae* family) have a homologous domain to SF_55464 [48]. Hopefully, the next release of SCOP (and SUPERFAMILY) will include up to date viral structural information. To avoid our subjective bias, SUPERFAMILY data and extended structural analysis data were interpreted separately. The quality of the data in databases in this kind of studies is very important. The amount of data used in our work is still comprehensible, allowing us to test correctness/quality of input data and our results, however, in larger-scale analyses it is not feasible or indeed possible. One PV related example is the network-like relationship studies of dsDNA viruses [49]. In the publication, data showed a connection (Figure 1 in [49]) between PV and polyomaviruses, which corresponds most likely to Bandicoot Papillomatosis virus; a chimera, containing capsid proteins from PV and a replication protein with a DnaJ domain from polyomaviruses. This connection was misinterpreted by the authors in the text. Therefore, to avoid or minimise misinterpretations in large scale studies, each scientific society should keep the data as correct as possible, to give confidence to large-scale analysis results.

4.2. Capsid Protein Connects PVs with a Rest of the Virosphere

The PV major capsid protein L1 has structural relatives at the SF level only in *Polyomaviridae*. In addition to L1, *Polyomaviridae* also codes domains structurally similar to E1 DBD and E1 helicase (including hexamerisation subdomain). PVs and *Polyomaviridae* are the only known viruses with nucleosomes inside virion [50,51]. The thirty-year-old statement by Favre et al. "The existence of a viral core containing DNA and cellular histones may be a further common structural characteristic of papovaviruses." (Papovaviruses—old name of PVs and polyomaviruses together) is still valid and this characteristic is not only common but also specific for these viruses [50]. Thus, there are several lines of evidence that PVs and *Polyomaviridae* are clearly evolutionary related.

According to published non-hierarchical structural analysis and supported by the common FOLD level in SCOP, PV L1 and *Polyomaviridae* major capsid protein VP1 belong to the "single jelly-roll" (eight-stranded beta barrel) capsid lineage also called "Picorna-like lineage". The single jelly-roll capsid lineage contains capsid proteins from a number of other viral families, including *Circoviridae*,

Geminiviridae, and *Parvoviridae* together with numerous families of RNA viruses [52,53]. Viral families in this lineage have different replication strategies and have host ranges both from Eukaryota and Bacteria. As noted earlier, SCOP classification to the same FOLD level does not guarantee a common ancestor, however, it also does not exclude it. Thus, most likely PVs are connected to the wider virosphere via their major capsid protein L1.

4.3. E2 DBD Most Likely Does Not Originate from Gammaherpesviruses

As summarized in Figure 4, E2 DBD domain has connection only with gammaherpesviruses. According to SUPERFAMILY results and HMMER searches only members of genus *Lymphocryptoviruses* gives significant hits. However, published structures of rhadinovirus (genus *Rhadinovirus* is another member of *Gammaherpesvirinae* subfamily) proteins (PDB codes 4blg, 2yq1, 4k2j and 5a76) prove that functionally and structurally homologous proteins are found also in rhadinoviruses [54–57]. The divergence time of gammaherpesviruses, where the SF_54957 domain is found have not been estimated explicitly; however it is possible to estimate their potential divergence time to no more than ~200 million years ago from published data [58,59]. PVs have existed at least ~315 million years [23] and assuming virus-host co-divergence also for fish viruses, the PVs are most likely more ancestral, at least ~415 million years [60]. Therefore, PV E2 DBD does not originate from gammaherpesviruses, at least not after their divergence.

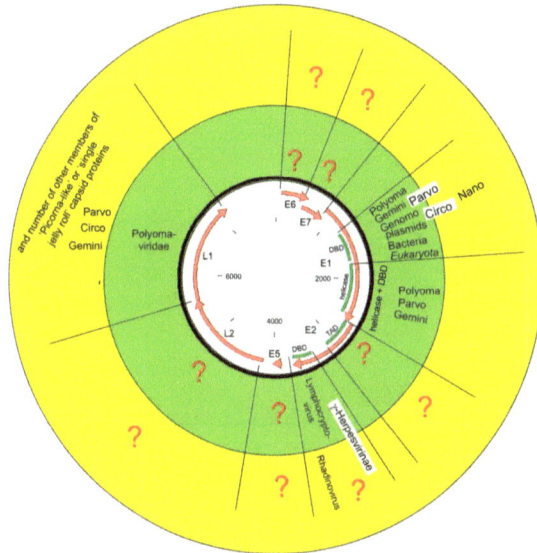

Figure 4. Summary figure of the relationship of PV domains with other parts of the biosphere. Virus family names are abbreviated without "*-viridae*" suffix. In the green circle, the relationships according to SCOP and SUPERFAMILY resource are shown. In the yellow circle, the relationships according to extended structural analysis from published articles and structures are shown. Genera *Lymphocryptovirus* and *Rhadinovirus* are subfamilies of γ-*Herpesvirinae*. For E1 helicase domain only evolutionary relationship via domain pair SF_55464:SF_52540 are shown.

4.4. Replication Protein Connects PVs with a Rest of Biosphere

SF_55464 is also found in more than hundred bacterial species (Table 3) with wide and sparse phylogenomic distribution. Most of the bacterial hits have best *e*-value for the relaxase HMM model (as in the case of bacterial plasmids). Non-relaxase hits in Bacteria are found only in 5 phytoplasma species. Extending to noncomplete genomes increases the number of bacterial hits to a few thousands of sequences. Thus, at least via the relationship with the relaxase domain, PVs have connections with bacteria and bacterial plasmids. The relationship of geminiviral replication proteins to plasmid have been published, however the direction of the transfer is not clear [61,62].

Phylogenetically closest genomes to currently known PV hosts (Vertebrates) where SF_55464 is found in SUPERFAMILY database are among Fungi (Supplementary Materials, File S1). Extending the search to non-complete genomes we also found SF_55464 in some Metazoa, however, closer examination shows, that they are all most likely misannotations. These sequences were almost identical to Bacterial ones and, if we exclude very recent "from Bacteria to Eukaryota" transfer (which is possible but very unlikely), these sequences are most likely contaminants or a part of the sequenced organism's microbiota. Detailed examination of true positive eukaryotic hits identified that in Stramenopiles this domain is coded by mitochondrial DNA, widening the phylogenetic distribution of RCR domains. All SF_55464 true positives in eukaryotes give best hit to geminiviral Rep HMM model. To test whether the eukaryotic hits are taxonomically restricted sequences or just a moderately divergent member of some other protein domain family we performed reciprocal sequence search using SF_55464 eukaryotic hits (parts of sequences corresponding to SF_55464) as a query in "phmmer" and "tblastn". Only sequences from three organisms belonging to *Basidiomycota* (*Serpula lacrymans var. lacrymans* S7.9, *Pisolithus tinctorius* Marx 270 v1.0 and *Laccaria bicolor* S238N-H82) recognised each other SF_55464 sequences (and after them viral sequences from *Geminiviridae* and *Genomoviridae*). All other eukaryotic sequences give hits only to viral sequences mostly from *Geminiviridae* and *Genomoviridae*. Thus, in the sequenced biosphere, these eukaryotic hits do not have close homologs in other organisms (even those not yet annotated as protein). This indicates that these sequences are taxonomically restricted and not on the periphery on some unidentified protein domain family. Therefore, PVs have connections to the sequenced eukaryotic world only via distant relatives in the virosphere.

SF_55464 is found in all or almost all members of PV, *Polyomaviridae*, and *Geminiviridae*. It is also found in 10 viruses currently classified into the new proposed family *Genomoviridae* and in a single member of *Circoviridae* (out of 45 in the database). In the phylogenetic tree of Rep proteins of circular single-stranded DNA (ssDNA) viruses, genomoviral Rep proteins form a well-supported monophyletic clade which branches as a sister group of *Geminiviridae* and they both are more distantly related to *Circo-* and *Nanoviridae* [44]. The Rep protein tree is supported by structural analyses showing that Rep proteins of *Geminiviridae*, *Circoviridae* and *Nanoviridae* are indeed structurally related [44,48]. Nanovirus and circovirus Rep protein structures are not yet classified in SCOP. Thus, in the virosphere, extended structural analyses of E1 DBD relatives connect PV with *Polyomaviridae*, *Parvoviridae*, *Geminiviridae*, *Circoviridae*, *Nanoviridae*, and *Genomoviridae*. However, connection outside the virosphere is still restricted to bacterial plasmids, bacteria, very few eukaryotes and few red algal plasmids.

4.5. Closest Domain Pair of E1 Protein Is Found Far from Known PV Hosts

Since SF_55464 and SF_52540 coexist as a domain pair in E1 protein, the existence of this pair in other genomes was studied. In Bacteria, this combination is found in 119 species in SUPERFAMILY "complete genomes", mostly in combination of "relaxase" domain with "Tandem AAA-ATPase domain". This domain pair was present in very few eukaryotes with very sparse phylogenomic distribution in three Fungi, one Alveolata and one Amoebozoa. In contrast to Bacterial sequences they have best fit to geminiviral "DNA-binding domain of REP protein" and "Extended AAA-ATPase domain" HMM models.

In the case of viruses, E1 domain pair was found in *Polyomaviridae*, *Parvoviridae*, *Geminiviridae*, and in 21 other viruses. Sharing SF_55464 and SF_52540 (including the hexamerisation subdomain) is true for *Polyomaviridae* and *Parvoviridae* (or at least for *Parvovirinae*). PVs and *Polyomaviridae* are dsDNA viruses; however, *Parvoviridae* belongs to ssDNA viruses encapsulating linear ssDNA. Two proteins coded by human herpesviruses (HHV) from this list belong to roseloviruses, namely HHV6A and HHV6B. They encode a protein with most likely parvovirus origin. Herpesviruses are known helperviruses for some parvoviruses [63,64] and according to phylogenetic distribution of this protein (and phylogenetic tree) there have been from virus-to virus transfer with direction from parvoviruses to roseloviruses (during the last 100 MYA as estimated from [58,59]).

In plasmids, SF_55464 and SF_52540 domain combination is also found (Table 4). In most of the plasmids, the sequence regions assigned to SF_55464 have the best *e*-value for "relaxase" HMM model and regions assigned to SF_52540 have the best hit to "Tandem AAA-ATPase" HMM (like in Bacteria). Only the sequences of the phytoplasma plasmids and red algal plasmids of *Porphyra pulchra* have best *e*-value for geminiviral "DNA-binding domain of REP protein" HMM and SF_52540 models other than "Tandem AAA-ATPase".

Considering "Virus to host" and "Host to virus" gene transfers and recombination of different viruses as well as accepting the statement by Rohwer and Barott "When considering the virosphere, extremely unlikely events become probabilistic certainties." it is very difficult to estimate the evolutionary history or trajectory of these domains [65]. It is possible to generate the phylogenetic tree of these sequences but it is much harder to find a root. Work on the age of some of these viral genera and families may give some information and restrictions, but this is beyond the scope of the current study.

5. Conclusions

Summarizing over all protein domains of PV, only domains coding less than half of total annotated coding sequences show confident evolutionary connection to the rest of biosphere. This half include ~1/5 of total amino acids (E1 DBD and helicase) showing connection to sequenced and annotated cellular proteins and less than 1/20 of total amino acids (E2 DBD) showing connection with gammaherpesviruses.

PVs are clearly related to *Polyomaviridae*, sharing structural homologs of capsid protein and two domains of replication protein at SCOP SF level. Both viral families have dsDNA viral genomes packed into nucleosomes inside the viral particle. *Parvoviridae* shares two replication related domains and, including extended structural similarity, also the capsid protein with PVs and with *Polyomaviridae*. As ssDNA viruses, *Parvoviridae* do not have nucleosomes in virions.

The relationship of PV, *Polyomaviridae* and *Parvoviridae* to *Geminiviridae*, *Circoviridae*, *Nanoviridae*, and *Genomoviridae* is not as clear and their exact relationship is out of the scope of this work. They all have SF_55464 and according to extended structural analysis, they all (except *Nano-* and *Genomoviridae*) have common capsid protein (there are no structural predictions for *Nano-* and *Genomoviridae* capsid protein).

The major capsid protein L1 and replication protein E1 connect PVs to the rest of the virosphere, E1 DBD also connects PVs to bacterial plasmids, bacteria and red algal plasmids. Excluding the E1 helicase domain, the connections to eukaryotic protein domains are almost non-existent, even including available structural information. There are clear connections with other parts of the biosphere but the exact evolutionary trajectory of PV proteins is not yet known. There are still almost no hints as to how PVs as a whole/entirety originate and how they become vertebrate and epithelium specific. The evolutionary history of other PV protein domains, which have not been found in cellular organisms, is as mysterious.

Most likely the last common ancestor of PV, *Polyomaviridae*, and *Parvoviridae*, or more precisely, the genome coding the ancestral replication and/or capsid protein of these viruses, inhabited a marine environment. Only very few non-fungal and non-vertebrate marine eukaryotic genomes are sequenced. Thus, most likely, we have an unexplored sequence and structure space in both cellular and viral taxons, as well as in other types of mobile elements in marine environments. Further characterisation (sequencing is only one part of the characterisations) of this and other biotopes will give more information and thus more hints on the origin of PV proteins. On the other hand, some connections between PVs and other viruses or cellular organisms may be lost forever due to gene loss events. For example, in the PV family, the E6 gene was lost at least twice in different virus clades [15].

To our current knowledge, PVs are connected to the rest of biosphere via replication and major capsid proteins. The origin and/or evolutionary history of other domains are still unknown. This makes the question "When and how did PV originate?" of continuing interest.

Supplementary Materials: The following are available online at www.mdpi.com/1999-4915/9/9/240/s1.

Acknowledgments: A.A. acknowledges support from Basic research financing to Estonian Biocentre.

Author Contributions: A.A. designed the experiments; M.P., H.K., and A.A. performed the experiments; M.P., H.K. and A.A. analysed the data; and M.P., H.K., K.G. and A.A. wrote the paper.

Conflicts of Interest: The authors declare no conflicts of interest.

References

1. Shope, R.E.; Hurst, E.W. Infectious Papillomatosis of Rabbits: With a Note on the Histopathology. *J. Exp. Med.* **1933**, *58*, 607–624. [CrossRef] [PubMed]
2. Papillomavirus Episteme. Available online: https://pave.niaid.nih.gov (accessed on 8 June 2017).
3. Van Doorslaer, K.; Li, Z.; Xirasagar, S.; Maes, P.; Kaminsky, D.; Liou, D.; Sun, Q.; Kaur, R.; Huyen, Y.; McBride, A.A. The Papillomavirus Episteme: A major update to the papillomavirus sequence database. *Nucleic Acids Res.* **2017**, *45*, 499–506. [CrossRef] [PubMed]
4. Danos, O.; Katinka, M.; Yaniv, M. Human papillomavirus 1a complete DNA sequence: A novel type of genome organization among papovaviridae. *EMBO J.* **1982**, *1*, 231–236. [PubMed]
5. Chen, E.Y.; Howley, P.M.; Levinson, A.D.; Seeburg, P.H. The primary structure and genetic organization of the bovine papillomavirus type 1 genome. *Nature* **1982**, *299*, 529–534. [CrossRef] [PubMed]
6. Danos, O.; Engel, L.W.; Chen, E.Y.; Yaniv, M.; Howley, P.M. Comparative analysis of the human type 1a and bovine type 1 papillomavirus genomes. *J. Virol.* **1983**, *46*, 557–566. [PubMed]
7. Clertant, P.; Seif, I. A common function for polyoma virus large-T and papillomavirus E1 proteins? *Nature* **1984**, *311*, 276–279. [CrossRef]
8. Karlin, S.; Ghandour, G.; Foulser, D.E.; Korn, L.J. Comparative analysis of human and bovine papillomaviruses. *Mol. Biol. Evol.* **1984**, *1*, 357–370. [PubMed]
9. Fuchs, P.G.; Iftner, T.; Weninger, J.; Pfister, H. Epidermodysplasia verruciformis-associated human papillomavirus 8: Genomic sequence and comparative analysis. *J. Virol.* **1986**, *58*, 626–634. [PubMed]
10. Campione-Piccardo, J.; Montpetit, M.L.; Grégoire, L.; Arella, M. A highly conserved nucleotide string shared by all genomes of human papillomaviruses. *Virus Genes* **1991**, *5*, 349–357. [CrossRef] [PubMed]
11. Shah, S.D.; Doorbar, J.; Goldstein, R.A. Analysis of host-parasite incongruence in papillomavirus evolution using importance sampling. *Mol. Biol. Evol.* **2010**, *27*, 1301–1314. [CrossRef] [PubMed]
12. Gottschling, M.; Stamatakis, A.; Nindl, I.; Stockfleth, E.; Alonso, A.; Bravo, I.G. Multiple evolutionary mechanisms drive papillomavirus diversification. *Mol. Biol. Evol.* **2007**, *24*, 1242–1258. [CrossRef] [PubMed]
13. Pimenoff, V.N.; de Oliveira, C.M.; Bravo, I.G. Transmission between archaic and modern human ancestors during the evolution of the oncogenic human papillomavirus 16. *Mol. Biol. Evol.* **2017**, *34*, 4–19. [CrossRef] [PubMed]
14. Van Doorslaer, K. Evolution of the Papillomaviridae. *Virology* **2013**, *445*, 11–20. [CrossRef] [PubMed]
15. Van Doorslaer, K.; McBride, A.A. Molecular archeological evidence in support of the repeated loss of a papillomavirus gene. *Sci. Rep.* **2016**, *6*, 33028. [CrossRef] [PubMed]
16. UniProt Rereference Proteomes. Available online: http://www.uniprot.org/help/reference_proteome (accessed on 8 June 2017).
17. UniProt Proteomes. Available online: http://www.uniprot.org/help/proteome (accessed on 8 June 2017).
18. Herbst, L.H.; Lenz, J.; van Doorslaer, K.; Chen, Z.; Stacy, B.A.; Wellehan, J.F.X.; Manire, C.A.; Burk, R.D. Genomic characterization of two novel reptilian papillomaviruses, *Chelonia mydas* papillomavirus 1 and *Caretta caretta* papillomavirus 1. *Virology* **2009**, *383*, 131–135. [CrossRef] [PubMed]
19. Rector, A.; van Ranst, M. Animal papillomaviruses. *Virology* **2013**, *445*, 213–223. [CrossRef] [PubMed]
20. Rector, A.; Lemey, P.; Tachezy, R.; Mostmans, S.; Ghim, S.-J.; van Doorslaer, K.; Roelke, M.; Bush, M.; Montali, R.J.; Joslin, J.; et al. Ancient papillomavirus-host co-speciation in Felidae. *Genome Biol.* **2007**, *8*, R57. [CrossRef] [PubMed]
21. Kumar, S.; Subramanian, S. Mutation rates in mammalian genomes. *Proc. Natl. Acad. Sci. USA.* **2002**, *99*, 803–808. [CrossRef] [PubMed]

22. Aiewsakun, P.; Katzourakis, A. Time-Dependent Rate Phenomenon in Viruses. *J. Virol.* **2016**, *90*, 7184–7195. [CrossRef] [PubMed]
23. Bravo, I.G.; Felez-Sanchez, M. Papillomaviruses: Viral evolution, cancer and evolutionary medicine. *Evol. Med. Public Heal.* **2015**, *2015*, 32–51. [CrossRef] [PubMed]
24. Chothia, C.; Lesk, A.M. The relation between the divergence of sequence and structure in proteins. *EMBO J.* **1986**, *5*, 823–826. [PubMed]
25. Illergård, K.; Ardell, D.H.; Elofsson, A. Structure is three to ten times more conserved than sequence—A study of structural response in protein cores. *Proteins Struct. Funct. Bioinform.* **2009**, *77*, 499–508. [CrossRef] [PubMed]
26. Challis, C.J.; Schmidler, S.C. A Stochastic Evolutionary Model for Protein Structure Alignment and Phylogeny. *Mol. Biol. Evol.* **2012**, *29*, 3575–3587. [CrossRef] [PubMed]
27. Herman, J.L.; Challis, C.J.; Novák, Á.; Hein, J.; Schmidler, S.C. Simultaneous Bayesian estimation of alignment and phylogeny under a joint model of protein sequence and structure. *Mol. Biol. Evol.* **2014**, *31*, 2251–2266. [CrossRef] [PubMed]
28. Finn, R.D.; Coggill, P.; Eberhardt, R.Y.; Eddy, S.R.; Mistry, J.; Mitchell, A.L.; Potter, S.C.; Punta, M.; Qureshi, M.; Sangrador-Vegas, A.; et al. The Pfam protein families database: Towards a more sustainable future. *Nucleic Acids Res.* **2016**, *44*, 279–285. [CrossRef] [PubMed]
29. Oates, M.E.; Stahlhacke, J.; Vavoulis, D.V.; Smithers, B.; Rackham, O.J.L.; Sardar, A.J.; Zaucha, J.; Thurlby, N.; Fang, H.; Gough, J. The SUPERFAMILY 1.75 database in 2014: A doubling of data. *Nucleic Acids Res.* **2015**, *43*, 227–233. [CrossRef] [PubMed]
30. HMMER: Biosequence Analysis Using Profile Hidden Markov Models. Available online: http://www.hmmer.org/ (accessed on 8 June 2017).
31. Finn, R.D.; Clements, J.; Eddy, S.R. HMMER web server: Interactive sequence similarity searching. *Nucleic Acids Res.* **2011**, *39*. [CrossRef] [PubMed]
32. Wu, S.; Zhang, Y. LOMETS: A local meta-threading-server for protein structure prediction. *Nucleic Acids Res.* **2007**, *35*, 3375–3382. [CrossRef] [PubMed]
33. Buck, C.B.; van Doorslaer, K.; Peretti, A.; Geoghegan, E.M.; Tisza, M.J.; An, P.; Katz, J.P.; Pipas, J.M.; McBride, A.A.; Camus, A.C.; et al. The Ancient Evolutionary History of Polyomaviruses. *PLoS Pathog.* **2016**, *12*. [CrossRef] [PubMed]
34. Alva, V.; Remmert, M.; Biegert, A.; Lupas, A.N.; Söding, J. A galaxy of folds. *Protein Sci.* **2010**, *19*, 124–130. [CrossRef] [PubMed]
35. Murzin, A.G.; Brenner, S.E.; Hubbard, T.; Chothia, C. SCOP: A structural classification of proteins database for the investigation of sequences and structures. *J. Mol. Biol.* **1995**, *247*, 536–540. [CrossRef]
36. Galaxy of Folds. Available online: https://toolkit.tuebingen.mpg.de/hhcluster/ (accessed on 27 May 2016).
37. Gough, J.; Karplus, K.; Hughey, R.; Chothia, C. Assignment of homology to genome sequences using a library of hidden Markov models that represent all proteins of known structure. *J. Mol. Biol.* **2001**, *313*, 903–919. [CrossRef] [PubMed]
38. Astell, C.R.; Mol, C.D.; Anderson, W.F. Structural and functional homology of parvovirus and papovavirus polypeptides. *J. Gen. Virol.* **1987**, *68*, 885–893. [CrossRef] [PubMed]
39. Berman, H.M.; Westbrook, J.; Feng, Z.; Gilliland, G.; Bhat, T.N.; Weissig, H.; Shindyalov, I.N.; Bourne, P.E. The protein data bank. *Nucleic Acids Res.* **2000**, *28*, 235–242. [CrossRef] [PubMed]
40. Abroi, A. A protein domain-based view of the virosphere-host relationship. *Biochimie* **2015**, *119*, 231–243. [CrossRef] [PubMed]
41. Sillitoe, I.; Lewis, T.E.; Cuff, A.; Das, S.; Ashford, P.; Dawson, N.L.; Furnham, N.; Laskowski, R.A.; Lee, D.; Lees, J.G.; et al. CATH: Comprehensive structural and functional annotations for genome sequences. *Nucleic Acids Res.* **2015**, *43*, 376–381. [CrossRef] [PubMed]
42. Gough, J. Convergent evolution of domain architectures (is rare). *Bioinformatics* **2005**, *21*, 1464–1471. [CrossRef] [PubMed]
43. Chandler, M.; de la Cruz, F.; Dyda, F.; Hickman, A.B.; Moncalian, G.; Ton-Hoang, B. Breaking and joining single-stranded DNA: The HUH endonuclease superfamily. *Nat. Rev. Microbiol.* **2013**, *11*, 525–538. [CrossRef] [PubMed]
44. Krupovic, M.; Ghabrial, S.A.; Jiang, D.; Varsani, A. Genomoviridae: A new family of widespread single-stranded DNA viruses. *Arch. Virol.* **2016**, *161*, 2633–2643. [CrossRef] [PubMed]

45. Abroi, A.; Gough, J. Are viruses a source of new protein folds for organisms?—Virosphere structure space and evolution. *BioEssays* **2011**, *33*, 626–635. [CrossRef] [PubMed]

46. Andreeva, A.; Howorth, D.; Chothia, C.; Kulesha, E.; Murzin, A.G. SCOP2 prototype: A new approach to protein structure mining. *Nucleic Acids Res.* **2014**, *42*. [CrossRef] [PubMed]

47. Fox, N.K.; Brenner, S.E.; Chandonia, J.-M. SCOPe: Structural Classification of Proteins—Extended, integrating SCOP and ASTRAL data and classification of new structures. *Nucleic Acids Res.* **2014**, *42*, 304–309. [CrossRef] [PubMed]

48. Vega-Rocha, S.; Gronenborn, B.; Gronenborn, A.; Campos-Olivas, R. Solution structure of the endonuclease domain from the master replication initiator protein of the nanovirus faba bean necrotic yellows virus and comparison with the corresponding geminivirus and circovirus structures. *Biochemistry* **2007**, *46*, 6201–6212. [CrossRef] [PubMed]

49. Iranzo, J.; Krupovic, M.; Koonin, E.V. The double-stranded DNA virosphere as a modular hierarchical network of gene sharing. *MBio* **2016**, *7*. [CrossRef] [PubMed]

50. Favre, M.; Breitburd, F.; Croissant, O.; Orth, G. Chromatin-like structures obtained after alkaline disruption of bovine and human papillomaviruses. *J. Virol.* **1977**, *21*, 1205–1209. [PubMed]

51. Friedmann, T.; David, D. Structural roles of polyoma virus proteins. *J. Virol.* **1972**, *10*, 776–782. [PubMed]

52. Krupovic, M.; Koonin, E.V. Multiple origins of viral capsid proteins from cellular ancestors. *Proc. Natl. Acad. Sci. USA* **2017**, *114*, E2401–E2410. [CrossRef] [PubMed]

53. Abrescia, N.G.A.; Bamford, D.H.; Grimes, J.M.; Stuart, D.I. Structure Unifies the Viral Universe. *Annu. Rev. Biochem.* **2012**, *81*, 795–822. [CrossRef] [PubMed]

54. Correia, B.; Cerqueira, S.A.; Beauchemin, C.; Pires de Miranda, M.; Li, S.; Ponnusamy, R.; Rodrigues, L.; Schneider, T.R.; Carrondo, M.A.; Kaye, K.M.; et al. Crystal Structure of the γ-2 Herpesvirus LANA DNA Binding Domain Identifies Charged Surface Residues Which Impact Viral Latency. *PLoS Pathog.* **2013**, *9*. [CrossRef] [PubMed]

55. Hellert, J.; Weidner-Glunde, M.; Krausze, J.; Richter, U.; Adler, H.; Fedorov, R.; Pietrek, M.; Rückert, J.; Ritter, C.; Schulz, T.F.; et al. A Structural Basis for BRD2/4-Mediated Host Chromatin Interaction and Oligomer Assembly of Kaposi Sarcoma-Associated Herpesvirus and Murine γherpesvirus LANA Proteins. *PLoS Pathog.* **2013**, *9*. [CrossRef] [PubMed]

56. Domsic, J.F.; Chen, H.S.; Lu, F.; Marmorstein, R.; Lieberman, P.M. Molecular Basis for Oligomeric-DNA Binding and Episome Maintenance by KSHV LANA. *PLoS Pathog.* **2013**, *9*. [CrossRef] [PubMed]

57. Ponnusamy, R.; Petoukhov, M.V.; Correia, B.; Custodio, T.F.; Juillard, F.; Tan, M.; Pires De Miranda, M.; Carrondo, M.A.; Simas, J.P.; Kaye, K.M.; et al. KSHV but not MHV-68 LANA induces a strong bend upon binding to terminal repeat viral DNA. *Nucleic Acids Res.* **2015**, *43*, 10039–10054. [CrossRef] [PubMed]

58. Grose, C. Pangaea and the Out-of-Africa Model of Varicella-Zoster Virus Evolution and Phylogeography. *J. Virol.* **2012**, *86*, 9558–9565. [CrossRef] [PubMed]

59. McGeoch, D.J.; Rixon, F.J.; Davison, A.J. Topics in herpesvirus genomics and evolution. *Virus Res.* **2006**, *117*, 90–104. [CrossRef] [PubMed]

60. López-Bueno, A.; Mavian, C.; Labella, A.M.; Castro, D.; Borrego, J.J.; Alcami, A.; Alejo, A. Concurrence of Iridovirus, Polyomavirus, and a Unique Member of a New Group of Fish Papillomaviruses in Lymphocystis Disease-Affected Gilthead Sea Bream. *J. Virol.* **2016**, *90*, 8768–8779. [CrossRef] [PubMed]

61. Saccardo, F.; Cettul, E.; Palmano, S.; Noris, E.; Firrao, G. On the alleged origin of geminiviruses from extrachromosomal DNAs of phytoplasmas. *BMC Evol. Biol.* **2011**, *11*, 185. [CrossRef] [PubMed]

62. Krupovic, M.; Ravantti, J.J.; Bamford, D.H. Geminiviruses: A tale of a plasmid becoming a virus. *BMC Evol. Biol.* **2009**, *9*, 112. [CrossRef] [PubMed]

63. Thomson, B.J.; Weindler, F.W.; Gray, D.; Schwaab, V.; Heilbronn, R. Human herpesvirus-6 (HHV-6) is a helper virus for adenoassociated virus type-2 (AAV-2) and the AAV-2 *Rep* gene homolog in HHV-6 can mediate AAV-2 DNA-replication and regulate gene-expression. *Virology* **1994**, *204*, 304–311. [CrossRef] [PubMed]

64. Thomson, B.J.; Efstathiou, S.; Honess, R.W. Acquisition of the Human Adenoassociated Virus Type-2 *Rep* Gene By Human Herpesvirus Type-6. *Nature* **1991**, *351*, 78–80. [CrossRef] [PubMed]

65. Rohwer, F.; Barott, K. Viral information. *Biol. Philos.* **2013**, *28*, 283–297. [CrossRef] [PubMed]

viruses

MDPI

Review

Structural Insights in Multifunctional Papillomavirus Oncoproteins

Irina Suarez and Gilles Trave *

Équipe Labellisée Ligue 2015, Department of Integrated Structural Biology, Institut de Génétique et de Biologie Moléculaire et Cellulaire (IGBMC), INSERM U1258/CNRS UMR 7104/Université de Strasbourg, 1 rue Laurent Fries, BP 10142, F-67404 Illkirch, France; suarezi@igbmc.fr
* Correspondence: gilles.trave@unistra.fr

Received: 3 December 2017; Accepted: 11 January 2018; Published: 15 January 2018

Abstract: Since their discovery in the mid-eighties, the main papillomavirus oncoproteins E6 and E7 have been recalcitrant to high-resolution structure analysis. However, in the last decade a wealth of three-dimensional information has been gained on both proteins whether free or complexed to host target proteins. Here, we first summarize the diverse activities of these small multifunctional oncoproteins. Next, we review the available structural data and the new insights they provide about the evolution of E6 and E7, their multiple interactions and their functional variability across human papillomavirus (HPV) species.

Keywords: papillomaviruses; oncoproteins; structure; X-ray; NMR; virus-host interactomics

1. Introduction

Papillomaviruses (PVs) constitute a large family of small oncogenic DNA viruses that infect mucosal or cutaneous epithelia [1,2]. PVs have been found in most vertebrate species investigated including human [3], other mammals, birds, reptiles [4], and very recently in fish [5,6]. PVs contain a very small double-stranded circular DNA genome of 7000 to 8000 base pairs [3]. Papillomaviruses are classified into distinct genera, species, types, or subtypes whenever the DNA sequences of their L1 gene share less than 60%, 70%, 90% or 98% identity, respectively. To date, 353 PV types have been identified and classified into 49 distinct genera [7,8] (see curated list at Papillomavirus Episteme (PaVE); https://pave.niaid.nih.gov/#home) [7]. They include more than 200 HPV types belonging to five distinct genera (alpha, beta, gamma, mu, and nu). Most alpha HPVs display a mucosal tropism while most beta, mu and nu HPVs infect the skin. However, HPV species alpha-2, alpha-4, and alpha-8 are found in skin warts [9] whereas HPV species beta-3 seems to be more commonly found in the nasal cavity than in skin [10], and therefore may display a mucosal tropism [11]. Gamma HPV appear to display both cutaneous and mucosal tropism [9].

All PVs induce proliferation of the infected cells. This proliferation mobilizes the cellular DNA replication and protein production machinery, which are hijacked by the virus to the benefit of its own replication [1–3,12]. In general, proliferation lasts only for a short time, allowing the production of virions ready to infect another individual. This activates an immunological response from the host, leading in most cases to virus clearance. Most HPV types that are designated as "low-risk" generate mild benign pathogenic effects, such as skin warts, mucosal lesions, or in the worst case, mucosal condylomas that require surgery [1–3]. However, for a subset of HPV types dubbed "high-risk", viral genes sometimes do not get fully eradicated by the host. All or part of the viral genome remains maintained in at least one host cell, either in an episomal form or inserted in the host cell genome [13]. These remaining viral oncogenes have the capacity to promote, over a long period that may last up to three decades, further changes to the infected cell, which may eventually lead to cancer [1–3,12,14].

Indeed, high-risk mucosal HPVs are responsible for practically 100% of cervical, anal, rectal, and penile cancers, as well as for an increasingly high proportion of oropharyngeal cancers [15].

The two main viral HPV oncogenes required to establish and maintain the tumorigenic phenotype encode two early expressed oncoproteins, called E6 and E7. Both E6 and E7 are very small proteins (in most mammalian PVs about 150 and 100 amino acids, respectively). Nonetheless, they are responsible for most of the proliferative and transforming events that lead to carcinogenesis [12,16–18]. Altogether, E6 and E7 display pro-proliferative and anti-apoptotic effects as well as cell adhesion, cell polarity, and cell differentiation-altering properties that facilitate a transient period of proliferation of the infected epithelial cells.

Remarkably, turning off the expression of E6 oncoprotein in HPV-positive cancer-derived cells by means of RNA interference induces growth arrest followed by either apoptosis or senescence [1,19–21]. This indicates that HPV-positive cancers are "addicted" to E6 expression for their survival, so that inhibition of key E6 functions may represent a promising strategy for counteracting the growth of HPV-positive tumors. Indeed, blocking E6 molecular activities by means of E6-binding recombinant proteins [22], peptides [23–25], or antibodies [26,27] has been shown to drive specifically growth arrest and/or death of HPV-positive tumor cells. These data indicate that well-designed small molecule ligands of E6 will represent promising avenues for therapy of HPV-positive cancers. Yet, a prerequisite for the ab initio design of such small molecule inhibitors, and for the rationale improvement of current low-affinity E6 inhibitors [28–30], is to obtain high-resolution information on the three-dimensional structure of E6 proteins alone or in complex with their cellular targets.

2. E6 and E7 Are Multifunctional Proteins

2.1. E6 and E7 Interact with Large Numbers of Host Target Proteins

For the best-studied mammalian PVs, both E6 and E7 have been found to interact with numerous distinct cellular target proteins involved in a variety of cellular functions [16,17,31–36]. Importantly, E6 proteins from distinct HPV species recognize distinct subsets of the full panel of potential E6 targets [31]. This exquisite capacity of different E6 proteins to recognize particular pools of targets likely contributes to the particular biological traits of each HPV type in terms of tropism, viral cycle, or pathogenicity. The same multifunctional character, undergoing variations across HPV species, has been observed for HPV E7 oncoproteins [31].

2.2. Viral Domain-Motif Hijacking Strategies Explain E6 and E7 Multifunctionality

The ability of the E6 and E7 proteins to recruit large numbers of different and often functionally unrelated proteins is at first thought surprising, since such small proteins cannot, in principle, present many distinct interaction surfaces. This can, however, be explained by considering the "domain-motif hijacking strategies" employed by many viral or bacterial pathogen proteins to disrupt or reprogram particular functions of the infected hosts [37]. In brief, a large part of cellular protein–protein interactions boil down to specific interactions between folded globular domains and short intrinsically unfolded linear motifs, which get folded upon binding to their target domains [38]. Viral and bacterial infectious agents produce proteins bearing the ability to divert these "domain-motif functional interaction networks" for the sake of their own lifecycle [37,39].

One can cite at least three well-documented examples of domain-motif network hijacking by the viral oncoproteins E6 and E7.

The E7 proteins from most papillomaviruses contain a conserved LxCxE motif, also found in many host proteins. The LxCxE motif is specialized in the recognition of the "pocket domain" of members of the retinoblastoma (Rb) protein family, including the Retinoblastoma tumor suppressor pRB (RB1), p107 (RBL1), and p130 (RB2) [1,17]. The binding of E7 via its LxCxE motif to Rb pocket proteins, and their subsequent proteasomal degradation (as discussed later in this review) disrupts complexes between Rb proteins and E2 transcription factors (E2F). Consequently, the active E2F factors released in this way facilitate cell division and its subsequent hijacking by the virus.

All E6 oncoproteins from High-Risk Mucosal HPVs (hrm-HPVs) present a conserved C-terminal PDZ-Binding Motif (PBM) that allows them to bind, and sometimes provoke the cellular degradation of, multidomain proteins containing PDZ domains. PDZ domains, named from the first letters of three proteins sharing such domains (post synaptic density protein (PSD95), Drosophila disc large tumor suppressor (Dlg1), and zonula occludens-1 protein (zo-1)) are involved in various processes including cell polarity, cell adhesion, and apoptosis [40–46].

In addition, most E6 proteins from mammalian papillomaviruses contain a charged hydrophobic pocket, which recognizes peptides prone to alpha helical structure that presents a conserved LxxLL consensus sequence interspaced with acidic residues [16,47–50]. This pocket allows E6 proteins to recruit a variety of host cellular proteins containing LxxLL motifs and involved in a variety of apparently unrelated functions, such as the ubiquitin ligase E6-Associated Protein (E6AP) [47], the focal adhesion protein paxillin [49], the cell fate-determining Mastermind-like transcriptional coactivator 1 (MAML1) [51–53], or the antiviral Interferon Regulatory transcriptional Factor 3 (IRF3) [54]. In all these proteins, the LxxLL motif is found in a region predicted as intrinsically unfolded, ensuring the accessibility of the motif for interaction with E6.

2.3. E6 and E7 Divert the Host Ubiquitination Machinery

Frequently, E6 and E7 not only bind to their targets but also provoke their accelerated destruction by the Ubiquitin Proteasome System (UPS) [55,56]. This is generally achieved via a tripartite interaction, in which the viral oncoprotein recruits, on the one hand, a UPS enzyme (generally, an E3 ubiquitin ligase) and on the other hand a target cellular protein. The target protein is subsequently poly-ubiquitinated, then degraded by the proteasome system, while the viral oncoprotein and the ubiquitin ligase are recycled for degrading the next molecule of target protein. The best documented case of such "UPS hijacking" by HPVs is performed by the E6 oncoprotein of high-risk mucosal (hrm) HPVs. The Hrm-HPV E6 binds to E6AP (also called UbE3A), a cellular E3 ubiquitin ligase containing a HECT (Homologous to E6AP C-Terminal) domain specialized in poly-ubiquitination of target proteins. The resulting E6/E6AP complex then recruits the p53 anti-apoptotic tumor suppressor protein, provoking its poly-ubiquitination and subsequent degradation by the proteasome [57,58]. In a somehow comparable process, the E7 oncoprotein of hrm-HPVs binds to the cullin 2 ubiquitin ligase, resulting in the proteasome-dependent degradation of pRb [59,60]. In a recent work, screening of a library expressing 590 proteins related to the UPS confirmed E6AP and members of cullin family as UPS targets of hrm HPV E6 and E7, respectively, and identified novel potential targets for both oncoproteins [61].

2.4. E6 and E7 and Nucleic Acids

While the best studied molecular activities of E6 and E7 proteins are related to their ability to interact with target cellular proteins, both E6 and E7 have early been suggested to interact also with nucleic acids [62–64]. This hypothesis was based on the fact that both E6 and E7 contain conserved zinc-binding domains with sequences distantly reminiscent of the zinc finger domains of transcription factors, including repeats of four conserved cysteine residues. Indeed, a subset of E6 proteins from high-risk mucosal HPV types was found to interact with high affinity and selectivity with 4-way DNA junctions [65] and the residues responsible for DNA binding were localized within the C-terminal zinc-binding domain of these E6 proteins [66–68]. The E6 proteins of high-risk mucosal HPVs have also been found to be RNA-binding proteins that can inhibit splicing of pre-mRNAs [69]. How these nucleic-acid binding properties are utilized during the virus lifecycle, and whether they contribute to the oncogenic phenotype, is still poorly understood.

2.5. The Multifunctionality Issue: How to Make Sense of Complexity?

The multifunctional character of PV oncoproteins raises the problem of how to rank/evaluate the individual contributions of their numerous interactions and of the subsequent biochemical reactions to viral tropism, lifecycle, and pathogenesis. At the limits, one may either consider that all observed interactions are equally and indistinctively relevant ("big-bag complexity" view point), or consider on the contrary that only a few interactions are relevant with all the rest being artefactual or fortuitous ("extreme reductionism" view point). A way to reconcile these opposite view points may be to consider that each identified interaction of E6 or E7 participates in viral tropism, lifecycle, and pathogenesis according to a given "weight", and that the observed phenotype emerges from the combination of all the differently weighted interactions and subsequent reactions. In such a "combinatorial-weighted" approach, the list of interactions taken into consideration remains the same for each E6 or E7 protein across the HPV phylogenetic tree, but the weight allocated to each individual interaction varies for each viral type considered.

How could one quantify the weight allocated to each viral-host interaction? One approach may be to use quantitative information derived from mass spectrometry (MS). In a remarkable work, White and collaborators [31,36] have used E6 proteins from 16 different HPV types belonging to eight different species and two different HPV genera (α and β), to identify 153 E6-binding cellular proteins. Each E6 protein considered was found to bind detectably to only a reduced subset of these 153 proteins. In addition, each E6 protein varied in its efficiency of pulling down each particular partner. This information was quantified in the form of a "Normalized Weighted D-score" that computed the uniqueness, abundance, and reproducibility of each identified E6-target interaction. In another study by Thomas et al. [70], resin beads pre-saturated with synthetic PDZ-Binding Motifs (PBMs) derived from 10 different HPV E6 proteins were incubated with keratinocyte extracts, leading to the pull-down and MS-based identification of 19 E6-binding PDZ domain-containing host cell proteins. Thanks to a reproducible protocol where only the PBM sequence varied, the normalized mean numbers of peptides of each pulled-down host protein identified by MS provided quantitative scores representative of the preferences of each E6 protein.

Another approach to allocating a quantitative weight to the different viral-host interactions could be to measure the affinity displayed by the viral protein towards each potential cellular partner. Recently, Vincentelli et al. [71] bacterially expressed 209 PDZ domains, representing 79% of the entire complement of human PDZ domains (the "PDZome") and used a high-throughput chromatographic approach (the holdup assay) to systematically measure in vitro the affinities of each expressed PDZ domain towards the PBMs of E6 proteins from the two highest-risk mucosal HPV types, 16 and 18. The data were represented in the form of "PDZome-binding specificity profiles", which allow visualizing and comparing at one glance the binding strengths of each E6 PBM towards all individual domains of the human PDZome.

3. Sequence and Structure of E6 and E7 Oncoproteins

3.1. Amino Acid Sequence Features of E6 and E7

The E6 and E7 sequences are present in the large majority of vertebrates (mammalian, avian, and reptilian) papillomaviruses characterized to date [6,7]. However, there are exceptions, including a few human papillomaviruses, in which the E6 sequences are missing [72,73]. The recently discovered fish papillomaviruses are devoid of both E6 and E7 sequences [6].

In most mammalian PVs, the E6 sequence spans about 150 amino acids comprising two repeated 70-residue zinc-binding domains called E6N and E6C [68,74], connected by a linker helix [75] (Figure 1A). The tandem repeat is surrounded by N-terminal and C-terminal extensions displaying higher sequence variability and a higher propensity to intrinsic disorder.

Mammalian E7 proteins span about 100 residues comprising a 50 residue-long N-terminal disordered region [76,77] followed by a 50-residue-long C-terminal zinc-binding domain, which folds as an obligate homodimer [76,78] (Figure 1A). The E7 zinc-binding domain, sometimes called CR3 in former publications, will be referred to as "E7-ZBD" from now on in this text.

In avian papillomaviruses, the E6 protein is constituted by a single zinc-binding 80-residue domain whereas the E7 protein contains a presumably disordered N-terminal region encompassing about 80 residues followed by a folded C-terminal zinc-binding domain encompassing about 50 residues [6,7,79] (Figure 1B).

Turtle papillomaviruses [6,7,80] also comprise an 80-residue single-domain E6 protein and a 110-residue E7. Curiously, in the two turtle papillomaviruses identified, the predicted 50-residue E7 zinc-binding domain is situated at the N-terminus and the presumably disordered region at the C-terminus (Figure 1C), in contrast to what is observed in both mammalian and avian E7.

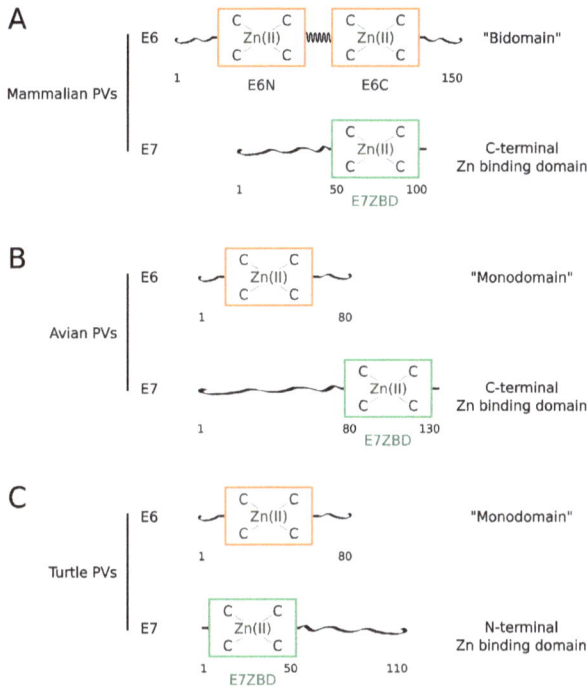

Figure 1. Schematics of E6 and E7 sequences across distinct vertebrates: (**A**) mammals; (**B**) birds; (**C**) turtles. Folded zinc-binding domains and unfolded regions are represented as rectangles and extended strings, respectively. Approximate amino-acid numbering is indicated below the schemes.

The presumably disordered regions, both in E6 and E7, harbor "mimics" of host short linear interaction motifs (sLiMs), allowing E6 and E7 to hijack cellular signaling networks involving such interaction motifs, as observed for many viral proteins [37,39]. This is the case of the PDZ-binding motif found at the C-terminus of E6 from high-risk mucosal (hrm) HPVs as well as the LxCxE Rb-binding motif found within the disordered region of most mammalian E7 proteins.

As will be discussed later in this review, structural data on two phylogenetically distant mammalian E6 proteins (from HPV16 and BPV1) [75] have shown that E6-LxxLL motif recognition is conserved across mammalian PVs, and that this property is related to the bi-domain structure of these E6 proteins. Since the E6 proteins of bird and turtle papillomavirus are constituted of a single domain, E6-LxxLL recognition is likely to be an acquired property that emerged for the mammalian papillomaviruses only.

3.2. History of Progress towards E6 and E7 Tri-Dimensional Structures

The E6 and E7 proteins were identified as major PV oncoproteins in the mid-80s [81–85], immediately promoting the first attempts to produce these proteins in recombinant form [86–89]. Nevertheless, the first structures of the isolated zinc-binding domains of either E6 [67] or E7 [76,78] were only released in 2006, whereas the first structures of full-length mammalian PV E6 proteins were released in 2013 [75], almost thirty years after the discovery of E6 and E7 oncoproteins. This was mainly due to difficulties in producing homogeneous soluble samples of recombinant E6 and E7 proteins. The search for soluble E6 samples amenable to structural analysis has been particularly arduous, going through the following steps: (i) experimental delimitation of the two zinc-binding domains [68,74,90,91]; (ii) solubilization of E6 by means of fusion to the highly soluble bacterial Maltose Binding Protein (MBP) [92–95]; (iii) biophysical characterization of E6 oligomers [92–94,96,97]; (iv) separation of soluble monomeric MBP-E6 fusions from soluble aggregated MBP-E6 fusions [92–94]; (v) solubilizing mutagenesis of E6 proteins to replace exposed non-conserved cysteine residues that promote intermolecular disulfide bridging during E6 purification and storage [92–94]; (vi) solubilizing mutagenesis of E6 proteins to replace surface exposed hydrophobic residues that promote oligomerization of purified E6 [91]; (vii) solubilization and stabilization of E6 proteins by binding to their cognate LxxLL motif [75,95,98]. Therefore, it was the interdependent progress in understanding the mechanisms of E6 self-association and in finding strategies to prevent these mechanisms that ultimately led to the resolution of structures of E6 proteins [67,75,91,98].

3.3. Structure of the Zinc-Binding Domains of E6

The folded regions of E6 and E7 contain zinc-binding folds that have not yet been observed elsewhere in the living kingdom, although some of their structural characteristics can be found within other proteins.

Two structures of distinct mammalian PV E6N domains are available to date, one from HPV16 and another one from the phylogenetically distant bovine PV, bovine papillomavirus 1 (BPV1). The structure of HPV16 E6N has been solved both by Nuclear Magnetic Resonance (NMR) in the form of an isolated fragment of E6 [91] and by crystallography in the frame of LxxLL motif-bound full-length HPV16 E6 [75,98]. The solution structure of the isolated E6N domain and its crystal structure within the crystallized full-length E6 are well superimposed (Figure 2B). The structure of the BPV1 E6N domain, as observed in the crystal structure of the BPV1 E6/paxillin complex [75] is homologous to that of HPV16 E6N, except for an N-terminal region of BPV1 E6N which was not visible in the BPV1 E6 crystal. The core of the HPV16 E6N fold consists of a three-stranded β-sheet and three α-helices, reinforced by a peripheral zinc-binding site. Two zinc-liganding cysteines are contributed by a knuckle situated at the junction between strand β1 and helix α2, while the other two zinc-liganding cysteines are contributed by the C-terminal α-helix α3 (Figure 2A,D,E). The BPV1 E6N lacks the N-terminal α-helix, which may exist as a part of the nonobservable region in the crystal (See Supplementary Figure S3 of [75]). In HPV16 E6N, the first ten N-terminal residues adopt an extended and rather flexible structure, which is, however, anchored to the hydrophobic core by phenylalanine F2 [75,91]. This explains why mutagenesis of residue F2 was previously found to alter various E6 activities [99,100], despite the fact that F2 is not involved in the interfaces with E6AP and p53 in the E6/E6AP and E6/E6AP/p53 complexes [75,98]. The conserved C-terminal helix of E6N domains constitutes the start of the "linker helix" that connects the E6N and E6C domains within the crystal structures of HPV16 E6 and BPV1 E6 [75] (Figure 2E).

Three structures of distinct mammalian PV E6C domains are available to date: from HPV16 [75,91], from the related high-risk mucosal HPV 51 [101], and from BPV1 [75]. As observed for HPV16 E6N, HPV16 E6C displays identical structures when solved either as an isolated fragment by NMR or as part of full-length HPV16 E6 by crystallography [75]. Moreover, the HPV51 E6C (solved in isolation by

NMR) [101] and the BPV1 E6C (as part of the crystal structure of full-length BPV1 E6 [75]) are perfectly superimposed on the structure of HPV16 E6C (Figure 2C).

Despite sharing only 10% sequence identity, the E6N and E6C domains have homologous structures, with the exception of the flanking N- and C-terminal regions (Figure 2A,D). Whereas the N-terminal region of HPV16 E6N is a flexible loop, the N-terminus of all solved E6C domains (from HPV16, HPV51 and BPV1) folds as an additional β-strand (β4), which extends the β-sheet. The C-terminal region of E6N corresponds to the start of the interdomain linker (see below), while the C-terminus of E6C harbors a PDZ-Binding Motif in high-risk mucosal HPVs and distinctive sequences in other types. The structural homology between E6N and E6C confirms sequence-based analyses [6,62], which suggests that they arose from duplication of a single-domain ancestor. This hypothesis is further supported by the fact that the E6 proteins of avian and turtle PVs are composed of a single zinc-binding domain. Indeed, the single-domain avian PV E6 [79] is similar, by its sequence and overall fold, to the E6C domains of mammalian PV E6.

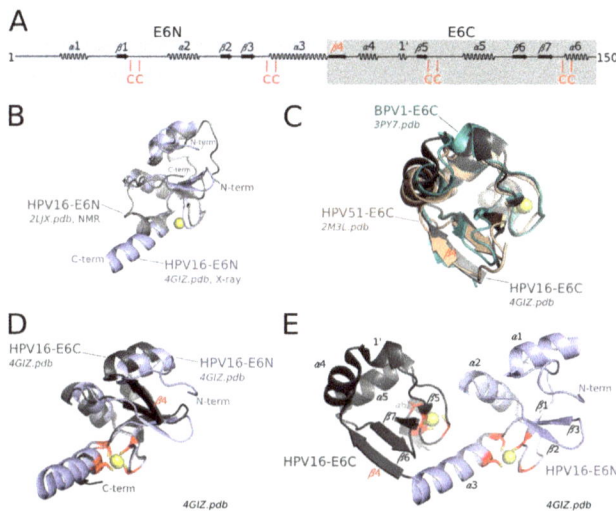

Figure 2. Architecture of mammalian human papillomavirus (HPV) E6. (**A**) Secondary structure elements of mammalian HPV16 E6. E6N is shaded in blue, E6C is shaded in gray; (**B**) Superimposition of HPV16 E6N domain solved by NMR [91] and HPV16 E6N domain from the crystal structure of the E6/E6AP LxxLL complex [75]; (**C**) Superimposition of HPV16 E6C domain from the crystal structure of the E6/E6AP LxxLL complex [75], HPV51 E6C domain solved by NMR [101] and bovine papillomavirus 1 (BPV1) E6C domain from the crystal structure of the BPV1 E6/paxillin LxxLL complex [75]; (**D**) Superimposition of HPV16 E6N and E6C domains from the crystal structure of the HPV16 E6/E6AP LxxLL complex [75]; (**E**) Structure of LxxLL-bound mammalian HPV16 E6 [75] (LxxLL motif not shown for clarity). Zinc-coordinating cysteines are highlighted in red. Secondary elements are numbered as in [75]. Spheres are Zinc (II) atoms. All structural views in this article were prepared using Pymol (http://www.pymol.org/).

Within full-length E6, the two domains E6N and E6C are connected by a linker. In the crystal structures of full-length E6 bound to LxxLL motifs [75], this linker forms a helix, in continuation of helix α3 at the exit of the E6N zinc binding site (Figure 2E).

3.4. Structure of the Zinc-Binding Domains of E7 and Comparison to E6 Domains

The structures of the C-terminal zinc-binding domain of E7 (E7 ZBD) from two phylogenetically distant HPVs, namely HPV 1 (low-risk cutaneous species μ-1) and HPV 45 (high-risk mucosal species α-7), have been solved by X-ray crystallography and solution NMR, respectively [76,78]. The backbone structures are quasi-identical for the two species, indicating the high conservation of this fold across mammalian PV species. In contrast to mammalian E6N and E6C, which are both monomeric in solution, the E7 ZBD presents a dimeric conformation (Figure 3B) that is observed both in solution and in the crystal [76,78]. The interface between the two monomers is highly hydrophobic, suggesting that this is an "obligate" dimer, which is highly favored in solution as compared to the monomeric form [102]. In solution, full-length E7 also forms a dimer since characteristic resonances of the dimeric E7 ZBD are preserved in NMR spectra of full-length E7 [76,103,104]. Indeed, the dimeric structure of high-risk HPV16 E7 favors the formation of a ternary complex between E7, pRb and CBP-p300 (CREB-Binding Protein-histone acetyltransferase protein 300) [103].

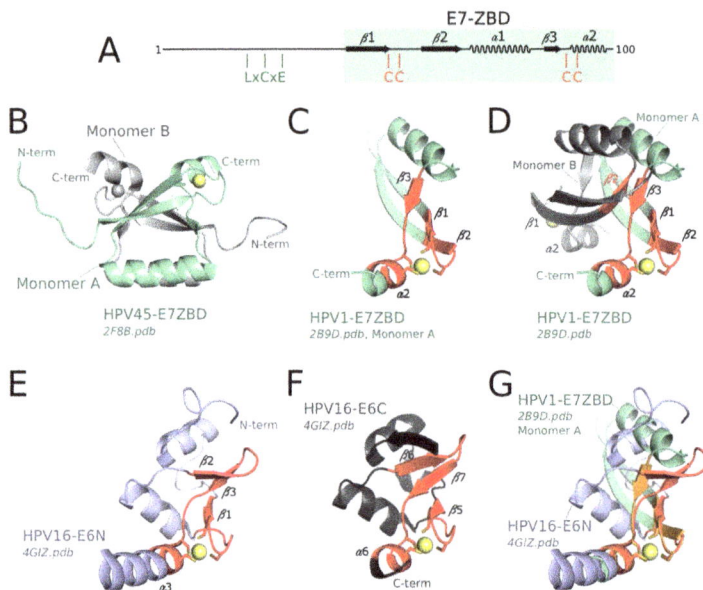

Figure 3. Structure of mammalian papillomavirus (PV) E7 ZBD and comparison to mammalian PV E6N and E6C. (**A**) Schematics of mammalian HPV E7 sequence. Mammalian PV E7 proteins contain an intrinsically unfolded N-terminus bearing a conserved LxCxE sequence (as indicated on the sheme) among other putative interaction motifs, followed by a folded zinc-binding domain (E7-ZBD) shaded in green, with four conserved cysteine (C) residues as indicated; (**B**) Solution structure of HPV45 E7 ZBD homodimer [76]. A comparable structure was also obtained by crystallography for the E7 ZBD of HPV1 E7 [78]; (**C**) Elements of the treble clef motif of E7 ZBD (in red) in context of the E7 ZBD monomer; (**D**) Elements of the treble clef motif of E7 ZBD (in red) in context of the E7 ZBD dimer; (**E**) Elements of the treble clef motif of E6N (in red); (**F**) Elements of the treble clef motif of E6C (in red); (**G**) Superimposition of HPV1 E7-ZBD and HPV16 E6N, enlighting common features of their treble clef motifs, indicated in orange and red color, respectively. Spheres are Zinc (II) atoms.

In addition to being dimeric, the fold of mammalian E7 ZBD is clearly distinct from that of E6N and E6C, with a different secondary structure ordering and topology (Figure 3A, compare to Figure 2A). Nevertheless, both folds probably originate from a very distant common ancestor. This is supported by the conservation of a common core structure comprising the four Zinc-coordinated cysteines and the secondary structure elements bearing these cysteines (Figure 3, compare panel C, E and F). Indeed, this core zinc-binding structural element has been noticed to exist in a variety of proteins including E6 and E7 [105] and was originally described as the "treble clef motif" [106,107]. The treble clef motif contains the following succession of elements: zinc knuckle (i.e., two short β-strands connected by a turn that bear two zinc-binding residues), loop, β-hairpin, and α-helix. Two zinc binding cysteines are contributed by the knuckle, a third cysteine is situated at the junction between the β-hairpin and the α-helix, and the fourth cysteine is contributed by the α-helix. The mammalian E6N and E6C folds contain the four secondary structure elements (Figure 3E,F) and therefore can be classified in this category. The mammalian E7 ZBD monomer contains the zinc knuckle, a loop, a single β-strand, and an α-helix (Figure 3C,D). Two zinc-binding cysteines are contributed by the knuckle, a third cysteine is situated at the junction between the β-strand and the α-helix, and the fourth cysteine is contributed by the α-helix. This organization also fits to the definition of the treble clef, with the difference that the β-hairpin is replaced by a single β-strand (β3, in contact with β2 from opposite monomer, see Figure 3D). Nonetheless, all the secondary structure elements that surround the zinc ion in mammalian E7, E6N, and E6C all superimpose very well (Figure 3G), supporting a common evolutionary origin for these three zinc-binding domains.

Remarkably, the avian E7 ZBD [79] is more similar, by its sequence and overall fold, to the mammalian E6N domain than to the mammalian E7 ZBD. Therefore, avian E7 ZBD and full-length avian E6 are structurally homologous to the mammalian E6N and E6C domains, respectively. This indicates that the mammalian "bidomain" E6 has arisen from the fusion of two ancestral "monodomain" E7 and E6 proteins, that were themselves derived from a single ancestral zinc-binding fold, presenting the common features shared by mammalian E6N and E6C domains. The distinctive dimeric fold of mammalian E7 ZBD must correspond to a very ancient duplicate of the common ancestor of E6N and E6C domain, that has later followed a divergent evolutionary process.

The cysteine-rich and dimeric mammalian E7 ZBD may contribute, depending on purification and/or storage conditions, to the redox-sensitivity and self-association properties of E7, which have been thoroughly analyzed under different aspects by the Prat-Gay group [108–111]. Interestingly, based on these studies the authors have proposed to use hyperstable E7 oligomers for vaccination against HPV-positive tumors [112]. This demonstrates that basic research focusing on the biophysical properties of HPV oncoprotein folds may also lead to practical applications of direct medical interest.

3.5. Conformation of Uncomplexed Mammalian E6 in Solution

To date, no crystal structure of uncomplexed full-length mammalian PV E6 could be obtained despite significant efforts. However, the NMR analysis of full-length E6 has provided extended insight on the conformation of free E6 in solution [91]. Most backbone amide signals of the full-length E6 construct could be identified and were found to overlay with signals in its separated E6N and E6C domains. This indicated that the structures of the domains were preserved in the context of the full-length protein. However, most of the resonances of the interdomain linker (residues 75–82) could not be observed, likely due to dynamic processes. NMR measurements of the tumbling correlation time (τc) indicated that the uncomplexed full-length E6 protein behaved as a rigid monomer rather than two independently tumbling domains [91]. This suggests that the interdomain linker may transiently adopt, in free E6 in solution, the helical conformation that is observed in the crystal structure of LxxLL-complexed E6 (Figure 2E, see also Figure 4A). This would represent a typical example of "conformational selection". Conformational selection is a theoretical framework, which proposes that unliganded proteins already adopt—or "sample"—, in a proportion that varies for each case considered, the conformation that they will adopt in the bound state [113].

While the conformation of free E6 represents an interesting and not fully resolved biophysical issue, it is rather the E6 proteins bound to their host targets that exert biological and pathogenic effects. In addition, full-length HPV E6 proteins appear to be expressed in very low amounts in tumor cells [97,114] so that they may essentially exist in the form of target-bound complexes. These considerations have motivated structural investigation of E6 proteins in complexes with cellular target proteins, as will be described in the next paragraphs.

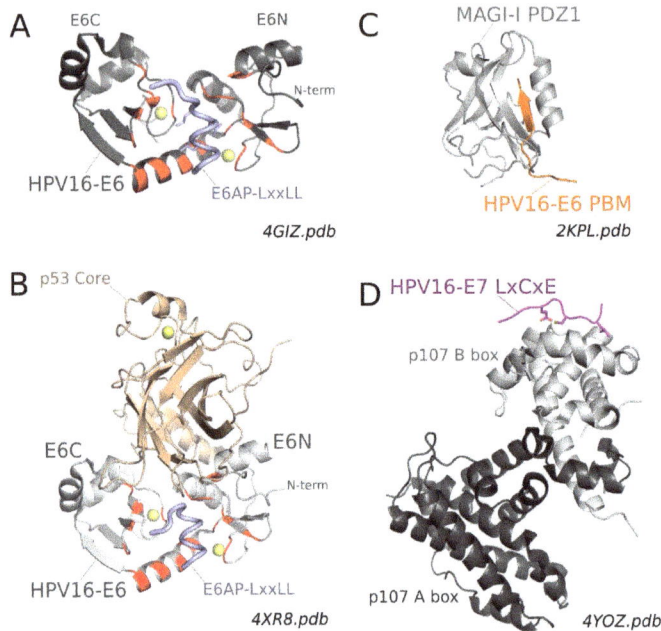

Figure 4. Main structural data on HPV16 E6 and E7 bound to cellular targets. (**A**) Crystal structure of full-length HPV16 E6 bound to the LxxLL motif of E6AP [75] LxxLL binding residues are highlighted in red; (**B**) Crystal structure of HPV16 E6 bound to the LxxLL motif of E6AP and the core domain of p53 [98]. (**C**) Solution structure of MAGI-1 (membrane associated guanylate kinase inverted 1) PDZ 2/6 bound to the C-terminal PDZ-Binding Motif (PBM) of HPV16 E6 [115]. LxxLL binding residues are highlighted in red; (**D**) Crystal structure of p107 pocket domain (composed of A-box and B-box) bound to the LxCxE motif of HPV16 E7 [116]. A comparable crystal structure was previously obtained for pRb pocket domain bound to the LxCxE motif of HPV16 E7 [117].

4. Structure and Specificity of E6-Target Complexes

4.1. Structure of Full-Length Mammalian E6 Proteins Bound to Target LxxLL Motifs

All mammalian PV E6 proteins recognize peptides presenting the LxxLL consensus sequence interspaced with acidic residues. To date, two X-ray structures of E6 proteins bound to such acidic LxxLL motifs have been solved [75]. First, the structure of a soluble mutant of HPV16 E6, complexed to a construct including the LxxLL motif of the E6AP ubiquitin ligase fused to the C-terminus of bacterial MBP, was solved at a 2.6 Å resolution (Figure 4A). Second, the structure of a triple fusion construct including bacterial MBP, the LxxLL motif of the cellular focal adhesion protein paxillin and the E6 protein of Bovine papillomavirus BPV1 was solved at a 2.3 Å resolution [75].

Despite the low sequence identity (30%) of the two E6-LxxLL complexes, their overall structures were very similar [75]. This strongly supports the notion that LxxLL recognition is a conserved structural property of E6 proteins throughout mammalian papillomaviruses [118], a view corroborated by recent proteomic data on a variety of HPV E6 proteins, that were all found to pull-down LxxLL motif-containing host proteins [31,32,34,36].

As discussed before, the BPV1 and HPV16 E6C domains have a similar fold while the two E6N domains have structurally resolved regions that are essentially superposable but differ in their N-terminal regions, partly due to the fact that the N-terminus of BPV1 E6 was unresolved in the crystal [75]. The LxxLL motif is bound in the form of a helix, which inserts within a pocket formed by the two domains and the linker helix of E6 (Figure 4A). This conserved general mode of recognition relies on the conserved secondary structure topology of the two E6 proteins, and on the conservation of a few key positions that could be identified [75]. Most logically, E6 residues conserved for general recognition were found to interact with either backbone atoms of the LxxLL peptides or side chain atoms of the three invariant Leucine residues.

Comparative examination of the complexes also allowed for identification of subtle amino acid differences at other positions, dubbed "reader" positions, that dictate the discriminative preferences of E6 from HPV16 and BPV1 for the LxxLL motifs of E6AP and Paxillin, respectively [75]. At least two regions of E6 were found to be critical for discriminative "reading" of LxxLL subsets. The C-terminus of the E6 linker helix and the N terminus of E6N domain were suggested to influence the selection of residues at the N or C terminus of the peptide, respectively [75]. Conversely, despite being conserved in HPV16 E6 and BPV1 E6, some reader arginine residues were found to establish contacts with acidic side chains belonging to different turns of the bound helix [75]. In other words, the same reader residue in two different E6 proteins can read a distinct position in the target LxxLL motif. It therefore appears that E6 proteins from different mammalian PVs have undergone sequence evolution mechanisms, which have allowed them to conserve the general capacity to interact with LxxLL motifs while specializing, through subtle variations in their "reader" positions, for the capture of different panels of target proteins bearing variations of the LxxLL motif.

4.2. Why Do Apparently Unrelated E6 Target Proteins Contain a Conserved E6-Binding LxxLL Motif? The CBP-P300 Hypothesis

Most mammalian E6 proteins recognize acidic LxxLL motifs [75], yet E6 proteins of distinct PV species recognize different subcategories of these motifs and hence recruit distinct LxxLL-containing host proteins [32,36,75]. The LxxLL motifs targetted by E6 proteins are highly conserved in their respective cellular proteins across mammals and sometimes even across the whole vertebrate lineage. These LxxLL motifs are unlikely to be conserved in diverse host proteins for the sole purpose of binding to viral E6. Rather, they may participate in a common host function, whose perturbation is useful to the viral lifecycle. LxxLL motifs are frequent in transcriptional co-activators, where they mediate crucial protein–protein interactions [119,120]. Remarkably, the best characterized LxxLL-containing E6 targets, i.e., E6AP [121], p53 [122], MAML1 [123], IRF3 [124], paxillin [125] and its close paralog ARA55 (Androgen Receptor Associated protein 55) [126], hADA3 (human Transcriptional Adapter 3) [127] and CCR4-Not complex [128], are all involved in transcriptional activation or co-activation. Moreover, most of them have been shown to interact with the CBP-p300 protein [121–124,126–128], which is itself a direct target of some E6 proteins, such as those of HPV16 and β1 HPVs [36,129]. CBP-p300 is a central transcriptional co-activator with acetyltransferase activity involved in numerous functions, including the host innate antiviral response [124,130]. Therefore perturbation of CBP-P300 activity is a consistent feature of many viruses [131]. Moreover, CBP-p300 is also involved in tumorigenic pathways [132]. Remarkably, CBP-p300 not only contains LxxLL motifs that bind to transcription factors [132,133], but it also recruits LxxLL motifs from co-activator proteins. In particular, the KIX domain of CBP-p300 modulates CBP-p300 activity by binding to acidic LxxLL motifs [134] reminiscent of those preferentially targeted by E6 proteins.

Based on the above-mentioned published observations, it is tempting to speculate that all mammalian E6 proteins share the ability to interfere with CBP-300 activity by directly interacting with CBP-300 or by capturing acidic LxxLL motifs within cellular partners of CBP-p300. This conserved ability to interfere with CBP-p300 activity would not only make sense for mammalian PVs to counteract innate immunity responses, but might also contribute to PV-induced oncogenesis. Indeed, the oncoprotein AML1-ETO, a fusion protein associated with acute myeloid leukemia (AML), competes with CBP-p300 for binding to the acidic LxxLL motif of E-proteins, a family of transcription factors that normally interact via their LxxLL motif with the KIX domain of CBP-p300 [134,135]. In a similar way, E6 oncoproteins might perform part of their tumorigenic action by competitively blocking the access of cellular acidic LxxLL motifs to CBP-p300. Besides, the acidic LxxLL motifs recognized by E6 proteins may also bind and regulate transcriptional activators or co-activators other than CBP-p300, that remain to be identified.

4.3. Structure of the Ternary E6/E6AP/p53 Complex

As discussed before, E6 of high-risk mucosal (hrm) HPVs binds to the LxxLL motif of cellular E3 ubiquitin ligase E6AP. The resulting E6/E6AP complex then recruits the p53 anti-apoptotic tumor suppressor protein, provoking its poly-ubiquitination and subsequent degradation by the proteasome [58]. The structural basis of the ternary E6/E6AP/p53 complex formation has been recently elucidated [98]. It was found that binding to a minimal 12-meric LxxLL E6AP-derived peptide was sufficient to render full-length E6 proficient for interaction with p53 [136]. Furthermore, the folded "core" domain (residues 94–292) of p53 was efficiently recruited by the pre-formed E6-LxxLL dimer. The crystal structure of the resulting E6-LxxLL-p53 core trimeric complex (involving bacterial Maltose Binding Protein fused to the LxxLL motif for solubility and crystallizability purposes) was solved at 2.25 Å resolution (Figure 4B). The structure of the E6-LxxLL subunit of the trimeric complex is practically superimposable to that of the E6-LxxLL complex previously solved [75] (Figure 4, compare panel A & B). The p53 core domain is recruited to this complex via a large interface essentially provided by E6. Whereas the presence of the LxxLL motif is required for E6 to recruit p53core, the LxxLL motif does not significantly interact with the p53core in the complex. This indicates that the main role of the LxxLL motif in p53 recruitment is to stabilize E6 in the LxxLL-bound conformation presenting the extended p53-binding interface. In light of our previous discussion of the solution structure of E6, this suggests that, while free E6 may transiently adopt in solution the conformation observed in the crystallized E6-LxxLL complexes, this conformation is not sufficiently sampled to allow for detectable p53 binding in the absence of the LxxLL motif of E6AP.

The p53-binding surface of E6 involves both E6N and E6C domains [98]. It is relatively large (1200 Å^2), in contrast with the rather weak affinity of the preformed E6-LxxLL complex for p53core (Kd = 22 μM). The weak affinity might be related to the predominance of polar contacts in the E6/p53 interface. In vivo, the affinity of p53 for E6/E6AP is likely to be enhanced by avidity effects, since full-length p53 and E6AP are known to form tetramers [137] and trimers [138], respectively. The E6-binding surface of p53 has not been seen before to be involved in binding to cellular p53 target molecules, including DNA. This suggests that E6 can recruit p53 and drive its degradation even when p53 is complexed to its natural targets.

The p53-binding surface of E6 comprises residues D44 and F47 of the E6N domain, which also mediate E6N dimerization and subsequent self-association of E6 [91]. Therefore, E6 self-association and E6 binding to p53 are two competing processes. However, this does not rule out the possibility that E6 might self-associate when it is not bound to p53.

While the assembly of E6, E6AP and p53 has been revealed, the structural mechanism of the subsequent poly-ubiquitination of p53 remains to be investigated. At least two possible models can be proposed. In one model, the sole formation of the triple complex is sufficient to place the E6AP HECT domain in atomic proximity of the ubiquitinable regions of p53, which thus becomes poly-ubiquitinated without further need of activating E6AP. In a second model, E6 not only induces the formation of the triple complex but also potentializes the ubiquitination activity of E6AP, for instance by inducing an activatory conformational change onto E6AP [139]. To decipher the full mechanism, further structural studies will be required, including the exploration of complexes involving larger parts of the 850 residues composing the E6AP protein. To date, high-resolution structural information on E6AP remains restricted to the N-terminal zinc-binding domain (~80 residues) [140], the central 12-meric LxxLL peptide in complex with E6 [75], and the C-terminal "HECT" domain (~350 residues) bearing the ubiquitin ligase activity [141].

4.4. Structural Basis of Hijacking of PDZ Domains and Rb Pocket Domains by High-Risk Mucosal HPVs

As discussed before, viral proteins often evolve mimics of host small linear interaction motifs, allowing them to competitively capture host domains and thereby perturb domain-motif networks carrying out particular biological functions [37,39]. The structural analysis of such interactions is facilitated by the fact that they do not necessarily require production of the full-length viral protein, since they can focus on complexes involving the isolated target domain(s) and short peptides corresponding to the viral motif.

As concerns the conserved PDZ-Binding Motif (PBM) found at the C-terminus of high-risk mucosal E6 proteins, several structures of the isolated E6 PBM have been solved by NMR and crystallography in complex with a variety of human PDZ domains: MAGI-1 PDZ2/6 domain [115,142], DLG1 (Disk Large homolog 1) PDZ2 and PDZ3 domains [78,101,142] and CAL (CFTR-associated ligand) PDZ domain [143]. Expectedly, all these structures demonstrated that, both in solution and in crystals, the viral E6 PBM interacts with the peptide-binding pocket of the PDZ domains following the "canonical binding mode" generally observed for host–host PDZ-PBM complexes (Figure 4C). For the strongest PDZ-E6 PBM complexes, binding affinities were in the micromolar (μM) range [22,71,78,101,115,144–147]. Subtle variations in binding thermodynamics and/or kinetics could be detected, that depended on the nature and length of amino acid extensions flanking either the N- and C-termini of the PDZ domains or the N-terminus of the PBM used for the interaction assays [101,115,146,147].

Of note, among the structures above mentioned, only the complexes involving PDZ domains from MAGI1 and DLG proteins correspond to interactions likely to be relevant in vivo. The E6-CAL complex [143] is probably physiologically irrelevant as the E6 PBM does not detectably interact with the CAL protein in in vitro assays. Indeed, this E6-CAL complex was investigated with the sole aim of observing at a high-resolution the unfavorable atomic interactions that are responsible for the very low affinity displayed by the E6 PBM for the CAL PDZ domains. Crystallization of this unnatural complex could, however, be achieved, thanks to the very high local concentrations reached during crystal formation. Therefore, the determination of a high-resolution crystal structure of a complex does not always warrant that this complex is relevant in vivo.

The human proteome comprises 266 human PDZ domains distributed over 152 proteins [71,148]. Due to the high degree of conservation of their structure and peptide-binding mode, PDZ domains can be relatively promiscuous, i.e., several PDZ domains may compete for binding to the same PBM-containing proteins. This illustrates the interest of recent studies [70,71] (see paragraph 2.5), that aimed to address in a quantitative way the preferences of E6 proteins towards the entire set of host PDZ domain-containing proteins. Such unbiased proteome-wide quantitative studies, combined with high-resolution structural data, will help understanding how subtle sequence variations of E6 proteins rule their preferences for different pools of target host proteins, thereby impacting the viral and pathogenic phenotypes.

As concerns the conserved LxCxE motif found in the N-terminal region of high-risk mucosal E7 proteins, two crystal structures of this isolated motif have been solved in complex with the Pocket domains of two different members of the Rb protein family: pRb [117] and p107 [116] (Figure 4D). These structures demonstrated that the viral E7 LxCxE interacts with the peptide binding groove of the pocket domains, in the same way as the equivalent LxCxE motifs from host proteins [116,117].

5. Exploitation of High-Resolution Structural Data for In Vivo Inhibition of E6 Oncogenic Activity

Despite their very recent release, the high-resolution structural data available on E6-target complexes have already started to be utilized for the exploration of potential therapeutic applications.

As discussed above, all high-risk mucosal E6 oncoproteins bind to PDZ domains and LxxLL motifs; both of these activities are crucial for HPV-induced oncogenesis; and high-resolution data are available on both E6-PDZ and E6-LxxLL complexes. This inspired two distinct research groups to design chimeric PDZ-LxxLL [22] and LxxLL-PDZ [149] fusion proteins. Both types of constructs were shown to act as strong bivalent E6 ligands displaying a nanomolar affinity. Furthermore, the PDZ-LxxLL chimera was shown to bind strongly and specifically to E6 proteins of all high-risk mucosal HPV types, and to provoke apoptotic death of HPV-positive cells derived from cervical tumors [22].

In another study, pep11**, a peptide displaying potent pro-apoptotic effects in HPV16-positive cells [24], has been probed by NMR for its specific interaction with full-length HPV16 E6 in solution [23]. The interaction was shown to be in the nanomolar affinity range, and to induce time-dependent aggregation and subsequent precipitation of E6. Prior knowledge of most backbone frequencies of HPV16 E6 [91] allowed for the mapping of the pep11**-binding surface on HPV16 E6. This surface was found to overlap with the E6AP-binding surface of E6, thereby providing a strong structural basis to the likely hypothesis that the pro-apoptotic effects of pep11** are mediated, at least in part, by its specific binding to HPV16 E6.

Finally, the structural data on E6/E6AP complexes have been utilized for in silico docking of small molecule E6 ligands displaying E6 inhibitory activity in the micromolar affinity range. It should be noted however, that the docking onto E6 3D structure did not serve the design of the small molecules, as it was only performed *a posteriori* to verify that the molecules characterized in this work could potentially be accommodated within the E6 pocket [28,30].

6. Conclusions

High-resolution structural analysis of the multifunctional E6 and E7 oncoproteins had been awaited for many years after their discovery. In the last decade, progress in production and solubilization of these proteins paved the way to breakthrough advances. The structures of zinc-binding domains of several E6 and E7 proteins have been solved by NMR and crystallography, revealing two zinc-binding folds probably derived from a common ancestor. Next, combined structural and biophysical data have depicted the strategies employed by E6 and E7 oncoproteins to hijack host domain-motif interaction networks. Several X-ray and NMR structures have described the mode of binding of short linear interaction motifs from both E6 and E7 to target host domains (PDZ and pocket domains, respectively). In parallel, quantitative proteomics approaches have been developed to identify and rank among the entire complement of human PDZ domains, those which bind best to each HPV E6 oncoprotein analyzed. Furthermore, X-ray structures of two E6-LxxLL complexes have described commonalities and differences in the recognition of host LxxLL motifs, a conserved property of most mammalian E6 proteins. Very recently, a trimer composed of HPV16 E6, the LxxLL motif of E6AP, and core domain of p53 has been solved by X-ray crystallography, depicting the mode of assembly of a viral p53 degradation complex.

Nonetheless, there are still exciting avenues to explore concerning the biophysical and structural characterization of HPV oncoprotein activities. It will be interesting to further investigate the molecular basis of papillomaviral motif/domain hijacking strategies by combining quantitative proteomic analysis of the LxxLL interactomes of diverse human or mammalian PV species, with the resolution of additional E6-LxxLL structures of particular interest, such as α-HPV E6/IRF3, β-HPV E6/MAML1 and β-HPV E6/CBP-p300. In these complexes, it will be important to extend the size of the host protein constructs involved by including, whenever possible, the full-length cellular proteins, rather than their sole LxxLL motifs. A particularly challenging aim will be to visualize at high resolution the p53 degradation complex including full-length E6 protein, full-length E6AP ligase (850 amino acids, possibly trimeric) and full-length p53 tetramer (4 × 393 amino acids). Last but not least, a research with high potential impact on human health will consist of exploiting the novel high-resolution structures of HPV oncoprotein–host protein complexes for the design and optimization of small molecule inhibitors of the pathogenic activities of HPVs.

Acknowledgments: The authors thank colleagues of the Travé team for helpful discussion, and former colleagues of the Oncoprotein team, particularly Katia Zanier, Sebastian Charbonnier and Yves Nominé, for their invaluable participation in the structural work cited in this review. The authors also thank the reviewers for careful reading, useful comments and expert editing of the manuscript. This work received institutional support from le Centre

Viruses **2018**, *10*, 37

National de la Recherche Scientifique (CNRS), Université de Strasbourg, Institut National de la Santé et de la Recherche Médicale (INSERM) and Région Alsace. The work was supported in part by grants from Ligue contre le Cancer (équipe labellisée 2015), National Institutes of Health (grant R01CA134737), Instruct (ESFRI), the French Infrastructure for Integrated Structural Biology (FRISBI) and Fondation pour La Recherche Medicale (fellowship to Irina Suarez). The authors declare that the content is solely their responsibility and does not represent the official views of the National Institutes of Health.

Conflicts of Interest: The authors declare no conflict of interest.

References

1. Harden, M.E.; Munger, K. Human papillomavirus molecular biology. *Mutat. Res. Rev. Mutat. Res.* **2017**, *772*, 3–12. [CrossRef] [PubMed]
2. Doorbar, J.; Quint, W.; Banks, L.; Bravo, I.G.; Stoler, M.; Broker, T.R.; Stanley, M.A. The biology and life-cycle of human papillomaviruses. *Vaccine* **2012**, *30*, F55–F70. [CrossRef] [PubMed]
3. Tommasino, M. The human papillomavirus family and its role in carcinogenesis. *Semin. Cancer Biol.* **2014**, *26*, 13–21. [CrossRef] [PubMed]
4. Rector, A.; Van Ranst, M. Animal papillomaviruses. *Virology* **2013**, *445*, 213–223. [CrossRef] [PubMed]
5. Lopez-Bueno, A.; Mavian, C.; Labella, A.M.; Castro, D.; Borrego, J.J.; Alcami, A.; Alejo, A. Concurrence of Iridovirus, Polyomavirus, and a Unique Member of a New Group of Fish Papillomaviruses in Lymphocystis Disease-Affected Gilthead Sea Bream. *J. Virol.* **2016**, *90*, 8768–8779. [CrossRef] [PubMed]
6. Van Doorslaer, K.; Ruoppolo, V.; Schmidt, A.; Lescroel, A.; Jongsomjit, D.; Elrod, M.; Kraberger, S.; Stainton, D.; Dugger, K.M.; Ballard, G.; et al. Unique genome organization of non-mammalian papillomaviruses provides insights into the evolution of viral early proteins. *Virus Evol.* **2017**, *3*, vex027. [CrossRef] [PubMed]
7. PaVE. PaVE: Papillomavirus Episteme. Available online: https://pave.niaid.nih.gov (accessed on 10 December 2017).
8. Van Doorslaer, K.; Li, Z.; Xirasagar, S.; Maes, P.; Kaminsky, D.; Liou, D.; Sun, Q.; Kaur, R.; Huyen, Y.; McBride, A.A. The Papillomavirus Episteme: A major update to the papillomavirus sequence database. *Nucleic Acids Res.* **2017**, *45*, D499–D506. [CrossRef] [PubMed]
9. Egawa, N.; Doorbar, J. The low-risk papillomaviruses. *Virus Res.* **2017**, *231*, 119–127. [CrossRef] [PubMed]
10. Forslund, O.; Johansson, H.; Madsen, K.G.; Kofoed, K. The nasal mucosa contains a large spectrum of human papillomavirus types from the Betapapillomavirus and Gammapapillomavirus genera. *J. Infect. Dis.* **2013**, *208*, 1335–1341. [CrossRef] [PubMed]
11. Tommasino, M. The biology of beta human papillomaviruses. *Virus Res.* **2017**, *231*, 128–138. [CrossRef] [PubMed]
12. Mittal, S.; Banks, L. Molecular mechanisms underlying human papillomavirus E6 and E7 oncoprotein-induced cell transformation. *Mutat. Res. Rev. Mutat. Res.* **2017**, *772*, 23–35. [CrossRef] [PubMed]
13. McBride, A.A. Playing with fire: Consequences of human papillomavirus DNA replication adjacent to genetically unstable regions of host chromatin. *Curr. Opin. Virol.* **2017**, *26*, 63–68. [CrossRef] [PubMed]
14. Moore, P.S.; Chang, Y. Why do viruses cause cancer? Highlights of the first century of human tumour virology. *Nat. Rev. Cancer* **2010**, *10*, 878–889. [CrossRef] [PubMed]
15. McBride, A.A. Oncogenic human papillomaviruses. *Philos. Trans. R. Soc. Lond. B Biol. Sci.* **2017**, *372*. [CrossRef] [PubMed]
16. Vande Pol, S.B.; Klingelhutz, A.J. Papillomavirus E6 oncoproteins. *Virology* **2013**, *445*, 115–137. [CrossRef] [PubMed]
17. Roman, A.; Munger, K. The papillomavirus E7 proteins. *Virology* **2013**, *445*, 138–168. [CrossRef] [PubMed]
18. Klingelhutz, A.J.; Roman, A. Cellular transformation by human papillomaviruses: Lessons learned by comparing high- and low-risk viruses. *Virology* **2012**, *424*, 77–98. [CrossRef] [PubMed]
19. Jiang, M.; Milner, J. Selective silencing of viral gene expression in HPV-positive human cervical carcinoma cells treated with siRNA, a primer of RNA interference. *Oncogene* **2002**, *21*, 6041–6048. [CrossRef] [PubMed]
20. Butz, K.; Ristriani, T.; Hengstermann, A.; Denk, C.; Scheffner, M.; Hoppe-Seyler, F. siRNA targeting of the viral E6 oncogene efficiently kills human papillomavirus-positive cancer cells. *Oncogene* **2003**, *22*, 5938–5945. [CrossRef] [PubMed]

21. Bonetta, A.C.; Mailly, L.; Robinet, E.; Trave, G.; Masson, M.; Deryckere, F. Artificial microRNAs against the viral E6 protein provoke apoptosis in HPV positive cancer cells. *Biochem. Biophys. Res. Commun.* **2015**, *465*, 658–664. [CrossRef] [PubMed]

22. Ramirez, J.; Poirson, J.; Foltz, C.; Chebaro, Y.; Schrapp, M.; Meyer, A.; Bonetta, A.; Forster, A.; Jacob, Y.; Masson, M.; et al. Targeting the Two Oncogenic Functional Sites of the HPV E6 Oncoprotein with a High-Affinity Bivalent Ligand. *Angew. Chem. Int. Ed. Engl.* **2015**, *54*, 7958–7962. [CrossRef] [PubMed]

23. Zanier, K.; Stutz, C.; Kintscher, S.; Reinz, E.; Sehr, P.; Bulkescher, J.; Hoppe-Seyler, K.; Trave, G.; Hoppe-Seyler, F. The E6AP binding pocket of the HPV16 E6 oncoprotein provides a docking site for a small inhibitory peptide unrelated to E6AP, indicating druggability of E6. *PLoS ONE* **2014**, *9*, e112514. [CrossRef] [PubMed]

24. Stutz, C.; Reinz, E.; Honegger, A.; Bulkescher, J.; Schweizer, J.; Zanier, K.; Trave, G.; Lohrey, C.; Hoppe-Seyler, K.; Hoppe-Seyler, F. Intracellular Analysis of the Interaction between the Human Papillomavirus Type 16 E6 Oncoprotein and Inhibitory Peptides. *PLoS ONE* **2015**, *10*, e0132339. [CrossRef] [PubMed]

25. Dymalla, S.; Scheffner, M.; Weber, E.; Sehr, P.; Lohrey, C.; Hoppe-Seyler, F.; Hoppe-Seyler, K. A novel peptide motif binding to and blocking the intracellular activity of the human papillomavirus E6 oncoprotein. *J. Mol. Med.* **2009**, *87*, 321–331. [CrossRef] [PubMed]

26. Courtete, J.; Sibler, A.P.; Zeder-Lutz, G.; Dalkara, D.; Oulad-Abdelghani, M.; Zuber, G.; Weiss, E. Suppression of cervical carcinoma cell growth by intracytoplasmic codelivery of anti-oncoprotein E6 antibody and small interfering RNA. *Mol. Cancer Ther.* **2007**, *6*, 1728–1735. [CrossRef] [PubMed]

27. Lagrange, M.; Boulade-Ladame, C.; Mailly, L.; Weiss, E.; Orfanoudakis, G.; Deryckere, F. Intracellular scFvs against the viral E6 oncoprotein provoke apoptosis in human papillomavirus-positive cancer cells. *Biochem. Biophys. Res. Commun.* **2007**, *361*, 487–492. [CrossRef] [PubMed]

28. Cherry, J.J.; Rietz, A.; Malinkevich, A.; Liu, Y.; Xie, M.; Bartolowits, M.; Davisson, V.J.; Baleja, J.D.; Androphy, E.J. Structure based identification and characterization of flavonoids that disrupt human papillomavirus-16 E6 function. *PLoS ONE* **2013**, *8*, e84506. [CrossRef] [PubMed]

29. Malecka, K.A.; Fera, D.; Schultz, D.C.; Hodawadekar, S.; Reichman, M.; Donover, P.S.; Murphy, M.E.; Marmorstein, R. Identification and Characterization of Small Molecule Human Papillomavirus E6 Inhibitors. *ACS Chem. Biol.* **2014**, *9*, 1603–1612. [CrossRef] [PubMed]

30. Rietz, A.; Petrov, D.P.; Bartolowits, M.; DeSmet, M.; Davisson, V.J.; Androphy, E.J. Molecular Probing of the HPV-16 E6 Protein Alpha Helix Binding Groove with Small Molecule Inhibitors. *PLoS ONE* **2016**, *11*, e0149845. [CrossRef] [PubMed]

31. White, E.A.; Howley, P.M. Proteomic approaches to the study of papillomavirus-host interactions. *Virology* **2013**, *435*, 57–69. [CrossRef] [PubMed]

32. Grace, M.; Munger, K. Proteomic analysis of the gamma human papillomavirus type 197 E6 and E7 associated cellular proteins. *Virology* **2017**, *500*, 71–81. [CrossRef] [PubMed]

33. Neveu, G.; Cassonnet, P.; Vidalain, P.O.; Rolloy, C.; Mendoza, J.; Jones, L.; Tangy, F.; Muller, M.; Demeret, C.; Tafforeau, L.; et al. Comparative analysis of virus-host interactomes with a mammalian high-throughput protein complementation assay based on Gaussia princeps luciferase. *Methods* **2012**, *58*, 349–359. [CrossRef] [PubMed]

34. Rozenblatt-Rosen, O.; Deo, R.C.; Padi, M.; Adelmant, G.; Calderwood, M.A.; Rolland, T.; Grace, M.; Dricot, A.; Askenazi, M.; Tavares, M.; et al. Interpreting cancer genomes using systematic host network perturbations by tumour virus proteins. *Nature* **2012**, *487*, 491–495. [CrossRef] [PubMed]

35. Wang, J.; Dupuis, C.; Tyring, S.K.; Underbrink, M.P. Sterile alpha Motif Domain Containing 9 Is a Novel Cellular Interacting Partner to Low-Risk Type Human Papillomavirus E6 Proteins. *PLoS ONE* **2016**, *11*, e0149859.

36. White, E.A.; Kramer, R.E.; Tan, M.J.; Hayes, S.D.; Harper, J.W.; Howley, P.M. Comprehensive analysis of host cellular interactions with human papillomavirus E6 proteins identifies new E6 binding partners and reflects viral diversity. *J. Virol.* **2012**, *86*, 13174–13186. [CrossRef] [PubMed]

37. Davey, N.E.; Trave, G.; Gibson, T.J. How viruses hijack cell regulation. *Trends Biochem. Sci.* **2011**, *36*, 159–169. [CrossRef] [PubMed]

38. Diella, F.; Haslam, N.; Chica, C.; Budd, A.; Michael, S.; Brown, N.P.; Trave, G.; Gibson, T.J. Understanding eukaryotic linear motifs and their role in cell signaling and regulation. *Front. Biosci.* **2008**, *13*, 6580–6603. [CrossRef] [PubMed]

39. Chemes, L.B.; de Prat-Gay, G.; Sanchez, I.E. Convergent evolution and mimicry of protein linear motifs in host-pathogen interactions. *Curr. Opin. Struct. Biol.* **2015**, *32*, 91–101. [CrossRef] [PubMed]
40. Lee, S.S.; Weiss, R.S.; Javier, R.T. Binding of human virus oncoproteins to hDlg/SAP97, a mammalian homolog of the *Drosophila* discs large tumor suppressor protein. *Proc. Natl. Acad. Sci. USA* **1997**, *94*, 6670–6675. [CrossRef] [PubMed]
41. Kiyono, T.; Hiraiwa, A.; Fujita, M.; Hayashi, Y.; Akiyama, T.; Ishibashi, M. Binding of high-risk human papillomavirus E6 oncoproteins to the human homologue of the *Drosophila* discs large tumor suppressor protein. *Proc. Natl. Acad. Sci. USA* **1997**, *94*, 11612–11616. [CrossRef] [PubMed]
42. Nakagawa, S.; Huibregtse, J.M. Human scribble (Vartul) is targeted for ubiquitin-mediated degradation by the high-risk papillomavirus E6 proteins and the E6AP ubiquitin-protein ligase. *Mol. Cell. Biol.* **2000**, *20*, 8244–8253. [CrossRef] [PubMed]
43. Glaunsinger, B.A.; Lee, S.S.; Thomas, M.; Banks, L.; Javier, R. Interactions of the PDZ-protein MAGI-1 with adenovirus E4-ORF1 and high-risk papillomavirus E6 oncoproteins. *Oncogene* **2000**, *19*, 5270–5280. [CrossRef] [PubMed]
44. Javier, R.T.; Rice, A.P. Emerging theme: Cellular PDZ proteins as common targets of pathogenic viruses. *J. Virol.* **2011**, *85*, 11544–11556. [CrossRef] [PubMed]
45. James, C.D.; Roberts, S. Viral Interactions with PDZ Domain-Containing Proteins-An Oncogenic Trait? *Pathogens* **2016**, *5*, 8. [CrossRef] [PubMed]
46. Banks, L.; Pim, D.; Thomas, M. Human tumour viruses and the deregulation of cell polarity in cancer. *Nat. Rev. Cancer* **2012**, *12*, 877–886. [CrossRef] [PubMed]
47. Huibregtse, J.M.; Scheffner, M.; Howley, P.M. Localization of the E6-AP regions that direct human papillomavirus E6 binding, association with p53, and ubiquitination of associated proteins. *Mol. Cell. Biol.* **1993**, *13*, 4918–4927. [CrossRef] [PubMed]
48. Chen, J.J.; Hong, Y.; Rustamzadeh, E.; Baleja, J.D.; Androphy, E.J. Identification of an alpha helical motif sufficient for association with papillomavirus E6. *J. Biol. Chem.* **1998**, *273*, 13537–13544. [CrossRef] [PubMed]
49. Vande Pol, S.B.; Brown, M.C.; Turner, C.E. Association of Bovine Papillomavirus Type 1 E6 oncoprotein with the focal adhesion protein paxillin through a conserved protein interaction motif. *Oncogene* **1998**, *16*, 43–52. [PubMed]
50. Bohl, J.; Das, K.; Dasgupta, B.; Vande Pol, S.B. Competitive binding to a charged leucine motif represses transformation by a papillomavirus E6 oncoprotein. *Virology* **2000**, *271*, 163–170. [CrossRef] [PubMed]
51. Tan, M.J.; White, E.A.; Sowa, M.E.; Harper, J.W.; Aster, J.C.; Howley, P.M. Cutaneous β-human papillomavirus E6 proteins bind Mastermind-like coactivators and repress Notch signaling. *Proc. Natl. Acad. Sci. USA* **2012**, *109*, E1473–E1480. [CrossRef] [PubMed]
52. Meyers, J.M.; Spangle, J.M.; Munger, K. The human papillomavirus type 8 E6 protein interferes with NOTCH activation during keratinocyte differentiation. *J. Virol.* **2013**, *87*, 4762–4767. [CrossRef] [PubMed]
53. Brimer, N.; Lyons, C.; Wallberg, A.E.; Vande Pol, S.B. Cutaneous papillomavirus E6 oncoproteins associate with MAML1 to repress transactivation and NOTCH signaling. *Oncogene* **2012**, *31*, 4639–4646. [CrossRef] [PubMed]
54. Ronco, L.V.; Karpova, A.Y.; Vidal, M.; Howley, P.M. Human papillomavirus 16 E6 oncoprotein binds to interferon regulatory factor-3 and inhibits its transcriptional activity. *Genes Dev.* **1998**, *12*, 2061–2072. [CrossRef] [PubMed]
55. Scheffner, M.; Whitaker, N.J. Human papillomavirus-induced carcinogenesis and the ubiquitin-proteasome system. *Semin. Cancer Biol.* **2003**, *13*, 59–67. [CrossRef]
56. Lou, Z.; Wang, S. E3 ubiquitin ligases and human papillomavirus-induced carcinogenesis. *J. Int. Med. Res.* **2014**, *42*, 247–260. [CrossRef] [PubMed]
57. Scheffner, M.; Huibregtse, J.M.; Vierstra, R.D.; Howley, P.M. The HPV-16 E6 and E6-AP complex functions as a ubiquitin-protein ligase in the ubiquitination of p53. *Cell* **1993**, *75*, 495–505. [CrossRef]
58. Scheffner, M.; Nuber, U.; Huibregtse, J.M. Protein ubiquitination involving an E1-E2-E3 enzyme ubiquitin thioester cascade. *Nature* **1995**, *373*, 81–83. [CrossRef] [PubMed]
59. Jones, D.L.; Thompson, D.A.; Münger, K. Destabilization of the RB tumor suppressor protein and stabilization of p53 contribute to HPV type 16 E7-induced apoptosis. *Virology* **1997**, *239*, 97–107. [CrossRef] [PubMed]

60. Boyer, S.N.; Wazer, D.E.; Band, V. E7 protein of human papilloma virus-16 induces degradation of retinoblastoma protein through the ubiquitin-proteasome pathway. *Cancer Res.* **1996**, *56*, 4620–4624. [PubMed]

61. Poirson, J.; Biquand, E.; Straub, M.L.; Cassonnet, P.; Nominé, Y.; Jones, L.; van der Werf, S.; Travé, G.; Zanier, K.; Jacob, Y.; et al. Mapping the interactome of HPV E6 and E7 oncoproteins with the ubiquitin-proteasome system. *FEBS J.* **2017**, *284*, 3171–3201. [CrossRef] [PubMed]

62. Cole, S.T.; Danos, O. Nucleotide sequence and comparative analysis of the human papillomavirus type 18 genome. Phylogeny of papillomaviruses and repeated structure of the E6 and E7 gene products. *J. Mol. Biol.* **1987**, *193*, 599–608. [CrossRef]

63. Mallon, R.G.; Wojciechowicz, D.; Defendi, V. DNA-binding activity of papillomavirus proteins. *J. Virol.* **1987**, *61*, 1655–1660. [PubMed]

64. Imai, Y.; Tsunokawa, Y.; Sugimura, T.; Terada, M. Purification and DNA-binding properties of human papillomavirus type 16 E6 protein expressed in *Escherichia coli*. *Biochem. Biophys. Res. Commun.* **1989**, *164*, 1402–1410. [CrossRef]

65. Ristriani, T.; Masson, M.; Nomine, Y.; Laurent, C.; Lefevre, J.F.; Weiss, E.; Trave, G. HPV oncoprotein E6 is a structure-dependent DNA-binding protein that recognizes four-way junctions. *J. Mol. Biol.* **2000**, *296*, 1189–1203. [CrossRef] [PubMed]

66. Ristriani, T.; Nomine, Y.; Masson, M.; Weiss, E.; Trave, G. Specific recognition of four-way DNA junctions by the C-terminal zinc-binding domain of HPV oncoprotein E6. *J. Mol. Biol.* **2001**, *305*, 729–739. [CrossRef] [PubMed]

67. Nomine, Y.; Masson, M.; Charbonnier, S.; Zanier, K.; Ristriani, T.; Deryckere, F.; Sibler, A.P.; Desplancq, D.; Atkinson, R.A.; Weiss, E.; et al. Structural and functional analysis of E6 oncoprotein: Insights in the molecular pathways of human papillomavirus-mediated pathogenesis. *Mol. Cell* **2006**, *21*, 665–678. [CrossRef] [PubMed]

68. Nomine, Y.; Charbonnier, S.; Ristriani, T.; Stier, G.; Masson, M.; Cavusoglu, N.; van Dorsselaer, A.; Weiss, E.; Kieffer, B.; Trave, G. Domain substructure of HPV E6 oncoprotein: Biophysical characterization of the E6 C-terminal DNA-binding domain. *Biochemistry* **2003**, *42*, 4909–4917. [CrossRef] [PubMed]

69. Bodaghi, S.; Jia, R.; Zheng, Z.M. Human papillomavirus type 16 E2 and E6 are RNA-binding proteins and inhibit in vitro splicing of pre-mRNAs with suboptimal splice sites. *Virology* **2009**, *386*, 32–43. [CrossRef] [PubMed]

70. Thomas, M.; Myers, M.P.; Massimi, P.; Guarnaccia, C.; Banks, L. Analysis of Multiple HPV E6 PDZ Interactions Defines Type-Specific PDZ Fingerprints that Predict Oncogenic Potential. *PLoS Pathog.* **2016**, *12*, e1005766. [CrossRef] [PubMed]

71. Vincentelli, R.; Luck, K.; Poirson, J.; Polanowska, J.; Abdat, J.; Blemont, M.; Turchetto, J.; Iv, F.; Ricquier, K.; Straub, M.L.; et al. Quantifying domain-ligand affinities and specificities by high-throughput holdup assay. *Nat. Methods* **2015**, *12*, 787–793. [CrossRef] [PubMed]

72. Nobre, R.J.; Herraez-Hernandez, E.; Fei, J.W.; Langbein, L.; Kaden, S.; Grone, H.J.; de Villiers, E.M. E7 oncoprotein of novel human papillomavirus type 108 lacking the E6 gene induces dysplasia in organotypic keratinocyte cultures. *J. Virol.* **2009**, *83*, 2907–2916. [CrossRef] [PubMed]

73. Van Doorslaer, K.; McBride, A.A. Molecular archeological evidence in support of the repeated loss of a papillomavirus gene. *Sci. Rep.* **2016**, *6*, 33028. [CrossRef] [PubMed]

74. Lipari, F.; McGibbon, G.A.; Wardrop, E.; Cordingley, M.G. Purification and biophysical characterization of a minimal functional domain and of an N-terminal Zn2+-binding fragment from the human papillomavirus type 16 E6 protein. *Biochemistry* **2001**, *40*, 1196–1204. [CrossRef] [PubMed]

75. Zanier, K.; Charbonnier, S.; Sidi, A.O.; McEwen, A.G.; Ferrario, M.G.; Poussin-Courmontagne, P.; Cura, V.; Brimer, N.; Babah, K.O.; Ansari, T.; et al. Structural Basis for Hijacking of Cellular LxxLL Motifs by Papillomavirus E6 Oncoproteins. *Science* **2013**, *339*, 694–698. [CrossRef] [PubMed]

76. Ohlenschlager, O.; Seiboth, T.; Zengerling, H.; Briese, L.; Marchanka, A.; Ramachandran, R.; Baum, M.; Korbas, M.; Meyer-Klaucke, W.; Durst, M.; et al. Solution structure of the partially folded high-risk human papilloma virus 45 oncoprotein E7. *Oncogene* **2006**, *25*, 5953–5959. [CrossRef] [PubMed]

77. Garcia-Alai, M.M.; Alonso, L.G.; de Prat-Gay, G. The N-terminal module of HPV16 E7 is an intrinsically disordered domain that confers conformational and recognition plasticity to the oncoprotein. *Biochemistry* **2007**, *46*, 10405–10412. [CrossRef] [PubMed]

78. Liu, X.; Clements, A.; Zhao, K.; Marmorstein, R. Structure of the human Papillomavirus E7 oncoprotein and its mechanism for inactivation of the retinoblastoma tumor suppressor. *J. Biol. Chem.* **2006**, *281*, 578–586. [CrossRef] [PubMed]

79. Van Doorslaer, K.; Sidi, A.O.; Zanier, K.; Rybin, V.; Deryckere, F.; Rector, A.; Burk, R.D.; Lienau, E.K.; van Ranst, M.; Trave, G. Identification of unusual E6 and E7 proteins within avian papillomaviruses: Cellular localization, biophysical characterization, and phylogenetic analysis. *J. Virol.* **2009**, *83*, 8759–8770. [CrossRef] [PubMed]

80. Herbst, L.H.; Lenz, J.; van Doorslaer, K.; Chen, Z.; Stacy, B.A.; Wellehan, J.F., Jr.; Manire, C.A.; Burk, R.D. Genomic characterization of two novel reptilian papillomaviruses, *Chelonia mydas* papillomavirus 1 and *Caretta caretta* papillomavirus 1. *Virology* **2009**, *383*, 131–135. [CrossRef] [PubMed]

81. Danos, O.; Georges, E.; Orth, G.; Yaniv, M. Fine structure of the cottontail rabbit papillomavirus mRNAs expressed in the transplantable VX2 carcinoma. *J. Virol.* **1985**, *53*, 735–741. [PubMed]

82. Georges, E.; Croissant, O.; Bonneaud, N.; Orth, G. Physical state and transcription of the cottontail rabbit papillomavirus genome in warts and transplantable VX2 and VX7 carcinomas of domestic rabbits. *J. Virol.* **1984**, *51*, 530–538. [PubMed]

83. Sarver, N.; Rabson, M.S.; Yang, Y.C.; Byrne, J.C.; Howley, P.M. Localization and analysis of bovine papillomavirus type 1 transforming functions. *J. Virol.* **1984**, *52*, 377–388. [PubMed]

84. Schiller, J.T.; Vass, W.C.; Lowy, D.R. Identification of a second transforming region in bovine papillomavirus DNA. *Proc. Natl. Acad. Sci. USA* **1984**, *81*, 7880–7884. [CrossRef] [PubMed]

85. Schwarz, E.; Freese, U.K.; Gissmann, L.; Mayer, W.; Roggenbuck, B.; Stremlau, A.; zur Hausen, H. Structure and transcription of human papillomavirus sequences in cervical carcinoma cells. *Nature* **1985**, *314*, 111–114. [CrossRef] [PubMed]

86. Androphy, E.J.; Schiller, J.T.; Lowy, D.R. Identification of the protein encoded by the E6 transforming gene of bovine papillomavirus. *Science* **1985**, *230*, 442–445. [CrossRef] [PubMed]

87. Matlashewski, G.; Banks, L.; Wu-Liao, J.; Spence, P.; Pim, D.; Crawford, L. The expression of human papillomavirus type 18 E6 protein in bacteria and the production of anti-E6 antibodies. *J. Gen. Virol.* **1986**, *67*, 1909–1916. [CrossRef] [PubMed]

88. Seedorf, K.; Oltersdorf, T.; Krammer, G.; Rowekamp, W. Identification of early proteins of the human papilloma viruses type 16 (HPV 16) and type 18 (HPV 18) in cervical carcinoma cells. *EMBO J.* **1987**, *6*, 139–144. [PubMed]

89. Smotkin, D.; Wettstein, F.O. Transcription of human papillomavirus type 16 early genes in a cervical cancer and a cancer-derived cell line and identification of the E7 protein. *Proc. Natl. Acad. Sci. USA* **1986**, *83*, 4680–4684. [CrossRef] [PubMed]

90. Liu, Y.; Cherry, J.J.; Dineen, J.V.; Androphy, E.J.; Baleja, J.D. Determinants of stability for the E6 protein of papillomavirus type 16. *J. Mol. Biol.* **2009**, *386*, 1123–1137. [CrossRef] [PubMed]

91. Zanier, K.; ould M'hamed ould Sidi, A.; Boulade-Ladame, C.; Rybin, V.; Chappelle, A.; Atkinson, A.; Kieffer, B.; Trave, G. Solution structure analysis of the HPV16 E6 oncoprotein reveals a self-association mechanism required for E6-mediated degradation of p53. *Structure* **2012**, *20*, 604–617. [CrossRef] [PubMed]

92. Nomine, Y.; Ristriani, T.; Laurent, C.; Lefevre, J.F.; Weiss, E.; Trave, G. Formation of soluble inclusion bodies by HPV E6 oncoprotein fused to maltose-binding protein. *Protein Expr. Purif.* **2001**, *23*, 22–32. [CrossRef] [PubMed]

93. Nomine, Y.; Ristriani, T.; Laurent, C.; Lefevre, J.F.; Weiss, E.; Trave, G. A strategy for optimizing the monodispersity of fusion proteins: Application to purification of recombinant HPV E6 oncoprotein. *Protein Eng.* **2001**, *14*, 297–305. [CrossRef] [PubMed]

94. Zanier, K.; Nominé, Y.; Charbonnier, S.; Ruhlmann, C.; Schultz, P.; Schweizer, J.; Travé, G. Formation of well-defined soluble aggregates upon fusion to MBP is a generic property of E6 proteins from various human papillomavirus species. *Protein Expr. Purif.* **2007**, *51*, 59–70. [CrossRef] [PubMed]

95. Sidi, A.O.; Babah, K.O.; Brimer, N.; Nomine, Y.; Romier, C.; Kieffer, B.; Pol, S.V.; Trave, G.; Zanier, K. Strategies for bacterial expression of protein-peptide complexes: Application to solubilization of papillomavirus E6. *Protein Expr. Purif.* **2011**, *80*, 8–16. [CrossRef] [PubMed]

96. García-Alai, M.M.; Dantur, K.I.; Smal, C.; Pietrasanta, L.; de Prat-Gay, G. High-risk HPV E6 oncoproteins assemble into large oligomers that allow localization of endogenous species in prototypic HPV-transformed cell lines. *Biochemistry* **2007**, *46*, 341–349. [CrossRef] [PubMed]

97. Zanier, K.; Ruhlmann, C.; Melin, F.; Masson, M.; Ould M'hamed Ould Sidi, A.; Bernard, X.; Fischer, B.; Brino, L.; Ristriani, T.; Rybin, V.; et al. E6 proteins from diverse papillomaviruses self-associate both in vitro and in vivo. *J. Mol. Biol.* **2010**, *396*, 90–104. [CrossRef] [PubMed]

98. Martinez-Zapien, D.; Ruiz, F.X.; Poirson, J.; Mitschler, A.; Ramirez, J.; Forster, A.; Cousido-Siah, A.; Masson, M.; Vande Pol, S.; Podjarny, A.; et al. Structure of the E6/E6AP/p53 complex required for HPV-mediated degradation of p53. *Nature* **2016**, *529*, 541–545. [CrossRef] [PubMed]

99. Dalal, S.; Gao, Q.; Androphy, E.J.; Band, V. Mutational analysis of human papillomavirus type 16 E6 demonstrates that p53 degradation is necessary for immortalization of mammary epithelial cells. *J. Virol.* **1996**, *70*, 683–688. [PubMed]

100. Cooper, B.; Schneider, S.; Bohl, J.; Jiang, Y.; Beaudet, A.; Vande Pol, S. Requirement of E6AP and the features of human papillomavirus E6 necessary to support degradation of p53. *Virology* **2003**, *306*, 87–99. [CrossRef]

101. Mischo, A.; Ohlenschläger, O.; Hortschansky, P.; Ramachandran, R.; Görlach, M. Structural insights into a wildtype domain of the oncoprotein E6 and its interaction with a PDZ domain. *PLoS ONE* **2013**, *8*, e62584. [CrossRef] [PubMed]

102. Alonso, L.G.; Garcia-Alai, M.M.; Nadra, A.D.; Lapena, A.N.; Almeida, F.L.; Gualfetti, P.; Prat-Gay, G.D. High-risk (HPV16) human papillomavirus E7 oncoprotein is highly stable and extended, with conformational transitions that could explain its multiple cellular binding partners. *Biochemistry* **2002**, *41*, 10510–10518. [CrossRef] [PubMed]

103. Jansma, A.L.; Martinez-Yamout, M.A.; Liao, R.; Sun, P.; Dyson, H.J.; Wright, P.E. The high-risk HPV16 E7 oncoprotein mediates interaction between the transcriptional coactivator CBP and the retinoblastoma protein pRb. *J. Mol. Biol.* **2014**, *426*, 4030–4048. [CrossRef] [PubMed]

104. Nogueira, M.O.; Hosek, T.; Calcada, E.O.; Castiglia, F.; Massimi, P.; Banks, L.; Felli, I.C.; Pierattelli, R. Monitoring HPV-16 E7 phosphorylation events. *Virology* **2017**, *503*, 70–75. [CrossRef] [PubMed]

105. De Souza, R.F.; Iyer, L.M.; Aravind, L. Diversity and evolution of chromatin proteins encoded by DNA viruses. *Biochim. Biophys. Acta* **2010**, *1799*, 302–318. [CrossRef] [PubMed]

106. Grishin, N.V. Treble clef finger—A functionally diverse zinc-binding structural motif. *Nucleic Acids Res.* **2001**, *29*, 1703–1714. [CrossRef] [PubMed]

107. Kaur, G.; Subramanian, S. Classification of the treble clef zinc finger: Noteworthy lessons for structure and function evolution. *Sci. Rep.* **2016**, *6*, 32070. [CrossRef] [PubMed]

108. Camporeale, G.; Lorenzo, J.R.; Thomas, M.G.; Salvatierra, E.; Borkosky, S.S.; Risso, M.G.; Sanchez, I.E.; de Prat Gay, G.; Alonso, L.G. Degenerate cysteine patterns mediate two redox sensing mechanisms in the papillomavirus E7 oncoprotein. *Redox Biol.* **2017**, *11*, 38–50. [CrossRef] [PubMed]

109. Chemes, L.B.; Camporeale, G.; Sanchez, I.E.; de Prat-Gay, G.; Alonso, L.G. Cysteine-rich positions outside the structural zinc motif of human papillomavirus E7 provide conformational modulation and suggest functional redox roles. *Biochemistry* **2014**, *53*, 1680–1696. [CrossRef] [PubMed]

110. Smal, C.; Alonso, L.G.; Wetzler, D.E.; Heer, A.; de Prat Gay, G. Ordered self-assembly mechanism of a spherical oncoprotein oligomer triggered by zinc removal and stabilized by an intrinsically disordered domain. *PLoS ONE* **2012**, *7*, e36457. [CrossRef] [PubMed]

111. Alonso, L.G.; Garcia-Alai, M.M.; Smal, C.; Centeno, J.M.; Iacono, R.; Castano, E.; Gualfetti, P.; de Prat-Gay, G. The HPV16 E7 viral oncoprotein self-assembles into defined spherical oligomers. *Biochemistry* **2004**, *43*, 3310–3317. [CrossRef] [PubMed]

112. Cerutti, M.L.; Alonso, L.G.; Tatti, S.; de Prat-Gay, G. Long-lasting immunoprotective and therapeutic effects of a hyperstable E7 oligomer based vaccine in a murine human papillomavirus tumor model. *Int. J. Cancer* **2012**, *130*, 1813–1820. [CrossRef] [PubMed]

113. Weikl, T.R.; Paul, F. Conformational selection in protein binding and function. *Protein Sci.* **2014**, *23*, 1508–1518. [CrossRef] [PubMed]

114. Masson, M.; Hindelang, C.; Sibler, A.P.; Schwalbach, G.; Trave, G.; Weiss, E. Preferential nuclear localization of the human papillomavirus type 16 E6 oncoprotein in cervical carcinoma cells. *J. Gen. Virol.* **2003**, *84*, 2099–2104. [CrossRef] [PubMed]

115. Charbonnier, S.; Nominé, Y.; Ramírez, J.; Luck, K.; Chapelle, A.; Stote, R.H.; Travé, G.; Kieffer, B.; Atkinson, R.A. The Structural and Dynamic Response of MAGI-1 PDZ1 with Noncanonical Domain Boundaries to the Binding of Human Papillomavirus E6. *J. Mol. Biol.* **2011**, *406*, 745–763. [CrossRef] [PubMed]

116. Guiley, K.Z.; Liban, T.J.; Felthousen, J.G.; Ramanan, P.; Litovchick, L.; Rubin, S.M. Structural mechanisms of DREAM complex assembly and regulation. *Genes Dev.* **2015**, *29*, 961–974. [CrossRef] [PubMed]

117. Lee, J.O.; Russo, A.A.; Pavletich, N.P. Structure of the retinoblastoma tumour-suppressor pocket domain bound to a peptide from HPV E7. *Nature* **1998**, *391*, 859–865. [CrossRef] [PubMed]

118. Vande Pol, S. Papillomavirus E6 Oncoproteins Take Common Structural Approaches to Solve Different Biological Problems. *PLoS Pathog.* **2015**, *11*, e1005138. [CrossRef] [PubMed]

119. Heery, D.M.; Kalkhoven, E.; Hoare, S.; Parker, M.G. A signature motif in transcriptional co-activators mediates binding to nuclear receptors. *Nature* **1997**, *387*, 733–736. [CrossRef] [PubMed]

120. Plevin, M.J.; Mills, M.M.; Ikura, M. The LxxLL motif: A multifunctional binding sequence in transcriptional regulation. *Trends Biochem. Sci.* **2005**, *30*, 66–69. [CrossRef] [PubMed]

121. Catoe, H.W.; Nawaz, Z. E6-AP facilitates efficient transcription at estrogen responsive promoters through recruitment of chromatin modifiers. *Steroids* **2011**, *76*, 897–902. [CrossRef] [PubMed]

122. Krois, A.S.; Ferreon, J.C.; Martinez-Yamout, M.A.; Dyson, H.J.; Wright, P.E. Recognition of the disordered p53 transactivation domain by the transcriptional adapter zinc finger domains of CREB-binding protein. *Proc. Natl. Acad. Sci. USA* **2016**, *113*, E1853–E1862. [CrossRef] [PubMed]

123. Wallberg, A.E.; Pedersen, K.; Lendahl, U.; Roeder, R.G. p300 and PCAF Act Cooperatively to Mediate Transcriptional Activation from Chromatin Templates by Notch Intracellular Domains In Vitro. *Mol. Cell. Biol.* **2002**, *22*, 7812–7819. [CrossRef] [PubMed]

124. Qin, B.Y.; Liu, C.; Srinath, H.; Lam, S.S.; Correia, J.J.; Derynck, R.; Lin, K. Crystal structure of IRF-3 in complex with CBP. *Structure* **2005**, *13*, 1269–1277. [CrossRef] [PubMed]

125. Sen, A.; de Castro, I.; Defranco, D.B.; Deng, F.M.; Melamed, J.; Kapur, P.; Raj, G.V.; Rossi, R.; Hammes, S.R. Paxillin mediates extranuclear and intranuclear signaling in prostate cancer proliferation. *J. Clin. Investig.* **2012**, *122*, 2469–2481. [CrossRef] [PubMed]

126. Bengtsen, M.; Sorensen, L.; Aabel, L.; Ledsaak, M.; Matre, V.; Gabrielsen, O.S. The adaptor protein ARA55 and the nuclear kinase HIPK1 assist c-Myb in recruiting p300 to chromatin. *Biochim. Biophys. Acta* **2017**, *1860*, 751–760. [CrossRef] [PubMed]

127. Germaniuk-Kurowska, A.; Nag, A.; Zhao, X.; Dimri, M.; Band, H.; Band, V. Ada3 requirement for HAT recruitment to estrogen receptors and estrogen-dependent breast cancer cell proliferation. *Cancer Res.* **2007**, *67*, 11789–11797. [CrossRef] [PubMed]

128. Sharma, S.; Poetz, F.; Bruer, M.; Ly-Hartig, T.B.; Schott, J.; Seraphin, B.; Stoecklin, G. Acetylation-Dependent Control of Global Poly(A) RNA Degradation by CBP/p300 and HDAC1/2. *Mol. Cell* **2016**, *63*, 927–938. [CrossRef] [PubMed]

129. Howie, H.L.; Koop, J.I.; Weese, J.; Robinson, K.; Wipf, G.; Kim, L.; Galloway, D.A. Beta-HPV 5 and 8 E6 promote p300 degradation by blocking AKT/p300 association. *PLoS Pathog.* **2011**, *7*, e1002211. [CrossRef] [PubMed]

130. Taylor, K.E.; Mossman, K.L. Recent advances in understanding viral evasion of type I interferon. *Immunology* **2013**, *138*, 190–197. [CrossRef] [PubMed]

131. Hottiger, M.O.; Nabel, G.J. Viral replication and the coactivators p300 and CBP. *Trends Microbiol.* **2000**, *8*, 560–565. [CrossRef]

132. Wang, F.; Marshall, C.B.; Ikura, M. Transcriptional/epigenetic regulator CBP/p300 in tumorigenesis: Structural and functional versatility in target recognition. *Cell. Mol. Life Sci.* **2013**, *70*, 3989–4008. [CrossRef] [PubMed]

133. Heery, D.M.; Hoare, S.; Hussain, S.; Parker, M.G.; Sheppard, H. Core LXXLL motif sequences in CREB-binding protein, SRC1, and RIP140 define affinity and selectivity for steroid and retinoid receptors. *J. Biol. Chem.* **2001**, *276*, 6695–6702. [CrossRef] [PubMed]

134. Denis, C.M.; Chitayat, S.; Plevin, M.J.; Wang, F.; Thompson, P.; Liu, S.; Spencer, H.L.; Ikura, M.; LeBrun, D.P.; Smith, S.P. Structural basis of CBP/p300 recruitment in leukemia induction by E2A-PBX1. *Blood* **2012**, *120*, 3968–3977. [CrossRef] [PubMed]

135. Zhang, J.; Kalkum, M.; Yamamura, S.; Chait, B.T.; Roeder, R.G. E protein silencing by the leukemogenic AML1-ETO fusion protein. *Science* **2004**, *305*, 1286–1289. [CrossRef] [PubMed]

136. Ansari, T.; Brimer, N.; Vande Pol, S.B. Peptide Interactions Stabilize and Restructure Human Papillomavirus Type 16 E6 to Interact with p53. *J. Virol.* **2012**, *86*, 11386–11391. [CrossRef] [PubMed]

137. Joerger, A.C.; Fersht, A.R. Structural biology of the tumor suppressor p53. *Annu. Rev. Biochem.* **2008**, *77*, 557–582. [CrossRef] [PubMed]

138. Ronchi, V.P.; Klein, J.M.; Edwards, D.J.; Haas, A.L. The active form of E6-associated protein (E6AP)/UBE3A ubiquitin ligase is an oligomer. *J. Biol. Chem.* **2014**, *289*, 1033–1048. [CrossRef] [PubMed]

139. Mortensen, F.; Schneider, D.; Barbic, T.; Sladewska-Marquardt, A.; Kuhnle, S.; Marx, A.; Scheffner, M. Role of ubiquitin and the HPV E6 oncoprotein in E6AP-mediated ubiquitination. *Proc. Natl. Acad. Sci. USA* **2015**, *112*, 9872–9877. [CrossRef] [PubMed]

140. Lemak, A.; Yee, A.; Bezsonova, I.; Dhe-Paganon, S.; Arrowsmith, C.H. Zn-binding AZUL domain of human ubiquitin protein ligase Ube3A. *J. Biomol. NMR* **2011**, *51*, 185–190. [CrossRef] [PubMed]

141. Huang, L.; Kinnucan, E.; Wang, G.; Beaudenon, S.; Howley, P.M.; Huibregtse, J.M.; Pavletich, N.P. Structure of an E6AP-UbcH7 Complex: Insights into Ubiquitination by the E2-E3 Enzyme Cascade. *Science* **1999**, *286*, 1321–1326. [CrossRef] [PubMed]

142. Zhang, Y.; Dasgupta, J.; Ma, R.Z.; Banks, L.; Thomas, M.; Chen, X.S. Structures of a human papillomavirus (HPV) E6 polypeptide bound to MAGUK proteins: Mechanisms of targeting tumor suppressors by a high-risk HPV oncoprotein. *J. Virol.* **2007**, *81*, 3618–3626. [CrossRef] [PubMed]

143. Amacher, J.F.; Cushing, P.R.; Brooks, L., 3rd; Boisguerin, P.; Madden, D.R. Stereochemical preferences modulate affinity and selectivity among five PDZ domains that bind CFTR: Comparative structural and sequence analyses. *Structure* **2014**, *22*, 82–93. [CrossRef] [PubMed]

144. Fournane, S.; Charbonnier, S.; Chapelle, A.; Kieffer, B.; Orfanoudakis, G.; Trave, G.; Masson, M.; Nomine, Y. Surface plasmon resonance analysis of the binding of high-risk mucosal HPV E6 oncoproteins to the PDZ1 domain of the tight junction protein MAGI-1. *J. Mol. Recognit.* **2011**, *24*, 511–523. [CrossRef] [PubMed]

145. Luck, K.; Fournane, S.; Kieffer, B.; Masson, M.; Nomine, Y.; Trave, G. Putting into practice domain-linear motif interaction predictions for exploration of protein networks. *PLoS ONE* **2011**, *6*, e25376. [CrossRef] [PubMed]

146. Chi, C.N.; Bach, A.; Engstrom, A.; Stromgaard, K.; Lundstrom, P.; Ferguson, N.; Jemth, P. Biophysical characterization of the complex between human papillomavirus E6 protein and synapse-associated protein 97. *J. Biol. Chem.* **2011**, *286*, 3597–3606. [CrossRef] [PubMed]

147. Ramirez, J.; Recht, R.; Charbonnier, S.; Ennifar, E.; Atkinson, R.A.; Trave, G.; Nomine, Y.; Kieffer, B. Disorder-to-order transition of MAGI-1 PDZ1 C-terminal extension upon peptide binding: Thermodynamic and dynamic insights. *Biochemistry* **2015**, *54*, 1327–1337. [CrossRef] [PubMed]

148. Luck, K.; Charbonnier, S.; Trave, G. The emerging contribution of sequence context to the specificity of protein interactions mediated by PDZ domains. *FEBS Lett.* **2012**, *586*, 2648–2661. [CrossRef] [PubMed]

149. Karlsson, O.A.; Ramirez, J.; Oberg, D.; Malmqvist, T.; Engstrom, A.; Friberg, M.; Chi, C.N.; Widersten, M.; Trave, G.; Nilsson, M.T.; et al. Design of a PDZbody, a bivalent binder of the E6 protein from human papillomavirus. *Sci. Rep.* **2015**, *5*, 9382. [CrossRef] [PubMed]

viruses

MDPI

Review

Subcellular Trafficking of the Papillomavirus Genome during Initial Infection: The Remarkable Abilities of Minor Capsid Protein L2

Samuel K. Campos [1,2,3,4]

1   The Department of Immunobiology, The University of Arizona, Tucson, AZ 85721-0240, USA;
    skcampos@email.arizona.edu; Tel.: +1-520-626-4842
2   The Department of Molecular & Cellular Biology, The University of Arizona, Tucson, AZ 85721-0240, USA
3   The Cancer Biology Graduate Interdisciplinary Program, The University of Arizona,
    Tucson, AZ 85721-0240, USA
4   The BIO5 Institute, Tucson, AZ 85721-0240, USA

Received: 31 October 2017; Accepted: 2 December 2017; Published: 3 December 2017

Abstract: Since 2012, our understanding of human papillomavirus (HPV) subcellular trafficking has undergone a drastic paradigm shift. Work from multiple laboratories has revealed that HPV has evolved a unique means to deliver its viral genome (vDNA) to the cell nucleus, relying on myriad host cell proteins and processes. The major breakthrough finding from these recent endeavors has been the realization of L2-dependent utilization of cellular sorting factors for the retrograde transport of vDNA away from degradative endo/lysosomal compartments to the Golgi, prior to mitosis-dependent nuclear accumulation of L2/vDNA. An overview of current models of HPV entry, subcellular trafficking, and the role of L2 during initial infection is provided below, highlighting unresolved questions and gaps in knowledge.

Keywords: human papillomavirus; HPV16; L2; subcellular trafficking; mitosis; transmembrane domain; translocation; membrane penetration; toxin; fusion peptide; gamma secretase; retromer

1. Introduction

HPVs infect and replicate in cutaneous and mucosal epithelium (skin and oral/genital mucosa). Of the hundreds of HPV types [1–3], a set of about 15 HPV "high-risk" types are associated with cervical, anogenital, and oropharyngeal cancers. An additional set of "low-risk" mucosal types cause benign anogenital warts. HPVs are currently the most common sexually transmitted infection, and collectively, these viruses account for 5% of cancers worldwide [4–6].

As for most other DNA viruses, a successful HPV infection requires that the viral genome (vDNA) be transported from an extracellular encapsidated state (i.e., viral particles) to a free unencapsidated state within the host cell nucleus, to allow for viral gene expression and vDNA replication. The non-enveloped HPV capsid, comprised of two proteins (L1 and L2), is the molecular machine that accomplishes this task. Seventy-two pentamers of the major capsid protein L1 form the 55 nm icosahedral particle, which together with L2, encapsidate the vDNA. The minor capsid protein L2 is present in variable, but low, amounts, with a maximal occupancy of 72 molecules per virion [7]. Most studies report a range between 12 and 60 molecules of L2 per virion [7,8], and my lab generally estimates 20–40 copies of L2 per particle as measured by Coomassie staining of SDS-PAGE gels and densitometry. Although CryoEM reconstructions indicate that the bulk of L2 density resides beneath the capsid surface underneath the L1 pentamers [7], it is important to remember that regions of L2 are known to be exposed on the surface of the virion [8–10]. Likewise, it is important to consider that individual L2 molecules can likely assume different conformations or configurations within the virion, although this has yet to be proven.

L2 is thought to be physically complexed to the vDNA within viral particles, and is responsible for the intracellular transport and nuclear accumulation of the vDNA during infection [11]. Although many studies have reported in vitro DNA-binding activity for the conserved, positively charged N- and C-termini of L2 [12–15], the structural nature of the L2/vDNA complex within the actual virion remains poorly understood. This review focuses on the remarkable actions of the L2 protein (Figure 1) and the molecular mechanisms and cellular pathways of subcellular trafficking of the L2/vDNA subviral complex. Recent progress will be summarized and outstanding questions and inconsistencies will be highlighted.

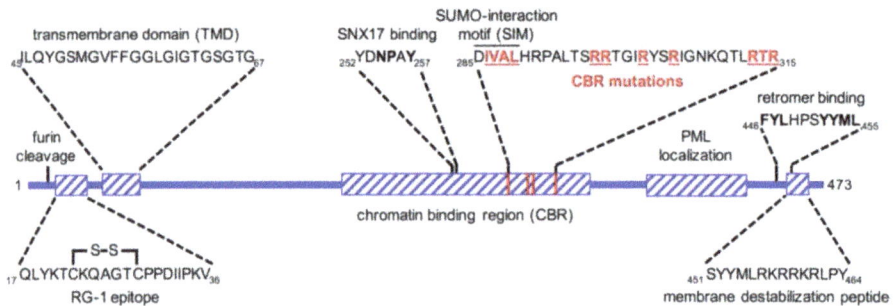

Figure 1. Diagram of the L2 protein. Positions of key components are illustrated. Relative distances and positions are to scale. Chromatin binding mutations are bolded and underlined in red to highlight the residues that were substituted.

It should be noted that many of the findings summarized in this review were found using differentiation-independent pseudovirus (PsV) or quasivirus (QsV) systems for generating infectious or reporter-containing HPV virions in 293TT cells [16]. While more "native" systems for the generation of HPV virions like organotypic raft systems might be ideal, the 293TT PsV/QsV system has been invaluable for basic research on HPV, as the system enables high titer production of highly purified infectious particles, as well as the means to generate virions that package L2 fusions, non-infectious L2 mutants, and 5-ethynyl-2′-deoxyuridine (EdU)-labeled vDNA for a variety of experimental systems. For a recent review on the differences between PsV/QsV and raft-derived HPV, see [17].

2. Viral Entry

Virion binding to extracellular heparin sulfate proteoglycans (HSPGs) induces conformational changes in both the L1 and L2 capsid proteins of the viral particle, and subsequent transfer of the virion to a cell surface entry receptor complex [18–20]. While bound, cell surface kallekrein-8 (KLK8) and furin cleave the L1 and L2 capsids, respectively, an important "priming" event that ensures the proper subsequent subcellular trafficking of L2/vDNA [21–23]. Endocytosis of the virion occurs through an actin-dependent process with similarities to macropinocytosis [24]. Tetraspanin CD151 and its associated $\alpha_3\beta_1$, $\alpha_6\beta_1$ and $\alpha_6\beta_4$ integrin partners, growth factor receptor tyrosine kinases, annexin A2, and the cytoskeletal adaptor obscurin-like 1 (OBSL1) have been implicated in tetraspanin-enriched microdomain (TEM)-dependent HPV16 entry [25–32] (Figure 2). While endocytosis of individual cell surface-bound particles can be quite rapid, overall bulk population-level internalization is asynchronous and slow, occurring on the time scale of many hours [24]. CD151 and the associated TEMs likely coordinate organization and assembly of entry receptor complexes; once assembled and bound to virion, these complexes facilitate rapid entry. Similar scaffolding roles for other tetraspanins have been reported for other enveloped and nonenveloped viruses [33,34].

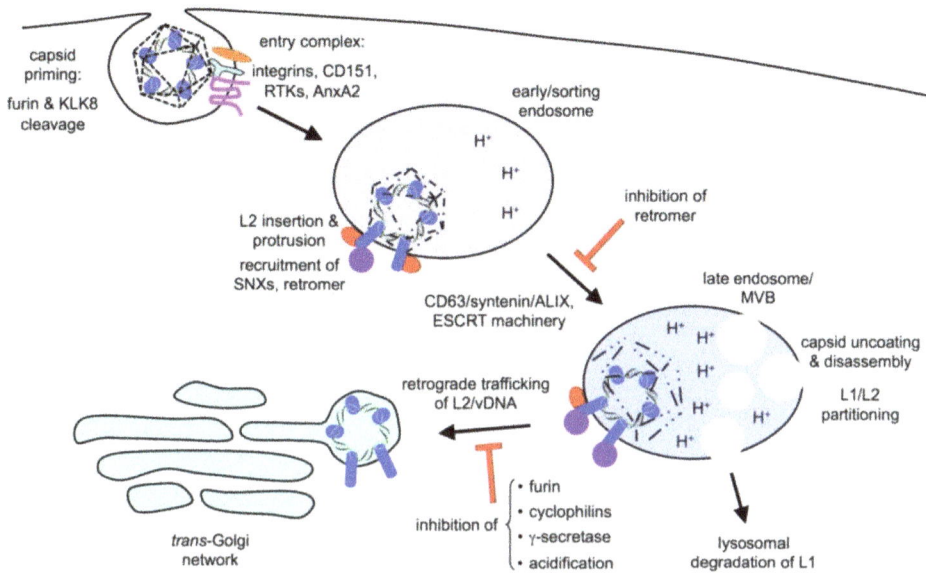

Figure 2. Early subcellular trafficking and uncoating. Internalized virions, primed by cleavage on the cell surface, enter the endolysosomal pathway and begin pH-dependent uncoating and L2 insertion/penetration. L2 recruitment of cytosolic sorting factors including sorting nexins (SNXs) and retromer modulates the trafficking pathway. Retromer binding is important for EE to LE/MVB transport. Retrograde transport of L2/vDNA from LE/MVBs to the *trans*-Golgi network (TGN) occurs in a furin-, cyclophilin- γ-sec-, and pH-dependent manner.

Cleavage of L2 by the host protease furin occurs on the surface of host cells and on the extracellular matrix (ECM) in response to binding of the virion to HSPGs [35]. This cleavage occurs C-terminal to the final arginine residue of a conserved consensus site (RTKR, residues 9–12 for HPV16), removing twelve N-terminal residues of HPV16 L2 [36] (Figure 1). The molecular basis for the requirement of this cleavage remains unknown, but inhibition of cleavage through mutation of the cleavage site or by biochemical inhibition of furin results in aberrant trafficking of the L2/vDNA complex and potent abrogation of infection [23]. Cleavage appears to trigger a conformational change in capsid and/or L2 structure, as the conserved and neutralizing RG-1 epitope (residues 17–36 for HPV16 [37]) becomes accessible to antibody staining shortly after virion binding in a furin-dependent manner [38]. Cell surface cyclophilins (peptidyl-prolyl isomerases, PPIs) also appear to modulate the conformation of L2, as RG-1 epitope exposure is sensitive to cyclosporine A, a broad PPI inhibitor [39]. RG-1 epitope exposure was initially believed to be a convenient marker for furin cleavage, and cyclophilins were believed to control L2 accessibility and susceptibility to furin, but recent work disfavors this idea, as furin cleavage still occurs despite PPI inhibition of RG-1 exposure [36]. Thus, while RG-1 staining is a convenient marker for an L2 conformational change that is both furin- and cyclophilin-dependent, it is not a direct readout for furin proteolysis of L2, as cleavage can occur without RG-1 exposure.

3. Subcellular Trafficking

Shortly after entry, early trafficking of HPV is modulated by the tetraspanin CD63 and its partners syntenin/ALIX [40]. These molecules are necessary for sorting of HPV virions from early endosomes (EE) into acidic late endosome (LE) and multivesicular bodies (MVBs), a prerequisite for capsid disassembly, uncoating, and segregation of L2/vDNA from L1 (Figure 2). When MVB trafficking of HPV was interrupted by knockdown of either CD63 or syntenin, subcellular transport of L2/vDNA was altered and infection was partially blocked, demonstrating a requirement for virion transport into MVBs [40]. Accordingly, components of the ESCRT machinery, a group of cytosolic multisubunit complexes that facilitate endosomal maturation and MVB biogenesis, are involved in efficient HPV infection [41,42]. Endosomal acidification is a strict requirement for HPV infection [24,43,44], but it remains unclear if it is simply needed for proper endosomal maturation and MVB biogenesis, or if acidification itself also triggers conformational changes in the HPV capsid or host proteins that are required for downstream processes like capsid uncoating and vDNA trafficking.

Acid-dependent cathepsin proteases further cleave and process the L1 capsid within the endolysosomal compartments. Capsid proteolysis and disassembly can be visualized by immunofluorescence with the monoclonal antibody L1-7 [45], specific for L1 residues 303–313, located in central cavities underneath each of the L1 pentamers. This region is only available for binding to L1-7 after capsid disassembly; around 8 h post infection [28]. While useful for marking capsid disassembly, L1-7 reactivity does not reveal true infectious uncoating, as staining is blocked by cathepsin inhibitors with no effect on infectivity [22]. Within LE/MVBs the L2/vDNA complex segregates away from the partially degraded L1 capsid in a cyclophilin-dependent manner [46]. The complex then traffics to the trans-Golgi network (TGN) in a retromer-dependent manner [23,47,48], where it resides until the onset of mitosis [49–51]. Recent work has shown that a fraction of the L1 capsid, in the form of conformationally intact pentamers, accompanies the L2/vDNA complex to the TGN and nucleus, but a functional role for these L1 pentamers remains unclear [52].

Colocalization of L2/vDNA with TGN markers is well established, but several groups have reported transport of vDNA to more distal retrograde sites. Partial colocalization of vDNA with *cis/medial*-Golgi markers like GM130 and giantin has been reported [23,53]. Likewise, sensitive techniques like the proximity ligation assay [54] have suggested that L2/vDNA retrograde traffics past the Golgi to the ER [55]. Whether these represent primary or alternative routes of infection, or even unproductive dead ends, is not clear. It is worth noting that the dynamic flux of proteins within the secretory compartments makes it difficult to precisely determine where colocalization is occurring by microscopy. Many ER proteins contain a C-terminal KDEL sequence, and although they are maintained within the ER at steady state, they are constantly trafficking into the Golgi, where they must be recycled back through KDEL cargo receptor [56,57]. Sensitive techniques like the proximity ligation assay using such KDEL-containing ER proteins must therefore be interpreted with caution. Since retrograde trafficking of L2/vDNA to more distal compartments has not been well established, this review will simply refer to the final retrograde destination of L2/vDNA as the "TGN".

4. Retromer and Sorting Factors

The retromer, a trimer of Vps26, Vps29, and Vps35, is a cytosolic sorting adaptor complex that binds to peptide motifs within the intracellular domains and cytosolic tails of membrane-bound receptors destined for the TGN. Retromer works in concert with molecules like Rab7b, Rab9a, and members of the sorting nexin family, including SNX3, SNX27, and the BAR-domain sorting nexins (SNX-BAR), to sort cargos from a variety of endosomal compartments to the TGN [58,59]. L2 contains conserved hydrophobic retromer-binding sites near the C-terminus (FYL at residues 446–448 and YYML at residues 452–455 for HPV16, see Figure 1). Mutation of these sites prevents association of L2 with retromer, and blocks the trafficking of L2/vDNA to the TGN, instead causing an accumulation within EEA1-positive endocytic compartments [48] suggesting a retromer-dependent sorting event away from EE compartments (Figure 2). Likewise, siRNA knockdown of retromer components also prevents L2/vDNA from reaching the TGN [48]. In addition to retromer, L2 is capable of interaction with SNX17 and SNX27 to direct endosomal and retrograde trafficking of the vDNA [60,61]. The interaction with SNX17 through a conserved motif (NPxY, residues 254–257 for HPV16, see Figure 1) is believed to occur very early after entry. One recent study observed recruitment of SNX17 to HPV positive endosomes by 2 h post infection, a phenotype that was dependent on the conserved NPxY motif within L2 [62]. The SNX17-L2 interaction likely promotes retention/recycling of the L2/vDNA complex within the endosomal compartment, preventing the rapid trafficking and degradation of L2/vDNA within lysosomal compartments [60]. Perhaps the virion requires a relatively long retention time in moderately acidic EE and LE/MVB environments for efficient uncoating, partitioning of the L2/vDNA subviral complex from L1, and/or recruitment of the retromer? Mutation of the NPxY motif or knockdown of SNX17 results in aberrant L2/vDNA trafficking and decreased infectivity [60].

SNX17 uses its FERM domain to bind to cargo harboring the NPxY motif. SNX27 is another FERM domain-containing SNX involved in L2/vDNA trafficking but, unlike SNX17, SNX27 does not interact with L2 through the conserved NPxY motif. In addition to a FERM domain, SNX27 also contains a PDZ domain, which mediates interaction with L2 through a non-canonical PDZ ligand located somewhere in between residues 192–292 of HPV16 L2 [61]. Notably, both SNX17 and SNX27 have been implicated in efficient retrograde trafficking through the retromer [59,63], raising the possibility that cooperative interactions of L2 with these SNXs may somehow promote retromer-dependent TGN localization.

Recent work has revealed the existence of an additional trimeric sorting complex called the retriever, which functions in concert with SNX17, the CCC, and WASH complexes to sort cargo from degradative to recycling compartments. Retriever consists of three subunits-DSCR3, C16orf62, and Vps29, a subunit in common with the retromer [64]. HPV infection is decreased upon knockdown of retriever components DSCR3 and C16orf62, as well as CCC components CCDC 22 and CCDC93, suggesting a role for this novel pathway in HPV infection [64].

5. γ-Secretase

ESCRT proteins, tetraspanins, SNXs, and retromer all have physiological roles in subcellular trafficking and protein transport, and thus many of these natural pathways and components are exploited and commandeered by different viruses for entry or assembly. Perhaps the most mysterious host factor necessary for HPV infection is the multisubunit intramembrane protease γ-secretase (γ-sec), which appears to be a unique requirement of papillomaviruses. The γ-sec complex is a transmembrane protease comprised of four subunits: presenilin1/2 (PS1/2), nicastrin (Nic), Aph1a/b, and PEN2. Two isoforms exist for both PS and Aph1, so there is heterogeneity among cellular γ-sec complexes [65]. γ-sec catalyzes the intramembrane cleavage of TMDs from a wide variety of membrane proteins [66], and is perhaps best known as an important component of the Notch signaling pathway and the biogenesis of Aβ peptides from amyloid precursor protein (APP) [67–70]. Biochemical inhibition of γ-sec or knockdown of any of the four subunits results in a potent block of HPV infection [55,71]. In a screen of a diverse panel of 34 different mucosal and cutaneous HPV types, sensitivity to γ-sec inhibition was the most conserved feature among the 29 alpha and 5 beta HPV types tested, even higher than sensitivity to furin inhibition [72]. The molecular basis for the γ-sec requirement is unknown, but inhibition of γ-sec activity results in a failure of L2/vDNA to reach the TGN, even though L2 appears to exit EEA1 positive endosomal compartments [55] (Figure 2). This is in contrast to retromer knockdown, which causes an accumulation of vDNA within EEA1 endosomes [48]. This may suggest that, in the absence of γ-sec activity, L2/vDNA never exits the MVB/LE compartments, and instead continues to lysosomes for degradation; or that γ-sec controls trafficking of L2/vDNA to a discrete intermediate compartment, between the MVB and the TGN. Consistent with the observed effects on L2/vDNA trafficking, HPV16 is only sensitive to γ-sec inhibition during the first 6–8 h of infection [55], and γ-sec inhibition has no effect on post-TGN trafficking of vDNA [51]. Failure of L2/vDNA to reach the TGN in the absence of γ-sec activity means that the retrograde trafficking pathway utilized by HPV involves more than just the canonical players like retromer and SNXs, suggesting that new γ-sec-dependent retrograde pathways may exist and are being exploited by papillomaviruses. The catalytic PS1/2 subunit of γ-sec are known to modulate protein trafficking, lysosomal maturation, and Ca^{2+} homeostasis independently of γ-sec-activity, but direct connections to retrograde trafficking pathways are scant [73]. Retromer has, however, been implicated in retrograde-dependent trafficking and cleavage of gamma secretase substrates, including APP [74,75].

6. L2 Is an "Inducible Transmembrane" Protein

How does L2, the minor capsid protein from a non-enveloped virus, complexed with the vDNA within the lumen of intracellular vesicular compartments, interact with a variety of cytosolic sorting molecules to direct its own transport to the TGN? Evidence from multiple laboratories suggests that L2 can interact with and span across vesicular membranes, thereby allowing L2 to gain access to the cytosol, to recruit cytosolic factors necessary for retrograde trafficking (Figure 2). Post-TGN transit, full translocation across the limiting membrane, and nuclear accumulation of the L2/vDNA complex requires mitosis and is summarized in greater detail in latter sections of this review. For the sake of consistency within the field, the following nomenclature is proposed for the 3-step process describing L2's remarkable ability to shift from being a soluble protein to a transmembrane protein and back again:

(1) **Insertion**—Within the lumen, part(s) of L2 insert(s) into the local membrane.
(2) **Protrusion**—L2 becomes a transmembrane protein. Part of L2 remains lumenal and is complexed with the vDNA, while other parts of L2 stick through the membrane and are accessible to cytosolic proteins.
(3) **Translocation**—L2/vDNA exits vesicular compartments, passing across the limiting membrane to establish infection within the cell nucleus.

7. Insertion and Protrusion of L2

How does the L2 protein initially interact with and insert into membranes? When naturally or ectopically expressed, L2 is a soluble nuclear protein, not a membrane protein [76,77]. Yet, during infection, L2 must somehow interact with and cross membranes; how is this accomplished? In 2006, a conserved "membrane destabilizing peptide" near the C-terminus of L2 (SYYMLRKRRKRLPY, residues 451–464 for HPV16, see Figure 1) was identified as having a role in the endosomal escape of L2/vDNA via membrane disruption or destabilization [78]. Using synthetic C-terminal peptides from HPV33 L2, containing the corresponding membrane destabilization moiety (SYFILRRRRKRFPYFFTDVRVAA, residues 445–467), in vitro cytotoxicity and propidium iodide uptake experiments showed a pH-dependent ability to disrupt cellular membranes. While it is possible that this C-terminal region aids in membrane insertion of L2 within the acidic endosomes, it is important to note that a direct role in L2 membrane insertion has yet to be demonstrated.

Regardless of how L2 initially inserts into membranes during infection, multiple groups have published indirect evidence that L2 interacts with cytosolic factors and thus must protrude into the cytosol across vesicular membranes. In 2013, my laboratory identified a glycine-rich transmembrane-like domain (TMD) towards the N-terminus of L2 (ILQYGSMGVFFGGLGIGTGSGTG, residues 45–67 of HPV16, see Figure 1) [79]. Taking advantage of TMD-flanking monoclonal antibody epitopes and using elegant immunofluorescence staining procedures, the Sapp laboratory has since demonstrated that L2 utilizes this TMD to span intracellular vesicular membranes with residues C-terminal of the TMD being cytosolic, consistent with a type-I transmembrane topology [80]. Moreover, additional data from the Sapp laboratory suggest the vDNA remains lumenal within these intracellular vesicles [81]. Exactly how L2 is able to insert into the membrane and span across remains unknown, but endosomal acidification seems to be required to adopt this conformation [80], although it remains to be determined if this may simply reflect a requirement for L1 capsid disassembly or L1/L2 partitioning, rather than low pH having a direct effect on L2 protein structure or conformation. Thus, virion-associated L2 appears to be an inducible transmembrane protein, with the ability to insert into membranes and adopt a transmembrane configuration, to drive vDNA subcellular trafficking by physically linking the lumenal vDNA to host cytosolic sorting proteins (Figure 2).

L2's function to facilitate vDNA delivery across the limiting membrane is analogous to that of many bacterial toxins, which penetrate intracellular membranes and deliver toxin domains to the cytosol. Many of these bacterial toxins including diphtheria toxin, anthrax toxin protective antigen, Shiga toxin, and *Pseudomonas* exotoxin A [82,83] also rely on proteolytic activation by furin and other proteases to trigger conformational changes and structural rearrangements that underlie toxin membrane insertion and penetration. Given the requirement for furin in TGN localization of L2/vDNA, it is very likely that cleavage triggers a structural change that enables L2 to insert and protrude into the local membrane via the TMD to recruit cytosolic SNXs and retromer. Until structural data on L2 is obtained, the nature of any cleavage-induced conformational changes will remain elusive.

Although direct evidence is lacking, the TMD itself may play a role in the initial insertion of L2 into membranes. It is noteworthy that the L2 TMD is quite similar to the fusion peptides (FPs) from many type-I fusogenic glycoproteins of enveloped viruses of the *Orthomyxoviridae*, *Paramyxoviridae*, and *Retroviridae* families [79]. These fusion peptides generally consist of ~20 apolar residues, and are typically enriched for glycine, a composition believed to impart conformational flexibility. The structurally dynamic nature of these fusion peptides is thought to be critical for their ability to partition into and destabilize local membranes [84–87]. It is also noteworthy that, like L2, these viral fusion proteins require "priming" by proteolytic cleavage and "activation" by environmental cues like low pH or receptor binding [88]. It is conceptually challenging to envision how L2 could achieve a type-I transmembrane state upon insertion of its N-terminal TMD—L2 would have to essentially drag 400 residues C-terminal to the TMD across the membrane. Perhaps cooperative interactions between the TMD, the C-terminal membrane disruption peptide, and the membrane are required for insertion and protrusion of L2.

How does γ-sec facilitate L2/vDNA trafficking? No interaction between L2 and the γ-sec complex has been reported, although it is tempting to envision that L2 could interact with the complex via its TMD while protruding through the membrane. Alternatively, γ-sec activity could somehow be required for L2 to initially insert into membranes to achieve membrane protrusion. This latter possibility would be consistent with the trafficking defects observed upon inhibition of gamma secretase [48], as failure to insert and protrude through the membrane would be expected to block TGN localization, and may instead cause trafficking away from EEA1 positive early endosomes into a degradative lysosomal pathway. Although no evidence exists supporting a role for γ-sec in HPV endocytosis, it has been found as part of a "tetraspanin interactome", associated with many of the same molecules believed to be part of the initial entry receptor complex, including tetraspanins CD9 and CD81, integrins α3 and β1, and annexin-A2 [89]. The γ-sec complex may therefore be present locally during uncoating when L2 presumably adopts the protruding conformation (Figure 2). In this scenario, one could also imagine L2 actually being a substrate for γ-sec cleavage, triggering a conformational change of some kind upon cleavage of the L2 TMD. While attractive, there is no published evidence supporting this notion. This begs the question- if L2 is not a substrate for γ-sec then how does inhibition of γ-sec catalysis affect L2/vDNA trafficking so drastically? Perhaps γ-sec cleavage of a cellular protein somehow modulates trafficking, or L2 may actually be a "pseudosubstrate", interacting with γ-sec without cleavage. TMD substrates of γ-sec are believed to first dock into a substrate binding site prior to transfer to the catalytic active site for proteolysis [90]. γ-sec inhibitors perturb the global structure of the γ-sec complex and may therefore prevent initial substrate docking [91,92].

As discussed above, many host proteins and pathways are expoited by HPV to facilitate virion trafficking and viral infection, but some proteins can restrict HPV infection suggesting an inherent anti-papillomaviral function. The endosomal protein stannin restricts HPV infection by rerouting virions away from the TGN to degradative compartments [93]. Similarly, the α-defensin HD5 alters L2/vDNA trafficking, accelerating the degradation of virions within LE/lysosomal compartments to restrict HPV infection [94]. Stannin appears to work by blocking association of L2 with retromer, to prevent retrograde trafficking of L2/vDNA, causing an increased accumulation of L2/vDNA in LAMP1-positive lysosomal compartments [93]. This is in contrast to retromer knockdown or mutation of retromer binding sites in L2, which instead cause a trafficking block within EEA1-positive endosomal compartments [48]. Rather than directly inhibiting the L2-retromer association, stannin likely blocks the insertion and protrusion of L2 within vesicular membranes to indirectly prevent binding and recruitment of retromer. Similarly, HD5 may induce aberrant trafficking of L2/vDNA by directly binding the virion to interfere with L2 insertion and protrusion [94]. Interestingly, interferon gamma has recently been found to restrict HPV infection in an L2-dependent and type-specific manner, decreasing the proteolytic degradation of L1 capsid and causing a block of L2/vDNA in LE/MVB/lysosomal compartments [95]. Cathepsin proteases also appear to limit HPV, as infection is increased upon genetic knockout, siRNA silencing, or biochemical inhibition [22,96]; this is contrary to other non-enveloped viruses, like reoviruses and adeno-associated viruses, which depend on these endosomal proteases for uncoating [97,98]. Vimentin is another recently reported inhibitory factor, found to limit infection at the level of viral entry [99]. Much will be gained from further mechanistic studies of these inhibitory host factors.

8. Post-TGN Transport- Mitosis, Translocation, and Chromatin Binding

Prior to 2013, the consensus view was that the L2/vDNA complex egressed from endosomal compartments into the cytosol, as is the case for many other non-enveloped viruses [100–102]. Thus, discovery of the TGN as an important stop in the retrograde route of incoming L2/vDNA was a major advance in the field [23,47]. Initially, the L2/vDNA was believed to penetrate the TGN directly and wait in the cytosol prior to nuclear entry. Cell cycle progression into mitosis is known to be important for HPV infection [103], and nuclear envelope breakdown was thought to facilitate L2/vDNA transfer from the cytosol into the nucleus like some retroviruses [50,104].

Recent work supports a model where, at the onset of mitosis, when the Golgi and TGN naturally begins to fragment and vesiculate, the vesicle-bound vDNA egresses from what was the interphase TGN (Figure 3). The Sapp laboratory pioneered an EdU/vDNA staining technique based on selective membrane permeabilization and sequential EdU labeling to demonstrate the lumenal state of the vDNA [81]. While the exact nature of these vesicles remains unknown, immunofluorescence microscopy reveals that these vDNA-containing vesicles stain negative for classical TGN markers like TGN46 and p230, and appear to migrate along microtubules, clustering around the centrioles during progression from G2/M to prometaphase [51,81]. L2 likely remains in the protruding conformation, spanning across the limiting membrane to coordinate microtubule-dependent traffic of these vDNA-containing vesicles along the mitotic spindle [81]. These vesicles eventually make their way to the condensed chromosomes, and by metaphase, vDNA can be seen to be associated with and presumably bound to the host chromosomes [49,51,81] (Figure 3). From there, the chromosome-bound vDNA is partitioned into daughter cells. In this manner, infection of both daughter cells is favored at an MOI > 2.

When does the L2/vDNA complex fully translocate across the limiting membrane? While useful for observing the trafficking of vDNA during HPV infection, standard subcellular localization of EdU-labeled vDNA by microscopy is insufficient to reveal the actual translocation event. Using their specialized immunofluorescence protocol for sequential fluorophore-azide conjugation of EdU-labeled vDNA in differentially permeabilized cells, the Sapp laboratory concluded that vDNA became cytosolic sometime during G1, well after mitosis [81]. In this model, L2/vDNA would reside within these unique mitotic transport vesicles, bound to chromosomes for an extended period of time (Figure 3), until translocation occurred sometime in G1 after completion of mitosis.

To better understand translocation, my laboratory has developed an alternative platform to detect and measure this elusive process. Our system is based on the biotin-protein ligase BirA, from *Escherichia coli*. The BirA enzyme will covalently attach a biotin molecule to a specific lysine residue of a short peptide substrate, termed the biotin acceptor peptide (BAP). The BAP is a specific substrate for bacterial BirA, and is not recognized by mammalian biotin-protein ligases [51,105]. By generating HPV pseudoviruses that encapsidate a functional L2-BirA fusion and a HaCaT keratinocyte line that stably expresses a cytosolic GFP-BAP fusion, we have set up a two-component compartmentalization assay to detect L2-BirA translocation. In this system, lumenal L2-BirA is separated from the cytosolic GFP-BAP substrate by limiting membranes. Only upon translocation of L2 will BirA encounter the BAP; thus, biotinylation of GFP-BAP is a readout for translocation. However, it should be noted that, since the assay relies on the fusion of BirA to the extreme C-terminus of L2, biotinylation of GFP-BAP could result from cytosolic exposure of just the C-terminus of L2, rather than full translocation of the entire L2/vDNA complex.

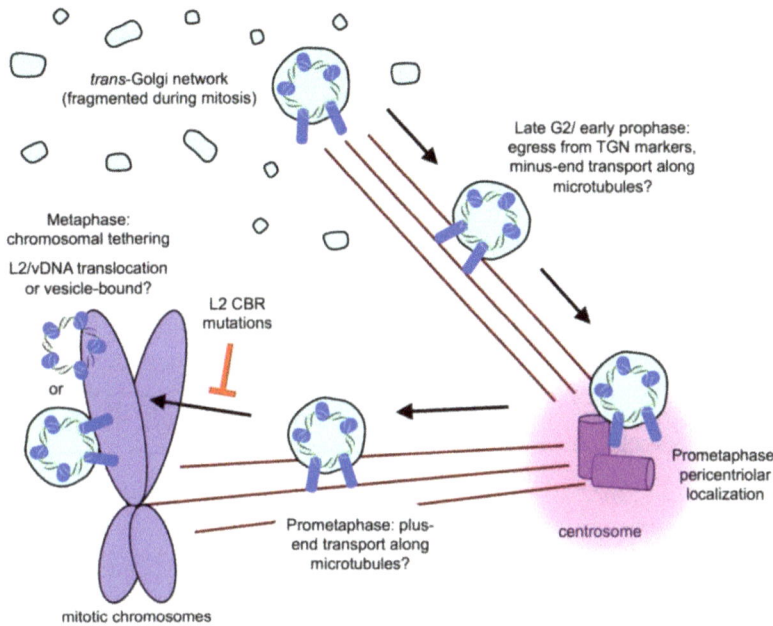

Figure 3. Post-TGN mitotic trafficking of L2/vDNA complex. Upon entry into mitosis, L2/vDNA remains vesicle-bound but loses coincidence with TGN markers. These L2/vDNA-containing vesicles likely travel along astral microtubules in the minus-end direction towards the centrosome, where they accumulate during prometaphase. The vesicles likely switch polarity and travel along the spindle microtubules in the plus-end direction to reach the host chromosomes by metaphase. Chromosome-bound L2/vDNA partitions with host chromosomes, eventually localizing to PML bodies of the daughter cells. Chromosome binding of L2/vDNA is through the chromatin binding region (CBR) of L2 and mutation of this region causes a block in translocation, with vesicular L2/vDNA becoming reabsorbed back into the nascent Golgi after mitosis. Chromosome-bound L2/vDNA may be in a membrane-bound vesiclular state, or may have penetrated the limiting membrane upon chromosome binding, further work is needed to clarify this stage of the HPV life cycle.

Using this system, we found that L2 translocation required TGN localization of L2/vDNA and cell cycle progression past G2/M. Timecourse experiments with synchronized HaCaT-GFP-BAP cells demonstrated that the earliest biotinylated GFP-BAP signal was detected at or just prior to the onset of mitosis. Although these experiments were performed with a bulk population, we believe it suggests that L2 translocation (or at least of the C-terminus of L2) begins during mitosis (Figure 3), well before transition into G1. Moreover, this timing of L2/vDNA translocation would be consistent with the visual "jump" of vDNA from a punctate pericentriolar distribution in prometaphase to being chromosome-bound by metaphase [51].

The chromosome binding ability of L2 was first reported in 2014 [50], and since then, the Schelhaas group has mapped a minimal chromatin binding region (CBR, residues 188–334 for HPV16) within L2 [49]. Interestingly, the ability of ectopically expressed L2-GFP fusion to associate with mitotic chromatin was found to require cell cycle progression into prometaphase. This finding is suggestive that either the interaction between L2 and mitotic chromatin is indirect, requiring a prometaphase-specific factor, or that L2 is post-translationally modified during prometaphase to somehow activate its chromatin binding ability. Substitution of specific residues (IVAL; 286–289, R302/305, and RTR; 313–315) were found to completely abrogate the chromatin binding activity while mutation of RR396/397 resulted in a partial-inhibition CBR function. When these same residues were mutated in reporter-expressing PsV, packaged with either L2 or L2-BirA, the same phenotypes were observed—infectivity and translocation were completely blocked for IVAL; 286–289, R302/305, and RTR; 313–315 and partially blocked for RR396/397 [49,51].

The striking correlation between the ability of L2 to bind chromatin and to translocate during infection supports a model whereby chromatin binding is required for L2 translocation [49,51]. A mechanistic linkage between these processes favors a model where translocation of L2/vDNA out of the mitotic vesicles occurs while these compartments encounter mitotic chromosomes during prometaphase, and is supported by appearance of translocation signal in timecourse experiments [51]. While further work is needed to validate one model over another, it should be noted that the two models do not have to be mutually exclusive. Translocation studies based on L2-BirA may in fact only be revealing exposure of the L2 C-terminus, and full translocation of vDNA could be occurring at a later time post-mitosis, as suggested by the Sapp laboratory. Alternatively, the nature of these vesicles is unknown, and if their lipid composition and detergent solubility changes during mitosis, it could affect sequential labeling efficiency or EdU-labeled vDNA availability to fluorophore-azides after differential detergent permeabilization.

9. A Topology Conundrum

As mentioned above, immunofluorescence staining with mAbs specific for L2 epitopes flanking the TMD suggest a type-I topology for L2 protrusion, with the N-terminal ~45 residues being lumenal, a ~25 residue TMD, and the C-terminal ~400 residues being cytosolic [80] (Figure 4). This topology is in agreement with the placement of known SNX17 and retromer binding motifs within L2, as well as the newly defined CBR (Figure 1). However, translocation studies with PsV encapsidating the L2-BirA fusion are suggestive of a different topology for L2. HPV infection in the presence of S-phase blockers like aphidicolin traps incoming L2/vDNA at the TGN, likely in the protruding conformation [51]. This block is reversible, as removal of the drug releases cell cycle inhibition and enables synchronized egress and translocation of L2/vDNA out of the TGN upon entry into mitosis [51]. This data, however, is not in agreement with a strict type-I membrane topology for L2. The lack of translocation signal in the presence of aphidicolin implies that the C-terminus of L2-BirA is not cytosolic, as it would be in a type-I topology. Rather, it suggests that BirA is either lumenal or is somehow obstructed from engaging the GFP-BAP substrate when L2 is protruding from endosomal and TGN compartments, only becoming accessible to the cytosol upon entry into mitosis. A double-pass topology of protruding L2-BirA would result in a lumenal C-terminal BirA (Figure 4).

It should be noted that the C-terminal membrane destabilization peptide bears no resemblance to a conventional TMD, and no other membrane-spanning regions of L2 have been identified, so it is unclear how L2 could span the membrane twice to keep the C-terminal BirA fusion lumenal. Membrane-spanning bacterial toxins, including anthrax toxin, Diphtheria toxin, botulinum toxin, tetanus toxin, and *Clostridium difficile* toxins, form pores through which they can extrude themselves into the cytosol by a variety of protein translocation mechanisms [106,107]. Oligomerization of individual L2 molecules, each with a single membrane-spanning TMD, could theoretically enable a double-pass topology by extrusion of the L2 C-termini back into the lumen through such a pore. This hypothetical configuration would place a central portion of L2 within the cytosol to recruit sorting factors and direct traffic of the associated vDNA (Figure 4). It should also be noted that such a double-pass topology would still be consistent with the immunofluorescence data supporting a type-I topology [80]. Clearly, much more work is needed to understand the protruding conformation of L2 during HPV infection.

A Type-I topology
- Agrees with L2 IF data and trypsin susceptibility
- Disagrees with L2-BirA studies
- Agrees with a single L2 TMD

B Double-pass topology
- Agrees with L2 IF data and trypsin susceptibility
- Agrees with L2-BirA studies
- Disagrees with single L2 TMD- unclear how this topology could be achieved

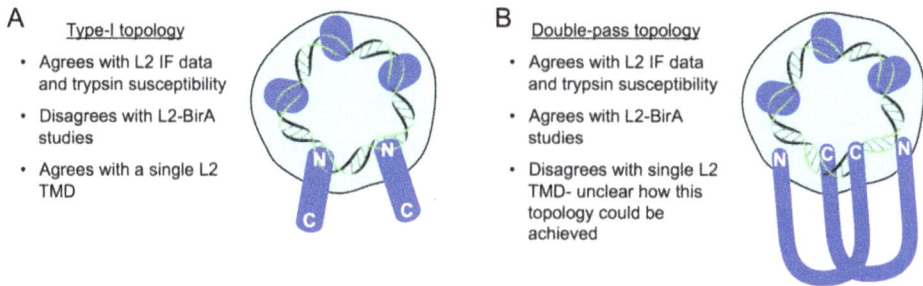

Figure 4. Topology models of L2 protrusion. Both models are consistent with published L2 immunofluorescence and trypsin susceptibility data [80]. (**A**) In the type-I model, the N-terminus remains lumenal with all ~400 residues downstream of the TMD being cytosolic to recruit sorting factors. L2-BirA would be expected to biotinylate substrate in this model, contradicting the actual data [51]. (**B**) In the double-pass model, both the N- and C-termini would be lumenal, with the bulk of L2 being cytosolic. L2-BirA would not be expected to biotinylate substrate as observed. However, the means by which L2 spans the membrane a second time is difficult to conceptualize, as the protein only has one TMD towards the N-terminus [79]. In vitro data suggest both the N- and C-termini are capable of non-speciific dsDNA binding through electrostatic interactions [12,14].

10. PML Bodies and Beyond

Regardless of the precise mechanisms of L2 translocation, the minor capsid protein eventually leaves vesicular compartments and is seen along with vDNA within interphase nuclei of infected cells, localized to punctate nuclear foci called promyelocytic leukemia (PML) nuclear domains, also known as PML oncogenic domains (PODS), or ND10 bodies [108]. PML bodies are small nuclear structures, organized by the PML protein for which they are named. These dynamic domains are present in most cells, and are assembled and remodeled in response to a variety of cellular stresses, including infection, innate immune triggers/interferon (IFN), heat shock, DNA damage pathways, and metabolic stress [109–111]. PML bodies modulate a wide variety of cellular responses via recruitment, retention, and modification of numerous proteins including the transcriptional repressor Daxx, tumor suppressor Sp100, transcriptional regulator ATRX, DNA helicase BLM, kinase HIPK2, and a multitude of other host proteins. The PML protein, which has many different isoforms, is critical to the assembly of PML bodies and recruitment of host proteins to these foci [112]. Many PML-associated proteins are either directly conjugated to small ubiquitin-like modifier (SUMO) proteins or contain short linear SUMO-interaction motifs (SIMs), or both. In addition to PML oligomerization, SUMOylation and SUMO-SIM networks are believed to be important to PML assembly and dynamics [113].

Given the role of PML bodies in innate antiviral responses, many viruses have been shown to target PML bodies or induce degradation or remodeling of specific PML components [114,115]. PML bodies have been shown to be important for efficient infection from reporter-expressing HPV16 pseudoviruses, as well as authentic BPV virions [108], suggesting that the vDNA is actively targeted to these sites by L2 upon infection. Ectopically expressed L2 can localize to PML bodies, remodeling them through recruitment of Daxx and depletion of Sp100 [116]. In older studies, the ability of GFP-L2 fusions to localize and induce remodeling of PML bodies was mapped to a C-terminal region of L2 (residues 360–420 for HPV33) [117]. L2 can itself be SUMOylated at a conserved lysine residue (K35 for HPV16) when ectopically expressed, but recent work suggests this modification is not important for PML body localization [118,119]. Rather, a moderately conserved SIM (DIVAL, residues 285–289 for HPV16, see Figure 1) has been implicated in PML localization of ectopically expressed untagged full length, L2 [118]. Precise mechanisms of L2-dependent PML body remodeling have yet to be worked out but ectopic overexpression studies must be interpreted with caution since the mode of L2 gene transfection/delivery has been shown to heavily influence nuclear/PML localization of L2 [120]. Recent work suggests that the PML component Sp100 restricts HPV transcription and vDNA replication [121], favoring a model whereby incoming L2 might promote a nuclear environment conducive for early HPV gene transcription and genome maintenance in basal cells.

During cell division, PML bodies show increased dynamics and disperse into the cytosol during open mitosis. Only after exit from mitosis and reformation of the nuclear envelope are PML bodies assembled de novo, and recruitment of Daxx and Sp100 is observed [122]. Whether L2 recruits PML to nucleate de novo assembly of PML bodies in the vicinity of the vDNA after mitosis or whether the L2/vDNA complex is targeted to newly formed PML bodies in early G1 remains to be determined. Likewise, much remains to be discovered regarding preferential remodeling of PML components like Daxx and Sp100 immediately after mitotic translocation of L2/vDNA, and the consequences of this remodeling for infection, immune evasion, and viral persistence.

11. Conclusions & Future Directions

In addition to role(s) in vDNA packaging, virion assembly, and particle stability [11], minor capsid protein L2 is tasked with ensuring nuclear delivery of the vDNA during HPV infection. This feat is accomplished via some remarkable means for a viral capsid protein present in low copy numbers. L2 is able to partition vDNA away from degradative endolysosomal compartments, instead diverting it to the TGN. L2 does this by possessing properties of an "inducible transmembrane" protein, with the ability to insert into and protrude across local vesicular membranes using a transmembrane-like domain. Portions of L2 containing conserved sorting motifs are exposed to the cytosol, recruiting cellular sorting factors that dictate retrograde trafficking of L2/vDNA to the TGN. Upon entry into mitosis, the vesicular L2/vDNA complex separates from the dispersed Golgi towards the pericentriolar region and by metaphase the vDNA can be seen to be associated with condensed chromosomes. Whether the visual association of vDNA with mitotic chromosomes represents full translocation of L2/vDNA across limiting membranes, or simply vDNA-filled post-Golgi vesicles bound to chromosomes, remains to be shown. Together with recent work on the chromatin-binding abilities of L2, translocation studies using a novel BirA-based approach suggest that chromatin binding is necessary for translocation. Timecourse experiments with synchronized cells suggest that translocation is concurrent with or slightly after the onset of mitosis. In contrast, sequential fluor-azide conjugation of EdU-labeled vDNA after differential detergent permeabilization suggests that translocation, as defined by the liberation of vDNA from membrane-bound compartments, occurs post-mitosis in G1. Regardless of the specific mechanisms and timing of translocation, L2/vDNA localizes to PML bodies of the daughter cells and likely functions to promote efficient viral gene expression.

Additional efforts are needed to further define the mechanisms of L2's remarkable abilities. Structural studies are needed to understand the nature of the L2/vDNA complex within viral particles, the molecular basis underlying the requirement for furin, the consequences of cleavage, and the mechanisms of L2-membrane interaction. Further work is needed to understand and define the nature of L2 protrusion through limiting membranes, specifically the topology of L2 within these membranes, and to identity any host proteins that may be necessary for L2-membrane insertion and protrusion. More work is necessary to identify the nature of the post-Golgi vesicles in which L2 resides upon entry into mitosis, and again to define the host proteins that may be interacting with L2 to enable subcellular transport of these compartments and eventual translocation. Elucidation of the timing and mechanisms of actual L2/vDNA translocation will require new approaches. Finally, the precise role(s) of virion-derived L2 in the remodeling of PML bodies, establishment of infection, initial viral gene expression, and potential immunoevasion all represent exciting avenues of future endeavors.

Acknowledgments: Sincere apologies to all whose work was neither discussed nor cited due to space limitations. I thank Koenraad Van Doorslaer, Brittany Forte, Shuaizhi Li, and Matthew Bronnimann for critical reading of the manuscript. Samuel K. Campos is supported by grant 1R01AI108751 from the National Institute for Allergy and Infectious Diseases.

Conflicts of Interest: The authors declare no conflict of interest.

References

1. Bzhalava, D.; Muhr, L.S.; Lagheden, C.; Ekstrom, J.; Forslund, O.; Dillner, J.; Hultin, E. Deep sequencing extends the diversity of human papillomaviruses in human skin. *Sci. Rep.* **2014**, *4*, 5807. [CrossRef] [PubMed]
2. Van Doorslaer, K. Evolution of the papillomaviridae. *Virology* **2013**, *445*, 11–20. [CrossRef] [PubMed]
3. Van Doorslaer, K.; Li, Z.; Xirasagar, S.; Maes, P.; Kaminsky, D.; Liou, D.; Sun, Q.; Kaur, R.; Huyen, Y.; McBride, A.A. The papillomavirus episteme: A major update to the papillomavirus sequence database. *Nucleic Acids Res.* **2017**, *45*, D499–D506. [CrossRef] [PubMed]
4. Doorbar, J.; Quint, W.; Banks, L.; Bravo, I.G.; Stoler, M.; Broker, T.R.; Stanley, M.A. The biology and life-cycle of human papillomaviruses. *Vaccine* **2012**, *30* (Suppl. 5), F55–F70. [CrossRef] [PubMed]
5. Forman, D.; de Martel, C.; Lacey, C.J.; Soerjomataram, I.; Lortet-Tieulent, J.; Bruni, L.; Vignat, J.; Ferlay, J.; Bray, F.; Plummer, M.; et al. Global burden of human papillomavirus and related diseases. *Vaccine* **2012**, *30* (Suppl. 5), F12–F23. [CrossRef] [PubMed]
6. Schiffman, M.; Doorbar, J.; Wentzensen, N.; de Sanjose, S.; Fakhry, C.; Monk, B.J.; Stanley, M.A.; Franceschi, S. Carcinogenic human papillomavirus infection. *Nat. Rev. Dis. Prim.* **2016**, *2*, 16086. [CrossRef] [PubMed]
7. Buck, C.B.; Cheng, N.; Thompson, C.D.; Lowy, D.R.; Steven, A.C.; Schiller, J.T.; Trus, B.L. Arrangement of L2 within the papillomavirus capsid. *J. Virol.* **2008**, *82*, 5190–5197. [CrossRef] [PubMed]
8. Bywaters, S.M.; Brendle, S.A.; Tossi, K.P.; Biryukov, J.; Meyers, C.; Christensen, N.D. Antibody competition reveals surface location of HPV L2 minor capsid protein residues 17–36. *Viruses* **2017**, *9*. [CrossRef] [PubMed]
9. Pastrana, D.V.; Gambhira, R.; Buck, C.B.; Pang, Y.Y.; Thompson, C.D.; Culp, T.D.; Christensen, N.D.; Lowy, D.R.; Schiller, J.T.; Roden, R.B. Cross-neutralization of cutaneous and mucosal papillomavirus types with anti-sera to the amino terminus of L2. *Virology* **2005**, *337*, 365–372. [CrossRef] [PubMed]
10. Kondo, K.; Ishii, Y.; Ochi, H.; Matsumoto, T.; Yoshikawa, H.; Kanda, T. Neutralization of HPV16, 18, 31, and 58 pseudovirions with antisera induced by immunizing rabbits with synthetic peptides representing segments of the HPV16 minor capsid protein L2 surface region. *Virology* **2007**, *358*, 266–272. [CrossRef] [PubMed]
11. Wang, J.W.; Roden, R.B. L2, the minor capsid protein of papillomavirus. *Virology* **2013**, *445*, 175–186. [CrossRef] [PubMed]
12. Bordeaux, J.; Forte, S.; Harding, E.; Darshan, M.S.; Klucevsek, K.; Moroianu, J. The L2 minor capsid protein of low-risk human papillomavirus type 11 interacts with host nuclear import receptors and viral DNA. *J. Virol.* **2006**, *80*, 8259–8262. [CrossRef] [PubMed]
13. Fay, A.; Yutzy, W.H., IV; Roden, R.B.S.; Moroianu, J. The positively charged termini of L2 minor capsid protein required for bovine papillomavirus infection function separately in nuclear import and DNA binding. *J. Virol.* **2004**, *78*, 13447–13454. [CrossRef] [PubMed]
14. Klucevsek, K.; Daley, J.; Darshan, M.S.; Bordeaux, J.; Moroianu, J. Nuclear import strategies of high-risk HPV18 L2 minor capsid protein. *Virology* **2006**, *352*, 200–208. [CrossRef] [PubMed]
15. Zhou, J.; Sun, X.Y.; Louis, K.; Frazer, I.H. Interaction of human papillomavirus (HPV) type 16 capsid proteins with HPV DNA requires an intact L2 N-terminal sequence. *J. Virol.* **1994**, *68*, 619–625. [PubMed]
16. Buck, C.B.; Thompson, C.D. Production of papillomavirus-based gene transfer vectors. *Curr. Protoc. Cell Biol.* **2007**. [CrossRef]
17. Biryukov, J.; Meyers, C. Papillomavirus infectious pathways: A comparison of systems. *Viruses* **2015**, *7*, 4303–4325. [CrossRef] [PubMed]
18. Aksoy, P.; Gottschalk, E.Y.; Meneses, P.I. HPV entry into cells. *Mutat. Res. Rev. Mutat. Res.* **2017**, *772*, 13–22. [CrossRef] [PubMed]
19. Day, P.M.; Schelhaas, M. Concepts of papillomavirus entry into host cells. *Curr. Opin. Virol.* **2014**, *4*, 24–31. [CrossRef] [PubMed]
20. DiGiuseppe, S.; Bienkowska-Haba, M.; Guion, L.G.; Sapp, M. Cruising the cellular highways: How human papillomavirus travels from the surface to the nucleus. *Virus Res.* **2017**, *231*, 1–9. [CrossRef] [PubMed]
21. Richards, R.M.; Lowy, D.R.; Schiller, J.T.; Day, P.M. Cleavage of the papillomavirus minor capsid protein, L2, at a furin consensus site is necessary for infection. *Proc. Natl. Acad. Sci. USA* **2006**, *103*, 1522–1527. [CrossRef] [PubMed]

22. Cerqueira, C.; Samperio Ventayol, P.; Vogeley, C.; Schelhaas, M. Kallikrein-8 proteolytically processes human papillomaviruses in the extracellular space to facilitate entry into host cells. *J. Virol.* **2015**, *89*, 7038–7052. [CrossRef] [PubMed]

23. Day, P.M.; Thompson, C.D.; Schowalter, R.M.; Lowy, D.R.; Schiller, J.T. Identification of a role for the trans-Golgi network in human papillomavirus 16 pseudovirus infection. *J. Virol.* **2013**, *87*, 3862–3870. [CrossRef] [PubMed]

24. Schelhaas, M.; Shah, B.; Holzer, M.; Blattmann, P.; Kuhling, L.; Day, P.M.; Schiller, J.T.; Helenius, A. Entry of human papillomavirus type 16 by actin-dependent, clathrin- and lipid raft-independent endocytosis. *PLoS Pathog.* **2012**, *8*, e1002657. [CrossRef] [PubMed]

25. Aksoy, P.; Abban, C.Y.; Kiyashka, E.; Qiang, W.; Meneses, P.I. HPV16 infection of HaCaTs is dependent on β4 integrin, and α6 integrin processing. *Virology* **2014**, *449*, 45–52. [CrossRef] [PubMed]

26. Dziduszko, A.; Ozbun, M.A. Annexin A2 and S100A10 regulate human papillomavirus type 16 entry and intracellular trafficking in human keratinocytes. *J. Virol.* **2013**, *87*, 7502–7515. [CrossRef] [PubMed]

27. Scheffer, K.D.; Gawlitza, A.; Spoden, G.A.; Zhang, X.A.; Lambert, C.; Berditchevski, F.; Florin, L. Tetraspanin CD151 mediates papillomavirus type 16 endocytosis. *J. Virol.* **2013**, *87*, 3435–3446. [CrossRef] [PubMed]

28. Spoden, G.; Freitag, K.; Husmann, M.; Boller, K.; Sapp, M.; Lambert, C.; Florin, L. Clathrin- and caveolin-independent entry of human papillomavirus type 16—Involvement of tetraspanin-enriched microdomains (tems). *PLoS ONE* **2008**, *3*, e3313. [CrossRef] [PubMed]

29. Spoden, G.; Kuhling, L.; Cordes, N.; Frenzel, B.; Sapp, M.; Boller, K.; Florin, L.; Schelhaas, M. Human papillomavirus types 16, 18, and 31 share similar endocytic requirements for entry. *J. Virol.* **2013**, *87*, 7765–7773. [CrossRef] [PubMed]

30. Surviladze, Z.; Dziduszko, A.; Ozbun, M.A. Essential roles for soluble virion-associated heparan sulfonated proteoglycans and growth factors in human papillomavirus infections. *PLoS Pathog.* **2012**, *8*, e1002519. [CrossRef] [PubMed]

31. Woodham, A.W.; da Silva, D.M.; Skeate, J.G.; Raff, A.B.; Ambroso, M.R.; Brand, H.E.; Isas, J.M.; Langen, R.; Kast, W.M. The S100A10 subunit of the annexin A2 heterotetramer facilitates L2-mediated human papillomavirus infection. *PLoS ONE* **2012**, *7*, e43519. [CrossRef] [PubMed]

32. Wustenhagen, E.; Hampe, L.; Boukhallouk, F.; Schneider, M.A.; Spoden, G.A.; Negwer, I.; Koynov, K.; Kast, W.M.; Florin, L. The cytoskeletal adaptor obscurin-like 1 interacts with the human papillomavirus 16 (HPV16) capsid protein L2 and is required for HPV16 endocytosis. *J. Virol.* **2016**, *90*, 10629–10641. [CrossRef] [PubMed]

33. Earnest, J.T.; Hantak, M.P.; Li, K.; McCray, P.B., Jr.; Perlman, S.; Gallagher, T. The tetraspanin CD9 facilitates mers-coronavirus entry by scaffolding host cell receptors and proteases. *PLoS Pathog.* **2017**, *13*, e1006546. [CrossRef] [PubMed]

34. Feneant, L.; Levy, S.; Cocquerel, L. CD81 and hepatitis c virus (HCV) infection. *Viruses* **2014**, *6*, 535–572. [CrossRef] [PubMed]

35. Day, P.M.; Schiller, J.T. The role of furin in papillomavirus infection. *Future Microbiol.* **2009**, *4*, 1255–1262. [CrossRef] [PubMed]

36. Bronnimann, M.P.; Calton, C.M.; Chiquette, S.F.; Li, S.; Lu, M.; Chapman, J.A.; Bratton, K.N.; Schlegel, A.M.; Campos, S.K. Furin cleavage of L2 during papillomavirus infection: Minimal dependence on cyclophilins. *J. Virol.* **2016**, *90*, 6224–6234. [CrossRef] [PubMed]

37. Gambhira, R.; Karanam, B.; Jagu, S.; Roberts, J.N.; Buck, C.B.; Bossis, I.; Alphs, H.H.; Culp, T.; Christensen, N.D.; Roden, R.B.S. A protective and broadly cross-neutralizing epitope of human papillomavirus L2. *J. Virol.* **2007**, *81*, 13927–13931. [CrossRef] [PubMed]

38. Day, P.M.; Gambhira, R.; Roden, R.B.; Lowy, D.R.; Schiller, J.T. Mechanisms of human papillomavirus type 16 neutralization by L2 cross-neutralizing and L1 type-specific antibodies. *J. Virol.* **2008**, *82*, 4638–4646. [CrossRef] [PubMed]

39. Bienkowska-Haba, M.; Patel, H.D.; Sapp, M. Target cell cyclophilins facilitate human papillomavirus type 16 infection. *PLoS Pathog.* **2009**, *5*, e1000524. [CrossRef] [PubMed]

40. Gräßel, L.; Fast, L.A.; Scheffer, K.D.; Boukhallouk, F.; Spoden, G.A.; Tenzer, S.; Boller, K.; Bago, R.; Rajesh, S.; Overduin, M.; et al. The CD63-syntenin-1 complex controls post-endocytic trafficking of oncogenic human papillomaviruses. *Sci. Rep.* **2016**, *6*, 32337. [CrossRef] [PubMed]

41. Broniarczyk, J.; Bergant, M.; Gozdzicka-Jozefiak, A.; Banks, L. Human papillomavirus infection requires the TSG101 component of the ESCRT machinery. *Virology* **2014**, *460–461*, 83–90. [CrossRef] [PubMed]

42. Broniarczyk, J.; Pim, D.; Massimi, P.; Bergant, M.; Gozdzicka-Jozefiak, A.; Crump, C.; Banks, L. The VPS4 component of the ESCRT machinery plays an essential role in hpv infectious entry and capsid disassembly. *Sci. Rep.* **2017**, *7*, 45159. [CrossRef] [PubMed]

43. Muller, K.H.; Spoden, G.A.; Scheffer, K.D.; Brunnhofer, R.; de Brabander, J.K.; Maier, M.E.; Florin, L.; Muller, C.P. Inhibition by cellular vacuolar ATPase impairs human papillomavirus uncoating and infection. *Antimicrob. Agents Chemother.* **2014**, *58*, 2905–2911. [CrossRef] [PubMed]

44. Smith, J.L.; Campos, S.K.; Wandinger-Ness, A.; Ozbun, M.A. Caveolin-1-dependent infectious entry of human papillomavirus type 31 in human keratinocytes proceeds to the endosomal pathway for pH-dependent uncoating. *J. Virol.* **2008**, *82*, 9505–9512. [CrossRef] [PubMed]

45. Sapp, M.; Kraus, U.; Volpers, C.; Snijders, P.J.; Walboomers, J.M.; Streeck, R.E. Analysis of type-restricted and cross-reactive epitopes on virus-like particles of human papillomavirus type 33 and in infected tissues using monoclonal antibodies to the major capsid protein. *J. Gen. Virol.* **1994**, *75 Pt 12*, 3375–3383. [CrossRef] [PubMed]

46. Bienkowska-Haba, M.; Williams, C.; Kim, S.M.; Garcea, R.L.; Sapp, M. Cyclophilins facilitate dissociation of the human papillomavirus type 16 capsid protein L1 from the L2/DNA complex following virus entry. *J. Virol.* **2012**, *86*, 9875–9887. [CrossRef] [PubMed]

47. Lipovsky, A.; Popa, A.; Pimienta, G.; Wyler, M.; Bhan, A.; Kuruvilla, L.; Guie, M.A.; Poffenberger, A.C.; Nelson, C.D.; Atwood, W.J.; et al. Genome-wide siRNA screen identifies the retromer as a cellular entry factor for human papillomavirus. *Proc. Natl. Acad. Sci. USA* **2013**, *110*, 7452–7457. [CrossRef] [PubMed]

48. Popa, A.; Zhang, W.; Harrison, M.S.; Goodner, K.; Kazakov, T.; Goodwin, E.C.; Lipovsky, A.; Burd, C.G.; DiMaio, D. Direct binding of retromer to human papillomavirus type 16 minor capsid protein L2 mediates endosome exit during viral infection. *PLoS Pathog.* **2015**, *11*, e1004699. [CrossRef] [PubMed]

49. Aydin, I.; Villalonga-Planells, R.; Greune, L.; Bronnimann, M.P.; Calton, C.M.; Becker, M.; Lai, K.Y.; Campos, S.K.; Schmidt, M.A.; Schelhaas, M. A central region in the minor capsid protein of papillomaviruses facilitates viral genome tethering and membrane penetration for mitotic nuclear entry. *PLoS Pathog.* **2017**, *13*, e1006308. [CrossRef] [PubMed]

50. Aydin, I.; Weber, S.; Snijder, B.; Samperio Ventayol, P.; Kuhbacher, A.; Becker, M.; Day, P.M.; Schiller, J.T.; Kann, M.; Pelkmans, L.; et al. Large scale RNAi reveals the requirement of nuclear envelope breakdown for nuclear import of human papillomaviruses. *PLoS Pathog.* **2014**, *10*, e1004162. [CrossRef] [PubMed]

51. Calton, C.M.; Bronnimann, M.P.; Manson, A.R.; Li, S.; Chapman, J.A.; Suarez-Berumen, M.; Williamson, T.R.; Molugu, S.K.; Bernal, R.A.; Campos, S.K. Translocation of the papillomavirus L2/vDNA complex across the limiting membrane requires the onset of mitosis. *PLoS Pathog.* **2017**, *13*, e1006200. [CrossRef] [PubMed]

52. DiGiuseppe, S.; Bienkowska-Haba, M.; Guion, L.G.M.; Keiffer, T.R.; Sapp, M. Human papillomavirus major capsid protein L1 remains associated with the incoming viral genome throughout the entry process. *J. Virol.* **2017**. [CrossRef] [PubMed]

53. Ishii, Y.; Nakahara, T.; Kataoka, M.; Kusumoto-Matsuo, R.; Mori, S.; Takeuchi, T.; Kukimoto, I. Identification of TrappC8 as a host factor required for human papillomavirus cell entry. *PLoS ONE* **2013**, *8*, e80297. [CrossRef] [PubMed]

54. Lipovsky, A.; Zhang, W.; Iwasaki, A.; DiMaio, D. Application of the proximity-dependent assay and fluorescence imaging approaches to study viral entry pathways. *Methods Mol. Biol.* **2015**, *1270*, 437–451. [PubMed]

55. Zhang, W.; Kazakov, T.; Popa, A.; DiMaio, D. Vesicular trafficking of incoming human papillomavirus 16 to the Golgi apparatus and endoplasmic reticulum requires γ-secretase activity. *mBio* **2014**, *5*, e01777-14. [CrossRef] [PubMed]

56. Gomez-Navarro, N.; Miller, E. Protein sorting at the ER-Golgi interface. *J. Cell Biol.* **2016**, *215*, 769–778. [CrossRef] [PubMed]

57. Villeneuve, J.; Duran, J.; Scarpa, M.; Bassaganyas, L.; van Galen, J.; Malhotra, V. Golgi enzymes do not cycle through the endoplasmic reticulum during protein secretion or mitosis. *Mol. Biol. Cell* **2017**, *28*, 141–151. [CrossRef] [PubMed]

58. Bonifacino, J.S.; Hurley, J.H. Retromer. *Curr. Opin. Cell Biol.* **2008**, *20*, 427–436. [CrossRef] [PubMed]

59. Burd, C.; Cullen, P.J. Retromer: A master conductor of endosome sorting. *Cold Spring Harb. Perspect. Biol.* **2014**, *6*. [CrossRef] [PubMed]

60. Bergant Marusic, M.; Ozbun, M.A.; Campos, S.K.; Myers, M.P.; Banks, L. Human papillomavirus L2 facilitates viral escape from late endosomes via sorting nexin 17. *Traffic* **2012**, *13*, 455–467. [CrossRef] [PubMed]

61. Pim, D.; Broniarczyk, J.; Bergant, M.; Playford, M.P.; Banks, L. A novel PDZ domain interaction mediates the binding between human papillomavirus 16 L2 and sorting nexin 27 and modulates virion trafficking. *J. Virol.* **2015**, *89*, 10145–10155. [CrossRef] [PubMed]

62. Bergant, M.; Peternel, S.; Pim, D.; Broniarczyk, J.; Banks, L. Characterizing the spatio-temporal role of sorting nexin 17 in human papillomavirus trafficking. *J. Gen. Virol.* **2017**, *98*, 715–725. [CrossRef] [PubMed]

63. Yin, W.; Liu, D.; Liu, N.; Xu, L.; Li, S.; Lin, S.; Shu, X.; Pei, D. SNX17 regulates notch pathway and pancreas development through the retromer-dependent recycling of jag1. *Cell Regen.* **2012**, *1*, 4. [CrossRef] [PubMed]

64. McNally, K.E.; Faulkner, R.; Steinberg, F.; Gallon, M.; Ghai, R.; Pim, D.; Langton, P.; Pearson, N.; Danson, C.M.; Nagele, H.; et al. Retriever is a multiprotein complex for retromer-independent endosomal cargo recycling. *Nat. Cell Biol.* **2017**, *19*, 1214–1225. [CrossRef] [PubMed]

65. De Strooper, B.; Iwatsubo, T.; Wolfe, M.S. Presenilins and γ-secretase: Structure, function, and role in alzheimer disease. *Cold Spring Harb. Perspect. Med.* **2012**, *2*, a006304. [CrossRef] [PubMed]

66. Beel, A.J.; Sanders, C.R. Substrate specificity of γ-secretase and other intramembrane proteases. *Cell. Mol. Life Sci.* **2008**, *65*, 1311–1334. [CrossRef] [PubMed]

67. Andrew, R.J.; Kellett, K.A.; Thinakaran, G.; Hooper, N.M. A greek tragedy: The growing complexity of alzheimer amyloid precursor protein proteolysis. *J. Biol. Chem.* **2016**, *291*, 19235–19244. [CrossRef] [PubMed]

68. De Strooper, B.; Annaert, W.; Cupers, P.; Saftig, P.; Craessaerts, K.; Mumm, J.S.; Schroeter, E.H.; Schrijvers, V.; Wolfe, M.S.; Ray, W.J.; et al. A presenilin-1-dependent gamma-secretase-like protease mediates release of notch intracellular domain. *Nature* **1999**, *398*, 518–522. [CrossRef] [PubMed]

69. Herreman, A.; Serneels, L.; Annaert, W.; Collen, D.; Schoonjans, L.; De Strooper, B. Total inactivation of gamma-secretase activity in presenilin-deficient embryonic stem cells. *Nat. Cell Biol.* **2000**, *2*, 461–462. [CrossRef] [PubMed]

70. Zhang, Z.; Nadeau, P.; Song, W.; Donoviel, D.; Yuan, M.; Bernstein, A.; Yankner, B.A. Presenilins are required for γ-secretase cleavage of β-APP and transmembrane cleavage of Notch-1. *Nat. Cell Biol.* **2000**, *2*, 463–465. [CrossRef] [PubMed]

71. Karanam, B.; Peng, S.; Li, T.; Buck, C.; Day, P.M.; Roden, R.B. Papillomavirus infection requires gamma secretase. *J. Virol.* **2010**, *84*, 10661–10670. [CrossRef] [PubMed]

72. Kwak, K.; Jiang, R.; Wang, J.W.; Jagu, S.; Kirnbauer, R.; Roden, R.B. Impact of inhibitors and L2 antibodies upon the infectivity of diverse alpha and beta human papillomavirus types. *PLoS ONE* **2014**, *9*, e97232. [CrossRef] [PubMed]

73. Duggan, S.P.; McCarthy, J.V. Beyond γ-secretase activity: The multifunctional nature of presenilins in cell signalling pathways. *Cell. Signal.* **2016**, *28*, 1–11. [CrossRef] [PubMed]

74. Choy, R.W.; Cheng, Z.; Schekman, R. Amyloid precursor protein (APP) traffics from the cell surface via endosomes for amyloid β (Aβ) production in the trans-Golgi network. *Proc. Natl. Acad. Sci. USA* **2012**, *109*, E2077–E2082. [CrossRef] [PubMed]

75. Small, S.A.; Gandy, S. Sorting through the cell biology of alzheimer's disease: Intracellular pathways to pathogenesis. *Neuron* **2006**, *52*, 15–31. [CrossRef] [PubMed]

76. Auvinen, E.; Kujari, H.; Arstila, P.; Hukkanen, V. Expression of the L2 and E7 genes of the human papillomavirus type 16 in female genital dysplasias. *Am. J. Pathol.* **1992**, *141*, 1217–1224. [PubMed]

77. Hagensee, M.E.; Yaegashi, N.; Galloway, D.A. Self-assembly of human papillomavirus type 1 capsids by expression of the L1 protein alone or by coexpression of the L1 and L2 capsid proteins. *J. Virol.* **1993**, *67*, 315–322. [PubMed]

78. Kämper, N.; Day, P.M.; Nowak, T.; Selinka, H.C.; Florin, L.; Bolscher, J.; Hilbig, L.; Schiller, J.T.; Sapp, M. A membrane-destabilizing peptide in capsid protein L2 is required for egress of papillomavirus genomes from endosomes. *J. Virol.* **2006**, *80*, 759–768. [CrossRef] [PubMed]

79. Bronnimann, M.P.; Chapman, J.A.; Park, C.K.; Campos, S.K. A transmembrane domain and GxxxG motifs within L2 are essential for papillomavirus infection. *J. Virol.* **2013**, *87*, 464–473. [CrossRef] [PubMed]

80. DiGiuseppe, S.; Keiffer, T.R.; Bienkowska-Haba, M.; Luszczek, W.; Guion, L.G.; Muller, M.; Sapp, M. Topography of the human papillomavirus minor capsid protein L2 during vesicular trafficking of infectious entry. *J. Virol.* **2015**, *89*, 10442–10452. [CrossRef] [PubMed]

81. DiGiuseppe, S.; Luszczek, W.; Keiffer, T.R.; Bienkowska-Haba, M.; Guion, L.G.; Sapp, M.J. Incoming human papillomavirus type 16 genome resides in a vesicular compartment throughout mitosis. *Proc. Natl. Acad. Sci. USA* **2016**, *113*, 6289–6294. [CrossRef] [PubMed]

82. Garred, O.; van Deurs, B.; Sandvig, K. Furin-induced cleavage and activation of shiga toxin. *J. Biol. Chem.* **1995**, *270*, 10817–10821. [CrossRef] [PubMed]

83. Gordon, V.M.; Klimpel, K.R.; Arora, N.; Henderson, M.A.; Leppla, S.H. Proteolytic activation of bacterial toxins by eukaryotic cells is performed by furin and by additional cellular proteases. *Infect. Immun.* **1995**, *63*, 82–87. [PubMed]

84. Epand, R.M. Fusion peptides and the mechanism of viral fusion. *Biochim. Biophys. Acta* **2003**, *1614*, 116–121. [CrossRef]

85. Tamm, L.K.; Han, X.; Li, Y.; Lai, A.L. Structure and function of membrane fusion peptides. *Biopolymers* **2002**, *66*, 249–260. [CrossRef] [PubMed]

86. Lorieau, J.L.; Louis, J.M.; Bax, A. The complete influenza hemagglutinin fusion domain adopts a tight helical hairpin arrangement at the lipid:Water interface. *Proc. Natl. Acad. Sci. USA* **2010**, *107*, 11341–11346. [CrossRef] [PubMed]

87. Hofmann, M.W.; Weise, K.; Ollesch, J.; Agrawal, P.; Stalz, H.; Stelzer, W.; Hulsbergen, F.; de Groot, H.; Gerwert, K.; Reed, J.; et al. De novo design of conformationally flexible transmembrane peptides driving membrane fusion. *Proc. Natl. Acad. Sci. USA* **2004**, *101*, 14776–14781. [CrossRef] [PubMed]

88. White, J.M.; Delos, S.E.; Brecher, M.; Schornberg, K. Structures and mechanisms of viral membrane fusion proteins: Multiple variations on a common theme. *Crit. Rev. Biochem. Mol. Biol.* **2008**, *43*, 189–219. [CrossRef] [PubMed]

89. Wakabayashi, T.; Craessaerts, K.; Bammens, L.; Bentahir, M.; Borgions, F.; Herdewijn, P.; Staes, A.; Timmerman, E.; Vandekerckhove, J.; Rubinstein, E.; et al. Analysis of the γ-secretase interactome and validation of its association with tetraspanin-enriched microdomains. *Nat. Cell Biol.* **2009**, *11*, 1340–1346. [CrossRef] [PubMed]

90. Wolfe, M.S. Structure, mechanism and inhibition of gamma-secretase and presenilin-like proteases. *Biol. Chem.* **2010**, *391*, 839–847. [CrossRef] [PubMed]

91. Kornilova, A.Y.; Das, C.; Wolfe, M.S. Differential effects of inhibitors on the γ-secretase complex. Mechanistic implications. *J. Biol. Chem.* **2003**, *278*, 16470–16473. [CrossRef] [PubMed]

92. Li, Y.; Bohm, C.; Dodd, R.; Chen, F.; Qamar, S.; Schmitt-Ulms, G.; Fraser, P.E.; St George-Hyslop, P.H. Structural biology of presenilin 1 complexes. *Mol. Neurodegener.* **2014**, *9*, 59. [CrossRef] [PubMed]

93. Lipovsky, A.; Erden, A.; Kanaya, E.; Zhang, W.; Crite, M.; Bradfield, C.; MacMicking, J.; DiMaio, D.; Schoggins, J.W.; Iwasaki, A. The cellular endosomal protein stannin inhibits intracellular trafficking of human papillomavirus during virus entry. *J. Gen. Virol.* **2017**. [CrossRef] [PubMed]

94. Wiens, M.E.; Smith, J.G. α-defensin HD5 inhibits human papillomavirus 16 infection via capsid stabilization and redirection to the lysosome. *mBio* **2017**, *8*, e02304-16. [CrossRef] [PubMed]

95. Day, P.M.; Thompson, C.D.; Lowy, D.R.; Schiller, J.T. Interferon gamma prevents infectious entry of human papillomavirus 16 via an L2-dependent mechanism. *J. Virol.* **2017**, *91*. [CrossRef] [PubMed]

96. Calton, C.M.; Schlegel, A.M.; Chapman, J.A.; Campos, S.K. Human papillomavirus type 16 does not require cathepsin L or B for infection. *J. Gen. Virol.* **2013**, *94*, 1865–1869. [CrossRef] [PubMed]

97. Akache, B.; Grimm, D.; Shen, X.; Fuess, S.; Yant, S.R.; Glazer, D.S.; Park, J.; Kay, M.A. A two-hybrid screen identifies cathepsins B and L as uncoating factors for adeno-associated virus 2 and 8. *Mol. Ther. J. Am. Soc. Gene Ther.* **2007**, *15*, 330–339. [CrossRef] [PubMed]

98. Ebert, D.H.; Deussing, J.; Peters, C.; Dermody, T.S. Cathepsin L and cathepsin B mediate reovirus disassembly in murine fibroblast cells. *J. Biol. Chem.* **2002**, *277*, 24609–24617. [CrossRef] [PubMed]

99. Schafer, G.; Graham, L.M.; Lang, D.M.; Blumenthal, M.J.; Bergant Marusic, M.; Katz, A.A. Vimentin modulates infectious internalization of human papillomavirus 16 pseudovirions. *J. Virol.* **2017**, *91*. [CrossRef] [PubMed]

100. Schiller, J.T.; Day, P.M.; Kines, R.C. Current understanding of the mechanism of HPV infection. *Gynecol. Oncol.* **2010**, *118*, S12–S17. [CrossRef] [PubMed]

101. Smith, A.E.; Helenius, A. How viruses enter animal cells. *Science* **2004**, *304*, 237–242. [CrossRef] [PubMed]
102. Tsai, B. Penetration of nonenveloped viruses into the cytoplasm. *Annu. Rev. Cell Dev. Biol.* **2007**, *23*, 23–43. [CrossRef] [PubMed]
103. Pyeon, D.; Pearce, S.M.; Lank, S.M.; Ahlquist, P.; Lambert, P.F. Establishment of human papillomavirus infection requires cell cycle progression. *PLoS Pathog.* **2009**, *5*, e1000318. [CrossRef] [PubMed]
104. Lewis, P.F.; Emerman, M. Passage through mitosis is required for oncoretroviruses but not for the human immunodeficiency virus. *J. Virol.* **1994**, *68*, 510–516. [PubMed]
105. Schatz, P.J. Use of peptide libraries to map the substrate specificity of a peptide-modifying enzyme: A 13 residue consensus peptide specifies biotinylation in *Escherichia coli*. *Nat. Biotechnol.* **1993**, *11*, 1138–1143. [CrossRef]
106. Murphy, J.R. Mechanism of diphtheria toxin catalytic domain delivery to the eukaryotic cell cytosol and the cellular factors that directly participate in the process. *Toxins* **2011**, *3*, 294–308. [CrossRef] [PubMed]
107. Pirazzini, M.; Azarnia Tehran, D.; Leka, O.; Zanetti, G.; Rossetto, O.; Montecucco, C. On the translocation of botulinum and tetanus neurotoxins across the membrane of acidic intracellular compartments. *Biochim. Biophys. Acta* **2016**, *1858*, 467–474. [CrossRef] [PubMed]
108. Day, P.M.; Baker, C.C.; Lowy, D.R.; Schiller, J.T. Establishment of papillomavirus infection is enhanced by promyelocytic leukemia protein (PML) expression. *Proc. Natl. Acad. Sci. USA* **2004**, *101*, 14252–14257. [CrossRef] [PubMed]
109. Lallemand-Breitenbach, V.; de The, H. PML nuclear bodies. *Cold Spring Harb. Perspect. Biol.* **2010**, *2*, a000661. [CrossRef] [PubMed]
110. Sahin, U.; Lallemand-Breitenbach, V.; de The, H. PML nuclear bodies: Regulation, function and therapeutic perspectives. *J. Pathol.* **2014**, *234*, 289–291. [CrossRef] [PubMed]
111. Scherer, M.; Stamminger, T. Emerging role of PML nuclear bodies in innate immune signaling. *J. Virol.* **2016**, *90*, 5850–5854. [CrossRef] [PubMed]
112. Weidtkamp-Peters, S.; Lenser, T.; Negorev, D.; Gerstner, N.; Hofmann, T.G.; Schwanitz, G.; Hoischen, C.; Maul, G.; Dittrich, P.; Hemmerich, P. Dynamics of component exchange at PML nuclear bodies. *J. Cell Sci.* **2008**, *121*, 2731–2743. [CrossRef] [PubMed]
113. Sahin, U.; Ferhi, O.; Jeanne, M.; Benhenda, S.; Berthier, C.; Jollivet, F.; Niwa-Kawakita, M.; Faklaris, O.; Setterblad, N.; de The, H.; et al. Oxidative stress-induced assembly of pml nuclear bodies controls sumoylation of partner proteins. *J. Cell Biol.* **2014**, *204*, 931–945. [CrossRef] [PubMed]
114. Everett, R.D. DNA viruses and viral proteins that interact with PML nuclear bodies. *Oncogene* **2001**, *20*, 7266–7273. [CrossRef] [PubMed]
115. Everett, R.D.; Chelbi-Alix, M.K. PML and PML nuclear bodies: Implications in antiviral defence. *Biochimie* **2007**, *89*, 819–830. [CrossRef] [PubMed]
116. Florin, L.; Schafer, F.; Sotlar, K.; Streeck, R.E.; Sapp, M. Reorganization of nuclear domain 10 induced by papillomavirus capsid protein L2. *Virology* **2002**, *295*, 97–107. [CrossRef] [PubMed]
117. Becker, K.A.; Florin, L.; Sapp, C.; Sapp, M. Dissection of human papillomavirus type 33 L2 domains involved in nuclear domains (ND) 10 homing and reorganization. *Virology* **2003**, *314*, 161–167. [CrossRef]
118. Bund, T.; Spoden, G.A.; Koynov, K.; Hellmann, N.; Boukhallouk, F.; Arnold, P.; Hinderberger, D.; Florin, L. An L2 sumo interacting motif is important for PML localization and infection of human papillomavirus type 16. *Cell. Microbiol.* **2014**, *16*, 1179–1200. [CrossRef] [PubMed]
119. Marusic, M.B.; Mencin, N.; Licen, M.; Banks, L.; Grm, H.S. Modification of human papillomavirus minor capsid protein L2 by sumoylation. *J. Virol.* **2010**, *84*, 11585–11589. [CrossRef] [PubMed]
120. Kieback, E.; Muller, M. Factors influencing subcellular localization of the human papillomavirus L2 minor structural protein. *Virology* **2006**, *345*, 199–208. [CrossRef] [PubMed]
121. Stepp, W.H.; Meyers, J.M.; McBride, A.A. Sp100 provides intrinsic immunity against human papillomavirus infection. *mBio* **2013**, *4*, e00845-13. [CrossRef] [PubMed]
122. Chen, Y.C.; Kappel, C.; Beaudouin, J.; Eils, R.; Spector, D.L. Live cell dynamics of promyelocytic leukemia nuclear bodies upon entry into and exit from mitosis. *Mol. Biol. Cell* **2008**, *19*, 3147–3162. [CrossRef] [PubMed]

viruses

MDPI

Review

Mechanisms by which HPV Induces a Replication Competent Environment in Differentiating Keratinocytes

Cary A. Moody [1,2]

[1] Department of Microbiology and Immunology, University of North Carolina at Chapel Hill,
     Chapel Hill, NC 27599, USA; camoody@med.unc.edu
[2] Lineberger Comprehensive Cancer Center, University of North Carolina at Chapel Hill,
     Chapel Hill, NC 27599, USA

Received: 31 August 2017; Accepted: 15 September 2017; Published: 19 September 2017

Abstract: Human papillomaviruses (HPV) are the causative agents of cervical cancer and are also associated with other genital malignancies, as well as an increasing number of head and neck cancers. HPVs have evolved their life cycle to contend with the different cell states found in the stratified epithelium. Initial infection and viral genome maintenance occurs in the proliferating basal cells of the stratified epithelium, where cellular replication machinery is abundant. However, the productive phase of the viral life cycle, including productive replication, late gene expression and virion production, occurs upon epithelial differentiation, in cells that normally exit the cell cycle. This review outlines how HPV interfaces with specific cellular signaling pathways and factors to provide a replication-competent environment in differentiating cells.

Keywords: virus; HPV; cell cycle; differentiation; replication; DNA damage response

1. Introduction

Human papillomaviruses (HPV) are non-enveloped, small DNA viruses that exhibit a strict tropism for epithelial cells. Over 200 types of HPVs have been identified and are classified into five evolutionary genera (α, β, γ, μ, ν) based on DNA sequence similarity [1]. The alpha group is the largest, containing approximately 64 HPV types, and is divided based on tropism of each type for cutaneous or mucosal epithelium. The cutaneous types cause common warts, which are rarely associated with malignancy [2]. About 40 alpha HPVs infect the mucosal epithelium and are categorized as high-risk or low-risk based on their association with cancer [1,3]. Low-risk types (e.g., HPV11 and HPV6) are most commonly associated with benign genital warts, but are also implicated in the development of laryngeal papillomas. The fifteen types termed high-risk (16, 18, 31, 33, 35, 39, 45, 51, 51, 56, 58, 59, 68, 73, 82) are classified as oncogenic based on their association with anogenital cancers [4]. In addition, certain high-risk types, particularly HPV16, infect the oropharyngeal mucosa and are associated with an increasing number of head and neck cancers [5]. Beta HPVs exhibit a tropism for the cutaneous epithelium, with infection occurring early in life and typically producing an asymptomatic infection [6]. However, persistent infection with certain types of beta HPVs are associated with the development of non-melanoma skin cancers at sun exposed sites, particularly in immunosuppressed patients and patients with the rare disease epidermodysplasia verruciformis [7]. The mu, nu and gamma HPVs infect the cutaneous epithelium and are most commonly associated with the formation of benign papillomas [8].

2. HPV Life Cycle

The life cycle of HPV is intimately linked to the differentiation status of the host cell keratinocyte and is characterized by three distinct phases of replication [9,10] (Figure 1). High-risk and low-risk HPVs initiate infection by gaining access to the proliferating basal cells of the stratified epithelium through a microwound [11]. Upon entry, HPV undergoes a transient round of replication referred to as "establishment replication", which results in a copy number of 50–100 viral genomes per cell. Viral episomes are subsequently maintained in the undifferentiated basal cells by replicating along with the host cell chromosomes. Only upon epithelial differentiation is the productive phase of the viral life cycle activated, resulting in the amplification of viral genomes to thousands of viral copies per cell in the suprabasal layers, as well as activation of late gene expression and virion assembly and release [10,12]. Regulation of the viral life cycle in this manner allows HPV to avoid detection by the immune response as high levels of viral gene expression as well as virion production are restricted to the uppermost layers of the epithelium, which are not under immune surveillance [4].

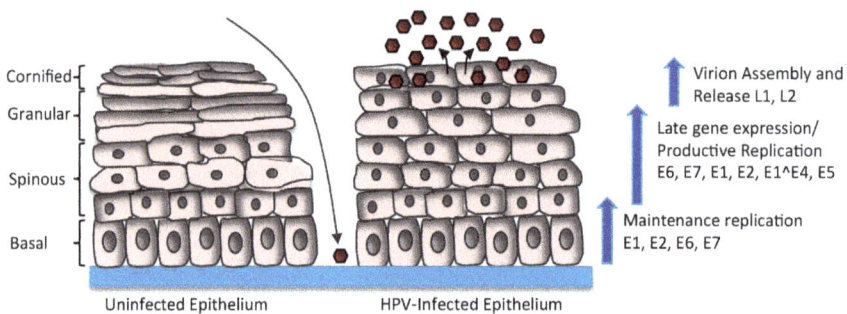

Figure 1. Human papillomavirus (HPV) Life Cycle. HPV infects the basal layer of the stratified epithelium through a microwound. Upon entry into the cell, the virus transiently amplifies to 50–100 copies per cell. HPV genomes are maintained at a stable copy number in undifferentiated basal cells by replicating along with cellular DNA. Upon differentiation, the productive phase of the life cycle is activated, resulting in late gene expression and amplification of viral genomes to thousands of copies per cell. The expression of E6 and E7 allows for cell cycle re-entry upon differentiation, providing cellular factors for productive replication. E4 and E5 also contribute to efficient productive replication. Expression of L1 and L2 promotes the encapsidation of newly replicated genomes, resulting in virion release from the uppermost layers of the epithelium (brown hexagons).

Due to the small coding capacity of the viral genome, HPV depends on the host DNA replication machinery to synthesize its DNA. While readily available in undifferentiated cells to stably maintain viral episomes, epithelial differentiation normally results in an exit from the cell cycle, limiting the availability of replication machinery in post-mitotic cells [13]. This provides a conundrum for HPV since differentiation is required to activate the productive phase of the life cycle, yet HPV also depends on cellular factors for replication. To support productive replication, HPV employs numerous mechanisms to subvert key regulatory pathways that regulate host cell replication, in turn maintaining differentiating cells active in the cell cycle. As such, HPV is able to reactivate cellular genes and signaling pathways necessary to support late gene expression and amplification of viral DNA. The majority of our insights into productive HPV replication have emerged from studying the alpha HPV types, primarily the high-risk types HPV16, HPV18 and HPV31. This review will focus on the mechanisms by which alpha HPVs renders post-mitotic, differentiating cells permissive for DNA synthesis during the productive phase of the viral life cycle.

3. HPV Genome Organization

The HPV genome exists as a covalently closed circle (episome) of approximately 8 kb [8]. HPV genomes are histone-associated in the virion as well as in infected cells, exhibiting a nucleosomal pattern similar to that of cellular DNA [14,15]. HPV genomes contain six to eight open reading frames (ORF) that are expressed as polycistronic transcripts that are then alternatively spliced to yield individual gene products [16,17] (Figure 2). High-risk HPV genomes contain two main promoters that are active at different stages in the viral life cycle [18–20]. In undifferentiated epithelial cells, viral gene expression is regulated by the early promoter, which is located adjacent to the E6 ORF in the upstream regulatory region (URR) and is referred to as p97 for HPV16 and HPV31, and p105 for HPV18. The early promoter directs expression of E1 and E2, which is necessary for viral replication. E1 is an ATP-dependent helicase that facilitates unwinding of the viral DNA and also recruits cellular factors to the viral origin of replication, located in the URR [21]. E2 is a sequence-specific DNA binding protein that has multiple binding sites in the URR. E2 binds and recruits E1 to a specific E1 binding site in the viral origin. E2 also regulates viral gene expression from the early promoter. In addition, E2 contributes to episomal maintenance in undifferentiated cells by tethering viral genomes to host mitotic chromosomes [22]. E6 and E7, which are the oncoproteins for the high-risk HPV types, are also expressed from the early promoter. E6 and E7 contribute to viral replication through their ability to modulate cell cycle control, cell survival, cellular differentiation, immune evasion, as well as DNA damage responses [23–27]. E1^E4 is encoded by a spliced RNA that fuses the first five amino acids of the E1 ORF with E4 [28]. While E1^E4 and E5 are expressed at low levels from the early promoter in undifferentiated cells, the high-risk E4 and E5 proteins seem to be primarily involved in facilitating efficient productive replication in differentiating cells [29–33]. Some HPV types express a fusion of E8 and the C-terminal half of the E2 ORF (E8^E2), which initiates from a promoter in the E1 ORF and functions to limit viral replication and transcription in undifferentiated and differentiated cells [34]. The late promoter is located in the E7 ORF (p742 HPV31, p811 HPV18, p670 HPV16) and is activated upon epithelial differentiation [35–37]. The late promoter is not regulated by E2 and drives high levels of expression of E1 and E2, as well as E1^E4 and E5 to facilitate productive viral replication. In addition, the late promoter directs expression of the L1 and L2 capsid genes to allow for encapsidation of viral genomes in the uppermost layers of the epithelium.

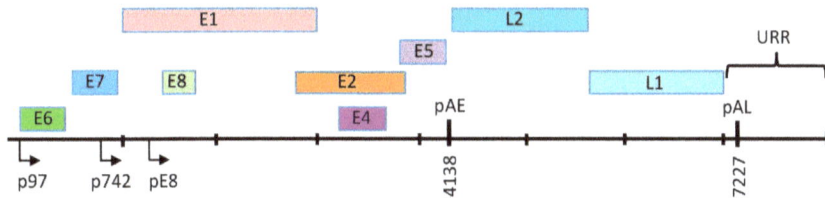

Figure 2. Linear depiction of the HPV31 genome. The open reading frames (ORF) are indicated by the color blocks. The early promoter is located upstream of the E6 ORF (p97) and the late promoter is located in the E7 ORF (p742). E8^E2 is expressed from a promoter located in the E1 ORF (pE8). The early polyadenylation site is located at the 3′ end of the E5 ORF (pAE) and the late polyadenylation site (pAL) is located in the URR (Upstream Regulatory Region). The origin of replication, as well as E1 and E2 binding sites are also located in the URR.

4. Regulation of Viral Gene Expression upon Keratinocyte Differentiation

Efficient amplification of HPV genomes upon differentiation requires activation of the late promoter to provide increased levels of E1, E2, E1^E4 and E5 [10]. The early promoter remains active upon differentiation, directing expression of E6 and E7, which is also necessary for late viral events. The tight link between differentiation and late gene expression suggests that differentiation-specific factors are required for late promoter activation. Chromatin rearrangements and histone modifications are detected at the late promoter upon differentiation, though how this is regulated remains unclear [38,39]. A variety of transcription factors have been shown to bind to the late promoter in the context of complete, episomal genomes, both in undifferentiated and differentiated cells, including c-Myb, C/EBPα, C/EBPβ, NFAT (Nuclear Factor of Activated T-cells), YY1 (Yin Yang 1), NF1 (Nuclear Factor 1), Oct-1 (Octamer-binding transcription factor 1), c-Jun, and Sp1 (Specificity Protein 1) [38,40]. However, only the LIP (Liver-enriched Inhibitory Protein) and LAP (Liver-enriched Activator Protein) isoforms of C/EBPα have been shown to regulate late promoter activity [41]. More recent studies have shown that transcription elongation regulates late promoter activity through the recruitment of elongation mediators (e.g., CDK8, BRD4) to viral genomes upon differentiation [42]. Late gene expression is also regulated by alternative splicing and changes in polyadenylation site usage [16]. Upon differentiation, read-through of the early polyadenylation site (pAE) located at the end of the E5 ORF allows late transcripts to be polyadenylated at the late polyadenylation (pAL) site located in the URR, facilitating expression of L1 and L2. Transcriptional read-through may be influenced by E2 expression, which increases in the mid to upper epithelial layers of the epithelium and has been shown to repress polyadenylation at the early site [43–45]. Polyadenylation is co-transcriptionally regulated with splicing, and certain splicing factors have been shown to influence polyadenylation site usage for HPV16 [46,47]. Splicing of HPV transcripts is positively regulated by Serine Arginine splicing factors (SRSF) (e.g., SRSF1, SRSF2, SRSF3), which increase upon differentiation and are regulated transcriptionally by E2 [46,48]. SRSF9 has also been shown to increase the efficiency of late RNA splicing [49]. Studies by the Parish lab recently demonstrated that CTCF insulator proteins regulate viral transcript splicing upon differentiation through binding to the HPV18 E2 ORF [50]. Mutation of the E2 CTCF binding site in the context of the HPV18 genome results in increased levels of E6 and E7 and increased proliferative capacity in suprabasal cells [50]. The E2 CTCF binding site is conserved across high-risk types, suggesting that HPV has evolved CTCF recruitment to viral genomes to control the levels of E6 and E7 upon differentiation, possibly to facilitate eventual exit from the cell cycle to allow for virion assembly and release.

5. Maintenance of Proliferative Potential in Differentiating Cells

5.1. Disruption of Rb/E2F Complexes

As normal, uninfected cells leave the basal layer, they lose proliferative potential and begin a terminal differentiation program [13]. However, a fundamentally important aspect of the HPV life cycle is to maintain cell cycle competence in differentiating epithelial cells to provide cellular factors for productive replication. E7 plays a critical role in this process though the binding and targeted degradation of the tumor suppressor pRb, as well as the related pocket proteins p107 and p130 [51] (Figure 3). Rb family members regulate the G1 to S-phase transition by controlling the activity of E2F transcription factors [52]. E7 binds to Rb family members through a conserved LXCXE domain located in the extreme C-terminus that disrupts the interaction between Rb and E2F transcription factors [53,54]. Disruption of the Rb/E2F interaction results in constitutive activation of E2F-resposive genes, allowing E7 to push differentiating cells back into S-phase, disrupting suprabasal quiescence and reactivating cellular DNA synthesis [55–58]. As a result, suprabasal cells exhibit markers of differentiation, as well as markers of cell cycle re-entry, including PCNA, cyclin A and cyclin E [55,59]. Low-risk E7 proteins also bind pRb, p107 and p130, but with much lower affinity, and only target p130 for degradation [53,60,61]. The loss of E7 expression in the context of HPV16 infection prevents the induction of host cell replication machinery and productive viral replication in suprabasal epithelial cells of organotypic raft cultures, which recapitulate the three-dimensional architecture of the stratified epithelium [62,63]. These studies highlight the importance of E7 in differentiation-dependent viral events. In addition to Rb family members, the interaction between E7 and type 1 histone deacetylases (HDAC1-3) is also important in maintaining E2F activation upon differentiation and facilitating viral replication [64–66]. HPV31 E7 specifically increases the levels of E2F2 by preventing HDAC binding to the *e2f2* promoter [65]. The increase in E2F2 is necessary for productive viral replication, though the downstream targets of E2F2 have not yet been identified.

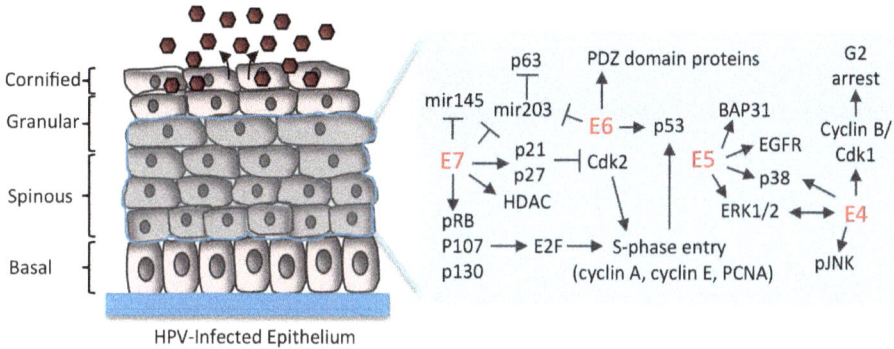

HPV-Infected Epithelium

Figure 3. Cellular factors targeted by HPV proteins to maintain cell cycle competency in suprabasal cells. E7 pushes differentiating cells back into the cell cycle by binding to pRb, p107 and p130, which regulate entry into S-phase by negatively regulating E2F transcription factors. Disruption of the pRb/E2F interaction by E7 allows for constitutive activation of E2F-responsive genes, allowing for S-phase re-entry by post-mitotic cells. Unscheduled S-phase entry induced by E7 results in increased p53 that is targeted for degradation by E6 to avoid apoptosis or cell cycle arrest in G1, as well as to block p53's negative effects on productive replication. E5 contributes to productive viral replication by maintaining cell cycle competency upon differentiation through interaction with BAP31, as well as through activation of epidermal growth factor receptor (EGFR), mitogen activated protein kinase (p38MAPK) and extracellular signal-regulated kinase (ERK)1/2. E4 may increase the efficiency of viral genome amplification by sustaining a G2-arrested environment upon differentiation, and through activation of MAPK signaling (p38, ERK1/2, pJNK). T bars indicate inhibition. Arrows indicate activation. HDAC: histone deacetylase; PCNA: proliferating cell nuclear antigen.

5.2. Uncoupling of Differentiation From Proliferation

Normal epithelial differentiation results in cell cycle arrest that is carried out by increased expression of the cyclin-dependent kinase inhibitors p21Cip1 and p27Kip1, which inhibits the activity of cyclin-dependent kinase 2 (Cdk2) [67]. Cdk2 facilitates G1 to S-phase entry and progression through interaction with cyclin E and cyclin A, respectively [68]. To circumvent this potential block, E7 targets cellular molecules that link differentiation with cell cycle exit (Figure 3). E7 does not affect the differentiation-dependent increase in p21, but rather binds to p21, at least in part through its Rb binding domain, in turn delaying differentiation and blocking the inhibitory effects on Cdk2 activity to establish a proliferative environment [57,69]. Low-risk E7 proteins do not bind as efficiently to p21 and are therefore not as successful at mitigating the inhibitory effects of p21 [57]. High-risk E7 proteins also maintain Cdk2 activity through direct interaction with cyclin E and cyclin A, as well as through maintaining high levels of the Cdc25a phosphatase, which removes inhibitory phosphorylation from Cdk2 [70–72]. Cdk2 activity may also contribute to productive viral replication by regulating the cellular localization of the E1 viral helicase. Cdk2-dependent phosphorylation of E1 prevents its nuclear export, leading to accumulation of E1 in the nucleus, which may allow for rapid amplification of viral genomes upon differentiation [73,74]. Proliferative potential in differentiating cells is also maintained by the E5 protein. E5 is expressed at high levels in suprabasal cells and contributes to efficient productive replication of HPV16 and HPV31 [33,75]. Loss of E5 expression in the context of the HPV31 genome results in decreased cyclin A and cyclin B levels upon methylcellulose-induced differentiation and reduces colony formation following differentiation [33]. Colony formation requires E5's ability to interact with B cell associated protein 31 (BAP31), an ER chaperone and regulator of apoptosis [76]. However, the mechanism by which this interaction maintains proliferative competence in differentiating cells is currently unclear. E5 also modulates several growth pathways that may contribute to viral genome amplification, including signaling through the epidermal growth factor receptor (EGFR), as well as activation of p38MAPK and ERK1/2 [77].

5.3. E6 Abrogation of p53 and Targeting of PDZ (PSD95/DLG1/ZO-1) Domain-Containing Proteins

To facilitate re-entry into the cell cycle and viral genome amplification in suprabasal cells, the activities of E7 coordinate with those of E6 [26] (Figure 3). One of the key functions of E6 is the inactivation of p53. For high-risk types, E6 promotes p53 ubiquitylation and proteasome-dependent degradation through interaction with the E6AP ubiquitin ligase [78–80]. E6-mediated p53 degradation is thought to protect cells from apoptosis or growth arrest due to E7-mediated cell cycle re-entry in the suprabasal layers. However, recent studies indicate that p53 negatively regulates viral genome amplification. E6 mutants in the context of the HPV18 genome that are unable to destabilize p53 result in fewer suprabasal cells supporting viral genome amplification in organotypic raft cultures [81]. The mechanism by which p53 negatively regulates productive viral genome amplification is unclear, but may be through interaction with the E2 origin binding protein [82,83]. E6 proteins also contribute to replication competence through the targeting of specific cellular proteins containing PDZ (PSD95/DLG1/ZO-1) domains [84]. E6 interacts with PDZ proteins through a C-terminal PDZ domain binding motif (PBM) that is found only in high-risk E6 proteins, suggesting this motif serves as a signature for oncogenic potential [85,86]. Most of the PDZ proteins that interact with E6 are targeted for proteasome-dependent degradation, or have an altered cellular localization [84]. PDZ proteins shown to associate with E6 are involved in the regulation of cell growth and polarity, as well as signal transduction pathways involved in cell proliferation, apoptosis, migration and intracellular trafficking [84]. The E6 PBM has been shown to play an essential role in viral genome amplification. Human foreskin keratinocytes transfected with HPV18 genomes containing a mutation in the E6 PBM exhibit a loss of productive viral replication and late gene expression in organotypic raft cultures, correlating with a decrease in the number of S-phase competent cells in the suprabasal layer [87]. A role for the E6 PBM in productive replication has also been observed for the high-risk types HPV31 and HPV16 [88,89]. More recent studies demonstrated that the E6 PBM protects the mitotic integrity of keratinocytes containing HPV18 episomes, with loss of the PBM domain leading to mitotic abnormalities that prevents the expansion of suprabasal cells to support vegetative viral replication [90]. What specific PDZ proteins are targeted by E6 to preserve mitotic integrity and to promote viral replication have yet to be defined.

5.4. Regulation of Differentiation-Induced microRNA Expression

While the HPV genome does not encode microRNAs, E6 and E7 of high-risk types have been shown to modulate the expression of cellular microRNAs to facilitate viral replication in differentiating cells [91] (Figure 3). microRNA-203 (mir203) is normally induced concomitantly with epithelial differentiation and restricts the proliferative potential of differentiating cells by repressing the expression of the p53 homolog p63 [92]. p63 is required for maintaining proliferative potential and acts as a switch between proliferation and differentiation [93]. Studies from the Laimins lab demonstrated that p63 is required for productive replication of HPV31 [94]. Expression of HPV31 E6 and E7 prevents upregulation of mir203 upon differentiation, which is necessary to maintain p63 in differentiating cells and presumably provide a proliferative environment for productive viral replication [95]. In support of this, knockdown of p63 expression in differentiating HPV31 positive keratinocytes using shRNAs results in decreased levels of cell cycle proteins, including cyclins A, B, and E, as well as Cdc25c, Cdk1 and Cdk2 [94]. mir145 is also normally induced upon differentiation and has been shown to negatively regulate the productive phase of the HPV31 life cycle [96]. mir145 regulates the levels of the transcription factor KLF4 (Kruppel-like factor 4), which is a target gene of p63 that plays a role in proliferation, differentiation, and maintenance of stem cells [97,98]. In the stratified epithelium, KLF4 also regulates expression of late epidermal differentiation markers and contributes to the formation of the cornified layer. KLF4 is present at high levels upon differentiation in HPV31 positive cells and is necessary for the productive phase of the viral life cycle [99]. HPV31 regulates KLF4 levels transcriptionally by p63, but also post-transcriptionally by E7-mediated suppression of differentiation-induced mir145 expression [96,99]. KLF4 levels are also regulated post-translationally by E6's ability to prevent inhibitory phosphorylation and sumoylation of KFL4 [99]. KLF4 directly activates late viral gene expression, and thus productive viral replication, by binding to the HPV31 URR in a complex with BLIMP1. Furthermore, KLF4 expression is necessary to maintain cyclin A and cyclin B in suprabasal cells [99]. KLF4 therefore has multiple functions in promoting the productive phase of the viral life cycle. In addition to KLF4, p63 also regulates expression of the DNA repair factors Rad51 and BRCA2, as well as activation of the checkpoint kinase Chk2 in HPV31 positive keratinocytes [94]. As described in more detail below, Chk2 kinase activity and Rad51 have been shown to be required for productive replication of HPV31 [100,101]. These studies suggest that upon cell cycle re-entry, HPV's ability to modulate differentiation-induced microRNAs results in maintenance of p63 levels, prolonging proliferative potential and ensuring the expression of a subset of cellular genes necessary for productive viral replication as well as late gene expression. In addition, p63 may contribute to activation of the DNA damage response that is necessary for viral DNA synthesis in differentiating cells.

6. Establishment of a G2-Arrested Environment

To provide a replication-competent environment upon differentiation, high-risk and low-risk E7 proteins push post-mitotic cells back into the cell cycle, rather than maintaining cells active in S-phase upon differentiation [102,103]. E7-induced cell cycle re-entry has traditionally been thought to result in an S-phase environment that provides HPV access to replication machinery that supports productive viral replication. However, more recent studies indicate that productive viral replication occurs post-cellular DNA synthesis in cells that are subsequently arrested in G2 [104–106]. Using organotypic raft cultures of HPV18 positive keratinocytes, Wang et al., demonstrated that cells undergoing viral genome amplification exhibit markers of G2/M arrest, including high levels of cytoplasmic cyclin B1 and inactive cyclin-dependent kinase 1 (Cdk1) [105]. Cdk1 normally forms a complex with cyclin B1 in the nucleus to stimulate entry into mitosis. In addition, these cells also contain the inactive form of the Cdc25C phosphatase, which functions to remove inhibitory phosphorylation from Cdk1 to allow entry into mitosis [107]. Overall, these studies indicate that HPV requires G2 arrest upon differentiation to support the productive phase of the viral life cycle.

The mechanism by which HPV induces G2 arrest upon differentiation is currently unclear. Arrest in G2 typically occurs in response to DNA damage or incomplete replication, which activates the ATM (Ataxia-Telangiectasia Mutated) and ATR (ATM and Rad3-related) DNA damage kinases [108]. ATM and ATR phosphorylate the checkpoint kinases Chk2 and Chk1, leading to their activation and the phosphorylation/inhibition of Cdc25C, preventing activation of the Cdk1/cyclin B1 complex [109]. As discussed below, high-risk HPV positive cells exhibit constitutive activation of ATM and ATR, with activation of both of these pathways necessary for productive viral replication [100,110,111]. Inhibition of Chk2 kinase activity in differentiating HPV31 positive cells results in decreased inhibitory phosphorylation of Cdc25C and Cdk1, offering support that activation of the ATM/ATR pathways contributes to the G2 arrest observed upon differentiation [100]. E7 expression alone is sufficient to induce ATM and ATR activation, as well as high levels of cytoplasmic cyclin B and Cdk1 in suprabasal cells of HPV18 organotypic raft cultures, suggesting that E7 is involved in facilitating cell cycle arrest upon differentiation [100,104,110]. However, several studies have shown that the overexpression of the E4 protein of multiple HPV types induces G2 arrest [112–114]. This is thought to occur through E4s ability to interact with cyclin B/Cdk1 complexes and to promote inhibitory phosphorylation of Cdk1 through the Wee1 kinase [114,115]. E1^E4 is the most abundantly expressed viral gene upon differentiation, occurring concomitantly with viral genome amplification due to activation of the late promoter [28]. In addition, HPV16 E4 protein stability is increased upon phosphorylation by ERK1/2, leading to high levels of E4 protein in differentiating cells [116]. E1^E4 expression has been shown to be necessary for efficient productive replication of HPV16, HPV18 and HPV31, but not for low-risk HPV11 [30–32,106]. Studies using normal immortalized keratinocytes containing HPV16 E4 mutants that no longer induce G2 arrest exhibit decreased viral genome amplification and L1 gene expression upon differentiation in methylcellulose, as well as in organotypic raft cultures [29]. These studies indicate that the G2 arrest function of E4 contributes to providing a replication-competent environment. The accumulation of E4 in G2 arrested cells may foster productive replication by enhancing the accumulation of E1 in the nucleus, possibly through activation of MAPK pathways that activate E1's nuclear localization sequence [29,117,118]. E4 has been proposed to induce G2 arrest to counteract E7-induced proliferation in order to establish an environment that allows for rapid amplification of viral genomes. It is possible that E7 initiates G2 arrest following cell cycle re-entry through activation of ATM and ATR, but increased E4 protein levels sustain G2 arrest, providing an environment conducive to productive viral replication. Productive replication in a G2 arrested environment is postulated to allow HPV to avoid competition with host DNA synthesis and appropriate necessary cellular factors for amplification of its genomes. E2 may contribute to this process through interaction with the cellular replication protein ORC2 (origin recognition complex), which promotes assembly of pre-replication (pre-RC) complexes on mammalian origins. Overexpression of HPV31 or HPV16 E2 decreases ORC2 occupancy at mammalian origins [119], raising the possibility that increased levels of E2 upon differentiation may serve to restrict pre-RC assembly at cellular origins that could compete with HPV for access to host replication machinery. This is important considering that increasing evidence supports a role for homologous recombination (HR) DNA repair pathways in the amplification of HPV genomes (discussed below) [25,120]. HR activity is restricted to the S- and G2-phases of the cell cycle [121]. By productively replicating post-cellular DNA synthesis in a G2 arrested environment, HPV has unfettered access to DNA repair factors, as well as other cellular factors, that are necessary for viral DNA synthesis.

7. Use of DNA Damage Response Pathways for Productive Replication

Numerous studies over the past several years have provided evidence to support a role for the DNA damage response (DDR) in productive replication of high-risk alpha HPV types [25]. The DDR is a complex series of signaling events that act to coordinate the cell cycle with DNA repair. There are three main kinases activated in response to DNA damage; ATM, ATR and DNA-PK (DNA-dependent Protein Kinase), all of which belong to the PIK-like kinase (Phosphatidyl inositol 3' kinase) family of serine/threonine kinases [122]. ATM and DNA-PK respond primarily to double-strand DNA breaks (DSBs) and promote repair through high fidelity homologous recombination (HR), or error prone non-homologous end joining (NHEJ), respectively [121] (Figure 4). In contrast, ATR facilitates repair of single-strand DNA that is generated in response to replication stress, or during the processing of DSBs [123] (Figure 4). However, due to the complexity of DNA repair, there is considerable cross-talk between these pathways to maintain genomic integrity. HPV requires activation of the ATM and ATR response pathways for productive viral replication, however whether the DNA-PK pathway also contributes to viral replication is not yet known. Activation of the DDR provides HPV access to the necessary repair factors that play a direct role in viral DNA synthesis. In addition, increasing evidence suggests that HPV utilizes these pathways to establish a G2 arrested environment that is amenable to recombination-directed amplification of viral genomes.

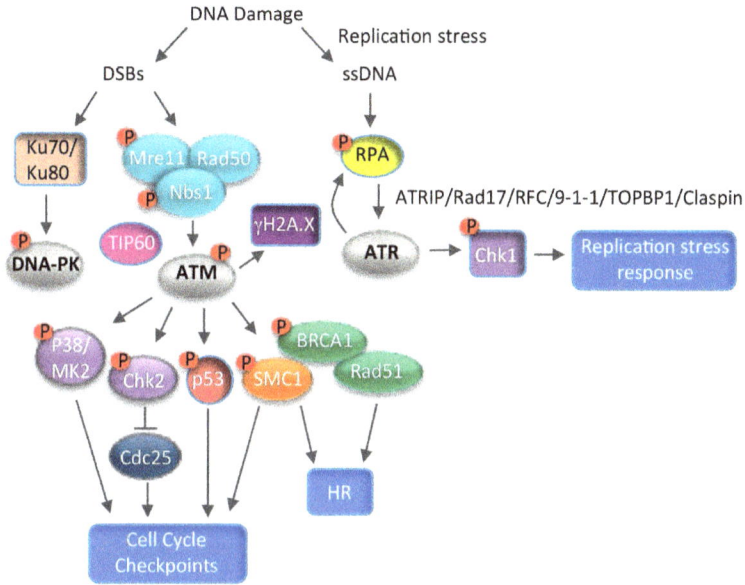

Figure 4. Schematic of the Ataxia-Telangiectasia Mutated (ATM), DNA-dependent Protein Kinase (DNA-PK), and ATM and Rad3-related (ATR) DNA damage response pathways. ATM and DNA-PK are activated in response to double strand DNA breaks (DSBs). ATM facilitates DNA repair through high-fidelity homologous recombination (HR), however, DNA-PK promotes repair through error-prone non-homologous end joining (NHEJ). DNA-PK is activated by the DNA damage sensor complex of Ku70/Ku80, while ATM is activated by the DNA damage sensor complex MRN (Mre11, Rad50, Nbs1) and the TIP60 acetyltransferase. ATM phosphorylates numerous downstream effectors, including H2A.X (gH2AX), Chk2, p38, p53, SMC1 and Breast Cancer Gene 1 (BRCA1) to induce cell cycle arrest and facilitate DNA repair, or to promote apoptosis in the case of severe DNA damage. ATR is activated by single-stranded DNA (ssDNA) generated by replication stress or the resection of DSBs. ssDNA is protected by the tripartite complex RPA, which promotes ATR activation through recruitment of ATRIP, a critical ATR regulator. The Rad17/RFC complex also binds to RPA-coated ssDNA and loads the 9-1-1 complex (Rad9-Hus1-Rad1). 9-1-1 recruits TOPBP1, which is necessary for ATR activation. Claspin mediates the activation of Chk1 by ATR, leading to the replication stress response. T bars indicates inhibition. Arrows indicate activation.

7.1. ATM Signaling and Productive Viral Replication

In response to DSBs, ATM is conically activated by the MRN (Mre11, Rad50, Nbs1) complex, which serves as a sensor of DNA damage, and by acetylation via the TIP60 acetyltransferase [122,124–126] (Figure 4). ATM then phosphorylates numerous downstream targets, including the histone variant H2A.X (histone 2A variant X), which initiates repair factor recruitment to sites of DNA damage in a highly ordered fashion [127]. ATM elicits its effects on cell cycle arrest and DNA repair through the activation of numerous kinases, including Chk2 and p38MAPK [122,128] (Figure 4). Chk2 phosphorylates many downstream effectors, including repair factors such as BRCA1 (Breast Cancer Gene 1), p53, and the Cdc25c family of phosphatases to mediate G2/M arrest [129]. p38MAPK (Mitogen Activated Protein Kinase) signaling is independent of Chk2 and induces the DDR through phosphorylation of MK2 (MAPK-activated protein kinase 2), which in turn phosphorylates downstream substrates to induce G2/M arrest [128]. The ATM effector SMC1 constitutes a third arm of the DDR, which along with Nbs1 induces cell cycle arrest and DNA repair [130,131] (Figure 4). A seminal study by the Laimins lab demonstrated that ATM is constitutively active in high-risk

HPV31 positive cells [100], and is characterized by the phosphorylation of multiple downstream targets, including H2A.X, Chk2, Nbs1, BRCA1, SMC1, p38MAPK and MK2 [100,132,133] (Figure 5). Subsequent studies demonstrated similar findings for HPV16 and HPV18 [104,134]. Inactivation of the MRN complex in HPV31 positive cells does not abrogate ATM activation [135], suggesting that HPV utilizes a non-canonical mechanism to induce the ATM DDR necessary for productive viral replication. Intriguingly, activation of the ATM pathway is specifically required for productive replication of HPV31 upon differentiation, with inhibition of ATM activity having no effect on episomal maintenance in undifferentiated cells [100]. Similar results were observed for the ATM effector Chk2 [100]. In addition to inactivation of Cdc25c, Chk2 activity is also necessary in differentiating HPV31 positive keratinocytes for activation of caspase-3/7, which is required for cleavage of the E1 viral helicase and viral genome amplification [100,136]. Interestingly, in contrast to Chk2, activation of the p38/MK2 axis of the ATM DDR is induced only upon differentiation [132]. The p38/MK2 complex is also necessary for productive replication of HPV31, though the downstream targets of this complex that drive viral DNA synthesis have not been defined [132].

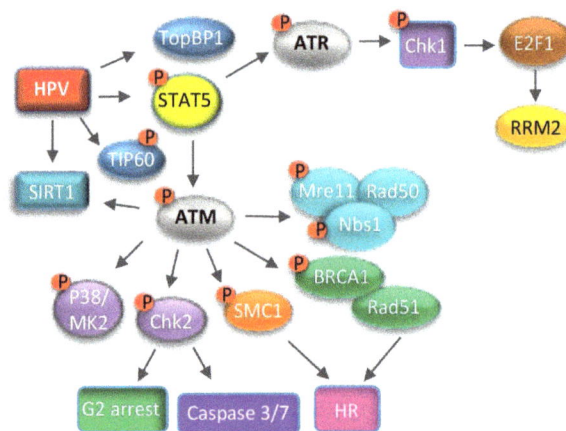

Figure 5. Modulation of the ATM and ATR DNA damage response pathways to promote productive viral replication. HPV-induced activation of ATM requires the STAT5 immune regulator, as well as TIP60, but not the MRN complex. Downstream effectors of ATM required for productive viral replication include the MRN complex, p38/MK2, Chk2, as well as factors involved in homologous recombination repair (Rad51, BRCA1, SMC1). HPV may utilize ATM activity to promote G2 arrest upon differentiation through activities of Chk2, as well as to direct repair to HR on viral genomes through epigenetic modifications and the recruitment of homologous recombination (HR) repair factors. ATR activation in HPV positive cells likely occurs through E7-induced replication stress and requires a STAT5-directed increase in TOPBP1. ATR/Chk1 activation leads to increased levels of E2F1, which drives expression of RRM2, resulting in increased dNTP pools to facilitate productive viral replication.

Multiple ATM signaling components are recruited to productively replicating viral genomes, including ATM, γH2A.X, Chk2, 53BP1, MRN, Rad51 and BRCA1, suggesting a direct role for DNA repair mechanisms in viral DNA synthesis [135,137–139]. Indeed, along with ATM, several of these factors, including the MRN complex, Rad51 and BRCA1, are necessary for DNA repair through homologous recombination (HR), and importantly, are also required for productive replication of HPV31 [101,122,135]. HR is a relatively error-free process, and HPV may preferentially use this method of repair to maintain the integrity of viral DNA during amplification. Structures consistent with recombination have been observed during productive replication of HPV31 and HPV16 that are not detected during maintenance replication in undifferentiated cells [140]. These observations suggest that amplification of viral genomes upon differentiation occurs in a distinct manner that may require ATM-driven HR. Initiation of HR requires resection of DSBs, which requires ATM activity, as well as BRCA1 and the MRN resection complex [121]. Resection is required for loading the Rad51 recombinase onto DNA, which then facilitates strand invasion into homologous sequences [122]. Rad51 binding to HPV31 DNA increases upon differentiation, and inhibition of Rad51's DNA binding ability blocks productive viral replication, suggesting that viral DNA resection is necessary for amplification of viral genomes [101]. In support of this, Anacker et al. demonstrated that the MRN complex is required for Rad51 localization to HPV31 replication foci, and that Mre11's nuclease activity is necessary for productive viral replication [135]. Recent studies have shown that SMC1 is also required for productive replication of HPV31 [133]. SMC1 is a member of the sister chromatid cohesion complex that is important for chromosome segregation during mitosis [141]. The role of SMC1 in productive viral replication is not clear, but SMC1 is recruited to the viral genome in a complex with CTCF insulator proteins [133]. SMC1 is postulated to promote HR by maintaining the close proximity of sister chromatids at DSBs [131], and may serve a similar role on HPV genomes to facilitate recombination-dependent replication.

In the context of the complete HPV31 genome, ATM activation occurs in a manner dependent on E7's Rb binding domain [142]. Expression of HPV18 E7 alone in organotypic raft cultures results in activation of ATM, Chk2 and Chk1 in the suprabasal layers, offering support that E7 contributes to productive viral replication through eliciting ATM activation in differentiating cells [104]. HPV31 E7 regulates the activation of ATM through STAT5, an immune regulator that is required for productive viral replication [143] (Figure 5). How STAT5 leads to ATM activation is currently unclear, but may involve STAT5-dependent activation of TIP60 [144]. In addition to ATM activation, HPV31 E7 contributes to productive viral replication by increasing the protein half-life of several DNA repair factors that are required for productive replication (e.g., ATM, Chk2, Chk1, Mre11, Rad50, Nbs1, Rad51 and BRCA1), ensuring high levels for efficient viral DNA synthesis [142]. Expression of the E1 viral helicase alone from high-risk and low-risk HPV types is sufficient to induce ATM activation, which may occur through the induction of DSBs due to E1's ability to non-specifically bind and unwind cellular DNA [134,145,146]. In the presence of E2, E1 is recruited to the viral origin of replication, along with multiple components of the ATM and ATR pathway [134,145,147]. How ATM activity is regulated by E7 versus E1 during the viral life cycle remains to be determined. In addition, whether activation of the ATM DDR occurs in the context of low-risk HPV infection, and if this response is required for productive replication is currently unknown. In contrast to the Alpha high-risk HPV types, beta HPV E6 and E7 proteins reduce expression of ATM and ATR, as well as the HR factors Rad51 and BRCA2, in turn delaying repair foci formation in response to UV exposure [148]. Whether inactivation of the DDR is necessary for the life cycle of beta HPVs is not yet known due to the lack of experimental systems to study replication.

The recruitment of DNA repair factors to sites of DNA damage requires alterations in chromatin structure orchestrated through ATP-dependent remodeling complexes and post-translational modifications of histones (e.g., acetylation, ubiquitylation, phosphorylation, methylation) [149,150]. ATM-induced phosphorylation of H2A.X (γH2A.X) is one of the key effectors in modulating chromatin dynamics in response to DSBs [127]. γH2A.X initiates the assembly of repair factors at DNA lesions in a highly regulated manner, including HR factors (MRN, Brca1 and Rad51) [151]. γH2A.X is bound to HPV31 DNA and binding increases during productive viral replication, suggesting that γH2A.X may serve to assemble HR repair factors at viral replication sites [138,152]. The DDR-associated histone deacetylase SIRT1 and the acetyltransferase TIP60 have also been linked to productive viral replication. SIRT1 channels repair to HR by recruiting Nbs1 and Rad51 to damaged DNA in an ATM- and γH2AX-dependent manner [153]. Interestingly, SIRT1 binds to HPV31 DNA and is necessary for productive viral replication, which may be mediated through the recruitment of Nbs1 and Rad51 to viral replication foci [139]. TIP60 is upregulated in HPV31 positive keratinocytes and is also necessary for productive viral replication [144]. While this presumably is due to TIP60's role in ATM activation, TIP60 can also influence repair to the HR pathway through the acetylation of histone H4 and attenuation of 53BP1 binding, which promotes repair through NHEJ [154]. SIRT1 and TIP60 may modify viral chromatin to ensure the recruitment of HR factors to productively replicating viral genomes. How the HPV life cycle may be epigenetically regulated through ATM activity is an interesting area of investigation.

7.2. ATR Signaling and Productive Viral Replication

Replication stress results in formation of single strand DNA (ssDNA) at stalled replication forks that activates the ATR kinase [155]. ATR and its downstream target Chk1 protect stalled replication forks and prevent excessive origin firing, maintaining genome integrity. High-risk HPV positive cells exhibit constitutive activation of the ATR pathway, indicating that replication stress is a chronic problem that HPV has to contend with [100,110,111]. Unscheduled cell cycle entry induced by high-risk HPV E6 and E7 proteins results in replication stress due to a disconnect between activation of cellular DNA synthesis and the availability of supplies required for replication [156,157]. This is thought to occur through E7's ability to target Rb for degradation. In support of this, mutation of E7's Rb binding domain in the context of the HPV31 genome prevents ATR signaling [142]. ATR activation requires recruitment to RPA-coated ssDNA by its regulator ATRIP [158] (Figure 4). ssDNA-RPA also recruits the RFC/Rad17 complex, which facilitates loading of the 9-1-1 complex at stalled replication forks [159]. The 9-1-1 complex then recruits TOPBP1 to activate ATR's kinase activity [160]. Intriguingly, HPV31 E7 ensures that infected cells can sufficiently respond to replication stress through ATR activation by increasing the levels of TOPBP1 in a STAT5-dependent manner [110] (Figure 5). Although E1 of high-risk and low-risk types can also independently activate ATR, it is unclear if this results from non-specific binding and unwinding of cellular DNA, or if increased E1 activity on viral DNA during productive viral replication results in in replication stress [134,145].

Inhibition of ATR, as well as its downstream target Chk1, blocks productive replication of HPV31, and also decreases HPV31 and HPV16 copy number in undifferentiated cells [110,111,161]. In response to replication stress, ATR phosphorylates RPA on Ser33 [162]. pRPA Ser33 localizes to HPV31 replication foci, suggesting that viral genomes are subject to replication stress during productive replication [138]. Activation of the ATR/Chk1 pathway may be important in repairing stalled forks that occur during amplification of viral genomes. Upon replication stress, activation of the ATR/Chk1 pathway is instrumental in maintaining E2F signaling, ensuring the expression of cellular genes that facilitate DNA repair and cell survival [163]. This is particularly important in cancer cells, which typically exhibit high levels of replication stress [164–166]. Recent studies from our lab demonstrated that HPV31 utilizes the ATR/Chk1/E2F1 arm of the DDR to increase levels of RRM2, the small subunit of the ribonucleotide reductase complex, in an E7-dependent manner [111] (Figure 5). RRM2, along with the large subunit RRM1, is necessary for the conversion of ribonucleotides to deoxyribonucleotides, providing dNTPs for replication, DNA repair and survival [167]. Knockdown of RRM2 reduced dNTP pools in differentiating HPV31 positive cells and blocked productive replication [111]. These studies indicate E7 induced cell-cycle re-entry upon differentiation results in replication stress that activates the ATR/Chk1 pathway to maintain E2F signaling. Importantly, these studies demonstrate that HPV exploits the ATR DNA damage response to ensure an adequate supply of dNTPs for productive replication, providing a replication competent environment in cells that are no longer dividing. Understanding the full extent of the ATR pathway throughout the viral life cycle is an important area of future investigation.

7.3. Consequences of Utilizing the DNA Damage Response for Replication

Studies have shown that HPV replication foci tend to form near common fragile sites, which are regions of the cellular genome that are prone to replication stress and recruit DNA repair factors to maintain genomic stability [168,169]. HPV may preferentially replicate adjacent to fragile sites to readily have access to DNA repair factors to facilitate recombination-directed replication. Interestingly, in cancers associated with oncogenic HPV types, viral DNA is often found integrated into host DNA at common fragile sites [170–173]. Integration is a dead-end for virus production and almost always results in increased expression of the E6 and E7 oncogenes [174]. Deregulated E6/E7 expression leads to a proliferative advantage and the clonal outgrowth of cells containing integrated viral DNA. While replicating near areas of cellular replication stress may be beneficial to viral persistence and productive viral replication, the close association of HPV replication foci with areas of the cellular DNA damage may increase the chance of accidental integration of the viral genome, and may explain the tendency for HPV to integrate into common fragile sites of host DNA [175]. Furthermore, recent studies from the Galloway lab demonstrated that high-risk E6 and E7 proteins attenuate the repair of cellular DSBs through the HR pathway. While this likely ensures HR factors are available for viral replication, the presence of persistent, unrepaired DNA breaks increases the opportunity for viral genome integration [176]. These integration events, in turn, may contribute to HPV oncogenesis through E6/E7-mediated genomic instability.

8. Conclusions

In order to provide a replication-competent environment, HPVs co-opt particular host cell pathways and interactions that regulate epithelial differentiation and cellular proliferation, as well facilitate repair of damaged DNA. Temporal regulation of viral gene expression is necessary to restrict high levels of viral gene expression, replication and virion production to the uppermost layers of the epithelium, protecting HPV-infected cells from detection by the immune response. This is achieved through differential usage of promoters and polyadenylation sites, as well as alternative splicing. In addition, E6 and E7 play critical roles in modulating innate immune responses to facilitate viral persistence and promote viral replication [27]. Cooperation between the activities of E6, E7, E1, E2, E4 and E5 upon differentiation allows HPV to establish an environment supportive of productive replication in non-dividing cells. Our understanding of how HPVs regulate the productive phase of the viral life cycle has increased dramatically over the past several years, particularly regarding how high-risk HPVs activate and utilize DNA repair pathways to amplify viral genomes. However, much remains to be learned regarding how alpha HPVs manipulate cellular pathways to facilitate viral replication, and in turn, how hijacking these pathways may affect the integrity of the cellular genome. Further understanding of the mechanisms by which HPV establishes a replication-competent environment throughout the viral life cycle is important to identify novel cellular targets that could be exploited therapeutically for the treatment of HPV-associated diseases.

Acknowledgments: This work was supported by the National Institutes of Health (1R01CA181581; to Cary A. Moody) and the American Cancer Society (A14-0113; to Cary A. Moody).

Conflicts of Interest: The authors declare no conflict of interest.

References

1. Van Doorslaer, K.; Tan, Q.; Xirasagar, S.; Bandaru, S.; Gopalan, V.; Mohamoud, Y.; Huyen, Y.; McBride, A.A. The papillomavirus episteme: A central resource for papillomavirus sequence data and analysis. *Nucleic Acids Res.* **2013**, *41*, 571–578. [CrossRef] [PubMed]
2. Bernard, H.U.; Burk, R.D.; Chen, Z.; van Doorslaer, K.; zur Hausen, H.; de Villiers, E.M. Classification of papillomaviruses (PVS) based on 189 PV types and proposal of taxonomic amendments. *Virology* **2010**, *401*, 70–79. [CrossRef] [PubMed]
3. Walboomers, J.M.; Jacobs, M.V.; Manos, M.M.; Bosch, F.X.; Kummer, J.A.; Shah, K.V.; Snijders, P.J.; Peto, J.; Meijer, C.J.; Munoz, N. Human papillomavirus is a necessary cause of invasive cervical cancer worldwide. *J. Pathol.* **1999**, *189*, 12–19. [CrossRef]
4. Stanley, M. Pathology and epidemiology of HPV infection in females. *Gynecol. Oncol.* **2010**, *117*, S5–S10. [CrossRef] [PubMed]
5. Gillison, M.L.; Chaturvedi, A.K.; Anderson, W.F.; Fakhry, C. Epidemiology of human papillomavirus-positive head and neck squamous cell carcinoma. *J. Clin. Oncol.* **2015**, *33*, 3235–3242. [CrossRef] [PubMed]

6. Tommasino, M. The biology of β human papillomaviruses. *Virus Res.* **2017**, *231*, 128–138. [CrossRef] [PubMed]
7. Howley, P.M.; Pfister, H.J. β genus papillomaviruses and skin cancer. *Virology* **2015**, *479–480*, 290–296. [CrossRef] [PubMed]
8. Egawa, N.; Egawa, K.; Griffin, H.; Doorbar, J. Human papillomaviruses; epithelial tropisms, and the development of neoplasia. *Viruses* **2015**, *7*, 3863–3890. [CrossRef] [PubMed]
9. McBride, A.A. Mechanisms and strategies of papillomavirus replication. *Biol. Chem.* **2017**, *398*, 919–927. [CrossRef] [PubMed]
10. Longworth, M.S.; Laimins, L.A. Pathogenesis of human papillomaviruses in differentiating epithelia. *Microbiol. Mol. Biol. Rev.* **2004**, *68*, 362–372. [CrossRef] [PubMed]
11. Pyeon, D.; Pearce, S.M.; Lank, S.M.; Ahlquist, P.; Lambert, P.F. Establishment of human papillomavirus infection requires cell cycle progression. *PLoS Pathog.* **2009**, *5*, e1000318. [CrossRef] [PubMed]
12. Maglennon, G.A.; McIntosh, P.; Doorbar, J. Persistence of viral DNA in the epithelial basal layer suggests a model for papillomavirus latency following immune regression. *Virology* **2011**, *414*, 153–163. [CrossRef] [PubMed]
13. Koster, M.I.; Roop, D.R. Mechanisms regulating epithelial stratification. *Annu. Rev. Cell Dev. Biol.* **2007**, *23*, 93–113. [CrossRef] [PubMed]
14. Favre, M.; Breitburd, F.; Croissant, O.; Orth, G. Chromatin-like structures obtained after alkaline disruption of bovine and human papillomaviruses. *J. Virol.* **1977**, *21*, 1205–1209. [PubMed]
15. Stunkel, W.; Bernard, H.U. The chromatin structure of the long control region of human papillomavirus type 16 represses viral oncoprotein expression. *J. Virol.* **1999**, *73*, 1918–1930. [PubMed]
16. Graham, S.V.; Faizo, A.A. Control of human papillomavirus gene expression by alternative splicing. *Virus Res.* **2017**, *231*, 83–95. [CrossRef] [PubMed]
17. Van Doorslaer, K.; Li, Z.; Xirasagar, S.; Maes, P.; Kaminsky, D.; Liou, D.; Sun, Q.; Kaur, R.; Huyen, Y.; McBride, A.A. The papillomavirus episteme: A major update to the papillomavirus sequence database. *Nucleic Acids Res.* **2017**, *45*, D499–D506. [CrossRef] [PubMed]
18. Geisen, C.; Kahn, T. Promoter activity of sequences located upstream of the human papillomavirus types of 16 and 18 late regions. *J. Gen. Virol.* **1996**, *77*, 2193–2200. [CrossRef] [PubMed]
19. Ozbun, M.A.; Meyers, C. Temporal usage of multiple promoters during the life cycle of human papillomavirus type 31b. *J. Virol.* **1998**, *72*, 2715–2722. [PubMed]
20. Braunstein, T.H.; Madsen, B.S.; Gavnholt, B.; Rosenstierne, M.W.; Johnsen, C.K.; Norrild, B. Identification of a new promoter in the early region of the human papillomavirus type 16 genome. *J. Gen. Virol.* **1999**, *80*, 3241–3250. [CrossRef] [PubMed]
21. Bergvall, M.; Melendy, T.; Archambault, J. The E1 proteins. *Virology* **2013**, *445*, 35–56. [CrossRef] [PubMed]
22. McBride, A.A. The papillomavirus E2 proteins. *Virology* **2013**, *445*, 57–79. [CrossRef] [PubMed]
23. Roman, A.; Munger, K. The papillomavirus E7 proteins. *Virology* **2013**, *445*, 138–168. [CrossRef] [PubMed]
24. Vande Pol, S.B.; Klingelhutz, A.J. Papillomavirus E6 oncoproteins. *Virology* **2013**, *445*, 115–137. [CrossRef] [PubMed]
25. Anacker, D.C.; Moody, C.A. Modulation of the DNA damage response during the life cycle of human papillomaviruses. *Virus Res.* **2017**, *231*, 41–49. [CrossRef] [PubMed]
26. Moody, C.A.; Laimins, L.A. Human papillomavirus oncoproteins: Pathways to transformation. *Nat. Rev. Cancer* **2010**, *10*, 550–560. [CrossRef] [PubMed]
27. Hong, S.; Laimins, L.A. Manipulation of the innate immune response by human papillomaviruses. *Virus Res.* **2017**, *231*, 34–40. [CrossRef] [PubMed]
28. Doorbar, J. The E4 protein; structure, function and patterns of expression. *Virology* **2013**, *445*, 80–98. [CrossRef] [PubMed]
29. Egawa, N.; Wang, Q.; Griffin, H.M.; Murakami, I.; Jackson, D.; Mahmood, R.; Doorbar, J. HPV16 and 18 genome amplification show different e4-dependence, with 16E4 enhancing E1 nuclear accumulation and replicative efficiency via its cell cycle arrest and kinase activation functions. *PLoS Pathog.* **2017**, *13*, e1006282. [CrossRef] [PubMed]
30. Fang, L.; Budgeon, L.R.; Doorbar, J.; Briggs, E.R.; Howett, M.K. The human papillomavirus type 11 E1^E4 protein is not essential for viral genome amplification. *Virology* **2006**, *351*, 271–279. [CrossRef] [PubMed]
31. Wilson, R.; Fehrmann, F.; Laimins, L.A. Role of the E1^E4 protein in the differentiation-dependent life cycle of human papillomavirus type 31. *J. Virol.* **2005**, *79*, 6732–6740. [CrossRef] [PubMed]

32. Wilson, R.; Ryan, G.B.; Knight, G.L.; Laimins, L.A.; Roberts, S. The full-length E1E4 protein of human papillomavirus type 18 modulates differentiation-dependent viral DNA amplification and late gene expression. *Virology* **2007**, *362*, 453–460. [CrossRef] [PubMed]

33. Fehrmann, F.; Klumpp, D.J.; Laimins, L.A. Human papillomavirus type 31 E5 protein supports cell cycle progression and activates late viral functions upon epithelial differentiation. *J. Virol.* **2003**, *77*, 2819–2831. [CrossRef] [PubMed]

34. Dreer, M.; van de Poel, S.; Stubenrauch, F. Control of viral replication and transcription by the papillomavirus E8^E2 protein. *Virus Res.* **2017**, *231*, 96–102. [CrossRef] [PubMed]

35. Grassmann, K.; Rapp, B.; Maschek, H.; Petry, K.U.; Iftner, T. Identification of a differentiation-inducible promoter in the E7 open reading frame of human papillomavirus type 16 (HPV-16) in raft cultures of a new cell line containing high copy numbers of episomal HPV-16 DNA. *J. Virol.* **1996**, *70*, 2339–2349. [PubMed]

36. Hummel, M.; Hudson, J.B.; Laimins, L.A. Differentiation-induced and constitutive transcription of human papillomavirus type 31B in cell lines containing viral episomes. *J. Virol.* **1992**, *66*, 6070–6080. [PubMed]

37. Klumpp, D.J.; Laimins, L.A. Differentiation-induced changes in promoter usage for transcripts encoding the human papillomavirus type 31 replication protein E1. *Virology* **1999**, *257*, 239–246. [CrossRef] [PubMed]

38. Wooldridge, T.R.; Laimins, L.A. Regulation of human papillomavirus type 31 gene expression during the differentiation-dependent life cycle through histone modifications and transcription factor binding. *Virology* **2008**, *374*, 371–380. [CrossRef] [PubMed]

39. Del Mar Pena, L.M.; Laimins, L.A. Differentiation-dependent chromatin rearrangement coincides with activation of human papillomavirus type 31 late gene expression. *J. Virol.* **2001**, *75*, 10005–10013. [CrossRef] [PubMed]

40. Carson, A.; Khan, S.A. Characterization of transcription factor binding to human papillomavirus type 16 DNA during cellular differentiation. *J. Virol.* **2006**, *80*, 4356–4362. [CrossRef] [PubMed]

41. Gunasekharan, V.; Hache, G.; Laimins, L. Differentiation-dependent changes in levels of C/EBPβ repressors and activators regulate human papillomavirus type 31 late gene expression. *J. Virol.* **2012**, *86*, 5393–5398. [CrossRef] [PubMed]

42. Songock, W.K.; Scott, M.L.; Bodily, J.M. Regulation of the human papillomavirus type 16 late promoter by transcriptional elongation. *Virology* **2017**, *507*, 179–191. [CrossRef] [PubMed]

43. Johansson, C.; Somberg, M.; Li, X.; Backstrom Winquist, E.; Fay, J.; Ryan, F.; Pim, D.; Banks, L.; Schwartz, S. HPV-16 E2 contributes to induction of HPV-16 late gene expression by inhibiting early polyadenylation. *EMBO J.* **2012**, *31*, 3212–3227. [CrossRef] [PubMed]

44. Maitland, N.J.; Conway, S.; Wilkinson, N.S.; Ramsdale, J.; Morris, J.R.; Sanders, C.M.; Burns, J.E.; Stern, P.L.; Wells, M. Expression patterns of the human papillomavirus type 16 transcription factor E2 in low- and high-grade cervical intraepithelial neoplasia. *J. Pathol.* **1998**, *186*, 275–280. [CrossRef]

45. Xue, Y.; Bellanger, S.; Zhang, W.; Lim, D.; Low, J.; Lunny, D.; Thierry, F. HPV16 E2 is an immediate early marker of viral infection, preceding E7 expression in precursor structures of cervical carcinoma. *Cancer Res.* **2010**, *70*, 5316–5325. [CrossRef] [PubMed]

46. Graham, S.V. Keratinocyte differentiation-dependent human papillomavirus gene regulation. *Viruses* **2017**, *9*, E245. [CrossRef] [PubMed]

47. Bentley, D.L. Coupling mRNA processing with transcription in time and space. *Nat. Rev. Genet.* **2014**, *15*, 163–175. [CrossRef] [PubMed]

48. Klymenko, T.; Hernandez-Lopez, H.; MacDonald, A.I.; Bodily, J.M.; Graham, S.V. Human papillomavirus E2 regulates SRSF3 (SRP20) to promote capsid protein expression in infected differentiated keratinocytes. *J. Virol.* **2016**, *90*, 5047–5058. [CrossRef] [PubMed]

49. Somberg, M.; Li, X.; Johansson, C.; Orru, B.; Chang, R.; Rush, M.; Fay, J.; Ryan, F.; Schwartz, S. Serine/arginine-rich protein 30c activates human papillomavirus type 16 L1 mRNA expression via a bimodal mechanism. *J. Gen. Virol.* **2011**, *92*, 2411–2421. [CrossRef] [PubMed]

50. Paris, C.; Pentland, I.; Groves, I.; Roberts, D.C.; Powis, S.J.; Coleman, N.; Roberts, S.; Parish, J.L. CCCTC-binding factor recruitment to the early region of the human papillomavirus 18 genome regulates viral oncogene expression. *J. Virol.* **2015**, *89*, 4770–4785. [CrossRef] [PubMed]

51. Dyson, N.; Howley, P.M.; Munger, K.; Harlow, E. The human papilloma virus-16 E7 oncoprotein is able to bind to the retinoblastoma gene product. *Science* **1989**, *243*, 934–937. [CrossRef] [PubMed]

52. Dyson, N. The regulation of E2F by pRB-family proteins. *Genes Dev.* **1998**, *12*, 2245–2262. [CrossRef] [PubMed]

53. Munger, K.; Werness, B.A.; Dyson, N.; Phelps, W.C.; Harlow, E.; Howley, P.M. Complex formation of human papillomavirus E7 proteins with the retinoblastoma tumor suppressor gene product. *EMBO J.* **1989**, *8*, 4099–4105. [PubMed]

54. Chellappan, S.; Kraus, V.B.; Kroger, B.; Munger, K.; Howley, P.M.; Phelps, W.C.; Nevins, J.R. Adenovirus E1A, simian virus 40 tumor antigen, and human papillomavirus E7 protein share the capacity to disrupt the interaction between transcription factor E2F and the retinoblastoma gene product. *Proc. Natl. Acad. Sci. USA* **1992**, *89*, 4549–4553. [CrossRef] [PubMed]

55. Cheng, S.; Schmidt-Grimminger, D.C.; Murant, T.; Broker, T.R.; Chow, L.T. Differentiation-dependent up-regulation of the human papillomavirus E7 gene reactivates cellular DNA replication in suprabasal differentiated keratinocytes. *Genes Dev.* **1995**, *9*, 2335–2349. [CrossRef] [PubMed]

56. Chen, H.Z.; Tsai, S.Y.; Leone, G. Emerging roles of E2Fs in cancer: An exit from cell cycle control. *Nat. Rev. Cancer* **2009**, *9*, 785–797. [CrossRef] [PubMed]

57. Jones, D.L.; Alani, R.M.; Munger, K. The human papillomavirus E7 oncoprotein can uncouple cellular differentiation and proliferation in human keratinocytes by abrogating p21CIP1-mediated inhibition of CDK2. *Genes Dev.* **1997**, *11*, 2101–2111. [CrossRef] [PubMed]

58. Demers, G.W.; Espling, E.; Harry, J.B.; Etscheid, B.G.; Galloway, D.A. Abrogation of growth arrest signals by human papillomavirus type 16 E7 is mediated by sequences required for transformation. *J. Virol.* **1996**, *70*, 6862–6869. [PubMed]

59. Demeter, L.M.; Stoler, M.H.; Broker, T.R.; Chow, L.T. Induction of proliferating cell nuclear antigen in differentiated keratinocytes of human papillomavirus-infected lesions. *Hum. Pathol.* **1994**, *25*, 343–348. [CrossRef]

60. Barrow-Laing, L.; Chen, W.; Roman, A. Low- and high-risk human papillomavirus E7 proteins regulate p130 differently. *Virology* **2010**, *400*, 233–239. [CrossRef] [PubMed]

61. Zhang, B.; Chen, W.; Roman, A. The E7 proteins of low- and high-risk human papillomaviruses share the ability to target the pRB family member p130 for degradation. *Proc. Natl. Acad. Sci. USA* **2006**, *103*, 437–442. [CrossRef] [PubMed]

62. Wilson, R.; Laimins, L.A. Differentiation of HPV-containing cells using organotypic "raft" culture or methylcellulose. *Methods Mol. Med.* **2005**, *119*, 157–169. [PubMed]

63. Flores, E.R.; Allen-Hoffmann, B.L.; Lee, D.; Lambert, P.F. The human papillomavirus type 16 E7 oncogene is required for the productive stage of the viral life cycle. *J. Virol.* **2000**, *74*, 6622–6631. [CrossRef] [PubMed]

64. Longworth, M.S.; Laimins, L.A. The binding of histone deacetylases and the integrity of zinc finger-like motifs of the E7 protein are essential for the life cycle of human papillomavirus type 31. *J. Virol.* **2004**, *78*, 3533–3541. [CrossRef] [PubMed]

65. Longworth, M.S.; Wilson, R.; Laimins, L.A. HPV31 E7 facilitates replication by activating E2F2 transcription through its interaction with HDACs. *EMBO J.* **2005**, *24*, 1821–1830. [CrossRef] [PubMed]

66. Brehm, A.; Nielsen, S.J.; Miska, E.A.; McCance, D.J.; Reid, J.L.; Bannister, A.J.; Kouzarides, T. The E7 oncoprotein associates with MI2 and histone deacetylase activity to promote cell growth. *EMBO J.* **1999**, *18*, 2449–2458. [CrossRef] [PubMed]

67. Missero, C.; Calautti, E.; Eckner, R.; Chin, J.; Tsai, L.H.; Livingston, D.M.; Dotto, G.P. Involvement of the cell-cycle inhibitor CIP1/WAF1 and the E1A-associated p300 protein in terminal differentiation. *Proc. Natl. Acad. Sci. USA* **1995**, *92*, 5451–5455. [CrossRef] [PubMed]

68. Deshpande, A.; Sicinski, P.; Hinds, P.W. Cyclins and CDKs in development and cancer: A perspective. *Oncogene* **2005**, *24*, 2909–2915. [CrossRef] [PubMed]

69. Funk, J.O.; Waga, S.; Harry, J.B.; Espling, E.; Stillman, B.; Galloway, D.A. Inhibition of CDK activity and pCNA-dependent DNA replication by p21 is blocked by interaction with the HPV-16 E7 oncoprotein. *Genes Dev.* **1997**, *11*, 2090–2100. [CrossRef] [PubMed]

70. Nguyen, C.L.; Munger, K. Direct association of the HPV16 E7 oncoprotein with cyclin A/CDK2 and cyclin E/CDK2 complexes. *Virology* **2008**, *380*, 21–25. [CrossRef] [PubMed]

71. Katich, S.C.; Zerfass-Thome, K.; Hoffmann, I. Regulation of the *CDC25A* gene by the human papillomavirus type 16 E7 oncogene. *Oncogene* **2001**, *20*, 543–550. [CrossRef] [PubMed]

72. Nguyen, D.X.; Westbrook, T.F.; McCance, D.J. Human papillomavirus type 16 E7 maintains elevated levels of the CDC25A tyrosine phosphatase during deregulation of cell cycle arrest. *J. Virol.* **2002**, *76*, 619–632. [CrossRef] [PubMed]

73. Deng, W.; Lin, B.Y.; Jin, G.; Wheeler, C.G.; Ma, T.; Harper, J.W.; Broker, T.R.; Chow, L.T. Cyclin/CDK regulates the nucleocytoplasmic localization of the human papillomavirus E1 DNA helicase. *J. Virol.* **2004**, *78*, 13954–13965. [CrossRef] [PubMed]

74. Fradet-Turcotte, A.; Moody, C.; Laimins, L.A.; Archambault, J. Nuclear export of human papillomavirus type 31 E1 is regulated by CDK2 phosphorylation and required for viral genome maintenance. *J. Virol.* **2010**, *84*, 11747–11760. [CrossRef] [PubMed]

75. Genther, S.M.; Sterling, S.; Duensing, S.; Munger, K.; Sattler, C.; Lambert, P.F. Quantitative role of the human papillomavirus type 16 E5 gene during the productive stage of the viral life cycle. *J. Virol.* **2003**, *77*, 2832–2842. [CrossRef] [PubMed]

76. Regan, J.A.; Laimins, L.A. Bap31 is a novel target of the human papillomavirus E5 protein. *J. Virol.* **2008**, *82*, 10042–10051. [CrossRef] [PubMed]

77. DiMaio, D.; Petti, L.M. The E5 proteins. *Virology* **2013**, *445*, 99–114. [CrossRef] [PubMed]

78. Huibregtse, J.M.; Scheffner, M.; Howley, P.M. A cellular protein mediates association of p53 with the E6 oncoprotein of human papillomavirus types 16 or 18. *EMBO J.* **1991**, *10*, 4129–4135. [PubMed]

79. Scheffner, M.; Werness, B.A.; Huibregtse, J.M.; Levine, A.J.; Howley, P.M. The E6 oncoprotein encoded by human papillomavirus types 16 and 18 promotes the degradation of p53. *Cell* **1990**, *63*, 1129–1136. [CrossRef]

80. Scheffner, M.; Huibregtse, J.M.; Vierstra, R.D.; Howley, P.M. The HPV-16 E6 and E6-AP complex functions as a ubiquitin-protein ligase in the ubiquitination of p53. *Cell* **1993**, *75*, 495–505. [CrossRef]

81. Kho, E.Y.; Wang, H.K.; Banerjee, N.S.; Broker, T.R.; Chow, L.T. HPV-18 E6 mutants reveal p53 modulation of viral DNA amplification in organotypic cultures. *Proc. Natl. Acad. Sci. USA* **2013**, *110*, 7542–7549. [CrossRef] [PubMed]

82. Massimi, P.; Pim, D.; Bertoli, C.; Bouvard, V.; Banks, L. Interaction between the HPV-16 E2 transcriptional activator and p53. *Oncogene* **1999**, *18*, 7748–7754. [CrossRef] [PubMed]

83. Brown, C.; Kowalczyk, A.M.; Taylor, E.R.; Morgan, I.M.; Gaston, K. p53 represses human papillomavirus type 16 DNA replication via the viral E2 protein. *Virol. J.* **2008**, *5*, 5. [CrossRef] [PubMed]

84. Ganti, K.; Broniarczyk, J.; Manoubi, W.; Massimi, P.; Mittal, S.; Pim, D.; Szalmas, A.; Thatte, J.; Thomas, M.; Tomaic, V.; et al. The human papillomavirus E6 PDZ binding motif: From life cycle to malignancy. *Viruses* **2015**, *7*, 3530–3551. [CrossRef] [PubMed]

85. Lee, S.S.; Weiss, R.S.; Javier, R.T. Binding of human virus oncoproteins to HDLG/Sap97, a mammalian homolog of the drosophila discs large tumor suppressor protein. *Proc. Natl. Acad. Sci. USA* **1997**, *94*, 6670–6675. [CrossRef] [PubMed]

86. Kiyono, T.; Hiraiwa, A.; Fujita, M.; Hayashi, Y.; Akiyama, T.; Ishibashi, M. Binding of high-risk human papillomavirus E6 oncoproteins to the human homologue of the drosophila discs large tumor suppressor protein. *Proc. Natl. Acad. Sci. USA* **1997**, *94*, 11612–11616. [CrossRef] [PubMed]

87. Delury, C.P.; Marsh, E.K.; James, C.D.; Boon, S.S.; Banks, L.; Knight, G.L.; Roberts, S. The role of protein kinase a regulation of the E6 PDZ-binding domain during the differentiation-dependent life cycle of human papillomavirus type 18. *J. Virol.* **2013**, *87*, 9463–9472. [CrossRef] [PubMed]

88. Lee, C.; Laimins, L.A. Role of the PDZ domain-binding motif of the oncoprotein E6 in the pathogenesis of human papillomavirus type 31. *J. Virol.* **2004**, *78*, 12366–12377. [CrossRef] [PubMed]

89. Nicolaides, L.; Davy, C.; Raj, K.; Kranjec, C.; Banks, L.; Doorbar, J. Stabilization of HPV16 E6 protein by PDZ proteins, and potential implications for genome maintenance. *Virology* **2011**, *414*, 137–145. [CrossRef] [PubMed]

90. Marsh, E.K.; Delury, C.P.; Davies, N.J.; Weston, C.J.; Miah, M.A.L.; Banks, L.; Parish, J.L.; Higgs, M.R.; Roberts, S. Mitotic control of human papillomavirus genome-containing cells is regulated by the function of the PDZ-binding motif of the E6 oncoprotein. *Oncotarget* **2017**, *8*, 19491–19506. [CrossRef] [PubMed]

91. Cai, X.; Li, G.; Laimins, L.A.; Cullen, B.R. Human papillomavirus genotype 31 does not express detectable microRNA levels during latent or productive virus replication. *J. Virol.* **2006**, *80*, 10890–10893. [CrossRef] [PubMed]

92. Yi, R.; Poy, M.N.; Stoffel, M.; Fuchs, E. A skin microrna promotes differentiation by repressing 'stemness'. *Nature* **2008**, *452*, 225–229. [CrossRef] [PubMed]

93. Melino, G.; Memmi, E.M.; Pelicci, P.G.; Bernassola, F. Maintaining epithelial stemness with p63. *Sci. Signal.* **2015**, *8*, 9. [CrossRef] [PubMed]

94. Mighty, K.K.; Laimins, L.A. P63 is necessary for the activation of human papillomavirus late viral functions upon epithelial differentiation. *J. Virol.* **2011**, *85*, 8863–8869. [CrossRef] [PubMed]
95. Melar-New, M.; Laimins, L.A. Human papillomaviruses modulate expression of microRNA 203 upon epithelial differentiation to control levels of p63 proteins. *J. Virol.* **2010**, *84*, 5212–5221. [CrossRef] [PubMed]
96. Gunasekharan, V.; Laimins, L.A. Human papillomaviruses modulate microrna 145 expression to directly control genome amplification. *J. Virol.* **2013**, *87*, 6037–6043. [CrossRef] [PubMed]
97. Sen, G.L.; Boxer, L.D.; Webster, D.E.; Bussat, R.T.; Qu, K.; Zarnegar, B.J.; Johnston, D.; Siprashvili, Z.; Khavari, P.A. Znf750 is a p63 target gene that induces KLF4 to drive terminal epidermal differentiation. *Dev. Cell.* **2012**, *22*, 669–677. [CrossRef] [PubMed]
98. Ghaleb, A.M.; Yang, V.W. Kruppel-like factor 4 (KLF4): What we currently know. *Gene* **2017**, *611*, 27–37. [CrossRef] [PubMed]
99. Gunasekharan, V.K.; Li, Y.; Andrade, J.; Laimins, L.A. Post-transcriptional regulation of KLF4 by high-risk human papillomaviruses is necessary for the differentiation-dependent viral life cycle. *PLoS Pathog.* **2016**, *12*, e1005747. [CrossRef] [PubMed]
100. Moody, C.A.; Laimins, L.A. Human papillomaviruses activate the ATM DNA damage pathway for viral genome amplification upon differentiation. *PLoS Pathog.* **2009**, *5*, e1000605. [CrossRef] [PubMed]
101. Chappell, W.H.; Gautam, D.; Ok, S.T.; Johnson, B.A.; Anacker, D.C.; Moody, C.A. Homologous recombination repair factors RAD51 and BRCA1 are necessary for productive replication of human papillomavirus 31. *J. Virol.* **2015**, *90*, 2639–2652. [CrossRef] [PubMed]
102. Banerjee, N.S.; Genovese, N.J.; Noya, F.; Chien, W.M.; Broker, T.R.; Chow, L.T. Conditionally activated E7 proteins of high-risk and low-risk human papillomaviruses induce S phase in postmitotic, differentiated human keratinocytes. *J. Virol.* **2006**, *80*, 6517–6524. [CrossRef] [PubMed]
103. Genovese, N.J.; Banerjee, N.S.; Broker, T.R.; Chow, L.T. Casein kinase II motif-dependent phosphorylation of human papillomavirus E7 protein promotes p130 degradation and S-phase induction in differentiated human keratinocytes. *J. Virol.* **2008**, *82*, 4862–4873. [CrossRef] [PubMed]
104. Banerjee, N.S.; Wang, H.K.; Broker, T.R.; Chow, L.T. Human papillomavirus (HPV) E7 induces prolonged G2 following S phase reentry in differentiated human keratinocytes. *J. Biol. Chem.* **2011**, *286*, 15473–15482. [CrossRef] [PubMed]
105. Wang, H.K.; Duffy, A.A.; Broker, T.R.; Chow, L.T. Robust production and passaging of infectious HPV in squamous epithelium of primary human keratinocytes. *Genes Dev.* **2009**, *23*, 181–194. [CrossRef] [PubMed]
106. Nakahara, T.; Peh, W.L.; Doorbar, J.; Lee, D.; Lambert, P.F. Human papillomavirus type 16 e1circumflexe4 contributes to multiple facets of the papillomavirus life cycle. *J. Virol.* **2005**, *79*, 13150–13165. [CrossRef] [PubMed]
107. Stark, G.R.; Taylor, W.R. Control of the G2/M transition. *Mol. Biotechnol.* **2006**, *32*, 227–248. [CrossRef]
108. Kousholt, A.N.; Menzel, T.; Sorensen, C.S. Pathways for genome integrity in G2 phase of the cell cycle. *Biomolecules* **2012**, *2*, 579–607. [CrossRef] [PubMed]
109. Zhou, B.B.; Elledge, S.J. The DNA damage response: Putting checkpoints in perspective. *Nature* **2000**, *408*, 433–439. [PubMed]
110. Hong, S.; Cheng, S.; Iovane, A.; Laimins, L.A. STAT-5 regulates transcription of the topoisomerase IIβ-binding protein 1 (*TOPBP1*) gene to activate the ATR pathway and promote human papillomavirus replication. *MBio* **2015**, *6*, e02006–02015. [CrossRef] [PubMed]
111. Anacker, D.C.; Aloor, H.L.; Shepard, C.N.; Lenzi, G.M.; Johnson, B.A.; Kim, B.; Moody, C.A. HPV31 utilizes the ATR-CHK1 pathway to maintain elevated RRM2 levels and a replication-competent environment in differentiating keratinocytes. *Virology* **2016**, *499*, 383–396. [CrossRef] [PubMed]
112. Davy, C.E.; Jackson, D.J.; Wang, Q.; Raj, K.; Masterson, P.J.; Fenner, N.F.; Southern, S.; Cuthill, S.; Millar, J.B.; Doorbar, J. Identification of a G(2) arrest domain in the E1 wedge E4 protein of human papillomavirus type 16. *J. Virol.* **2002**, *76*, 9806–9818. [CrossRef] [PubMed]
113. Nakahara, T.; Nishimura, A.; Tanaka, M.; Ueno, T.; Ishimoto, A.; Sakai, H. Modulation of the cell division cycle by human papillomavirus type 18 E4. *J. Virol.* **2002**, *76*, 10914–10920. [CrossRef] [PubMed]
114. Knight, G.L.; Turnell, A.S.; Roberts, S. Role for wee1 in inhibition of G2-to-M transition through the cooperation of distinct human papillomavirus type 1 E4 proteins. *J. Virol.* **2006**, *80*, 7416–7426. [CrossRef] [PubMed]

115. Davy, C.E.; Jackson, D.J.; Raj, K.; Peh, W.L.; Southern, S.A.; Das, P.; Sorathia, R.; Laskey, P.; Middleton, K.; Nakahara, T.; et al. Human papillomavirus type 16 E1∧E4-induced G2 arrest is associated with cytoplasmic retention of active CDK1/cyclin B1 complexes. *J. Virol.* **2005**, *79*, 3998–4011. [CrossRef] [PubMed]
116. Wang, Q.; Kennedy, A.; Das, P.; McIntosh, P.B.; Howell, S.A.; Isaacson, E.R.; Hinz, S.A.; Davy, C.; Doorbar, J. Phosphorylation of the human papillomavirus type 16 E1∧E4 protein at t57 by ERK triggers a structural change that enhances keratin binding and protein stability. *J. Virol.* **2009**, *83*, 3668–3683. [CrossRef] [PubMed]
117. Yu, J.H.; Lin, B.Y.; Deng, W.; Broker, T.R.; Chow, L.T. Mitogen-activated protein kinases activate the nuclear localization sequence of human papillomavirus type L1 E1 DNA helicase to promote efficient nuclear import. *J. Virol.* **2007**, *81*, 5066–5078. [CrossRef] [PubMed]
118. McIntosh, P.B.; Laskey, P.; Sullivan, K.; Davy, C.; Wang, Q.; Jackson, D.J.; Griffin, H.M.; Doorbar, J. E1∧E4-mediated keratin phosphorylation and ubiquitylation: A mechanism for keratin depletion in HPV16-infected epithelium. *J. Cell. Sci.* **2010**, *123*, 2810–2822. [CrossRef] [PubMed]
119. DeSmet, M.; Kanginakudru, S.; Rietz, A.; Wu, W.H.; Roden, R.; Androphy, E.J. The replicative consequences of papillomavirus E2 protein binding to the origin replication factor ORC2. *PLoS Pathog.* **2016**, *12*, e1005934. [CrossRef] [PubMed]
120. Sakakibara, N.; Chen, D.; McBride, A.A. Papillomaviruses use recombination-dependent replication to vegetatively amplify their genomes in differentiated cells. *PLoS Pathog.* **2013**, *9*, e1003321. [CrossRef] [PubMed]
121. Ceccaldi, R.; Rondinelli, B.; D'Andrea, A.D. Repair pathway choices and consequences at the double-strand break. *Trends Cell Biol.* **2016**, *26*, 52–64. [CrossRef] [PubMed]
122. Ciccia, A.; Elledge, S.J. The DNA damage response: Making it safe to play with knives. *Mol. Cell* **2010**, *40*, 179–204. [CrossRef] [PubMed]
123. Cimprich, K.A.; Cortez, D. ATR: An essential regulator of genome integrity. *Nat. Rev. Mol. Cell Biol.* **2008**, *9*, 616–627. [CrossRef] [PubMed]
124. Lee, J.H.; Paull, T.T. ATM activation by DNA double-strand breaks through the MRE11-RAD50-NBS1 complex. *Science* **2005**, *308*, 551–554. [CrossRef] [PubMed]
125. Sun, Y.; Jiang, X.; Chen, S.; Fernandes, N.; Price, B.D. A role for the TIP60 histone acetyltransferase in the acetylation and activation of ATM. *Proc. Natl. Acad. Sci. USA* **2005**, *102*, 13182–13187. [CrossRef] [PubMed]
126. Williams, R.S.; Williams, J.S.; Tainer, J.A. MRE11-RAD50-NBS1 is a keystone complex connecting DNA repair machinery, double-strand break signaling, and the chromatin template. *Biochem. Cell Biol.* **2007**, *85*, 509–520. [CrossRef] [PubMed]
127. Bakkenist, C.J.; Kastan, M.B. Chromatin perturbations during the DNA damage response in higher eukaryotes. *DNA Repair (Amst)* **2015**, *36*, 8–12. [CrossRef] [PubMed]
128. Reinhardt, H.C.; Yaffe, M.B. Kinases that control the cell cycle in response to DNA damage: CHK1, CHK2, and MK2. *Curr. Opin. Cell Biol.* **2009**, *21*, 245–255. [CrossRef] [PubMed]
129. Donzelli, M.; Draetta, G.F. Regulating mammalian checkpoints through CDC25 inactivation. *EMBO Rep.* **2003**, *4*, 671–677. [CrossRef] [PubMed]
130. Yazdi, P.T.; Wang, Y.; Zhao, S.; Patel, N.; Lee, E.Y.; Qin, J. SMC1 is a downstream effector in the ATM/NBS1 branch of the human S-phase checkpoint. *Genes Dev.* **2002**, *16*, 571–582. [CrossRef] [PubMed]
131. Lehmann, A.R. The role of SMC proteins in the responses to DNA damage. *DNA Repair (Amst)* **2005**, *4*, 309–314. [CrossRef] [PubMed]
132. Satsuka, A.; Mehta, K.; Laimins, L. p38MAPK and MK2 pathways are important for the differentiation-dependent human papillomavirus life cycle. *J. Virol.* **2015**, *89*, 1919–1924. [CrossRef] [PubMed]
133. Mehta, K.; Gunasekharan, V.; Satsuka, A.; Laimins, L.A. Human papillomaviruses activate and recruit SMC1 cohesin proteins for the differentiation-dependent life cycle through association with CTCF insulators. *PLoS Pathog.* **2015**, *11*, e1004763. [CrossRef] [PubMed]
134. Sakakibara, N.; Mitra, R.; McBride, A.A. The papillomavirus E1 helicase activates a cellular DNA damage response in viral replication foci. *J. Virol.* **2011**, *85*, 8981–8995. [CrossRef] [PubMed]
135. Anacker, D.C.; Gautam, D.; Gillespie, K.A.; Chappell, W.H.; Moody, C.A. Productive replication of human papillomavirus 31 requires DNA repair factor NBS1. *J. Virol.* **2014**, *88*, 8528–8544. [CrossRef] [PubMed]
136. Moody, C.A.; Fradet-Turcotte, A.; Archambault, J.; Laimins, L.A. Human papillomaviruses activate caspases upon epithelial differentiation to induce viral genome amplification. *Proc. Natl. Acad. Sci. USA* **2007**, *104*, 19541–19546. [CrossRef] [PubMed]

137. Sakakibara, N.; Chen, D.; Jang, M.K.; Kang, D.W.; Luecke, H.F.; Wu, S.Y.; Chiang, C.M.; McBride, A.A. Brd4 is displaced from HPV replication factories as they expand and amplify viral DNA. *PLoS Pathog.* **2013**, *9*, e1003777. [CrossRef] [PubMed]

138. Gillespie, K.A.; Mehta, K.P.; Laimins, L.A.; Moody, C.A. Human papillomaviruses recruit cellular DNA repair and homologous recombination factors to viral replication centers. *J. Virol.* **2012**, *86*, 9520–9526. [CrossRef] [PubMed]

139. Langsfeld, E.S.; Bodily, J.M.; Laimins, L.A. The deacetylase sirtuin 1 regulates human papillomavirus replication by modulating histone acetylation and recruitment of DNA damage factors NBS1 and RAD51 to viral genomes. *PLoS Pathog.* **2015**, *11*, e1005181. [CrossRef] [PubMed]

140. Flores, E.R.; Lambert, P.F. Evidence for a switch in the mode of human papillomavirus type 16 DNA replication during the viral life cycle. *J. Virol.* **1997**, *71*, 7167–7179. [PubMed]

141. Hirano, T. SMC proteins and chromosome mechanics: From bacteria to humans. *Philos. Trans. R. Soc. Lond. B Biol. Sci.* **2005**, *360*, 507–514. [CrossRef] [PubMed]

142. Johnson, B.A.; Aloor, H.L.; Moody, C.A. The RB binding domain of Hpv31 E7 is required to maintain high levels of DNA repair factors in infected cells. *Virology* **2017**, *500*, 22–34. [CrossRef] [PubMed]

143. Hong, S.; Laimins, L.A. The JAK-STAT transcriptional regulator, STAT-5, activates the ATM DNA damage pathway to induce Hpv 31 genome amplification upon epithelial differentiation. *PLoS Pathog.* **2013**, *9*, e1003295. [CrossRef] [PubMed]

144. Hong, S.; Dutta, A.; Laimins, L.A. The acetyltransferase tip60 is a critical regulator of the differentiation-dependent amplification of human papillomaviruses. *J. Virol.* **2015**, *89*, 4668–4675. [CrossRef] [PubMed]

145. Reinson, T.; Toots, M.; Kadaja, M.; Pipitch, R.; Allik, M.; Ustav, E.; Ustav, M. Engagement of the ATR-dependent DNA damage response at the human papillomavirus 18 replication centers during the initial amplification. *J. Virol.* **2013**, *87*, 951–964. [CrossRef] [PubMed]

146. Fradet-Turcotte, A.; Bergeron-Labrecque, F.; Moody, C.A.; Lehoux, M.; Laimins, L.A.; Archambault, J. Nuclear accumulation of the papillomavirus E1 helicase blocks S-phase progression and triggers an ATM-dependent DNA damage response. *J. Virol.* **2011**, *85*, 8996–9012. [CrossRef] [PubMed]

147. Gauson, E.J.; Donaldson, M.M.; Dornan, E.S.; Wang, X.; Bristol, M.; Bodily, J.M.; Morgan, I.M. Evidence supporting a role for TOPBP1 and BRD4 in the initiation but not continuation of human papillomavirus 16 E1/E2-mediated DNA replication. *J. Virol.* **2015**, *89*, 4980–4991. [CrossRef] [PubMed]

148. Galloway, D.A.; Laimins, L.A. Human papillomaviruses: Shared and distinct pathways for pathogenesis. *Curr. Opin. Virol.* **2015**, *14*, 87–92. [CrossRef] [PubMed]

149. Papamichos-Chronakis, M.; Peterson, C.L. Chromatin and the genome integrity network. *Nat. Rev. Genet.* **2013**, *14*, 62–75. [CrossRef] [PubMed]

150. Polo, S.E.; Jackson, S.P. Dynamics of DNA damage response proteins at DNA breaks: A focus on protein modifications. *Genes Dev.* **2011**, *25*, 409–433. [CrossRef] [PubMed]

151. Van Attikum, H.; Gasser, S.M. Crosstalk between histone modifications during the DNA damage response. *Trends Cell Biol.* **2009**, *19*, 207–217. [CrossRef] [PubMed]

152. Gautam, D.; Moody, C.A. Impact of the DNA damage response on human papillomavirus chromatin. *PLoS Pathog.* **2016**, *12*, e1005613. [CrossRef] [PubMed]

153. Oberdoerffer, P.; Michan, S.; McVay, M.; Mostoslavsky, R.; Vann, J.; Park, S.K.; Hartlerode, A.; Stegmuller, J.; Hafner, A.; Loerch, P.; et al. SIRT1 redistribution on chromatin promotes genomic stability but alters gene expression during aging. *Cell* **2008**, *135*, 907–918. [CrossRef] [PubMed]

154. Tang, J.; Cho, N.W.; Cui, G.; Manion, E.M.; Shanbhag, N.M.; Botuyan, M.V.; Mer, G.; Greenberg, R.A. Acetylation limits 53BP1 association with damaged chromatin to promote homologous recombination. *Nat. Struct. Mol. Biol.* **2013**, *20*, 317–325. [CrossRef] [PubMed]

155. Saldivar, J.C.; Cortez, D.; Cimprich, K.A. The essential kinase ATR: Ensuring faithful duplication of a challenging genome. *Nat. Rev. Mol. Cell Biol.* **2017**. [CrossRef] [PubMed]

156. Spardy, N.; Covella, K.; Cha, E.; Hoskins, E.E.; Wells, S.I.; Duensing, A.; Duensing, S. Human papillomavirus 16 E7 oncoprotein attenuates DNA damage checkpoint control by increasing the proteolytic turnover of claspin. *Cancer Res.* **2009**, *69*, 7022–7029. [CrossRef] [PubMed]

157. Bester, A.C.; Roniger, M.; Oren, Y.S.; Im, M.M.; Sarni, D.; Chaoat, M.; Bensimon, A.; Zamir, G.; Shewach, D.S.; Kerem, B. Nucleotide deficiency promotes genomic instability in early stages of cancer development. *Cell* **2011**, *145*, 435–446. [CrossRef] [PubMed]

158. Ellison, V.; Stillman, B. Biochemical characterization of DNA damage checkpoint complexes: Clamp loader and clamp complexes with specificity for 5′ recessed DNA. *PLoS Biol.* **2003**, *1*, E33. [CrossRef] [PubMed]

159. Zou, L.; Liu, D.; Elledge, S.J. Replication protein a-mediated recruitment and activation of RAD17 complexes. *Proc. Natl. Acad. Sci. USA* **2003**, *100*, 13827–13832. [CrossRef] [PubMed]

160. Mordes, D.A.; Glick, G.G.; Zhao, R.; Cortez, D. TOPBP1 activates ATR through ATRIP and a PIKK regulatory domain. *Genes Dev.* **2008**, *22*, 1478–1489. [CrossRef] [PubMed]

161. Edwards, T.G.; Helmus, M.J.; Koeller, K.; Bashkin, J.K.; Fisher, C. Human papillomavirus episome stability is reduced by aphidicolin and controlled by DNA damage response pathways. *J. Virol.* **2013**, *87*, 3979–3989. [CrossRef] [PubMed]

162. Liu, S.; Opiyo, S.O.; Manthey, K.; Glanzer, J.G.; Ashley, A.K.; Amerin, C.; Troksa, K.; Shrivastav, M.; Nickoloff, J.A.; Oakley, G.G. Distinct roles for DNA-PK, ATM and ATR in RPA phosphorylation and checkpoint activation in response to replication stress. *Nucleic Acids Res.* **2012**, *40*, 10780–10794. [CrossRef] [PubMed]

163. Bertoli, C.; Klier, S.; McGowan, C.; Wittenberg, C.; de Bruin, R.A. CHK1 inhibits E2F6 repressor function in response to replication stress to maintain cell-cycle transcription. *Curr. Biol.* **2013**, *23*, 1629–1637. [CrossRef] [PubMed]

164. Buisson, R.; Boisvert, J.L.; Benes, C.H.; Zou, L. Distinct but concerted roles of ATR, DNA-PK, and CHK1 in countering replication stress during S phase. *Mol. Cell* **2015**, *59*, 1011–1024. [CrossRef] [PubMed]

165. Murga, M.; Campaner, S.; Lopez-Contreras, A.J.; Toledo, L.I.; Soria, R.; Montana, M.F.; D'Artista, L.; Schleker, T.; Guerra, C.; Garcia, E.; et al. Exploiting oncogene-induced replicative stress for the selective killing of Myc-driven tumors. *Nat. Struct. Mol. Biol.* **2011**, *18*, 1331–1335. [CrossRef] [PubMed]

166. Toledo, L.I.; Murga, M.; Zur, R.; Soria, R.; Rodriguez, A.; Martinez, S.; Oyarzabal, J.; Pastor, J.; Bischoff, J.R.; Fernandez-Capetillo, O. A cell-based screen identifies ATR inhibitors with synthetic lethal properties for cancer-associated mutations. *Nat. Struct. Mol. Biol.* **2011**, *18*, 721–727. [CrossRef] [PubMed]

167. Nordlund, P.; Reichard, P. Ribonucleotide reductases. *Annu. Rev. Biochem.* **2006**, *75*, 681–706. [CrossRef] [PubMed]

168. Jang, M.K.; Shen, K.; McBride, A.A. Papillomavirus genomes associate with BRD4 to replicate at fragile sites in the host genome. *PLoS Pathog.* **2014**, *10*, e1004117. [CrossRef] [PubMed]

169. Sarni, D.; Kerem, B. The complex nature of fragile site plasticity and its importance in cancer. *Curr. Opin. Cell Biol.* **2016**, *40*, 131–136. [CrossRef] [PubMed]

170. Gao, G.; Johnson, S.H.; Vasmatzis, G.; Pauley, C.E.; Tombers, N.M.; Kasperbauer, J.L.; Smith, D.I. Common fragile sites (CFS) and extremely large *CFS* genes are targets for human papillomavirus integrations and chromosome rearrangements in oropharyngeal squamous cell carcinoma. *Genes Chromosomes Cancer* **2017**, *56*, 59–74. [CrossRef] [PubMed]

171. Bodelon, C.; Untereiner, M.E.; Machiela, M.J.; Vinokurova, S.; Wentzensen, N. Genomic characterization of viral integration sites in HPV-related cancers. *Int. J. Cancer* **2016**, *139*, 2001–2011. [CrossRef] [PubMed]

172. Thorland, E.C.; Myers, S.L.; Persing, D.H.; Sarkar, G.; McGovern, R.M.; Gostout, B.S.; Smith, D.I. Human papillomavirus type 16 integrations in cervical tumors frequently occur in common fragile sites. *Cancer Res.* **2000**, *60*, 5916–5921. [PubMed]

173. Choo, K.B.; Chen, C.M.; Han, C.P.; Cheng, W.T.; Au, L.C. Molecular analysis of cellular loci disrupted by papillomavirus 16 integration in cervical cancer: Frequent viral integration in topologically destabilized and transcriptionally active chromosomal regions. *J. Med. Virol.* **1996**, *49*, 15–22. [CrossRef]

174. McBride, A.A.; Warburton, A. The role of integration in oncogenic progression of HPV-associated cancers. *PLoS Pathog.* **2017**, *13*, e1006211. [CrossRef] [PubMed]

175. McBride, A.A. Playing with fire: Consequences of human papillomavirus DNA replication adjacent to genetically unstable regions of host chromatin. *Curr. Opin. Virol.* **2017**, *26*, 63–68. [CrossRef] [PubMed]

176. Wallace, N.A.; Khanal, S.; Robinson, K.L.; Wendel, S.O.; Messer, J.J.; Galloway, D.A. High risk α papillomavirus oncogenes impair the homologous recombination pathway. *J. Virol.* **2017**. [CrossRef] [PubMed]

viruses

MDPI

Review

Keratinocyte Differentiation-Dependent Human Papillomavirus Gene Regulation

Sheila V. Graham

MRC-University of Glasgow Centre for Virus Research, Institute of Infection, Immunity and Inflammation, College of Medical, Veterinary and Life Sciences, University of Glasgow, Garscube Estate, Glasgow G61 1QH, UK; sheila.graham@gla.ac.uk; Tel.: +44-141-330-6256

Academic Editors: Alison A. McBride and Karl Munger
Received: 7 August 2017; Accepted: 25 August 2017; Published: 30 August 2017

Abstract: Human papillomaviruses (HPVs) cause diseases ranging from benign warts to invasive cancers. HPVs infect epithelial cells and their replication cycle is tightly linked with the differentiation process of the infected keratinocyte. The normal replication cycle involves an early and a late phase. The early phase encompasses viral entry and initial genome replication, stimulation of cell division and inhibition of apoptosis in the infected cell. Late events in the HPV life cycle include viral genome amplification, virion formation, and release into the environment from the surface of the epithelium. The main proteins required at the late stage of infection for viral genome amplification include E1, E2, E4 and E5. The late proteins L1 and L2 are structural proteins that form the viral capsid. Regulation of these late events involves both cellular and viral proteins. The late viral mRNAs are expressed from a specific late promoter but final late mRNA levels in the infected cell are controlled by splicing, polyadenylation, nuclear export and RNA stability. Viral late protein expression is also controlled at the level of translation. This review will discuss current knowledge of how HPV late gene expression is regulated.

Keywords: human papillomavirus; infection; epithelial differentiation; gene regulation; RNA processing

1. Introduction

Over 210 human papillomavirus (HPV) genotypes have been characterized. Most HPVs do not cause any symptoms or disease. However, some can cause benign lesions such as warts on the hands or verrucas on the soles of the feet. Importantly, around 40 can infect the anogenital tract and cause benign lesions such as genital warts, or precancerous lesions such as cervical intraepithelial neoplasia [1]. Anogenital HPV infection is very common and affects around 80% of the population during their lifetime. Infection is usually transient because the virus is eventually recognized by the immune system and the infection is cleared without causing significant disease [2]. However, if infection with one of the so-called "high risk" (HR) anogenital-infective HPV genotypes becomes persistent, it can cause cellular changes that can lead to cancer formation [3]. HR-HPVs all belong to the alpha-papillomaviridae. There are 12 HR-HPVs for which the International Agency for Research on Cancer considers that there is sufficient evidence of carcinogenicity to humans (HPV16, HPV18, HPV 31, HPV 33, HPV35, HPV 39, HPV45, HPV51, HPV52, HPV56, HPV58 and HPV59). HPV68 is considered probably carcinogenic and HPV26, HPV53, HPV66, HPV67, HPV68, HPV70 and HPV73 are considered possibly carcinogenic [4]. The causative association of HPV infection with cervical cancer is well studied but more recently, HR-HPV has been found to be increasingly associated with oropharyngeal cancer, especially in younger men [5]. Understanding the infectious life cycle of HPVs that cause benign diseases—and of those HR-HPVs that can cause cancer—is essential for elucidating how persistent infection occurs and what molecular changes support cancer progression.

HPVs infect mucosal or cutaneous epithelia and there is a tight regulatory linkage between the viral life cycle and epithelial differentiation. Initial infection is targeted to dividing cells in the basal epithelial layer. Upon cell entry, the HPV genome is deposited in the nucleus and begins to express viral early proteins, which carry out an initial round of viral genome replication to yield around 50–100 genome copies per cell [6]. Once an infected basal epithelial cell divides, these viral genome copies are replicated and segregated equally into daughter cells. Continued presence of the viral genome over a period of several years in actively dividing epithelial cells results in persistent infection [6]. However, in a normal infection upon basal cell division, an infected daughter cell will become a transit amplifying cell that is destined to complete epithelial differentiation and move up through the various epithelial layers (Figure 1). During this process, there is a carefully orchestrated pattern of viral gene expression in response to epithelial differentiation such that different viral gene expression events are specifically linked to different epithelial layers (Figure 1). The viral genome within the cell nucleus responds to keratinocyte differentiation by activating viral late gene expression. Viral late proteins are required to accomplish productive viral genome amplification and virion production. These late events in the viral life cycle occur specifically in the differentiating upper layers of the epithelium (Figure 1) [7]. Therefore, cellular events linked to cellular gene expression make a significant impact on viral replication. It seems clear that persistent HPV infection is associated with a loss of the virus's ability to carry out productive viral replication and synthesize new viral particles [8]. This review article will explore current knowledge of the viral and cellular molecular mechanisms of gene regulation that are required for the virus to accomplish late events in its life cycle.

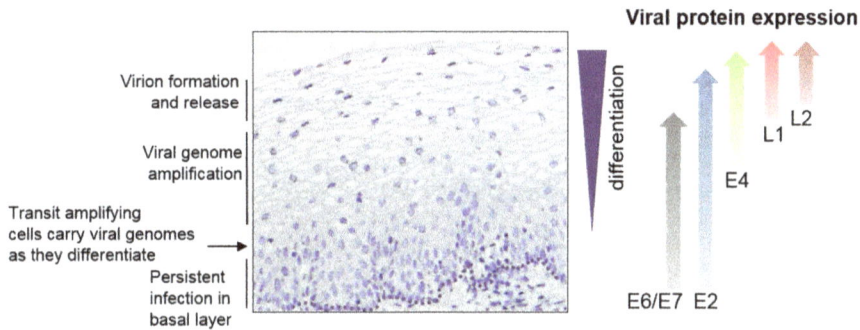

Figure 1. Eosin-stained HPV16-infected cervical epithelium (nuclei are stained purple). Key events in the viral replication cycle are noted on the left hand side. The approximate region of expression of viral proteins is shown on the right hand side. It is expected that E1 follows the pattern of E2 while E5 follows the pattern of E4. However, lack of suitable antibodies against these proteins have precluded confirmation of their expression pattern.

2. Viral Genome Organization

Non-enveloped icosahedral HPV virions are around 55 nm in diameter and each contains a double stranded circular DNA genome of around 8 kb in length. The genome can be divided into three regions, the early region containing open reading frames for at least seven viral regulatory proteins, the late region containing open reading frames for the two viral structural proteins, the capsid proteins, and an approximately 1 kb regulatory region called the Long Control Region (LCR) or Upstream Regulatory Region (URR) (Figure 2A). This non-coding region contains the viral early promoter and transcriptional enhancer, the viral origin of replication, the late polyadenylation site and the late (or negative) regulatory element (LRE/NRE) that controls late gene expression at various post-transcriptional levels (Figure 2A). The early viral promoter (HPV16: P97) is located at the 5′ end of the E6 open reading frame. Other start sites just upstream of this have been identified for HPV16 late mRNA in clinical samples (Figure 2B P7437) [9] and for HPV18 in raft cultured infected epithelial cells [10]. The viral late promoter (HPV16: P670) that is specifically activated in differentiated epithelial cells is located within the E7 open reading frame (Figure 2B). A third promoter, the E8 promoter, that is active both early and late in infection is located in the E1 open reading frame. Other possible promoters have been described but have not yet been fully tested for functionality (Figure 2B) [9,11–14].

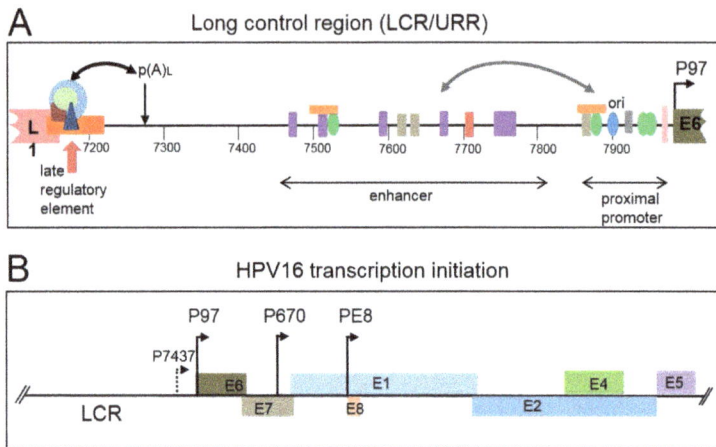

Figure 2. Regulatory elements on the HPV16 genome. (**A**) Details of the *cis*-acting elements, and the proteins that bind, located in the HPV16 long control region (LCR), otherwise called the upstream regulatory region (URR). The 3′ end of the L1 open reading frame is shown in pink. The 5′ end of the E6 open reading frame is shown in olive. The approximate positions of the proximal promoter and transcriptional enhancer are shown with double headed arrows. The origin of replication (ori) is marked with a blue oval. The late polyadenylation site is indicated with a downward facing arrow and p(A)$_L$. The late regulatory element at the end of the L1 open reading frame is indicated with a red upward arrow and an orange box. The light blue circle indicates U1 snRNP, light green circle; SRSF1, dark red shape; U2AF, dark blue triangle; CUG-BP1 (see Table 1). The curved double headed black arrow indicates interactions between these factors and the polyadenylation complex (not shown). The viral transcription/replication factor E2 is shown as green ovals. Transcription factors are shown as rectangles. NF1; purple, Oct1; red, AP1; beige, Sp1; gray, YY1; orange. This is not an exhaustive list of transcription factors that bind the LCR. The curved double headed gray arrow indicates interactions between the enhancer and proximal promoter; (**B**) illustration of the LCR and early region of the HPV16 genome. Early open reading frames are shown as coloured boxes. The location on the genome of characterised promoters (P) is shown with arrows and associated numbers.

2.1. Early Gene Expression

The early promoter is responsible for expression of the early gene region. The viral replication factor E1 and its auxiliary protein E2, which is also the viral transcription factor, are probably the first proteins expressed during infection [15]. The viral E6 and E7 oncoproteins are also expressed early in a normal infection, but at very low levels to limit their oncogenic activity. However, in differentiating epithelial cells where cell division would normally be repressed, they act to stimulate cell cycle progression [16,17]. Moreover, because inappropriately dividing cells in the upper epithelial layers would normally be subject to apoptosis, E6 degrades p53 by targeting it for proteasome-mediated degradation, thus allowing HPV-infected cells to survive and support viral replication [18]. Other open reading frames in the early region encode E8^E2, an antagonist of the viral E2 replication and transcription factor [19], E4, which can remodel differentiated keratinocytes to allow release of progeny viral particles [20] and regulate the cell cycle [21], and E5, which has roles in keratinocyte signaling and immune evasion [22]. The proteins encoded by these open reading frames are expressed at highest levels in differentiating virus-infected cells [23] and they are required for productive viral genome amplification [24–28]. mRNAs encoding these proteins have been found to be expressed from both viral early and late promoters and this suggests that they may have roles at both the early and late stages of the viral life cycle. Early region mRNAs are polyadenylated at the early polyadenylation site which is situated just downstream of the E5 open reading frame (Figure 3A) [29].

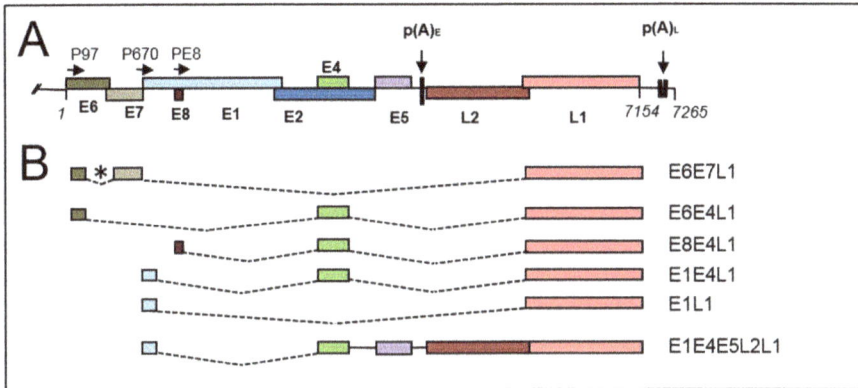

Figure 3. Splicing and polyadenylation of HPV16 late RNAs. (**A**) Diagram of the HPV16 genome. Coloured boxes indicate viral open reading frames. Promoters are indicated with arrows and (P). The positions of the early (pA$_E$) and late (pA$_L$) polyadenylation sites and indicated with downward arrows; (**B**) diagram of the main splicing events for generation of HPV16 late mRNAs. Boxes indicate open reading frames. Dotted lines indicate introns that are spliced out. (*), alternative splice acceptor sites can be used in the E6E7 gene region, including, for HPV16, sites at 409, 526 and 743. The putative coding potential of each mRNA is indicated to the right-hand side. This is a limited list of possible late mRNAs. A fuller list can be viewed at https://pave.niaid.nih.gov/#explore/transcript_maps.

2.2. Late Gene Expression and the Late Promoter

Late gene expression is initiated by the viral late promoter, terminated at the late polyadenylation site (Figure 3A) and occurs in the suprabasal layers (Figure 1) [11]. Viral late proteins E1, E2, E4 and E5 are required for vegetative viral genome amplification where the ~100 initially replicated viral genome copies are amplified to hundreds or thousands of copies. This is thought to take place using a rolling circle mode of replication, or recombination-dependent replication that is specific to this late life cycle stage [30,31] and is dependent upon the viral replication proteins E1 and E2. Expression of E2 appears to peak in the spinous layer of infected epithelia, co-incident with initiation of genome amplification [32–34]. Effective antibodies against E1 are lacking but it is likely that E1 expression follows a similar expression pattern to E2. HR-HPV E4, the most abundant viral protein, is expressed in the upper epithelial layers but cutaneous HPVs can express E4 in the basal layer [35]. E5 expression is likely also restricted to differentiated epithelial cells. Expression of major and minor capsid proteins L1 and L2 is always subsequent to E4 expression. However, each HPV genotype can display a distinct late gene expression program [35]. For example, mucosal HR-HPVs 16 and 31 express their capsid proteins only in the uppermost granular layer, but cutaneous HPVs such as HPV1 and HPV63 carry out late events much closer to the basal layer [35]. Importantly, HR-HPVs appear to restrict capsid protein synthesis to the uppermost layers [36–38]. This allows the virus to avoid activating the humoral immune response [2] and may facilitate the release of newly formed virions from disintegrating dead squames at the top of the epithelium.

Like many viral genomes, the genome organization of HPVs is complex. There are overlapping open reading frames (e.g., E2 and E4 or E1 and E8) and multicistronic transcription. Moreover, alternative promoter and polyadenylation site usage and alternative splicing yields many more mRNAs than there are open reading frames [39] and these seem to be essential for encoding the full complement of viral proteins. Therefore, although transcriptional regulation is important for the HPV life cycle and is a key factor in initiation of viral late gene expression in differentiating keratinocytes, much of the control of viral gene expression occurs at a post-transcriptional level.

3. Viral Late Promoter Activity

Viral late promoter activation is a key mechanism behind the initiation of late gene expression. The mucosal oncogenic virus late promoters (HPV16 P670 [36], HPV18 P811 [40] and HPV31 P742 [41]) are the best characterized but similarly positioned promoters exist for other HPVs including the cutaneous HPV5 (P840) [42], and HPV8 (P7535) [10,43,44]. The HPV31 late promoter has over 30 possible initiation sites stretching from genome positions 605 to 779 [41,45] while for HPV16, transcription initiation sites have been mapped to various nucleotides within 200 nucleotides surrounding P670 [11,36]. HPV31 possesses two differentiation-responsive elements in the late promoter region [41] and a number of transcription factors have been shown to bind and regulate the HPV16 and 31 promoter regions [46,47]. The HPV18 late promoter is likely also regulated by similar differentiation-dependent transcription factors but this has not been fully elucidated. A recent intriguing observation is that HPV18 late promoter activity is controlled by the orientation of viral DNA replication, thus providing a clear link between vegetative viral DNA amplification and late transcription [40]. Moreover, this study revealed that hnRNP proteins A/B and DOB bind a transcriptional repressor to regulate late gene expression. As well as promoter control by proximal elements, it has been shown that the HPV31 late promoter is also controlled by the viral enhancer located in the LCR [41]. Linked to this observation, recently it has been shown that the promoter is mainly controlled at the level of transcription elongation. Although RNA polymerase II is bound at the late promoter in undifferentiated keratinocytes, elongation is inefficient and does not progress to the late region. In contrast, in differentiated cells, the enhancer recruits members of the BET family proteins—including Brd4 and its binding partner CDK8—to the Mediator complex to stimulate transcription elongation [48]. Changes in signaling during keratinocyte differentiation must also regulate the late promoter, for example PKCδ [49] has been shown to activate it. Finally, the E8 promoter, responsible for expression of the replication and transcriptional repressor E8^E2, is active at both early and late phases of the life cycle but its regulation has not been extensively characterized as yet [50].

There is evidence of a reciprocal coordination between activity of the viral early and late promoters, because the early E7 protein has been shown to control the HPV16 late promoter [51]. Although it has been proposed that early oncoprotein activity wanes as the late phase of the life cycle gets underway [23,52], transcripts encoding E6 and E7 have been detected in the cytoplasm of cells in the mid-upper epithelial layers [53–56]. Indeed, it has been reported that transcription initiation at the HPV31 early promoter is not down-regulated by differentiation [13] and HPV31 late transcript mapping in differentiated keratinocytes detected transcription initiation in the vicinity of the early promoter giving rise to several RNAs encoding E6 and E7 [57]. It is unlikely that RNA polymerase could load simultaneously onto the three closely spaced promoters (Figure 2B) that are active at late stages in the viral life cycle. Although it is possible that a stochastic choice of promoters takes place, there must be differentiation-specific control exerted. Therefore, polymerase loading and/or rate of elongation, controlled by transcription complexes recruited in a differentiation-specific manner to the viral enhancer and promoter regions, is a more likely mechanism to ensure inter-dependent promoter activity at late times during the viral replication cycle.

4. Differentiation-Dependent Regulation of Polyadenylation

Polyadenylation is a two-step process involving cleavage at the cleavage site and subsequent addition of 200–250 A residues to the 3′ end of the mRNA [58]. Polyadenylation is essential for mRNA export from the nucleus to the cytoplasm, mRNA stability and translation. Mammalian polyadenylation sites comprise two cis-acting regulatory elements, an upstream A(A/U)UAAA element, located 10–30 nucleotides upstream of the cleavage site, that binds the cleavage and poyadenylation specificity factor (CPSF). A 73 kDa subunit of CPSF possesses endonuclease activity for the cleavage reaction, but CPSF is also required for poly(A) addition. A GU-rich element, followed by a UUU motif, located 10–30 nucleotides downstream of the cleavage site binds the cleavage stimulatory factor (CstF). CstF is required for cleavage and it stabilises the binding of CPSF upstream through protein-protein interactions between its 77 kDa subunit and the 160 kDa subunit of CPSF.

The exact sequence of the cis-acting elements to which these protein complexes bind determines whether polyadenylation is efficient or not by controlling the stability of the polyadenylation complex [58]. HPV genomes possess at least two polyadenylation sites—the early site ($p(A)_E$) situated in the early 3′ untranslated region (UTR) (Figure 3A) downstream of the E5 open reading frame, and the late site, $p(A)_L$, located in the 5′ end of the LCR [59]. Early trancription terminates at the early site, while late mRNA synthesis uses the late site [36,37,60,61]. There is good evidence for differentiation-dependent control of both sites. Because late mRNA production initiates from the late promoter located in the early region, transcription must proceed through the early polyadenylation site to synthesize late mRNAs. This suggests that either rate of RNA polymerase passage through the early polyadenylation site is up-regulated or use of the early polyadenylation site is repressed at late times of infection. The HPV31 early polyadenylation site contains a consensus CPSF binding site, but each of three possible CstF binding sites downstream are of weak consensus and lead to heterogeneity in polyadenylation site usage [62,63]. The HPV16 early polyadenylation site has a similar organisation but also contains a long U-rich stretch upstream of the CPSF-binding motif that binds auxilliary polyadenylation factors such as hFip 1, to enhance early polyadenylation [64]. Other HPV early polyadenylation sites are also of weak consensus [65], however, the HPV18 early site appears to contain good consensus polyadenylation signals in addition to several upstream U/GU-rich motifs that could regulate efficient polyadenylation [10]. The HPV31 late polyadenylation site has a simple AAUAAA and downstream G/U rich element organisation [66]. HPV16 uses two tandem late polyadenylation sites [11], the first of which is of weak conensus and used less frequently, while the second is of strong consensus [11,67]. HPV18 has a single late AAUAAA site and a downstream GU-rich motif but transcript cleavage was detected at a range of nucleotides over a 35 nt region [10].

4.1. Repression of the Early Polyadenylation Site and Late Gene Expression

Most studies have arrived at the consensus view that late mRNA expression is controlled largely through negative control of the early polyadenylation site, thus allowing transcription to proceed into the late region. This may be the reason why the early polyadenylation site has to be inherently weak to allow fine control by RNA binding proteins. The role of the early polyadenylation site in the control of late gene expression was first demonstrated for HPV31. Mutational analysis of the early polyadenylation site revealed that the early CPSF-binding site was a key inhibitor of late transcript production [63]. However, this study also uncovered an element in the 5′ end of the downstream L2 open reading frame that repressed capsid mRNA production, probably via positive enhancement of the early polyadenylation site through binding the 64 kDa subunit of CstF (CstF64) [62,63]. Subsequently, a similarly located element was discovered for HPV16, but in this case the element bound not only CstF64 but also hnRNP H through multiple GGG motifs [65,68]. It has been proposed that these motifs cause formation of a specific RNA secondary structure that is optimal for interaction of RNA binding proteins with the upstream polyadenylation complex. Whether these would be stimulatory or inhibitory interactions in vivo is not known.

These important findings indicate that the "weak" early viral polyadenylation site could be enhanced by upstream and downstream sequences that could bind polyadenylation-stimulatory factors and thus facilitate polyadenylation of early mRNAs. Alternatively, cellular factors binding to RNA motifs in the vicinity of the polyadenylation site could create secondary structures that are suboptimal for efficient early polyadenylation leading to stimulation of late gene expression. Such cellular factors could be expressed in a differentiation-dependent manner. Polyadenylation regulation, via the carboxyl terminal domain of RNA polymerase II, is coupled to other post-transcriptional events, including RNA capping and splicing, all of which take place co-transcriptionally [69]. Therefore, HPV RNA splicing could be expected to regulate polyadenylation. As shown in Figure 3B, HPV mRNAs are the products of a range of splicing events. The splicing regulatory protein Serine Arginine Splicing Factor 3 (SRSF3) binds a splicing enhancer element in the E4 open reading frame to facilitate polyadenylation at the early polyadenylation site [70]. Another E4 element binds SRSF1 and appears to favour synthesis of HPV16 mRNA early polyadenylation [71,72]. Splicing factors are known to control polyadenylation by forming protein-protein interactions across exons that link to downstream polyadenylation complexes [58,73]. Differential expression of factors controlling the switch from early to late polyadenylation during keratinocyte differentiation would be necessary to induce repression of the early polyadenylation site. However, there is controversy regarding differentiation-stage-specific expression of splicing and polyadenylation factors because different patterns are observed in uninfected versus HPV-infected or HPV-associated tumour tissue [70,74,75]. Future studies should analyse protein expression in normal human cervical keratinocytes and the same cells transfected with episomal HPV genomes. Finally, HPV E2 has been implicated in a general, potentially low level, inhibition of polyadenylation by inhibiting assembly of the CPSF complex [76]. Rising levels of E2 in the mid to upper epithelial layers could lead to selective repression of early gene expression because the majority of mRNAs are polyadenylated at the early polyadenylation site. Early polyadenylation repression would lead to increased transcriptional read-through from the early region to the late region and a corresponding increase (despite some inhibition of late polyadenylation) in production of HPV16 late mRNAs [76]. It remains to be tested whether other E2-interacting proteins, both viral (e.g., E4, E1, E8) and cellular (e.g., SR proteins) [77,78], might modulate the effects of E2 on polyadenylation.

4.2. Control of Late Gene Expression Occurs at Multiple Post-Transcriptional Levels

Several different mechanisms of regulating capsid protein expression have been reported. These include transcription initiation and elongation—and polyadenylation, as discussed above—but also splicing, mRNA stability and translation. These processes are all linked in the cell (Figure 4). Transcription and RNA processing is linked through the carboxyl terminal domain of RNA polymerase II [69], while mRNA stability and translation on the ribosomes are also intimately connected [79]. Cell signaling regulates each of these pathways in order that cells can direct expression of the appropriate set of mRNAs and proteins to ensure normal function. An early study revealed that protein kinase C (PKC) signalling post-transcriptionally controlled late protein expression in HPV31-infected keratinocytes [61]. The mechanism of action could be through polyadenylation, splicing or RNA stability control. The following sections will discuss each of these post-transcriptional mechanisms in turn. However, it is important to remember that changes in one process will impact on the others and alterations in differentiation-specific expression levels of cellular proteins will also play a major role.

4.3. Control of Late Polyadenylation

The HPV late polyadenylation sites that have been analysed thus far are predicted to be used efficiently because they possess excellent consensus *cis*-acting control sequences. However, polyadenylation efficiency can be positively or negatively controlled through upstream sequence elements (USEs) or downstream sequence elements (DSEs). These elements are usually U-rich and are found in the vicinity of some cellular—e.g., COX-2 [80], human complement factor C2 [81] and collagen genes [82]), and many viral, polyadenylation sites (e.g., HIV [83], SV40 [84,85] and adenovirus [86]. They work by recruiting RNA-binding proteins that can stimulate or stabilise the polyadenylation complexes perhaps by inducing secondary structures that facilitates appropriate protein-proteins interactions. The upstream and downstream *cis*-acting elements that regulate HPV16 and HPV31 early polyadenylation site usage appear to work in this way. Not all USE/DSEs stimulate polyadenylation. For example, U1A RNAs possess a 3'UTR USE that binds U1A protein to negatively autoregulate its expression [87].

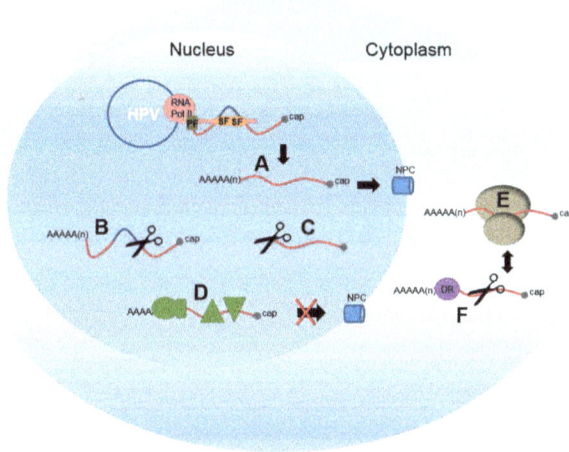

Figure 4. Cartoon of HPV gene regulation in the nucleus and cytoplasm. Transcription of the HPV genome (blue circle) is carried out by RNA polymerase II (pink oval). The carboxyl terminal domain of the polymerase (pink thin rectangle) acts to integrate transcription and RNA processing by binding splicing factors (SF; e.g., SRSFs and hnRNPs, orange ovals) and polyadenylation factors (PF, olive rectangle). These are then able to bind and process the nascent RNA molecule as it emerges from the RNA polymerase exit channel. A. Fully processed mRNAs are able to be exported from the nucleus to the cytoplasm through nuclear pore complexes (NPC, blue tubes) to access the ribosomes (beige ovoids) for translation (E.). Aberrantly formed mRNAs are targeted for decay. B. mRNAs that aberrantly retain introns. C. mRNAs that are not properly polyadenylated, or capped. D. RNAs that bind protein complexes (olive shapes) that inhibit nuclear export. F. RNA decay factors (DF, purple circle), especially those that bind 3' untranslated region motifs, can control levels of mRNAs in the cytoplasm linked to translation. This is not an exhaustive list of possible control mechanisms.

Upstream negative regulatory elements that control late gene expression have been identified in a wide range of HPV genomes [88]. These cis-acting U-rich elements are between 50 and 100 nts in length and are positioned at the 3' end of the L1 coding region and span into the late 3' untranslated region (3'UTR) (Figure 2A) [89]. They can bind a number of RNA-binding proteins that have known functions in splicing, polyadenylation and mRNA stability (Table 1). In particular, the HPV16, HPV18 and 31 elements contain 5' splice sites that could bind U1 snRNP [10,66,90]. U1snRNP is a large RNA-protein complex which, when bound upstream of the polyadenylation site, could repress formation of the polyadenylation complex. Indeed, there is direct evidence that the HPV16 element inhibts HPV late polyadenylation in HeLa cells that possess a basal epithelial cell phenotype [91], and this could be another mechanism of ensuring restriction of capsid protein expression to differentiated kertatinocytes. Other HPV late elements (including HPV16) are more similar to AU-rich elements (AREs) that control mRNA stability [88] and theiractivity may be mediated by differentially-expressed cellular proteins such as the mRNA stability regulator, HuR [89,92].

4.4. Regulation of Late RNA Stability

RNA stability is a very significant layer of control in gene expression because degradation or stabilisation of a transcript can rapidly alter its cellular levels (Figure 4) [93]. In the nucleus, incorrectly processed or mutant RNAs, detected by virtue of the RNA-binding proteins that are loaded upon them, are destroyed by the exosome and therefore are not exported to the cytoplasm for translation [94]. In the cytoplasm, mRNAs can be stabilised or destroyed by an AU-rich element (ARE)-mediated pathway or by nonsense-mediated decay (NMD) [93,95].

Table 1. RNA-binding proteins that interact with RNA regulatory elements in the late 3′UTR of different HPVs [96].

HPV Genotype	Protein	Function
HPV1	HuR	mRNA stability, polyadenylation, nuclear export
	hnRNPC1/C2	Splicing, polyadenylation, mRNA stability, nuclear export
	PABP	Polyadenylation, translation
HPV16	U1snRNP	splicing
	HuR [1]	mRNA stability, polyadenylation, nuclear export
	SRSF1 [3]	Splicing regulation
	hnRNP A1	Splicing regulation
	U2AF	Auxiliary splicing factor
	CstF-64	Polyadenylation
	CUG-BP	Alternative splicing, translation
HPV31	U1snRNP [2]	splicing
	HuR	mRNA stability, polyadenylation, nuclear export
	U2AF	Auxiliary splicing factor
	CstF-64	Polyadenylation
HPV18	U1snRNP[2]	splicing

[1] HuR binding was found by our group [92,97], but not by others [88,91]; [2] U1snRNP binding is inferred due to the presence of 5′ splice sites, but not proven; [3] Binding is indirect.

Evidence from early in situ hybridisation studies of mucosally-infective low risk HPV6, 11, and high risk 16, 18 and 31-positive lesions revealed that some late region transcripts were detected in mid-epithelial layers but that cytoplasmic late mRNAs, i.e., fully formed mRNAs that are able to be exported from the nucleus to the cytoplasm, were detected only in the upper epithelial layers [53–56]. This suggests that although the late region can be transcribed in less differentiated cells, RNA processing or export from the nucleus is inhibited, or cytoplasmic late mRNAs are highly unstable. Late pre-mRNAs may be synthesised in less differentiated keratinocytes, at least in the W12 and NIKS16 models of the virus life cycle, because it is possible to induce L1 protein expression by manipulating levels of certain RNA-binding proteins [92,98]. Moreover, expression of reporter constructs containing the entire wild type HPV16 late regulatory element gave significantly increased RNA nuclear retention compared to similar constructs with mutations affecting key protein-binding regions of the element [97]. This suggests that RNA-protein interactions on the late regulatory element in undifferentiated cells could inhibit late pre-mRNA processing or nuclear export, correlating with the previous in vivo observations [53–56]. Therefore, nuclear retention of late transcripts synthesised in undifferentiated keratinocytes could be a direct result of polyadenylation inhibition via the late regulatory element.

The cutaneous HPV1 and mucosal HPV16 late regulatory elements have been shown to bind the RNA stability regulator HuR [92,97,99–101]. The HPV1 element also binds the RNA stability regulator hnRNP C1/C2 [97,102]. Regulation of mRNA turnover could control capsid protein expression. For example, HuR overexpression in undifferentiated keratinocytes caused the induction of L1 protein expression, while siRNA-mediated inhibition in differentiated keratinocytes caused loss of L1 expresion [92]. Similarly, the HPV16 L2 open reading frame was found to contain a 5′and a 3′ regulatory element that acts at the level of mRNA stability. The 5′ element negatively regulated

CAT reporter mRNA stability by 3-fold in the cytoplasm of HeLa cells expressing the HIV *tat* protein, while the 3′ element was only weakly active [103]. Interestingly, the HPV1 L2 gene does not appear to contain such an element, or indeed, any major regulatory element in the late gene region, apart from the 3′UTR element [103].

4.5. Late RNA Splicing

Late mRNAs are polycistronic transcripts containing the 1.5 kb L1 open reading frame with several different possible upstream spliced exons (Figure 3B). Other late mRNAs contain both L1 and L2 (1.4 kb) open reading frames in a read-through mRNA that also contains the short E4 (274 bps) and E5 (236 bps) open reading frames (Figure 3B). Given that the average size of a eukaryotic terminal exon is 627 nts [104], this means that the terminal exons of these mRNAs are unusually long at either 1.5 or 2.9 kb. A process called exon definition is important for efficient splicing, especially of long exons [73] and exons are defined by protein complexes formed at the 5′ and 3′ exon-intron boundaries and by splicing regulatory factors such as SR proteins and hnRNP proteins bound to exon- and intron-internal cis-acting sequences [105]. Chromatin conformation and RNA secondary structure also play a role in determining efficiency of splicing and which exons are spliced out or retained [105]. Moreover, due to coupling of transcription elongation and RNA processing—including splicing (Figure 4)—the promoter used to initiate gene transcription can influence splicing efficiency [106]. Although exon size is often greatest in genes with few introns (late mRNAs have only one or two introns) [107], the cellular RNA processing machinery would be predicted to process the late RNAs inefficiently. The late mRNAs are generated by splicing from four possible splice donor sites, one in the E6 region (HPV16 nt 226, HPV18, nt 233, HPV31 nt 210), one in the E8 gene region (HPV16 nt 1301, HPV18 nt 1202) and one at the end of each of the E1 and E4 open reading frame (HPV16 nt 3632, HPV18, nt 3696, HPV31, nt 3590) (Figure 3B). The latter two splice events are by far the most frequently used. At least for HPVs 16, 18 and 31, late RNAs can be initiated at both the early and late promoters, although the relative abundance of early promoter-initiated late RNAs versus late, or E8, promoter-initiated late RNAs has not been determined in the different layers of the infected epithelium.

4.6. Control of Late mRNA Splicing

In general, cellular and viral splicing is positively controlled by SR splicing factors [108], and negatively by hnRNP proteins [109,110]. HPV controls expression of SRSF1, 2 and 3 through the viral E2 transcription factor [98] and SRSF levels peak in the mid to upper layers of HPV-infected epithelial [75,98] where late gene expression commences [23], and E2 is most highly expressed [32,33]. Recently, we reported that SRSF3 controls L1 protein expression in a differentiation-dependent manner. siRNA depletion of SRSF3 in differentiated HPV16-positive keratinocytes inhibited expression of the capsid mRNA and protein while SRSF3 overexpression in undifferentiated epithelial cells caused the induction of L1 protein. Levels of the E1^E4^L1 mRNA were undetectable in differentiated keratinocytes treated with siRNA agaisnt SRSF3. While there was a corresponding increase in levels of the read-through L2L1 mRNA, it is likely that loss of SRSF3 affects other RNA processing events as well as splicing. We also found that SRSF1 contributed to control of capsid protein expression [98]. This is in agreement with a previous study, which demonstrated that SRSF1 contributed to regulation of the major alternatively spliced E4^L1 mRNA and the L2L1 read-through late mRNA (Figure 3B) [72]. A third SR protein, SRSF9 has also been shown to enhance splicing of late RNAs [111].In HeLa cells, SRSF9 inhibited splicing at the splice acceptor site at the 5′ end of the E4 open reading frame. This led to use of the next downstream splice acceptor site located at the 5′ end of the L1 open reading frame resulting in enhanced L1 mRNA production.

The L1 open reading frame contains a splicing enhancer element that can be repressed by hnRNP A1 in HeLa cells [112,113]. However, hnRNP A1 appears to be upregulated during differentiation of HPV-infected keratinocytes [114]. Although this could be counter-balanced by increased expression of splicing enhancer proteins—such as SRSF1—that negate the effects of hnRNP A1, it will be important in future to try to understand the complexity of late RNA splicing regulation in differentiated keratinocytes. hnRNP H can bind a G-rich element in the L2 coding region and, as discussed in Section 4.1, can control early polyadenylation [65]. The importance of the early polyadenylation site usage in preventing late mRNA expression means, in effect, that hnRNP H can also inhibit late mRNA expression. Other sequences in the viral early region can control late mRNA expression. hnRNPs A2/B1 and D inhibit L1 mRNA production by binding an element in the E4 open reading frame that likely suppresses use of its 3′ splice donor site to inhibit splicing to the L1 splice acceptor site [115]. Similarly, hnRNP I, (polypyrimidine tract binding protein (PTB)) can induce late transcript expression by inhibiting the same splice donor site [116]. Conversely, overexpression of hnRNP C1, and another hnRNP protein, RALYL, induced late mRNA production in C33A cervical cancer cells. hnRNP C1 binding to the early 3′UTR was shown to activate L1 mRNA expression by activating the E4 splice donor site [117]. No doubt the work published thus far only scratches the surface of the possible SR and hnRNP regulatory interactions that can control HPV late gene expression. Most studies have examined HPV16. Therefore, it will be of interest to compare the data from this HR-HPV with that from other HPV types, for example another HR-HPV such as HPV31 or 18, the low risk anogenital infective HPVs 6 and 11, or the cutaneous infective HPV1. In fact, some elements that control late gene expression have been shown to be present in a range of HPV types [88,118] suggesting that such an important regulatory event is likely to be conserved across different HPV types. Finally, it is important that differentiation is taken into account in such studies due to possible HPV infection-associated changes in the various proteins that can control splicing.

4.7. Control of Capsid Protein Translation

Control of the translation of viral late proteins is still poorly understood. HPV polycistronic late mRNAs are predicted to be inefficiently translated to yield capsid proteins. This is because of the Kozak rules of translation initation that predict that subsequent open reading frames in polycistronic transcripts will be poorly translated [119]. In the scanning model of translation initiation, the ribosome recognises the first AUG in an mRNA as the start codon. Moreover, translation is inhibited when an intron, that often contains pseudo splice sites, is retained in the mRNA, as it will be for L2L1 readthrough RNAs. The first translation start codon in most late mRNAs is at the start of the E1 open reading frame because the E4 open reading frame does not possess an AUG codon; this is donated by splicing. E4 is the most abundant viral protein [23], suggesting that it is very efficiently translated. The L2 and L1 open reading frames are downstream of E4 in all mRNAs indicating inefficient translation. L1 protein is indeed much less abundant than E4 in HPV-infected tissues [120], and although levels of L2 relative to E4 have not been directly observed, it is likely that L2 is expressed at an even lower level than L1 because only one molecule of L2 is required for every 5 molecules of L1 in the virus capsid [120]. If the capsid protein mRNAs were particularly abundant, this might facilitate capsid protein expression, and they do seem to be readily detected in HPV-positive lesions [53–56]. Alternatively, it has been proposed that viral translation efficiency could be assured through use of rare codons that are less frequently used for cellular mRNA translation [121,122]. HPV genomes possess a high A+T frequency at the 3rd position in codons in the L2 and L1, but not the E4, open reading frames [122]. At least in experiments using reporter gene constructs, this seems to determine the keratinocyte differentiation-specific expression of L1 protein [123,124]. Undifferentiated and differentiated keratinocytes were found to have a different tRNA profile whereby only differentiated keratinocyte tRNAs allowed efficient L1 translation [123,125]. This regulatory mechanism was favoured in G2/M-like growth arrested cells that expressed keratinocyte differentiation markers [126]. Therefore, the tRNA pools available in differentiated keratinocytes may greatly favour translation of HPV late mRNAs.

Other studies have examined whether late mRNA-binding proteins control translation. The HPV16 L2 open reading frames contains a 3′ element that binds hnRNP K and poly (rC) binding protein (CBP) [127]. Depletion of these proteins in an in vitro system caused a reduction in L2 capsid protein production but a definitive role for the RNA binding proteins in translation, as opposed to other events—for example mRNA stability—was not tested. Indeed, the L2 open reading frame has a second inhibitory element located within the first 845 nucleotides that acts to destabilise the mRNA, and therefore to inhibit its translation [103]. Any of the late regulatory elements discussed above that may regulate late mRNA processing are also possible regulators of late mRNA translation because they bind cytoplasmic proteins that can affect mRNA stability [89]. For example, the HPV1 late regulatory element has five U-rich motifs that render the late mRNAs unstable in HeLa cells and this results in inefficient translation [128].

The above mechanisms may all play a part in facilitating viral capsid protein expression from late mRNAs. However, as stated previously, these mRNAs are unusual in structure due to their polycistronic nature and this could exacerbate recognition of appropriate start codons for translation initiaiton. For example, in the case of the L1 open reading frame, there is always at least one AUG start codon prior to the L1 start. For L2 in the L2L1 read-through RNA there are at least three start sites that the ribosome must pass through in order to translate L2. Perhaps RNA-binding proteins obscure the upstream intiation codons in some way during translation initiation. Alternatively, other putative promoters have been described for a number of HPVs that could give rise to monocistronic mRNAs encoding each of the late proteins separately [7,11]. However, the activities of these have not been examined in the context of the viral life cycle.

5. Conclusions

This review has aimed to highlight the various regulatory mechanisms that control late gene expression of human papillomaviruses. We have most information on the regulation of late gene expression for HR-HPVs and it cannot always be assumed that other HPVs will undergo similar regulatory controls. Over the last decade, the extent of the association of the HPV replication cycle with keratinocyte differentiation has become clearer. Future studies on HPV gene regulation should continue to focus on how differentiated keratinocytes control viral capsid production. In particular, what differentiation-specific changes in transcription factors and RNA-binding proteins might be relevant to facilitating viral late mRNA production and late protein translation should be worked out. Understanding the mechanism of control of capsid protein expression could facilitate the design of antivirals to induce expression of the highly immunogenic capsid proteins inappropriately in the lower epithelial layer to enhanced immune recognition and viral clearance.

Acknowledgments: Recent work in the Graham lab was funded by a grant to SVG from the Wellcome Trust (088848/Z/09/Z). She gratefully acknowledges support from the Medical Research Council as core funding for the MRC University of Glasgow Centre for Virus Research. Fund to cover publication of this article have been provided by the University of Glasgow.

Conflicts of Interest: The author declares no conflict of interest.

References

1. Van Doorslaer, K. Evolution of the papillomaviridae. *Virology* **2013**, *445*, 11–20. [CrossRef] [PubMed]
2. Stanley, M.A. Epithelial cell responses to infection with human papillomavirus. *Clin. Microbiol. Rev.* **2012**, *25*, 215–222. [CrossRef] [PubMed]
3. Doorbar, J.; Egawa, N.; Griffin, H.; Kranjec, C.; Murakami, I. Human papillomavirus molecular biology and disease association. *Rev. Med. Virol.* **2015**, *25*, 2–23. [CrossRef] [PubMed]
4. Arbyn, M.; Tommasino, M.; Depuydt, C.; Dillner, J. Are 20 human papillomavirus types causing cervical cancer? *J. Pathol.* **2014**, *234*, 431–435. [CrossRef] [PubMed]
5. Gillison, M.L.; Chaturvedi, A.K.; Anderson, W.F.; Fakhry, C. Epidemiology of human papillomavirus—Positive head and neck squamous cell carcinoma. *J. Clin. Oncol.* **2015**, *33*, 3235–3242. [CrossRef] [PubMed]
6. Egawa, N.; Egawa, K.; Griffin, H.; Doorbar, J. Human papillomaviruses; epithelial tropisms, and the development of neoplasia. *Viruses* **2015**, *7*, 3863–3890. [CrossRef] [PubMed]

7. Graham, S.V. Human papillomavirus: Gene expression, regulation and prospects for novel diagnostic methods and antiviral therapies. *Future Microbiol.* **2010**, *5*, 1493–1505. [CrossRef] [PubMed]

8. Groves, I.J.; Coleman, N. Pathogenesis of human papillomavirus-associated mucosal disease. *J. Pathol.* **2015**, *235*, 527–538. [CrossRef] [PubMed]

9. Chen, J.; Xue, Y.; Poidinger, M.; Lim, T.; Chew, S.H.; Pang, C.L.; Abastado, J.-P.; Thierry, F. Mapping of HPV transcripts in four human cervical lesions using RNAseq suggests quantitative rearrangements during carcinogenic progression. *Virology* **2014**, *462–463*, 14–24. [CrossRef] [PubMed]

10. Wang, X.; Meyers, C.; Wang, H.-K.; Chow, L.T.; Zheng, Z.-M. Construction of a full transcription map of human papillomavirus type 18 during productive viral infection. *J. Virol.* **2011**, *85*, 8080–8092. [CrossRef] [PubMed]

11. Milligan, S.G.; Veerapraditsin, T.; Ahamat, B.; Mole, S.; Graham, S.V. Analysis of novel human papillomavirus type 16 late mRNAs in differentiated W12 cervical epithelial cells. *Virology* **2007**, *360*, 172–181. [CrossRef] [PubMed]

12. Hansen, C.N.; Nielsen, L.; Norrild, B. Activities of E7 promoters in the human papillomavirus type 16 genome during cell differentiation. *Virus Res.* **2010**, *150*, 34–42. [CrossRef] [PubMed]

13. Ozbun, M.A.; Meyers, C. Temporal usage of multiple promoters during the life cycle of human papillomavirus type 31b. *J. Virol.* **1998**, *72*, 2715–2722. [PubMed]

14. Ozbun, M.A.; Meyers, C. Two novel promoters in the upstream regulatory region of human papillomavirus type 31b are negatively regulated by epithelial differentiation. *J. Virol.* **1999**, *73*, 3505–3510. [PubMed]

15. Ozbun, M.A. Human papillomavirus type 31b infection of human keratinocytes and the onset of early transcription. *J. Virol.* **2002**, *76*, 11291–11300. [CrossRef] [PubMed]

16. Vande Pol, S.B.; Klingelhutz, A.J. Papillomavirus E6 oncoproteins. *Virology* **2013**, *445*, 115–137. [CrossRef] [PubMed]

17. Roman, A.; Munger, K. The papillomavirus E7 proteins. *Virology* **2013**, *445*, 138–168. [CrossRef] [PubMed]

18. Moody, C.A.; Laimins, L.A. Human papillomavirus oncoproteins: Pathways to transformation. *Nat. Rev. Cancer* **2010**, *10*, 550–560. [CrossRef] [PubMed]

19. Stubenrauch, F.; Hummel, M.; Iftner, T.; Laimins, L.A. The E8^E2C protein, a negative regulator of viral transcription and replication, is required for extrachromosomal maintenance of human papillomavirus type 31 in keratinocytes. *J. Virol.* **2000**, *74*, 1178–1186. [CrossRef] [PubMed]

20. Wang, Q.; Griffin, H.M.; Southern, S.; Jackson, D.; Martin, A.; McIntosh, P.; Davy, C.; Masterson, P.J.; Walker, P.A.; Laskey, P.; et al. Functional analysis of the human papillomavirus type 16E1^E4 protein provides a mechanism for in vivo and in vitro keratin filament reorganization. *J. Virol.* **2004**, *78*, 821–833. [CrossRef] [PubMed]

21. Davy, C.E.; Jackson, D.J.; Raj, K.; Peh, W.L.; Southern, S.A.; Das, P.; Sorathia, R.; Laskey, P.; Middleton, K.; Nakahara, T.; et al. Human Papillomavirus type 16 E1^E4-induced G2 arrest is associated with cytoplasmic retention of active Cdk1/Cyclin B1 complexes. *J. Virol.* **2005**, *79*, 3998–4011. [CrossRef] [PubMed]

22. DiMaio, D.; Petti, L.M. The E5 proteins. *Virology* **2013**, *445*, 99–114. [CrossRef] [PubMed]

23. Middleton, K.; Peh, W.; Southern, S.A.; Griffin, H.M.; Sotlar, K.; Nakahara, T.; El-Sherif, A.; Morris, L.; Seth, R.; Hibma, M.; et al. Organisation of the human papillomavirus productive cycle during neoplastic progression provides a basis for the selection of diagnostic markers. *J. Virol.* **2003**, *77*, 10186–10201. [CrossRef] [PubMed]

24. Fehrmann, F.; Klumpp, D.J.; Laimins, L.A. Human papillomavirus type 31 E5 protein supports cell cycle progression and activates late viral functions upon epithelial differnetiation. *J. Virol.* **2003**, *77*, 2819–2831. [CrossRef] [PubMed]

25. Wilson, R.; Ryan, G.B.; Knight, G.L.; Laimins, L.A.; Roberts, S. The full-length E1^E4 protein of human papillomavirus type 18 modulates differentiation-dependent viral DNA amplification and late gene expression. *Virology* **2007**, *362*, 453–460. [CrossRef] [PubMed]

26. Genther, S.M.; Sterling, S.; Duensing, S.; Munger, K.; Sattler, C.; Lambert, P.F. Quantitative role of the human papillomavirus type 16 E5 gene during the productive stage of the viral life cycle. *J. Virol.* **2003**, *77*, 2832–2842. [CrossRef] [PubMed]

27. Nakahara, T.; Peh, W.L.; Doorbar, J.; Lee, D.; Lambert, P.F. Human papillomavirus type 16 E1^E4 contributes to multiple facets of the papillomavirus life cycle. *J. Virol.* **2005**, *79*, 13150–13165. [CrossRef] [PubMed]

28. Nakahara, T.; Nishimura, A.; Tanaka, M.; Ueno, T.; Ishimoto, A.; Sakai, H. Modulation of the cell division cycle by human papillomavirus type 18 E4. *J. Virol.* **2002**, *76*, 10914–10920. [CrossRef] [PubMed]

29. Johansson, C.; Schwartz, S. Regulation of human papillomavirus gene expression by splicing and polyadenylation. *Nat. Rev. Microbiol.* **2013**, *11*, 239–251. [CrossRef] [PubMed]
30. Flores, E.R.; Lambert, P.F. Evidence for a switch in the mode of human papillomavirus type 16 DNA replication during the viral life cycle. *J. Virol.* **1997**, *71*, 7167–7179. [PubMed]
31. Sakakibara, N.; Chen, D.; McBride, A.A. Papillomaviruses use recombination-dependent replication to vegetatively amplify their genomes in differentiated cells. *PLoS Pathog.* **2013**, *9*, e1003321. [CrossRef] [PubMed]
32. Maitland, N.J.; Conway, S.; Wilkinson, N.S.; Ramsdale, J.; Morris, J.R.; Sanders, C.M.; Burns, J.E.; Stern, P.L.; Wells, M. Expression patterns of the human papillomavirus type 16 transcription factor E2 in low- and high-grade cervical intraepithelial neoplasia. *J. Pathol.* **1998**, *186*, 275–280. [CrossRef]
33. Xue, Y.; Bellanger, S.; Zhang, W.; Lim, D.; Low, J.; Lunny, D.; Thierry, F. HPV16 E2 is an immediate early marker of viral infection, preceding E7 expression in precursor structures of cervical carcinoma. *Cancer Res.* **2010**, *70*, 5316–5325. [CrossRef] [PubMed]
34. Paris, C.; Pentland, I.; Groves, I.; Roberts, D.C.; Powis, S.J.; Coleman, N.; Roberts, S.; Parish, J.L. CCCTC-Binding factor recruitment to the early region of the human papillomavirus 18 genome regulates viral oncogene expression. *J. Virol.* **2015**, *89*, 4770–4785. [CrossRef] [PubMed]
35. Peh, W.L.; Middleton, K.; Christensen, N.; Nicholls, P.; Egawa, K.; Sotlar, K.; Brandsma, J.; Percival, A.; Lewis, J.; Liu, W.J.; et al. Life cycle heterogeneity in animal models of human papillomavirus-associated disease. *J. Virol.* **2002**, *76*, 10411–10416. [CrossRef]
36. Grassman, K.; Rapp, B.; Maschek, H.; Petry, K.U.; Iftner, T. Identification of a differentiation-inducible promoter in the E7 open reading frame of human papillomavirus type 16 (HPV-16) in raft cultures of a new cell line containing high copy numbers of episomal HPV-16 DNA. *J. Virol.* **1996**, *70*, 2339–2349.
37. Hummel, M.; Hudson, J.B.; Laimins, L.A. Differentiation-induced and constitutive transcription of human papillomavirus type 31b in cell lines containing viral episomes. *J. Virol.* **1992**, *66*, 6070–6080. [PubMed]
38. Ruesch, M.N.; Stubenrauch, F.; Laimins, L.A. Activation of papillomavirus late gene transcription and genome amplification upon differentiation in semisolid medium is coincident with expression of involucrin and transglutaminase but not keratin. *J. Virol.* **1998**, *72*, 5016–5024. [PubMed]
39. Van Doorslaer, K.; Li, Z.; Xirasagar, S.; Maes, P.; Kaminsky, D.; Liou, D.; Sun, Q.; Kaur, R.; Huyen, Y.; McBride, A.A. The papillomavirus episteme: A major update to the papillomavirus sequence database. *Nucleic Acids Res.* **2017**, *45*, D499–D506. [CrossRef] [PubMed]
40. Wang, X.; Liu, H.; Ge, H.; Ajiro, M.; Sharma, N.R.; Meyers, C.; Morozov, P.; Tuschl, T.; Klar, A.; Court, D.; et al. Viral DNA replication orientation and hnRNPs regulate transcription of the human papillomavirus 18 late promoter. *Microbiology* **2017**, *8*, 3309–3321. [CrossRef] [PubMed]
41. Bodily, J.M.; Meyers, C. Genetic analysis of the human papillomavirus type 31 differentiation-dependent late promoter. *J. Virol.* **2005**, *79*, 3309–3321. [CrossRef] [PubMed]
42. Sankovski, E.; Männik, A.; Geimanen, J.; Ustav, E.; Ustav, M. Mapping of betapapillomavirus human papillomavirus 5 transcription and characterization of viral-genome replication function. *J. Virol.* **2014**, *88*, 961–973. [CrossRef] [PubMed]
43. Stubenrauch, F.; Malejczyk, J.; Fuchs, P.G.; Pfister, H.J. Late promoter of human papillomavirus type 8 and its regualtion. *J. Virol.* **1992**, *64*, 3144–3149.
44. Meyers, C.; Mayer, T.J.; Ozbun, M.A. Synthesis of infectious human papillomavirus type 18 in differentiating epithelium transfected with viral DNA. *J. Virol.* **1997**, *71*, 7381–7386. [PubMed]
45. Del Mar Pena, L.; Laimins, L.A. Differentiation-dependent chromatin rearrangement coincides with activation of human papillomavirus type 31 late gene expression. *J. Virol.* **2001**, *75*, 10005–10013. [CrossRef] [PubMed]
46. Carson, A.; Khan, S.A. Characterisation of transcription factor binding to human papillomavirus type 16 DNA during cellular differentiation. *J. Virol.* **2006**, *80*, 4356–4362. [CrossRef] [PubMed]
47. Wooldridge, T.R.; Laimins, L.A. Regulation of human papillomavirus type 31 gene expression during the differentiation-dependent life cycle through histone modifications and transcription factor binding. *Virology* **2008**, *374*, 371–380. [CrossRef] [PubMed]
48. Songock, W.K.; Scott, M.L.; Bodily, J.M. Regulation of the human papillomavirus type 16 late promoter by transcriptional elongation. *Virology* **2017**, *507*, 179–191. [CrossRef] [PubMed]

49. Bodily, J.M.; Alam, S.; Meyers, C. Regulation of human papillomavirus type 31 late promoter activation and genome amplification by protein kinase C. *Virology* **2006**, *348*, 328–340. [CrossRef] [PubMed]
50. Straub, E.; Fertey, J.; Dreer, M.; Iftner, T.; Stubenrauch, F. Characterization of the human papillomavirus 16 E8 promoter. *J. Virol.* **2015**, *89*, 7304–7313. [CrossRef] [PubMed]
51. Bodily, J.M.; Hennigan, C.; Wrobel, G.A.; Rodriguez, C.M. Regulation of the human papillomavirus type 16 late promoter by E7 and the cell cycle. *Virology* **2013**, *443*, 11–19. [CrossRef] [PubMed]
52. Weschler, E.I.; Wang, Q.; Roberts, I.; Pagliarulo, E.; Jackson, D.; Untersperger, C.; Coleman, N.; Griffin, H.; Doorbar, J. Reconstruction of human papillomavirus type 16-mediated early-stage neoplasia implicated E6/E7 deregulation and the loss of contact inhibition in neoplastic progression. *J. Virol.* **2012**, *86*, 6358–6364.
53. Stoler, M.H.; Wolinsky, S.M.; Whitbeck, A.; Broker, T.R.; Chow, L.T. Differentiation-linked human papillomavirus types 6 and 11 transcription in genital condylomata revealed by in situ hybridisation with message-specific RNA probes. *Virology* **1989**, *172*, 331–340. [CrossRef]
54. Stoler, M.H.; Rhodes, C.R.; Whitbeck, A.; Wolinsky, S.M.; Chow, L.T.; Broker, T.R. Human papillomavirus type 16 and 18 gene expression in cervical neoplasia. *Hum. Pathol.* **1992**, *23*, 117–128. [CrossRef]
55. Beyer-Finkler, E.; Stoler, M.H.; Girardi, F.; Pfister, H.J. Cell differentiation-related gene expression of human papillomavirus 33. *Med. Microbiol. Immunol.* **1990**, *179*, 185–192. [CrossRef] [PubMed]
56. Crum, C.P.; Nuovo, G.; Freidman, D.; Silverstein, S.J. Accumulation of RNA homologous to human papillomavirus type 16 open reading frames in genital precancers. *J. Virol.* **1988**, *62*, 84–90. [PubMed]
57. Ozbun, M.A.; Meyers, C. Characterisation of late gene transcripts expressed during vegetative replication of human papillomavirus type 31b. *J. Virol.* **1997**, *71*, 5161–5172. [PubMed]
58. Neve, J.; Patel, R.; Wang, Z.; Louey, A.; Furger, A.M. Cleavage and polyadenylation: Ending the message expands gene regulation. *RNA Biol.* **2017**, *14*, 865–890. [CrossRef] [PubMed]
59. Rohlfs, M.; Winkenbach, S.; Meyer, S.; Rupp, T.; Dürst, M. Viral transcription in human keratinocyte cell lines immortalized by human papillomvirus type-16. *Virology* **1991**, *183*, 331–342. [CrossRef]
60. Higgins, G.D.; Uzelin, D.M.; Phillips, G.E.; McEvoy, P.; Burrel, C.J. Transcription patterns of human papillomavirus type 16 in genital intraepithelia neoplasia: Evidence for promoter usage within the E7 open reading frame during epithelial differentiation. *J. Gen. Virol.* **1992**, *73*, 2047–2057. [CrossRef] [PubMed]
61. Hummel, M.; Lim, H.B.; Laimins, L.A. Human papillomavirus type 31b late gene expression is regulated through protein kinase C-mediated changes in RNA processing. *J. Virol.* **1995**, *69*, 3381–3388. [PubMed]
62. Terhune, S.S.; Milcarek, C.; Laimins, L.A. Regulation of human papillomavirus type 31 polyadenylation during the differentiation-dependent life cycle. *J. Virol.* **1999**, *73*, 7185–7192. [PubMed]
63. Terhune, S.S.; Hubert, W.G.; Thomas, J.T.; Laimins, L.A. Early polyadenylation signals of human papillomavirus type 31 negatively regulates capsid gene expression. *J. Virol.* **2001**, *75*, 8147–8157. [CrossRef] [PubMed]
64. Zhao, X.; Oberg, D.; Rush, M.; Fay, J.; Lambkin, H.; Schwartz, S. A 57-nucleotide upstream early polyadenylation element in human papillomavirus type 16 interacts with hFip1, CstF-64, hnRNP C1/C2 and polypyrimidine tract binding protein. *J. Virol.* **2005**, *79*, 4270–4288. [CrossRef] [PubMed]
65. Öberg, D.; Fay, J.; Lambkin, H.; Schwartz, S. A downstream polyadenylation element in human papillomavirus type 16 L2 encodes multiple GGG motifs and interacts with hnRNP H. *J. Virol.* **2005**, *79*, 9254–9269. [CrossRef] [PubMed]
66. Cumming, S.A.; Repellin, C.E.; McPhillips, M.; Radford, J.C.; Clements, J.B.; Graham, S.V. The human papillomavirus type 31 late 3′ untranslated region contains a complex bipartite negative regulatory element. *J. Virol.* **2002**, *76*, 5993–6003. [CrossRef] [PubMed]
67. Kennedy, I.M.; Haddow, J.K.; Clements, J.B. Analysis of human papillomavirus type 16 late mRNA 3′ processing signals in vitro and in vivo. *J. Virol.* **1990**, *64*, 1825–1829. [PubMed]
68. Öberg, D.; Collier, B.; Zhao, X.; Schwartz, S. Mutational inactivation of two distinct negative RNA elements in the human papillomavirus type 16 L2 coding region induces production of high levels of L2 in human cells. *J. Virol.* **2003**, *77*, 11674–11684. [CrossRef] [PubMed]
69. Bentley, D.L. Coupling mRNA processing with transcription in time and space. *Nat. Rev. Genet.* **2014**, *15*, 163–175. [CrossRef] [PubMed]
70. Jia, R.; Liu, X.; Tao, M.; Kruhlak, M.; Guo, M.; Meyers, C.; Baker, C.C.; Zheng, Z.M. Control of the papillomavirus early-to-late switch by differentially expressed SRp20. *J. Virol.* **2009**, *83*, 167–180. [CrossRef] [PubMed]

71. Rush, M.; Zhao, X.; Schwartz, S. A splicing enhancer in the E4 coding region of human papillomavirus type 16 is required for early mRNA splicing and polyadenylation as well as inhibition of premature late gene expression. *J. Virol.* **2005**, *79*, 12002–12015. [CrossRef] [PubMed]

72. Somberg, M.; Schwartz, S. Multiple ASF/SF2 sites in the human papillomavirus type 16 (HPV-16) E4-coding region promote splicing to the most commonly Used 3'-splice site on the HPV-16 genome. *J. Virol.* **2010**, *84*, 8219–8230. [CrossRef] [PubMed]

73. Berget, S. Exon recognition in vertebrate splicing. *J. Biol. Chem.* **1995**, *270*, 2411–2414. [CrossRef] [PubMed]

74. Fay, J.; Kelehan, P.; Lambkin, H.; Schwartz, S. Increased expression of cellular RNA-binding proteins in HPV-induced neoplasia and cervical cancer. *J. Med. Virol.* **2009**, *81*, 897–907. [CrossRef] [PubMed]

75. Mole, S.; McFarlane, M.; Chuen-Im, T.; Milligan, S.G.; Millan, D.; Graham, S.V. RNA splicing factors regulated by HPV16 during cervical tumour progression. *J. Pathol.* **2009**, *219*, 383–391. [CrossRef] [PubMed]

76. Johannson, C.; Somberg, M.; Li, X.; Winquist, E.B.; Fay, J.; Ryan, F.; Pim, D.; Banks, L.; Schwartz, S. HPV-16 E2 contributes to induction of HPV-16 late gene expression by inhibiting early polyadenylation. *EMBO J.* **2012**, *31*, 3212–3227. [CrossRef] [PubMed]

77. McBride, A.A. The papillomavirus E2 proteins. *Virology* **2013**, *445*, 57–79. [CrossRef] [PubMed]

78. Jang, M.K.; Anderson, D.E.; van Doorslaer, K.; McBride, A.A. A proteomic approach to discover and compare interacting partners of papillomavirus E2 proteins from diverse phylogenetic groups. *Proteomics* **2015**, *15*, 2038–2050. [CrossRef] [PubMed]

79. Radhakrishnan, A.; Green, R. Connections underlying translation and mRNA stability. *J. Mol. Biol.* **2016**, *428*, 3558–3564. [CrossRef] [PubMed]

80. Hall-Pogar, T.; Liang, S.; Hague, L.K.; Lutz, C.S. Specific trans-acting proteins interact with auxiliary RNA polyadenylation elements in the COX-2 3'-UTR. *RNA* **2007**, *13*, 1103–1115. [CrossRef] [PubMed]

81. Moriera, A.; Takagaki, Y.; Brackenridge, S.; Wollerton, M.; Manley, J.L.; Proudfoot, N.J. The upstream sequence element of the C2 complement poly(A) signal activates mRNA 3' end formation by two distinct mechanisms. *Genes Dev.* **1998**, *12*, 2522–2534. [CrossRef]

82. Natalizio, B.J.; Muñiz, L.C.; Arhin, G.K.; Wilusz, J.; Lutz, C.S. Upstream elements present in the 3'-untranslated region of collagen genes influence the processing efficiency of overlapping polyadenylation signals. *J. Biol. Chem.* **2002**, *277*, 42733–42740. [CrossRef] [PubMed]

83. Gilmartin, G.M.; Fleming, E.S.; Oetjen, J.; Graveley, B.R. CPSF recognition of an HIV-1 mRNA 3' processing enhancer: Multiple sequence contacts involved in poly(A) site definition. *Genes Dev.* **1995**, *9*, 72–83. [CrossRef] [PubMed]

84. Lutz, C.S.; Alwine, J.C. Direct interaction of the U1 snRNP-A protein with the upstream efficiency element of the SV40 late polyadenylation signal. *Genes Dev.* **1994**, *8*, 576–586. [CrossRef] [PubMed]

85. Lutz, C.S.; Murthy, K.G.K.; Schek, N.; O'Connor, J.P.; Manley, J.L.; Alwine, J.C. Interaction between the U1 snRNP-A protein and the 160 kD subunit of cleavage-polyadenylation specificity factor increases polyadenylation efficiency in vitro. *Genes Dev.* **1996**, *10*, 325–337. [CrossRef] [PubMed]

86. DeZazzo, J.D.; Imperiale, M.J. Sequences upstream of AAUAAA influence poly(A) site selection in a complex transcription unit. *Mol. Cell. Biol.* **1989**, *9*, 4951–4961. [CrossRef] [PubMed]

87. Boelens, W.C.; Jansen, E.J.R.; van Venrooij, W.J.; Stripecke, R.; Mattaj, I.M.; Gunderson, S.I. The human U1 snRNP-specific U1A protein inhibits polyadenylation of its own pre-mRNA. *Cell* **1993**, *72*, 881–892. [CrossRef]

88. Zhao, X.; Rush, M.; Carlsson, A.; Schwartz, S. The presence of inhibitory RNA elements in the late 3'-untranslated region is a conserved property of human papillomaviruses. *Virus Res.* **2007**, *125*, 135–144. [CrossRef] [PubMed]

89. Graham, S.V. Papillomavirus 3'UTR regulatory elements. *Front. Biosci.* **2008**, *13*, 5646–5663. [CrossRef] [PubMed]

90. Cumming, S.A.; McPhillips, M.G.; Veerapraditsin, T.; Milligan, S.G.; Graham, S.V. Activity of the human papillomavirus type 16 late negative regulatory element is partly due to four weak consensus 5' splice sites that bind a U1 snRNP-like complex. *J. Virol.* **2003**, *77*, 5167–5177. [CrossRef] [PubMed]

91. Goraczniak, R.; Gunderson, S.I. The regulatory element in the 3'-untranslated region of human papillomavirus 16 inhibits expression by binding CUG-binding protein 1. *J. Biol. Chem.* **2008**, *283*, 2286–2296. [CrossRef] [PubMed]

92. Cumming, S.A.; Chuen-Im, T.; Zhang, J.; Graham, S.V. The RNA stability regulator HuR regulates L1 protein expression in vivo in differentiating cervical epithelial cells. *Virology* **2009**, *383*, 142–149. [CrossRef] [PubMed]

93. Garneau, N.L.; Wilusz, J.; Wilusz, C.J. The highways and byways of mRNA decay. *Nat. Rev. Mol. Cell Biol.* **2007**, *8*, 113–126. [CrossRef] [PubMed]

94. Eberle, A.B.; Visa, N. Quality control of mRNP biogenesis: Networking at the transcription site. *Semin. Cell Dev. Biol.* **2014**, *32*, 37–46. [CrossRef] [PubMed]

95. Celik, A.; He, F.; Jacobson, A. NMD monitors translational fidelity 24/7. *Curr. Genet.* **2017**, 1–4. [CrossRef] [PubMed]

96. Kajitani, N.; Schwartz, S. RNA binding proteins that control human papillomavirus gene expression. *Biomolecules* **2015**, *5*, 758–774. [CrossRef] [PubMed]

97. Koffa, M.D.; Graham, S.V.; Takagaki, Y.; Manley, J.L.; Clements, J.B. The human papillomavirus type 16 negative regulatory element interacts with three proteins that act at different posttranscriptional levels. *Proc. Natl. Acad. Sci. USA* **2000**, *97*, 4677–4682. [CrossRef] [PubMed]

98. Klymenko, T.; Hernandez-Lopez, H.; MacDonald, A.I.; Bodily, J.M.; Graham, S.V. Human papillomavirus E2 regulates SRSF3 (SRp20) to promote capsid protein expression in infected differentiated keratinocytes. *J. Virol.* **2016**, *90*, 5047–5058. [CrossRef] [PubMed]

99. Zhao, C.; Sokolowski, M.; Tan, W.; Schwartz, S. Characterisation and partial purification of cellular factors interacting with a negative element on human papillomavirus type 1 late mRNAs. *Virus Res.* **1998**, *55*, 1–13. [CrossRef]

100. Sokolowski, M.; Zhao, C.; Tan, W.; Schwartz, S. AU-rich mRNA instability elements on human papillomavirus type 1 late mRNAs and c-*fos* mRNAs interact with the same cellular factors. *Oncogene* **1997**, *15*, 2303–2319. [CrossRef] [PubMed]

101. Sokolowski, M.; Furneaux, H.; Schwartz, S. The inhibitory activity of the AU-rich RNA element in the human papillomavirus type 1 late 3' untranslated region correlates with its affinity for the elav-like HuR protein. *J. Virol.* **1999**, *73*, 1080–1091. [PubMed]

102. Sokolowski, M.; Schwartz, S. Heterogeneous nuclear ribonucleoprotein C binds exclusively to the functionally important UUUUU-motifs in the human papillomavirus type-1 AU-rich inhibitory element. *Virus Res.* **2001**, *73*, 163–175. [CrossRef]

103. Sokolowski, M.; Tan, W.; Jellne, M.; Schwartz, S. mRNA instability elements in the human papillomavirus type 16 L2 coding region. *J. Virol.* **1998**, *72*, 1504–1515. [PubMed]

104. Hawkins, J.D. A survey on intron and exon lengths. *Nucleic Acids Res.* **1988**, *16*, 9893–9908. [CrossRef] [PubMed]

105. De Conti, L.; Baralle, M.; Buratti, E. Exon and intron definition in pre-mRNA splicing. *Wiley Interdiscip. Rev. RNA* **2013**, *4*, 49–60. [CrossRef] [PubMed]

106. Dujardin, G.; Lafaille, C.; Petrillo, E.; Buggiano, V.; Gómez Acuña, L.I.; Fiszbein, A.; Godoy Herz, M.A.; Nieto Moreno, N.; Muñoz, M.J.; Alló, M.; et al. Transcriptional elongation and alternative splicing. *Biochim. Biophys. Acta Gene Regul. Mech.* **2013**, *1829*, 134–140. [CrossRef] [PubMed]

107. Atambayeva, S.A.; Khailenko, V.A.; Ivashchenko, A.T. Intron and exon length variation in Arabidopsis, rice, nematode, and human. *Mol. Biol.* **2008**, *42*, 352–361.

108. Howard, J.M.; Sanford, J.R. The RNAissance family: SR proteins as multifaceted regulators of gene expression. *Wiley Interdiscip. Rev. RNA* **2015**, *6*, 93–110. [CrossRef] [PubMed]

109. Busch, A.; Hertel, K.J. Evolution of SR protein and hnRNP splicing regulatory factors. *Wiley Interdiscip. Rev. RNA* **2012**, *3*, 1–12. [CrossRef] [PubMed]

110. Eperon, I.C.; Makarova, O.V.; Mayeda, A.; Munroe, S.H.; Caceres, J.F.; Hayward, D.G.; Krainer, A.R. Selection of alternative 5' splice sites: Role of U1 snRNP and models for the antagonistic effects of SF2/ASF and hnRNP A1. *Mol. Cell. Biol.* **2000**, *20*, 8303–8318. [CrossRef] [PubMed]

111. Somberg, M.; Li, X.; Johansson, C.; Orru, B.; Chang, R.; Rush, M.; Fay, J.; Ryan, F.; Schwartz, S. Serine/arginine-rich protein 30c activates human papillomavirus type 16 L1 mRNA expression via a bimodal mechanism. *J. Gen. Virol.* **2011**, *92*, 2411–2421. [CrossRef] [PubMed]

112. Zhao, X.; Fay, J.; Lambkin, H.; Schwartz, S. Identification of a 17-nucleotide splicing enhancer in HPV-16 L1 that counteracts the effect of multiple hnRNP A1-binding splicing silencers. *Virology* **2007**, *369*, 351–363. [CrossRef] [PubMed]

113. Zhao, X.; Rush, M.; Schwartz, S. Identification of an hnRNP A1-dependent splicing silencer in the human papillomavirus type 16 L1 coding region that prevents premature expression of the late L1 gene. *J. Virol.* **2004**, *78*, 10888–10905. [CrossRef] [PubMed]

114. Chuen-Im, T.; Zhang, J.; Milligan, S.G.; McPhillips, M.G.; Graham, S.V. The alternative splicing factor hnRNP A1 is up-regulated during virus-infected epithelial cell differentiation and binds the human papillomavirus type 16 late regulatory element. *Virus Res.* **2008**, *131*, 189–198.

115. Li, X.; Johansson, C.; Glahder, J.; Mossberg, A.-K.; Schwartz, S. Suppression of HPV-16 late L1 5′-splice site SD3632 by binding of hnRNP D proteins and hnRNP A2/B1 to upstream AUAGUA RNA motifs. *Nucleic Acids Res.* **2013**, *41*, 10488–10508. [CrossRef] [PubMed]

116. Somberg, M.; Zhao, X.; Fröhlich, M.; Evander, M.; Schwartz, S. Polypyrimidine tract binding protein induces human papillomavirus type 16 late gene expression by interfering with splicing inhibitory elements at the major late 5′ splice site, SD3632. *J. Virol.* **2008**, *82*, 3665–3678. [CrossRef] [PubMed]

117. Dhanjal, S.; Kajitani, N.; Glahder, J.; Mossberg, A.-K.; Johansson, C.; Schwartz, S. Heterogeneous nuclear ribonucleoprotein C proteins interact with the human papillomavirus type 16 (HPV16) early 3′-untranslated region and alleviate suppression of HPV16 late L1 mRNA splicing. *J. Biol. Chem.* **2015**, *290*, 13354–13371. [CrossRef] [PubMed]

118. Collier, B.; Oberg, D.; Zhao, X.; Schwartz, S. Specific inactivation of inhibitory sequences in the 5′ end of the human papillomavirus type 16 L1 open reading frame results in production of high levels of L1 protein in human epithelial cells. *J. Virol.* **2002**, *76*, 2739–2752. [CrossRef] [PubMed]

119. Kozak, M. Some thoughts about translational regulation: Forward and backward glances. *J. Cell. Biochem.* **2007**, *102*, 280–290. [CrossRef] [PubMed]

120. Buck, C.; Trus, B. The papillomavirus virion: A machine built to hide molecular achilles' heels. In *Viral Molecular Machines*; Rossmann, M.G., Rao, V.B., Eds.; Springer: New York, NY, USA, 2012; Volume 726, pp. 403–422.

121. Zhou, J.; Lui, W.J.; Peng, S.W.; Sun, X.Y.; Frazer, I.H. Papillomavirus capsid protein expression levels depends on the match between codon usage and tRNA availability. *J. Virol.* **1999**, *73*, 4972–4982. [PubMed]

122. Zhao, K.N.; Lui, W.J.; Frazer, I.H. Codon usage bias and A+T content variation in human papillomavirus genomes. *Virus Res.* **2003**, *98*, 95–104. [CrossRef] [PubMed]

123. Zhao, K.-N.; Gu, W.; Fang, N.X.; Saunders, N.A.; Frazer, I.H. Gene codon composition determines differentiation-dependent expression of a viral capsid gene in keratinocytes in vitro and in vivo. *Mol. Cell. Biol.* **2005**, *25*, 8643–8655. [CrossRef] [PubMed]

124. Fang, N.-X.; Gu, W.; Ding, J.; Saunders, N.A.; Frazer, I.H.; Zhao, K.-N. Calcium enhances mouse keratinocyte differentiation in vitro to differentially regulate expression of papillomavirus authentic and codon modified L1 genes. *Virology* **2007**, *365*, 187–197. [CrossRef] [PubMed]

125. Gu, W.; Ding, J.; Wang, X.; de Kluyver, R.L.; Saunders, N.A.; Frazer, I.H.; Zhao, K.-N. Generalized substitution of isoencoding codons shortens the duration of papillomavirus L1 protein expression in transiently gene-transfected keratinocytes due to cell differentiation. *Nucleic Acids Res.* **2007**, *35*, 4820–4832. [CrossRef] [PubMed]

126. Ding, J.; Doorbar, J.; Li, B.; Zhou, F.; Gu, W.; Zhao, L.; Saunders, N.A.; Frazer, I.H.; Zhao, K.-N. Expression of papillomavirus L1 proteins regulated by authentic gene codon usage is favoured in G2/M-like cells in differentiating keratinocytes. *Virology* **2010**, *399*, 46–58. [CrossRef] [PubMed]

127. Collier, B.; Goobar, L.; Sokolowski, M.; Schwartz, S. Translational inhibition in vitro of human papillomavirus type 16 L2 mRNA mediated through interaction with heterogeneous ribonucleoprotein K and poly (rC)-binding proteins 1 and 2. *J. Biol. Chem.* **1998**, *273*, 22648–22656. [CrossRef] [PubMed]

128. Wiklund, L.; Sokolowski, M.; Carlsson, A.; Rush, M.; Schwartz, S. Inhibition of translation by UAUUUAU and UAUUUUUAU motifs of the AU-rich RNA instability element in the HPV-1 late 3′ untranslated region. *J. Biol. Chem.* **2002**, *277*, 40462–40471. [CrossRef] [PubMed]

Review

The Mouse Papillomavirus Infection Model

Jiafen Hu [1,2,*] ![ORCID], Nancy M. Cladel [1,2], Lynn R. Budgeon [1,2], Karla K. Balogh [1,2] and
Neil D. Christensen [1,2,3,*]

[1] The Jake Gittlen Laboratories for Cancer Research, Hershey, PA 17033, USA; ncladel@gmail.com (N.M.C.);
lrb11@psu.edu (L.R.B.); kkb2@psu.edu (K.K.B.)
[2] Department of Pathology, Pennsylvania State University College of Medicine, Hershey, PA 17033, USA
[3] Department of Microbiology and Immunology, Pennsylvania State University College of Medicine, Hershey,
PA 17033, USA
* Correspondence: fjh4@psu.edu (J.H.); ndc1@psu.edu (N.D.C.); Tel.: +1-717-531-4700 (J.H. & N.D.C.);
Fax: +1-717-531-5634 (J.H. & N.D.C.)

Academic Editors: Alison A. McBride and Karl Munger
Received: 7 August 2017; Accepted: 24 August 2017; Published: 30 August 2017

Abstract: The mouse papillomavirus (MmuPV1) was first reported in 2011 and has since become a powerful research tool. Through collective efforts from different groups, significant progress has been made in the understanding of molecular, virological, and immunological mechanisms of MmuPV1 infections in both immunocompromised and immunocompetent hosts. This mouse papillomavirus provides, for the first time, the opportunity to study papillomavirus infections in the context of a small common laboratory animal for which abundant reagents are available and for which many strains exist. The model is a major step forward in the study of papillomavirus disease and pathology. In this review, we summarize studies using MmuPV1 over the past six years and share our perspectives on the value of this unique model system. Specifically, we discuss viral pathogenesis in cutaneous and mucosal tissues as well as in different mouse strains, immune responses to the virus, and local host-restricted factors that may be involved in MmuPV1 infections and associated disease progression.

Keywords: the mouse papillomavirus; tissue tropism; anogenital; oral infection; skin carcinoma; pathogenesis; innate immunity; adaptive immunity; RNA sequencing; host defense

---

1. Introduction

The papillomavirus research community has been searching for a mouse papillomavirus model since the identification of the cottontail rabbit papillomavirus in 1933 [1]. Over the years, a number of rodent papillomaviruses have been isolated, cloned and sequenced but none of them infected laboratory mouse strains [2]. In 1989, Tilbrook et al., identified papillomavirus DNA in hairless mouse tumors resulting from ultraviolet irradiation exposure [3]. They subsequently inoculated this mouse with the cell-free extract of these skin tumors and observed increased tumor incidence and degree of malignancy upon irradiation [4]. They demonstrated that the viral DNA shared homology with *Mastomys natalensis* papillomavirus DNA as well as with HPV11, -13, -16 and -18 by hybridization. The viral genome, however, was not isolated or sequenced [3,4]. In 2011, a mouse papillomavirus (subsequently labeled MmuPV1) was identified in a colony of nude (NMRI-Foxn1^{nu}/Foxn1^{nu}) mice in India [5]. The DNA sequence was reported in a subsequent publication [6]. A variant of MmuPV1 was later identified in a house mouse [7]. In the few years since it was first reported, MmuPV1 has become a valuable animal papillomavirus because it provides, for the first time, the opportunity to study papillomavirus infections in the context of a small common laboratory animal for which abundant reagents are available and for which many strains exist. Several groups, including our own, have established this mouse model system and have made significant progress in understanding molecular, virological, and immunological mechanisms of MmuPV1 infections in both immunocompromised and immunocompetent hosts [5–23]. In this review, we summarize studies using this mouse papillomavirus model over the past six years and share our perspectives on the value of this unique model system. Specifically, we discuss viral pathogenesis in cutaneous and mucosal tissues as well as in different mouse strains, immune responses to the virus, and local host-restricted factors that may be involved in MmuPV1 infections and associated disease progression.

2. Mouse Papillomavirus Exhibits both Cutaneous and Mucosal Tropism

First reports of MmuPV1 identified the virus as strictly cutaneous [5]. Indeed, the lesions first observed were florid muzzle tumors [5]. Most laboratories have focused their work on cutaneous sites including the tail, the muzzle, the back, and the ear [11,13–18]. When compared with muzzle and tail sites, the back skin was the least susceptible site for the primary infections ([14]. Our recent work has supported this observation and has shown significantly less encapsidated DNA in back skin vis a vis in muzzle and tail sites (manuscript in press). Work in our laboratory showed, however, that the vaginal, anal and oral mucosae of immunocompromised Hsd:NU Foxn1^{nu}, NU/J-Foxn1^{nu}, and B6.Cg-Foxn1^{nu} mice are also easily infected [20–23]. Figure 1 is representative of mucosal lesions in Hsd:NU (Foxn1^{nu}) nude mice.

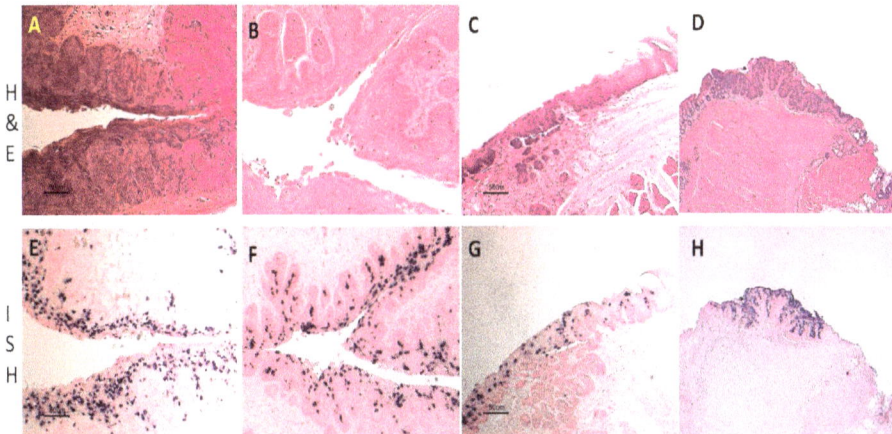

Figure 1. Histology of infected mucosal sites in Hsd:NU Foxn1^{nu} mice: the lower genital tract (**A**,**B**, 20×); the tongue (**C**, 20×); and the anal tract (**D**, 10×). Viral DNA was detected at the corresponding sites by in situ hybridization. Vaginal tract (**E**, 20×) and cervix (**F**, 20×) were positive for viral DNA. The Circumvallate papilla was the primary target for tongue infection (**G**, 20×). The transition zone of the anal tract was the most susceptible site for viral infection (**H**, 10×).

Vaginal infections were later confirmed by other groups [8,11]. We are able to readily establish infections in these disparate mucosal tissues and follow the progress of disease by QPCR examination of viral DNA in lavage samples [21]. Cytology was also very useful to monitor infections in the lower genital tract [21,22]. These noninvasive methods have proved to be powerful tools in our laboratory as they allow not only for disease progression to be tracked longitudinally but also for maximal data to be obtained from a small number of animals. Histological analyses were also conducted following sacrifice of the animals. Using these tools, we have demonstrated that:

(1) The single circumvallate papilla at the back of the mouse tongue is preferentially targeted by the virus [20]. This observation provides opportunities to study viral infections at the back of the tongue, the site of an increasing number of papillomavirus-associated human tumors [24].

(2) The vaginal and anal tracts are highly susceptible to virus infections [21,22]. This observation provides a novel in vivo model to study both anal and genital infections concurrently in the same host. These studies may reveal information that could lead to better understanding and control of corresponding human infections and diseases.

(3) Primary infections at cutaneous sites can lead to secondary infections at mucosal sites [23]. In a similar manner, secondary cutaneous infections often follow primary infections in immunocompromised mice (manuscript in press). Thus, primary infection in the oral cavity or vagina can lead to secondary infections at skin sites on the back, muzzle or tail.

3. Certain Strains of Immunocompetent Mice Are Susceptible to MmuPV1 Infection

MmuPV1 was identified in a colony of NMRI-Foxn1^{nu}/Foxn1^{nu} nude mice in India [5]. These mice were immunocompromised and there was interest in determining whether immunocompetent strains could also maintain infections. In this early report, the authors noted the development of small cutaneous papillomas on the back skin of immunocompetent S/RV/Cri-ba/ba (bare) mice [11,17]. These lesions regressed by eight weeks and no further analyses were conducted. Subsequently, Jiang et al. tested SKH1 (Crl: SKH1-Hrhr) mice and found that tail lesions in a subset of these animals persisted over time [18]. Uberoi et al. subjected FVB/NJ immunocompetent mice to UVB radiation and noted that, following treatment, ear lesions persisted although the mice were resistant to infection prior to irradiation [13]. In work submitted for publication from our laboratory, we observed that several immunocompetent mouse strains (e.g., C57BL6 and hairless SKH-1) mounted transient mucosal infections, which quickly regressed. However, the immunocompetent heterozygotes of inbred NU/J mice (Foxn1$^{nu/+}$), outbred Hsd (Foxn1$^{nu/+}$), and C57BL/6 (Foxn1$^{nu/+}$) mice maintained persistent vaginal infections. Thus, it is clear that MmuPV1 infections are not restricted to immunocompromised mouse strains.

4. MmuPV1 Has Malignant Potential

4.1. Mouse Papillomavirus Oncogenes E6 and E7 Are Tumorigenic

As in many other papillomaviruses, the mouse papillomavirus also contains two putative viral oncogenes, E6 and E7 [6]. We have tested the tumorigenicity of these proteins in vitro and in vivo using standard methods that we have previously described [25]. Both gene products showed significantly higher proliferative activity in vitro (Figure 2A) and tumorigenicity in vivo (Figure 2C,D). A recent study showed that MmuPV1E6 shared some biochemical and functional characteristics with cutaneous HPV8 E6 including inhibiting NOTCH and TGF-β signaling as well as contributing to delayed differentiation and prolonged survival of differentiated keratinocytes [16]. MmuPV1 E6 mutants also failed to induce tumor growth in nude mice indicating that E6 contributes to papillomavirus pathogenesis and carcinogenesis [16]. White et al. noted that MmuPV1 E7 bound PTPN14, a classical nontransmembrane protein tyrosine phosphatase (PTP), as did numerous oncogenic HPV E7s [26]. These observations suggest that E7 may share oncogenic properties with the high-risk human papillomavirus E7. Much work remains to be done to further clarify the roles of MmuPV1 E6 and E7.

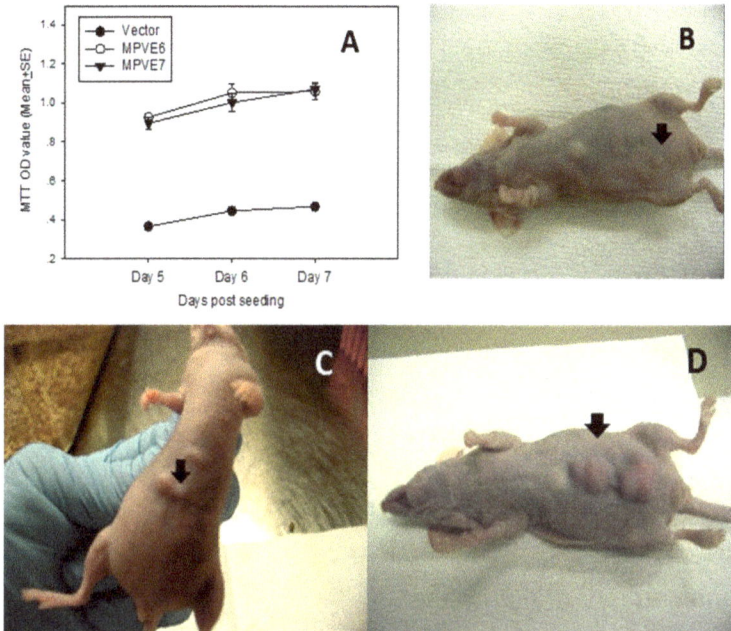

Figure 2. MmuPV1 E6 and E7 cloned into the expression vector (PCR3) were transfected into NIH3T3 cells under the selection of G418. The stably transfected cells were tested for proliferation in vitro. Both E6 and E7 showed significantly higher proliferative activity (**A**); E6 and E7 stably transfected NIH3T3 cells also showed tumorigenicity in vivo ((**C,D**), respectively, see arrows); and the vector control showed minimal disease (**B**).

4.2. Cutaneous Lesions Can Develop into Cancers

Sundberg et al. reported the development of locally invasive poorly differentiated carcinomas on the dorsal skin of B6.Cg-Foxn1^{nu}/Foxn1^{nu} mice [11]. They noted that the tumors resembled trichoblastomas seen in humans [27]. Uberoi et al. subjected FVB/NJ immunocompetent mice to UVB radiation and subsequently infected the mice at ear and tail sites [13]. Many of the mice developed ear lesions, some of which progressed to squamous cell carcinomas. This was the first report of an MmuPV1 cancer in an immunocompetent mouse. In our laboratory, we have detected two different and spontaneously developing cutaneous skin carcinomas as secondary skin infections in Hsd:NU Foxn1^{nu} nude mice. Taken together, these findings clearly demonstrate the malignant potential of MmuPV1 at cutaneous sites.

4.3. Mucosal Lesions Can Develop into Cancers

In data under review for publication, our laboratory has shown that both homozygous NU/J mice (immunocompromised) and their immunocompetent heterozygous counterparts develop carcinoma in situ in MmuPV1-infected vaginal tissues. These carcinomas were detected at 7.5 months post infection. These findings and those above suggest that MmuPV1 may prove to be a useful model to study papillomavirus-associated malignant progression at both cutaneous and mucosal sites in a tractable animal model.

5. Immune Responses to MmuPV1 Infections

Both adaptive and innate immunity play a role in MmuPV1 infections at both cutaneous and mucosal sites. Studies of adaptive immune responses have been focused on T-cell mediated immune responses in cutaneous infections [13,15,17,18]. Initial innate immune responses have been studied in both cutaneous and mucosal infections in our laboratory.

5.1. T- and B-Cell Mediated Immunity in the Control of MmuPV1 Infections

Several immunocompetent mouse strains including C56BL/6, FVB/NJ, and SKH-1 have been tested for susceptibility to MmuPV1 infections at cutaneous sites [8,11,15,17]. Both CD4 and CD8 T cells have been found to play a crucial role in the control of papillomavirus infection at these sites, although neither CD4 nor CD8-knockout or -depletion led to visible disease in these immunocompetent mice [15,17,18]. Further studies demonstrated that a strong E6- and E7-specific CD8+ T cell response is correlated with viral clearance and tumor regression in vaccinated mice [18]. Specifically, transferred E6/90-99 specific CD8 T cells can prevent the development of tumor growth in MmuPV1-infected athymic mice [17]. We have observed persistence and delayed regression of anal infections in C57BL/6 mice depleted of both CD4 and CD8 T cells (Figure 3A). Passive immunization with serum from virus-particle immunized mice provided strong protection against primary viral infection [8]. Our recent studies also demonstrated complete protection at both cutaneous and mucosal sites as a result of passive immunization with a neutralizing monoclonal antibody in athymic mice (unpublished observations). These findings suggest that both T- and B-cell mediated immune responses play a critical role in the clinical outcome of MmuPV1 infections.

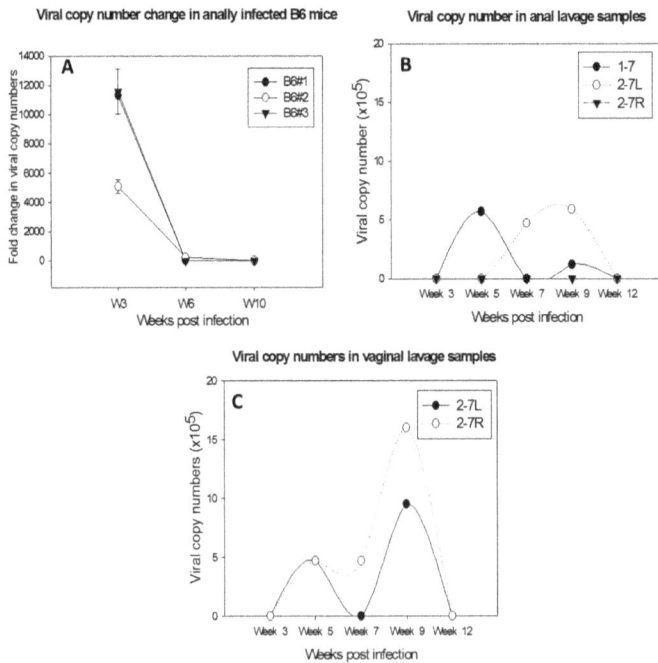

Figure 3. Delayed regression of anal infections was observed in C57BL/6 mice depleted of both CD4 and CD8 T cells (**A**). Viral DNA was detected in lavage samples of the anal tract of one male (1–7) and two females (2–7L and 2–7R) (**B**); and the lower genital tract of 2–7L and 2–7R of Ifnar$^{-/-}$ mice (**C**). In contrast, viral infections were cleared in the wild type B6 mice before Week 5 post-infection.

5.2. Innate Immunity in the Control of MmuPV1 Infection at Mucosal Sites

Type I interferon pathways are the first defense against pathogen invasion [28–30]. A previous study failed to detect visible disease at cutaneous sites in ifnar$^{-/-}$ mice [17]. In contrast, we found a prolonged time to regression at the anogenital (vaginal and anal) sites in these mice compared to wild-type mice (Figure 3B,C). The infections were detectable up to three months post infection in contrast to infections in wild type animals in which the viral DNA usually became undetectable by week four post infection (our unpublished observations).

5.2.1. Neutrophils and NK Cells Are Associated with Decreased Local MmuPV1 Mucosal Infection in Immunocompetent Heterozygous (Foxn1$^{nu/+}$) NU/J Mice

Innate immune cells including NK cells and neutrophils are important in host defense against the early stages of viral infections. Previous studies demonstrated that adaptive immunity is sufficient to eliminate skin MmuPV1 infections in immunocompetent mice [15,17]. We examined both neutrophils and NK cells in situ in infected mucosal and cutaneous tissues of both immunocompetent heterozygous NU/J mice (Foxn1$^{nu/+}$) and immunocompromised homozygous (Foxn1$^{nu/nu}$) NU/J mice. We found that increased numbers of NK cells were observed in the tail tissues of the heterozygous NU/J mice (with minimal infections) relative to those detected in the homozygotes with visible lesions (data not shown). Tissue resident NK cells are reported to be very different from conventional splenic NK cells in that they produce cytokines TNFα and GM-CSF rather than of IFNγ [31]. Although cytokines are key players in numerous inflammatory processes and the production of cytokines is under tight genetic control [32], this difference in resident NK cells may result in differential local cytokine production leading to tissue-specific disease outcome.

Immunocompetent (Foxn1$^{nu/+}$) and immunocompromised (Foxn1$^{nu/nu}$) NU/J mice showed persistent infections in vaginal tissues. We detected higher numbers of infiltrating neutrophils in the vaginal tissues of the immunocompetent (Foxn1$^{nu/+}$) NU/J mice than in those of immunocompromised (Foxn1$^{nu/nu}$) mice. The homozygous (Foxn1$^{nu/nu}$) mice had more severe disease indicating that neutrophils may have contributed to the control of disease in the immunocompetent (Foxn1$^{nu/+}$) mice (manuscript in preparation). Collectively, our findings suggest that neutrophils and NK cells may be associated with reduced MmuPV1 infection in immunocompetent mice. The role of these and other immune cells in MmuPV1 infection and persistence needs further investigation.

5.2.2. RNA Sequencing Data Support the Involvement of NK Cells and Neutrophils as Well as Type I IFNs in MmuPV1-Infected Tissues

Recently, we conducted genome-wide RNA sequencing on different tissues with or without MmuPV1 infection. In agreement with findings in HPV-associated human studies, many molecules in different signaling pathways associated with antiviral activities, cell growth and differentiation, cancer development, and inflammation were dysregulated in the infected tissues [33–37]. Host defense factors including β-defensins [36,38–41] and TLR5 [42,43] were associated with MmuPV1 infection (Table 1). Type I interferon (IFNs) are the most potent known antiviral factors limiting the replication and spread of most viruses [30,44,45]. Significant changes were found in type I IFNs as well as IL-12, IL-15, and IL-18 (Table 1). These molecules are produced by macrophages and dendritic cells and are critical for NK-cell maturation and regulation of NK-cell function [46–49]. CXCR2 expressed by circulating neutrophils was upregulated in infected muzzle tissues (Table 1). We have also detected significant changes of several interferon-stimulated genes (ISGs). For example, Stat3 and Stat6, which promote innate antiviral responses and contribute to the detrimental effects of viral infection, were upregulated in MmuPV1-infected muzzle and tongue tissues (Table 1). Other ISGs including Trim proteins (Trim 23 and Trim 29) were dysregulated in infected tissues (Table 1). We also detected downregulation of CD53, a pan-leukocyte surface glycoprotein proposed to play an important role in thymopoiesis and leukocyte signal transduction in all infected tissues [50].

Table 1. Some significantly changed molecules in MmuPV1-infected tissues.

Transcripts	MmuPV1 Infected Tissues		
	Muzzle	Tongue	Vagina
IL15	Down	Down	N.S.
Il1rn	UP	UP	N.S.
Il4ra	N.S.	UP	UP
IFNar1	N.S.	UP	N.S.
Ifi2712b	N.S.	Down	Down
Ifi27	Down	N.S.	N.S.
Ifit2	N.S.	N.S.	Down
TLR5	N.S.	UP	N.S.
CXCR2	UP	N.S.	N.S.
CD53	Down	Down	Down
Stat3	UP	UP	N.S.
Stat6	UP	UP	N.S.
Trim23	Down	Down	N.S.
Trim29	UP	N.S.	UP
Defb4	Down	Down	N.S.
Defb6	Down	Down	N.S.

N.S. Not significant between the infected vs. non-infected tissues.

6. Other Host-Restricted Factors in Local MmuPV1 Infections

Besides the involvement of innate and adaptive immunity in papillomavirus infections, other host-restricted factors may play a role in disease outcome. A previous study showed that NOD/SCID mice, a strain that is deficient in T-, B-, and NK cells, were resistant to cutaneous MmuPV1 infections [11]. In agreement with this finding, we detected a single small tail lesion on one of seven infected NOD/SCID mice although all infected sites were positive for viral DNA (manuscript under review). This finding suggests that a latent or subclinical infection may have been established in these infected tail tissues and that additional host defenses may have played a role in the control of cutaneous infections in this strain.

We and others have observed site specific infections in other mouse strains. For example, the inbred NU/J nude mice, which manifested minimal cutaneous disease relative to that in outbred Hsd:NU mice, showed advanced mucosal infections. We and others observed that the back skin is less susceptible to viral infection when compared with the muzzle and tail in the same animal [14]. In addition, we have shown that viral encapsidation is significantly reduced in back skin relative to other cutaneous sites (manuscript in press). To understand the site-specificity, Sundberg et al. conducted transcriptome assays in different skin papillomas [11]. While they found dysregulation of several skin cancer-associated genes in papilloma tissues, they also found a significant difference in gene expression in different skin sites supporting the concept that the local environment may contribute to the disease outcome at these sites [11]. In another study, the authors noted that the tail showed less disease than the ear skin in UV-irradiated mice [13]. All these findings suggest local host defense factors may have contributed to disease outcome in these different mouse strains. More thorough molecular and genetic studies should provide new evidence of the site-specific host control in viral infections.

7. Conclusions and Future Directions

The discovery of MmuPV1 is very recent. In the short time since it was first reported, several groups have made significant findings with respect to the immunology, molecular biology, and tropism of the disease [5–23]. The findings have set the stage for some very useful new models to investigate papillomavirus disease in vivo. They also introduce new opportunities to better refine our understanding of tissue-specific immune responses to papillomavirus infections. Among the most important findings and their implications are the following.

(1) The first small animal model to study papillomavirus infections and associated diseases in anogenital and oral tissues:

More than 66% of cervical cancers are associated with PV. Cervical cancers create an enormous medical burden for the world's women. More than 250,000 individuals die each year of the disease [51,52]. Current prophylactic vaccines have no effect on existing disease and so a therapeutic vaccine would represent an important advancement [53]. With the finding that certain immunocompetent strains of mice support persistent vaginal MmuPV1 infections, the model is well-placed to test potential therapeutic vaccines [54].

PV-related oral cancers are on the rise in younger men [55,56]. They present a major challenge to the medical community and result in considerable morbidity for the patients. A suitable preclinical model of oral PV-associated disease is therefore in great demand. The finding that the circumvallate papilla of the mouse tongue is a preferred site for infection suggests that the MmuPV1/mouse model could be ideal for the study of papillomavirus-related oral cancers in humans, in which cancers tend to occur at the back of the tongue and in the tonsillar region [57].

Anal cancers are on the rise in women [58]. The reasons are poorly understood and few interventions are available. MmuPV1 infects the anal canal, especially the transition zone. The MmuPV-1 mouse model thus could provide a much-needed model to study these enigmatic infections.

(2) A small animal model to study host defense against papillomavirus infections:

T- and B-cells have been shown to be critical factors in generating immunity to MmuPV1. However, the fact that not all mouse strains deficient in T- and B-cells develop MmuPV1 lesions supports the idea that other immune components and pathways can play a role. Our early work has shown possible roles for neutrophils and NK cells and both RNAseq data generated in our laboratory and the microarray data of Sundberg et al. support this hypothesis as well [11]. We expect that a more complete analysis of the RNA sequencing and array data will suggest other avenues for investigation and validation. The availability of many genetically-modified mice will also allow for the expanded investigation of host factors in viral infections.

(3) A small animal model to study papillomavirus-associated tissue tropism:

Tissue tropism has always been of interest to researchers in the papillomavirus field. Most papillomaviruses display either cutaneous or mucosal tropism and even within those broad categories only specific tissues are commonly targeted [51]. MmuPV1 is different in that it displays both cutaneous and mucosal tropism [11,23]. The tissues of different strains can be differentially susceptible to the virus infections and this provides a tool to plumb local immune and other factors contributing to tissue tropism. We anticipate that high throughput analyses such as RNAseq will be instrumental in helping to elucidate the molecular and cellular components of PV tissue tropism.

(4) A small animal model to study papillomavirus-associated skin cancers:

Cutaneous cancers are sometimes associated with papillomaviruses [59]. MmuPV1 lesions have been shown to undergo malignant transformation in immunocompromised animals [11] as well as in immunocompetent irradiated mice [13] and so the MmuPV1 model may become a useful tool to study cutaneous skin cancer disease and progression.

(5) A small animal model to study papillomavirus-associated transmission:

Our recent studies demonstrated that genital tissues of both males and females are susceptible to viral infections. In addition, our studies have shown that the development of secondary lesions resulted from virus shedding from primary infections [21,23]. These observations suggest that the MmuPV1 model will be useful to study both horizontal (between partners) and vertical (mother to child) transmission of virus. HPV-induced Recurrent Respiratory Papillomatosis (RRP) is believed to result from HPV transmission from mothers to newborns as a result of passage through the birth canal [60]. RRP is a devastating pathological condition in children. It is characterized by the recurrent appearance of wart-like lesions in the respiratory tract, particularly at the larynx and vocal cords [61]. These patients must undergo repeated surgery or other invasive treatment to manage the disease. There is a great need for model systems to study papillomavirus-associated vertical transmission.

(6) A small animal model to study the role of the menstrual cycle and contraceptives in papillomavirus-associated diseases:

The influence of the menstrual cycle and contraception on other viruses such as genital herpes and zika has been reported [62–65]. Whether the menstrual cycle and contraception play a role in genital papillomavirus infections is of great practical interest. We demonstrated in our previous studies that highest viral titers was detected at the estrus stage [21]. Whether contraception plays a role in viral susceptibility and persistence is under investigation in the laboratory. The MmuPV1 vaginal model will be an excellent tool to study the interplay between viral infection, the menstrual cycle, and contraceptive use. Figure 4 illustrates in graphic form the potential of the MmuPV1 model for multiple studies in situ. The model is a major step forward in the study of papillomavirus disease and pathology.

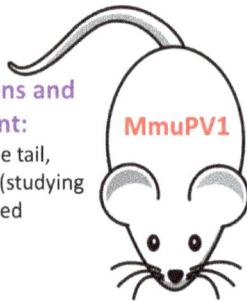

Immunology studies:
Innate immunity (Interferons and cytokines)
Adaptive immunity (T and B-mediated immune responses)
Vaccine development

Genetic studies:
Host restricted factors for viral infections using different gene-modified mice, microarray and RNAseq analyses

Cutaneous infections and cancer development:
Several sites include the tail, muzzle, back, and ear (studying UV-irradiation associated malignancies)

MmuPV1

Mucosal infections and cancer development:
Several sites include the vagina, anus, and tongue. These mucosal sites parallel to the corresponding HPV-associated infections in humans.

Viral transmissions:
Horizontal transmissions and infertility (between partners); Vertical transmissions and associated childhood diseases (mother-to-newborn)

Hormone studies:
The menstrual cycle on viral infections; Long-term contraceptive on viral infections and persistence

Figure 4. The potential application of the MmuPV1/mouse model.

Acknowledgments: We thank Ziaur Rahman for providing ifnar$^{-/-}$ mice. Research reported in this publication was supported by the National Institute of Allergy and Infectious Diseases of the National Institutes of Health under Award Number R21AI121822 (Christensen and Hu) and the Jake Gittlen Memorial Golf Tournament.

Conflicts of Interest: The authors declare no conflict of interest.

References

1. Shope, R.E.; Hurst, E.W. Infectious Papillomatosis of Rabbits: With a Note on the Histopathology. *J. Exp. Med.* **1933**, *58*, 607–624. [CrossRef] [PubMed]
2. Rector, A.; van Ranst, M. Animal papillomaviruses. *Virology* **2013**, *445*, 213–223. [CrossRef] [PubMed]
3. Tilbrook, P.A.; Greenoak, G.E.; Reeve, V.E.; Canfield, P.J.; Gissmann, L.; Gallagher, C.H.; Kulski, J.K. Identification of papillomaviral DNA sequences in hairless mouse tumours induced by ultraviolet irradiation. *J. Gen. Virol.* **1989**, *70*, 1005–1009. [CrossRef] [PubMed]
4. Reeve, V.E.; Greenoak, G.E.; Canfield, P.J.; Boehm-Wilcox, C.; Tilbrook, P.A.; Kulski, J.K.; Gallagher, C.H. Enhancement of UV-induced skin carcinogenesis in the hairless mouse by inoculation with cell-free extracts of skin tumours. *Immunol. Cell Biol.* **1989**, *67*, 421–427. [CrossRef] [PubMed]
5. Ingle, A.; Ghim, S.; Joh, J.; Chepkoech, I.; Bennett Jenson, A.; Sundberg, J.P. Novel laboratory mouse papillomavirus (MusPV) infection. *Vet. Pathol.* **2011**, *48*, 500–505. [CrossRef] [PubMed]
6. Joh, J.; Jenson, A.B.; King, W.; Proctor, M.; Ingle, A.; Sundberg, J.P.; Ghim, S.J. Genomic analysis of the first laboratory-mouse papillomavirus. *J. Gen. Virol.* **2011**, *92*, 692–698. [CrossRef] [PubMed]
7. Schulz, E.; Gottschling, M.; Ulrich, R.G.; Richter, D.; Stockfleth, E.; Nindl, I. Isolation of three novel rat and mouse papillomaviruses and their genomic characterization. *PLoS ONE* **2012**, *7*, e47164. [CrossRef] [PubMed]
8. Joh, J.; Ghim, S.J.; Chilton, P.M.; Sundberg, J.P.; Park, J.; Wilcher, S.A.; Proctor, M.L.; Bennett Jenson, A. MmuPV1 infection and tumor development of T cell-deficient mice is prevented by passively transferred hyperimmune sera from normal congenic mice immunized with MmuPV1 virus-like particles (VLPs). *Exp. Mol. Pathol.* **2016**, *100*, 212–219. [CrossRef] [PubMed]
9. Everts, H.B.; Suo, L.; Ghim, S.; Bennett Jenson, A.; Sundberg, J.P. Retinoic acid metabolism proteins are altered in trichoblastomas induced by mouse papillomavirus 1. *Exp. Mol. Pathol.* **2015**, *99*, 546–551. [CrossRef] [PubMed]
10. Joh, J.; Jenson, A.B.; Ingle, A.; Sundberg, J.P.; Ghim, S.J. Searching for the initiating site of the major capsid protein to generate virus-like particles for a novel laboratory mouse papillomavirus. *Exp. Mol. Pathol.* **2014**, *96*, 155–161. [CrossRef] [PubMed]
11. Sundberg, J.P.; Stearns, T.M.; Joh, J.; Proctor, M.; Ingle, A.; Silva, K.A.; Dadras, S.S.; Jenson, A.B.; Ghim, S.J. Immune status, strain background, and anatomic site of inoculation affect mouse papillomavirus (MmuPV1) induction of exophytic papillomas or endophytic trichoblastomas. *PLoS ONE* **2014**, *9*, e113582. [CrossRef] [PubMed]
12. Joh, J.; Jenson, A.B.; Proctor, M.; Ingle, A.; Silva, K.A.; Potter, C.S.; Sundberg, J.P.; Ghim, S.J. Molecular diagnosis of a laboratory mouse papillomavirus (MusPV). *Exp. Mol. Pathol.* **2012**, *93*, 416–421. [CrossRef] [PubMed]
13. Uberoi, A.; Yoshida, S.; Frazer, I.H.; Pitot, H.C.; Lambert, P.F. Role of Ultraviolet Radiation in Papillomavirus-Induced Disease. *PLoS Pathog.* **2016**, *12*, e1005664. [CrossRef] [PubMed]
14. Handisurya, A.; Day, P.M.; Thompson, C.D.; Buck, C.B.; Pang, Y.Y.; Lowy, D.R.; Schiller, J.T. Characterization of Mus musculus papillomavirus 1 infection in situ reveals an unusual pattern of late gene expression and capsid protein localization. *J. Virol.* **2013**, *87*, 13214–13225. [CrossRef] [PubMed]
15. Handisurya, A.; Day, P.M.; Thompson, C.D.; Bonelli, M.; Lowy, D.R.; Schiller, J.T. Strain-Specific Properties and T Cells Regulate the Susceptibility to Papilloma Induction by Mus musculus Papillomavirus 1. *PLoS Pathog.* **2014**, *10*, e1004314. [CrossRef] [PubMed]
16. Meyers, J.M.; Uberoi, A.; Grace, M.; Lambert, P.F.; Munger, K. Cutaneous HPV8 and MmuPV1 E6 Proteins Target the NOTCH and TGF-β Tumor Suppressors to Inhibit Differentiation and Sustain Keratinocyte Proliferation. *PLoS Pathog.* **2017**, *13*, e1006171. [CrossRef] [PubMed]
17. Wang, J.W.; Jiang, R.; Peng, S.; Chang, Y.N.; Hung, C.F.; Roden, R.B. Immunologic Control of Mus musculus Papillomavirus Type 1. *PLoS Pathog.* **2015**, *11*, e1005243. [CrossRef] [PubMed]
18. Jiang, R.T.; Wang, J.W.; Peng, S.; Huang, T.C.; Wang, C.; Cannella, F.; Chang, Y.N.; Viscidi, R.P.; Best, S.R.A.; Hung, C.F.; et al. Spontaneous and vaccine-induced clearance of Mus musculus Papillomavirus type 1 (MmuPV1/MusPV1) infection. *J. Virol.* **2017**, *91*. [CrossRef] [PubMed]
19. Christensen, N.D.; Budgeon, L.R.; Cladel, N.M.; Hu, J. Recent advances in preclinical model systems for papillomaviruses. *Virus Res.* **2017**, *231*, 108–118. [CrossRef] [PubMed]

20. Cladel, N.M.; Budgeon, L.R.; Balogh, K.K.; Cooper, T.K.; Hu, J.; Christensen, N.D. Mouse papillomavirus MmuPV1 infects oral mucosa and preferentially targets the base of the tongue. *Virology* **2016**, *488*, 73–80. [CrossRef] [PubMed]
21. Hu, J.; Budgeon, L.R.; Cladel, N.M.; Balogh, K.; Myers, R.; Cooper, T.K.; Christensen, N.D. Tracking vaginal, anal and oral infection in a mouse papillomavirus infection model. *J. Gen. Virol.* **2015**, *96*, 3554–3565. [CrossRef] [PubMed]
22. Cladel, N.M.; Budgeon, L.R.; Balogh, K.K.; Cooper, T.K.; Hu, J.; Christensen, N.D. A novel pre-clinical murine model to study the life cycle and progression of cervical and anal papillomavirus infections. *PLoS ONE* **2015**, *10*, e0120128. [CrossRef] [PubMed]
23. Cladel, N.M.; Budgeon, L.R.; Cooper, T.K.; Balogh, K.K.; Hu, J.; Christensen, N.D. Secondary infections, expanded tissue tropism, and evidence for malignant potential in immunocompromised mice infected with Mus musculus papillomavirus 1 DNA and virus. *J. Virol.* **2013**, *87*, 9391–9395. [CrossRef] [PubMed]
24. Woods, R., Sr.; O'Regan, E.M.; Kennedy, S.; Martin, C.; O'Leary, J.J.; Timon, C. Role of human papillomavirus in oropharyngeal squamous cell carcinoma: A review. *World J. Clin. Cases* **2014**, *2*, 172–193. [PubMed]
25. Hu, J.; Cladel, N.M.; Budgeon, L.R.; Christensen, N.D. Characterization of three rabbit oral papillomavirus oncogenes. *Virology* **2004**, *325*, 48–55. [CrossRef] [PubMed]
26. White, E.A.; Munger, K.; Howley, P.M. High-Risk Human Papillomavirus E7 Proteins Target PTPN14 for Degradation. *MBio* **2016**, *7*. [CrossRef] [PubMed]
27. Battistella, M.; Peltre, B.; Cribier, B. Composite tumors associating trichoblastoma and benign epidermal/follicular neoplasm: Another proof of the follicular nature of inverted follicular keratosis. *J. Cutan. Pathol.* **2010**, *37*, 1057–1063. [CrossRef] [PubMed]
28. Fensterl, V.; Sen, G.C. Interferons and viral infections. *BioFactors* **2009**, *35*, 14–20. [CrossRef] [PubMed]
29. Stanley, M.A. Epithelial cell responses to infection with human papillomavirus. *Clin. Microbiol. Rev.* **2012**, *25*, 215–222. [CrossRef] [PubMed]
30. Durbin, R.K.; Kotenko, S.V.; Durbin, J.E. Interferon induction and function at the mucosal surface. *Immunol. Rev.* **2013**, *255*, 25–39. [CrossRef] [PubMed]
31. Sojka, D.K.; Plougastel-Douglas, B.; Yang, L.; Pak-Wittel, M.A.; Artyomov, M.N.; Ivanova, Y.; Zhong, C.; Chase, J.M.; Rothman, P.B.; Yu, J.; et al. Tissue-resident natural killer (NK) cells are cell lineages distinct from thymic and conventional splenic NK cells. *eLife* **2014**, *3*, e01659. [CrossRef] [PubMed]
32. De Craen, A.J.; Posthuma, D.; Remarque, E.J.; van den Biggelaar, A.H.; Westendorp, R.G.; Boomsma, D.I. Heritability estimates of innate immunity: An extended twin study. *Genes Immun.* **2005**, *6*, 167–170. [CrossRef] [PubMed]
33. Zhang, J.; Zhu, L.; Feng, P. Dissecting innate immune signaling in viral evasion of cytokine production. *J. Vis. Exp.* **2014**. [CrossRef] [PubMed]
34. Schneider, W.M.; Chevillotte, M.D.; Rice, C.M. Interferon-stimulated genes: A complex web of host defenses. *Annu. Rev. Immunol.* **2014**, *32*, 513–545. [CrossRef] [PubMed]
35. Jin, L.; Sturgis, E.M.; Cao, X.; Song, X.; Salahuddin, T.; Wei, Q.; Li, G. Interleukin-10 promoter variants predict HPV-positive tumors and survival of squamous cell carcinoma of the oropharynx. *FASEB J.* **2013**, *27*, 2496–2503. [CrossRef] [PubMed]
36. Ding, J.; Chou, Y.Y.; Chang, T.L. Defensins in viral infections. *J. Innate Immun.* **2009**, *1*, 413–420. [CrossRef] [PubMed]
37. Gregorczyk, K.P.; Krzyzowska, M. Innate immunity to infection in the lower female genital tract. *Postepy Hig. Med. Dosw.* **2013**, *67*, 388–401. [CrossRef]
38. Wilson, S.S.; Wiens, M.E.; Smith, J.G. Antiviral mechanisms of human defensins. *J. Mol. Biol.* **2013**, *425*, 4965–4980. [CrossRef] [PubMed]
39. Abe, S.; Miura, K.; Kinoshita, A.; Mishima, H.; Miura, S.; Yamasaki, K.; Hasegawa, Y.; Higashijima, A.; Jo, O.; Sasaki, K.; et al. Copy number variation of the antimicrobial-gene, defensin β4, is associated with susceptibility to cervical cancer. *J. Hum. Genet.* **2013**, *58*, 250–253. [CrossRef] [PubMed]
40. Erhart, W.; Alkasi, O.; Brunke, G.; Wegener, F.; Maass, N.; Arnold, N.; Arlt, A.; Meinhold-Heerlein, I. Induction of human β-defensins and psoriasin in vulvovaginal human papillomavirus-associated lesions. *J. Infect. Dis.* **2011**, *204*, 391–399. [CrossRef] [PubMed]

41. Kreuter, A.; Skrygan, M.; Gambichler, T.; Brockmeyer, N.H.; Stucker, M.; Herzler, C.; Potthoff, A.; Altmeyer, P.; Pfister, H.; Wieland, U. Human papillomavirus-associated induction of human β-defensins in anal intraepithelial neoplasia. *Br. J. Dermatol.* **2009**, *160*, 1197–1205. [CrossRef] [PubMed]

42. Daud, I.I.; Scott, M.E.; Ma, Y.; Shiboski, S.; Farhat, S.; Moscicki, A.B. Association between toll-like receptor expression and human papillomavirus type 16 persistence. *Int. J. Cancer* **2011**, *128*, 879–886. [CrossRef] [PubMed]

43. Sasagawa, T.; Takagi, H.; Makinoda, S. Immune responses against human papillomavirus (HPV) infection and evasion of host defense in cervical cancer. *J. Infect. Chemother.* **2012**, *18*, 807–815. [CrossRef] [PubMed]

44. Ma, W.; Tummers, B.; van Esch, E.M.; Goedemans, R.; Melief, C.J.; Meyers, C.; Boer, J.M.; van der Burg, S.H. Human Papillomavirus Downregulates the Expression of IFITM1 and RIPK3 to Escape from IFNγ- and TNFα-Mediated Antiproliferative Effects and Necroptosis. *Front Immunol.* **2016**, *7*, 496. [CrossRef] [PubMed]

45. Behbahani, H.; Walther-Jallow, L.; Klareskog, E.; Baum, L.; French, A.L.; Patterson, B.K.; Garcia, P.; Spetz, A.L.; Landay, A.; Andersson, J. Proinflammatory and type 1 cytokine expression in cervical mucosa during HIV-1 and human papillomavirus infection. *J. Acquir. Immune. Defic. Syndr.* **2007**, *45*, 9–19. [CrossRef] [PubMed]

46. Tummers, B.; Burg, S.H. High-risk human papillomavirus targets crossroads in immune signaling. *Viruses* **2015**, *7*, 2485–2506. [CrossRef] [PubMed]

47. Amador-Molina, A.; Hernandez-Valencia, J.F.; Lamoyi, E.; Contreras-Paredes, A.; Lizano, M. Role of innate immunity against human papillomavirus (HPV) infections and effect of adjuvants in promoting specific immune response. *Viruses* **2013**, *5*, 2624–2642. [CrossRef] [PubMed]

48. Fernandes, J.V.; DE Medeiros Fernandes, T.A.; DE Azevedo, J.C.; Cobucci, R.N.; DE Carvalho, M.G.; Andrade, V.S.; DE Araújo, J.M. Link between chronic inflammation and human papillomavirus-induced carcinogenesis (Review). *Oncol. Lett.* **2015**, *9*, 1015–1026. [CrossRef] [PubMed]

49. Torres-Poveda, K.; Bahena-Roman, M.; Madrid-Gonzalez, C.; Burguete-Garcia, A.I.; Bermudez-Morales, V.H.; Peralta-Zaragoza, O.; Madrid-Marina, V. Role of IL-10 and TGF-β1 in local immunosuppression in HPV-associated cervical neoplasia. *World J. Clin. Oncol.* **2014**, *5*, 753–763. [CrossRef] [PubMed]

50. Bos, S.D.; Lakenberg, N.; van der Breggen, R.; Houwing-Duistermaat, J.J.; Kloppenburg, M.; de Craen, A.J.; Beekman, M.; Meulenbelt, I.; Slagboom, P.E. A genome-wide linkage scan reveals CD53 as an important regulator of innate TNF-alpha levels. *Eur. J. Hum. Genet.* **2010**, *18*, 953–959. [CrossRef] [PubMed]

51. Martinez, G.G.; Troconis, J.N. Natural history of the infection for human papillomavirus: An actualization. *Investig. Clin.* **2014**, *55*, 82–91.

52. Ting, J.; Rositch, A.F.; Taylor, S.M.; Rahangdale, L.; Soeters, H.M.; Sun, X.; Smith, J.S. Worldwide incidence of cervical lesions: A systematic review. *Epidemiol. Infect.* **2015**, *143*, 225–241. [CrossRef] [PubMed]

53. Dochez, C.; Bogers, J.J.; Verhelst, R.; Rees, H. HPV vaccines to prevent cervical cancer and genital warts: An update. *Vaccine* **2014**, *32*, 1595–1601. [CrossRef] [PubMed]

54. Scheinfeld, N. Update on the treatment of genital warts. *Dermatol. Online J.* **2013**, *19*, 18559. [PubMed]

55. Liu, H.; Li, J.; Zhou, Y.; Hu, Q.; Zeng, Y.; Mohammadreza, M.M. Human papillomavirus as a favorable prognostic factor in a subset of head and neck squamous cell carcinomas: A meta-analysis. *J. Med. Virol.* **2017**, *89*, 710–725. [CrossRef] [PubMed]

56. Petrelli, F.; Sarti, E.; Barni, S. Predictive value of human papillomavirus in oropharyngeal carcinoma treated with radiotherapy: An updated systematic review and meta-analysis of 30 trials. *Head Neck* **2014**, *36*, 750–759. [CrossRef] [PubMed]

57. Dalianis, T. Human papillomavirus and oropharyngeal cancer, the epidemics, and significance of additional clinical biomarkers for prediction of response to therapy (Review). *Int. J. Oncol.* **2014**, *44*, 1799–1805. [CrossRef] [PubMed]

58. Assi, R.; Reddy, V.; Einarsdottir, H.; Longo, W.E. Anorectal Human Papillomavirus: Current Concepts. *Yale J. Biol. Med.* **2014**, *87*, 537–547. [PubMed]

59. Varada, S.; Posnick, M.; Alessa, D.; Ramirez-Fort, M.K. Management of cutaneous human papillomavirus infection in immunocompromised patients. *Curr. Probl. Dermatol.* **2014**, *45*, 197–215. [PubMed]

60. Mammas, I.N.; Sourvinos, G.; Spandidos, D.A. The paediatric story of human papillomavirus (Review). *Oncol. Lett.* **2014**, *8*, 502–506. [CrossRef] [PubMed]

61. Chan, Y.H.; Lo, C.M.; Lau, H.Y.; Lam, T.H. Vertically transmitted nasopharyngeal infection of the human papillomavirus: Does it play an aetiological role in nasopharyngeal cancer? *Oral Oncol.* **2014**, *50*, 326–329. [CrossRef] [PubMed]

62. Kaushic, C.; Ashkar, A.A.; Reid, L.A.; Rosenthal, K.L. Progesterone increases susceptibility and decreases immune responses to genital herpes infection. *J. Virol.* **2003**, *77*, 4558–4565. [CrossRef] [PubMed]

63. Gallichan, W.S.; Rosenthal, K.L. Effects of the estrous cycle on local humoral immune responses and protection of intranasally immunized female mice against herpes simplex virus type 2 infection in the genital tract. *Virology* **1996**, *224*, 487–497. [CrossRef] [PubMed]

64. Teepe, A.G.; Allen, L.B.; Wordinger, R.J.; Harris, E.F. Effect of the estrous cycle on susceptibility of female mice to intravaginal inoculation of herpes simplex virus type 2 (HSV-2). *Antivir. Res.* **1990**, *14*, 227–235. [CrossRef]

65. Tang, W.W.; Young, M.P.; Mamidi, A.; Regla-Nava, J.A.; Kim, K.; Shresta, S. A Mouse Model of Zika Virus Sexual Transmission and Vaginal Viral Replication. *Cell Rep.* **2016**, *17*, 3091–3098. [CrossRef] [PubMed]

viruses

MDPI

Review

Natural History of HPV Infection across the Lifespan: Role of Viral Latency

Patti E. Gravitt [1],* and Rachel L. Winer [2]

[1]   Department of Global Health, George Washington University Milken Institute School of Public Health, Washington, DC 20052, USA
[2]   Department of Epidemiology, University of Washington School of Public Health, Seattle, WA 98195, USA; rlw@uw.edu
*   Correspondence: pgravitt@gwu.edu; Tel.: +1-202-994-8939

Academic Editors: Alison A. McBride and Karl Munger
Received: 24 August 2017; Accepted: 19 September 2017; Published: 21 September 2017

Abstract: Large-scale epidemiologic studies have been invaluable for elaboration of the causal relationship between persistent detection of genital human papillomavirus (HPV) infection and the development of invasive cervical cancer. However, these studies provide limited data to adequately inform models of the individual-level natural history of HPV infection over the course of a lifetime, and particularly ignore the biological distinction between HPV-negative tests and lack of infection (i.e., the possibility of latent, undetectable HPV infection). Using data from more recent epidemiological studies, this review proposes an alternative model of the natural history of genital HPV across the life span. We argue that a more complete elucidation of the age-specific probabilities of the alternative transitions is highly relevant with the expanded use of HPV testing in cervical cancer screening. With routine HPV testing in cervical cancer screening, women commonly transition in and out of HPV detectability, raising concerns for the patient and the provider regarding the source of the positive test result, its prognosis, and effective strategies to prevent future recurrence. Alternative study designs and analytic frameworks are proposed to better understand the frequency and determinants of these transition pathways.

Keywords: papillomavirus; latency; cervical cancer

1. Introduction

Prospective epidemiologic studies conducted in the late 1990s and 2000s established the temporal association between exposure to high-risk (HR) human papillomavirus (HPV) and the subsequent development of cervical intraepithelial neoplasia (CIN) and cervical cancer. These data, combined with strong biological plausibility derived from the basic sciences, led to acceptance of HR-HPV as a necessary, but insufficient cause of nearly 100% of cervical cancers [1]. Not only did these data consistently fulfill the Bradford–Hill criteria for causality, but the well-accepted temporal pathway from HPV infection to invasive cervical cancer (ICC) that was derived from these studies has also translated into important changes to cervical cancer screening guidelines worldwide, increasing both the impact and cost effectiveness of secondary cervical cancer prevention.

Because these studies collected data on HPV repeatedly over time, typically once every 4–6 months, a natural analytic extension was to model the HPV measures as outcomes, rather than exposures, to establish incidence and clearance rates—the basic natural history parameters of the infection. These incidence and clearance estimates have formed the basis for mathematical models of HPV infection, used in health policy analyses to predict the impact of interventions such as HPV vaccination and screening. In this review, we reflect on the accuracy of these estimates in representing the natural history of viral infection within an individual over the course of the life span by reviewing the data through the lens of an infectious disease epidemiologist. We further highlight the clinical and public health relevance for enhancing our understanding of the within-woman HPV infection natural history across the life span.

2. Overview of Well-Established Aspects of HPV Natural History

Figure 1 represents the current paradigm of HPV natural history from infection to cervical cancer, and highlights several uncertainties in the interpretation of natural history estimates derived from the original prospective studies of HPV and cervical cancer. In this conceptual model, HPV infections are acquired via sexual exposures, with newly sexually active adolescent and young adult women at highest risk of acquisition [2]. During productive HPV infection, low-grade cervical abnormalities may be clinically detectable in screening (e.g., low grade squamous intraepithelial lesions (LSIL) or CIN grade 1 (CIN1)), but are usually transient and resolve without intervention within 1–2 years [3]. The majority (~90%) of newly acquired HPV infections similarly become undetectable within 1–2 years [3], a phenomenon routinely described as "viral clearance," but which may also represent immune control below detectable levels or viral latency [4–6]. A detectable immune response is generated approximately 60% of the time [7], evidenced by the presence of serum antibodies specific to the HPV type causing infection, with uncertain ability to provide immunity against re-infection [8]. A minority of HPV infections are persistently detected beyond 12 months, increasing the risk of carcinogenic progression to cervical pre-cancer (high grade squamous intraepithelial lesions (HSIL) or CIN grade 2 or 2 (CIN2/3) and potentially cancer if untreated [3].

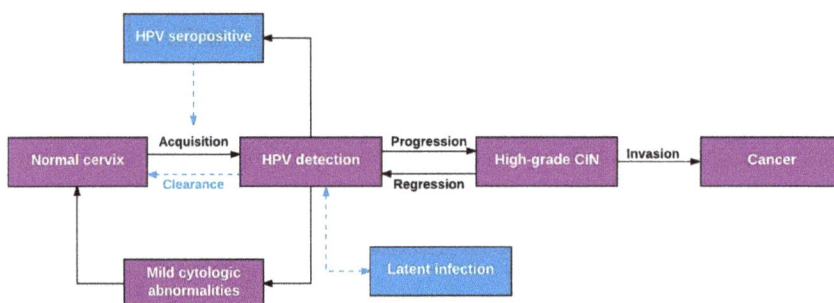

Figure 1. Schematic model of the population-level natural history of human papillomavirus infection and cervical cancer. Purple boxes indicate well-accepted natural history model parameters; blue boxes represent uncertainties.

3. Expanded View of HPV Natural History

The paradigm in Figure 1 is based on our understanding of the population-level natural history of HPV over the course of 5–10 years (the typical duration of prospective natural history studies). Because of the practical limitations prohibiting longer duration studies, extrapolating this population-level model to the within-woman natural history of HPV infections over an entire life span requires several explicit assumptions. In most applications of HPV "incidence and clearance" estimates derived from this model, two critical assumptions prevail: (1) new HPV detection reflects recent acquisition either as a new infection or a re-infection; and (2) loss of HPV detection reflects viral clearance, or eradication. To more fully explore the validity of these assumptions and evaluate the evidence to support them, we have elaborated a more nuanced natural history of an HPV infection within an individual woman (Figure 2). In our model, we posit that each HPV infection may follow a number of non-linear, non-mutually exclusive pathways over a woman's life span. Specifically, new HPV detection can result not only from a recent sexual acquisition or re-infection, but also from recurrent detection of a controlled or latent infection [9], auto-inoculation from other epithelial sites (e.g., anus) [10], or transient deposition of viral nucleic acid from a recent sex act [11]. Similarly, loss of HPV detection (aka clearance) may reflect viral eradication with or without acquired immunity against re-infection or viral control below limits of detection (aka viral latency) [12].

We emphasize that the short duration of typical HPV natural history studies (usually no more than 4 years of follow-up) and the infrequent sampling to measure HPV outcomes (most often only once every 4–6 months), has limited our ability to observe the full spectrum of natural history infection transitions depicted in Figure 2. A more complete elucidation of the age-specific probabilities of the alternative transitions is becoming more relevant with the expanded use of HPV testing in cervical cancer screening. Women are receiving HPV test results over a decade or more of screening, and even in the context of infrequent screening intervals (3–5 years), transitioning in and out of HPV detectability is increasingly more common, raising concerns for the patient and the provider regarding the source of the positive test result and its prognosis. While definitive answers are elusive, alternative study designs and analytic frameworks provide critical evidence to understand the frequency and determinants of these transitions.

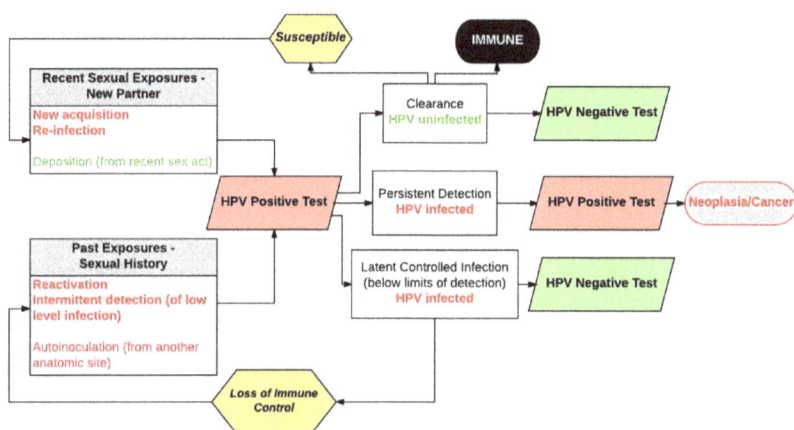

Figure 2. Schematic model of individual-level natural history of female genital HPV infection across the life span. The model assumes two pathways to type-specific HPV positivity after an HPV-negative test result—new acquisition or reinfection due to current sexual activity, or reactivation/recurrent detection of controlled, latent HPV infection. Red boxes indicate positive HPV molecular test results and green boxes indicate negative HPV molecular test results. Colored fonts represent the true underlying infectious status independent of concurrent molecular test results from exfoliated samples; red = HPV infection, green = HPV uninfected.

4. Evidence Supporting Re-Detection as an Important Transition from Negative to Positive HPV Tests

A key uncertainty in the natural history of HPV infection within an individual is whether an HPV infection that becomes undetectable on repeat testing has truly cleared, or whether the virus persists at low, undetectable levels or has entered a latent state. While distinctions between the latter two scenarios are controversial, studies suggest that re-detection of the same HPV type is relatively common, occurring in at least 10–20% of women observed to have "cleared" the virus [8]. Furthermore, convincing data from multiple studies of immune compromised, sexually abstinent [9,13], older, less sexually active populations [4,14], and adolescents with long-term intensive follow-up [5] support the phenomena of immunologically controlled re-detection or reactivation of a previously acquired type-specific HPV infection. The human observational data is largely consistent with animal models of papillomavirus latency, which have quite elegantly shown latent detection of low copy papillomavirus DNA with minimal gene expression in basal cells following clinical resolution of initial viral infection [15]. Upon iatrogenic immunosuppression, papillomavirus copy number increased in the epithelium of the latently infected animals, though no recurrence of warts (the consequence of initial infection) was observed [16].

5. New Acquisition versus Re-Detection of Prior Infection: Evidence from Studies of Mid-Adult Women

The likelihood that detectable HPV is due to new acquisition is high in newly sexually active young women [2], and decreases with age [4,14,17,18]. Although sex with new partners remains a risk factor for new infection, rates of acquiring new partners decline with age [19–21] and the ratio of detectable HPV that is attributable to prior versus new infection increases. These concepts are supported by data from three recent U.S. cohort studies in mid-adult women, including women aged 35–60 years in Baltimore, Maryland [14], women aged 30–50 years in Seattle, Washington [4], and female online daters aged 25–65 years recruited from various U.S. cities [17]. These studies all used repeat HPV DNA testing and sexual behavior data to estimate risks of incident HPV detection associated with recent versus past sexual exposures [4,14,17], with the Seattle study adding baseline HPV serology testing as a biomarker of prior infection [4]. Although HPV serology is an imperfect marker of prior infection (due to limited assay sensitivity [22] and the fact that antibody responses to natural infection are neither uniformly detected [7,23–25] nor lifelong [7,23]), the addition of serology data to analyses of HPV DNA genotyping and sexual behavior offer a unique contribution to our understanding of HPV natural history.

In these mid-adult cohort studies, reporting sex with recent new, or otherwise high-risk male sex partners (e.g., casual partners or partners with other concurrent partners) increased the risk of incident HPV detection [4,14,17]. Among women with recent new, or high-risk partners, the fraction of new HPV detection that could be attributed to one of these recent partners was estimated to be between 64 [17] and 82% [14], with the remainder likely due to re-detection of prior infection. It is important to note that these estimates that measure attributable risk in the "exposed" (i.e., those with recent new partners) are distinct from population attributable risk. The latter measure is dependent not just on the relative strength of new partners as a risk factor for new HPV detection, but also on the proportion of women in a study population who report new partners. Notably, in the Baltimore cohort, although incident HPV detection was 5.6-fold more likely in women reporting recent new partners than in women reporting no recent sexual activity (a strong relative risk), the fraction of all incident HPV detections attributable to recent partners was low (27%) because only 10% of women reported a new sex partner during study follow-up. Conversely, 72% of apparent new HPV detection in this cohort was attributed to a lifetime number of ≥5 male sex partners. In addition, the relative risk of detection associated with lifetime partners increased with age, further suggesting a relative shift from new acquisition to reactivation as the more likely source of new HPV detection in older women. In the Seattle cohort, recent sexual behaviors were also associated with increased risk of incident HPV detection, but only in the absence of serologic evidence of prior infection with the same type [4]. Compared to the Baltimore cohort, reports of new partners and other recent high-risk sexual behaviors were more common in the Seattle cohort; thus, the fraction of incident HPV detections attributable to recent sexual risk behavior was higher. A combination of sexual behavior and baseline serology data were used to create a composite variable to reflect risk categories for type-specific incident HPV detection: seronegative with no recent high-risk sexual behavior, seronegative with recent high-risk sexual behavior, and seropositive (irrespective of recent sexual behavior). With this approach, 40% of incident HPV in the Seattle cohort was attributed to probable new infection, due to a lack of serologic evidence of prior infection coupled with recent sexual risk behavior. On the other hand, 30% of incident detections were attributed to likely re-detection of prior infection due to the presence of natural antibodies to the same type. Of note, while the proportion of new detections attributable to prior infection remained constant with age, the proportion attributable to probable new infection was higher in women aged 30–39 years (48%) than in women aged 40–50 years (21%). Declining rates of new acquisition with age (as seen in the other mid-adult cohorts) combined with possible waning antibodies with age (such that a larger proportion of new detections due to prior infection could not be classified as such in the absence of natural antibodies) are likely explanations for these findings.

Results from these three cohort studies indicate that, while new partners remain a strong risk factor for new HPV infection into mid-adulthood, the likelihood that newly detected HPV is due to new infection versus re-detection of prior infection declines with age. In these studies, it is also notable that recent sex with non-new or otherwise low-risk partners did not appreciably increase the risk for incident HPV detection compared to no recent sexual activity. This information may be reassuring for women in monogamous relationships who test HPV-positive in clinical settings.

Natural history studies in older women using similar strategies for attribution of new HR-HPV detection to acquisition vs. recurrent detection, but with a longer duration than the Baltimore and Seattle studies, are required to better estimate whether there is similar or differential risk for cervical precancer or cancer. Since the studies cited above consistently show an increasing attributable fraction of new HPV detection due to prior infection relative to recent acquisition with age, precancer risk by age provides a reasonable surrogate for risk difference by the source of HPV infection. In a large US study of women receiving routine co-testing, the risk of CIN3+ in women with newly detected HR-HPV was largely similar across age groups [26], suggesting no difference in risk in women with newly acquired vs. recurrent detection of HR-HPV. From the perspective of clinical management of newly detected HR-HPV in screening, similar risks independent of the transition pathway to HPV detection as depicted in Figure 2 is reassuring since unequivocal differentiation of the transition path for any given woman is currently impractical. However, this does not obviate the need to develop a more complete understanding and differentiation of these natural history pathways, because while patients with transiently detected HR-HPV are reassured by a negative test as it relates to their immediate precancer risk, they remain concerned about how to protect themselves from re-infection.

6. Protection against Re-Infection: Role of Naturally Acquired Antibodies and Vaccine

A key area of controversy is whether serum antibodies from natural HPV infection protect against re-infection with the same HPV type. To date, studies addressing this issue have produced mixed results [27–33]. In a study of the control arm of a bivalent HPV vaccine trial, Safaeian and colleagues reported a 50% and 64% reduction in incidence of HPV16 and HPV18 detection, respectively, in women with the highest levels of HPV antibodies [27]. It should be noted, however, that HPV antibody titers are generally not bimodal, and selection of a cutpoint to define seropositives is somewhat arbitrary, often relying on statistical definitions of 3–5 standard deviations over the mean optical density (OD) of virginal (presumably HPV-negative) girls [32]. Thus, it is possible that a lack of protection associated with lower HPV antibody titers may be the result of misclassification of baseline HPV serostatus, rather than a lack of protection associated with lower antibody titers. Other studies suggest that antibodies may offer protection against type-specific re-infection that wanes with age [8,12]. In the placebo arm of the quadrivalent HPV vaccine trial in mid-adult women, protection against type-specific re-infection was observed for women aged 24–34 years, as demonstrated by a higher incidence of vaccine-type HPV in seronegative versus seropositive women (5.7 versus 1.0 per 100 person-years) [34]. In contrast, in women aged 35–45 years, the incidence of vaccine-type HPV was observed to be slightly higher in seropositive versus seronegative women (2.8 versus 2.1 per 100 person-years), suggesting a possible lack of protection from natural infection in older women. An important consideration, however, is that reactivation of previously acquired, latent HPV infection, rather than re-infection, may explain the apparent lack of protection against re-infection from natural antibodies in the older age group. In the Seattle mid-adult cohort, type-specific HPV incidence was also observed to be higher in seropositive versus seronegative women (with a 6-month cumulative incidence of 2.9% versus 1.2%, respectively), a trend observed both in 30–39 and 40–50 year old women. Furthermore, as noted above, recent sexual behaviors were unassociated with new HPV detection in the presence of serologic evidence of prior infection with the same type (suggesting a low likelihood of re-infection versus re-detection of prior infection) [4]. These results suggest that, in the presence of naturally acquired antibodies, reactivation or intermittent viral shedding of a previously acquired infection is a more probable source of new HPV detection than new acquisition. This theory is further supported by earlier studies showing that it is rare for an individual woman to be infected with more than one variant of a specific HPV genotype [35].

HPV vaccine trials in mid-adult populations (aged 26–45 years) show similar efficacy in women naïve for the vaccine types at the time of immunization (e.g., both DNA and seronegative) [36,37], suggesting that women at risk for new HPV exposures may benefit from HPV vaccination at any age (though we note that new partner acquisition declines precipitously with increasing age [21] and thus the value of routine immunization of mid-adult women may have little benefit at the population level). In contrast, women with currently detected vaccine-targeted DNA at the time of immunization are not protected from progression to CIN3+, nor do they control their infections more quickly compared with control arm women [38]. Previously infected control arm women who were seropositive but DNA-negative at the time of immunization (i.e., women with prior, but not current infection) had a lower risk of new vaccine-type detection compared with seronegative controls. Yet, the rate of new vaccine type detection in the seropositive vaccine arm participants was significantly lower than seropositive controls [36], suggesting the possibility that vaccination may reduce risk of reactivation of controlled infection. It has been shown that B-cell memory is substantially boosted in vaccinated women with prior exposure to vaccine types, and that neutralizing antibodies were elicited in all women with a single dose vaccine boost, but only in a limited number of women with natural memory B-cell-derived antibodies [39]. These data provides some biological plausibility for enhancement of immunologic control of previously acquired HPV infections to prevent, or reduce, the frequency or duration of reactivation (see future research needs below).

7. Rationale for Resolving Remaining Uncertainties

Why is it important to resolve the remaining uncertainties in our understanding of HPV natural history? From a clinical perspective, clarifying the unknowns has critical implications for patient psychosocial counseling. As women participating in cervical cancer screening accumulate HPV testing histories, there is a strong need for comprehensive, accurate, and reassuring information to guide clinician–patient interactions for complex, yet common scenarios such as non-consecutive HPV-positive results. While the increased risk of carcinogenic progression associated with persistent HPV detection is clear [3] and risk associated with newly detected HPV is similar across age [26], it remains uncertain whether an intermittent viral detection or reactivation history increases the risk of progression compared to a history of consistent negative HPV test results. Several health systems in the United States have accumulated over a decade of HPV co-testing data (e.g., Kaiser Permanente, Northern California, CA, USA) and are well suited to evaluate risk associated with decades of negative HPV testing vs. persistent detection vs. intermittent detection. A recent report from the POBASCAM study of HPV testing in the Netherlands showed a higher risk of new HPV detection and CIN3+ development in women with intermittent negative HPV tests in screening compared with women remaining persistently HPV-negative. Recurrent positive results following apparent clearance were most often due to the same type, suggesting reactivation rather than new infection [40].

From a public health perspective, accurate natural history models are necessary for developing evidence-based HPV vaccination and cervical cancer screening strategies for older populations of women. For example, while data from vaccine trials in mid-adult women suggest that the protective benefits of prophylactic HPV vaccines may extend beyond newly acquired infection—with clear efficacy demonstrated in the subgroups of women with baseline antibodies to vaccine-type HPV compared to seronegative women [36]—the underlying mechanism of protection is unclear. The mathematical models used to estimate vaccine effectiveness and cost-effectiveness in older populations of women are sensitive to assumptions regarding natural immunity, viral latency, and re-infection [8,33,41]. More accurate parameters are needed to inform the validity of bold strategies such as HPV-FASTER [42] that propose integrated screening, treatment, and vaccination programs in women up to 45–50 years of age.

However, elucidating a more accurate picture of the natural history of HPV infection within women over time will require different approaches to study design and analysis. For example, Liu et al. reported on the temporal dynamics of type-specific HPV infection in women with bi-weekly sample collection over 4 months, and found that many types were repeatedly detected with very short duration (<2 weeks) over the course of the study [6]. Recurrent detection was not associated with sexual activity, but was associated with stage of the menstrual cycle [43] and microbiome transitions [44], suggesting a much more dynamic reactivation/control of virus in the lower genital tract than has been previously acknowledged. The evolving understanding of viral natural history when employing more frequent and intensive sampling designs has strong analogies with herpes simplex virus type 2 (HSV-2), where a paradigm of lifelong latency with infrequent reactivation was replaced with a model of very frequent bursts of viral reactivation and shedding, with local immunologic memory keeping the duration of these shedding episodes to a few hours (reviewed in [45]). Using data collected from HSV-2 seropositive women with multiple genital sample collections per day (up to every 6 h while awake), mathematical simulation models were generated to both explore new hypotheses about the mechanisms of HSV-2 reactivation/control cycles and the impact of interventions [46]. Similar studies are envisioned to provide critical evidence for the mechanisms of HPV recurrent detection, as well as the possible impact of vaccine and probiotic or immunologic interventions aimed at minimizing the frequency and/or duration of recurrent HPV detection. This could translate broadly in terms of public health impact on cervical precancer/cancer risk in adult women, as well as potentially reducing the newly recognized risk of HPV infection (and "clearance") on increasing the susceptibility to HIV infection in populations with high co-infection rates.

Viewed from the lens of an infectious disease epidemiologist, the epidemiology of female genital HPV infection across the life span is likely more complex than previously appreciated. Evidence supporting an important transition pathway from HPV detection to non-detection, which represents virologic control or latency, rather than viral clearance or eradication, is rapidly accumulating. As women across the world begin to accumulate their own personal HPV testing histories, understanding the risks of precancer/cancer incidence associated with intermittently positive vs. repeatedly negative tests results becomes imperative for setting rational clinical management guidelines for screening intervals and safe exiting from screening in older women. In addition, the notion of prophylactic vaccination in older, less sexually active women is gaining in popularity. However, the individual- and population-level benefits of vaccination in women well past their sexual debut is dependent on understanding the proportion of new HPV detection resulting from recent acquisition vs. recurrent detection/reactivation, as well as the potential role for HPV vaccines in preventing or controlling reactivated infection. In other words, the time to act on improving our understanding of the causes and consequences of HPV latency and prevention of reactivation is now.

Acknowledgments: Funding support from National Cancer Institute CA123467 (P.E.G.) and National Institute of Allergy and Infectious Diseases AI083224 (R.L.W.).

Author Contributions: P.E.G. and R.L.W. equally conceived the alternative individual-level HPV natural history model and wrote the paper.

Conflicts of Interest: The authors declare no conflict of interest.

References

1. Bosch, F.X.; Lorincz, A.; Munoz, N.; Meijer, C.J.; Shah, K.V. The causal relation between human papillomavirus and cervical cancer. *J. Clin. Pathol.* **2002**, *55*, 244–265. [CrossRef] [PubMed]
2. Burchell, A.N.; Winer, R.L.; de Sanjose, S.; Franco, E.L. Chapter 6: Epidemiology and Transmission Dynamics of Genital HPV Infection. *Vaccine* **2006**, *24* (Suppl. 3), S52–S61. [CrossRef] [PubMed]
3. Schiffman, M.; Castle, P.E.; Jeronimo, J.; Rodriguez, A.C.; Wacholder, S. Human papillomavirus and cervical cancer. *Lancet* **2007**, *370*, 890–907. [CrossRef]

4. Fu, T.C.; Carter, J.J.; Hughes, J.P.; Feng, Q.; Hawes, S.E.; Schwartz, S.M.; Xi, L.F.; Lasof, T.; Stern, J.E.; Galloway, D.A.; et al. Re-detection vs. new acquisition of high-risk human papillomavirus in mid-adult women. *Int. J. Cancer* **2016**, *139*, 2201–2212. [CrossRef] [PubMed]
5. Shew, M.L.; Ermel, A.C.; Tong, Y.; Tu, W.; Qadadri, B.; Brown, D.R. Episodic detection of human papillomavirus within a longitudinal cohort of young women. *J. Med. Virol.* **2015**, *87*, 2122–2129. [CrossRef] [PubMed]
6. Liu, S.H.; Cummings, D.A.; Zenilman, J.M.; Gravitt, P.E.; Brotman, R.M. Characterizing the temporal dynamics of human papillomavirus DNA detectability using short-interval sampling. *Cancer Epidemiol. Biomark. Prev.* **2014**, *23*, 200–208. [CrossRef] [PubMed]
7. Carter, J.J.; Koutsky, L.A.; Hughes, J.P.; Lee, S.K.; Kuypers, J.; Kiviat, N.; Galloway, D.A. Comparison of human papillomavirus types 16, 18, and 6 capsid antibody responses following incident infection. *J. Infect. Dis.* **2000**, *181*, 1911–1919. [CrossRef] [PubMed]
8. Gravitt, P.E. Evidence and impact of human papillomavirus latency. *Open Virol. J.* **2012**, *6*, 198–203. [CrossRef] [PubMed]
9. Strickler, H.D.; Burk, R.D.; Fazzari, M.; Anastos, K.; Minkoff, H.; Massad, L.S.; Hall, C.; Bacon, M.; Levine, A.M.; Watts, D.H.; et al. Natural history and possible reactivation of human papillomavirus in human immunodeficiency virus-positive women. *J. Natl. Cancer Inst.* **2005**, *97*, 577–586. [CrossRef] [PubMed]
10. Fu, T.C.; Hughes, J.P.; Feng, Q.; Hulbert, A.; Hawes, S.E.; Xi, L.F.; Schwartz, S.M.; Stern, J.E.; Koutsky, L.A.; Winer, R.L. Epidemiology of Human Papillomavirus Detected in the Oral Cavity and Fingernails of Mid-Adult Women. *Sex. Transm. Dis.* **2015**, *42*, 677–685. [CrossRef] [PubMed]
11. Baay, M.F.; Francois, K.; Lardon, F.; Van Royen, P.; Pauwels, P.; Vermorken, J.B.; Peeters, M.; Verhoeven, V. The presence of Y chromosomal deoxyribonucleic acid in the female vaginal swab: Possible Implications for Human Papillomavirus Testing. *Cancer Epidemiol.* **2011**, *35*, 101–103. [CrossRef] [PubMed]
12. Gravitt, P.E. The known unknowns of HPV natural history. *J. Clin. Investig.* **2011**, *121*, 4593–4599. [CrossRef] [PubMed]
13. Theiler, R.N.; Farr, S.L.; Karon, J.M.; Paramsothy, P.; Viscidi, R.; Duerr, A.; Cu-Uvin, S.; Sobel, J.; Shah, K.; Klein, R.S.; et al. High-risk human papillomavirus reactivation in human immunodeficiency virus-infected women: Risk Factors for Cervical Viral Shedding. *Obstet. Gynecol.* **2010**, *115*, 1150–1158. [CrossRef] [PubMed]
14. Rositch, A.F.; Burke, A.E.; Viscidi, R.P.; Silver, M.I.; Chang, K.; Gravitt, P.E. Contributions of recent and past sexual partnerships on incident human papillomavirus detection: Acquisition and Reactivation in Older Women. *Cancer Res.* **2012**, *72*, 6183–6190. [CrossRef] [PubMed]
15. Maglennon, G.A.; McIntosh, P.; Doorbar, J. Persistence of viral DNA in the epithelial basal layer suggests a model for papillomavirus latency following immune regression. *Virology* **2011**, *414*, 153–163. [CrossRef] [PubMed]
16. Maglennon, G.A.; McIntosh, P.B.; Doorbar, J. Immunosuppression facilitates the reactivation of latent papillomavirus infections. *J. Virol.* **2014**, *88*, 710–716. [CrossRef] [PubMed]
17. Winer, R.L.; Hughes, J.P.; Feng, Q.; Stern, J.E.; Xi, L.F.; Koutsky, L.A. Incident Detection of High-Risk Human Papillomavirus Infections in a Cohort of High-Risk Women Aged 25–65 Years. *J. Infect. Dis.* **2016**, *214*, 665–675. [CrossRef] [PubMed]
18. Gravitt, P.E.; Rositch, A.F.; Silver, M.I.; Marks, M.A.; Chang, K.; Burke, A.E.; Viscidi, R.P. A cohort effect of the sexual revolution may be masking an increase in human papillomavirus detection at menopause in the United States. *J. Infect. Dis.* **2013**, *207*, 272–280. [CrossRef] [PubMed]
19. Herbenick, D.; Reece, M.; Schick, V.; Sanders, S.A.; Dodge, B.; Fortenberry, J.D. Sexual behavior in the United States: Results from a National Probability Sample of Men and Women Ages 14–94. *J. Sex. Med.* **2010**, *7* (Suppl. 5), 255–265. [CrossRef] [PubMed]
20. Mercer, C.H.; Tanton, C.; Prah, P.; Erens, B.; Sonnenberg, P.; Clifton, S.; Macdowall, W.; Lewis, R.; Field, N.; Datta, J.; et al. Changes in sexual attitudes and lifestyles in Britain through the life course and over time: Findings from the National Surveys of Sexual Attitudes and Lifestyles (Natsal). *Lancet* **2013**, *382*, 1781–1794. [CrossRef]
21. Ryser, M.D.; Rositch, A.; Gravitt, P.E. Age and sexual behavior indicates an increasing trend of HPV infection following the sexual revolution. *J. Infect. Dis.* **2017**, *3*, 46–49.

22. Strickler, H.D.; Schiffman, M.H.; Shah, K.V.; Rabkin, C.S.; Schiller, J.T.; Wacholder, S.; Clayman, B.; Viscidi, R.P. A survey of human papillomavirus 16 antibodies in patients with epithelial cancers. *Eur. J. Cancer Prev.* **1998**, *7*, 305–313. [CrossRef] [PubMed]

23. Carter, J.J.; Koutsky, L.A.; Wipf, G.C.; Christensen, N.D.; Lee, S.K.; Kuypers, J.; Kiviat, N.; Galloway, D.A. The natural history of human papillomavirus type 16 capsid antibodies among a cohort of university women. *J. Infect. Dis.* **1996**, *174*, 927–936. [CrossRef] [PubMed]

24. Wang, S.S.; Schiffman, M.; Herrero, R.; Carreon, J.; Hildesheim, A.; Rodriguez, A.C.; Bratti, M.C.; Sherman, M.E.; Morales, J.; Guillen, D.; et al. Determinants of human papillomavirus 16 serological conversion and persistence in a population-based cohort of 10 000 women in Costa Rica. *Br. J. Cancer* **2004**, *91*, 1269–1274. [CrossRef] [PubMed]

25. Ho, G.Y.; Studentsov, Y.Y.; Bierman, R.; Burk, R.D. Natural history of human papillomavirus type 16 virus-like particle antibodies in young women. *Cancer Epidemiol. Biomark. Prev.* **2004**, *13*, 110–116. [CrossRef]

26. Gage, J.C.; Katki, H.A.; Schiffman, M.; Fetterman, B.; Poitras, N.E.; Lorey, T.; Cheung, L.C.; Castle, P.E.; Kinney, W.K. Age-stratified 5-year risks of cervical precancer among women with enrollment and newly detected HPV infection. *Int. J. Cancer* **2015**, *136*, 1665–1671. [CrossRef] [PubMed]

27. Safaeian, M.; Porras, C.; Schiffman, M.; Rodriguez, A.C.; Wacholder, S.; Gonzalez, P.; Quint, W.; Van Doorn, L.J.; Sherman, M.E.; Xhenseval, V.; et al. Epidemiological study of anti-HPV16/18 seropositivity and subsequent risk of HPV16 and -18 infections. *J. Natl. Cancer Inst.* **2010**, *102*, 1653–1662. [CrossRef] [PubMed]

28. Ho, G.Y.; Studentsov, Y.; Hall, C.B.; Bierman, R.; Beardsley, L.; Lempa, M.; Burk, R.D. Risk factors for subsequent cervicovaginal human papillomavirus (HPV) infection and the protective role of antibodies to HPV-16 virus-like particles. *J. Infect. Dis.* **2002**, *186*, 737–742. [CrossRef] [PubMed]

29. Malik, Z.A.; Hailpern, S.M.; Burk, R.D. Persistent antibodies to HPV virus-like particles following natural infection are protective against subsequent cervicovaginal infection with related and unrelated HPV. *Viral Immunol.* **2009**, *22*, 445–449. [CrossRef] [PubMed]

30. Schiffman, M.; Wentzensen, N.; Wacholder, S.; Kinney, W.; Gage, J.C.; Castle, P.E. Human papillomavirus testing in the prevention of cervical cancer. *J. Natl. Cancer Inst.* **2011**, *103*, 368–383. [CrossRef] [PubMed]

31. Trottier, H.; Ferreira, S.; Thomann, P.; Costa, M.C.; Sobrinho, J.S.; Prado, J.C.; Rohan, T.E.; Villa, L.L.; Franco, E.L. Human papillomavirus infection and reinfection in adult women: The Role of Sexual Activity and Natural Immunity. *Cancer Res.* **2010**, *70*, 8569–8577. [CrossRef] [PubMed]

32. Viscidi, R.P.; Schiffman, M.; Hildesheim, A.; Herrero, R.; Castle, P.E.; Bratti, M.C.; Rodriguez, A.C.; Sherman, M.E.; Wang, S.; Clayman, B.; et al. Seroreactivity to human papillomavirus (HPV) types 16, 18, or 31 and risk of subsequent HPV infection: Results from a Population-based Study in Costa Rica. *Cancer Epidemiol. Biomark. Prev.* **2004**, *13*, 324–327. [CrossRef]

33. Beachler, D.C.; Jenkins, G.; Safaeian, M.; Kreimer, A.R.; Wentzensen, N. Natural Acquired Immunity Against Subsequent Genital Human Papillomavirus Infection: A Systematic Review and Meta-analysis. *J. Infect. Dis.* **2016**, *213*, 1444–1454. [CrossRef] [PubMed]

34. Velicer, C.; Zhu, X.; Vuocolo, S.; Liaw, K.L.; Saah, A. Prevalence and incidence of HPV genital infection in women. *Sex. Transm. Dis.* **2009**, *36*, 696–703. [CrossRef] [PubMed]

35. Xi, L.F.; Koutsky, L.A.; Castle, P.E.; Edelstein, Z.R.; Hulbert, A.; Schiffman, M.; Kiviat, N.B. Human papillomavirus type 16 variants in paired enrollment and follow-up cervical samples: Implications for a Proper Understanding of Type-specific Persistent Infections. *J. Infect. Dis.* **2010**, *202*, 1667–1670. [CrossRef] [PubMed]

36. Wheeler, C.M.; Skinner, S.R.; Del Rosario-Raymundo, M.R.; Garland, S.M.; Chatterjee, A.; Lazcano-Ponce, E.; Salmerón, J.; McNeil, S.; Stapleton, J.T.; Bouchard, C.; et al. Efficacy, safety, and immunogenicity of the human papillomavirus 16/18 AS04-adjuvanted vaccine in women older than 25 years: 7-year Follow-up of the Phase 3, Double-blind, Randomised Controlled Viviane Study. *Lancet Infect. Dis.* **2016**, *16*, 1154–1168. [CrossRef]

37. Castellsague, X.; Munoz, N.; Pitisuttithum, P.; Ferris, D.; Monsonego, J.; Ault, K.; Luna, J.; Myers, E.; Mallary, S.; Bautista, O.M.; et al. End-of-study safety, immunogenicity, and efficacy of quadrivalent HPV (types 6, 11, 16, 18) recombinant vaccine in adult women 24–45 years of age. *Br. J. Cancer* **2011**, *105*, 28–37. [CrossRef] [PubMed]

38. Hildesheim, A.; Herrero, R.; Wacholder, S.; Rodriguez, A.C.; Solomon, D.; Bratti, M.C.; Schiller, J.T.; Gonzalez, P.; Dubin, G.; Porras, C.; et al. Effect of human papillomavirus 16/18 L1 viruslike particle vaccine among young women with preexisting infection: A Randomized Trial. *JAMA* **2007**, *298*, 743–753. [CrossRef] [PubMed]

39. Scherer, E.M.; Smith, R.A.; Gallego, D.F.; Carter, J.J.; Wipf, G.C.; Hoyos, M.; Stern, M.; Thurston, T.; Trinklein, N.D.; Wald, A.; et al. A Single Human Papillomavirus Vaccine Dose Improves B Cell Memory in Previously Infected Subjects. *EBioMedicine* **2016**, *10*, 55–64. [CrossRef] [PubMed]

40. Polman, N.J.; Veldhuijzen, N.J.; Heideman, D.A.M.; Snijders, P.J.F.; Meijer, C.J.L.M.; Berkhof, J. HPV-positive women with normal cytology remain at increased risk of CIN3 after a negative repeat HPV test. *Br. J. Cancer* **2017**. [CrossRef] [PubMed]

41. Franceschi, S.; Baussano, I. Naturally acquired immunity against human papillomavirus (HPV): Why It Matters in the HPV Vaccine Era. *J. Infect. Dis.* **2014**, *210*, 507–509. [CrossRef] [PubMed]

42. Bosch, F.X.; Robles, C.; Diaz, M.; Arbyn, M.; Baussano, I.; Clavel, C.; Ronco, G.; Dillner, J.; Lehtinen, M.; Petry, K.U.; et al. HPV-FASTER: Broadening the Scope for Prevention of HPV-related Cancer. *Nat. Rev. Clin. Oncol.* **2016**, *13*, 119–132. [CrossRef] [PubMed]

43. Liu, S.H.; Brotman, R.M.; Zenilman, J.M.; Gravitt, P.E.; Cummings, D.A. Menstrual cycle and detectable human papillomavirus in reproductive-age women: A Time Series Study. *J. Infect. Dis.* **2013**, *208*, 1404–1415. [CrossRef] [PubMed]

44. Brotman, R.M.; Shardell, M.D.; Gajer, P.; Tracy, J.K.; Zenilman, J.M.; Ravel, J.; Gravitt, P.E. Interplay between the temporal dynamics of the vaginal microbiota and human papillomavirus detection. *J. Infect. Dis.* **2014**, *210*, 1723–1733. [CrossRef] [PubMed]

45. Johnston, C.; Corey, L. Current Concepts for Genital Herpes Simplex Virus Infection: Diagnostics and Pathogenesis of Genital Tract Shedding. *Clin. Microbiol. Rev.* **2016**, *29*, 149–161. [CrossRef] [PubMed]

46. Schiffer, J.T.; Abu-Raddad, L.; Mark, K.E.; Zhu, J.; Selke, S.; Koelle, D.M.; Wald, A.; Corey, L.; et al. Mucosal host immune response predicts the severity and duration of herpes simplex virus-2 genital tract shedding episodes. *Proc. Natl. Acad. Sci. USA* **2010**, *107*, 18973–18978. [CrossRef] [PubMed]

Review

Why Human Papillomavirus Acute Infections Matter

Samuel Alizon * , **Carmen Lía Murall** and **Ignacio G. Bravo**

MIVEGEC (UMR CNRS 5290, UR IRD 224, UM), 911 avenue Agropolis, 34394 Montpellier CEDEX 5, France; carmenlia.murall@outlook.com (C.L.M.); ignacio.bravo@ird.fr (I.G.B.)
* Correspondence: samuel.alizon@cnrs.fr; Tel.: +33-4-48-19-18-67

Academic Editors: Alison McBride and Karl Munger
Received: 9 July 2017; Accepted: 2 October 2017; Published: 10 October 2017

Abstract: Most infections by human papillomaviruses (HPVs) are 'acute', that is non-persistent. Yet, for HPVs, as for many other oncoviruses, there is a striking gap between our detailed understanding of chronic infections and our limited data on the early stages of infection. Here we argue that studying HPV acute infections is necessary and timely. Focusing on early interactions will help explain why certain infections are cleared while others become chronic or latent. From a molecular perspective, descriptions of immune effectors and pro-inflammatory pathways during the initial stages of infections have the potential to lead to novel treatments or to improved handling algorithms. From a dynamical perspective, adopting concepts from spatial ecology, such as meta-populations or meta-communities, can help explain why HPV acute infections sometimes last for years. Furthermore, cervical cancer screening and vaccines impose novel iatrogenic pressures on HPVs, implying that anticipating any viral evolutionary response remains essential. Finally, hints at the associations between HPV acute infections and fertility deserve further investigation given their high, worldwide prevalence. Overall, understanding asymptomatic and benign infections may be instrumental in reducing HPV virulence.

Keywords: clearance; persistence; latency; chronic; meta-population; fertility; virome; warts; cancer; evolution

1. Introduction

The most oncogenic viruses to humans are a group of around 20, closely related, human papillomavirus (HPV) types. All of them are classified in the *Alphapapillomavirus* genus and classically referred to as 'high risk types' [1]. HPV-induced cancers typically occur after several years of infection (Figure 1). The importance of viral persistence in the natural history of these cancers has driven most research to focus on chronic infections and to relatively neglect acute infections (sensu Virgin et al. [2]). For instance, when studying the duration of HPV infections in young women, infections that clear within two to three years tend to be referred to only indirectly, i.e., without a qualifying adjective [3]. Here we aim at clarifying what acute HPV infections are and to summarize the current understanding about them, in the context of the recent progress in the fight against HPVs. Finally, we identify the main gaps in our knowledge about such acute infections, which, if filled, would have direct implications for preventing, controlling and treating infections by HPVs.

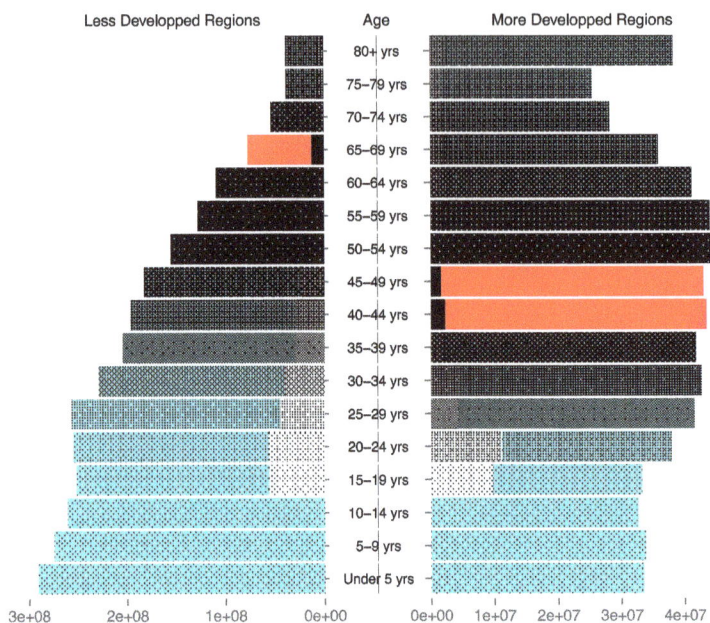

Figure 1. Number of women with asymptomatic cervical human papillomavirus (HPV) infections (black bars) as a function of age class (red bars) and of country development status. Bar color intensity reflects the prevalence of cervical cancer in the corresponding age class (and hence the focus of current research). Data should be read as follows, using the 30–34 years of age as an example: in less developed regions, among the 228 million women in the age class, 40 million (17.8%) display a normal cytology but are actually infected by HPVs, and 30 thousand (0.013%) suffer from cervical cancer, while in more developed regions among the 42 million women in the age class, 4.3 million (10.2%) display a normal cytology but are actually infected by HPVs, and six thousand (0.015%) suffer from cervical cancer. Data correspond to HPV prevalence in women with normal cytology, as estimated by the HPV information center (http://hpvcentre.net/, [4]). Overall HPV prevalence values will actually be larger, as they will include women with abnormal cytology. No data were available for HPV prevalence in women below 15 years of age. Demographic data correspond to the UN projections for 2015 (note the different scales between the less and more developed regions).

2. State of the Fight Against Human Papillomavirus

Infection-driven cancers are distinctive because they can be fought using the arsenal developed against infectious diseases: identification of risk factors, prevention of transmission and early detection of infected individuals. Identification of risk factors has led to the recognition of a few, closely related oncogenic HPVs as necessary etiologic agents of several cancers [5]. Contagion can now be prevented by the use of safe and effective vaccines targeting the most oncogenic HPVs along with certain non-oncogenic HPVs that cause anogenital warts [6,7]. Screening programs for early detection of (pre)neoplastic lesions caused by HPVs infections have also been successful in decreasing the burden of cervical cancer in rich countries [8]. However, their differential implementation has also increased the inequality between countries [9,10].

In spite of the primary and secondary prevention measurements available, HPVs will continue to infect millions of people in the foreseeable future, thereby causing significant morbidity and mortality worldwide [11]. Indeed, vaccine coverage varies widely both within and between countries [6], as does access to screening programs [12]. Beyond socio-economical factors, screening effectiveness is hampered by the fact that certain forms of cancer are more difficult to detect than others. This is the case for glandular forms of cervical cancer compared to the more common squamous carcinoma. They are often overlooked during standard screening procedures and their incidence is increasing [13,14]. Furthermore, cancers induced by HPVs in anatomical locations other than the cervix (e.g., anal [15] or oropharyngeal [16]) are on the rise in many countries, albeit in different populations [5]. These cancers are particularly worrying—either because they are detected once the carcinogenic process is more advanced, as is the case of head and neck cancer [16], or because they affect populations at increased risk, as is the case of anal cancer in HIV-infected men having sex with men [15]. Finally, from an economical perspective, non-carcinogenic HPVs should not be overlooked since the total health care cost linked to treating genital warts can exceed that of treating HPV-induced cancers [17], despite the obvious differences in severity and indirect impact of both diseases.

3. HPV Acute Infections

3.1. A Definition Challenge

HPV infections that are not chronic or latent have been referred to as 'acute' [18,19], 'non-persistent' [20], 'transient' [21], or 'cleared' [22] infections. Here, following Virgin et al. [2], we define acute infections as a non-equilibrium process that results either in infection clearance, host death or chronic infection.

As illustrated in Figure 2, clinical detection patterns may often lead to ambiguities. First, after sexual intercourse with an infected partner, viral genetic material may be detected for several days, even in the absence of an infection. We refer to these as 'transitory infections', although 'transitory detection' might be more accurate. Second, some infections successfully establish, replicate the viral genome, produce virions, and are eventually cleared (Figure 2). We refer to these as 'acute infections'. Third, some acute infections only appear to clear, but the viral genome remains in the infected cell without detectable activity [18,23]. We refer to these as 'latent infections'. The viral genetic material may occasionally be detected during latency. Reactivation of the viral activity in latent infections may occur much later, for instance triggered by immunesupression [24], but often also without any obvious reason. Finally, some acute infections are not cleared and maintain viral activity over time. We refer to these as 'chronic infections'. Clinically relevant chronic infections may still resolve naturally, sometimes in a matter of years [8,21]. Acute, latent and chronic infections most likely differ in terms of viral activity, e.g., viral and cellular gene expression patterns, effects on cell replication dynamics, or induced local immunosuppression [25]. Nevertheless, in the absence of a proper follow-up study design, characterizing the stage of an infection remains difficult, as the detection of viral genetic material associated to latent or to chronic infections [2] can bias our estimates about the prevalence of acute infections, as explained below.

One important reason for the blurry line between acute and chronic infections is that HPVs are very diverse in terms of genetics as well as in the clinical presentation of the infection. For instance, papillomaviruses colonize the skin and mucosa of virtually all humans from very early in life [26–28] and replicate at very low levels without any apparent clinical or cellular damage. In this respect, humans are continuously being infected (and reinfected) by HPVs, and most humans host a number of latent or chronic infections by HPVs. Certain viral genotypes cause proliferative infections with clinical manifestations such as warts or pre-neoplastic lesions. As we will see below, most of these proliferative lesions resolve naturally albeit in a matter of years, meaning that acute infections by HPVs can be of long duration. Overall, only a detailed understanding of the role of the immune response and clearance, of the nature and permissiveness for latency/chronicity of the cellular targets, and of the timing and repertoire of viral gene expression, will allow us to clarify the frontier between acute and chronic HPV infections.

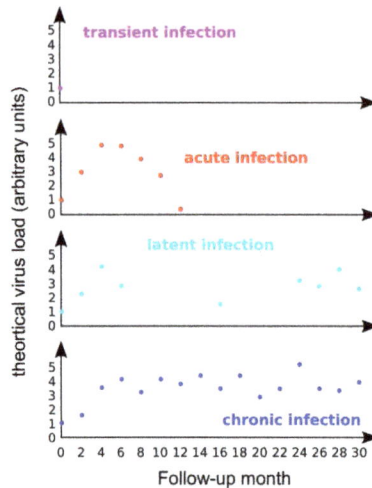

Figure 2. Kinetics of HPV virus load in different genital infections. The panels present made-up data with four imaginary patients monitored every two months from the onset of the infection (month 0). Currently most data describe viral load in precancerous lesions and in cancers [29], while very little data focus on how virus loads vary over the course of an acute infection.

3.2. Most Infections by HPVs Are Acute

For most viruses the acute, clinically prominent phases of viral infections are generally better understood than the chronic ones [2]. Research on oncogenic HPVs is an exception because of its focus on chronicity and on virus-induced cell transformation. Nevertheless, epidemiology data strongly suggest that the vast majority of anogenital infections by oncogenic HPVs never become chronic [8]. In women, the incidence of novel anogenital infections by oncogenic HPVs decreases with age, while persistence increases with age [30,31]. In men, this risk for novel infections is stable with age [32]. For infections in the female genital tract, the prevalence is U-shaped with age [4,33]. For both men and women, young adults exhibit the highest prevalence, which often rises above 25% (black bars in Figure 1). HPV16 is the most persistent type, but by 12 months 40% of infections clear or are treated because of diagnosis of a (pre)neoplastic lesion (Figure 3). This proportion reaches 85% by 36 months [3]. For HPV6, which is only very rarely associated with cancer [34] and instead most often with genital warts, these numbers are 66% and 98%, respectively. These figures should nevertheless be taken with caution as they directly depend on our ability to detect latent infections [18,23].

Besides studies on the natural history of the infection, the placebo arms of vaccine trials have been instrumental in increasing our understanding of the epidemiology of these infections by following thousands of young adults over time (e.g., [35–37]). In order to detect viral persistence of anogenital infections, the typical interval between two visits is usually six months. Studies with denser sampling (e.g., twice per week, as in [38]) often come at the expense of duration of the follow-up (16 weeks in this case), and therefore may provide accurate information about incidence but less information about persistence.

Some longitudinal studies performed more frequent sampling over a long period of time. For instance, two studies sampled women every three months. One of these followed unvaccinated girls aged 15–16 years in Tanzania and found that the median time from reported sexual debut to first HPV infection was five months, while median infection duration was six months [22]. Another study followed 150 adolescent women (aged 14 to 17) visiting primary care clinics for a median of six years [39] and found that 21% of HPV type-specific infections consisted of re-detection after apparent clearance (two negative visits). Note, however, that these girls were simultaneously infected by many HPVs, which could have generated some interference in the detection with the technique used. Furthermore, the population studied exhibited high rates of sexually transmitted infections (71.2% for chlamydia and 49.3% for gonorrhea) and a mean of above 10 sexual partners during the study. In spite of their limitations, these two studies suggest that the little we know about the duration and the actual clearance of acute infections could be challenged by studies with deeper resolution.

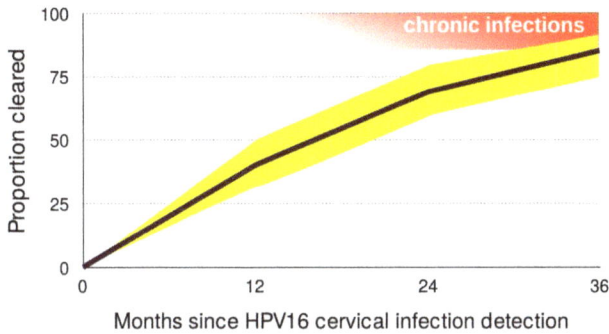

Figure 3. Proportion of HPV16 infections cleared 12, 24 and 36 months after the first detection. 95% confidence intervals are shown in yellow. Acute infections result in clearance or chronic infections, shown in red (note that chronic infections may clear as well). Data on clearance proportions originate from Table 4 in [3] and were obtained for 895 women in the United States of America, aged 16 to 23. Notice that chronic infections (in red) can also clear.

3.3. The Immunology of Clearance

Why most infections clear, while only few persist and progress to cancer remains largely unresolved (see Box 1 at the end of this article). Overall, we know that some cofactors such as HPV genotype, host genetic background, age of sexual debut, or coinfections, have an effect on clearance time [40]. However, our mechanistic understanding is still lagging behind.

The immunology of HPV infections has been thoroughly reviewed elsewhere (e.g., [41]). We know that HPV infections concur with a local anti-inflammatory environment, and that although the adaptive immune response is very efficient at clearing the infection, its activation is variable and sometimes insufficient to prevent future re-infections [36,41,42].

Recent research into the role of innate effectors (e.g., natural killer T cells) and Th-17 responses (e.g., $\delta\gamma$ T cells) in cancerous lesions has provided new insights into chronic infections [43], but the implications for acute infection clearance is not obvious. More work, then, into innate and adaptive immunity activities during acute infections would help better elucidate mechanisms of clearance, as they are likely to be several. This is particularly true for anatomical locations outside the cervix, where the specific immunity microenvironment is less understood.

4. Open Challenges

4.1. Deciphering HPV Kinetics: The Meta-Population Hypothesis

The extant genetic diversity of HPVs is enormous, with many different viral life styles. For anogenital infections alone, there is a large variance in infection duration, with values ranging from a few months [22] to years [3]. We propose here that the meta-population framework used in ecology can provide strong explanatory power and insight into this variance.

In ecology, a meta-population is a set of populations of the same species that are connected through dispersal [44,45]. Each population displays its own dynamics but the processes and patterns that emerge at a higher (meta-population) level can be very different from what happens within each population. This ecological framework has relevance for host-pathogen interactions. For instance, genetic data suggest that HIV infections may exhibit a meta-population structure in the spleen [46]. Also, within-host viral genetic structure such as this has been recently put forward as an underlying explanation for the great variance in set-point viral load observed between patients [47]. A meta-population framework was first proposed and applied to HPVs in a study of multiple type infections under natural and vaccinated immunological scenarios [48]. Here, we broaden this as a way to understand the variation in dynamics of HPV infections.

The modular nature of mammalian skin and the individual proliferation of cells set the scene for a meta-population scenario, one where discrete viral populations can establish in various 'patches' within the same anatomical 'site' and/or in various sites of the same host. We further argue that this framework can help explain the thin line between acute and chronic HPV infections.

Viral colonization of an individual patch is largely a rare event, provided that the common assumption of access to basal cells through microlesions holds true. The ability of the virus to access other patches increases with infection duration, virion productivity and weakening of epithelium integrity. Following Ryser et al. [49], we argue that local extinction/success are stochastic processes. Some initially infected patches may not successfully infect a new patch before going extinct, while others might. To further complicate the picture, heterogeneity of cell types and skin structures leads to patch differences in susceptibility to infection, in permissivity to virion production, or in intensity of immune surveillance. Thus, certain patches will act as reservoirs (known as 'source patches' in ecology) while others will act as 'sink patches'. In such source–sink dynamics, infection duration is driven by the extinction rate in the reservoir patches but the virus load detected depends on the number and productivity of the sink patches infected as well [47].

Regarding infection duration, the meta-population hypothesis allows the differentiation between genuine and apparent chronic infections. Genuinely chronic infections would refer to sustained and characteristic viral activity in a given patch for a long period of time, therefore allowing for the accumulation of molecular damage that may lead to malignization [50,51]. Apparent chronic infections would be sequential transmission chains of infections from one patch to another within the same anatomical location (Figure 4). In this case, each individual patch infection may eventually die out, so that there is no sustained virus–cell interaction, even if the patient remains positive for the infection by the same viral type during a long time. Obviously, the repetitive infection events in these apparent chronic infections would increase the chances that one of them becomes genuinely chronic, if the virus-cell-environment interactions allow for it.

The meta-population framework, thus, may provide a powerful ecological explanation for the poorly understood differential progression of certain chronic infections towards cancer. Also, viral and cellular genomes will not accumulate similar genetic and/or epigenetic modifications during genuine and apparent chronic infections, thus allowing for molecular tests to differentiate between them. Testing the meta-population hypothesis will require HPV research to engage in spatial ecology studies, sampling between various macro/microscopic sites within the same individual (e.g., [52–55]), to study 'auto-inoculation' [56], and to systematically resort to microdissection studies to investigate the details of infected patches within sites [34,57–59].

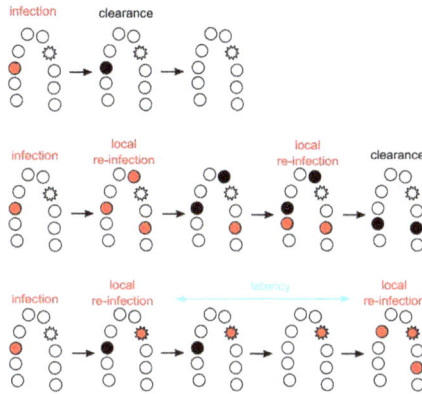

Figure 4. The meta-population hypothesis for HPV kinetics, exemplified in the cervix. Each circle represents a population (i.e., a cellular patch or a site) that can be susceptible (white), infected (red), or recovered from infection (black). The star-shape indicates a cellular patch that can remain infected for a longer time than the regular sites and can act as a source site if infected. For simplicity, recovered sites are here assumed to transitorily exhibit higher local immunity and resource depletion, thus making it unable to allow for re-infection. The top line shows a case with a local acute infection that is cleared without reaching other patches. The middle one shows the meta-population dynamics of local re-infection from an initial patch, without cell heterogeneity. The bottom line shows a case with patch heterogeneity (source-sink dynamics).

4.2. Cellular, Viral and Environmental Heterogeneity: Towards Meta-Communities

The mammalian skin is a complex environment with an intricate tri-dimensional structure. The most distinctive feature is the presence of hairs and the expression of keratins, keratin-associated proteins, and the epidermal differentiation complex [60–62], as well as other hair evolutionary-related structures such as the mammary, sebaceous, and sweat glands [63]. Different sites of the body have specific cell types, with unique composition, arrangement and density of skin appendages. Particular anatomical features are often observed at transitions between specialized epithelia, as in the tonsillar crypta, the transition zone in the cervix, the perianal region or the periungual area. Stratified skin renewal dynamics are strongly regulated and proceed vertically, with a gradient from the dividing cells in the basal layer to the differentiating and cornified keratinocytes in the outermost layer. The skin structure is, thus, conceivably modular, with columnar functional 'patches', which we will define here as the continuum of locations in the epithelium. The 'vertical' nature of the proliferating/desquamating keratinocyte patches is complemented by the 'horizontal' immune surveillance by Langerhans cells, dendritic cells and macrophages [64,65].

This heterogeneity goes beyond the strong assumption of the meta-population concept, which is that it only applies to populations, that is individuals from the same species. To account for different target cell types and for more 'trophic interactions' (here immune cells), we need to invoke the meta-community concept [66]. This has already proven to be appropriate to analyze within-host dynamics, because various parasite species often co-infect the same host and thus compete for space, energy, and resources [67,68].

A first advantage of a meta-community approach for HPVs is that it can factor in different target cell types. Papillomaviruses have been infecting amniotes since their first appearance [69] and have evolved strategies to successfully thrive across the diversity of structures and specializations of present-day mammalian skin [70]. For instance, cells in hair follicles seem to act as reservoirs for chronic infection by cutaneous HPVs, while in stratified mucosal epithelia viral infections can become chronic in the basal cell layer. The columnar nature of the skin influences the natural history of the proliferative lesions induced by HPVs, which derive from the clonal expansion of basal/parabasal cells [70,71].

The heterogeneous nature of the skin underlies also the different propensity for chronic viral infections to lead to cancer in different anatomical locations. Many squamous cervical carcinomas associated with HPVs are thought to arise from a discrete population of cuboidal epithelial cells located in the transformation zone between the endo- and the ectocervix [72]. The absence of such transformation zones in the vagina, vulva or penis might explain why the cervix has a much higher burden of infection, higher incidence and younger age-at-diagnosis for cancers compared to other anatomical locations [73]. We can only speculate on why HPV-related cancers also occur in absence of a transformation zone, but a likely explanation is that while all infected epithelial stem cells can potentially end up being carcinogenic, the probability of such event depends on the cell phenotype and the local environment. Finally, heterogeneity between squamous and glandular patches results also in differential susceptibility towards chronification and malignisation by different HPVs and even by closely related viral variants, possibly also modulated by the host genetic background [74]. These known effects of cellular heterogeneity on chronic infections and cancer are likely to also affect acute infections, but our knowledge about the latter is very limited.

More generally, the underlying hypothesis in our meta-population and meta-community approaches is that acute, latent and chronic infections exhibit differential specific features at the virocellular level, which further vary as a function of the infected cell type and differentiation status. Markers of acute versus true persisting infections will most likely be different for different anatomical sites and skin structures. The overall dynamics of the virus–host interaction is thus an integration of the interactions between viral genotype, host genotype, cellular phenotype and environment.

4.3. HPV Latency

Papillomavirus latent infections are still very poorly characterized and understood. The prevalence of latency remains largely unexplored, and its contribution to the oncogenic process in chronic infections is obscure. Establishment of latent infections is a life-history trait shared by very divergent viruses. For large DNA viruses, such as herpesviruses, the infection often starts with an acute phase, and then viral gene expression changes towards a different profile where the virus enters a latent state, with limited or even no genome replication, and no cellular damage, until reactivation is triggered. In small DNA viruses, such as Torque-Teno viruses, the initial infection goes unnoticed (the acute infection stage has actually never been documented) and the viral genomes remain in the host and replicate chronically at very low levels without apparent damage [2]. For papillomaviruses, most of the evidence on latency originates from animal models, where latency is defined as low-level viral genome maintenance in the basal layers without a productive viral life cycle [24,75].

For HPVs, two ways for latency to arise have been proposed: infections may directly enter latency without going through an acute phase, or it may instead arise after a productive phase without successful clearance [23]. In both cases, latency is directly linked to acute infection dynamics. In the meta-communities context, latency can additionally be conceived to arise from heterogeneity in the interactions between virus and host cells. The molecular decision for cell division in the basal layer is stochastic and given that the viral genome requires cell division to replicate [76], latency-reactivation episodes could be mechanistically understood to reflect stochasticity in the time lapses of basal cell mitotic activity, without necessarily requiring viral manipulation of the host cell. Again, anatomical heterogeneity and viral genetic diversity may render certain anatomical locations combined with particular viral linages more prone to such latency-reactivation cycles.

Overall, in addition to molecular virology investigations into latency mechanisms, studies combining a detailed initial follow-up (to demonstrate clearance), a long follow-up (to demonstrate chronic infection) and sequencing (to demonstrate that the virus causing the acute infection actually persisted) will help generate a complete picture on papillomavirus infection latency. For instance, a recent analysis of HPV16 complete genomes in 57 persistent infections revealed one case of reinfection by the same HPV type, which could have been mistaken for a persitent infection [77].

4.4. Immunotherapies

Since vaccination will not be widespread for many years [78] and given that screening interventions are often expensive and difficult to implement, the development of treatments remains urgent. Treatment development remains historically the less successful front for HPV research. Understanding the mechanisms of HPV clearance promises aiding the development of immunotherapies, which consist in treating a disease by stimulating or suppressing the immune system.

Currently, the bulk of clinical and animal model research into HPV immunotherapies is understandably focused on cervical cancer treatment. Indeed, there are numerous approaches to developing these treatments, such as protein/peptide vaccines, bacteria-and-viral based vectors, and immunomodulators [79]. One of the most promising examples to date is the therapeutic vaccine VGX-3100, a plasmid containing synthetic versions of *E6* and *E7* genes of HPV16 and HPV18, which showed efficiency in a controlled trial at improving the regression of high-grade lesions (CIN2/3) [80]. The iatrogenic exposure to viral oncoproteins triggers a strong cellular immune response against the infected cells that the natural infection is not able to initiate. An alternative approach is to reverse the anti-inflammatory microenvironment that the virus creates during infections, in order to alert the immune system and boost its functioning (reviewed in [81]). While several of these therapies have reached clinical trial stages, it has become clear that one mechanism alone (e.g., augmenting CTL infiltration and function) will not suffice, and combinations of mechanisms are needed. Studying how these mechanisms work in natural acute infections might help.

In practice, immunotherapies cannot be envisaged to treat all acute HPVs infections given their prevalence and their often sub-clinical presentations. However, knowledge on the acute stage could be transferable to fight chronic stage infections, for instance, by identifying any immune stimulants or immune cell subtypes that help clear natural acute infections. However, it may be possible that several treatments will be needed for different lesion grades (e.g., for low to high-grade neoplastic lesions than for late-stage cancers), given that the heightened immune suppression microenvironment in advanced malignancies is particularly complex and strong. Therefore, insights from acute infection clearance could be particularly well suited for development of therapies against premalignant lesions that are usually removed surgically.

More focused applications could arise from studying HPV infections in sites other than the cervix, for instance in the case of respiratory recurrent papillomatosis. This is a rare condition caused by chronic infection by HPV6 or HPV11 that shares features of both acute and chronic infections. The chronic disease imposes a recurrent burden to the patients to control the acute, benign clinical presentations of the disease, which can progress fatally to the lungs [82]. Furthermore, given the unique microenvironments of non-cervical sites, studies of infection dynamics at these sites are greatly needed. In particular, anal intraepithelial neoplasias and anal squamous cell carcinomas are increasing in prevalence worldwide yet studies of HPV-immunity interactions in anal infections are few [83].

4.5. Fertility

The Zika epidemic has reminded us of the risk viral infections have on pregnancies [84]. Anogenital infections by HPVs deserve to be studied in this context because rare deleterious effects could translate into an important burden, given their high prevalence. There is a well-established link between cervical disease and pregnancy complications [85], but we are only beginning to understand the connection between clinically asymptomatic HPVs infections and complications such as pre-eclampsia, fetal growth restriction or pre-term delivery [86].

In men, anogenital asymptomatic infections by mucosal and cutaneous HPVs are very common [37] and some are associated with penile cancer [5]. Surprisingly, HPV DNA can also be found in human semen [87] and although this viral DNA could originate from desquamating epithelial cells, data suggest the presence of viral DNA directly associated to sperm cells [29]. Evidence supporting a correlation between infection by HPVs in men and infertility is still controversial [88] and the mechanisms involved remain largely unknown [89].

Finally, in utero transmission of HPVs is rare [90], but asymptomatic cervical viral infections can reduce barrier integrity [91] and viral DNA can be detected in the placentae [92]. Further, there is a sound body of evidence describing transmission of HPVs during vaginal birth, as illustrated by the case of infantile respiratory recurrent papillomatosis described above [82].

Overall, the presence of mucosal HPVs in placenta and semen suggests non-classical tropisms, and thus, viral life cycle in these particular cellular environments needs to be elucidated. Clinical studies investigating acute anogenital infections in pregnant women (along the various stages of pregnancy) as well as in their partners are required to unravel how HPVs (independently of their oncogenic potential) may be either directly or indirectly increasing the risk of infertility, spontaneous abortions, pre-term labor, pre-eclampsia or other complications.

4.6. Vaccine Escape

Effective vaccines with high coverage rates exert major selective pressures on pathogens [93]. The risk of an evolutionary response from the pathogen relies on its genetic diversity, and does not necessarily require de novo mutations if it is already diverse. Extant diversity in papillomaviruses is immense, with hundreds of viral lineages, which themselves harbor significant genetic variation [94].

While vaccines against HPVs are effective and safe, they may be leaving the door open for an evolutionary response [48]. This can occur if some vaccinated individuals are infected by HPV vaccine types (due to infection before vaccination, immunosuppression or failure to mount a sufficiently strong immune response). In the clinical trial for the nonavalent vaccine, even in the per-protocol efficacy population, where the $n = 5812$ participants received all three doses of vaccine within one year and were HPV-uninfected at inclusion, the median number of infections by one of the targeted viruses was 3.6 for 1000 person-years at risk (data from Supplementary Table S5 in [95]). In the intention-to-treat population, which did not exclude 988 additional participants who received at least one vaccine dose, this number increased to 36. Despite the outstanding vaccine efficacy, when millions of women are vaccinated, this will still represent thousands of infections by vaccine-targeted viral genotypes that would be eventually cleared and not result in malignant disease. These acute infections would thus occur in vaccinated women, with a strong immune response, and the question arises whether they may last long enough to allow specific viral lineages (either pre-existing or de novo generated viral variants) to be differentially transmitted, thus paving the way for viral adaptation to this special environment.

HPV vaccines can generate off-target immune activity and offer protection against closely related viruses not targeted by the vaccine formulation [96,97]. Yet viral diversity remains so large that in the few regions where vaccination has reduced the prevalence of vaccine-targeted types, most other HPVs continue to circulate [6,98]. Overall, vaccination against HPVs will undoubtedly have a strong and highly desirable impact in disease prevention, but it will create novel host environments to which viruses may adapt.

Thanks to next generation sequencing, we now have access to an increasing number of full genomes. For instance, very recently, an analysis of more than 5500 HPV16 genomes identified shared feature between viruses isolated from precancers and cancers compared to viruses from case-control samples [99]. However, evaluating the potential for HPV evolution will require genomic data of intra- and inter-individual viral populations collected through time. Since HPVs are double-stranded DNA viruses replicated by host polymerases, mutation rates are expected to be low. Nevertheless, we know little about polymerase fidelity in somatic cells, both in general and during a viral infections. Investigation of long follow-up data sets with short sampling intervals are lacking, especially since evolutionary rates may exhibit periods of rapid increase followed by long periods of stasis, as demonstrated for Influenza A virus [100]. Further, in the case of HPVs, displaying acute, chronic and latent stages of the infection, the evolutionary dynamics may strongly depend on the presence of a latent phase and on the epidemiological transmission patterns, as described for Varicella-zoster virus [101]. For HPVs, focusing on the acute infections could therefore yield novel insights since this is where the viral life cycle is most productive.

4.7. Scars That Matter Long after Clearance?

Even when viral clearance does occur, recent work shows that acute infections can impair the immune system causing chronic inflammation or 'immunological scaring' [102]. Certain oncoviruses are believed to leave behind molecular damage in their host cells, which can lead to cancer several years later [103]. Cervical cancers are a clear example of direct carcinogenesis, since chronic infection by oncogenic HPVs is a necessary cause for virtually all of these cancers. However, even acute infections can induce modifications in the cellular (epi)genome, creating the stage for pre-cancerous lesions [103,104]. Although an old hypothesis, the 'hit-and-run' effects of acute infections are poorly understood for bacterial or viral infections, and exploring how HPVs may cause this kind of damage remains an important research direction [105].

Speculatively, such a long-term impact could concern not only anogenital but also infections at cutaneous sites. Indeed, virtually all humans become infected by very diverse cutaneous HPVs, chiefly beta- and gammapapillomaviruses, which are proposed to act as cofactors for the risk of developing non-melanoma skin cancers in certain human populations [106]. Given that such HPVs infections are extremely prevalent (Figure 5), often causing subclinical infections and interact extensively with the immune system, they could be an ideal model for studying 'under-the-radar' viral infections and their potential side-effects on immune functioning.

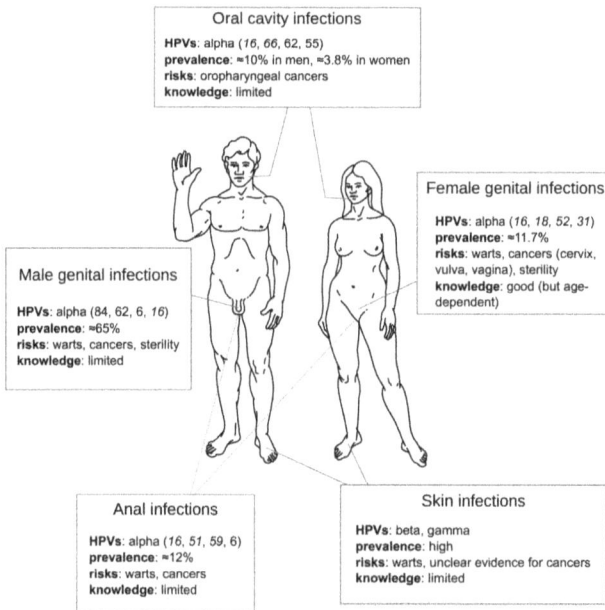

Figure 5. Challenges of HPV acute infections per anatomical location. For each location, HPVs are ordered per prevalence and oncogenic ones are in italic. Most prevalences are global estimates [1,4,107], except for oral (US [108]), genital in men (Brazil, Mexico, US [109]) and anal (Brazil, Mexico, US [110]) sites. Detailed data are available in Supplementary Table S1.

5. Conclusions

Acute infections by HPVs are the rule rather than the exception, and a better understanding of such infections is urgent because of their enormous fundamental and public health implications. We propose that metapopulation and metacommunity approaches borrowed from ecology can provide a strong explanatory framework to the study of the course of HPV infections and better distinguish between acute, latent and chronic infections. In addition to the promise of identifying early markers of infection chronicity, understanding viral-immunity interactions can help design new treatments to boost natural immunity against novel infections. Such therapies may reduce the chances of an acute infection to chronify and to reach (pre)cancer stages. Detailed studies with long follow-up and short time intervals are further needed to precisely assess the role acute HPV infections could have on fertility or on the long-term 'immune scars' these infections may leave behind.

Finally, as with most infectious diseases, it is important to remain humble and to accept that elimination [111] of virulent HPVs is unlikely. In fact, it is not clear whether such selective removal should be desirable, as most HPVs are part of our virome and might be playing important ecological roles. This is why the development of vaccines should not stop us from improving our characterization of the natural history of HPVs so that we can effectively shift from an elimination to a control perspective.

Box 1. Open Questions about HPV Acute Infections

- Do some innate immunity evasion mechanisms lead to delayed clearance? [43]
- How common is immunological tolerance to HPVs?
- What role, if any, do natural B-cell responses play in clearance?
- Are there differences in immunological clearance mechanisms for different HPVs or for different infected cell types, tissue structures or anatomical sites?
- Are there long term effects after HPV infection clearance (immunological scaring)?
- How do HPV infections affect male and female fertility?
- How do HPV infection kinetics affect transmission? [112]
- Are HPVs infections structured in host tissues (the meta-population hypothesis)?
- What is the role of the microbiota in HPV infection clearance, persistence and progression to cancer? [113,114]

Supplementary Materials: The following are available online at www.mdpi.com/1999-4915/9/10/293/s1, Table S1: HPV epidemiological data for Figure 5.

Acknowledgments: Samuel Alizon is supported by the European Research Council (ERC) under the European Union's Horizon 2020 research and innovation program (EVOLPROOF, grant agreement No 648963). Ignacio G. Bravo is supported by the European Research Council (ERC) under the European Union's Horizon 2020 research and innovation program (CODOVIREVOL, grant agreement No 647916). All authors acknowledge further support from the CNRS and the IRD.

Author Contributions: All authors conceived and wrote the review.

Conflicts of Interest: The authors declare no conflict of interest. The funding sponsors had no role in the writing of the manuscript, and in the decision to publish it.

Abbreviations

The following abbreviations are used in this manuscript:

CIN	Cervical Intraepithelial Neoplasia
CTL	Cytotoxic T Lymphocytes
HPVs	Human Papillomaviruses
UN	United Nations

References

1. *IARC Monographs on the Evaluation of Carcinogenic Risks to Human*, v. 100B; International Agency for Research on Cancer: Lyon, France, 2012.
2. Virgin, H.W.; Wherry, E.J.; Ahmed, R. Redefining chronic viral infection. *Cell* **2009**, *138*, 30–50.
3. Insinga, R.P.; Dasbach, E.J.; Elbasha, E.H.; Liaw, K.L.; Barr, E. Incidence and duration of cervical human papillomavirus 6, 11, 16, and 18 infections in young women: An evaluation from multiple analytic perspectives. *Cancer Epidemiol. Biomark. Prev.* **2007**, *16*, 709–715.
4. Bruni, L.; Diaz, M.; Castellsagué, X.; Ferrer, E.; Bosch, F.X.; de Sanjosé, S. Cervical human papillomavirus prevalence in 5 continents: Meta-analysis of 1 million women with normal cytological findings. *J. Infect. Dis.* **2010**, *202*, 1789–1799.
5. Plummer, M.; de Martel, C.; Vignat, J.; Ferlay, J.; Bray, F.; Franceschi, S. Global burden of cancers attributable to infections in 2012: A synthetic analysis. *Lancet Glob. Health* **2016**, *4*, e609–e616.
6. Drolet, M.; Bénard, É.; Boily, M.C.; Ali, H.; Baandrup, L.; Bauer, H.; Beddows, S.; Brisson, J.; Brotherton, J.M.L.; Cummings, T.; et al. Population-level impact and herd effects following human papillomavirus vaccination programmes: a systematic review and meta-analysis. *Lancet Infect. Dis.* **2015**, *15*, 565–580.
7. Herrero, R.; González, P.; Markowitz, L.E. Present status of human papillomavirus vaccine development and implementation. *Lancet Oncol.* **2015**, *16*, e206–e216.
8. Schiffman, M.; Solomon, D. Cervical-cancer screening with human papillomavirus and cytologic cotesting. *N. Engl. J. Med.* **2013**, *369*, 2324–2331.
9. Gakidou, E.; Nordhagen, S.; Obermeyer, Z. Coverage of Cervical Cancer Screening in 57 Countries: Low Average Levels and Large Inequalities. *PLoS Med.* **2008**, *5*, e132.
10. Palència, L.; Espelt, A.; Rodríguez-Sanz, M.; Puigpinós, R.; Pons-Vigués, M.; Pasarín, M.I.; Spadea, T.; Kunst, A.E.; Borrell, C. Socio-economic inequalities in breast and cervical cancer screening practices in Europe: Influence of the type of screening program. *Int. J. Epidemiol.* **2010**, *39*, 757–765.
11. Bruni, L.; Diaz, M.; Barrionuevo-Rosas, L.; Herrero, R.; Bray, F.; Bosch, F.X.; de Sanjosé, S.; Castellsagué, X. Global estimates of human papillomavirus vaccination coverage by region and income level: A pooled analysis. *Lancet Glob. Health* **2016**, *4*, e453–e463.
12. Ronco, G.; Dillner, J.; Elfström, K.M.; Tunesi, S.; Snijders, P.J.F.; Arbyn, M.; Kitchener, H.; Segnan, N.; Gilham, C.; Giorgi-Rossi, P.; et al. Efficacy of HPV-based screening for prevention of invasive cervical cancer: Follow-up of four European randomised controlled trials. *Lancet* **2014**, *383*, 524–532.
13. Pettersson, B.F.; Hellman, K.; Vaziri, R.; Andersson, S.; Hellström, A.C. Cervical cancer in the screening era: Who fell victim in spite of successful screening programs? *J. Gynecol. Oncol.* **2011**, *22*, 76–82.
14. Castanon, A.; Landy, R.; Sasieni, P.D. Is cervical screening preventing adenocarcinoma and adenosquamous carcinoma of the cervix? *Int. J. Cancer* **2016**, *139*, 1040–1045.
15. Bertisch, B.; Franceschi, S.; Lise, M.; Vernazza, P.; Keiser, O.; Schöni-Affolter, F.; Bouchardy, C.; Dehler, S.; Levi, F.; Jundt, G.; et al. Risk Factors for Anal Cancer in Persons Infected with HIV: A Nested Case-Control Study in the Swiss HIV Cohort Study. *Am. J. Epidemiol.* **2013**, *178*, 877–884.
16. Gillison, M.L.; Chaturvedi, A.K.; Anderson, W.F.; Fakhry, C. Epidemiology of Human Papillomavirus-Positive Head and Neck Squamous Cell Carcinoma. *J. Clin. Oncol.* **2015**, *33*, 3235–3242.
17. Herse, F.; Reissell, E. The annual costs associated with human papillomavirus types 6, 11, 16, and 18 infections in Finland. *Scand. J. Infect. Dis.* **2011**, *43*, 209–215.
18. Gravitt, P.E. The known unknowns of HPV natural history. *J. Clin. Investig.* **2011**, *121*, 4593–4599.
19. Schiffman, M.; Wentzensen, N.; Wacholder, S.; Kinney, W.; Gage, J.C.; Castle, P.E. Human Papillomavirus Testing in the Prevention of Cervical Cancer. *J. Natl. Cancer Inst.* **2011**, *103*, 368–383.
20. Hazard, K.; Karlsson, A.; Andersson, K.; Ekberg, H.; Dillner, J.; Forslund, O. Cutaneous Human Papillomaviruses Persist on Healthy Skin. *J. Investig. Dermatol.* **2007**, *127*, 116–119.
21. Depuydt, C.E.; Criel, A.M.; Benoy, I.H.; Arbyn, M.; Vereecken, A.J.; Bogers, J.J. Changes in type-specific human papillomavirus load predict progression to cervical cancer. *J. Cell. Mol. Med.* **2012**, *16*, 3096–3104.
22. Houlihan, C.F.; Baisley, K.; Bravo, I.G.; Kapiga, S.; de Sanjosé, S.; Changalucha, J.; Ross, D.A.; Hayes, R.J.; Watson-Jones, D. Rapid acquisition of HPV around the time of sexual debut in adolescent girls in Tanzania. *Int. J. Epidemiol.* **2016**, *45*, 762–773.
23. Doorbar, J. Latent papillomavirus infections and their regulation. *Curr. Opin. Virol.* **2013**, *3*, 416–421.

24. Maglennon, G.A.; McIntosh, P.B.; Doorbar, J. Immunosuppression Facilitates the Reactivation of Latent Papillomavirus Infections. *J. Virol.* **2014**, *88*, 710–716.
25. Doorbar, J.; Quint, W.; Banks, L.; Bravo, I.G.; Stoler, M.; Broker, T.R.; Stanley, M.A. The biology and life-cycle of human papillomaviruses. *Vaccine* **2012**, *30* (Suppl. 5), F55–F70.
26. Antonsson, A.; Karanfilovska, S.; Lindqvist, P.G.; Hansson, B.G. General Acquisition of Human Papillomavirus Infections of Skin Occurs in Early Infancy. *J. Clin. Microbiol.* **2003**, *41*, 2509–2514.
27. Rintala, M.A.M.; Grénman, S.E.; Puranen, M.H.; Isolauri, E.; Ekblad, U.; Kero, P.O.; Syrjänen, S.M. Transmission of High-Risk Human Papillomavirus (HPV) between Parents and Infant: A Prospective Study of HPV in Families in Finland. *J. Clin. Microbiol.* **2005**, *43*, 376–381.
28. Michael, K.M.; Waterboer, T.; Sehr, P.; Rother, A.; Reidel, U.; Boeing, H.; Bravo, I.G.; Schlehofer, J.; Gärtner, B.C.; Pawlita, M. Seroprevalence of 34 Human Papillomavirus Types in the German General Population. *PLoS Pathog.* **2008**, *4*, e1000091.
29. Depuydt, C.E.; Jonckheere, J.; Berth, M.; Salembier, G.M.; Vereecken, A.J.; Bogers, J.J. Serial type-specific human papillomavirus (HPV) load measurement allows differentiation between regressing cervical lesions and serial virion productive transient infections. *Cancer Med.* **2015**, *4*, 1294–1302.
30. Muñoz, N.; Méndez, F.; Posso, H.; Molano, M.; van den Brule, A.J.; Ronderos, M.; Meijer, C.; Muñoz, A. Incidence, Duration, and Determinants of Cervical Human Papillomavirus Infection in a Cohort of Colombian Women with Normal Cytological Results. *J. Infect. Dis.* **2004**, *190*, 2077–2087.
31. Castle, P.E.; Schiffman, M.; Herrero, R.; Hildesheim, A.; Rodriguez, A.C.; Bratti, M.C.; Sherman, M.E.; Wacholder, S.; Tarone, R.; Burk, R.D. A Prospective Study of Age Trends in Cervical Human Papillomavirus Acquisition and Persistence in Guanacaste, Costa Rica. *J. Infect. Dis.* **2005**, *191*, 1808–1816.
32. Giuliano, A.R.; Lee, J.H.; Fulp, W.; Villa, L.L.; Lazcano, E.; Papenfuss, M.R.; Abrahamsen, M.; Salmeron, J.; Anic, G.M.; Rollison, D.E.; et al. Incidence and clearance of genital human papillomavirus infection in men (HIM): A cohort study. *Lancet* **2011**, *377*, 932–940.
33. Sanjosé, S.D.; Diaz, M.; Castellsagué, X.; Clifford, G.; Bruni, L.; Muñoz, N.; Bosch, F.X. Worldwide prevalence and genotype distribution of cervical human papillomavirus DNA in women with normal cytology: a meta-analysis. *Lancet Infect. Dis.* **2007**, *7*, 453–459.
34. Guimerà, N.; Lloveras, B.; Lindeman, J.; Alemany, L.; Sandt, M.V.D.; Alejo, M.; Hernandez-suarez, G.; Bravo, I.G.; Molijn, A.; Jenkins, D.; et al. The Occasional Role of Low-risk Human Papillomaviruses 6, 11, 42, 44, and 70 in Anogenital Carcinoma Defined by Laser Capture Microdissection/PCR Methodology: Results From a Global Study. *Am. J. Surg. Pathol.* **2013**, *37*, 1299–1310.
35. Herrero, R.; Wacholder, S.; Rodríguez, A.C.; Solomon, D.; González, P.; Kreimer, A.R.; Porras, C.; Schussler, J.; Jiménez, S.; Sherman, M.E.; et al. Prevention of persistent human papillomavirus infection by an HPV16/18 vaccine: A community-based randomized clinical trial in Guanacaste, Costa Rica. *Cancer Discov.* **2011**, *1*, 408–419.
36. Castellsagué, X.; Naud, P.; Chow, S.N.; Wheeler, C.M.; Germar, M.J.V.; Lehtinen, M.; Paavonen, J.; Jaisamrarn, U.; Garland, S.M.; Salmerón, J.; et al. Risk of newly detected infections and cervical abnormalities in women seropositive for naturally acquired human papillomavirus type 16/18 antibodies: Analysis of the control arm of PATRICIA. *J. Infect. Dis.* **2014**, *210*, 517–534.
37. Moreira, E.D.; Giuliano, A.R.; Palefsky, J.; Flores, C.A.; Goldstone, S.; Ferris, D.; Hillman, R.J.; Moi, H.; Stoler, M.H.; Marshall, B.; et al. Incidence, Clearance, and Disease Progression of Genital Human Papillomavirus Infection in Heterosexual Men. *J. Infect. Dis.* **2014**, *210*, 192–199.
38. Liu, S.H.; Cummings, D.A.T.; Zenilman, J.M.; Gravitt, P.E.; Brotman, R.M. Characterizing the Temporal Dynamics of Human Papillomavirus DNA Detectability Using Short-Interval Sampling. *Cancer Epidemiol. Biomark. Prev.* **2014**, *23*, 200–208.
39. Shew, M.L.; Ermel, A.C.; Tong, Y.; Tu, W.; Qadadri, B.; Brown, D.R. Episodic detection of human papillomavirus within a longitudinal cohort of young women. *J. Med. Virol.* **2015**, *87*, 2122–2129.
40. Ferenczy, A.; Franco, E. Persistent human papillomavirus infection and cervical neoplasia. *Lancet Oncol.* **2002**, *3*, 11–16.
41. Stanley, M. Immunology of HPV Infection. *Curr. Obstet. Gynecol. Rep.* **2015**, *4*, 195–200.
42. Stanley, M. Immune responses to human papillomavirus. *Vaccine* **2006**, *24* (Suppl. 1), S16–S22.
43. Van Hede, D.; Langers, I.; Delvenne, P.; Jacobs, N. Origin and immunoescape of uterine cervical cancer. *Presse Med.* **2014**, *43*, e413–e421.

44. Levins, R. Some Demographic and Genetic Consequences of Environmental Heterogeneity for Biological Control. *Bull. Entomol. Soc. Am.* **1969**, *15*, 237–240.
45. Hanski, I. *Metapopulation Ecology*; Oxford University Press: Oxford, UK, 1999.
46. Frost, S.D.W.; Dumaurier, M.J.; Wain-Hobson, S.; Brown, A.J.L. Genetic drift and within-host metapopulation dynamics of HIV-1 infection. *Proc. Natl. Acad. Sci. USA* **2001**, *98*, 6975–6980.
47. Lythgoe, K.A.; Blanquart, F.; Pellis, L.; Fraser, C. Large Variations in HIV-1 Viral Load Explained by Shifting-Mosaic Metapopulation Dynamics. *PLoS Biol.* **2016**, *14*, e1002567.
48. Murall, C.L.; Bauch, C.T.; Day, T. Could the human papillomavirus vaccines drive virulence evolution? *Proc. Biol. Sci.* **2015**, *282*, 20141069.
49. Ryser, M.D.; Myers, E.R.; Durrett, R. HPV clearance and the neglected role of stochasticity. *PLoS Comput. Biol.* **2015**, *11*, e1004113.
50. Moore, P.S.; Chang, Y. Why do viruses cause cancer? Highlights of the first century of human tumour virology. *Nat. Rev. Cancer* **2010**, *10*, 878–889.
51. Hanahan, D.; Weinberg, R.A. Hallmarks of cancer: The next generation. *Cell* **2011**, *144*, 646–674.
52. Smith, E.M.; Ritchie, J.M.; Yankowitz, J.; Wang, D.; Turek, L.P.; Haugen, T.H. HPV prevalence and concordance in the cervix and oral cavity of pregnant women. *Infect. Dis. Obstet. Gynecol.* **2004**, *12*, 45–56.
53. De Vuyst, H.; Chung, M.H.; Baussano, I.; Mugo, N.R.; Tenet, V.; van Kemenade, F.J.; Rana, F.S.; Sakr, S.R.; Meijer, C.J.; Snijders, P.J.; et al. Comparison of HPV DNA testing in cervical exfoliated cells and tissue biopsies among HIV-positive women in Kenya. *Int. J. Cancer* **2013**, *133*, 1441–1446.
54. Nunes, E.M.; López, R.V.M.; Sudenga, S.L.; Gheit, T.; Tommasino, M.; Baggio, M.L.; Ferreira, S.; Galan, L.; Silva, R.C.; Lazcano-Ponce, E.; et al. Concordance of Beta-papillomavirus across anogenital and oral anatomic sites of men: The HIM Study. *Virology* **2017**, *510*, 55–59.
55. Hampras, S.S.; Rollison, D.E.; Giuliano, A.R.; McKay-Chopin, S.; Minoni, L.; Sereday, K.; Gheit, T.; Tommasino, M. Prevalence and Concordance of Cutaneous Beta Human Papillomavirus Infection at Mucosal and Cutaneous Sites. *J. Infect. Dis.* **2017**, *216*, 92–96.
56. Hernandez, B.Y.; Shvetsov, Y.B.; Goodman, M.T.; Wilkens, L.R.; Thompson, P.J.; Zhu, X.; Tom, J.; Ning, L. Genital and extra-genital warts increase the risk of asymptomatic genital human papillomavirus infection in men. *Sex. Transm. Infect.* **2011**, *87*, 391–395.
57. Kalantari, M.; Garcia-Carranca, A.; Morales-Vazquez, C.D.; Zuna, R.; Montiel, D.P.; Calleja-Macias, I.E.; Johansson, B.; Andersson, S.; Bernard, H.U. Laser capture microdissection of cervical human papillomavirus infections: Copy number of the virus in cancerous and normal tissue and heterogeneous DNA methylation. *Virology* **2009**, *390*, 261–267.
58. Quint, W.; Jenkins, D.; Molijn, A.; Struijk, L.; van de Sandt, M.; Doorbar, J.; Mols, J.; Van Hoof, C.; Hardt, K.; Struyf, F.; et al. One virus, one lesion—Individual components of CIN lesions contain a specific HPV type. *J. Pathol.* **2012**, *227*, 62–71.
59. Lechner, M.; Fenton, T.; West, J.; Wilson, G.; Feber, A.; Henderson, S.; Thirlwell, C.; Dibra, H.K.; Jay, A.; Butcher, L.; et al. Identification and functional validation of HPV-mediated hypermethylation in head and neck squamous cell carcinoma. *Genome Med.* **2013**, *5*, 15.
60. Candi, E.; Schmidt, R.; Melino, G. The cornified envelope: a model of cell death in the skin. *Nat. Rev. Mol. Cell Biol.* **2005**, *6*, 328–340.
61. Eckhart, L.; Valle, L.D.; Jaeger, K.; Ballaun, C.; Szabo, S.; Nardi, A.; Buchberger, M.; Hermann, M.; Alibardi, L.; Tschachler, E. Identification of reptilian genes encoding hair keratin-like proteins suggests a new scenario for the evolutionary origin of hair. *Proc. Natl. Acad. Sci. USA* **2008**, *105*, 18419–18423.
62. Henry, J.; Toulza, E.; Hsu, C.Y.; Pellerin, L.; Balica, S.; Mazereeuw-Hautier, J.; Paul, C.; Serre, G.; Jonca, N.; Simon, M. Update on the epidermal differentiation complex. *Front. Biosci.* **2012**, *17*, 1517–1532.
63. Dhouailly, D. A new scenario for the evolutionary origin of hair, feather, and avian scales. *J. Anat.* **2009**, *214*, 587–606.
64. Gilliam, A.C.; Kremer, I.B.; Yoshida, Y.; Stevens, S.R.; Tootell, E.; Teunissen, M.B.; Hammerberg, C.; Cooper, K.D. The human hair follicle: A reservoir of CD40+ B7-deficient Langerhans cells that repopulate epidermis after UVB exposure. *J. Investig. Dermatol.* **1998**, *110*, 422–427.
65. Malissen, B.; Tamoutounour, S.; Henri, S. The origins and functions of dendritic cells and macrophages in the skin. *Nat. Rev. Immunol.* **2014**, *14*, 417–428.

66. Wilson, D.S. Complex Interactions in Metacommunities, with Implications for Biodiversity and Higher Levels of Selection. *Ecology* **1992**, *73*, 1984–2000.
67. Murall, C.L.; McCann, K.S.; Bauch, C.T. Food webs in the human body: Linking ecological theory to viral dynamics. *PLoS ONE* **2012**, *7*, e48812.
68. Rynkiewicz, E.C.; Pedersen, A.B.; Fenton, A. An ecosystem approach to understanding and managing within-host parasite community dynamics. *Trends Parasitol.* **2015**, *31*, 212–221.
69. Bravo, I.G.; Félez-Sánchez, M. Papillomaviruses: Viral evolution, cancer and evolutionary medicine. *Evol. Med. Public Health* **2015**, *2015*, 32–51.
70. Egawa, N.; Egawa, K.; Griffin, H.; Doorbar, J. Human Papillomaviruses; Epithelial Tropisms, and the Development of Neoplasia. *Viruses* **2015**, *7*, 3863–3890.
71. Murray, R.F.; Hobbs, J.; Payne, B. Possible clonal origin of common warts (*Verruca vulgaris*). *Nature* **1971**, *232*, 51–52.
72. Herfs, M.; Yamamoto, Y.; Laury, A.; Wang, X.; Nucci, M.R.; McLaughlin-Drubin, M.E.; Münger, K.; Feldman, S.; McKeon, F.D.; Xian, W.; et al. A discrete population of squamocolumnar junction cells implicated in the pathogenesis of cervical cancer. *Proc. Natl. Acad. Sci. USA* **2012**, *109*, 10516–10521.
73. Nicolás-Párraga, S.; Gandini, C.; Pimenoff, V.N.; Alemany, L.; de Sanjosé, S.; Xavier Bosch, F.; Bravo, I.G.; The RIS HPV TT and HPV VVAP study groups. HPV16 variants distribution in invasive cancers of the cervix, vulva, vagina, penis, and anus. *Cancer Med.* **2016**, *5*, 2909–2919.
74. Nicolás-Párraga, S.; Alemany, L.; de Sanjosé, S.; Bosch, F.; Bravo, I.; RIS HPV TT and HPV VVAP study groups. Differential HPV16 variant distribution in squamous cell carcinoma, adenocarcinoma and adenosquamous cell carcinoma. *Int. J. Cancer* **2017**, *140*, 2092–2100.
75. Maglennon, G.A.; McIntosh, P.; Doorbar, J. Persistence of viral DNA in the epithelial basal layer suggests a model for papillomavirus latency following immune regression. *Virology* **2011**, *414*, 153–163.
76. Reinson, T.; Henno, L.; Toots, M., Jr.; Ustav, M. The Cell Cycle Timing of Human Papillomavirus DNA Replication. *PLoS ONE* **2015**, *10*, e0131675.
77. Van der Weele, P.; Meijer, C.J.L.M.; King, A.J. Whole-Genome Sequencing and Variant Analysis of Human Papillomavirus 16 Infections. *J. Virol.* **2017**, *91*, e00844-17.
78. Bosch, F.X.; Robles, C.; Díaz, M.; Arbyn, M.; Baussano, I.; Clavel, C.; Ronco, G.; Dillner, J.; Lehtinen, M.; Petry, K.U.; et al. HPV-FASTER: Broadening the scope for prevention of HPV-related cancer. *Nat. Rev. Oncol.* **2015**, *13*, 119–132.
79. Skeate, J.G.; Woodham, A.W.; Einstein, M.H.; Silva, D.M.D.; Kast, W.M. Current therapeutic vaccination and immunotherapy strategies for HPV-related diseases. *Hum. Vaccines Immunother.* **2016**, *12*, 1418–1429.
80. Trimble, C.L.; Morrow, M.P.; Kraynyak, K.A.; Shen, X.; Dallas, M.; Yan, J.; Edwards, L.; Parker, R.L.; Denny, L.; Giffear, M.; et al. Safety, efficacy, and immunogenicity of VGX-3100, a therapeutic synthetic DNA vaccine targeting human papillomavirus 16 and 18 E6 and E7 proteins for cervical intraepithelial neoplasia 2/3: A randomised, double-blind, placebo-controlled phase 2b trial. *Lancet* **2015**, *386*, 2078–2088.
81. Amador-Molina, A.; Hernández-Valencia, J.F.; Lamoyi, E.; Contreras-Paredes, A.; Lizano, M. Role of innate immunity against human papillomavirus (HPV) infections and effect of adjuvants in promoting specific immune response. *Viruses* **2013**, *5*, 2624–2642.
82. Derkay, C.S.; Wiatrak, B. Recurrent Respiratory Papillomatosis: A Review. *Laryngoscope* **2008**, *118*, 1236–1247.
83. Martin, D.; Rödel, F.; Balermpas, P.; Rödel, C.; Fokas, E. The immune microenvironment and HPV in anal cancer: Rationale to complement chemoradiation with immunotherapy. *BBA Rev. Cancer* **2017**, *1868*, 221–230.
84. Silasi, M.; Cardenas, I.; Kwon, J.Y.; Racicot, K.; Aldo, P.; Mor, G. Viral Infections During Pregnancy. *Am. J. Reprod. Immunol.* **2015**, *73*, 199–213.
85. Kyrgiou, M.; Athanasiou, A.; Paraskevaidi, M.; Mitra, A.; Kalliala, I.; Martin-Hirsch, P.; Arbyn, M.; Bennett, P.; Paraskevaidis, E. Adverse obstetric outcomes after local treatment for cervical preinvasive and early invasive disease according to cone depth: systematic review and meta-analysis. *BMJ* **2016**, *354*, i3633.
86. Zuo, Z.; Goel, S.; Carter, J.E. Association of Cervical Cytology and HPV DNA Status During Pregnancy With Placental Abnormalities and Preterm Birth. *Am. J. Clin. Pathol.* **2011**, *136*, 260–265.
87. Laprise, C.; Trottier, H.; Monnier, P.; Coutlée, F.; Mayrand, M.H. Prevalence of human papillomaviruses in semen: A systematic review and meta-analysis. *Hum. Reprod.* **2014**, *29*, 640–651.

88. Luttmer, R.; Dijkstra, M.G.; Snijders, P.J.F.; Hompes, P.G.A.; Pronk, D.T.M.; Hubeek, I.; Berkhof, J.; Heideman, D.A.M.; Meijer, C.J.L.M. Presence of human papillomavirus in semen in relation to semen quality. *Hum. Reprod.* **2016**, *31*, 280–286.
89. Foresta, C.; Noventa, M.; De Toni, L.; Gizzo, S.; Garolla, A. HPV-DNA sperm infection and infertility: From a systematic literature review to a possible clinical management proposal. *Andrology* **2015**, *3*, 163–173.
90. Castellsagué, X.; Drudis, T.; Cañadas, M.P.; Goncé, A.; Ros, R.; Pérez, J.M.; Quintana, M.J.; Muñoz, J.; Albero, G.; de Sanjosé, S.; et al. Human Papillomavirus (HPV) infection in pregnant women and mother-to-child transmission of genital HPV genotypes: a prospective study in Spain. *BMC Infect. Dis.* **2009**, *9*, 74.
91. Racicot, K.; Cardenas, I.; Wünsche, V.; Aldo, P.; Guller, S.; Means, R.E.; Romero, R.; Mor, G. Viral Infection of the Pregnant Cervix Predisposes to Ascending Bacterial Infection. *J. Immunol.* **2013**, *191*, 934–941.
92. Slatter, T.L.; Hung, N.G.; Clow, W.M.; Royds, J.A.; Devenish, C.J.; Hung, N.A. A clinicopathological study of episomal papillomavirus infection of the human placenta and pregnancy complications. *Mod. Pathol.* **2015**, *28*, 1369–1382.
93. Gandon, S.; Day, T. Evidences of parasite evolution after vaccination. *Vaccine* **2008**, *26S*, C4–C7.
94. Bernard, H.U.; Burk, R.D.; Chen, Z.; van Doorslaer, K.; Hausen, H.Z.; de Villiers, E.M. Classification of papillomaviruses (PVs) based on 189 PV types and proposal of taxonomic amendments. *Virology* **2010**, *401*, 70–79.
95. Joura, E.A.; Giuliano, A.R.; Iversen, O.E.; Bouchard, C.; Mao, C.; Mehlsen, J.; Moreira, E.D., Jr.; Ngan, Y.; Petersen, L.K.; Lazcano-Ponce, E.; et al. A 9-valent HPV vaccine against infection and intraepithelial neoplasia in women. *N. Engl. J. Med.* **2015**, *372*, 711–723.
96. Draper, E.; Bissett, S.L.; Howell-Jones, R.; Edwards, D.; Munslow, G.; Soldan, K.; Beddows, S. Neutralization of non-vaccine human papillomavirus pseudoviruses from the A7 and A9 species groups by bivalent HPV vaccine sera. *Vaccine* **2011**, *29*, 8585–8590.
97. Kemp, T.J.; Hildesheim, A.; Safaeian, M.; Dauner, J.G.; Pan, Y.; Porras, C.; Schiller, J.T.; Lowy, D.R.; Herrero, R.; Pinto, L.A. HPV16/18 L1 VLP vaccine induces cross-neutralizing antibodies that may mediate cross-protection. *Vaccine* **2011**, *29*, 2011–2014.
98. Markowitz, L.E.; Liu, G.; Hariri, S.; Steinau, M.; Dunne, E.F.; Unger, E.R. Prevalence of HPV After Introduction of the Vaccination Program in the United States. *Pediatrics* **2016**, *137*, e20151968.
99. Mirabello, L.; Yeager, M.; Yu, K.; Clifford, G.M.; Xiao, Y.; Zhu, B.; Cullen, M.; Boland, J.F.; Wentzensen, N.; Nelson, C.W.; et al. HPV16 E7 Genetic Conservation Is Critical to Carcinogenesis. *Cell* **2017**, *170*, 1164.e6.
100. Worobey, M.; Han, G.Z.; Rambaut, A. A synchronized global sweep of the internal genes of modern avian influenza virus. *Nature* **2014**, *508*, 254–257.
101. Weinert, L.A.; Depledge, D.P.; Kundu, S.; Gershon, A.A.; Nichols, R.A.; Balloux, F.; Welch, J.J.; Breuer, J. Rates of Vaccine Evolution Show Strong Effects of Latency: Implications for Varicella Zoster Virus Epidemiology. *Mol. Biol. Evol.* **2015**, *32*, 1020–1028.
102. Da Fonseca, D.M.; Hand, T.W.; Han, S.J.; Gerner, M.Y.; Glatman Zaretsky, A.; Byrd, A.L.; Harrison, O.J.; Ortiz, A.M.; Quinones, M.; Trinchieri, G.; et al. Microbiota-Dependent Sequelae of Acute Infection Compromise Tissue-Specific Immunity. *Cell* **2015**, *163*, 354–366.
103. Paschos, K.; Allday, M.J. Epigenetic reprogramming of host genes in viral and microbial pathogenesis. *Trends Microbiol.* **2010**, *18*, 439–447.
104. Hattori, N.; Ushijima, T. Epigenetic impact of infection on carcinogenesis: Mechanisms and applications. *Genome Med.* **2016**, *8*, 10.
105. Johannsen, E.; Lambert, P.F. Epigenetics of human papillomaviruses. *Virology* **2013**, *445*, 205–212.
106. Quint, K.D.; Genders, R.E.; de Koning, M.N.; Borgogna, C.; Gariglio, M.; Bouwes Bavinck, J.N.; Doorbar, J.; Feltkamp, M.C. Human Beta-papillomavirus infection and keratinocyte carcinomas. *J. Pathol.* **2015**, *235*, 342–354.
107. Bzhalava, D.; Johansson, H.; Ekström, J.; Faust, H.; Möller, B.; Eklund, C.; Nordin, P.; Stenquist, B.; Paoli, J.; Persson, B.; et al. Unbiased Approach for Virus Detection in Skin Lesions. *PLoS ONE* **2013**, *8*, e65953.
108. Gillison, M.L.; Broutian, T.; Pickard, R.K.L.; Tong, Z.Y.; Xiao, W.; Kahle, L.; Graubard, B.I.; Chaturvedi, A.K. Prevalence of Oral HPV Infection in the United States, 2009–2010. *JAMA* **2012**, *307*, 693–703.

109. Giuliano, A.R.; Lazcano-Ponce, E.; Villa, L.L.; Flores, R.; Salmeron, J.; Lee, J.H.; Papenfuss, M.R.; Abrahamsen, M.; Jolles, E.; Nielson, C.M.; et al. The Human Papillomavirus Infection in Men Study: Human Papillomavirus Prevalence and Type Distribution among Men Residing in Brazil, Mexico, and the United States. *Cancer Epidemiol. Biomark. Prev.* **2008**, *17*, 2036–2043.

110. Nyitray, A.G.; Smith, D.; Villa, L.; Lazcano-Ponce, E.; Abrahamsen, M.; Papenfuss, M.; Giuliano, A.R. Prevalence of and Risk Factors for Anal Human Papillomavirus Infection in Men Who Have Sex with Women: A Cross-National Study. *J. Infect. Dis.* **2010**, *201*, 1498–1508.

111. Brisson, M.; Bénard, E.; Drolet, M.; Bogaards, J.A.; Baussano, I.; Vänskä, S.; Jit, M.; Boily, M.C.; Smith, M.A.; Berkhof, J.; et al. Population-level impact, herd immunity, and elimination after human papillomavirus vaccination: A systematic review and meta-analysis of predictions from transmission-dynamic models. *Lancet Public Health* **2016**, *1*, e8–e17.

112. Grabowski, M.K.; Kong, X.; Gray, R.H.; Serwadda, D.; Kigozi, G.; Gravitt, P.E.; Nalugoda, F.; Reynolds, S.J.; Wawer, M.J.; Redd, A.D.; et al. Partner Human Papillomavirus Viral Load and Incident Human Papillomavirus Detection in Heterosexual Couples. *J. Infect. Dis.* **2016**, *213*, 948–956.

113. Brotman, R.M.; Shardell, M.D.; Gajer, P.; Tracy, J.K.; Zenilman, J.M.; Ravel, J.; Gravitt, P.E. Interplay between the temporal dynamics of the vaginal microbiota and human papillomavirus detection. *J. Infect. Dis.* **2014**, *210*, 1723–1733.

114. Mitra, A.; MacIntyre, D.A.; Marchesi, J.R.; Lee, Y.S.; Bennett, P.R.; Kyrgiou, M. The vaginal microbiota, human papillomavirus infection and cervical intraepithelial neoplasia: what do we know and where are we going next? *Microbiome* **2016**, *4*, 58.

Review

Deciphering the Multifactorial Susceptibility of Mucosal Junction Cells to HPV Infection and Related Carcinogenesis

Michael Herfs [1,*], Thing R. Soong [2], Philippe Delvenne [1] and Christopher P. Crum [2]

[1]   Laboratory of Experimental Pathology, GIGA-Cancer, University of Liege, 4000 Liege, Belgium;
      P.Delvenne@ulg.ac.be
[2]   Division of Women's and Perinatal Pathology, Department of Pathology, Brigham and Women's Hospital,
      Harvard Medical School, Boston, MA 02115, USA; tsoong@partners.org (T.R.S.);
      ccrum@bwh.harvard.edu (C.P.C.)
*   Correspondence: M.Herfs@ulg.ac.be; Tel.: +32-4-366-4282

Academic Editors: Alison A. McBride and Karl Munger
Received: 27 March 2017; Accepted: 18 April 2017; Published: 20 April 2017

Abstract: Human papillomavirus (HPV)-induced neoplasms have long been considered to originate from viral infection of the basal cell layer of the squamous mucosa. However, this paradigm has been recently undermined by accumulating data supporting the critical role of a discrete population of squamo-columnar (SC) junction cells in the pathogenesis of cervical (pre)cancers. The present review summarizes the current knowledge on junctional cells, discusses their high vulnerability to HPV infection, and stresses the potential clinical/translational value of the novel dualistic model of HPV-related carcinogenesis.

Keywords: human papillomavirus; squamo-columnar junctions; (pre)neoplastic lesions

---

1. Introduction

Cancer of the uterine cervix is generally observed near the squamo-columnar (SC) junction and is almost always caused by carcinogenic human papillomaviruses (HPV). However, the nature of cervical epithelial cells initially infected by HPV has been a matter of debate/speculation for a long time. Classically, the HPV-target cells are presumed to be the basal keratinocytes of the pluristratified mucosa lining the outer part of the cervix (ectocervix) and the region where the endocervical columnar epithelium, sensitive to the acidic vaginal pH, has undergone a squamous metaplastic transformation during puberty (transformation zone (TZ)). According to this theory, microtrauma, ulcerative lesions, or abrasions into the squamous epithelium would render the proliferating basal cell layer uniquely vulnerable to viral infection [1]. This hypothesis was emphasized by (1) the detection of HPV transcripts in basal cells; (2) the basal location of dysplastic cells in the initial phase of malignancy (low-grade cervical intraepithelial neoplasia (CIN1)); and (3) the observation that chemical disruption of the integrity of the stratified epithelium is required for pseudovirion infection of the murine genital tract [2]. Obviously, it is indisputable that transcriptionally-active HPV infection can occur in basal keratinocytes and result in the subsequent development of (pre)neoplastic lesions in both the ectocervix and the lower genital tract (vagina/vulva). This mechanism is also likely to explain the HPV infections occurring in both the oral cavity and the anal margin/perianal skin (microtrauma resulting from mastication and anal intercourse, respectively) because these two sites are entirely lined by a pluristratified epithelium. However, this hypothesis is incompatible with the historical observation that HPV infection causes cervical cancer and its precursor lesions mainly within the SC junction microenvironment [3,4]. In addition, this theory does not explain the discrepancy in terms of cancer risk between the uterine cervix and other HPV-infected mucosa. Indeed, with approximately 520,000 new cases diagnosed each year worldwide, cervical squamous cell carcinoma (SCC) is approximately 20-fold more prevalent than vaginal/vulvar/penile neoplasms [5]. Finally, considering that routine cervical screening allows for the annual detection of several millions of CIN, no biological or physical factor is likely to induce traumatic lesions in the cervical os of such a large number of women.

In the early 2000s, endocervical reserve cells were also proposed as the cell of origin for cervical cancer [6]. However, the broad distribution of reserve cells in the cervical canal [7], the extremely low reported incidence (less than 1/1000) of CIN in endocervical polyps [8], and the absence of convincing evidence demonstrating HPV infection in normal monolayered reserve cells do not support a major role of these latter cells in cervical carcinogenesis.

Consequently, for years, it has been thought that the ecto-endocervical junction microenvironment contains unique epithelial cells highly susceptible to high-risk HPV infection and related (pre)cancer development. However, the topographical preference of cervical neoplasia for the SC junction remained unexplained. In 2012, a discrete population of cells residing within, or in close proximity to, the cervical SC junction was identified [9] (Figure 1). Displaying a cuboidal/immature columnar phenotype, these cells were shown to have an embryonic origin similar to their counterparts discovered one year earlier at the gastroesophageal junction [10,11]. In addition to being actively involved in adult adaptive processes (metaplasia, hyperplasia) [10], SC junction cells exhibit intriguing phenotypic similarities with approximately 90% of cervical SCC and high-grade precursors [9,12]. These findings underlying the instrumental role of junctional cells in cervical carcinogenesis were recently confirmed by several immunohistochemical and/or transcriptional studies analyzing the expression of SC junction-overexpressed biomarkers (i.e., cytokeratin 7 (Krt7), anterior gradient protein 2 (AGR2) or cystic fibrosis transmembrane conductance regulator (CFTR)) in large cohorts of cervical (pre)neoplastic lesions [13–17]. Moreover, the evidence that normal-appearing SC junction cells harbor both HPV transcripts (E6*I/II) and early viral proteins (HPV E2) was reported and sustains the possibility that these cuboidal cells could serve as a reservoir for latent infections and subsequent CIN development in asymptomatic HPV-positive patients [18].

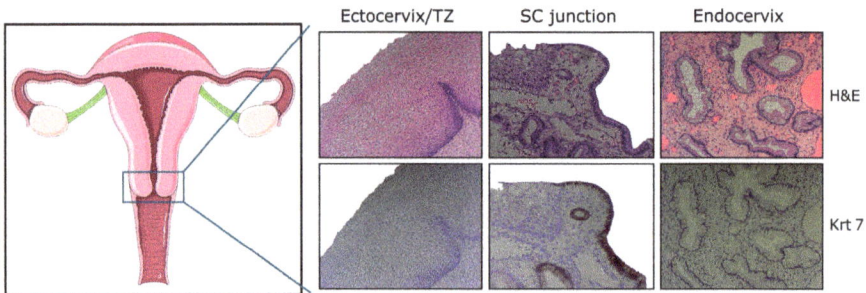

Figure 1. Schematic representation of the female genital tract and histology of the adult cervix with ectocervical (squamous), junctional (cuboidal), and endocervical (columnar) cells. Note the uniform keratin 7 (Krt7) immunoreactivity displayed by cuboidal cells observed within, or in close proximity to, the squamo-columnar (SC) junction. H&E: hematoxylin and eosin; TZ: transformation zone.

2. High Vulnerability of SC Junctions to HPV Infection and (Pre)Cancer Development: A Multifaceted Process

2.1. Altered Secretion of Antimicrobial Peptides

Similar to the skin or other mucosal surfaces, the gynecological tract expresses a distinct set of host–defense peptides adapted specifically for limiting pathogen invasion and replication within this unique environment [19]. Involved in several processes beyond their antimicrobial activity (i.e., wound healing, angiogenesis, etc.) [20,21], several innate peptides, especially the lysozyme, the secretory leukocyte protease inhibitor (SLPI), as well as the defensin superfamily, have been recently the subject of extensive mechanistic studies. Using pseudoviruses of both cutaneous and mucosal (non-carcinogenic (low-risk) and carcinogenic (high-risk)) genotypes, Buck and colleagues first demonstrated that alpha defensins 1–3 (also called human neutrophil peptides (HNP) 1–3) and 5 (known as human defensin 5 (HD5)) strongly inhibit HPV infection in vitro [22]. In contrast, beta defensins, lyzosyme, or SLPI exhibited very little or no antagonist activity. Based on these important findings and the strong chemotactic activity exerted by alpha defensins for antigen presenting cells, these latter peptides were considered as interesting candidates for the non-surgical management of HPV-positive lesions in young women [23]. Despite encouraging results [24], the therapeutic effect of these antimicrobial peptides was never precisely determined in vivo. Recently, interest in the defensin superfamily in the context of HPV infection was reactivated by the observed absence of HD5 expression in both normal SC junction cells and (pre)neoplastic lesions arising from these latter cells [25]. Moreover, the HD5-related anti-HPV activity was recently deciphered. By preventing proteolytic processing (furin cleavage) of the minor capsid protein L2, Wiens and Smith showed that HD5 does not block virus internalization, but significantly alters the viral entry pathway [26]. The same research team further investigated the molecular mechanism and demonstrated that, similarly to the inhibitory effect reported for both human adenoviruses and polyomaviruses [27–29], HD5 alters HPV intracellular trafficking via capsid stabilization and redirection of the incoming viral particles to the lysosome [30]. Altogether, these data suggest that SC junction cells may display an increased vulnerability to HPV infection through a deficient expression of innate molecules inhibiting the intracellular steps of virus processing.

2.2. Overexpression of Key Proteins Implicated in Post-Endocytic HPV Trafficking

Although some unclear points remain (i.e., the identity of the internalization receptor(s) or the pathway(s) involved in HPV entry), the mechanisms governing HPV entry into host cells is now relatively well-characterized [31]. After initial binding to heparan sulfate proteoglycans, several conformational changes and, as mentioned above, a furin-dependent L2 cleavage have been described. Viral particles would then interact with a complex of proteins including alpha 6 integrin and tetraspanin CD151 before their endocytosis in a clathrin- and caveolin-independent manner. Following HPV entry, virions are transported via the endosomal system where capsid disassembly occurs. Significantly, a recent article demonstrated the crucial requirement of tetraspanin CD63, a SC junction-overexpressed protein [9], during post-endocytic steps [32]. Accordingly, CD63 was shown to interact with L1 capsid protein and virus uncoating was dramatically decreased in CD63-depleted cells. In contrast, this junctional-overexpressed molecule was not involved in cell surface interactions or endocytosis. Given that post-endocytic trafficking strongly determines the success of the infection, the high expression of some key proteins controlling both vesicular trafficking and HPV disassembly could render the SC junction cells more vulnerable to HPV infection.

2.3. Possible Translational Regulation of HPV mRNAs by the Cytokeratin Filaments 7 and 19

All mucosal HPVs are characterized by a circular DNA genome of approximately 8000 base pairs coding for eight major proteins (early region: E1, E2, E4, E5, E6, E7; late region: L1 and L2). Viral gene expression has been shown to be regulated at the level of transcription (by HPV E2 protein [33], as well as host cell factors, such as Specificity protein 1 (Sp1) and Activator protein 1 (AP-1) [34]), polyadenylation, and RNA splicing. The latter is evidenced by the full transcriptional maps reported for several HPV genotypes [35–38]. In the early 2000s, several studies analyzed the interactions occurring between viral transcripts and intermediate filaments and reported exciting effects of Krt7 and 19 on E7 oncoprotein expression. Indeed, by interacting with a 6-mer amino acid peptide SEQIKA present at position 91–96 in the human Krt7 sequence, HPV E7 messenger RNA (mRNA) was shown to be protected/stabilized [39]. In addition, viral mRNA translation was presumed to be increased by Krt19 [40], another member of the keratin family overexpressed in both cervical and anal SC junctions [17,41]. Although these intriguing findings need to be confirmed, they support the hypothesis of a specific regulation of viral transcripts in SC junction cells leading to increased viral oncoprotein levels and, subsequently, to (pre)cancer development/progression.

2.4. Altered Adaptive Immune Responses in SC Junction Microenvironment

The junction between the ectocervical squamous and the endocervical muco-secreting epithelia appears to be dynamic. Indeed, this junction moves to the ectocervix until the age of about 14. Then, a metaplastic process occurs due to an estrogen-dependent acidification of the vaginal luminal pH and this subsequent replacement of the columnar epithelium by metaplastic squamous foci traces the SC junction into the endocervical canal, where it is often highly situated in menopause. The regenerative metaplastic response (TZ) between the former and the new SC junction has been shown to be driven by the junctional cell-dependent production of squamous metaplastic (reserve) cells [10]. In general, the metaplastic epithelium does not exhibit the same structure throughout the transformation zone. Depending on the persistence, or not (exfoliation), of the remaining precursor SC junction cells at the top of the metaplastic epithelium, the degree of metaplasia is considered as immature or mature, respectively. This histological feature is probably related to vaginal pH fluctuations during the menstrual cycle as, in healthy women, pH ranges between 4.5 and 5.5 with alkalinization before ovulation [42]. Therefore, a chronic, but inconstant, irritation of the mucosa occurs inducing a chronic inflammation in the cervical os.

Prior to and following the discovery of SC junction cells, we, and others, reported altered local immune responses within the SC junction or in metaplastic areas (TZ) located in close proximity. Focusing on $CD1a^+$ antigen-presenting cells (Langerhans cells (LC) and dendritic cells (DC)), several studies showed that LC/DC density is largely decreased in both the SC junction microenvironment and the cervical (pre)neoplastic lesions compared to the surrounding HPV-uninfected ectocervical/vaginal squamous mucosa [43–45]. By inducing the expression of epithelial to mesenchymal transition regulators, the strong secretion of transforming growth factor-β (TGF-β) observed in TZ (especially in immature metaplastic patches) was shown to inhibit E-cadherin expression [43]. The resulting disruption of E-cadherin-mediated LC-epithelial cell adhesion was shown to alter antigen presenting cell maturation/differentiation promoting T regulatory (Treg) cell development [46,47]. Moreover, an inverse correlation between LC/DC density and the expression of prostaglandin E2 (PGE2) enzymatic pathways was reported [48]. The significance of these observations in terms of antigen presenting cell trafficking was clearly demonstrated. Indeed, PGE2 decreased the migratory capacity of immature LC/DC as well as induced the tolerogenic phenotype of these cells by altering accessory molecule expression and by modifying Interleukin (IL)-10/IL-12 secretion ratio [48]. Similarly, receptor activator of nuclear factor κ-B ligand (RANKL), which is strongly secreted in both the cervical SC junction and HPV-related (pre)neoplastic microenvironment, was also shown to promote the emergence of tolerogenic DC [49]. The altered density of LC/DC within, or in close proximity of, the SC junction could be further exacerbated by the absence of HD5 expression, which has been shown to have a strong chemotactic activity on several cell types involved in the immune responses [25]. In addition to LC/DC, the functionality of both T lymphocytes (higher number of Treg cells) and plasmacytoid dendritic cells (pDC) is also altered within the cervical SC junction microenvironment and the high mobility group box 1 (HMGB1) was recently identified as a key soluble factor involved in the acquisition of tolerogenic pDC [50]. Altogether, the modified expression of soluble/adhesion molecules within the SC junction microenvironment could promote the immunoescape of infected/transformed cells. The possible mechanisms underlying the mucosal junctional cells' unique susceptibility to HPV-related carcinogenesis are summarized in Figure 2.

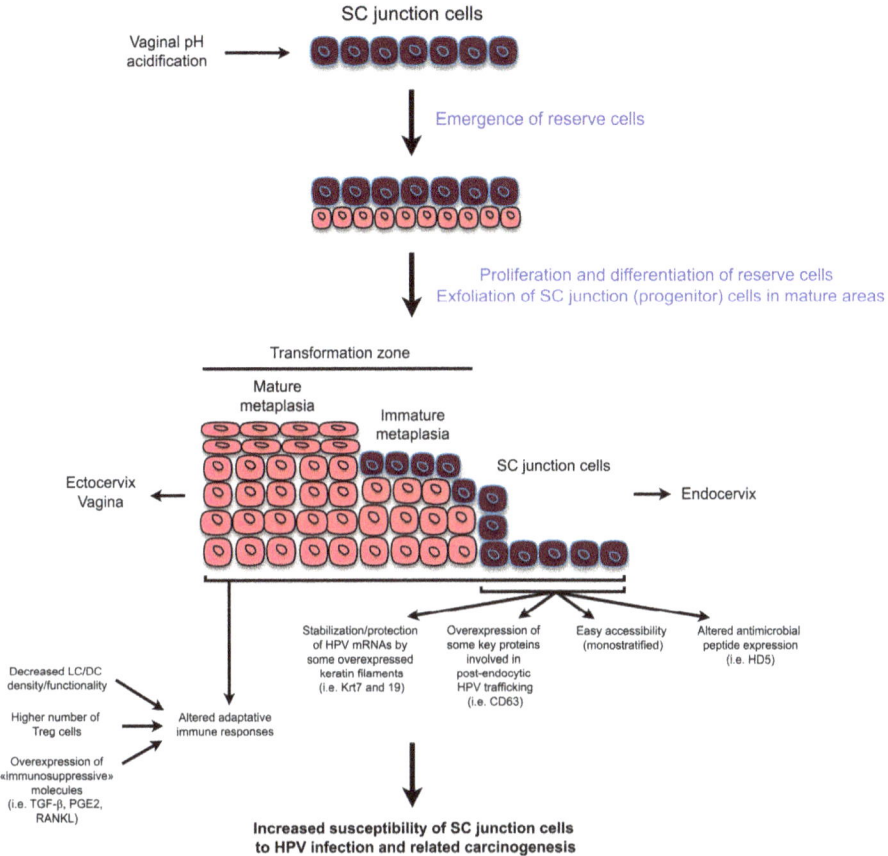

Figure 2. Schematic representation highlighting both the tissue remodeling observed in the SC junction microenvironment and the possible mechanisms explaining the high susceptibility of the SC junction cells to human papillomavirus (HPV)infection and related carcinogenesis. mRNA: messenger RNA; HD5: human defensin 5; LC/DC: langerhans cells/dendritic cells; PGE2: prostaglandin E2; RANKL: receptor activator of nuclear factor κ-B ligand.

3. Dualistic Model of HPV-Related Carcinogenesis

3.1. Explanation for Multiple Neoplastic Phenotypes

In pathology practical examination, it is well established that HPV genotypes impact the morphology of (pre)neoplastic lesions. Low-risk HPVs (mainly HPV6 and 11) are typically related to condylomata acuminata (genital warts) development and viral cytopathic effect [51,52]. In contrast, carcinogenic (high-risk) strains are less frequently associated to koilocytic morphology or virion production [52,53]. In addition to the viral component, recent findings support that the nature of the epithelial cells (basal keratinocytes versus SC junction cells) originally infected by HPV also considerably influences the appearance of CIN. While the infections occurring in the ectocervix (and, in general, in the lower genital tract) lead, traditionally, to well-differentiated squamous lesions, an immature phenotype is frequently observed when high-risk HPV-mediated transformation appears within the endocervical canal [12]. These morphological differences governed by the expression profile/intrinsic features of "the cell of origin" critically affect the pathologists' interpretation and subsequent classification of HPV-related lesions. Indeed, diagnostic reproducibility of CIN is well-known to be problematic leading to inappropriate management of patients [54]. The inter-observer agreement is especially low (~50%) for Krt7-positive low-grade dysplasias (arising from the SC junction) and CIN2 supporting that this latter "considered high-risk" lesion cannot be consistently distinguished in the spectrum of cervical (pre)cancers [12,55]. This has led to proposals that this uncertainty be conveyed in the diagnostic report (i.e., (pre)neoplastic lesions of uncertain grade (CIN1-2)) and managed accordingly [56].

In addition to lesions demonstrating a squamous phenotype (~85% of all cervical neoplasms), adenocarcinoma, as well as intriguing lesions that display both squamous and columnar differentiation are observed within the SC junction microenvironment [57–59]. These latter may range from individuated and adjacent squamous (CIN) and columnar (adenocarcinoma in situ) lesions to cases in which both components are closely blended. Such a mixed phenotype can also be detected in benign lesions, such as microglandular hyperplasia [10,60] (Figure 3). Although both the stemness properties and differentiation potential of SC junction cells are still not clearly defined, the expression of SC junction-specific/overexpressed biomarkers exhibited by all types of cervical lesions (squamous, columnar, adenosquamous) represents the strongest argument for the multipotent potential of junctional cells. The immuno-phenotypic homology between cervical SC junction cells observed in adult tissues and embryonic Müllerian epithelium also supports this innovative hypothesis [10].

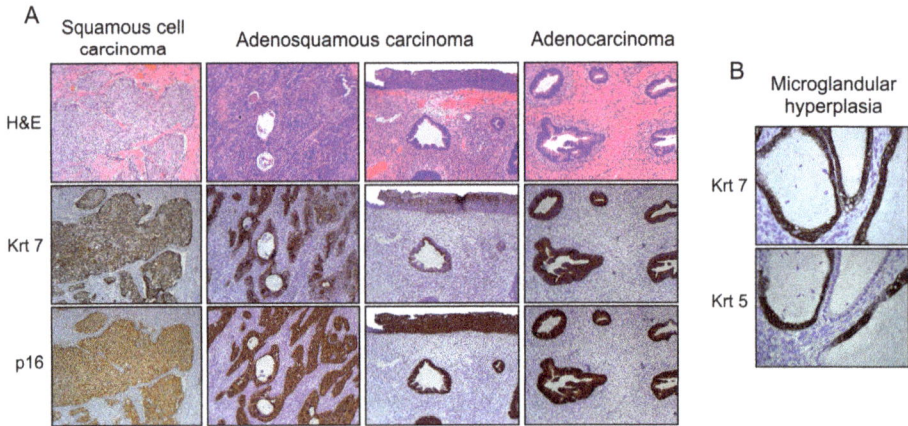

Figure 3. Phenotypic variants among cervical malignant epithelial tumors (**A**). Note the Krt7 immunoreactivity displayed by squamous, adenosquamous (blended or individuated/adjacent lesions), and columnar neoplasms supporting their similar cell of origin (SC junction); and (**B**) under benign conditions (i.e., microglandular hyperplasia), a mixed phenotype can also be observed. Note the Krt5 (squamous biomarker) expression in Krt7-positive cuboidal cells without evidence of reserve cells.

3.2. From the Bench to Cervical Cancer Prevention

Although bi- or quadrivalent HPV vaccines have been on the market for a decade and have been shown to be highly efficient to prevent HPV16/18-related high-grade anal/cervical lesions [61–64], HPV is still the most common sexually transmitted infection worldwide according to the World Health Organization (WHO). Based on estimations, up to 500 million individuals could be carriers of HPV DNA [65]. Although the large majority of infected patients are asymptomatic and will never develop cancer, every year, billions of euros/dollars are spent for both screening (Pap smear and/or HPV testing) and follow-up of patients at very low-risk of tumor development [66]. In an era of personalized medicine, the identification of one or, more realistically, several reliable biomarkers (used in combination) and allowing to accurately predict the outcome of HPV-related lesions, might substantially relieve the current ponderous and costly management of infected patients [67]. In the last decade, this area of research has been the subject of numerous studies and several candidate (viral and cellular) biomarkers have been highlighted. Among these latter, p16^{ink4}, the HPV viral load and genotyping were the most investigated. Although, all gave some indications on high-risk patients, a lack of concordance has been extensively reported and all failed to precisely predict CIN, which will finally progress to cancer [68–75].

Supported by the very large discrepancy, in terms of cancer risk, between the uterine cervix and the vagina/vulva, it is now very likely that the nature of the HPV-infected cells strongly influences disease outcome. Recently, four studies analyzed the predictive value of junctional biomarkers (especially Krt7) and, interestingly, all showed that low-grade CIN arising from the SC junction have a significantly higher risk to progress to CIN2/3 compared to their counterparts observed in the transformation zone/ectocervix [12–15]. Furthermore, when compared to other risk factors, such as HPV16 infection and diffuse p16^{ink4} expression, full-thickness Krt7 immunoreactivity demonstrated the highest correlation with lesion progression [14]. Although these conclusions were shown to be reproducible and in agreement with the high percentage (~90%) of CIN3 and cervical cancers displaying immunophenotypic similarities with SC junction cells, unfortunately, Krt7 or another SC junction-specific/overexpressed protein is unlikely to be sufficiently specific to provide actionable information in the clinical setting. Indeed, whatever the HPV genotype or the cellular origin of (pre)neoplastic lesions, the majority of HPV infections clear spontaneously (without any medical action) within 1–2 year(s) [76,77]. In addition, predicting disease progression is extremely challenging and subject to several biases, such as the tissue sampling, over/underdiagnosis, and the age of patients. Therefore, despite cumulative efforts for discovering reliable predictive biomarkers, the most cost-effective management of HPV-related lesions is to avoid relying on these costly and imprecise "false prophets".

HPV vaccines, recommended for young sexually-naive women, hold promise to make a major impact on cervical cancer mortality in the future. However, a large number of older and/or vulnerable women are in need of a cancer preventive. The discovery of SC junction cells might point the way to a different, low-cost, and simple way to prevent cervical cancer in underserved populations: by removing the vulnerable cells that are its source. This recently-proposed clinical perspective is supported by historical data [78–80]. Over 50 years ago, Paul Younge (Boston Lying-in Hospital) routinely cauterized the uterine cervices of his patients postpartum. He apparently treated approximately 6000 cervices in this manner and had never witnessed a subsequent cancer in his patients [81]. Several years later, another study noted a profound reduction in cancer risk with cauterization of the cervix [82]. These single practice results were further confirmed by a population-based program conducted in Finland during the 1960s and 1970s [83]. A recent study in South Africa analyzed the occurrence of HPV infections following cervical cryotherapy in healthy (human immunodeficiency virus (HIV)/HPV negative) women. Importantly, a 55% reduction rate was detected in the treated group compared to control individuals [84]. Although prophylactic SC junction (cryo)ablation is unlikely to completely prevent the development of cervical (pre)cancers, this procedure could significantly reduce the risk of subsequent (pre)cancerous lesions similar to what is extensively reported in women who underwent conization [85]. Indeed, these latter patients are characterized by an impressively low recurrence rate given the risk of reinfection by new HPV genotypes, suggesting that the SC junction excision induces a protective effect extending the type-specific immunity potentially acquired by these women. However, this concept still needs to be validated in a controlled clinical trial. The potential value of SC junction ablation in lowering the risk of CIN2/3 in HPV-positive women is another concept worthy of exploration. Repetitive HPV testing is highly inefficient and prone to the vagaries of patient attitudes towards return visits. A strategy in which a single positive HPV test would generate an SC junction intervention of low morbidity might be an interesting alternative in cancer prevention. In addition, by reducing the rate of HPV infections, SC junction removal might also impact on HIV acquisition, which has been shown to be higher in HPV-positive individuals [86–90].

4. Conclusions and Perspectives

With the development and commercialization of two vaccines, as well as the Nobel Prize attributed to Harald Zur Hausen, experimental research on HPV-related carcinogenesis was supposed to be (almost) over in the late 2000s. Three major findings/confirmations have recently raised the interest of the whole HPV community: (1) the involvement of beta HPV genotypes in skin cancer development [91,92]; (2) the significance of HPV variants in terms of cancer risk [93]; and (3) the detection of discrete cell populations topographically located in SC junctions and displaying a high susceptibility to HPV infection [9]. The existence of these latter non-squamous cells was assumed for a long time due, mainly, to the observation of most cervical (pre)cancers in the SC junction, as well as the high percentage (~80%) of columnar neoplasms (adenocarcinoma) etiologically linked to carcinogenic HPV [3,4,94]. However, these well-described results were frequently not taken into account by both virologists and epidemiologists. Moving forward, there is little doubt that additional features explaining the high vulnerability of SC junction cells to HPV infection and related carcinogenesis will be discovered in the early future. Therefore, the description of results highlighted in the present review is likely to be incomplete. One additional matter of investigations is the HPV life cycle in SC junction cells and/or adenocarcinoma. Still unknown in these latter cases, it is undoubted that both squamous differentiation and cellular stratification, considered as crucial for the occurrence of a productive infection [95], will not play a role.

Acknowledgments: The authors thank their colleagues at the University of Liege and Brigham and Women's Hospital for their helpful discussions. This work was supported in part by the Belgian Fund for Medical Scientific Research (FNRS/Televie), by the Centre Anti-Cancereux près l'Université de Liège, by the Fonds Léon Frédéricq, and by the Seventh Framework Program for Research and Technological Development (European Commission: Infect-ERA 2015 (HPV-Motiva)).

Author Contributions: M.H. reviewed the literature and wrote the manuscript. M.H. and T.R.S. collected tissue specimens. M.H. generated the figures. P.D. and C.P.C. provided edits and approved the final version.

Conflicts of Interest: M.H. served as member of one advisory board for GlaxoSmithKline (GSK). The other authors declare no conflict of interest.

References

1. Woodman, C.B.; Collins, S.I.; Young, L.S. The natural history of cervical HPV infection: Unresolved issues. *Nat. Rev. Cancer* **2007**, *7*, 11–22. [CrossRef] [PubMed]
2. Roberts, J.N.; Buck, C.B.; Thompson, C.D.; Kines, R.; Bernardo, M.; Choyke, P.L.; Lowy, D.R.; Schiller, J.T. Genital transmission of HPV in a mouse model is potentiated by nonoxynol-9 and inhibited by carrageenan. *Nat. Med.* **2007**, *13*, 857–861. [CrossRef] [PubMed]
3. Marsh, M. Original site of cervical carcinoma; topographical relationship of carcinoma of the cervix to the external os and to the squamocolumnar junction. *Obstet. Gynecol.* **1956**, *7*, 444–452. [CrossRef] [PubMed]
4. Richart, R.M. Cervical intraepithelial neoplasia. *Pathol. Ann.* **1973**, *8*, 301–328.
5. Forman, D.; de Martel, C.; Lacey, C.J.; Soerjomataram, I.; Lortet-Tieulent, J.; Bruni, L.; Vignat, J.; Ferlay, J.; Bray, F.; Plummer, M.; Franceschi, S. Global burden of human papillomavirus and related diseases. *Vaccine* **2012**, *30* (Suppl. 5), F12–F23. [CrossRef] [PubMed]
6. Martens, J.E.; Arends, J.; Van der Linden, P.J.; De Boer, B.A.; Helmerhorst, T.J. Cytokeratin 17 and p63 are markers of the HPV target cell, the cervical stem cell. *Anticancer Res.* **2004**, *24*, 771–775. [PubMed]
7. Martens, J.E.; Smedts, F.M.; Ploeger, D.; Helmerhorst, T.J.; Ramaekers, F.C.; Arends, J.W.; Hopman, A.H. Distribution pattern and marker profile show two subpopulations of reserve cells in the endocervical canal. *Int. J. Gynecol. Pathol.* **2009**, *28*, 381–388. [CrossRef] [PubMed]
8. Chin, N.; Platt, A.B.; Nuovo, G.J. Squamous intraepithelial lesions arising in benign endocervical polyps: A report of 9 cases with correlation to the Pap smears, HPV analysis, and immunoprofile. *Int. J. Gynecol. Pathol.* **2008**, *27*, 582–590. [CrossRef] [PubMed]

9. Herfs, M.; Yamamoto, Y.; Laury, A.; Wang, X.; Nucci, M.R.; McLaughlin-Drubin, M.E.; Munger, K.; Feldman, S.; McKeon, F.D.; Xian, W.; et al. A discrete population of squamocolumnar junction cells implicated in the pathogenesis of cervical cancer. *Proc. Natl. Acad. Sci. USA* **2012**, *109*, 10516–10521. [CrossRef] [PubMed]

10. Herfs, M.; Vargas, S.O.; Yamamoto, Y.; Howitt, B.E.; Nucci, M.R.; Hornick, J.L.; McKeon, F.D.; Xian, W.; Crum, C.P. A novel blueprint for 'top down' differentiation defines the cervical squamocolumnar junction during development, reproductive life, and neoplasia. *J. Pathol.* **2013**, *229*, 460–468. [CrossRef] [PubMed]

11. Wang, X.; Ouyang, H.; Yamamoto, Y.; Kumar, P.A.; Wei, T.S.; Dagher, R.; Vincent, M.; Lu, X.; Bellizzi, A.M.; Ho, K.Y.; et al. Residual embryonic cells as precursors of a Barrett's-like metaplasia. *Cell* **2011**, *145*, 1023–1035. [CrossRef] [PubMed]

12. Herfs, M.; Parra-Herran, C.; Howitt, B.E.; Laury, A.R.; Nucci, M.R.; Feldman, S.; Jimenez, C.A.; McKeon, F.D.; Xian, W.; Crum, C.P. Cervical squamocolumnar junction-specific markers define distinct, clinically relevant subsets of low-grade squamous intraepithelial lesions. *Am. J. Surg. Pathol.* **2013**, *37*, 1311–1318. [CrossRef] [PubMed]

13. Paquette, C.; Mills, A.M.; Stoler, M.H. Predictive Value of Cytokeratin 7 Immunohistochemistry in Cervical Low-grade Squamous Intraepithelial Lesion as a Marker for Risk of Progression to a High-grade Lesion. *Am. J. Surg. Pathol.* **2016**, *40*, 236–243. [CrossRef] [PubMed]

14. Mills, A.M.; Paquette, C.; Terzic, T.; Castle, P.E.; Stoler, M.H. CK7 Immunohistochemistry as a Predictor of CIN1 Progression: A Retrospective Study of Patients From the Quadrivalent HPV Vaccine Trials. *Am. J. Surg. Pathol.* **2017**, *41*, 143–152. [CrossRef] [PubMed]

15. Huang, E.C.; Tomic, M.M.; Hanamornroongruang, S.; Meserve, E.E.; Herfs, M.; Crum, C.P. p16ink4 and cytokeratin 7 immunostaining in predicting HSIL outcome for low-grade squamous intraepithelial lesions: A case series, literature review and commentary. *Mod. Pathol.* **2016**, *29*, 1501–1510. [CrossRef] [PubMed]

16. Peng, X.; Wu, Z.; Yu, L.; Li, J.; Xu, W.; Chan, H.C.; Zhang, Y.; Hu, L. Overexpression of cystic fibrosis transmembrane conductance regulator (CFTR) is associated with human cervical cancer malignancy, progression and prognosis. *Gynecol. Oncol.* **2012**, *125*, 470–476. [CrossRef] [PubMed]

17. Lee, H.; Lee, H.; Cho, Y.K. Cytokeratin7 and cytokeratin19 expression in high grade cervical intraepithelial neoplasm and squamous cell carcinoma and their possible association in cervical carcinogenesis. *Diagn. Pathol.* **2017**, *12*, 18. [CrossRef] [PubMed]

18. Mirkovic, J.; Howitt, B.E.; Roncarati, P.; Demoulin, S.; Suarez-Carmona, M.; Hubert, P.; McKeon, F.D.; Xian, W.; Li, A.; Delvenne, P.; et al. Carcinogenic HPV infection in the cervical squamo-columnar junction. *J. Pathol.* **2015**, *236*, 265–271. [CrossRef] [PubMed]

19. Yarbrough, V.L.; Winkle, S.; Herbst-Kralovetz, M.M. Antimicrobial peptides in the female reproductive tract: A critical component of the mucosal immune barrier with physiological and clinical implications. *Hum. Reprod. Updat.* **2015**, *21*, 353–377. [CrossRef] [PubMed]

20. Suarez-Carmona, M.; Hubert, P.; Delvenne, P.; Herfs, M. Defensins: "Simple" antimicrobial peptides or broad-spectrum molecules? *Cytokine Growth Factor Rev.* **2015**, *26*, 361–370. [CrossRef] [PubMed]

21. Hancock, R.E.; Haney, E.F.; Gill, E.E. The immunology of host defence peptides: Beyond antimicrobial activity. *Nat. Rev. Immunol.* **2016**, *16*, 321–334. [CrossRef] [PubMed]

22. Buck, C.B.; Day, P.M.; Thompson, C.D.; Lubkowski, J.; Lu, W.; Lowy, D.R.; Schiller, J.T. Human α-defensins block papillomavirus infection. *Proc. Natl. Acad. Sci. USA* **2006**, *103*, 1516–1521. [CrossRef] [PubMed]

23. Hubert, P.; Herman, L.; Maillard, C.; Caberg, J.H.; Nikkels, A.; Pierard, G.; Foidart, J.M.; Noel, A.; Boniver, J.; Delvenne, P. Defensins induce the recruitment of dendritic cells in cervical human papillomavirus-associated (pre)neoplastic lesions formed in vitro and transplanted in vivo. *FASEB J.* **2007**, *21*, 2765–2775. [CrossRef] [PubMed]

24. Zhao, S.; Zhou, H.Y.; Li, H.; Yi, T.; Zhao, X. The therapeutic impact of HNP-1 in condyloma acuminatum. *Int. J. Dermatol.* **2015**, *54*, 1205–1210. [CrossRef] [PubMed]

25. Hubert, P.; Herman, L.; Roncarati, P.; Maillard, C.; Renoux, V.; Demoulin, S.; Erpicum, C.; Foidart, J.M.; Boniver, J.; Noel, A.; et al. Altered α-defensin 5 expression in cervical squamocolumnar junction: Implication in the formation of a viral/tumour-permissive microenvironment. *J. Pathol.* **2014**, *234*, 464–477. [CrossRef] [PubMed]

26. Wiens, M.E.; Smith, J.G. α-defensin HD5 inhibits furin cleavage of human papillomavirus 16 L2 to block infection. *J. Virol.* **2015**, *89*, 2866–2874. [CrossRef] [PubMed]

27. Nguyen, E.K.; Nemerow, G.R.; Smith, J.G. Direct evidence from single-cell analysis that human α-defensins block adenovirus uncoating to neutralize infection. *J. Virol.* **2010**, *84*, 4041–4049. [CrossRef] [PubMed]

28. Zins, S.R.; Nelson, C.D.; Maginnis, M.S.; Banerjee, R.; O'Hara, B.A.; Atwood, W.J. The human α-defensin HD5 neutralizes JC polyomavirus infection by reducing endoplasmic reticulum traffic and stabilizing the viral capsid. *J. Virol.* **2014**, *88*, 948–960. [CrossRef] [PubMed]

29. Smith, J.G.; Nemerow, G.R. Mechanism of adenovirus neutralization by Human α-defensins. *Cell Host Microbe* **2008**, *3*, 11–19. [CrossRef] [PubMed]

30. Wiens, M.E.; Smith, J.G. α-Defensin HD5 Inhibits Human Papillomavirus 16 Infection via Capsid Stabilization and Redirection to the Lysosome. *mBio* **2017**, *8*, e02304–e02316. [CrossRef] [PubMed]

31. Day, P.M.; Schelhaas, M. Concepts of papillomavirus entry into host cells. *Curr. Opin. Virol.* **2014**, *4*, 24–31. [CrossRef] [PubMed]

32. Grassel, L.; Fast, L.A.; Scheffer, K.D.; Boukhallouk, F.; Spoden, G.A.; Tenzer, S.; Boller, K.; Bago, R.; Rajesh, S.; Overduin, M.; et al. The CD63-Syntenin-1 Complex Controls Post-Endocytic Trafficking of Oncogenic Human Papillomaviruses. *Sci. Rep.* **2016**, *6*, 32337. [CrossRef] [PubMed]

33. McBride, A.A. The papillomavirus E2 proteins. *Virology* **2013**, *445*, 57–79. [CrossRef] [PubMed]

34. Thierry, F. Transcriptional regulation of the papillomavirus oncogenes by cellular and viral transcription factors in cervical carcinoma. *Virology* **2009**, *384*, 375–379. [CrossRef] [PubMed]

35. Wang, X.; Meyers, C.; Wang, H.K.; Chow, L.T.; Zheng, Z.M. Construction of a full transcription map of human papillomavirus type 18 during productive viral infection. *J. Virol.* **2011**, *85*, 8080–8092. [CrossRef] [PubMed]

36. Chen, J.; Xue, Y.; Poidinger, M.; Lim, T.; Chew, S.H.; Pang, C.L.; Abastado, J.P.; Thierry, F. Mapping of HPV transcripts in four human cervical lesions using RNAseq suggests quantitative rearrangements during carcinogenic progression. *Virology* **2014**, *462–463*, 14–24. [CrossRef] [PubMed]

37. Taguchi, A.; Nagasaka, K.; Kawana, K.; Hashimoto, K.; Kusumoto-Matsuo, R.; Plessy, C.; Thomas, M.; Nakamura, H.; Bonetti, A.; Oda, K.; et al. Characterization of novel transcripts of human papillomavirus type 16 using cap analysis gene expression technology. *J. Virol.* **2015**, *89*, 2448–2452. [CrossRef] [PubMed]

38. Schmitt, M.; Dalstein, V.; Waterboer, T.; Clavel, C.; Gissmann, L.; Pawlita, M. The HPV16 transcriptome in cervical lesions of different grades. *Mol. Cell. Probes* **2011**, *25*, 260–265. [CrossRef] [PubMed]

39. Kanduc, D. Translational regulation of human papillomavirus type 16 E7 mRNA by the peptide SEQIKA, shared by rabbit α_1-globin and human cytokeratin 7. *J. Virol.* **2002**, *76*, 7040–7048. [CrossRef] [PubMed]

40. Favia, G.; Kanduc, D.; Lo Muzio, L.; Lucchese, A.; Serpico, R. Possible association between HPV16 E7 protein level and cytokeratin 19. *Int. J. Cancer* **2004**, *111*, 795–797. [CrossRef] [PubMed]

41. Herfs, M.; Longuespee, R.; Quick, C.M.; Roncarati, P.; Suarez-Carmona, M.; Hubert, P.; Lebeau, A.; Bruyere, D.; Mazzucchelli, G.; Smargiasso, N.; et al. Proteomic signatures reveal a dualistic and clinically relevant classification of anal canal carcinoma. *J. Pathol.* **2017**, *241*, 522–533. [CrossRef] [PubMed]

42. Gorodeski, G.I.; Hopfer, U.; Liu, C.C.; Margles, E. Estrogen acidifies vaginal pH by up-regulation of proton secretion via the apical membrane of vaginal-ectocervical epithelial cells. *Endocrinology* **2005**, *146*, 816–824. [CrossRef] [PubMed]

43. Herfs, M.; Hubert, P.; Kholod, N.; Caberg, J.H.; Gilles, C.; Berx, G.; Savagner, P.; Boniver, J.; Delvenne, P. Transforming growth factor-β1-mediated Slug and Snail transcription factor up-regulation reduces the density of Langerhans cells in epithelial metaplasia by affecting E-cadherin expression. *Am. J. Pathol.* **2008**, *172*, 1391–1402. [CrossRef] [PubMed]

44. Caberg, J.H.; Hubert, P.M.; Begon, D.Y.; Herfs, M.F.; Roncarati, P.J.; Boniver, J.J.; Delvenne, P.O. Silencing of E7 oncogene restores functional E-cadherin expression in human papillomavirus 16-transformed keratinocytes. *Carcinogenesis* **2008**, *29*, 1441–1447. [CrossRef] [PubMed]

45. Hubert, P.; Caberg, J.H.; Gilles, C.; Bousarghin, L.; Franzen-Detrooz, E.; Boniver, J.; Delvenne, P. E-cadherin-dependent adhesion of dendritic and Langerhans cells to keratinocytes is defective in cervical human papillomavirus-associated (pre)neoplastic lesions. *J. Pathol.* **2005**, *206*, 346–355. [CrossRef] [PubMed]

46. Jiang, A.; Bloom, O.; Ono, S.; Cui, W.; Unternaehrer, J.; Jiang, S.; Whitney, J.A.; Connolly, J.; Banchereau, J.; Mellman, I. Disruption of E-cadherin-mediated adhesion induces a functionally distinct pathway of dendritic cell maturation. *Immunity* **2007**, *27*, 610–624. [CrossRef] [PubMed]

47. Mayumi, N.; Watanabe, E.; Norose, Y.; Watari, E.; Kawana, S.; Geijtenbeek, T.B.; Takahashi, H. E-cadherin interactions are required for Langerhans cell differentiation. *Eur. J. Immunol.* **2013**, *43*, 270–280. [CrossRef] [PubMed]

48. Herfs, M.; Herman, L.; Hubert, P.; Minner, F.; Arafa, M.; Roncarati, P.; Henrotin, Y.; Boniver, J.; Delvenne, P. High expression of PGE2 enzymatic pathways in cervical (pre)neoplastic lesions and functional consequences for antigen-presenting cells. *Cancer Immunol. Immunother. CII* **2009**, *58*, 603–614. [CrossRef] [PubMed]

49. Demoulin, S.A.; Somja, J.; Duray, A.; Guenin, S.; Roncarati, P.; Delvenne, P.O.; Herfs, M.F.; Hubert, P.M. Cervical (pre)neoplastic microenvironment promotes the emergence of tolerogenic dendritic cells via RANKL secretion. *Oncoimmunology* **2015**, *4*, e1008334. [CrossRef] [PubMed]

50. Demoulin, S.; Herfs, M.; Somja, J.; Roncarati, P.; Delvenne, P.; Hubert, P. HMGB1 secretion during cervical carcinogenesis promotes the acquisition of a tolerogenic functionality by plasmacytoid dendritic cells. *Int. J. Cancer* **2015**, *137*, 345–358. [CrossRef] [PubMed]

51. Brown, D.R.; Schroeder, J.M.; Bryan, J.T.; Stoler, M.H.; Fife, K.H. Detection of multiple human papillomavirus types in Condylomata acuminata lesions from otherwise healthy and immunosuppressed patients. *J. Clin. Microbiol.* **1999**, *37*, 3316–3322. [PubMed]

52. Alves de Sousa, N.L.; Alves, R.R.; Martins, M.R.; Barros, N.K.; Ribeiro, A.A.; Zeferino, L.C.; Dufloth, R.M.; Rabelo-Santos, S.H. Cytopathic effects of human papillomavirus infection and the severity of cervical intraepithelial neoplasia: A frequency study. *Diagn. Cytopathol.* **2012**, *40*, 871–875. [CrossRef] [PubMed]

53. Vrdoljak-Mozetic, D.; Krasevic, M.; Versa Ostojic, D.; Stemberger-Papic, S.; Rubesa-Mihaljevic, R.; Bubonja-Sonje, M. HPV16 genotype, p16/Ki-67 dual staining and koilocytic morphology as potential predictors of the clinical outcome for cervical low-grade squamous intraepithelial lesions. *Cytopathology* **2015**, *26*, 10–18. [CrossRef] [PubMed]

54. Ceballos, K.M.; Chapman, W.; Daya, D.; Julian, J.A.; Lytwyn, A.; McLachlin, C.M.; Elit, L. Reproducibility of the histological diagnosis of cervical dysplasia among pathologists from 4 continents. *Int. J. Gynecol. Pathol.* **2008**, *27*, 101–107. [CrossRef] [PubMed]

55. Dalla Palma, P.; Giorgi Rossi, P.; Collina, G.; Buccoliero, A.M.; Ghiringhello, B.; Gilioli, E.; Onnis, G.L.; Aldovini, D.; Galanti, G.; Casadei, G.; et al. The reproducibility of CIN diagnoses among different pathologists: Data from histology reviews from a multicenter randomized study. *Am. J. Clin. Pathol.* **2009**, *132*, 125–132. [CrossRef] [PubMed]

56. Herfs, M.; Crum, C.P. Laboratory management of cervical intraepithelial neoplasia: Proposing a new paradigm. *Adv. Anat. Pathol.* **2013**, *20*, 86–94. [CrossRef] [PubMed]

57. Park, J.J.; Sun, D.; Quade, B.J.; Flynn, C.; Sheets, E.E.; Yang, A.; McKeon, F.; Crum, C.P. Stratified mucin-producing intraepithelial lesions of the cervix: Adenosquamous or columnar cell neoplasia? *Am. J. Surg. Pathol.* **2000**, *24*, 1414–1419. [CrossRef] [PubMed]

58. Crum, C.P. Contemporary theories of cervical carcinogenesis: The virus, the host, and the stem cell. *Mod. Pathol.* **2000**, *13*, 243–251. [CrossRef] [PubMed]

59. Smotkin, D.; Berek, J.S.; Fu, Y.S.; Hacker, N.F.; Major, F.J.; Wettstein, F.O. Human papillomavirus deoxyribonucleic acid in adenocarcinoma and adenosquamous carcinoma of the uterine cervix. *Obstet. Gynecol.* **1986**, *68*, 241–244. [PubMed]

60. Witkiewicz, A.K.; Hecht, J.L.; Cviko, A.; McKeon, F.D.; Ince, T.A.; Crum, C.P. Microglandular hyperplasia: A model for the de novo emergence and evolution of endocervical reserve cells. *Hum. Pathol.* **2005**, *36*, 154–161. [CrossRef] [PubMed]

61. Palefsky, J.M.; Giuliano, A.R.; Goldstone, S.; Moreira, E.D., Jr.; Aranda, C.; Jessen, H.; Hillman, R.; Ferris, D.; Coutlee, F.; Stoler, M.H.; et al. HPV vaccine against anal HPV infection and anal intraepithelial neoplasia. *N. Engl. J. Med.* **2011**, *365*, 1576–1585. [CrossRef] [PubMed]

62. Lehtinen, M.; Paavonen, J.; Wheeler, C.M.; Jaisamrarn, U.; Garland, S.M.; Castellsague, X.; Skinner, S.R.; Apter, D.; Naud, P.; Salmeron, J.; et al. Overall efficacy of HPV-16/18 AS04-adjuvanted vaccine against grade 3 or greater cervical intraepithelial neoplasia: 4-year end-of-study analysis of the randomised, double-blind PATRICIA trial. *Lancet Oncol.* **2012**, *13*, 89–99. [CrossRef]

63. Munoz, N.; Kjaer, S.K.; Sigurdsson, K.; Iversen, O.E.; Hernandez-Avila, M.; Wheeler, C.M.; Perez, G.; Brown, D.R.; Koutsky, L.A.; Tay, E.H.; et al. Impact of human papillomavirus (HPV)-6/11/16/18 vaccine on all HPV-associated genital diseases in young women. *J. Natl. Cancer Inst.* **2010**, *102*, 325–339. [CrossRef] [PubMed]

64. Schiller, J.T.; Castellsague, X.; Garland, S.M. A review of clinical trials of human papillomavirus prophylactic vaccines. *Vaccine* 2012, *30* (Suppl. 5), F123–F138. [CrossRef] [PubMed]
65. De Sanjose, S.; Diaz, M.; Castellsague, X.; Clifford, G.; Bruni, L.; Munoz, N.; Bosch, F.X. Worldwide prevalence and genotype distribution of cervical human papillomavirus DNA in women with normal cytology: A meta-analysis. *Lancet Infect. Dis.* 2007, *7*, 453–459. [CrossRef]
66. Chesson, H.W.; Ekwueme, D.U.; Saraiya, M.; Watson, M.; Lowy, D.R.; Markowitz, L.E. Estimates of the annual direct medical costs of the prevention and treatment of disease associated with human papillomavirus in the United States. *Vaccine* 2012, *30*, 6016–6019. [CrossRef] [PubMed]
67. Massad, L.S.; Einstein, M.H.; Huh, W.K.; Katki, H.A.; Kinney, W.K.; Schiffman, M.; Solomon, D.; Wentzensen, N.; Lawson, H.W.; Conference, A.C.G. 2012 updated consensus guidelines for the management of abnormal cervical cancer screening tests and cancer precursors. *Obstet. Gynecol.* 2013, *121*, 829–846. [CrossRef] [PubMed]
68. Sagasta, A.; Castillo, P.; Saco, A.; Torne, A.; Esteve, R.; Marimon, L.; Ordi, J.; Del Pino, M. p16 staining has limited value in predicting the outcome of histological low-grade squamous intraepithelial lesions of the cervix. *Mod. Pathol.* 2016, *29*, 51–59. [CrossRef] [PubMed]
69. Liao, G.D.; Sellors, J.W.; Sun, H.K.; Zhang, X.; Bao, Y.P.; Jeronimo, J.; Chen, W.; Zhao, F.H.; Song, Y.; Cao, Z.; et al. p16INK4A immunohistochemical staining and predictive value for progression of cervical intraepithelial neoplasia grade 1: A prospective study in China. *Int. J. Cancer* 2014, *134*, 1715–1724. [CrossRef] [PubMed]
70. Matsumoto, K.; Oki, A.; Furuta, R.; Maeda, H.; Yasugi, T.; Takatsuka, N.; Mitsuhashi, A.; Fujii, T.; Hirai, Y.; Iwasaka, T.; et al. Predicting the progression of cervical precursor lesions by human papillomavirus genotyping: A prospective cohort study. *Int. J. Cancer* 2011, *128*, 2898–2910. [CrossRef] [PubMed]
71. Pajtler, M.; Milicic-Juhas, V.; Milojkovic, M.; Topolovec, Z.; Curzik, D.; Mihaljevic, I. Assessment of HPV DNA test value in management women with cytological findings of ASC-US, CIN1 and CIN2. *Coll. Antropol.* 2010, *34*, 81–86. [PubMed]
72. Gravitt, P.E.; Kovacic, M.B.; Herrero, R.; Schiffman, M.; Bratti, C.; Hildesheim, A.; Morales, J.; Alfaro, M.; Sherman, M.E.; Wacholder, S.; et al. High load for most high risk human papillomavirus genotypes is associated with prevalent cervical cancer precursors but only HPV16 load predicts the development of incident disease. *Int. J. Cancer* 2007, *121*, 2787–2793. [CrossRef] [PubMed]
73. Dalstein, V.; Riethmuller, D.; Pretet, J.L.; Le Bail Carval, K.; Sautiere, J.L.; Carbillet, J.P.; Kantelip, B.; Schaal, J.P.; Mougin, C. Persistence and load of high-risk HPV are predictors for development of high-grade cervical lesions: A longitudinal French cohort study. *Int. J. Cancer* 2003, *106*, 396–403. [CrossRef] [PubMed]
74. Hesselink, A.T.; Berkhof, J.; Heideman, D.A.; Bulkmans, N.W.; van Tellingen, J.E.; Meijer, C.J.; Snijders, P.J. High-risk human papillomavirus DNA load in a population-based cervical screening cohort in relation to the detection of high-grade cervical intraepithelial neoplasia and cervical cancer. *Int. J. Cancer* 2009, *124*, 381–386. [CrossRef] [PubMed]
75. Xi, L.F.; Koutsky, L.A.; Castle, P.E.; Wheeler, C.M.; Galloway, D.A.; Mao, C.; Ho, J.; Kiviat, N.B. Human papillomavirus type 18 DNA load and 2-year cumulative diagnoses of cervical intraepithelial neoplasia grades 2–3. *J. Natl. Cancer Inst.* 2009, *101*, 153–161. [CrossRef] [PubMed]
76. Holowaty, P.; Miller, A.B.; Rohan, T.; To, T. Natural history of dysplasia of the uterine cervix. *J. Natl. Cancer Inst.* 1999, *91*, 252–258. [CrossRef] [PubMed]
77. Schlecht, N.F.; Platt, R.W.; Duarte-Franco, E.; Costa, M.C.; Sobrinho, J.P.; Prado, J.C.; Ferenczy, A.; Rohan, T.E.; Villa, L.L.; Franco, E.L. Human papillomavirus infection and time to progression and regression of cervical intraepithelial neoplasia. *J. Natl. Cancer Inst.* 2003, *95*, 1336–1343. [CrossRef] [PubMed]
78. Herfs, M.; Crum, C.P. Cervical cancer: Squamocolumnar junction ablation—Tying up loose ends? *Nat. Rev. Clin. Oncol.* 2015, *12*, 378–380. [CrossRef] [PubMed]
79. Herfs, M.; Somja, J.; Howitt, B.E.; Suarez-Carmona, M.; Kustermans, G.; Hubert, P.; Doyen, J.; Goffin, F.; Kridelka, F.; Crum, C.P.; et al. Unique recurrence patterns of cervical intraepithelial neoplasia after excision of the squamocolumnar junction. *Int. J. Cancer* 2015, *136*, 1043–1052. [CrossRef] [PubMed]
80. Franceschi, S. Past and future of prophylactic ablation of the cervical squamocolumnar junction. *Ecancermedicalscience* 2015, *9*, 527. [PubMed]
81. Younge, P.A. Cancer of the uterine cervix; a preventable disease. *Obstet. Gynecol.* 1957, *10*, 469–481. [PubMed]

82. Peyton, F.W.; Peyton, R.R.; Anderson, V.L.; Pavnica, P. The importance of cauterization to maintain a healthy cervix. Long-term study from a private gynecologic practice. *Am. J. Obstet. Gynecol.* **1978**, *131*, 374–380. [CrossRef]

83. Kauraniemi, T.; Rasanen-Virtanen, U.; Hakama, M. Risk of cervical cancer among an electrocoagulated population. *Am. J. Obstet. Gynecol.* **1978**, *131*, 533–538. [CrossRef]

84. Taylor, S.; Wang, C.; Wright, T.C.; Denny, L.; Tsai, W.Y.; Kuhn, L. Reduced acquisition and reactivation of human papillomavirus infections among older women treated with cryotherapy: Results from a randomized trial in South Africa. *BMC Med.* **2010**, *8*, 40. [CrossRef] [PubMed]

85. Kocken, M.; Helmerhorst, T.J.; Berkhof, J.; Louwers, J.A.; Nobbenhuis, M.A.; Bais, A.G.; Hogewoning, C.J.; Zaal, A.; Verheijen, R.H.; Snijders, P.J.; et al. Risk of recurrent high-grade cervical intraepithelial neoplasia after successful treatment: A long-term multi-cohort study. *Lancet Oncol.* **2011**, *12*, 441–450. [CrossRef]

86. Herfs, M.; Hubert, P.; Moutschen, M.; Delvenne, P. Mucosal junctions: Open doors to HPV and HIV infections? *Trends Microbiol.* **2011**, *19*, 114–120. [CrossRef] [PubMed]

87. Auvert, B.; Lissouba, P.; Cutler, E.; Zarca, K.; Puren, A.; Taljaard, D. Association of oncogenic and nononcogenic human papillomavirus with HIV incidence. *J. Acquir. Immune Defic. Syndr.* **2010**, *53*, 111–116. [CrossRef] [PubMed]

88. Lissouba, P.; Van de Perre, P.; Auvert, B. Association of genital human papillomavirus infection with HIV acquisition: A systematic review and meta-analysis. *Sex. Transm. Infect.* **2013**, *89*, 350–356. [CrossRef] [PubMed]

89. Averbach, S.H.; Gravitt, P.E.; Nowak, R.G.; Celentano, D.D.; Dunbar, M.S.; Morrison, C.S.; Grimes, B.; Padian, N.S. The association between cervical human papillomavirus infection and HIV acquisition among women in Zimbabwe. *Aids* **2010**, *24*, 1035–1042. [CrossRef] [PubMed]

90. Smith-McCune, K.K.; Shiboski, S.; Chirenje, M.Z.; Magure, T.; Tuveson, J.; Ma, Y.; Da Costa, M.; Moscicki, A.B.; Palefsky, J.M.; Makunike-Mutasa, R.; et al. Type-specific cervico-vaginal human papillomavirus infection increases risk of HIV acquisition independent of other sexually transmitted infections. *PLoS ONE* **2010**, *5*, e10094. [CrossRef] [PubMed]

91. Tommasino, M. The biology of beta human papillomaviruses. *Virus Res.* **2017**, *231*, 128–138. [CrossRef] [PubMed]

92. Viarisio, D.; Mueller-Decker, K.; Kloz, U.; Aengeneyndt, B.; Kopp-Schneider, A.; Grone, H.J.; Gheit, T.; Flechtenmacher, C.; Gissmann, L.; Tommasino, M. E6 and E7 from beta HPV38 cooperate with ultraviolet light in the development of actinic keratosis-like lesions and squamous cell carcinoma in mice. *PLoS Pathog.* **2011**, *7*, e1002125. [CrossRef] [PubMed]

93. Mirabello, L.; Yeager, M.; Cullen, M.; Boland, J.F.; Chen, Z.; Wentzensen, N.; Zhang, X.; Yu, K.; Yang, Q.; Mitchell, J.; et al. HPV16 Sublineage Associations With Histology-Specific Cancer Risk Using HPV Whole-Genome Sequences in 3200 Women. *J. Natl. Cancer Inst.* **2016**, *108*. [CrossRef] [PubMed]

94. Pirog, E.C.; Kleter, B.; Olgac, S.; Bobkiewicz, P.; Lindeman, J.; Quint, W.G.; Richart, R.M.; Isacson, C. Prevalence of human papillomavirus DNA in different histological subtypes of cervical adenocarcinoma. *Am. J. Pathol.* **2000**, *157*, 1055–1062. [CrossRef]

95. Kajitani, N.; Satsuka, A.; Kawate, A.; Sakai, H. Productive Lifecycle of Human Papillomaviruses that Depends Upon Squamous Epithelial Differentiation. *Front. Microbiol.* **2012**, *3*, 152. [CrossRef] [PubMed]

viruses

Review

Human Papillomavirus and the Stroma: Bidirectional Crosstalk during the Virus Life Cycle and Carcinogenesis

Megan E. Spurgeon and Paul F. Lambert *

McArdle Laboratory for Cancer Research, Department of Oncology, University of Wisconsin-Madison, Madison, WI 53705, USA; megan.spurgeon@wisc.edu
* Correspondence: plambert@wisc.edu; Tel.: +1-608-262-8533

Academic Editors: Alison A. McBride and Karl Munger
Received: 30 June 2017; Accepted: 4 August 2017; Published: 9 August 2017

Abstract: Human papillomaviruses (HPVs) are double-stranded DNA (dsDNA) tumor viruses that are causally associated with human cancers of the anogenital tract, skin, and oral cavity. Despite the availability of prophylactic vaccines, HPVs remain a major global health issue due to inadequate vaccine availability and vaccination coverage. The HPV life cycle is established and completed in the terminally differentiating stratified epithelia, and decades of research using in vitro organotypic raft cultures and in vivo genetically engineered mouse models have contributed to our understanding of the interactions between HPVs and the epithelium. More recently, important and emerging roles for the underlying stroma, or microenvironment, during the HPV life cycle and HPV-induced disease have become clear. This review discusses the current understanding of the bidirectional communication and relationship between HPV-infected epithelia and the surrounding microenvironment. As is the case with other human cancers, evidence suggests that the stroma functions as a significant partner in tumorigenesis and helps facilitate the oncogenic potential of HPVs in the stratified epithelium.

Keywords: human papillomavirus; stroma; tumor microenvironment; cervical cancer; paracrine signaling; epithelial–stromal interactions

---

1. Introduction

Viruses are the etiological agents of approximately 15% of human cancers worldwide [1]. There are currently seven identified human tumor viruses, and techniques to discover new viruses associated with human malignancies are continuously emerging and evolving [2]. High-risk human papillomaviruses (HPVs), a group of small DNA tumor viruses that infect the stratified squamous epithelia (referred to hereafter as the stratified epithelium or stratified epithelia), are alone responsible for nearly 5% of worldwide cancers [3]. There are more than 150 types of HPV, and around 40 HPV types are considered mucosotropic in that they preferentially infect the mucosal stratified epithelia of the anogenital tract (cervix, vagina, and anus) and the oral cavity [4,5]. HPV infections are the most common sexually transmitted infection in the U.S., and sexually active individuals are usually infected with HPV at least once during their lifetime [6,7].

However, not all HPV infections cause cancer. HPVs are considered low-risk or high-risk based on their oncogenic potential [8]. Low-risk mucosotropic HPVs cause benign genital warts, whereas those HPV types considered high-risk are causally associated with nearly all cases of cervical cancer and other cancers of the lower female reproductive tract and anus, as well as an increasing proportion of oropharyngeal cancers. The high-risk HPV types HPV16 and HPV18 are particularly formidable, causing the majority of HPV-associated cervical, anal, and oral cancers [9–11]. The introduction of prophylactic vaccines that prevent infection by those high-risk HPVs most highly linked to cancers (e.g., HPV16 and HPV18) was a major breakthrough in controlling HPV-associated carcinogenesis [12]. Unfortunately, suboptimal vaccine availability and lagging vaccination coverage continue to make HPV-associated cancers a significant global health issue [13]. A more complete understanding of HPV virus–host interactions during infection and carcinogenesis therefore remains key to identifying new preventative and therapeutic options.

More than 90% of HPV infections in the cervix are naturally cleared [14]. Thus, the factors and mechanisms involved in cases that do ultimately progress to cancer remain the focus of intense research. In the absence of clearance, HPV infections can persist for years or even decades, and these persistent high-risk HPV infections are a major risk factor for subsequent cancer development (Reviewed in [15,16]). While the mechanisms that govern viral persistence and progression to cancer are still unclear, both HPV type and viral load appear to be involved [17,18]. Integration of the HPV viral genome into host DNA is a frequent event during persistent HPV infection and occurs at random sites unique to each cancer [19,20], although a preference for chromosomal fragile sites has been reported [21]. Viral genome integration increases and stabilizes transcription of the viral oncogenes E6 and E7 [22,23] and provides a growth advantage to epithelial cells [24]. HPV E6 and E7 are highly multifunctional proteins that contribute to carcinogenesis, primarily through their ability to inactivate the p53 and retinoblastoma protein (pRb) tumor suppressor pathways [25,26]. Despite its demonstrated role in carcinogenesis, persistent HPV infection is not sufficient for human cervical cancer development [27,28], and thus other co-factors likely contribute to this process.

An increasingly recognized co-factor in many types of cancer is the adjacent stroma, or 'microenvironment'. The stroma is a supportive scaffold upon which epithelial cells reside, and is composed of connective tissue, vasculature, and various cell types (Figure 1). The role of the tumor microenvironment (TME) in carcinogenesis is a relatively nascent, yet growing, area of cancer biology research. Fibroblasts are a major component of the stroma, and tumor-associated or cancer-associated fibroblasts (TAFs or CAFs, respectively) have become a major focus of TME research [29]. Communication between epithelial cells and the TME has been reported to affect processes ranging from tumor initiation and neoplastic progression to metastasis and therapeutic response [30]. HPV-associated cervical carcinogenesis is accompanied by changes in stromal gene expression throughout neoplastic progression, which reflect the likely role of the stroma in supporting angiogenesis and epithelial invasion [31]. However, an in-depth understanding of the interaction between HPV-infected tissue and the surrounding microenvironment has not been achieved. In this review, we explore the current knowledge of the interplay between HPV-positive epithelia and the stroma during viral infection and carcinogenesis. Emerging evidence suggests that these tissue compartments engage in bidirectional crosstalk to facilitate HPV infection and HPV-associated cancers.

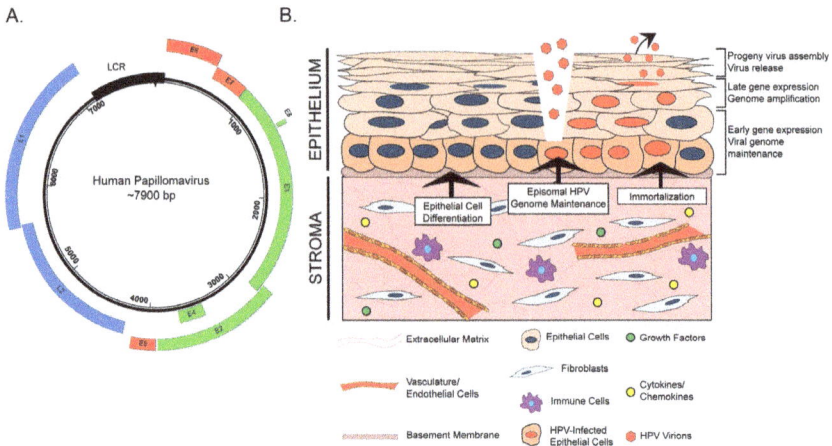

Figure 1. The human papillomavirus (HPV) life cycle and the stroma. (**A**) Map of the HPV genome. Shown is the circular, ~7900 base pair (bp) double-stranded DNA (dsDNA) genome of HPV18. The boxes represent translational open reading frames that encode HPV proteins. Early (E) genes are shown in green and late genes (L) are shown in blue. The long control region (LCR), which regulates transcription and viral DNA replication, is shown by a black box. Three HPV oncogenes, E5, E6, and E7, are shown in red. (**B**) A schematic representation of the stratified epithelium, underlying stroma, and the HPV life cycle is shown. Basal cells, which are adjacent to the basement membrane and underlying stroma, are shown in dark tan. HPV virions infect the basal cells of the stratified epithelium, presumably through a break or wound in the epithelial layer. The virus life cycle proceeds throughout the epithelium and is tied to differentiation, ending with progeny virus production and release from the terminally differentiated cells. Labels indicating the spatiotemporal regulation of key events in the HPV life cycle are shown to the right of the stratified epithelium. The stroma, which is composed of various cell types, vasculature, and connective tissue, has been shown to promote epithelial cell immortalization, maintenance of episomal HPV genomes, and epithelial cell differentiation in support of the HPV life cycle (indicated by black arrows). Please refer to the legend to identify common components of the stroma, and to the text for further detail.

2. The HPV Life Cycle and the Stroma

HPVs are icosahedral, non-enveloped viruses approximately 50–60 nm in diameter. These viruses contain a double-stranded, circular or episomal DNA genome of approximately 8000 base pairs (Figure 1A). A temporal pattern of viral gene expression gives rise to 'early' and 'late' genes, which are transcribed using a complex system of promoter usage and splicing patterns [32,33]. HPVs exhibit tropism for the stratified epithelia, and the viral life cycle is intimately tied to the process of cellular differentiation in this tissue (Reviewed in [34]). HPVs infect the poorly differentiated, basal keratinocytes of the stratified epithelium, to which they are believed to gain access through wounds or breaks in the epithelial layer (Figure 1B). Naturally, the basal cells are the only cells actively engaged in cell division within the stratified epithelium. This characteristic of basal keratinocytes helps explain their being targeted by HPV for infection, as both viral genome nuclear entry and genome maintenance require active cell division [35]. Episomal viral genomes are then maintained in basal cells at a low copy number. As infected daughter cells enter the process of terminal differentiation, early viral genes reprogram the parabasal and suprabasal cells to re-enter the cell cycle. A process of viral genome amplification, which is dependent upon host DNA synthesis/repair machinery [36,37], is followed by late gene expression and viral progeny production in and release from the superficial layers of the terminally differentiated stratified epithelium.

The role of the stroma during initial HPV infection and the life cycle is an underexplored area of research. There is evidence that stromal fibroblasts and/or feeder layers, which are comprised of mitotically inactive fibroblasts, affect various stages of the HPV virus life cycle through both direct and indirect mechanisms (Figure 1B). Rheinwald and Green first identified feeder layers as an important component of in vitro keratinocyte culture and propagation [38]. Fibroblasts feeder layers promote epithelial homeostasis and proliferation by providing critical soluble factors, such as growth factors and cytokines, and extracellular matrix components that function in juxtacrine and paracrine manners [39]. There is evidence that feeder layers are necessary for maintenance of the HPV genome, which is replicated and maintained as an extrachromosomal circular plasmid (or episome) in HPV-infected cells. In W12 keratinocytes, which were isolated from an HPV16-infected low-grade cervical lesion, the HPV genome exists in extrachromosomal episomal form. Removal of co-cultured feeder cells from several W12 clonal populations resulted in the gradual loss of HPV episomes and promoted viral genome integration in all clones [40]. Consistent with this finding, unpublished work from our laboratory indicates that high-risk HPV genomes are not maintained in oral keratinocytes in the absence of a feeder layer [41]. Another role for the stroma in the HPV life cycle involves its critical function to facilitate differentiation and stratification of the epithelium through paracrine interactions [42,43], which is essential for completion of the HPV life cycle [44]. To study the complete HPV life cycle, in vitro methods such as three-dimensional organotypic raft culture methods are often used [45,46], which are composed of keratinocytes cultured at the air–liquid interface on top of a dermal equivalent containing fibroblasts [47]. The proper architecture, differentiation, and proliferation of the stratified epithelium in these biomimetic in vitro cultures requires the physical presence of fibroblasts, the reciprocal exchange of diffusible factors, and crosstalk between the two compartments through growth factors and cytokines [42,48–53]. Other stromal cell types, such as endothelial cells, also support epithelial differentiation [50,51]. Thus, while the underlying mechanistic understanding is very limited to date, interactions between the stroma and the epithelial compartment clearly contribute to HPV genome maintenance and the differentiation-dependent completion of the HPV life cycle.

3. HPV and the Stroma: Bidirectional Paracrine Effects

The stromal microenvironment is a critical, if not wholly essential, participant in several facets of both normal and cancer epithelial tissue biology [29,42]. In the previous section, we discussed the limited knowledge about how stromal cells contribute to HPV infection and the viral life cycle. In the following sections, we will review our current understanding of how HPV-positive epithelia and the stromal microenvironment communicate with and influence one another during HPV-associated disease and cancer development. While, below, we describe the epithelial-to-stromal communication separately from stromal-to-epithelial communication, it should be kept in mind that there likely are complex feedback loops in which signals emanating from one compartment inform on how the other compartment responds back, and that this bidirectional cross talk is likely key to how the viruses manipulate their environment, and in the case of tumor-associated viruses, contribute to carcinogenesis.

3.1. Effects of the HPV-Positive Epithelium on the Stromal Microenvironment

It is increasingly clear that virus-infected cells communicate with the surrounding microenvironment [54]. In the section below, we describe key examples of effects of HPV-positive epithelial cells on stromal architecture, angiogenesis, inflammation, and immune cell recruitment (Figure 2). This epithelial-to-stromal communication likely initiates a process of bidirectional crosstalk and interdependent signaling pathways that persists during HPV-associated disease progression.

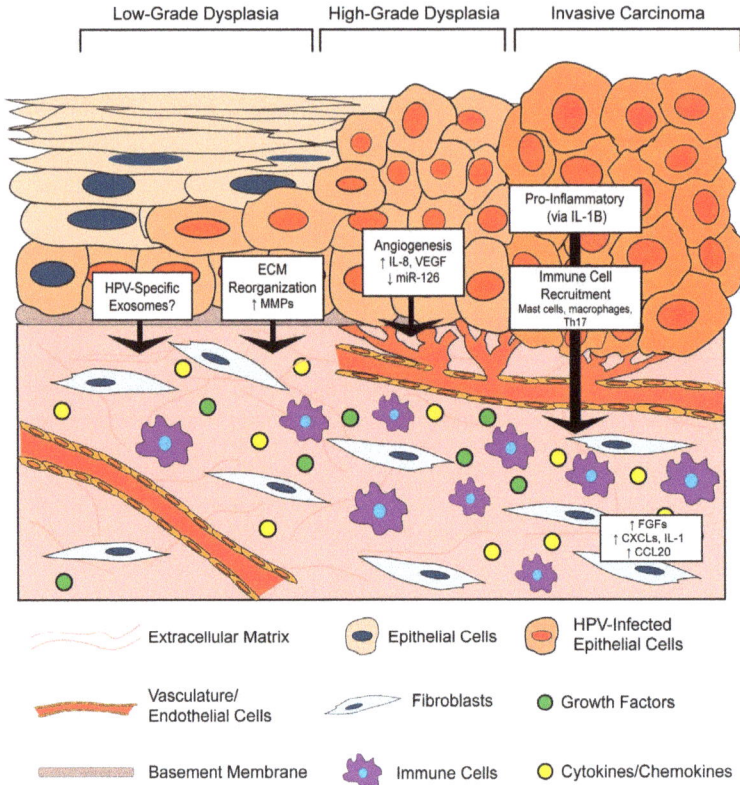

Figure 2. Effects of the HPV-positive epithelium on the stroma. The HPV-infected epithelium undergoes neoplastic progression from low-grade to high-grade dysplasia and eventually invasive carcinoma. Key epithelial-to-stromal communication events between the HPV-positive epithelium and the stroma, as well as the reported paracrine factors involved in these processes, are depicted. Large black arrows indicate events in the HPV-positive epithelium that affect the underlying stroma. Small black arrows represent an increase (up arrow) or decrease (down arrow) in the indicated factor. The location of signaling events within the process of neoplastic progression is not necessarily reflective of their temporal activities. ECM: extracellular matrix; CCL20: C-C Motif Chemokine Ligand 20; CXCLs: C-X-C Motif Chemokine Ligands; FGFs: Fibroblast growth factors; IL-1: Interleukin-1; IL-1B: Interleukin-1B; IL-8: Interleukin-8; miR-126: microRNA-126; MMP: matrix metalloproteinase; Th17: T helper 17 cells; VEGF: vascular endothelial growth factor.

3.1.1. HPV Effects on Stromal Architecture

HPV-positive epithelial cells influence the composition and organization of the microenvironment in both in vitro and in vivo models. This paracrine effect of HPV on the underlying stroma may, in part, be due to the ability of E6 and E7 to alter the secretory profile of epithelial cells, which has been shown to regulate the invasive potential of in vitro organotypic raft epithelial cells via signaling between the epithelium and stroma [55]. For instance, HPV16-positive CSCC27 cervical cancer cells promote changes in co-cultured cervical fibroblasts, including the production of a laminin-rich matrix and reduced fibronectin and collagen [56]. Mechanistically, these changes were attributed to reciprocal actions between the HPV-positive epithelial cells and fibroblasts involving matrix metalloproteinases (MMPs): the presence of fibroblasts promoted MMP-7 expression in the epithelial cells, whereas CSCC27 epithelial cells induced MMP-2 expression in the fibroblasts. MMPs are extracellular proteinases with well-known functions in modifying TME architecture, and their enzymatic activities can directly contribute to carcinogenesis by affecting processes such as extracellular matrix reorganization, angiogenesis, and inflammation [57]. Several additional lines of evidence support a role for HPVs in modulating MMP expression. In another in vitro study, HPV18-positive SKG-II cells increased pro-MMP-1 and pro-MMP-3 in co-cultured cervical fibroblasts, presumably through tumor-cell derived soluble factors [58]. Similar findings were reported for HPV16- and HPV18-immortalized cells and HPV-positive cervical cancer cells, where an increase in the membrane-type 1 MMP (MT-1 MMP) as well as MMP-2 was observed, and some evidence suggests that this is a unique function of E7 proteins from high-risk HPV types [59]. Other groups showed that E6 cooperates with E7 to increase MT-1 MMP, MMP-2 and MMP-9 expression in order to enhance epithelial cell stromal invasion and migration [60,61]. Therefore, HPV-positive epithelial cells increase MMP expression, which can contribute to stromal reorganization.

The influence of HPV on the stroma in vivo has been largely investigated through the study of genetically engineered HPV transgenic mice. These in vivo models allow observations within a complete biological system, and are therefore especially critical to understanding the interplay between the epithelium and tumor microenvironment. There are two main HPV transgenic murine models, both of which target HPV genes to the basal cells of the stratified epithelium using the basal keratinocyte-specific keratin 14 (K14) promoter. One model is the *K14-HPV16* model in which the entire HPV16 early region is expressed [62]. In the other model, transgenic mice singly expressing the HPV16 E6 or E7 oncogenes, *K14E6* [63] and *K14E7* [64] mice, which can be crossed to generate bitransgenic *K14E6/E7* mice. In addition to studying cervical cancer, *K14E6*, *K14E7*, and *K14E6/E7* transgenic mice have been used to study HPV-associated cancers at other anatomical sites, including the skin [65], oral cavity [66], and anus [67].

In the skin of *K14HPV16* transgenic mice, the underlying stromal architecture was extensively remodeled during the course of neoplastic progression [68,69]. Architectural changes arose even in premalignant lesions, in the absence of epithelial dysplasia and malignancy, indicating that HPV-positive epithelia can induce reorganization of the microenvironment beginning during the early stages of neoplastic progression. These structural changes included thinning of the basement membrane, apparent degradation and disruption of the collagen fibril network, and additional disintegration of the extracellular matrix [68]. Much of this reorganization was attributed to an infiltration of inflammatory cells, primarily mast cells, and their associated protease activities [69]. Thus, data support a role for HPV in facilitating epithelial-to-stromal signals that result in extracellular matrix reorganization at least in part through HPV-induced MMP expression.

3.1.2. HPV Effects on Angiogenesis in the Stroma

In both the human cervix and the murine cervix of *K14HPV16* mice, angiogenesis and vascular density increases during progression to cancer [70]. Increased vascularity is observed even in early cervical lesions, which implies that HPV infection itself or early consequences of infection promote angiogenesis [71]. HPV-mediated angiogenesis has been directly linked to the functions of the HPV oncoproteins in a variety of in vitro and in vivo studies. In work by Chen et al. [72], conditioned media was collected from human foreskin keratinocytes (HFKs) either transduced with HPV16 E6/E7 or stably transfected with the entire HPV16 genome, or media from the HPV31-positive, cervical intraepithelial neoplasia (CIN) derived cell line, CIN612. Application of conditioned media from these HPV positive cells to endothelial cells in vitro increased their proliferation and migration. This conditioned media was also analyzed in an in vivo Matrigel plug assay, which showed remarkably enhanced vascularization at seven days post-implantation in those plugs composed of HPV-positive media compared to HPV-negative controls. Interestingly, there was a much greater response in vivo, leading the authors to speculate that multiple stromal cell types contribute to this HPV-dependent angiogenic response. Analysis of conditioned media from cells expressing HPV16 E6 identified a significant increase in the pro-angiogenic factors vascular endothelial growth factor (VEGF) compared to that of parental cells [73]. Others observed an increase in VEGF and interleukin (IL)-8 along with reduced expression of angiogenesis inhibitors, thrombospondin-1 and maspin, in human keratinocytes expressing both HPV16 E6 and E7 [72,74] and that expression of both HPV16 E6 and E7 together was necessary to induce angiogenesis [75].

In addition to the secretion of pro-angiogenic factors from HPV-positive epithelial cells that function in a paracrine manner, there is also evidence that HPV-positive cells can stimulate pro-angiogenic gene expression in cells within the adjacent stroma. For instance, CAFs isolated from the stroma of a cervical cancer secreted more VEGF than cervical cancer epithelial cells under both normal and hypoxic conditions [76]. More recently, an intriguing mechanism was reported in which HPV16-positive CaSki cells were found to reduce expression of a micro-RNA (miRNA), miR-126, in endothelial cells [77]. This observation was made using an in vitro tri-culture system composed of CaSki cancer epithelial cells, endothelial cells, and fibroblasts. The reduction of miR-126 in endothelial cells required the presence of both epithelial cells and fibroblasts in the tri-culture, suggesting that a complex network of paracrine interactions is involved in miR-126 regulation. Interestingly, miR-126 was previously identified as a miRNA downregulated in HPV16-positive cervical epithelial cells compared to normal epithelial cells [78], perhaps suggesting that this miRNA is specifically targeted by HPVs. The effect of this miRNA was then investigated using xenografts, which are cells injected subcutaneously into live mice and allowed to generate tumors in vivo. Increased microvasculature density and tube formation was observed in xenografts composed of CaSki cells and CAFs, and this effect was associated with a decrease in miR-126 in host-derived endothelial cells recruited to the tumor. The pro-angiogenic gene for adrenomedullin (*ADM*) and several other pro-angiogenic genes are targets of miR-126 [79]. Adrenomedullin expression was highly increased in the cervical cancer stroma, and its increased expression correlated with miR-126 downregulation in the tri-culture system. Therefore, this fascinating study reveals a cell non-autonomous role of HPV-positive cervical cancer epithelial cells in promoting angiogenesis in the stroma through the regulation of expression of a miRNA in endothelial cells. Exploration of the role of epithelial–stromal crosstalk and the influence of HPV in regulating miRNA expression is only now beginning, but promises to reveal important mechanisms involved in HPV-associated carcinogenesis.

The pro-angiogenic effects of HPV-positive epithelial cells on the stroma have also been observed using in vivo murine models. One such example involves the release of platelet-derived growth factor (PDGF) from the *K14HPV16* cervical epithelium, which leads to expression of fibroblast growth factors 2 and 7 (FGF-2 and FGF-7) in the stroma to promote angiogenesis and tumor cell growth [80]. Fibroblasts isolated from the dermis adjacent to *K14HPV16* murine skin produce several pro-inflammatory cytokines with well-known roles in angiogenesis [81]. When these pro-inflammatory fibroblasts were mixed and co-injected with PDSC5 HPV16-positive skin carcinoma epithelial cells, there was a higher level of vascularization and microvasculature density compared to xenografts with normal fibroblasts. Similar results were observed using an in vivo Matrigel plug assay. Interestingly, this study also found that HPV-positive epithelial cells could 'educate' normal fibroblasts in vitro to become pro-inflammatory, and found that this paracrine education is mediated by epithelial-derived interleukin-1β (IL-1B). In a detailed and descriptive study by Coussens et al., important insight is provided into how pro-inflammatory stromal cells function as 'co-conspirators' in angiogenesis in the *K14HPV16* epidermis [69]. Beginning early in neoplastic progression, the stroma is characterized by increased capillary density, corroborating other reports that HPV may promote angiogenesis prior to malignant conversion [71]. Throughout neoplastic progression, dermal capillaries become more dilated and enlarged, their density increases, and they become increasingly localized adjacent to the basement membrane. These changes are associated with an increasing pro-inflammatory environment and infiltration of mast cells [69] and macrophages [81]. The mast-cell protease monocyte chemotactic protein-4 (MCP-4), or chymase, was found to be key in promoting an angiogenic switch, as angiogenesis was severely reduced in mast-cell deficient *K14HPV16* mice. Therefore, there are several mechanisms by which the HPV-positive epithelium can instruct changes in fibroblasts, endothelial cells, and other stromal cells that enhance angiogenesis and cancer progression.

3.1.3. HPV Effects on Inflammation and Immune Cell Recruitment in the Stroma

HPVs are notorious for their ability to evade the host immune response during infection, mainly due to the nature of their intraepithelial life cycle that provides ample cover from host detection [82,83]. However, during the course of progression, the microenvironment becomes increasingly associated with an immune or inflammatory infiltrate [71]. Inflammation is involved in several stages of HPV-associated carcinogenesis [84]. While the mechanisms are not completely clear, several studies indicate that HPV-positive epithelial cells direct pro-inflammatory gene expression in fibroblasts or other stromal cell types, which results in immune cell recruitment. For a more comprehensive review of the interaction between HPVs, the immune system, and the stroma, please refer to the review by Woodby et al. [85].

As described in previous sections, a strong pro-inflammatory gene expression signature was measured in fibroblasts isolated from dysplastic skin of *K14HPV16* transgenic mice and was induced by epithelial cell-secreted IL-1B [81]. A prominently represented group of genes within this inflammatory signature included several ELR+ C-X-C family of chemokines (*CXCL1, CXCL2, CXCL5*), interleukins (*IL-1B, IL-6*), the cyclooxygenase-2 gene *COX2*, and several other genes with well-known pro-inflammatory roles. This group of C-X-C motif ligand chemokines (CXCLs), which are all ligands for the CXCR2 receptor, are considered pro-angiogenic and also function as chemoattractants for neutrophils and macrophages [86,87]. Indeed, there was an increase in macrophage recruitment to xenografts and Matrigel plugs containing these pro-inflammatory fibroblasts [81]. A similar pro-inflammatory signature was observed in fibroblasts isolated from human cervical cancers, and antiestrogen treatment of these cervical CAFs revealed that the inflammation-associated gene expression is partially mediated by estrogen [88]. Our own unpublished results have identified a strong pro-inflammatory signature in the cervical stroma of estrogen-treated *K14E6/E7* mice [89]. It will be important to further define the role of the HPV-positive epithelia in inducing pro-inflammatory cytokines and chemokines, as well as their role in the stroma during HPV infection and carcinogenesis.

In addition to epithelial-derived IL-1B as described above [81], several other mechanisms have been proposed for induction of inflammation and immune cell recruitment by HPV-positive epithelial cells. For instance, it was reported that HPV-positive epithelial cells create a pro-tumorigenic and inflammatory microenvironment by secreting IL-6 [90,91]. The IL-6 secreted from HPV-positive epithelial cells induces expression of the chemokine C-C motif chemokine ligand 20 (CCL20) in stromal fibroblasts to attract pro-inflammatory T helper 17 (Th17) cells [92]. Xenograft studies comparing cervical cell lines concluded that the HPV-positive HeLa and SiHa cell xenografts recruited significantly more inflammatory cells, particularly macrophages, than the HPV-negative C33A cell xenografts in vivo, and this was due to increased IL-6 and IL-8 secretion from HPV-positive epithelial cells [93]. Tumor-associated macrophages in the stroma of HPV-positive tumors have been shown to secrete IL-10 in order to suppress antitumor T cell responses and create a pro-tumorigenic microenvironment [94,95].

Inflammatory cell recruitment by the HPV-positive epithelium was also observed in HPV transgenic mice. In *K14HPV16* mice, a significant influx of mast cells that secrete MMP-9 was observed in the microenvironment, which was necessary for promoting angiogenesis and epithelial proliferation in the skin [69,96]: Interestingly, the MMP-9 appeared to be mainly supplied by infiltrating immune cells, including mast cells, neutrophils, and macrophages, and not the epithelial cells [96]. While the individual contributions of the HPV viral proteins to inflammation and immune cell recruitment have not been fully elucidated, the HPV16 E7 protein alone in *K14E7* mice is sufficient for leukocyte trafficking to the skin and this function depends on its interaction with the pRb tumor suppressor [97,98]. An independent study found that the HPV16 E7 expressed in *K14E7* mouse skin is sufficient to recruit mast cells through elicitation of CCL cytokines, CCL2 and CCL5, and that this contributes to the oncogene driving establishment of an immunosuppressive environment within the skin, in a manner that is dependent upon the induction of epithelial hyperplasia by E7 [97]. Therefore, HPV-positive cells coordinate with the surrounding stroma to elicit a pro-inflammatory microenvironment, which in turn facilitates the immune and inflammatory infiltrate observed in the stroma of HPV-associated lesions and cancers [71]. Further elucidation of the role of HPV in promoting epithelial-to-stromal crosstalk in these processes will enhance our understanding of HPV-driven carcinogenesis.

3.2. Effects of the Stromal Microenvironment on HPV-Positive Epithelia

Upon HPV infection, it is clear that the stratified epithelium initiates communication with the underlying stroma. Presumably in response to this communication, the microenvironment reciprocates contact through various stromal-to-epithelial signaling events. While these processes are still not well understood, there is evidence that the stroma plays a role in HPV-positive epithelial cell growth and disease initiation and maintenance.

3.2.1. Stromal Effects on Growth and Differentiation of HPV-Positive Epithelia

Given its role in normal epithelial cell biology, it is not surprising that the stromal compartment also affects the growth and differentiation of HPV-infected epithelial cells (Figure 3). One such effect is HPV-induced cellular immortalization. High-risk HPVs can extend the life span of epithelial cells and ultimately induce cell immortalization via the functions of the E6 and E7 oncoproteins (Reviewed in [99]), and fibroblasts appear to cooperate in epithelial cell immortalization. For instance, feeder layers of growth-arrested murine 3T3 fibroblasts were required for epithelial cell immortalization by Rho kinase (ROCK) inhibitors [100,101]. Moreover, the presence of a feeder layer greatly enhanced the ability of the HPV E6/E7 proteins to immortalize cells in vitro [102]. Stromal cells also affect the growth properties of HPV-positive epithelial cells. Dermal fibroblasts induced anchorage-independent growth of HPV16-immortalized cervical epithelial cells, and this likely involved paracrine-acting factors since the effect did not require direct cell-to-cell contact [103]. In this same set of experiments, the presence of fibroblasts enhanced epithelial cell growth in organotypic rafts and increased expression of IL-1 in the co-cultured HPV16-positive epithelia. IL-1 is a cytokine implicated in epithelial–stromal interactions. In an illuminating body of work using organotypic rafts, Maas-Szabowski and colleagues defined a 'double-paracrine' epithelial–stromal signaling mechanism involving the IL-1 proteins, IL-1α (IL-1A) and IL-1B, that regulates epithelial differentiation and growth [43,49,104]. This work demonstrates that release of IL-1 by epithelial cells causes increased expression of keratinocyte growth factor (KGF; also known as FGF-7) in co-cultured fibroblasts, which subsequently feeds back in a paracrine manner to increase epithelial proliferation and tissue formation. Interestingly, IL-1A was reported to provide a selective growth advantage specifically to HPV16- and HPV18-positive cervical epithelial cells, and inhibited the growth of normal epithelial cells [105]. This paracrine signaling loop was further implicated in stromal regulation of growth and differentiation of HPV-positive epithelia in a series of studies by the McCance laboratory. In organotypic raft cultures composed of primary human foreskin keratinocytes immortalized with HPV16 E6 and E7 proteins and primary human foreskin fibroblasts, fibroblast-specific depletion of pRb increased KGF expression to facilitate epithelial invasion [106,107]. They also found that AKT kinase activation in the stromal fibroblasts increases the invasiveness of HPV-positive epithelia in a KGF-dependent manner [108]. These in vitro experiments indicate that both pRb and AKT in fibroblasts regulate the invasion of HPV-positive epithelia in a cell non-autonomous manner, and provide direct experimental evidence that the stroma regulates growth properties of the HPV-infected stratified epithelia.

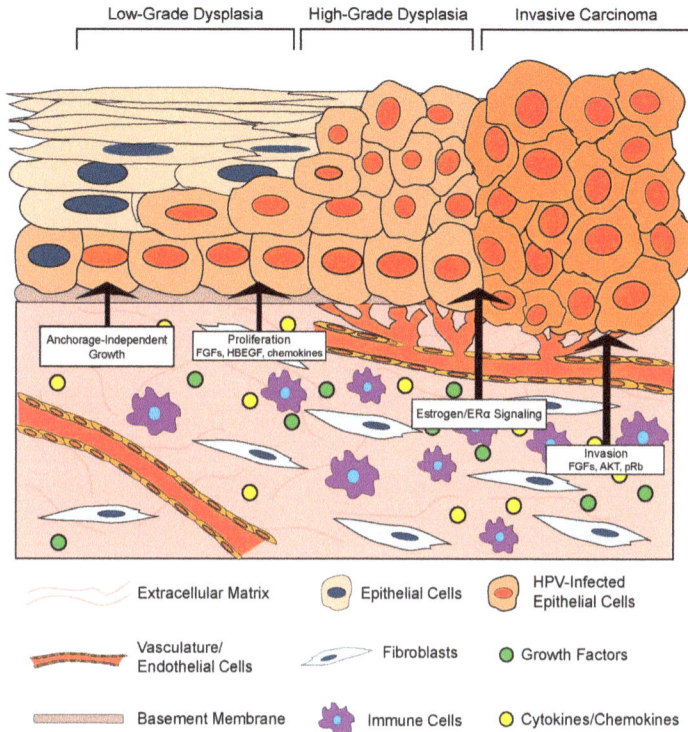

Figure 3. Effects of the stroma on HPV-positive epithelium. The HPV-infected epithelium undergoes neoplastic progression from low-grade to high-grade dysplasia and eventually invasive carcinoma. Key stromal-to-epithelial communication events between the stroma and the HPV-positive epithelium, as well as the reported paracrine factors involved in these processes, are depicted (large black arrows). The location of signaling events within the process of neoplastic progression is not necessarily reflective of their temporal activities. ERα: estrogen receptor α; HBEGF: heparin-binding epidermal growth factor-like growth factor; pRb: retinoblastoma protein.

Fibroblast growth factors are a family of secreted factors that are significant players in communication between the stroma and epithelium and are involved in processes ranging from development to cancer [109]. There is evidence that FGFs and their receptors may facilitate HPV-mediated carcinogenesis through a variety of mechanisms involving stromal-to-epithelial crosstalk [110]. As outlined above, FGF-7 (or KGF) has been implicated in stromal regulation of HPV-positive epithelial invasion. There is also evidence that another FGF, FGF-2 (or bFGF) correlates with the invasive potential of HPV-positive epithelial cells. The effects of primary human fibroblasts isolated from cervical normal or cancer tissue on the invasive properties of normal keratinocytes or HPV16-immortalized keratinocytes were analyzed using an in vitro Matrigel invasion assay [111]. In this study, conditioned media from CAFs stimulated invasivity of epithelial cells and this was associated with increased expression of FGF-2. Interestingly, the pro-invasion effect of the CAF-conditioned media and more specifically FGF-2 was only observed with HPV-positive epithelial cells. Conversely, conditioned media from normal fibroblasts inhibited penetration, and this inhibitory effect was mediated by transforming growth factor-β (TGF-β). From these studies, the authors concluded that the paracrine effects mediated by fibroblasts were due to an increased sensitivity of HPV-positive epithelial cells to FGF-2 and TGF-β, thus increasing their mobility and invasive phenotype. Other factors besides FGFs are also implicated. Human cervical fibroblasts also increased the invasiveness of the HPV18-postive cervical cancer cell line SKG-II by increasing MMP expression in in vitro Matrigel invasion assays [58]. Another group has also reported that CAFs induce migration of HPV-positive cervical cancer cells more than normal fibroblasts, and attributed this effect to differences in extracellular matrix remodeling [56]. Co-culture of HPV+ cervical ME180 and CaSki cancer epithelial cells with CAFs in vitro also enhanced epithelial cell proliferation, and this was dependent on production of an epidermal growth factor receptor (EGFR) ligand, the heparin-binding epidermal growth factor-like growth factor (HBEGF) [112]. Therefore, while a complete understanding of the specific interactions between the stroma and HPV-positive cells is still lacking, it is clear that the microenvironment and cells contained within can have a unique effect on the growth and proliferation of the epithelium when HPV or HPV oncogenes are present.

3.2.2. Stromal Effects on Disease in HPV-Associated Cancer Models

Virus-associated carcinogenesis can be studied in vivo using xenografts as well as genetically engineered murine models [113]. As introduced in previous sections, there are two main HPV transgenic mouse models in which expression of HPV genes; primarily, the E6 and E7 oncogenes are directed specifically to epithelial cells by placing the viral genes under the transcriptional control of K14 transcriptional promoter. Initial studies of HPV transgenic mice revealed that expression of HPV16 E6 and E7 in the cervical epithelium is necessary but not sufficient to cause cervical cancer, which mirrors human data. Subsequent studies identified the female hormone estrogen (17β-estradiol) as a necessary co-factor for HPV-associated cervical carcinogenesis in both murine models [114,115]. In transgenic mice, treatment with 17-β-estradiol and expression of its receptor ERα are required for the onset, maintenance, and progression of neoplastic disease [114–117]. Furthermore, continuous treatment with estrogen signaling inhibitors promotes regression of cervical cancer and precancerous lesions, and prevents their recurrence, in HPV transgenic mice [118–121]. Human epidemiological data also correlate higher estrogen levels and an increased risk factor for cervical cancer [122,123]. The following section focuses on observations of stroma-to-epithelium interactions in vivo using both HPV-positive cervical cancer xenografts and the above referenced animal models of HPV-associated cancers. Emerging evidence linking the requirement for estrogen in cervical carcinogenesis to the stromal compartment will also be discussed.

Several experiments using xenografts have identified positive effects of fibroblasts on HPV-positive epithelial cells. In one set of experiments, xenografts of the HPV-positive cervical cancer cell line ME180 were significantly larger when the epithelial cells were co-inoculated with cervical cancer CAFs or mouse embryonic fibroblasts compared to epithelial cells alone [112]. The positive effect of fibroblasts on epithelial cell growth was mediated by the EGFR ligand, HB-EGF, expressed by the fibroblasts in response to PDGF secreted by HPV-positive epithelial cells [80]. Validating the HB-EGF-dependent mechanism, fibroblasts isolated from HB-EGF-null mice failed to positively affect ME180 xenograft growth. Notably, ME180 xenografts in nude mice rarely metastasize. However, lymph node metastases were observed in mice injected with xenografts composed of both ME180 cells and cervical cancer CAFs [124], suggesting that the fibroblasts provided a microenvironment favorable to HPV-positive epithelial cell migration and metastasis.

Another set of xenograft experiments analyzed the effect of stromal cells on the growth of PDSC5 cells, which are a murine HPV16-positive squamous cell carcinoma cell line derived from the skin of *K14HPV16* mice. In a study by Erez and colleagues, dermal fibroblasts were isolated from areas adjacent to normal skin in non-transgenic mice and dysplastic skin lesions spontaneously arising in *K14HPV16* transgenic mice, and their gene expression transcriptomes compared [81]. The CAFs isolated from the stroma adjacent to HPV-transgenic skin exhibited a significant pro-inflammatory signature composed of several pro-angiogenic cytokines and chemokines. Xenografts containing PDSC5 cells and CAFs or normal dermal fibroblasts were generated in mice. Similar to the results in ME180 cervical cancer cell xenografts, the presence of the pro-inflammatory CAFs increased the growth rate, size, and vascularization of PDSC5 skin cancer cell xenografts. In the PDSC5 xenograft model, expression of the cysteine protease cathepsin C by fibroblasts and leukocytes in the stromal compartment was necessary for tumor growth and angiogenesis [125]. Collectively, these results suggest that stromal cells can actively influence the in vivo growth of HPV-positive epithelial cells derived from various anatomical sites.

There is also evidence that the stroma influences de novo cervical cancers in transgenic mouse models of HPV-associated cervical cancer. In estrogen-treated *K14HPV16* mice, a paracrine loop between the epithelia and stroma involving PDGF and FGFs was identified [80]. In this study, both FGF-2 and FGF-7 were expressed by fibroblasts in response to PDGF secreted by the HPV-positive epithelia. When estrogen-treated *K14HPV16* mice were treated with imatinib to inhibit PDGF signaling, FGF-2 and FGF-7 expression was decreased in the stroma. Imatinib treatment reduced proliferation and increased apoptosis in the epithelium and effectively delayed progression and reduced in vivo cervical cancer growth. These effects were linked to a reduction in angiogenesis driven by fibroblast-derived FGFs. Expression of FGF-2 was similarly increased in the stroma of human cervical cancers. These experiments clearly define a role for the microenvironment in promoting HPV-associated cervical cancer growth through paracrine mechanisms.

In our own studies, we have made the important observation that estrogen contributes to cervical carcinogenesis by signaling through the underlying stroma. An extensive body of work by Cunha and colleagues had previously established that stromal–epithelial interactions mediate the effects of estrogen on morphogenesis of the female reproductive tract, and more specifically that stromal estrogen receptor α (ERα) facilitates the proliferative effects of estrogen on the epithelium [126–129]. In line with these observations, we found that neoplastic cervical disease in *K14E7* mice requires the continuous expression of ERα in the cervical stroma [130]. Stroma-specific deletion of ERα in the cervices of estrogen-treated HPV transgenic mice resulted in a hypoplastic epithelium with significantly reduced proliferation, despite the continued expression of ERα in the epithelial compartment. Moreover, the status of stromal ERα highly correlated with cervical disease, and ERα ablation in the stroma of estrogen-treated HPV transgenic mice significantly reduced cervical disease in the epithelia, and most mice had no signs of disease. These results revealed that the ability of estrogen to induce cervical disease in the HPV-positive epithelium involves estrogen signaling through the stromal compartment.

In a parallel study analyzing human cervical tissue samples spanning neoplastic disease progression from normal cervix to cervical cancers, ERα expression became progressively confined to the stromal compartment but was lost in epithelial cancer cells [131]. Fibroblasts were subsequently identified as being the predominant ERα-positive cell type in the stroma of human cervical cancers, a finding validated by other studies [88]. Similar to previous observations reporting a pro-inflammatory signature in fibroblasts adjacent to *K14HPV16* skin [81], human cervical cancer-associated fibroblasts likewise showed increased pro-tumorigenic and inflammation-associated gene expression [88]. Interestingly, the expression of several pro-tumorigenic genes in ex vivo cultured cervical CAFs was decreased upon treatment antiestrogen drugs, suggesting that stromal fibroblasts can indeed mediate estrogen-dependent signaling in the human cervix [88]. Collectively, these in vivo studies highlight that estrogen signaling in the cervical stroma drives cervical carcinogenesis in the HPV-positive cervical epithelium. We hypothesize that a unique repertoire of paracrine-acting factors is elicited by estrogen in the stroma of HPV transgenic mice that promote cervical cancer development in the adjacent epithelium, and current studies are underway to identify these factors. Nonetheless, there is increasing evidence that the stroma plays a pivotal role in the dynamics and progression of HPV-driven cervical cancer development.

4. Potential Mechanisms of Bidirectional Crosstalk

The functional consequences of HPV-induced epithelial–stromal crosstalk have been discussed throughout this review, and include influences on the HPV life cycle and virus-induced disease. Many of the factors implicated in the crosstalk are proteins secreted into the extracellular space (e.g., growth factors, cytokines/chemokines, proteases). In many cases, however, the mechanism for crosstalk has not been elucidated. One emerging mechanism to explain the communication between tissue compartments is the reciprocal secretion and uptake of extracellular vesicles by epithelial and stromal cells. Exosomes are one particular type of vesicle that has been extensively studied in the context of cancer biology [132]. Exosomes are 40–150 nm-sized vesicles composed of a lipid bilayer and containing intracellular cargo such as proteins, RNA, and/or DNA, all of which can be delivered to recipient cells. The exchange of genetic material, in the form of mRNA and miRNAs, is a particularly potent mechanism to modulate gene expression and function in recipient cells [133]. Importantly, viruses can modulate the vesicle profile of infected cells [54]. For instance, nasopharyngeal carcinoma cells harboring the oncogenic gammaherpesvirus Epstein–Barr virus (EBV) release exosomes that contain viral proteins and miRNAs, as well as signal transduction molecules, and were shown to activate pro-tumorigenic signaling pathways in recipient cells [134]. Several recent reports indicate that HPVs modulate exosome secretion and content. Exosomes secreted from keratinocytes expressing HPV E6 and E7 oncogenes contained several antiapoptotic proteins [135] and unique miRNA expression patterns [136–138]. While one report did not find evidence for E6 or E7 protein in exosomes [135], both E6 and E7 transcripts were detected in exosomes secreted from HPV-positive keratinocytes and cervical cancer cells in a more recent study [136]. The possibility that E6 and E7 transcripts and/or proteins may be transferred to cells via microvesicles would not only provide a potential mechanistic basis for epithelial–stromal crosstalk, but would also represent a major paradigm shift in our understanding of HPV-associated disease. In this regard, reports from one lab that E6 and E7 can be detected in the extracellular space and purified protein can be internalized by endothelial cells is particularly intriguing [139–141]. Furthermore, our own recent unpublished studies indicate that in vivo expression of the HPV oncogenes in the cervical epithelium has a profound, cell non-autonomous effect on global gene expression in nearby stromal cells [89]. It is not currently known whether or how microvesicles secreted from HPV-positive epithelial cells affect the function of surrounding cells, and such experiments will be crucial to dissecting their role in epithelial–stromal communication. A complete dissection of the mechanisms that underlie HPV-induced crosstalk between the stratified epithelium and the stroma will be important in fully understanding how HPV contributes to disease.

5. Conclusions and Future Perspectives

In this review, we have outlined the current knowledge of key interactions between human papillomaviruses and the underlying stroma. Established and emerging evidence indicate that bidirectional crosstalk between HPV-infected epithelial cells and the microenvironment contribute to the HPV life cycle (Figure 1), stromal architecture, angiogenesis, and inflammation (Figure 2), as well as growth of the epithelium and disease progression (Figure 3). While the role of the stroma in HPV infection and disease is increasingly appreciated, future studies are necessary to provide a complete picture of the reciprocal interactions between the two tissue compartments and the role of HPV in these processes. Organotypic cultures have historically been an important tool to study epithelial–stromal interactions and HPV infection in vitro [45,46,104] and will continue to provide a relevant experimental platform for future studies. Likewise, genetically engineered mouse models of HPV-associated disease [113] will allow interrogation of compartment-specific cell types and signaling pathways and their role in carcinogenesis in various anatomical sites. In addition to the animal model in which the rodent *Mastomys coucha* is naturally infected with *Mastomys natalensis* papillomavirus (MnPV) [142], the exciting discovery of a murine papillomavirus (MmuPV1) [143] will allow the ability to study the natural course of papillomavirus infections and disease progression in tractable and genetically modifiable hosts, a traditionally and historically elusive endeavor in the field. Our laboratory has already used the murine MmuPV1 model to study papillomavirus-associated skin disease [144], and future studies are underway to study cervical and oral disease induced by MmuPV1. The MmuPV1-based infection models, together with pre-existing HPV transgenic mouse models, will allow future studies necessary to understanding in vivo mechanisms of epithelial–stromal crosstalk during papillomavirus-induced disease.

Acknowledgments: The National Institutes of Health (NIH) grants CA022443 and CA211246 support Paul F. Lambert and Megan E. Spurgeon.

Author Contributions: M.E.S. and P.F.L. wrote the manuscript.

Conflicts of Interest: The authors declare no conflict of interest.

References

1. De Martel, C.; Ferlay, J.; Franceschi, S.; Vignat, J.; Bray, F.; Forman, D.; Plummer, M. Global burden of cancers attributable to infections in 2008: A review and synthetic analysis. *Lancet Oncol.* **2012**, *13*, 607–615. [CrossRef]
2. Mirvish, E.D.; Shuda, M. Strategies for human tumor virus discoveries: From microscopic observation to digital transcriptome subtraction. *Front. Microbiol.* **2016**, *7*, 676. [CrossRef] [PubMed]
3. Zur Hausen, H. Papillomaviruses in the causation of human cancers—A brief historical account. *Virology* **2009**, *384*, 260–265. [CrossRef] [PubMed]
4. Bernard, H.U.; Burk, R.D.; Chen, Z.; van Doorslaer, K.; zur Hausen, H.; de Villiers, E.M. Classification of papillomaviruses (PVs) based on 189 PV types and proposal of taxonomic amendments. *Virology* **2010**, *401*, 70–79. [CrossRef] [PubMed]
5. Van Doorslaer, K.; Li, Z.; Xirasagar, S.; Maes, P.; Kaminsky, D.; Liou, D.; Sun, Q.; Kaur, R.; Huyen, Y.; McBride, A.A. The papillomavirus episteme: A major update to the papillomavirus sequence database. *Nucleic Acids Res.* **2017**, *45*, D499–D506. [CrossRef] [PubMed]
6. Satterwhite, C.L.; Torrone, E.; Meites, E.; Dunne, E.F.; Mahajan, R.; Ocfemia, M.C.; Su, J.; Xu, F.; Weinstock, H. Sexually transmitted infections among us women and men: Prevalence and incidence estimates, 2008. *Sex. Transm. Dis.* **2013**, *40*, 187–193. [CrossRef] [PubMed]
7. Baseman, J.G.; Koutsky, L.A. The epidemiology of human papillomavirus infections. *J. Clin. Virol.* **2005**, *32* (Suppl. S1), S16–S24. [CrossRef] [PubMed]
8. Cogliano, V.; Baan, R.; Straif, K.; Grosse, Y.; Secretan, B.; El Ghissassi, F.; WHO International Agency for Research on Cancer. Carcinogenicity of human papillomaviruses. *Lancet Oncol.* **2005**, *6*, 204. [CrossRef]
9. D'Souza, G.; Kreimer, A.R.; Viscidi, R.; Pawlita, M.; Fakhry, C.; Koch, W.M.; Westra, W.H.; Gillison, M.L. Case-control study of human papillomavirus and oropharyngeal cancer. *N. Engl. J. Med.* **2007**, *356*, 1944–1956. [CrossRef] [PubMed]
10. Hoots, B.E.; Palefsky, J.M.; Pimenta, J.M.; Smith, J.S. Human papillomavirus type distribution in anal cancer and anal intraepithelial lesions. *Int. J. Cancer* **2009**, *124*, 2375–2383. [CrossRef] [PubMed]
11. Munoz, N.; Bosch, F.X.; Castellsague, X.; Diaz, M.; de Sanjose, S.; Hammouda, D.; Shah, K.V.; Meijer, C.J. Against which human papillomavirus types shall we vaccinate and screen? The international perspective. *Int. J. Cancer* **2004**, *111*, 278–285. [CrossRef] [PubMed]
12. Frazer, I.H.; Leggatt, G.R.; Mattarollo, S.R. Prevention and treatment of papillomavirus-related cancers through immunization. *Annu. Rev. Immunol.* **2011**, *29*, 111–138. [CrossRef] [PubMed]
13. Jemal, A.; Simard, E.P.; Dorell, C.; Noone, A.M.; Markowitz, L.E.; Kohler, B.; Eheman, C.; Saraiya, M.; Bandi, P.; Saslow, D.; et al. Annual report to the nation on the status of cancer, 1975–2009, featuring the burden and trends in human papillomavirus (HPV)-associated cancers and HPV vaccination coverage levels. *J. Natl. Cancer Inst.* **2013**, *105*, 175–201. [CrossRef] [PubMed]
14. Rodriguez, A.C.; Schiffman, M.; Herrero, R.; Hildesheim, A.; Bratti, C.; Sherman, M.E.; Solomon, D.; Guillen, D.; Alfaro, M.; Morales, J.; et al. Longitudinal study of human papillomavirus persistence and cervical intraepithelial neoplasia grade 2/3: Critical role of duration of infection. *J. Natl. Cancer Inst.* **2010**, *102*, 315–324. [CrossRef] [PubMed]
15. Bodily, J.; Laimins, L.A. Persistence of human papillomavirus infection: Keys to malignant progression. *Trends Microbiol.* **2011**, *19*, 33–39. [CrossRef] [PubMed]
16. Moscicki, A.B.; Schiffman, M.; Burchell, A.; Albero, G.; Giuliano, A.R.; Goodman, M.T.; Kjaer, S.K.; Palefsky, J. Updating the natural history of human papillomavirus and anogenital cancers. *Vaccine* **2012**, *30* (Suppl. S5), F24–F33. [CrossRef] [PubMed]

17. Munoz, N.; Hernandez-Suarez, G.; Mendez, F.; Molano, M.; Posso, H.; Moreno, V.; Murillo, R.; Ronderos, M.; Meijer, C.; Munoz, A.; et al. Persistence of HPV infection and risk of high-grade cervical intraepithelial neoplasia in a cohort of colombian women. *Br. J. Cancer* **2009**, *100*, 1184–1190. [CrossRef] [PubMed]

18. Schiffman, M.; Rodriguez, A.C.; Chen, Z.; Wacholder, S.; Herrero, R.; Hildesheim, A.; Desalle, R.; Befano, B.; Yu, K.; Safaeian, M.; et al. A population-based prospective study of carcinogenic human papillomavirus variant lineages, viral persistence, and cervical neoplasia. *Cancer Res.* **2010**, *70*, 3159–3169. [CrossRef] [PubMed]

19. Luft, F.; Klaes, R.; Nees, M.; Durst, M.; Heilmann, V.; Melsheimer, P.; von Knebel Doeberitz, M. Detection of integrated papillomavirus sequences by ligation-mediated PCR (DIPS-PCR) and molecular characterization in cervical cancer cells. *Int. J. Cancer* **2001**, *92*, 9–17. [CrossRef]

20. Wentzensen, N.; Ridder, R.; Klaes, R.; Vinokurova, S.; Schaefer, U.; Doeberitz, M. Characterization of viral-cellular fusion transcripts in a large series of HPV16 and 18 positive anogenital lesions. *Oncogene* **2002**, *21*, 419–426. [CrossRef] [PubMed]

21. Jang, M.K.; Shen, K.; McBride, A.A. Papillomavirus genomes associate with BRD4 to replicate at fragile sites in the host genome. *PLoS Pathog.* **2014**, *10*, e1004117. [CrossRef] [PubMed]

22. Jeon, S.; Lambert, P.F. Integration of human papillomavirus type 16 DNA into the human genome leads to increased stability of E6 and E7 mRNAs: Implications for cervical carcinogenesis. *Proc. Natl. Acad. Sci. USA* **1995**, *92*, 1654–1658. [CrossRef] [PubMed]

23. Ziegert, C.; Wentzensen, N.; Vinokurova, S.; Kisseljov, F.; Einenkel, J.; Hoeckel, M.; von Knebel Doeberitz, M. A comprehensive analysis of HPV integration loci in anogenital lesions combining transcript and genome-based amplification techniques. *Oncogene* **2003**, *22*, 3977–3984. [CrossRef] [PubMed]

24. Jeon, S.; Allen-Hoffmann, B.L.; Lambert, P.F. Integration of human papillomavirus type 16 into the human genome correlates with a selective growth advantage of cells. *J. Virol.* **1995**, *69*, 2989–2997. [PubMed]

25. Munger, K.; Baldwin, A.; Edwards, K.M.; Hayakawa, H.; Nguyen, C.L.; Owens, M.; Grace, M.; Huh, K. Mechanisms of human papillomavirus-induced oncogenesis. *J. Virol.* **2004**, *78*, 11451–11460. [CrossRef] [PubMed]

26. Zur Hausen, H. Papillomaviruses causing cancer: Evasion from host-cell control in early events in carcinogenesis. *J. Natl. Cancer Inst.* **2000**, *92*, 690–698. [CrossRef] [PubMed]

27. Schiffman, M.; Castle, P.E.; Jeronimo, J.; Rodriguez, A.C.; Wacholder, S. Human papillomavirus and cervical cancer. *Lancet* **2007**, *370*, 890–907. [CrossRef]

28. Walboomers, J.M.; Jacobs, M.V.; Manos, M.M.; Bosch, F.X.; Kummer, J.A.; Shah, K.V.; Snijders, P.J.; Peto, J.; Meijer, C.J.; Muñoz, N. Human papillomavirus is a necessary cause of invasive cervical cancer worldwide. *J. Pathol.* **1999**, *189*, 12–19. [CrossRef]

29. Bhowmick, N.A.; Neilson, E.G.; Moses, H.L. Stromal fibroblasts in cancer initiation and progression. *Nature* **2004**, *432*, 332–337. [CrossRef] [PubMed]

30. Quail, D.F.; Joyce, J.A. Microenvironmental regulation of tumor progression and metastasis. *Nat. Med.* **2013**, *19*, 1423–1437. [CrossRef] [PubMed]

31. Gius, D.; Funk, M.C.; Chuang, E.Y.; Feng, S.; Huettner, P.C.; Nguyen, L.; Bradbury, C.M.; Mishra, M.; Gao, S.; Buttin, B.M.; et al. Profiling microdissected epithelium and stroma to model genomic signatures for cervical carcinogenesis accommodating for covariates. *Cancer Res.* **2007**, *67*, 7113–7123. [CrossRef] [PubMed]

32. De Villiers, E.M. Cross-roads in the classification of papillomaviruses. *Virology* **2013**, *445*, 2–10. [CrossRef] [PubMed]

33. Zheng, Z.M.; Baker, C.C. Papillomavirus genome structure, expression, and post-transcriptional regulation. *Front. Biosci.* **2006**, *11*, 2286–2302. [CrossRef] [PubMed]

34. Doorbar, J.; Egawa, N.; Griffin, H.; Kranjec, C.; Murakami, I. Human papillomavirus molecular biology and disease association. *Rev. Med. Virol.* **2015**, *25* (Suppl. S1), 2–23. [CrossRef] [PubMed]

35. Pyeon, D.; Pearce, S.M.; Lank, S.M.; Ahlquist, P.; Lambert, P.F. Establishment of human papillomavirus infection requires cell cycle progression. *PLoS Pathog.* **2009**, *5*, e1000318. [CrossRef] [PubMed]

36. Flores, E.R.; Allen-Hoffmann, B.L.; Lee, D.; Lambert, P.F. The human papillomavirus type 16 E7 oncogene is required for the productive stage of the viral life cycle. *J. Virol.* **2000**, *74*, 6622–6631. [CrossRef] [PubMed]

37. Moody, C.A.; Laimins, L.A. Human papillomaviruses activate the ATM DNA damage pathway for viral genome amplification upon differentiation. *PLoS Pathog.* **2009**, *5*, e1000605. [CrossRef] [PubMed]

38. Rheinwald, J.G.; Green, H. Serial cultivation of strains of human epidermal keratinocytes: The formation of keratinizing colonies from single cells. *Cell* **1975**, *6*, 331–343. [CrossRef]

39. Llames, S.; Garcia-Perez, E.; Meana, A.; Larcher, F.; del Rio, M. Feeder layer cell actions and applications. *Tissue Eng. Part B Rev.* **2015**, *21*, 345–353. [CrossRef] [PubMed]

40. Dall, K.L.; Scarpini, C.G.; Roberts, I.; Winder, D.M.; Stanley, M.A.; Muralidhar, B.; Herdman, M.T.; Pett, M.R.; Coleman, N. Characterization of naturally occurring HPV16 integration sites isolated from cervical keratinocytes under noncompetitive conditions. *Cancer Res.* **2008**, *68*, 8249–8259. [CrossRef] [PubMed]

41. Lee, D.; Lambert, P. Maintenance of HPV genomes in NOKs cells requires fibroblast feeder layer. McArdle Laboratory for Cancer Research, University of Wisconsin-Madison, Madison, WI, USA. Unpublished work, 2017.

42. Werner, S.; Smola, H. Paracrine regulation of keratinocyte proliferation and differentiation. *Trends Cell Biol.* **2001**, *11*, 143–146. [CrossRef]

43. Maas-Szabowski, N.; Shimotoyodome, A.; Fusenig, N.E. Keratinocyte growth regulation in fibroblast cocultures via a double paracrine mechanism. *J. Cell Sci.* **1999**, *112 Pt 12*, 1843–1853. [PubMed]

44. Longworth, M.S.; Laimins, L.A. Pathogenesis of human papillomaviruses in differentiating epithelia. *Microbiol. Mol. Biol. Rev.* **2004**, *68*, 362–372. [CrossRef] [PubMed]

45. Meyers, C.; Laimins, L.A. In vitro systems for the study and propagation of human papillomaviruses. *Curr. Top. Microbiol. Immunol.* **1994**, *186*, 199–215. [PubMed]

46. Lambert, P.F.; Ozbun, M.A.; Collins, A.; Holmgren, S.; Lee, D.; Nakahara, T. Using an immortalized cell line to study the HPV life cycle in organotypic "raft" cultures. *Methods Mol. Med.* **2005**, *119*, 141–155. [PubMed]

47. Lee, D.; Norby, K.; Hayes, M.; Chiu, Y.F.; Sugden, B.; Lambert, P.F. Using organotypic epithelial tissue culture to study the human papillomavirus life cycle. *Curr. Protoc. Microbiol.* **2016**, *41*, 14B.8.1–14B.8.19. [PubMed]

48. El Ghalbzouri, A.; Ponec, M. Diffusible factors released by fibroblasts support epidermal morphogenesis and deposition of basement membrane components. *Wound Repair Regen.* **2004**, *12*, 359–367. [CrossRef] [PubMed]

49. Maas-Szabowski, N.; Stark, H.J.; Fusenig, N.E. Keratinocyte growth regulation in defined organotypic cultures through IL-1-induced keratinocyte growth factor expression in resting fibroblasts. *J. Investig. Dermatol.* **2000**, *114*, 1075–1084. [CrossRef] [PubMed]

50. Smola, H.; Stark, H.J.; Thiekotter, G.; Mirancea, N.; Krieg, T.; Fusenig, N.E. Dynamics of basement membrane formation by keratinocyte-fibroblast interactions in organotypic skin culture. *Exp. Cell Res.* **1998**, *239*, 399–410. [CrossRef] [PubMed]

51. Smola, H.; Thiekotter, G.; Fusenig, N.E. Mutual induction of growth factor gene expression by epidermal-dermal cell interaction. *J. Cell Biol.* **1993**, *122*, 417–429. [CrossRef] [PubMed]

52. Schumacher, M.; Schuster, C.; Rogon, Z.M.; Bauer, T.; Caushaj, N.; Baars, S.; Szabowski, S.; Bauer, C.; Schorpp-Kistner, M.; Hess, J.; et al. Efficient keratinocyte differentiation strictly depends on JNK-induced soluble factors in fibroblasts. *J. Investig. Dermatol.* **2014**, *134*, 1332–1341. [CrossRef] [PubMed]

53. Werner, S.; Krieg, T.; Smola, H. Keratinocyte-fibroblast interactions in wound healing. *J. Investig. Dermatol.* **2007**, *127*, 998–1008. [CrossRef] [PubMed]

54. Schorey, J.S.; Harding, C.V. Extracellular vesicles and infectious diseases: New complexity to an old story. *J. Clin. Investig.* **2016**, *126*, 1181–1189. [CrossRef] [PubMed]

55. Pickard, A.; McDade, S.S.; McFarland, M.; McCluggage, W.G.; Wheeler, C.M.; McCance, D.J. HPV16 down-regulates the insulin-like growth factor binding protein 2 to promote epithelial invasion in organotypic cultures. *PLoS Pathog.* **2015**, *11*, e1004988. [CrossRef] [PubMed]

56. Fullar, A.; Dudas, J.; Olah, L.; Hollosi, P.; Papp, Z.; Sobel, G.; Karaszi, K.; Paku, S.; Baghy, K.; Kovalszky, I. Remodeling of extracellular matrix by normal and tumor-associated fibroblasts promotes cervical cancer progression. *BMC Cancer* **2015**, *15*, 256. [CrossRef] [PubMed]

57. Kessenbrock, K.; Plaks, V.; Werb, Z. Matrix metalloproteinases: Regulators of the tumor microenvironment. *Cell* **2010**, *141*, 52–67. [CrossRef] [PubMed]

58. Sato, T.; Sakai, T.; Noguchi, Y.; Takita, M.; Hirakawa, S.; Ito, A. Tumor-stromal cell contact promotes invasion of human uterine cervical carcinoma cells by augmenting the expression and activation of stromal matrix metalloproteinases. *Gynecol. Oncol.* **2004**, *92*, 47–56. [CrossRef] [PubMed]

59. Smola-Hess, S.; Pahne, J.; Mauch, C.; Zigrino, P.; Smola, H.; Pfister, H.J. Expression of membrane type 1 matrix metalloproteinase in papillomavirus-positive cells: Role of the human papillomavirus (HPV) 16 and HPV8 E7 gene products. *J. Gen. Virol.* **2005**, *86*, 1291–1296. [CrossRef] [PubMed]

60. Kaewprag, J.; Umnajvijit, W.; Ngamkham, J.; Ponglikitmongkol, M. HPV16 oncoproteins promote cervical cancer invasiveness by upregulating specific matrix metalloproteinases. *PLoS ONE* **2013**, *8*, e71611. [CrossRef] [PubMed]

61. Zhu, D.; Ye, M.; Zhang, W. E6/E7 oncoproteins of high risk HPV-16 upregulate MT1-MMP, MMP-2 and MMP-9 and promote the migration of cervical cancer cells. *Int. J. Clin. Exp. Pathol.* **2015**, *8*, 4981–4989. [PubMed]

62. Arbeit, J.M.; Munger, K.; Howley, P.M.; Hanahan, D. Progressive squamous epithelial neoplasia in K14-human papillomavirus type 16 transgenic mice. *J. Virol.* **1994**, *68*, 4358–4368. [PubMed]

63. Herber, R.; Liem, A.; Pitot, H.; Lambert, P.F. Squamous epithelial hyperplasia and carcinoma in mice transgenic for the human papillomavirus type 16 E7 oncogene. *J. Virol.* **1996**, *70*, 1873–1881. [PubMed]

64. Song, S.; Pitot, H.C.; Lambert, P.F. The human papillomavirus type 16 E6 gene alone is sufficient to induce carcinomas in transgenic animals. *J. Virol.* **1999**, *73*, 5887–5893. [PubMed]

65. Lambert, P.F.; Pan, H.; Pitot, H.C.; Liem, A.; Jackson, M.; Griep, A.E. Epidermal cancer associated with expression of human papillomavirus type 16 E6 and E7 oncogenes in the skin of transgenic mice. *Proc. Natl. Acad. Sci. USA* **1993**, *90*, 5583–5587. [CrossRef] [PubMed]

66. Strati, K.; Pitot, H.C.; Lambert, P.F. Identification of biomarkers that distinguish human papillomavirus (HPV)-positive versus HPV-negative head and neck cancers in a mouse model. *Proc. Natl. Acad. Sci. USA* **2006**, *103*, 14152–14157. [CrossRef] [PubMed]

67. Stelzer, M.K.; Pitot, H.C.; Liem, A.; Schweizer, J.; Mahoney, C.; Lambert, P.F. A mouse model for human anal cancer. *Cancer Prev. Res.* **2010**, *3*, 1534–1541. [CrossRef] [PubMed]

68. Coussens, L.M.; Hanahan, D.; Arbeit, J.M. Genetic predisposition and parameters of malignant progression in K14-HPV16 transgenic mice. *Am. J. Pathol.* **1996**, *149*, 1899–1917. [PubMed]

69. Coussens, L.M.; Raymond, W.W.; Bergers, G.; Laig-Webster, M.; Behrendtsen, O.; Werb, Z.; Caughey, G.H.; Hanahan, D. Inflammatory mast cells up-regulate angiogenesis during squamous epithelial carcinogenesis. *Genes Dev.* **1999**, *13*, 1382–1397. [CrossRef] [PubMed]

70. Smith-McCune, K.; Zhu, Y.H.; Hanahan, D.; Arbeit, J. Cross-species comparison of angiogenesis during the premalignant stages of squamous carcinogenesis in the human cervix and K14-HPV16 transgenic mice. *Cancer Res.* **1997**, *57*, 1294–1300. [PubMed]

71. Mazibrada, J.; Ritta, M.; Mondini, M.; De Andrea, M.; Azzimonti, B.; Borgogna, C.; Ciotti, M.; Orlando, A.; Surico, N.; Chiusa, L.; et al. Interaction between inflammation and angiogenesis during different stages of cervical carcinogenesis. *Gynecol. Oncol.* **2008**, *108*, 112–120. [CrossRef] [PubMed]

72. Chen, W.; Li, F.; Mead, L.; White, H.; Walker, J.; Ingram, D.A.; Roman, A. Human papillomavirus causes an angiogenic switch in keratinocytes which is sufficient to alter endothelial cell behavior. *Virology* **2007**, *367*, 168–174. [CrossRef] [PubMed]

73. Lopez-Ocejo, O.; Viloria-Petit, A.; Bequet-Romero, M.; Mukhopadhyay, D.; Rak, J.; Kerbel, R.S. Oncogenes and tumor angiogenesis: The HPV-16 E6 oncoprotein activates the vascular endothelial growth factor (VEGF) gene promoter in a p53 independent manner. *Oncogene* **2000**, *19*, 4611–4620. [CrossRef] [PubMed]

74. Toussaint-Smith, E.; Donner, D.B.; Roman, A. Expression of human papillomavirus type 16 E6 and E7 oncoproteins in primary foreskin keratinocytes is sufficient to alter the expression of angiogenic factors. *Oncogene* **2004**, *23*, 2988–2995. [CrossRef] [PubMed]

75. Walker, J.; Smiley, L.C.; Ingram, D.; Roman, A. Expression of human papillomavirus type 16 E7 is sufficient to significantly increase expression of angiogenic factors but is not sufficient to induce endothelial cell migration. *Virology* **2011**, *410*, 283–290. [CrossRef] [PubMed]

76. Pilch, H.; Schlenger, K.; Steiner, E.; Brockerhoff, P.; Knapstein, P.; Vaupel, P. Hypoxia-stimulated expression of angiogenic growth factors in cervical cancer cells and cervical cancer-derived fibroblasts. *Int. J. Gynecol. Cancer* **2001**, *11*, 137–142. [CrossRef] [PubMed]

77. Huang, T.H.; Chu, T.Y. Repression of mir-126 and upregulation of adrenomedullin in the stromal endothelium by cancer-stromal cross talks confers angiogenesis of cervical cancer. *Oncogene* **2014**, *33*, 3636–3647. [CrossRef] [PubMed]

78. Martinez, I.; Gardiner, A.S.; Board, K.F.; Monzon, F.A.; Edwards, R.P.; Khan, S.A. Human papillomavirus type 16 reduces the expression of microRNA-218 in cervical carcinoma cells. *Oncogene* **2008**, *27*, 2575–2582. [CrossRef] [PubMed]

79. Zhu, N.; Zhang, D.; Xie, H.; Zhou, Z.; Chen, H.; Hu, T.; Bai, Y.; Shen, Y.; Yuan, W.; Jing, Q.; et al. Endothelial-specific intron-derived miR-126 is down-regulated in human breast cancer and targets both VEGFA and PIK3R2. *Mol. Cell. Biochem.* **2011**, *351*, 157–164. [CrossRef] [PubMed]

80. Pietras, K.; Pahler, J.; Bergers, G.; Hanahan, D. Functions of paracrine pdgf signaling in the proangiogenic tumor stroma revealed by pharmacological targeting. *PLoS Med.* **2008**, *5*, e19. [CrossRef] [PubMed]

81. Erez, N.; Truitt, M.; Olson, P.; Arron, S.T.; Hanahan, D. Cancer-associated fibroblasts are activated in incipient neoplasia to orchestrate tumor-promoting inflammation in an NF-κB-dependent manner. *Cancer Cell* **2010**, *17*, 135–147. [CrossRef] [PubMed]

82. Kanodia, S.; Fahey, L.M.; Kast, W.M. Mechanisms used by human papillomaviruses to escape the host immune response. *Curr. Cancer Drug Targets* **2007**, *7*, 79–89. [CrossRef] [PubMed]

83. Stanley, M.A. Epithelial cell responses to infection with human papillomavirus. *Clin. Microbiol. Rev.* **2012**, *25*, 215–222. [CrossRef] [PubMed]

84. Mangino, G.; Chiantore, M.V.; Iuliano, M.; Fiorucci, G.; Romeo, G. Inflammatory microenvironment and human papillomavirus-induced carcinogenesis. *Cytokine Growth Factor Rev.* **2016**, *30*, 103–111. [CrossRef] [PubMed]

85. Woodby, B.; Scott, M.; Bodily, J. The interaction between human papillomaviruses and the stromal microenvironment. *Prog. Mol. Biol. Transl. Sci.* **2016**, *144*, 169–238. [PubMed]

86. Chow, M.T.; Luster, A.D. Chemokines in cancer. *Cancer Immunol. Res.* **2014**, *2*, 1125–1131. [CrossRef] [PubMed]

87. Keeley, E.C.; Mehrad, B.; Strieter, R.M. CXC chemokines in cancer angiogenesis and metastases. *Adv. Cancer Res.* **2010**, *106*, 91–111. [PubMed]

88. Kumar, M.M.; Davuluri, S.; Poojar, S.; Mukherjee, G.; Bajpai, A.K.; Bafna, U.D.; Devi, U.K.; Kallur, P.P.; Kshitish, A.K.; Jayshree, R.S. Role of estrogen receptor α in human cervical cancer-associated fibroblasts: A transcriptomic study. *Tumour Biol.* **2016**, *37*, 4409–4420. [CrossRef] [PubMed]

89. Spurgeon, M.E.; Horswill, M.; den Boon, J.A.; Barthakur, S.; Forouzan, O.; Beebe, D.J.; Roopra, A.; Ahlquist, P.; Lambert, P. Human papillomavirus oncogenes reprogram the cervical cancer microenvironment independently of and synergistically with estrogen. 2017, submitted for publication.

90. Pahne-Zeppenfeld, J.; Schroer, N.; Walch-Ruckheim, B.; Oldak, M.; Gorter, A.; Hegde, S.; Smola, S. Cervical cancer cell-derived interleukin-6 impairs CCR7-dependent migration of MMP-9-expressing dendritic cells. *Int. J. Cancer* **2014**, *134*, 2061–2073. [CrossRef] [PubMed]

91. Schroer, N.; Pahne, J.; Walch, B.; Wickenhauser, C.; Smola, S. Molecular pathobiology of human cervical high-grade lesions: Paracrine STAT3 activation in tumor-instructed myeloid cells drives local MMP-9 expression. *Cancer Res.* **2011**, *71*, 87–97. [CrossRef] [PubMed]

92. Walch-Ruckheim, B.; Mavrova, R.; Henning, M.; Vicinus, B.; Kim, Y.J.; Bohle, R.M.; Juhasz-Boss, I.; Solomayer, E.F.; Smola, S. Stromal fibroblasts induce CCL20 through IL6/C/EBPβ to support the recruitment of Th17 cells during cervical cancer progression. *Cancer Res.* **2015**, *75*, 5248–5259. [CrossRef] [PubMed]

93. Stone, S.C.; Rossetti, R.A.; Lima, A.M.; Lepique, A.P. HPV associated tumor cells control tumor microenvironment and leukocytosis in experimental models. *Immun. Inflamm. Dis.* **2014**, *2*, 63–75. [CrossRef] [PubMed]

94. Lepique, A.P.; Daghastanli, K.R.; Cuccovia, I.M.; Villa, L.L. HPV16 tumor associated macrophages suppress antitumor T cell responses. *Clin. Cancer Res.* **2009**, *15*, 4391–4400. [CrossRef] [PubMed]

95. Bolpetti, A.; Silva, J.S.; Villa, L.L.; Lepique, A.P. Interleukin-10 production by tumor infiltrating macrophages plays a role in human papillomavirus 16 tumor growth. *BMC Immunol.* **2010**, *11*, 27. [CrossRef] [PubMed]

96. Coussens, L.M.; Tinkle, C.L.; Hanahan, D.; Werb, Z. MMP-9 supplied by bone marrow-derived cells contributes to skin carcinogenesis. *Cell* **2000**, *103*, 481–490. [CrossRef]

97. Bergot, A.S.; Ford, N.; Leggatt, G.R.; Wells, J.W.; Frazer, I.H.; Grimbaldeston, M.A. HPV16-E7 expression in squamous epithelium creates a local immune suppressive environment via CCL2- and CCL5-mediated recruitment of mast cells. *PLoS Pathog.* **2014**, *10*, e1004466. [CrossRef] [PubMed]

98. Choyce, A.; Yong, M.; Narayan, S.; Mattarollo, S.R.; Liem, A.; Lambert, P.F.; Frazer, I.H.; Leggatt, G.R. Expression of a single, viral oncoprotein in skin epithelium is sufficient to recruit lymphocytes. *PLoS ONE* **2013**, *8*, e57798. [CrossRef] [PubMed]

99. Munger, K.; Howley, P.M. Human papillomavirus immortalization and transformation functions. *Virus Res.* **2002**, *89*, 213–228. [CrossRef]

100. Chapman, S.; McDermott, D.H.; Shen, K.; Jang, M.K.; McBride, A.A. The effect of Rho kinase inhibition on long-term keratinocyte proliferation is rapid and conditional. *Stem Cell Res. Ther.* **2014**, *5*, 60. [CrossRef] [PubMed]

101. Liu, X.; Ory, V.; Chapman, S.; Yuan, H.; Albanese, C.; Kallakury, B.; Timofeeva, O.A.; Nealon, C.; Dakic, A.; Simic, V.; et al. ROCK inhibitor and feeder cells induce the conditional reprogramming of epithelial cells. *Am. J. Pathol.* **2012**, *180*, 599–607. [CrossRef] [PubMed]

102. Fu, B.; Quintero, J.; Baker, C.C. Keratinocyte growth conditions modulate telomerase expression, senescence, and immortalization by human papillomavirus type 16 E6 and E7 oncogenes. *Cancer Res.* **2003**, *63*, 7815–7824. [PubMed]

103. Zheng, J.; Vaheri, A. Human skin fibroblasts induce anchorage-independent growth of HPV-16-DNA-immortalized cervical epithelial cells. *Int. J. Cancer* **1995**, *61*, 658–665. [CrossRef] [PubMed]

104. Maas-Szabowski, N.; Szabowski, A.; Stark, H.J.; Andrecht, S.; Kolbus, A.; Schorpp-Kistner, M.; Angel, P.; Fusenig, N.E. Organotypic cocultures with genetically modified mouse fibroblasts as a tool to dissect molecular mechanisms regulating keratinocyte growth and differentiation. *J. Investig. Dermatol.* **2001**, *116*, 816–820. [CrossRef] [PubMed]

105. Woodworth, C.D.; McMullin, E.; Iglesias, M.; Plowman, G.D. Interleukin 1α and tumor necrosis factor α stimulate autocrine amphiregulin expression and proliferation of human papillomavirus-immortalized and carcinoma-derived cervical epithelial cells. *Proc. Natl. Acad. Sci. USA* **1995**, *92*, 2840–2844. [CrossRef] [PubMed]

106. Pickard, A.; Cichon, A.C.; Barry, A.; Kieran, D.; Patel, D.; Hamilton, P.; Salto-Tellez, M.; James, J.; McCance, D.J. Inactivation of Rb in stromal fibroblasts promotes epithelial cell invasion. *EMBO J.* **2012**, *31*, 3092–3103. [CrossRef] [PubMed]

107. Pickard, A.; Cichon, A.C.; Menges, C.; Patel, D.; McCance, D.J. Regulation of epithelial differentiation and proliferation by the stroma: A role for the retinoblastoma protein. *J. Investig. Dermatol.* **2012**, *132*, 2691–2699. [CrossRef] [PubMed]

108. Cichon, A.C.; Pickard, A.; McDade, S.S.; Sharpe, D.J.; Moran, M.; James, J.A.; McCance, D.J. AKT in stromal fibroblasts controls invasion of epithelial cells. *Oncotarget* **2013**, *4*, 1103–1116. [CrossRef] [PubMed]

109. Ornitz, D.M.; Itoh, N. The fibroblast growth factor signaling pathway. *Wiley Interdiscip. Rev. Dev. Biol.* **2015**, *4*, 215–266. [CrossRef] [PubMed]

110. Arbeit, J.M.; Olson, D.C.; Hanahan, D. Upregulation of fibroblast growth factors and their receptors during multi-stage epidermal carcinogenesis in K14-HPV16 transgenic mice. *Oncogene* **1996**, *13*, 1847–1857. [PubMed]

111. Turner, M.A.; Darragh, T.; Palefsky, J.M. Epithelial-stromal interactions modulating penetration of matrigel membranes by HPV 16-immortalized keratinocytes. *J. Investig. Dermatol.* **1997**, *109*, 619–625. [CrossRef] [PubMed]

112. Murata, T.; Mizushima, H.; Chinen, I.; Moribe, H.; Yagi, S.; Hoffman, R.M.; Kimura, T.; Yoshino, K.; Ueda, Y.; Enomoto, T.; et al. HB-EGF and PDGF mediate reciprocal interactions of carcinoma cells with cancer-associated fibroblasts to support progression of uterine cervical cancers. *Cancer Res.* **2011**, *71*, 6633–6642. [CrossRef] [PubMed]

113. Lambert, P.F. Transgenic mouse models of tumor virus action. *Annu. Rev. Virol.* **2016**, *3*, 473–489. [CrossRef] [PubMed]

114. Arbeit, J.M.; Howley, P.M.; Hanahan, D. Chronic estrogen-induced cervical and vaginal squamous carcinogenesis in human papillomavirus type 16 transgenic mice. *Proc. Natl. Acad. Sci. USA* **1996**, *93*, 2930–2935. [CrossRef] [PubMed]

115. Riley, R.R.; Duensing, S.; Brake, T.; Munger, K.; Lambert, P.F.; Arbeit, J.M. Dissection of human papillomavirus E6 and E7 function in transgenic mouse models of cervical carcinogenesis. *Cancer Res.* **2003**, *63*, 4862–4871. [PubMed]

116. Brake, T.; Lambert, P.F. Estrogen contributes to the onset, persistence, and malignant progression of cervical cancer in a human papillomavirus-transgenic mouse model. *Proc. Natl. Acad. Sci. USA* **2005**, *102*, 2490–2495. [CrossRef] [PubMed]

117. Chung, S.H.; Wiedmeyer, K.; Shai, A.; Korach, K.S.; Lambert, P.F. Requirement for estrogen receptor alpha in a mouse model for human papillomavirus-associated cervical cancer. *Cancer Res.* **2008**, *68*, 9928–9934. [CrossRef] [PubMed]

118. Chung, S.H.; Lambert, P.F. Prevention and treatment of cervical cancer in mice using estrogen receptor antagonists. *Proc. Natl. Acad. Sci. USA* **2009**, *106*, 19467–19472. [CrossRef] [PubMed]

119. Mehta, F.F.; Baik, S.; Chung, S.H. Recurrence of cervical cancer and its resistance to progestin therapy in a mouse model. *Oncotarget* **2017**, *8*, 2372–2380. [CrossRef] [PubMed]

120. Spurgeon, M.E.; Chung, S.H.; Lambert, P.F. Recurrence of cervical cancer in mice after selective estrogen receptor modulator therapy. *Am. J. Pathol.* **2014**, *184*, 530–540. [CrossRef] [PubMed]

121. Yoo, Y.A.; Son, J.; Mehta, F.F.; DeMayo, F.J.; Lydon, J.P.; Chung, S.H. Progesterone signaling inhibits cervical carcinogenesis in mice. *Am. J. Pathol.* **2013**, *183*, 1679–1687. [CrossRef] [PubMed]

122. Bronowicka-Klys, D.E.; Lianeri, M.; Jagodzinski, P.P. The role and impact of estrogens and xenoestrogen on the development of cervical cancer. *Biomed. Pharmacother.* **2016**, *84*, 1945–1953. [CrossRef] [PubMed]

123. Chung, S.H.; Franceschi, S.; Lambert, P.F. Estrogen and ERα: Culprits in cervical cancer? *Trends Endocrinol. Metab.* **2010**, *21*, 504–511. [CrossRef] [PubMed]

124. Murata, T.; Mekada, E.; Hoffman, R.M. Reconstitution of a metastatic-resistant tumor microenvironment with cancer-associated fibroblasts enables metastasis. *Cell Cycle* **2017**, *16*, 533–535. [CrossRef] [PubMed]

125. Ruffell, B.; Affara, N.I.; Cottone, L.; Junankar, S.; Johansson, M.; DeNardo, D.G.; Korets, L.; Reinheckel, T.; Sloane, B.F.; Bogyo, M.; et al. Cathepsin C is a tissue-specific regulator of squamous carcinogenesis. *Genes Dev.* **2013**, *27*, 2086–2098. [CrossRef] [PubMed]

126. Cooke, P.S.; Buchanan, D.L.; Young, P.; Setiawan, T.; Brody, J.; Korach, K.S.; Taylor, J.; Lubahn, D.B.; Cunha, G.R. Stromal estrogen receptors mediate mitogenic effects of estradiol on uterine epithelium. *Proc. Natl. Acad. Sci. USA* **1997**, *94*, 6535–6540. [CrossRef] [PubMed]

127. Cunha, G.R. Stromal induction and specification of morphogenesis and cytodifferentiation of the epithelia of the mullerian ducts and urogenital sinus during development of the uterus and vagina in mice. *J. Exp. Zool.* **1976**, *196*, 361–370. [CrossRef] [PubMed]

128. Cunha, G.R.; Cooke, P.S.; Kurita, T. Role of stromal-epithelial interactions in hormonal responses. *Arch. Histol. Cytol.* **2004**, *67*, 417–434. [CrossRef] [PubMed]

129. Kurita, T.; Cooke, P.S.; Cunha, G.R. Epithelial-stromal tissue interaction in paramesonephric (Mullerian) epithelial differentiation. *Dev. Biol.* **2001**, *240*, 194–211. [CrossRef] [PubMed]

130. Chung, S.H.; Shin, M.K.; Korach, K.S.; Lambert, P.F. Requirement for stromal estrogen receptor α in cervical neoplasia. *Horm. Cancer* **2013**, *4*, 50–59. [CrossRef] [PubMed]

131. den Boon, J.A.; Pyeon, D.; Wang, S.S.; Horswill, M.; Schiffman, M.; Sherman, M.; Zuna, R.E.; Wang, Z.; Hewitt, S.M.; Pearson, R.; et al. Molecular transitions from papillomavirus infection to cervical precancer and cancer: Role of stromal estrogen receptor signaling. *Proc. Natl. Acad. Sci. USA* **2015**, *112*, E3255–E3264. [CrossRef] [PubMed]

132. Kalluri, R. The biology and function of exosomes in cancer. *J. Clin. Investig.* **2016**, *126*, 1208–1215. [CrossRef] [PubMed]

133. Valadi, H.; Ekstrom, K.; Bossios, A.; Sjostrand, M.; Lee, J.J.; Lotvall, J.O. Exosome-mediated transfer of mRNAs and microRNAs is a novel mechanism of genetic exchange between cells. *Nat. Cell Biol.* **2007**, *9*, 654–659. [CrossRef] [PubMed]

134. Meckes, D.G., Jr.; Shair, K.H.; Marquitz, A.R.; Kung, C.P.; Edwards, R.H.; Raab-Traub, N. Human tumor virus utilizes exosomes for intercellular communication. *Proc. Natl. Acad. Sci. USA* **2010**, *107*, 20370–20375. [CrossRef] [PubMed]

135. Honegger, A.; Leitz, J.; Bulkescher, J.; Hoppe-Seyler, K.; Hoppe-Seyler, F. Silencing of human papillomavirus (HPV) E6/E7 oncogene expression affects both the contents and the amounts of extracellular microvesicles released from HPV-positive cancer cells. *Int. J. Cancer* **2013**, *133*, 1631–1642. [CrossRef] [PubMed]

136. Chiantore, M.V.; Mangino, G.; Iuliano, M.; Zangrillo, M.S.; De Lillis, I.; Vaccari, G.; Accardi, R.; Tommasino, M.; Columba Cabezas, S.; Federico, M.; et al. Human papillomavirus E6 and E7 oncoproteins affect the expression of cancer-related microRNAs: Additional evidence in HPV-induced tumorigenesis. *J. Cancer Res. Clin. Oncol.* **2016**, *142*, 1751–1763. [CrossRef] [PubMed]

137. Harden, M.E.; Munger, K. Human papillomavirus 16 E6 and E7 oncoprotein expression alters microRNA expression in extracellular vesicles. *Virology* **2017**, *508*, 63–69. [CrossRef] [PubMed]

138. Honegger, A.; Schilling, D.; Bastian, S.; Sponagel, J.; Kuryshev, V.; Sultmann, H.; Scheffner, M.; Hoppe-Seyler, K.; Hoppe-Seyler, F. Dependence of intracellular and exosomal micrornas on viral *E6/E7* oncogene expression in HPV-positive tumor cells. *PLoS Pathog.* **2015**, *11*, e1004712. [CrossRef] [PubMed]

139. Le Buanec, H.; D'Anna, R.; Lachgar, A.; Zagury, J.F.; Bernard, J.; Ittele, D.; d'Alessio, P.; Hallez, S.; Giannouli, C.; Burny, A.; et al. HPV-16 E7 but not E6 oncogenic protein triggers both cellular immunosuppression and angiogenic processes. *Biomed. Pharmacother.* **1999**, *53*, 424–431. [CrossRef]

140. Le Buanec, H.; Lachgar, A.; D'Anna, R.; Zagury, J.F.; Bizzini, B.; Bernard, J.; Ittele, D.; Hallez, S.; Giannouli, C.; Burny, A.; et al. Induction of cellular immunosuppression by the human papillomavirus type 16 E7 oncogenic protein. *Biomed. Pharmacother.* **1999**, *53*, 323–328. [CrossRef]

141. D'Anna, R.; Le Buanec, H.; Alessandri, G.; Caruso, A.; Burny, A.; Gallo, R.; Zagury, J.F.; Zagury, D.; D'Alessio, P. Selective activation of cervical microvascular endothelial cells by human papillomavirus 16-E7 oncoprotein. *J. Natl. Cancer Inst.* **2001**, *93*, 1843–1851. [CrossRef] [PubMed]

142. Vinzon, S.E.; Braspenning-Wesch, I.; Muller, M.; Geissler, E.K.; Nindl, I.; Grone, H.J.; Schafer, K.; Rosl, F. Protective vaccination against papillomavirus-induced skin tumors under immunocompetent and immunosuppressive conditions: A preclinical study using a natural outbred animal model. *PLoS Pathog.* **2014**, *10*, e1003924. [CrossRef] [PubMed]

143. Ingle, A.; Ghim, S.; Joh, J.; Chepkoech, I.; Bennett Jenson, A.; Sundberg, J.P. Novel laboratory mouse papillomavirus (MusPV) infection. *Vet. Pathol.* **2011**, *48*, 500–505. [CrossRef] [PubMed]

144. Uberoi, A.; Yoshida, S.; Frazer, I.H.; Pitot, H.C.; Lambert, P.F. Role of ultraviolet radiation in papillomavirus-induced disease. *PLoS Pathog.* **2016**, *12*, e1005664. [CrossRef] [PubMed]

Review

Changing Stem Cell Dynamics during Papillomavirus Infection: Potential Roles for Cellular Plasticity in the Viral Lifecycle and Disease

Katerina Strati

Department of Biological Sciences, University of Cyprus, 1 University Avenue, Nicosia, 2109, Cyprus; strati@ucy.ac.cy; Tel.: +357 22 892884

Academic Editors: Alison A. McBride and Karl Munger
Received: 7 July 2017; Accepted: 8 August 2017; Published: 12 August 2017

Abstract: Stem cells and cellular plasticity are likely important components of tissue response to infection. There is emerging evidence that stem cells harbor receptors for common pathogen motifs and that they are receptive to local inflammatory signals in ways suggesting that they are critical responders that determine the balance between health and disease. In the field of papillomaviruses stem cells have been speculated to play roles during the viral life cycle, particularly during maintenance, and virus-promoted carcinogenesis but little has been conclusively determined. I summarize here evidence that gives clues to the potential role of stem cells and cellular plasticity in the lifecycle papillomavirus and linked carcinogenesis. I also discuss outstanding questions which need to be resolved.

Keywords: HPV; stem cells; cellular plasticity

1. Introduction

Human papillomavirus (HPV) infection occurs via micro wounds which allow the virus access to the basal layer of stratified epithelia. This target site of infection has been the subject of intense scrutiny, leading to significant progress regarding the molecular mechanisms governing the attachment and eventual entry of the virus into initially infected cells [1,2]. However, the basal layer of stratified epithelia has a relatively heterogeneous composition containing committed undifferentiated cells, progenitors, and also the tissue stem cells [3]. It is not understood whether these are subject to differential infection and whether the fate of infected cells varies based on cell identity prior to infection. Nevertheless, it has been proposed that the virus may be more successfully maintained [4,5], and indeed infected cells are more likely to proceed to tumorigenesis when the initially infected cell is a tissue stem cell [6,7]. To this day there is little evidence to conclusively resolve this question. Yet as progress is being made to better understand the role of "stemness" in carcinogenesis overall, the question of its role in HPV-mediated carcinogenesis and potentially the viral lifecycle, grows more intriguing. Furthermore, our evolving understanding of cellular plasticity raises new possibilities which may be at play during the HPV lifecycle and pathogenesis. In this review, I will discuss the available evidence and its implications and propose significant questions which remain to be addressed.

2. The Target Cell of Papillomavirus Infection

Papillomaviruses can productively infect both mucosal and cutaneous stratified epithelia. Infection and the ensuing life cycle of the virus are intimately linked to the differentiation program of the tissue [8,9] thus our reconstruction of these events relies on the solid understanding of tissue biology. The skin is an easily accessible model tissue for stratified epithelial differentiation, homeostasis, and regeneration due to its rapid turnover and ability to regenerate quickly upon mechanical injury.

Much of our knowledge pertaining to the role of stem cell populations in these processes derives from studying the cutaneous epithelia in mouse models [3]. The characteristics of cutaneous epithelia and their stem cell populations may or may not extrapolate to the mucosal tissues. However, since the cervical stem cells have not been conclusively characterized the comparison of the cervical and cutaneous epithelia is useful and necessary. In stratified epithelia, the stem cells orchestrate tissue homeostasis as well as acute regeneration [3]. From work done in mouse models we know that different stem cell pools are primarily responsible for replacing lost cells during homeostasis and regeneration [10,11]. Genetic ablation experiments have also provided evidence that a certain amount of plasticity exists (for example if bulge stem cells are ablated, neighboring cells can replenish the stem cell niche) [11,12]. Stem cells in cutaneous epithelia are typically slow-cycling and can perform asymmetric division, dependent on the cellular niche, generating one stem and one transient amplifying/progenitor cell. Of course, during wound repair the balance of asymmetric division may be shifted to replenish stem cell populations to homeostatic levels [11]. Progenitors can then undergo large numbers of cell divisions and gradual differentiation helping to replenish the tissue.

Studies using HPV virus like particles (VLPs), or infectious virions for attachment and viral entry argue against the exclusive targeting of a small subpopulation such as the tissue stem cells. Rather, the virus associates with the exposed basement membrane at a site of wounding, and furin cleavage leads to conformational changes on the capsid which allow it to attach to nearby basal cells via a receptor which recognizes L1 capsid protein [1,2]. The high prevalence of infection with HPVs [13] seen in populations worldwide also provides an argument against a model in which the virus exclusively targets a rare subpopulation. It is likely that both stem and non-stem cells adjacent to a site of wounding can be infected. If there are smaller differences of susceptibilities to infection between cell types, those would be technically difficult to quantify at least in vivo. However, it is conceivable that the outcome of infection differs depending on the cell type infected.

3. A Cell Reservoir for Long-Term Viral Maintenance

If one takes for granted the more likely scenario that both stem and committed cells can be the targets of infection, then a vital question concerns the fate of infection in a stem as opposed to a committed cell. A popular hypothesis has been that long-term maintenance of the viral DNA can take place within infected tissue stem cells [4,5]. This is particularly relevant to low-level, asymptomatic, persistent infection which has been argued to be the source of future reactivation. [14,15]. During the long period between initial infection and disease development the viral DNA may be able to persist in a small subset of cells and is clinically undetectable. The reasons for lack of detection are likely linked to the small number of cells harboring genomes and low levels of viral replication. Such infections can escape immunological control later in life and lead to disease. Reactivation is marked by higher levels of viral replication and transcription which facilitate clinical detection. While reactivation has not been conclusively determined to occur in humans it is strongly supported by evidence obtained using rabbit papillomaviruses as a model [5,16]. In a cottontail rabbit papillomavirus (CRPV) infection model, viral genomes were detected in an area of the hair follicle which coincides with cells with in vitro clonogenic activity [16]. In a rabbit oral papillomavirus (ROPV) infection model it has been shown that the virus DNA can persist for long periods of time in a subset of the basal epithelial cells [5]. The authors speculate that these cells harboring viral DNA represent the epithelial stem cells. This scenario is not unlikely when one considers the contrast between the timeline of disease reactivation (over a year following infection) compared to the quick regenerative pace of the epithelium. One would predict that upon infection of a committed cell, the progeny cells harboring the virus would be cleared from the tissue in a matter of days/months. Thus, it is reasonable to hypothesize that cells which are maintained and harbor genomes for long periods of time represent a type of tissue stem cell. However, little is known about the molecular characteristics of a stem cell in the rabbit oral epithelium. Much less can be inferred about the identity of the cell prior to infection, thus it is difficult to assess this claim conclusively.

Hair follicle stem cells have also been proposed as the reservoirs for human cutaneous HPV infections. DNA of β-papillomaviruses is frequently isolated from plucked eyebrow hairs [17,18] suggesting that a cell type at this location is the long-term reservoir of the virus. β-Papillomaviruses are emerging as potential co-factors in a subset of non-melanoma skin cancers [19,20]. Definitively pinpointing the exact nature of the cell reservoir may be important for predicting the possibility of viral reactivation in at-risk populations (*Epidermodysplasia verruciformis* (EV) patients, transplant recipients etc.) [20].

The inability to define the concrete characteristics of the human cervical stem cells has complicated studies aiming to understand the maintenance of the human mucosotropic viruses. It is thought that the transformation zone which is characterized by a transition from columnar to squamous epithelium is the site of the so-called reserve cells which may act as the cervical stem cells. Some markers have been proposed for reserve cells (e.g., Keratin 17 (K17), p63, Keratin 7 (K7), etc.) [21,22] but the dearth of healthy human biopsy material and the loose anatomic equivalence of the mouse cervix have hindered functional studies of stemness on such putative stem cell populations. Importantly, a subset of cells in the transformation zone have been shown to be susceptible to HPV infection, and high-grade lesions stemming from this area are more likely to progress to carcinoma in situ [22]. Furthermore, lesions share the expression of markers of this area—e.g., K7, matrix metalloproteinase-7 (MMP-7), cluster of differentiation 63 (CD63)—and this immunophenotype was not regenerated after removal, in other sites, or by HPV oncogene expression in keratinocytes. It is likely that these junctional cells represent the source of at least some cervical malignancies and may represent a cervical stem cell population. Reserve cells can likely serve as a site for infection and potentially a viral reservoir. There are however HPV lesions which can be detected in other mucosal sites (e.g., the vagina) which do not share this anatomic feature thus it is unlikely that these cells are the unique targets of infection, maintenance or transformation.

4. Changes in Tissue Stem Cell Dynamics during Infection

Infected tissue stem cells are of interest due to their potential links to carcinogenesis. However, more recently stem cells have also been implicated in the tissue response to infection. There is an emerging understanding that tissue stem cells have evolved to respond directly both to commensal and pathogenic microbes as evidenced by the expression of pattern recognition receptors (PRRs) in tissue stem cells [23–25]. In addition to inflammatory signals (discussed in a later section of this review), tissue stem cells have been shown to respond to the presence of microbes in ways which define the balance between maintaining tissue health or disease development. The paradigm has been set by studies in the gut where expression of nucleotide-binding oligomerization domain-containing protein 2 (Nod2) [24] and Toll-like receptor 4 (TLR4) [25] receptors in intestinal stem cells has provided a direct link for the interaction of the stem cells with tissue commensals via the recognition of peptidoglycan and lipopolysaccharide (LPS), respectively. This interaction has been shown to be critical to tissue regeneration and homeostasis suggesting a direct link between microbes and tissue stem cells as essential to tissue health. Of course, tissue stem cell dynamics have also been shown to be perturbed by pathogenic bacteria in the gut [26,27] and other tissues such the urogenital tract where pathogenic *Escherichia coli* [28] mobilize tissue stem cells and progenitors during pathogenesis.

While the effects of infection on tissue stem cell dynamics are less well understood in cutaneous and mucosal epithelia compared to the gut, studies investigating the expression of viral gene products on skin stem cell populations suggest that important changes occur. Compelling evidence regarding the changes in stem cell dynamics during papillomavirus infection comes from studies using transgenic animals for both mucosotropic [6,29] and cutaneous HPVs [30]. The available evidence for HPV16 converges towards a model where the expression of early gene products pushes the tissue stem cells towards a hybrid state: one which retains typical markers of stem cells (e.g., K15) [6,29], but also expresses atypical markers (e.g., P-cadherin) [6] and loses key functional characteristics such as quiescence. Loss of quiescence and increased mobilization of the stem cells has been reported both upon individual expression of HPV16 E6 and HPV16 E7 likely through different pathways [29]. This change in stem cell dynamics may represent a critical aspect in the process of viral carcinogenesis. Stem cell quiescence is a tumor refractory state and its absence may render the tissue more vulnerable to additional carcinogenic insult [31]. Interestingly, HPV-associated tumorigenesis has also been linked to non-quiescent, skin stem cell populations [32] likely to be hierarchically linked to quiescent populations [33]. One study showed that in mice, tumors induced by HPV16 oncogenes are derived from descendants of leucine-rich repeat-containing G-protein coupled receptor 5 (LGR5)-positive stem cells [6]. These are long-lived, non-quiescent cells in the hair follicle, which have been shown to actively contribute to hair-follicle growth. Combined, these findings may represent a common pathway in which HPV infection can lead to carcinogenesis by increasing stem cell mobilization and promoting a stem cell state which is receptive to further oncogenic changes. These studies were performed in mice, using transgenic animals which express the viral oncogenes throughout the basal layer of the epithelium. Thus, it is difficult to precisely extrapolate these findings to what happens during human infection. But the results support the notion that expression of viral oncogenes in cells with stem cell properties has the potential to profoundly alter the behavior of these cells, their susceptibility to carcinogenic insult, and rendering them more likely to contribute to carcinogenesis.

5. Changes in Cellular Plasticity during Infection

Initial focus has been on uncovering how the behavior of stem cells may change upon papillomavirus infection. There is however also evidence which supports an alternate, not mutually exclusive scenario: that during infection it is possible that reprogramming events can contribute to the emergence of stem-cell like cells.

The increased stem cell mobilization seen in transgenic animals is also concurrent with the expression of stem cell markers such as Keratin15 (K15) outside the typical stem cell niche [6,29]. The expression of early genes from the cutaneous HPV8 has also been shown to lead to an expansion of stem cell markers, specifically leucine-rich repeats and immunoglobulin-like domains 1 (Lrig1) [30] in a model expressing the early genes of HPV8 in the skin epithelium of mice. Interestingly similar Lrig1 expression pattern was also seen in biopsies from EV patients. This may represent a field cancerization effect critical to β-papillomavirus-induced carcinogenesis which does not implicate the integration of the virus in developing carcinomas as seen in alpha-papillomaviruses.

In light of technologies describing cellular reprogramming developed after Yamanaka's seminal discoveries in 2006 [34], the re-expression of stem cell markers in differentiated cells has been re-evaluated as perhaps more meaningful than just mere dedifferentiation. It is now thought that increased cellular plasticity may have functional significance in the pathogen lifecycle and disease. In vitro cellular reprogramming of differentiated cells to pluripotency has been hailed as a way to derive pluripotent cells which has initiated the development of better research models in the lab and is under study to provide solutions in the field of regenerative medicine. Reprogramming is an epigenetic process which gradually shifts the transcriptional program of a differentiated "reprogramming" cell to that of a pluripotent one. Despite the concern for side-effects such as teratoma formation, the field of reprogramming has made progress with in vivo [35] approaches, particularly strategies to achieve tissue regeneration and rejuvenation. Studies aimed at understanding the mechanism of in vivo reprogramming, have revealed that physiological stimuli such as tissue damage, inflammation and senescence in a tissue can potentiate induced pluripotency in vivo [36]. Tissue damage or senescence triggers the release of interleukin-6 which is critical in facilitating reprogramming in neighboring cells [36]. This introduces the possibility that there may be evolutionarily conserved physiological importance for in vivo cellular reprogramming-like processes which remain to be understood. It is important to note that no findings to date suggest that such stimuli (tissue damage, senescence) can trigger pluripotency independently. However, they add to accumulating credible evidence that they may create a permissive environment for reprogramming events leading to intermediate and likely transitory stem-like states. Critically, both tissue damage and senescence are relevant to the papillomavirus lifecycle. Stem-like states do not adhere exactly to the characteristics of isolatable tissue or embryonic stem cells. However, such transitory states, are understood to form a continuum spanning between differentiated and stem cells [37,38]. They are amenable to directed differentiation in vitro, and likely represent safer avenues for transdifferentiation strategies than the use of pluripotent cells [39,40]. However, the potential roles of transitory stem-like states in physiological processes, including infection are poorly understood.

One of the most intriguing and compelling reported examples of reprogramming in vivo implicates the intracellular *Mycobacterium leprae* [41]. The bacterium has been shown to reprogram infected Schwann cells into stem cell-like cells. The reprogramming has significant implications for the course of pathogenesis as well as the life cycle of the pathogen [42]. While Schwann cells have high retention for the bacteria, the reprogrammed cells provide a route of dissemination into other cell types which may be critical for neuropathogenesis during leprosy. While the molecular pathways through which *M. leprae* infection leads to such events are incompletely understood, initial evidence suggests that they involve in part the innate immune response and inflammation which precede such reprogramming events [43].

Inflammation is naturally of relevance to the HPV lifecycle as well, particularly if one takes into account the mode of infection. Both infection-associated and sterile inflammation are understood to contribute to a regenerative response and inflammation is emerging as an evolutionarily conserved mechanism of tissue regeneration [23]. Regenerative inflammation can be mediated both via native signals to the tissue stem cells as well as to differentiated cells which may undergo profound dedifferentiation or experience increased "stemness". In fact, in the drosophila and the mammalian gut, where regenerative inflammation has been most extensively studied, it has been shown to be important both in the maintenance of tissue homeostasis in healthy tissue, as well as the promotion of disease [23,24,26,27,44,45]. This regenerative inflammatory response is thought to extend to other epithelial tissues and is likely an important aspect of epithelial recovery following papillomavirus infection [23,46]. However, its implications in the viral lifecycle and pathogenesis have not been addressed. Newer model systems involving a murine papillomavirus which infects via a site of wounding may shed light to this aspect of papillomavirus biology as they most closely mimic the conditions of real life infection [47–49].

Other than leading intracellular lifestyles HPVs do not have many commonalities with mycobacteria at a first glance. However, upon closer inspection of the available evidence there is good reason to suspect that the virus may be contributing to similar events. The viral oncogenes have prominent targets which are implicated in stem cell biology and may contribute to epigenetic reprogramming via their inactivation.

6. Mechanisms of Enhancing Cellular Plasticity

Work aimed at illuminating the critical steps during the process of reprogramming cells to pluripotency, clearly implicated prominent tumor suppressors in stem cell biology. p53, retinoblastoma protein (pRb) and other key players in tumor suppression have been shown to control checkpoints during cellular reprogramming [50–54]. Their loss or inhibition has been shown to facilitate the reprogramming process. Furthermore, they have been shown to impact the function of stem cell related factors: p53 can control the expression of the stem cell related protein Nanog [55], while pRb can directly bind sex determining region Y-box 2 (Sox2),and octamer-binding transcription factor 4 (Oct4) [56] thus controlling pluripotency networks via these key transcription factors. Since both p53 and pRb are critical targets of papillomavirus oncogenes it is important to interrogate whether in addition to removing critical barriers for cell cycle control, their targeting during viral infection creates a tissue environment which is more conducive to increased cellular plasticity (Figure 1). Consistent with this E6 contributed to reprogramming of cells from Fanconi anemia patients which are difficult to reprogram via its action on p53 [57]. The proteins encoded by the *INK4A/ARF* locus, some of which are implicated in papillomavirus pathogenesis are also important during reprogramming [52].

The extent to which in vivo inactivation of such tumor suppressors enables increased cellular plasticity and phenomena of natural reprogramming remains to be seen.

In addition to targeting tumor suppressors which have been linked to cellular reprogramming there is also evidence to suggest that the viral oncogenes may contribute to cellular reprogramming in ways which are independent of their ability to target p53 or pRb. High-risk HPVs (16, 31) have recently been shown to upregulate Kruppel-like factor 4 (Klf4) and contribute to its hypoSUMOylation [58]. This upregulation was necessary for the differentiation dependent lifecycle of the virus however it is also critical to note that the functions of Klf4 cells harboring viral genomes were markedly different to those seen in control keratinocytes. Biochemical evidence which dates back to the initial understanding of the function of Oct4 demonstrated the ability of high-risk HPV E7 (similar oncoproteins such as adenovirus E1A) to directly interact and synergize with Oct4 for the activation of its target genes [59,60]. More recent evidence from transgenic animals suggests that E7 may also contribute to the transcriptional upregulation of Oct4 [61]. Furthermore, E7 was shown to epigenetically reprogram cells via the transcriptional activation of histone demethylases lysine (K)-specific demethylase 6A and 6B (KDM6A and KDM6B, respectively) [62]. This upregulation led to a decrease in Histone3 Lysine27 (H3K27) trimethylation, and downstream transcriptional changes such as the activation of *Hox* genes. Critically this expression has been linked to the high p16 expression characteristic of HPV positive cancers and may represent a way of targeting HPV-positive cancer cells since they are dependent on its continued expression [63]. The E7 protein of cutaneous HPVs has been shown to lead to the upregulation of several stem cell markers genes in infected keratinocytes and to the formation of cells with stem-like properties—e.g., epithelial cell adhesion molecule (EpCaM) [64]. High-risk mucosal as well as cutaneous HPVs and murine papillomaviruses also inhibit the Notch pathway which is known to play important roles in commitment to differentiation [65,66]. Of course, the role of such reprogramming to the viral lifecycle and pathogenesis is not well understood. It may be linked to important aspects of the viral life cycle.

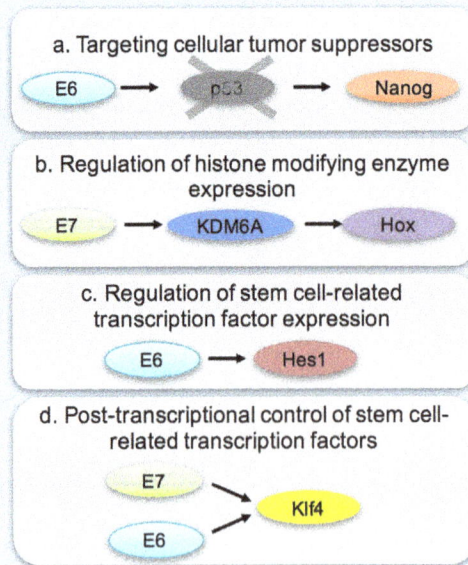

Figure 1. Potential mechanisms through which papillomavirus infection may contribute to the development of stem-like cells and cancer stem cells. (**a**) Targeting cellular tumor suppressors for

degradation: retinoblastoma protein (pRb) has been shown to bind both sex determining region Y-box 2 (Sox2), and octamer-binding transcription factor 4 (Oct4) leading to repression of pluripotency [56]. p53 Has been shown to directly bind and suppress transcription from the Nanog promoter [55]. Both pRb and p53 have been shown to be important gatekeepers during cellular reprogramming, and their absence significantly facilitates the process [50,51,53,56]. (b) Transcriptional upregulation of histone modifying enzymes: Upregulation of lysine (K)-specific demethylase 6A and 6B (KDM6A and KDM6B, respectively) mediated by the *E7* oncogene of HPV16 leads to a reduction of repressive H3K27 chromatin marks and downstream activation of targets such as *Hox* genes [62,63]; (c) Transcriptional upregulation of stem cell-related transcription factors: the viral oncogenes *E6* and *E7* of high-risk types have been linked to the upregulation of pluripotency associated transcription factors—Oct4 [61], Hes family basic helix-loop-helix transcription factor 1 (Hes1) [67]. Infection with cutaneous papillomaviruses has also been linked to the upregulation of stem cell related genes [30,64]. (d) Post-transcriptional control of stem cell related transcription factors has also been demonstrated: E7 has been reported to bind Oct4 and act as a transcriptional co-activator [60]. Both E6 and E7 have been shown to transcriptionally, post-transcriptionally and post-translationally regulate Kruppel-like factor 4 (Klf4) (e.g., via hyposumoylation) leading to modified Klf4 activity in infected keratinocytes [58]. The upregulation of stemness related genes has been most frequently attributed to the viral oncogenes *E6* and *E7* but the full mechanisms underlying some of these effects have yet to be elucidated. Furthermore, the impact of the re-expression or modulation of stemness related genes in the viral lifecycle and carcinogenesis is still poorly understood.

7. Potential Links of Cellular Plasticity to Disease

There is a dearth of studies which would allow definitive conclusions about the upregulation of stem cell related proteins in the viral life cycle. However, there is an accumulating body of evidence suggesting they may be linked to disease. The upregulation of genes related to stem cells may be a way in which the virus contributes to the formation and maintenance of cancer stem cell populations (Figure 1). Recently E6 has been shown to be upregulated in cancer stem cell-like cells isolated from primary tumors cervical cell lines based on their immunophenotype and sphere forming capabilities [67]. These cells were also shown to express Oct4, Sox2, Nanog, and Lrig1 and were dependent on E6-mediated expression of Hes1 for continued self-renewal.

The expression of stem cell related markers is now increasingly reported in HPV-related cancers. Proteins such as Sox2, Nanog, and Oct4 are now thought to serve not only as biomarkers tumor stage and therapy response in cancer but also may be directly implicated in the process of carcinogenesis. In HPV-associated cancers stem cell markers Oct4, Sox2 (in cervical cancers) [68], and Lrig1 (in head and neck squamous cell carcinomas) [69] have been reported and proposed as potential biomarkers. Given the established role of $p16^{INK4A}$ as a useful biomarker it remains to be seen whether the use of these are of added prognostic benefit. Nevertheless, the expression of these proteins in cancer biopsies and cervical cancer cell lines may provide insights into potential roles in the carcinogenic process. Critically, the links of early gene products, particularly the viral oncogenes of human papillomaviruses to the upregulation of such markers lends credibility to this notion.

A source of skepticism for studies supporting a changing cellular state during infection, derives from the refractoriness of keratinocytes to cellular reprogramming and perceived lack of plasticity. It is clear that no condition has been described thus far where a viral gene product or set of products have been shown to independently lead to an isolatable stem cell. However, it is critical to bear in mind that infection occurs in the context of a wound. It is an aspect of infection which has been poorly accounted for particularly since many experimental models used in papillomavirus research do not take it into consideration. In virion-infected monolayer keratinocytes, organotypic cultures of stably transfected keratinocytes or transgenic animals expressing viral gene products, the wound environment is not routinely modelled. However, during real-life infection, in addition to the viral invasion which triggers an inflammatory response via the presence of pathogen associated molecular patterns (PAMPs) there are also damage associated molecular patterns (DAMPs), and reactive oxygen species (ROS) released as a result of the wounding. Thus, both infection-associated, as well as sterile inflammation should be kept in mind as additional important factors which likely remove constraints towards increased cellular plasticity [36].

8. Conclusions

The role of stem cells and cellular stemness has been proposed to be important in the lifecycle and disease promotion by HPVs. Viral genomes have been detected in anatomical locations which are consistent with those of stem cells yet conclusive evidence as to whether infected stem cells are the sites of viral maintenance of all HPVs or the cancer initiating cells of HPV-related cancers is to this day tenuous. Early gene products, and particularly the viral oncogenes have been shown to modify stem cell dynamics and cellular stemness, however the extent to which this is critical for the viral lifecycle or for ensuing disease remains elusive. As HPV-related cancers account for about 5% of human cancers [13] (and potentially more if one considers the accumulating evidence for the role of β-papillomaviruses in cutaneous carcinogenesis) understanding the viral reservoir and mechanisms of carcinogenesis remains imperative. At the same time, emerging tools and concepts from the booming field of stem cell biology facilitate the task at hand. Current evidence suggests that viral maintenance occurs in cells whose characteristics are consistent to those of stem cells (Figure 2). Critical questions remain in tracing the origins of such cells: do they derive from infected tissue stem cells, or do they develop stem-like characteristics subsequent to infection? It is also clear that early gene products, and particularly the viral oncogenes can change the behavior of stem cells in a tissue, as well as reprogram cells towards states which resemble stem cells in some aspects. The next frontier would be to uncover direct links for these phenomena to aspects of the lifecycle on carcinogenesis.

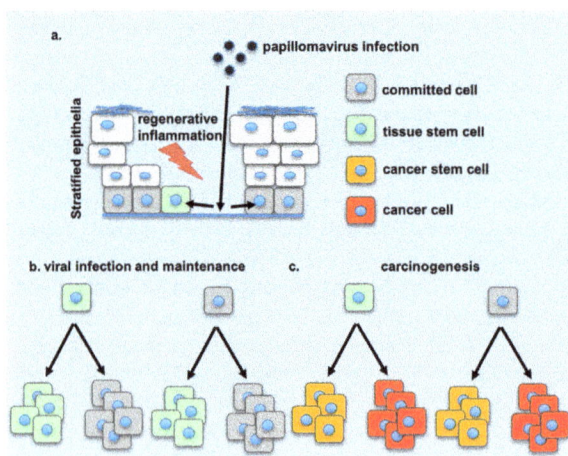

Figure 2. Model of changes in stem cells during papillomavirus infection and carcinogenesis. (a) Papillomaviruses can gain access to the basal layer of stratified epithelia via microwounds. Following attachment to the basement membrane, the virus can infect committed or stem cells in the basal layer of the epithelium. As the epithelium heals, infected cells are subject to changes due to the expression of early gene products and local inflammatory signals linked to infection and the regenerative response. (b) Stem cells or stem-like cells (with at least some atypical features) detected during infection and viral maintenance may be derived from infected stem cells or committed cells which have been reprogrammed. (c) Cancer stem cells may likewise be derived from infected tissue stem cells or from drastically de-differentiated committed cells.

Acknowledgments: The author acknowledges Stella Michael for proofreading and suggestions on the manuscript. This study has been funded by the University of Cyprus, Nicosia, 2109, Cyprus.

Conflicts of Interest: The author declares no conflict of interest.

References

1. Kines, R.C.; Thompson, C.D.; Lowy, D.R.; Schiller, J.T.; Day, P.M. The initial steps leading to papillomavirus infection occur on the basement membrane prior to cell surface binding. *Proc. Natl. Acad. Sci. USA* **2009**, *106*, 20458–20463. [CrossRef] [PubMed]

2. Schiller, J.T.; Day, P.M.; Kines, R.C. Current understanding of the mechanism of HPV infection. *Gynecol. Oncol.* **2010**, *118*, S12–S17. [CrossRef] [PubMed]

3. Mascre, G.; Dekoninck, S.; Drogat, B.; Youssef, K.K.; Brohee, S.; Sotiropoulou, P.A.; Simons, B.D.; Blanpain, C. Distinct contribution of stem and progenitor cells to epidermal maintenance. *Nature* **2012**, *489*, 257–262. [CrossRef] [PubMed]

4. Doorbar, J. Latent papillomavirus infections and their regulation. *Curr. Opin. Virol.* **2013**, *3*, 416–421. [CrossRef] [PubMed]

5. Maglennon, G.A.; McIntosh, P.; Doorbar, J. Persistence of viral DNA in the epithelial basal layer suggests a model for papillomavirus latency following immune regression. *Virology* **2011**, *414*, 153–163. [CrossRef] [PubMed]

6. Da Silva-Diz, V.; Sole-Sanchez, S.; Valdes-Gutierrez, A.; Urpi, M.; Riba-Artes, D.; Penin, R.M.; Pascual, G.; Gonzalez-Suarez, E.; Casanovas, O.; Vinals, F.; et al. Progeny of Lgr5-expressing hair follicle stem cell contributes to papillomavirus-induced tumor development in epidermis. *Oncogene* **2013**, *32*, 3732–3743. [CrossRef] [PubMed]

7. Kranjec, C.; Doorbar, J. Human papillomavirus infection and induction of neoplasia: A matter of fitness. *Curr. Opin. Virol.* **2016**, *20*, 129–136. [CrossRef] [PubMed]

8. Fehrmann, F.; Laimins, L.A. Human papillomaviruses: Targeting differentiating epithelial cells for malignant transformation. *Oncogene* **2003**, *22*, 5201–5207. [CrossRef] [PubMed]

9. Hong, S.; Laimins, L.A. Regulation of the life cycle of hpvs by differentiation and the DNA damage response. *Future Microbiol.* **2013**, *8*, 1547–1557. [CrossRef] [PubMed]

10. Ito, M.; Liu, Y.; Yang, Z.; Nguyen, J.; Liang, F.; Morris, R.J.; Cotsarelis, G. Stem cells in the hair follicle bulge contribute to wound repair but not to homeostasis of the epidermis. *Nat. Med.* **2005**, *11*, 1351–1354. [CrossRef] [PubMed]

11. Blanpain, C.; Fuchs, E. Stem cell plasticity. Plasticity of epithelial stem cells in tissue regeneration. *Science* **2014**, *344*, 1242281. [CrossRef] [PubMed]

12. Rompolas, P.; Mesa, K.R.; Greco, V. Spatial organization within a niche as a determinant of stem-cell fate. *Nature* **2013**, *502*, 513–518. [CrossRef] [PubMed]

13. Human papillomaviruses. In *IARC Monographs on the Evaluation of Carcinogenic Risks to Humans*; WHO: Geneva, Switzerland, 2007; Volume 90, pp. 1–636.

14. Brown, D.R.; Weaver, B. Human papillomavirus in older women: New infection or reactivation? *J. Infect. Dis.* **2013**, *207*, 211–212. [CrossRef] [PubMed]

15. Gravitt, P.E.; Rositch, A.F.; Silver, M.I.; Marks, M.A.; Chang, K.; Burke, A.E.; Viscidi, R.P. A cohort effect of the sexual revolution may be masking an increase in human papillomavirus detection at menopause in the united states. *J. Infect. Dis.* **2013**, *207*, 272–280. [CrossRef] [PubMed]

16. Schmitt, A.; Rochat, A.; Zeltner, R.; Borenstein, L.; Barrandon, Y.; Wettstein, F.O.; Iftner, T. The primary target cells of the high-risk cottontail rabbit papillomavirus colocalize with hair follicle stem cells. *J. Virol.* **1996**, *70*, 1912–1922. [PubMed]

17. Boxman, I.L.; Berkhout, R.J.; Mulder, L.H.; Wolkers, M.C.; Bouwes Bavinck, J.N.; Vermeer, B.J.; Ter Schegget, J. Detection of human papillomavirus dna in plucked hairs from renal transplant recipients and healthy volunteers. *J. Investig. Dermatol.* **1997**, *108*, 712–715. [CrossRef] [PubMed]

18. De Koning, M.N.; Struijk, L.; Bavinck, J.N.; Kleter, B.; Ter Schegget, J.; Quint, W.G.; Feltkamp, M.C. β-papillomaviruses frequently persist in the skin of healthy individuals. *J. Gen. Virol.* **2007**, *88*, 1489–1495. [CrossRef] [PubMed]

19. Galloway, D.A.; Laimins, L.A. Human papillomaviruses: Shared and distinct pathways for pathogenesis. *Curr. Opin. Virol.* **2015**, *14*, 87–92. [CrossRef] [PubMed]

20. Tommasino, M. The biology of β-human papillomaviruses. *Virus Res.* **2017**, *231*, 128–138. [CrossRef] [PubMed]

21. Martens, J.E.; Arends, J.; Van der Linden, P.J.; De Boer, B.A.; Helmerhorst, T.J. Cytokeratin 17 and p63 are markers of the HPV target cell, the cervical stem cell. *Anticancer Res.* **2004**, *24*, 771–775. [PubMed]

22. Herfs, M.; Yamamoto, Y.; Laury, A.; Wang, X.; Nucci, M.R.; McLaughlin-Drubin, M.E.; Munger, K.; Feldman, S.; McKeon, F.D.; Xian, W.; et al. A discrete population of squamocolumnar junction cells implicated in the pathogenesis of cervical cancer. *Proc. Natl. Acad. Sci. USA* **2012**, *109*, 10516–10521. [CrossRef] [PubMed]

23. Karin, M.; Clevers, H. Reparative inflammation takes charge of tissue regeneration. *Nature* **2016**, *529*, 307–315. [CrossRef] [PubMed]

24. Nigro, G.; Rossi, R.; Commere, P.H.; Jay, P.; Sansonetti, P.J. The cytosolic bacterial peptidoglycan sensor Nod2 affords stem cell protection and links microbes to gut epithelial regeneration. *Cell Host Microbe* **2014**, *15*, 792–798. [CrossRef] [PubMed]

25. Neal, M.D.; Sodhi, C.P.; Jia, H.; Dyer, M.; Egan, C.E.; Yazji, I.; Good, M.; Afrazi, A.; Marino, R.; Slagle, D.; et al. Toll-like receptor 4 is expressed on intestinal stem cells and regulates their proliferation and apoptosis via the p53 up-regulated modulator of apoptosis. *J. Biol. Chem.* **2012**, *287*, 37296–37308. [CrossRef] [PubMed]

26. Apidianakis, Y.; Pitsouli, C.; Perrimon, N.; Rahme, L. Synergy between bacterial infection and genetic predisposition in intestinal dysplasia. *Proc. Natl. Acad. Sci. USA* **2009**, *106*, 20883–20888. [CrossRef] [PubMed]

27. Pitsouli, C.; Apidianakis, Y.; Perrimon, N. Homeostasis in infected epithelia: Stem cells take the lead. *Cell Host Microbe* **2009**, *6*, 301–307. [CrossRef] [PubMed]

28. Mysorekar, I.U.; Isaacson-Schmid, M.; Walker, J.N.; Mills, J.C.; Hultgren, S.J. Bone morphogenetic protein 4 signaling regulates epithelial renewal in the urinary tract in response to uropathogenic infection. *Cell Host Microbe* **2009**, *5*, 463–475. [CrossRef] [PubMed]

29. Michael, S.; Lambert, P.F.; Strati, K. The HPV16 oncogenes cause aberrant stem cell mobilization. *Virology* **2013**, *443*, 218–225. [CrossRef] [PubMed]

30. Lanfredini, S.; Olivero, C.; Borgogna, C.; Calati, F.; Powell, K.; Davies, K.J.; De Andrea, M.; Harries, S.; Tang, H.K.C.; Pfister, H.; et al. HPV8 field cancerization in a transgenic mouse model is due to Lrig1+ keratinocyte stem cell expansion. *J. Investig. Dermatol.* **2017**. [CrossRef] [PubMed]

31. White, A.C.; Khuu, J.K.; Dang, C.Y.; Hu, J.; Tran, K.V.; Liu, A.; Gomez, S.; Zhang, Z.; Yi, R.; Scumpia, P.; et al. Stem cell quiescence acts as a tumour suppressor in squamous tumours. *Nat. Cell. Biol.* **2014**, *16*, 99–107. [CrossRef] [PubMed]

32. Jaks, V.; Barker, N.; Kasper, M.; Van Es, J.H.; Snippert, H.J.; Clevers, H.; Toftgård, R. Lgr5 marks cycling, yet long-lived, hair follicle stem cells. *Nat. Genet.* **2008**, *40*, 1291–1299. [CrossRef] [PubMed]

33. Lin, K.K.; Andersen, B. Have hair follicle stem cells shed their tranquil image? *Cell Stem Cell* **2008**, *3*, 581–582. [CrossRef] [PubMed]

34. Takahashi, K.; Yamanaka, S. Induction of pluripotent stem cells from mouse embryonic and adult fibroblast cultures by defined factors. *Cell* **2006**, *126*, 663–676. [CrossRef] [PubMed]

35. Abad, M.; Mosteiro, L.; Pantoja, C.; Canamero, M.; Rayon, T.; Ors, I.; Grana, O.; Megias, D.; Dominguez, O.; Martinez, D.; et al. Reprogramming in vivo produces teratomas and iPS cells with totipotency features. *Nature* **2013**, *502*, 340–345. [CrossRef] [PubMed]

36. Mosteiro, L.; Pantoja, C.; Alcazar, N.; Marión, R.M.; Chondronasiou, D.; Rovira, M.; Fernandez-Marcos, P.J.; Muñoz-Martin, M.; Blanco-Aparicio, C.; Pastor, J.; et al. Tissue damage and senescence provide critical signals for cellular reprogramming in vivo. *Science* **2016**, *354*, aaf4445. [CrossRef] [PubMed]

37. Stadtfeld, M.; Maherali, N.; Breault, D.T.; Hochedlinger, K. Defining molecular cornerstones during fibroblast to iPS cell reprogramming in mouse. *Cell Stem Cell* **2008**, *2*, 230–240. [CrossRef] [PubMed]

38. Hochedlinger, K.; Plath, K. Epigenetic reprogramming and induced pluripotency. *Development* **2009**, *136*, 509–523. [CrossRef] [PubMed]

39. Kelaini, S.; Cochrane, A.; Margariti, A. Direct reprogramming of adult cells: Avoiding the pluripotent state. *Stem Cells Cloning* **2014**, *7*, 19–29. [PubMed]

40. Margariti, A.; Winkler, B.; Karamariti, E.; Zampetaki, A.; Tsai, T.N.; Baban, D.; Ragoussis, J.; Huang, Y.; Han, J.D.; Zeng, L.; et al. Direct reprogramming of fibroblasts into endothelial cells capable of angiogenesis

and reendothelialization in tissue-engineered vessels. *Proc. Natl. Acad. Sci. USA* **2012**, *109*, 13793–13798. [CrossRef] [PubMed]

41. Masaki, T.; Qu, J.; Cholewa-Waclaw, J.; Burr, K.; Raaum, R.; Rambukkana, A. Reprogramming adult schwann cells to stem cell-like cells by leprosy bacilli promotes dissemination of infection. *Cell* **2013**, *152*, 51–67. [CrossRef] [PubMed]

42. Masaki, T.; McGlinchey, A.; Tomlinson, S.R.; Qu, J.; Rambukkana, A. Reprogramming diminishes retention of mycobacterium leprae in schwann cells and elevates bacterial transfer property to fibroblasts. *F1000Research* **2013**, *2*, 198. [CrossRef] [PubMed]

43. Masaki, T.; McGlinchey, A.; Cholewa-Waclaw, J.; Qu, J.; Tomlinson, S.R.; Rambukkana, A. Innate immune response precedes *Mycobacterium leprae*-induced reprogramming of adult Schwann cells. *Cell. Reprogram.* **2014**, *16*, 9–17. [CrossRef] [PubMed]

44. Panayidou, S.; Apidianakis, Y. Regenerative inflammation: Lessons from *Drosophila* intestinal epithelium in health and disease. *Pathogens* **2013**, *2*, 209–231. [CrossRef] [PubMed]

45. Taniguchi, K.; Wu, L.W.; Grivennikov, S.I.; De Jong, P.R.; Lian, I.; Yu, F.X.; Wang, K.; Ho, S.B.; Boland, B.S.; Chang, J.T.; et al. A gp130-Src-YAP module links inflammation to epithelial regeneration. *Nature* **2015**, *519*, 57–62. [CrossRef] [PubMed]

46. Michael, S.; Achilleos, C.; Panayiotou, T.; Strati, K. Inflammation shapes stem cells and stemness during infection and beyond. *Front. Cell Dev. Biol.* **2016**, *4*, 118. [CrossRef] [PubMed]

47. Ingle, A.; Ghim, S.; Joh, J.; Chepkoech, I.; Bennett Jenson, A.; Sundberg, J.P. Novel laboratory mouse papillomavirus (MusPV) infection. *Vet. Pathol.* **2011**, *48*, 500–505. [CrossRef] [PubMed]

48. Handisurya, A.; Day, P.M.; Thompson, C.D.; Buck, C.B.; Pang, Y.Y.; Lowy, D.R.; Schiller, J.T. Characterization of *Mus musculus* papillomavirus 1 infection in situ reveals an unusual pattern of late gene expression and capsid protein localization. *J. Virol.* **2013**, *87*, 13214–13225. [CrossRef] [PubMed]

49. Uberoi, A.; Yoshida, S.; Frazer, I.H.; Pitot, H.C.; Lambert, P.F. Role of ultraviolet radiation in papillomavirus-induced disease. *PLoS Pathog.* **2016**, *12*, e1005664. [CrossRef] [PubMed]

50. Kawamura, T.; Suzuki, J.; Wang, Y.V.; Menendez, S.; Morera, L.B.; Raya, A.; Wahl, G.M.; Izpisua Belmonte, J.C. Linking the p53 tumour suppressor pathway to somatic cell reprogramming. *Nature* **2009**, *460*, 1140–1144. [CrossRef] [PubMed]

51. Marion, R.M.; Strati, K.; Li, H.; Murga, M.; Blanco, R.; Ortega, S.; Fernandez-Capetillo, O.; Serrano, M.; Blasco, M.A. A p53-mediated DNA damage response limits reprogramming to ensure iPS cell genomic integrity. *Nature* **2009**, *460*, 1149–1153. [CrossRef] [PubMed]

52. Li, H.; Collado, M.; Villasante, A.; Strati, K.; Ortega, S.; Canamero, M.; Blasco, M.A.; Serrano, M. The Ink4/Arf locus is a barrier for ips cell reprogramming. *Nature* **2009**, *460*, 1136–1139. [CrossRef] [PubMed]

53. Marion, R.M.; Strati, K.; Li, H.; Tejera, A.; Schoeftner, S.; Ortega, S.; Serrano, M.; Blasco, M.A. Telomeres acquire embryonic stem cell characteristics in induced pluripotent stem cells. *Cell Stem Cell* **2009**, *4*, 141–154. [CrossRef] [PubMed]

54. Utikal, J.; Polo, J.M.; Stadtfeld, M.; Maherali, N.; Kulalert, W.; Walsh, R.M.; Khalil, A.; Rheinwald, J.G.; Hochedlinger, K. Immortalization eliminates a roadblock during cellular reprogramming into iPS cells. *Nature* **2009**, *460*, 1145–1148. [CrossRef] [PubMed]

55. Lin, T.; Chao, C.; Saito, S.; Mazur, S.J.; Murphy, M.E.; Appella, E.; Xu, Y. P53 induces differentiation of mouse embryonic stem cells by suppressing nanog expression. *Nat. Cell. Biol.* **2005**, *7*, 165–171. [CrossRef] [PubMed]

56. Kareta, M.S.; Gorges, L.L.; Hafeez, S.; Benayoun, B.A.; Marro, S.; Zmoos, A.F.; Cecchini, M.J.; Spacek, D.; Batista, L.F.; O'Brien, M.; et al. Inhibition of pluripotency networks by the Rb tumor suppressor restricts reprogramming and tumorigenesis. *Cell Stem Cell* **2015**, *16*, 39–50. [CrossRef] [PubMed]

57. Chlon, T.M.; Hoskins, E.E.; Mayhew, C.N.; Wikenheiser-Brokamp, K.A.; Davies, S.M.; Mehta, P.; Myers, K.C.; Wells, J.M.; Wells, S.I. High-risk human papillomavirus E6 protein promotes reprogramming of fanconi anemia patient cells through repression of p53 but does not allow for sustained growth of induced pluripotent stem cells. *J. Virol.* **2014**, *88*, 11315–11326. [CrossRef] [PubMed]

58. Gunasekharan, V.K.; Li, Y.; Andrade, J.; Laimins, L.A. Post-transcriptional regulation of Klf4 by high-risk human papillomaviruses is necessary for the differentiation-dependent viral life cycle. *PLoS Pathog.* **2016**, *12*, e1005747. [CrossRef] [PubMed]

59. Brehm, A.; Ohbo, K.; Scholer, H. The carboxy-terminal transactivation domain of Oct-4 acquires cell specificity through the POU domain. *Mol. Cell. Biol.* **1997**, *17*, 154–162. [CrossRef] [PubMed]
60. Brehm, A.; Ohbo, K.; Zwerschke, W.; Botquin, V.; Jansen-Durr, P.; Scholer, H.R. Synergism with germ line transcription factor Oct-4: Viral oncoproteins share the ability to mimic a stem cell-specific activity. *Mol. Cell. Biol.* **1999**, *19*, 2635–2643. [CrossRef] [PubMed]
61. Organista-Nava, J.; Gómez-Gómez, Y.; Ocadiz-Delgado, R.; García-Villa, E.; Bonilla-Delgado, J.; Lagunas-Martínez, A.; Tapia, J.S.; Lambert, P.F.; García-Carrancá, A.; Gariglio, P. The HPV16 E7 oncoprotein increases the expression of Oct3/4 and stemness-related genes and augments cell self-renewal. *Virology* **2016**, *499*, 230–242. [CrossRef] [PubMed]
62. McLaughlin-Drubin, M.E.; Crum, C.P.; Munger, K. Human papillomavirus E7 oncoprotein induces KDM6A and KDM6B histone demethylase expression and causes epigenetic reprogramming. *Proc. Natl. Acad. Sci. USA* **2011**, *108*, 2130–2135. [CrossRef] [PubMed]
63. McLaughlin-Drubin, M.E.; Park, D.; Munger, K. Tumor suppressor p16^{INK4A} is necessary for survival of cervical carcinoma cell lines. *Proc. Natl. Acad. Sci. USA* **2013**, *110*, 16175–16180. [CrossRef] [PubMed]
64. Hufbauer, M.; Biddle, A.; Borgogna, C.; Gariglio, M.; Doorbar, J.; Storey, A.; Pfister, H.; Mackenzie, I.; Akgül, B. Expression of β-papillomavirus oncogenes increases the number of keratinocytes with stem cell-like properties. *J. Virol.* **2013**, *87*, 12158–12165. [CrossRef] [PubMed]
65. Meyers, J.M.; Uberoi, A.; Grace, M.; Lambert, P.F.; Munger, K. Cutaneous HPV8 and MmuPV1 E6 proteins target the Notch and TGF-β tumor suppressors to inhibit differentiation and sustain keratinocyte proliferation. *PLoS Pathog.* **2017**, *13*, e1006171. [CrossRef] [PubMed]
66. Kranjec, C.; Holleywood, C.; Libert, D.; Griffin, H.; Mahmood, R.; Isaacson, E.; Doorbar, J. Modulation of basal cell fate during productive and transforming HPV16 infection is mediated by progressive E6-driven depletion of Notch. *J. Pathol.* **2017**, *242*, 448–462. [CrossRef] [PubMed]
67. Tyagi, A.; Vishnoi, K.; Mahata, S.; Verma, G.; Srivastava, Y.; Masaldan, S.; Roy, B.G.; Bharti, A.C.; Das, B.C. Cervical cancer stem cells selectively overexpress hpv oncoprotein E6 that controls stemness and self-renewal through upregulation of Hes1. *Clin. Cancer Res.* **2016**, *22*, 4170–4184. [CrossRef] [PubMed]
68. Kim, B.W.; Cho, H.; Choi, C.H.; Ylaya, K.; Chung, J.Y.; Kim, J.H.; Hewitt, S.M. Clinical significance of Oct4 and Sox2 protein expression in cervical cancer. *BMC Cancer* **2015**, *15*, 1015. [CrossRef] [PubMed]
69. Lindquist, D.; Näsman, A.; Tarján, M.; Henriksson, R.; Tot, T.; Dalianis, T.; Hedman, H. Expression of LRIG1 is associated with good prognosis and human papillomavirus status in oropharyngeal cancer. *Br. J. Cancer* **2014**, *110*, 1793–1800. [CrossRef] [PubMed]

viruses

MDPI

Review

Telomerase Induction in HPV Infection and Oncogenesis

Rachel Katzenellenbogen [1,2] (iD)

[1] Center for Global Infectious Disease Research, Seattle Children's Research Institute, Seattle, WA 98101, USA; rkatzen@uw.edu; Tel.: +1-206-884-1082

[2] Department of Pediatrics, Division of Adolescent Medicine, University of Washington, Seattle, WA 98195, USA

Academic Editors: Alison A. McBride and Karl Munger
Received: 15 June 2017; Accepted: 7 July 2017; Published: 10 July 2017

Abstract: Telomerase extends the repetitive DNA at the ends of linear chromosomes, and it is normally active in stem cells. When expressed in somatic diploid cells, it can lead to cellular immortalization. Human papillomaviruses (HPVs) are associated with and high-risk for cancer activate telomerase through the catalytic subunit of telomerase, human telomerase reverse transcriptase (hTERT). The expression of hTERT is affected by both high-risk HPVs, E6 and E7. Seminal studies over the last two decades have identified the transcriptional, epigenetic, and post-transcriptional roles high-risk E6 and E7 have in telomerase induction. This review will summarize these findings during infection and highlight the importance of telomerase activation as an oncogenic pathway in HPV-associated cancer development and progression.

Keywords: human papillomavirus; papillomaviruses; oncogenic virus; cancer; HPV E6; HPV E7; hTERT; telomerase

---

1. Introduction: Human Papillomavirus Infections

Human Papillomavirus Infection and Life Cycle in Stratified Squamous Epithelium

Human papillomaviruses (HPVs) are small, non-envelope, double stranded DNA viruses, and there are more than 200 papillomavirus types identified to date (see curated list at Papillomavirus Episteme (PaVE); https://pave.niaid.nih.gov/#home). All HPVs complete their life cycle in stratified squamous epithelium, either cutaneous or mucosal dependent on their tropism (reviewed in [1,2]). The HPV life cycle begins with infection of basal cells in stratified squamous epithelium. These cells are reached either through microabrasions or at anatomic sites where monolayer, columnar epithelium transition to stratified squamous epithelium. These sites of transition are the found at the cervical transformation zone, the anal verge, and crypts in the oral mucosa, and intriguingly these are also where many HPV-associated cancers occur [3]. There is also evidence the cervix and anal verge contain specific cells that support HPV-associated cancers and have signature gene expression profiles [4,5]. Whether or not these cells represent a preferred host for the virus or more narrowly for the initiation of cancer, we broadly understand that a productive HPV infection begins in the bottom layer of stratified squamous epithelium.

In the basal layer, the HPV genome escapes its viral capsid and is maintained at 50–60 copies per cells (ranging from 10 to 200 copies) [2,6], while the early (*E*) viral genes are expressed. E1 and E2 support HPV DNA replication and the measured expression of E6 and E7 [2,6]. One function of E6 and E7, as well as E5, is to reduce activation of the innate immune system and avoid viral clearance [7–9]. Evading immune sensing and maintaining copies of the viral genome are both critical to establishing an infection.

As cells leave the basal layer, they rise through the suprabasal and spinous layers and progress through differentiation. HPV requires this cellular differentiation program to complete its own viral life cycle. Without it, HPV has an abortive infection. Therefore, as HPV-infected host cells rise through the differentiating layers, HPV progresses through its own life cycle. In these differentiating layers, HPV DNA becomes amplified to several thousand copies per cell [2], and late (*L*) gene expression is activated. L1 and L2 form the protein shell of HPV, and they incorporate the HPV episomal DNA into new infectious virions. Those released as epithelial cells are sloughed off at the top of stratified squamous epithelium, completing the viral life cycle.

Although HPV requires its host cell to differentiate in order to complete its life cycle, it also requires it host cell to continue to grow when normally it would not. HPV dysregulates the balance of growth and differentiation found in stratified squamous epithelium to do so. In low-risk HPV types, this is manifest as warts. In high-risk HPV types, this leads to dysplasias and carcinoma in situ [10]. This dysregulation of growth and differentiation is driven primarily by the *E6* and *E7* genes. E6 and E7 drive cells to continue to grow and divide when they otherwise would not, and, to that end, E6 and E7 are expressed in the differentiating layers of stratified squamous epithelium [2,10]. By E6 and E7 disrupting the typical segregation of cell cycle and growth from differentiation, more HPV DNA can be copied and expressed, and more cells infected by HPV can grow.

There are at least 15 HPV types that are defined as high-risk (HR) by their association with cervical cancer [11]. HPV-associated cancers universally express the HR *E6* and *E7* genes, thus are considered to be HPV's viral oncogenes. If HR E6 and E7 are introduced into normal diploid cells, they become immortalized [12,13]. If HR E6 and E7 expression is reduced in HPV-positive cervical cancer cell lines, the cells growth arrest [14,15]. This implies that not only are the HR *E6* and *E7* genes required for oncogenesis, but they are also required for the maintenance of malignant phenotype. There are critical oncogenic pathways that HR E6 and E7 affect. HR E7 targets the retinoblastoma protein (Rb) for degradation in epithelial cells [16,17]. This allows infected epithelial cells to proceed through S phase and the cell cycle. HR E6 targets p53 for degradation to avoid apoptosis [18,19]. It similarly targets PSD-Dlg-ZO-1/2 (PDZ) containing proteins for degradation, disrupting cellular apicobasal orientation and cell-to-cell adhesion in the epithelium, leading to hyperplasia [20–24]. HR E6 also activates gene expression; its most critical gene to activate is human telomerase reverse transcriptase (hTERT), the catalytic subunit of telomerase. It is the degradation of Rb by HR E7 and the activation of hTERT by HR E6 that drives normal keratinocytes to immortalization [12,13].

In this review article, we will describe the roles E6 and E7 have in telomerase induction during HPV infection and in oncogenesis. We will first define telomeric DNA, its role in DNA protection, and the enzymatic function of telomerase. Then, we will highlight the multiple ways HR E6 and E7 derepress hTERT to activate and accelerate telomerase activity. Finally, we will discuss how E6, E7, and hTERT expression changes during oncogenesis.

2. Telomeric DNA and Telomerase

Telomeric DNA caps the ends of linear chromosomes, is repetitive, and is approximately 5000 to 15,000 nucleotides in length in humans [25,26]. No genetic material is found within telomeric DNA itself. Rather, it is bound by the shelterin protein complex to block double strand (dsDNA) repair signaling [27], protecting against non-homologous end joining and erroneous chromosomal break repair [27]. Telomerase, a ribonucleoprotein enzyme complex, extends this repetitive telomeric DNA. The holoenzyme includes the catalytic subunit hTERT that is expressed at rate-determining levels [28–30], the telomerase RNA component, TERC or TR, used to extend the six nucleotide repeat 5′ TTAGGG 3′ found in telomeric DNA, and the protein dyskerin [26,31]. Telomerase is typically active during embryonic and fetal development [32] and in stem cells [33]. It is not active in normal somatic cells. However, telomerase activity has been detected in almost all human tumors and immortalized cells in culture [29,30,34].

Without telomerase activity, the linear chromosomes of cellular DNA are serially shortened with every cell cycle and division by 100 to 200 nucleotides [35]. This DNA loss is called the "end replication problem". As telomeric DNA becomes critically shortened over time, normal somatic diploid cells enter mortality stage one (M1) and undergo either replicative senescence or apoptosis [35–38]. If these cells continue to cycle beyond stage M1, they lose the protective shelterin protein complexes and enter mortality stage two (M2) or crisis. In crisis, cells signal that there are dsDNA breaks at the ends of chromosomes requiring repair. This genomic instability creates the "end protection problem". It leads to anaphase bridges and chromosomal breaks that are catastrophic to the cell [39–41]. Only clonal cells survive that have had enormous chromosomal rearrangements [41]. Consequently, the extension of telomeric DNA by telomerase allows diploid cells to grow over time, avoiding apoptosis, senescence, and chromosomal rearrangements. It is because of this allowance that telomerase and its rate-determining catalytic subunit hTERT are expressed in nearly all cancers [30,34,42,43].

3. Telomerase and hTERT Activity Driven by HR E6 and E7

Studies in the late 1980s defined the roles HR E6 and E7 played in cellular immortalization, cancer development, and cancer progression [12,13]. In a seminal paper by Kiyono et al. HR E6 and E7 were found to collaborate in the immortalization of both primary fibroblasts and keratinocytes, specifically by dysregulating Rb/p16INK4A and telomerase [44]. HR E7 was important for immortalization, but it did not directly affect telomerase [44]. Rather, HR E6 did. HR E7 was described to increase hTERT driven expression of luciferase and augment telomerase activity driven by HR HPV E6 [45]. In HeLa cells, re-expression of either HR E6 or E7 after their removal led to increased hTERT [46]. Although E7 could synergize the E6 regulation of telomerase and cellular immortalization, HR E6 is the principal trigger and regulator of hTERT expression and telomerase activity (Table 1). Other studies built on these foundational reports.

Table 1. HR E6 and E7 regulation of hTERT and cellular protein targets for that regulation.

HPV Gene	Effect on hTERT	Cellular Protein Target
	Chromatin Effects	
E6 and E7	Promoter methylation changes	
E6	Increase promoter acetylation	HATs and HDACs, mSin3A
	Transcription Effects	
E6	Increase transcriptional activators	c-Myc/Max, Sp1
E6	Decrease transcriptional repressors	c-Myc/Mad, Maz, USF1, NFX1-91
E7	Increase expression with E6	
	RNA Effects	
E6	Increase transcript stability	NFX1-123, PABPCs
E6	Increase active spliced isoform of hTERT	c-Myc
	Protein Effects	
E6	Binds hTERT	hTERT

HR: high-risk; hTERT: human telomerase reverse transcriptase; HPV: human papillomavirus; HATs: histone acetyltransferases; HDACs: histone deacetylases; mSin3A: SIN3 transcription regulator family member A; c-Myc: MYC proto-oncogene; Max: MYC associated factor X; Maz: MYC associated zinc finger protein; USF1: Upstream transcription factor 1; NFX1-91: Nuclear transcription factor, X-box binding 1, isoform 3; NFX1-123: Nuclear transcription factor, X-box binding 1, isoform 1; PABPCs: cytoplasmic poly(A) binding proteins.

Recent studies have confirmed that low-risk (LR) E6 does not activate telomerase [47] while HR E6 is necessary and sufficient for telomerase activation in keratinocytes [48–51]. Without HR E6, telomerase is not detected in epithelial cells, and the catalytic subunit of telomerase, hTERT, is not expressed [48,50]. In addition to HR E6 regulating the activity of telomerase, HR E6 was found to bind hTERT itself and repetitive telomeric DNA [52]. Therefore, the role HR E6 has in hTERT, telomerase, and telomeric DNA regulation is multilayered, demonstrated by its redundant actions to drive immortalization.

The E3 Ubiquitin Ligase E6 Associated Protein (E6AP) is important for the activation of hTERT expression and telomerase activity by HR E6 [53–55]. E6AP partners with HR E6 to polyubiquitinate and degrade p53 and PDZ-containing proteins [18,19,21,56,57], but this partnership does not lead to the degradation of hTERT or telomerase; instead, it increases hTERT and telomerase. Decreasing HR E6 and E6AP by microRNA (miR375) indirectly reduced hTERT and telomerase activity in cells [58], and the E6 motifs needed to bind E6AP were also required for telomerase activation and immortalization in fibroblasts and keratinocytes [59]. Hence, HR E6 and E6AP (E6/E6AP) function as principal inductors of telomerase, and its catalytic subunit hTERT.

3.1. hTERT: Promoter Regulation

Most research on telomerase regulation has focused on the expression of its catalytic subunit, hTERT. The *hTERT* gene is constitutively repressed in somatic epithelial cells; this repression occurs at its promoter. The hTERT promoter is approximately 1100 nucleotides in length, with its core promoter being only 200 to 300 nucleotides long [60–62]. Normally, transcriptional repressors of hTERT are bound to *cis* elements in its core promoter, blocking transcription [48,60,61,63–69]. These *cis* elements are E boxes, GC-rich sites, and X boxes.

Two E box *cis* elements flank the transcriptional start site of hTERT [61,62,66], and if these E boxes are mutated or deleted, hTERT expression and telomerase activity are dramatically reduced [50,54]. These E boxes are normally bound by c-Myc as a heterodimer with either Max or Mad. These c-Myc/Max or c-Myc/Mad heterodimers are important for hTERT transcriptional activation or repression, respectively [65,66,70–72]. Upstream transcription factor 1 (USF1) also binds to E boxes, competitively and sterically repressing hTERT expression by c-Myc/Max [50,73,74] Although the amount of c-Myc that binds to the hTERT promoter does not correlate to hTERT expression, the presence of c-Myc at the promoter is important [51,72,75]. E6/E6AP are also bound at E boxes in the hTERT promoter [51,54,74,75], and they interact with c-Myc to drive gene expression [51]. The requirement for E6AP to drive hTERT expression at the promoter with HR E6 is controversial [53–55,76,77], but the *cis* E boxes within the hTERT promoter are required for its transcriptional activation with or without HR E6.

There are five GC-rich *cis* elements in the hTERT promoter 5′ of the transcriptional start site [50,71,74]. Sp1 binds to these elements and transcriptionally activates hTERT expression [50,71]. Maz also is bound at these sites but is a transcriptional repressor [71]. Like deletion of the E boxes, deletion of GC-rich *cis* elements leads to loss of hTERT promoter-driven transcriptional activation [50].

Finally, there are two X boxes in the hTERT promoter [48]. One is downstream of the hTERT transcriptional start site, lies within the 5′ UTR of hTERT, and overlaps with the downstream E box to which c-Myc/Max binds [48]. The second is upstream of the hTERT core promoter in an inverted position [48]. Nuclear transcription factor X-box binding 1, isoform 3 (NFX1-91) is a repressor of hTERT transcription and is bound constitutively at the hTERT downstream X box [48,78]. NFX1-91 is polyubiquitinated by E6/E6AP and targeted for proteasomal degradation [48]. Its removal from the hTERT promoter leads to transcriptional activation of hTERT [48].

3.2. hTERT: Epigenetic Regulation

Beyond studies of the hTERT promoter *cis* elements and the transcriptional proteins that bind those elements, epigenetic studies of the hTERT promoter demonstrate important structural chromatin changes that affect transcriptional activation of hTERT [78]. Several studies document the importance of E6/E6AP in opening the hTERT promoter chromatin structure as they change histone acetyltransferase (HAT) and histone deacetylase (HDAC) recruitment to the hTERT promoter [53,78]. The hTERT repressor NFX1-91 not only binds the promoter X box *cis* element but also binds SIN3 transcription regulator family member A (mSin3A), a transcriptional co-repressor that recruits HDACs to promoters [78]. When NFX1-91 is degraded by E6/E6AP, HDAC activity at the hTERT promoter is lost and HAT activity increases [78], and with this, histone acetylation increases further over time [53].

DNA methylation patterns at the hTERT promoter also shift during an HPV infection and in tissue culture studies of HPV positive cells. Specific regions of the promoter become hypermethylated, while other regions become hypomethylated, during long-term tissue culture of cells with HR E6 and E7 [79–82]. Although a direct causal association between hTERT promoter methylation and cancer development has not been seen [83], there are changes that parallel increases in hTERT expression, and these changes in methylation patterns correlate with HR and probable HR E6 expression [84].

3.3. hTERT: Post-Transcriptional Regulation

Post-transcriptional regulation of hTERT by alternative mRNA splicing and mRNA stabilization is important for telomerase activity [32,85,86]. In non-HPV studies, c-Myc shifts hTERT mRNA expression from a non-active splice variant to an active form [87]. RNA processing proteins, such as Serine-Arginine Rich Splicing Factors, are also expressed at increased levels in high-grade cervical dysplasias [88], pointing indirectly to HR HPV manipulating RNA processing proteins during oncogenesis.

We found that hTERT and telomerase activity are upregulated post-transcriptionally by HR E6 through the host cellular protein NFX1-123 [86]. NFX1-123 is the longer splice variant of the *NFX1* gene (the hTERT transcriptional repressor NFX1-91 is the shorter splice variant) [48,89]. Greater expression of NFX1-123 leads to increased hTERT and telomerase activity with HR E6, and knock down of endogenous NFX1-123 reduces the ability of HR E6 to increase hTERT and telomerase [86,89]. The mechanism by which NFX1-123 augments hTERT expression is through stabilization of the hTERT mRNA, and the 5′ UTR of the hTERT transcript is necessary for this stabilization [86].

NFX1-123 contains two protein motifs important for binding, stabilizing, and augmenting hTERT expression. The R3H domain of NFX1-123 has putative single-stranded nucleic acid binding capabilities [86,89,90], and when this motif is deleted, the stabilization and increased expression of hTERT seen in HR E6 expressing cells is lost [86,89]. Second, the poly(A) binding protein interacting motif (PAM2) of NFX1-123 directs binding of cytoplasmic poly(A) binding proteins (PABPCs) to NFX1-123, and PABPCs increase the stability and translation of genes with poly (A) tailed mRNA [89,91]. Like the R3H domain, when the PAM2 motif of NFX1-123 is mutated or deleted, its ability to augment hTERT expression and telomerase activity by HR E6 is also lost [86,89].

Cytoplasmic poly(A) binding proteins themselves are important in hTERT expression and telomerase activity in HR E6 positive cells. When PABPC types 1 and 4 are knocked down, hTERT and telomerase activity are reduced [92]. Conversely, when PABPC type 4 is overexpressed, hTERT and telomerase are augmented, and cells with either more hTERT or more PABPC type 4 grow better in culture [92].

Collectively, these research findings highlight multiple ways hTERT mRNA is post-transcriptionally regulated. Again, the duplicative mechanisms, from DNA, chromatin, and RNA regulation, that HR E6 uses to increase hTERT and telomerase emphasizes its importance to HPV and oncogenesis.

3.4. hTERT: Beta HPV E6

Most studies examining the regulation of telomerase by HPV have focused on HR HPVs from the α genus. More recent work has examined the role β genus HPVs play in nonmelanomatous squamous cell carcinoma, and how the beta E6 and E7 proteins may also activate oncogenic pathways, whether similar or disparate to α HR E6 and E7 proteins. Work by Galloway et al. has determined the oncogenic potential of β HPV types through direct analysis of their E6 and E7 protein functionality, and specifically how different E6 types activate hTERT expression, telomerase activity, and immortalization [40,93]. β E6 proteins with greater effect on hTERT activation and telomerase activity have improved cellular growth and longevity in culture [93]. This improvement is not only proportional to telomerase activity but also depends on the presence of E6AP [93]. Therefore, like α HR HPV types, several β genus *E6* genes drive hTERT expression and telomerase activity.

4. Telomerase in HPV-Associated Cancers

During cervical cancer initiation and progression, the expression of hTERT and the activity of telomerase parallels worsening disease [94–97]. Approximately half of HPV positive squamous intraepithelial lesions and cervical intraepithelial grade III lesions have detectable telomerase activity and that increases to over 90% in HPV positive cervical cancer samples [94,98]. The level of hTERT expression and telomerase activity found in cervical lesions is proportional to the pathologic severity of disease detected [94,96,98]. In HPV-positive cancers, telomerase is universally expressed (modeled in Figure 1) [34]. Telomerase is increasingly identified as having both canonical and non-canonical functions, and each is important to HPV-induced cellular immortalization and oncogenesis [99].

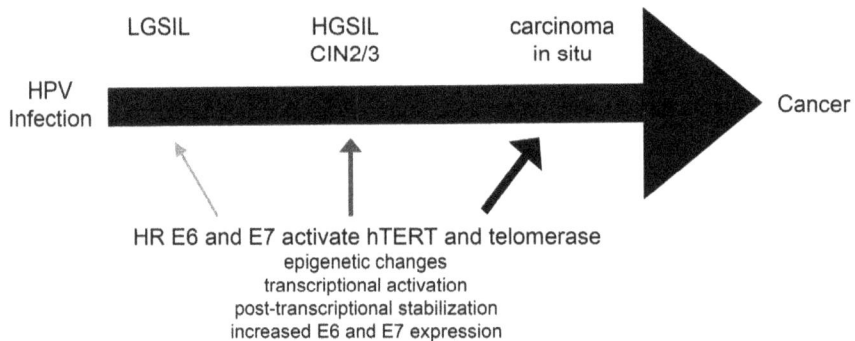

Figure 1. HPV infection and telomerase induction. Telomerase, and its rate determining catalytic subunit, hTERT, is normally not expressed in somatic cells. With a HR HPV infection, E6 and E7 activate the *hTERT* gene. With disease progression, *hTERT* activation and telomerase activity increases (demonstrated by darker, larger arrows), and the expression of HR E6 and E7 also increases with the integration of HPV DNA into the host cell chromosomal DNA or loss of E2 regulation. LGSIL (low-grade squamous intraepithelial lesion) is typical for an active HPV infection. HGSIL (high-grade squamous intraepithelial lesion) is typical for a HR HPV infection with worsening cytologic changes and parallel greater histologic involvement, with multiple layers of the stratified squamous epithelium. CIN2/3 (cervical intraepithelial neoplasia 2 or 3) shows histologic changes due to an active HPV infection that involved most (2) or all (3) of the stratified squamous epithelium. Carcinoma in situ is the full thickness involvement of stratified squamous epithelium without breakdown of the basement membrane.

Interestingly, during the transition from HR HPV infection, to dysplasia, to frank cancer, HPV DNA typically no longer remains episomal. It becomes integrated in the cell's chromosomes. This happens within the context of genomic instability, created and supported by the functions of HR E6 and E7 themselves. This, by definition, means HPV can no longer form infectious virions; it also means HPV gene expression itself is dysregulated. With HPV DNA integration, the HR E6 and E7 genes are universally preserved, but the regulatory *E1* and *E2* genes are often lost. Even in non-integrated HPV driven cancers, the binding sites for E2 often become methylated. These changes allow for greater expression of E6 and E7, as E2 moderates the expression level of these viral oncogenes [100,101]. Although not required, with increased E6 and E7, there is a parallel increase in telomerase activity [102]. During HR HPV infection and its associated cancer development and progression, HR E6, with E7, activates telomerase. This activation is augmented over time by changes that support cellular immortalization and growth and by the acceleration of viral and cellular genomic instability that was first initiated by HR E6 and E7.

High-risk HPV infections are associated with cancers in other sites besides the cervix [3]. These include vulvar, vaginal, anal, penile, and the head-and-neck. Each of these HPV-associated cancers are also associated with upregulated telomerase activity [34]. Therefore, the anatomic location of a HR HPV infection is not the singular instigator of immortalization—the commonality among these cancers is HR HPV, and HR E6 specifically, driving telomerase.

5. Conclusions

High-risk E6 hijacks host cell proteins from their usual function (E6AP, c-Myc, HDAC, HAT, mSin3A, NFX1-91, NFX1-123, PABPCs, and mRNA splicing factors) to activate hTERT and telomerase activity. This supports cellular immortalization. These viral-cellular protein partnerships primarily control the derepression of telomerase's catalytic subunit, hTERT. They increase hTERT through the promoter's *cis* and *trans* elements, the chromatin structure, the mRNA product, and associated RNA regulatory proteins. There are still many unanswered questions in the dysregulation of telomerase activation by HPV during infection and oncogenesis. However, its universality implies it is critical to the core function of HR HPV types and to induction of cancers caused by HPV.

Acknowledgments: The author would like to recognize all of the researchers' work that was not recognized due to space limitations. This review did not receive any specific grant from funding agencies in the public, commercial, or not-for-profit sectors.

Conflicts of Interest: The author declares no conflict of interest.

References

1. Doorbar, J.; Quint, W.; Banks, L.; Bravo, I.G.; Stoler, M.; Broker, T.R.; Stanley, M.A. The biology and life-cycle of human papillomaviruses. *Vaccine* **2012**, *30*, F55–F70. [CrossRef] [PubMed]

2. Doorbar, J. The papillomavirus life cycle. *J. Clin. Virol.* **2005**, *32*, S7–S15. [CrossRef] [PubMed]

3. De Martel, C.; Plummer, M.; Vignat, J.; Franceschi, S. Worldwide burden of cancer attributable to HPV by site, country and HPV type. *Int. J. Cancer* **2017**, *141*, 664–670. [CrossRef] [PubMed]

4. Herfs, M.; Yamamoto, Y.; Laury, A.; Wang, X.; Nucci, M.R.; McLaughlin-Drubin, M.E.; Munger, K.; Feldman, S.; McKeon, F.D.; Xian, W.; et al. A discrete population of squamocolumnar junction cells implicated in the pathogenesis of cervical cancer. *Proc. Natl. Acad. Sci. USA* **2012**, *109*, 10516–10521. [CrossRef] [PubMed]

5. Herfs, M.; Longuespee, R.; Quick, C.M.; Roncarati, P.; Suarez-Carmona, M.; Hubert, P.; Lebeau, A.; Bruyere, D.; Mazzucchelli, G.; Smargiasso, N.; et al. Proteomic signatures reveal a dualistic and clinically relevant classification of anal canal carcinoma. *J. Pathol.* **2017**, *241*, 522–533. [CrossRef] [PubMed]

6. Doorbar, J. Molecular biology of human papillomavirus infection and cervical cancer. *Clin. Sci.* **2006**, *110*, 525–541. [CrossRef] [PubMed]

7. Westrich, J.A.; Warren, C.J.; Pyeon, D. Evasion of host immune defenses by human papillomavirus. *Virus Res.* **2017**, *231*, 21–33. [CrossRef] [PubMed]

8. Hong, S.; Laimins, L.A. Manipulation of the innate immune response by human papillomaviruses. *Virus Res.* **2017**, *231*, 34–40. [CrossRef] [PubMed]

9. Grabowska, A.K.; Riemer, A.B. The invisible enemy—How human papillomaviruses avoid recognition and clearance by the host immune system. *Open Virol. J.* **2012**, *6*, 249–256. [CrossRef] [PubMed]

10. Doorbar, J.; Egawa, N.; Griffin, H.; Kranjec, C.; Murakami, I. Human papillomavirus molecular biology and disease association. *Rev. Med. Virol.* **2015**, *25*, 2–23. [CrossRef] [PubMed]

11. Munoz, N.; Bosch, F.X.; de Sanjose, S.; Herrero, R.; Castellsague, X.; Shah, K.V.; Snijders, P.J.; Meijer, C.J. Epidemiologic classification of human papillomavirus types associated with cervical cancer. *N. Engl. J. Med.* **2003**, *348*, 518–527. [CrossRef] [PubMed]

12. Hawley-Nelson, P.; Vousden, K.H.; Hubbert, N.L.; Lowy, D.R.; Schiller, J.T. HPV16 E6 and E7 proteins cooperate to immortalize human foreskin keratinocytes. *EMBO J.* **1989**, *8*, 3905–3910. [PubMed]

13. Munger, K.; Phelps, W.C.; Bubb, V.; Howley, P.M.; Schlegel, R. The E6 and E7 genes of the human papillomavirus type 16 together are necessary and sufficient for transformation of primary human keratinocytes. *J. Virol.* **1989**, *63*, 4417–4421. [PubMed]

14. Wells, S.I.; Francis, D.A.; Karpova, A.Y.; Dowhanick, J.J.; Benson, J.D.; Howley, P.M. Papillomavirus E2 induces senescence in HPV-positive cells via pRB- and p21^{CIP}-dependent pathways. *EMBO J.* **2000**, *19*, 5762–5771. [CrossRef] [PubMed]

15. Francis, D.A.; Schmid, S.I.; Howley, P.M. Repression of the integrated papillomavirus E6/E7 promoter is required for growth suppression of cervical cancer cells. *J. Virol.* **2000**, *74*, 2679–2686. [CrossRef] [PubMed]

16. Dyson, N.; Howley, P.M.; Munger, K.; Harlow, E. The human papilloma virus-16 E7 oncoprotein is able to bind to the retinoblastoma gene product. *Science* **1989**, *243*, 934–937. [CrossRef] [PubMed]

17. Boyer, S.N.; Wazer, D.E.; Band, V. E7 protein of human papilloma virus-16 induces degradation of retinoblastoma protein through the ubiquitin-proteasome pathway. *Cancer Res.* **1996**, *56*, 4620–4624. [PubMed]

18. Huibregtse, J.M.; Scheffner, M.; Howley, P.M. A cellular protein mediates association of p53 with the E6 oncoprotein of human papillomavirus types 16 or 18. *EMBO J.* **1991**, *10*, 4129–4135. [PubMed]

19. Scheffner, M.; Huibregtse, J.M.; Vierstra, R.D.; Howley, P.M. The HPV-16 E6 and E6-AP complex functions as a ubiquitin-protein ligase in the ubiquitination of p53. *Cell* **1993**, *75*, 495–505. [CrossRef]

20. Kiyono, T.; Hiraiwa, A.; Fujita, M.; Hayashi, Y.; Akiyama, T.; Ishibashi, M. Binding of high-risk human papillomavirus E6 oncoproteins to the human homologue of the Drosophila discs large tumor suppressor protein. *Proc. Natl. Acad. Sci. USA* **1997**, *94*, 11612–11616. [CrossRef] [PubMed]

21. Nakagawa, S.; Huibregtse, J.M. Human scribble (vartul) is targeted for ubiquitin-mediated degradation by the high-risk papillomavirus E6 proteins and the E6AP ubiquitin-protein ligase. *Mol. Cell. Biol.* **2000**, *20*, 8244–8253. [CrossRef] [PubMed]

22. Lee, S.S.; Glaunsinger, B.; Mantovani, F.; Banks, L.; Javier, R.T. Multi-PDZ domain protein MUPP1 is a cellular target for both adenovirus E4-ORF1 and high-risk papillomavirus type 18 E6 oncoproteins. *J. Virol.* **2000**, *74*, 9680–9693. [CrossRef] [PubMed]

23. Watson, R.A.; Thomas, M.; Banks, L.; Roberts, S. Activity of the human papillomavirus E6 PDZ-binding motif correlates with an enhanced morphological transformation of immortalized human keratinocytes. *J. Cell Sci.* **2003**, *116*, 4925–4934. [CrossRef] [PubMed]

24. Nguyen, M.L.; Nguyen, M.M.; Lee, D.; Griep, A.E.; Lambert, P.F. The PDZ ligand domain of the human papillomavirus type 16 E6 protein is required for E6's induction of epithelial hyperplasia in vivo. *J. Virol.* **2003**, *77*, 6957–6964. [CrossRef] [PubMed]

25. Cech, T.R. Beginning to understand the end of the chromosome. *Cell* **2004**, *116*, 273–279. [CrossRef]

26. Schmidt, J.C.; Cech, T.R. Human telomerase: Biogenesis, trafficking, recruitment, and activation. *Genes Dev.* **2015**, *29*, 1095–1105. [CrossRef] [PubMed]

27. De Lange, T. Shelterin: The protein complex that shapes and safeguards human telomeres. *Genes Dev.* **2005**, *19*, 2100–2110. [CrossRef] [PubMed]

28. Counter, C.M.; Hahn, W.C.; Wei, W.; Caddle, S.D.; Beijersbergen, R.L.; Lansdorp, P.M.; Sedivy, J.M.; Weinberg, R.A. Dissociation among in vitro telomerase activity, telomere maintenance, and cellular immortalization. *Proc. Natl. Acad. Sci. USA* **1998**, *95*, 14723–14728. [CrossRef] [PubMed]

29. Counter, C.M.; Meyerson, M.; Eaton, E.N.; Ellisen, L.W.; Caddle, S.D.; Haber, D.A.; Weinberg, R.A. Telomerase activity is restored in human cells by ectopic expression of hTERT (hEST2), the catalytic subunit of telomerase. *Oncogene* **1998**, *16*, 1217–1222. [CrossRef] [PubMed]

30. Meyerson, M.; Counter, C.M.; Eaton, E.N.; Ellisen, L.W.; Steiner, P.; Caddle, S.D.; Ziaugra, L.; Beijersbergen, R.L.; Davidoff, M.J.; Liu, Q.; et al. hEST2, the putative human telomerase catalytic subunit gene, is up-regulated in tumor cells and during immortalization. *Cell* **1997**, *90*, 785–795. [CrossRef]

31. Cohen, S.B.; Graham, M.E.; Lovrecz, G.O.; Bache, N.; Robinson, P.J.; Reddel, R.R. Protein composition of catalytically active human telomerase from immortal cells. *Science* **2007**, *315*, 1850–1853. [CrossRef] [PubMed]

32. Ulaner, G.A.; Hu, J.F.; Vu, T.H.; Giudice, L.C.; Hoffman, A.R. Telomerase activity in human development is regulated by human telomerase reverse transcriptase (hTERT) transcription and by alternate splicing of hTERT transcripts. *Cancer Res.* **1998**, *58*, 4168–4172. [PubMed]

33. Nandakumar, J.; Cech, T.R. Finding the end: Recruitment of telomerase to telomeres. *Nat. Rev. Mol. Cell Biol.* **2013**, *14*, 69–82. [CrossRef] [PubMed]

34. Shay, J.W.; Bacchetti, S. A survey of telomerase activity in human cancer. *Eur. J. Cancer* **1997**, *33*, 787–791. [CrossRef]

35. Levy, M.Z.; Allsopp, R.C.; Futcher, A.B.; Greider, C.W.; Harley, C.B. Telomere end-replication problem and cell aging. *J. Mol. Biol.* **1992**, *225*, 951–960. [CrossRef]

36. Hayflick, L. The limited in vitro lifetime of human diploid cell strains. *Exp. Cell Res.* **1965**, *37*, 614–636. [CrossRef]

37. Hayflick, L.; Moorhead, P.S. The serial cultivation of human diploid cell strains. *Exp. Cell Res.* **1961**, *25*, 585–621. [CrossRef]

38. Bodnar, A.G.; Ouellette, M.; Frolkis, M.; Holt, S.E.; Chiu, C.P.; Morin, G.B.; Harley, C.B.; Shay, J.W.; Lichtsteiner, S.; Wright, W.E. Extension of life-span by introduction of telomerase into normal human cells. *Science* **1998**, *279*, 349–352. [CrossRef] [PubMed]

39. Plug-DeMaggio, A.W.; Sundsvold, T.; Wurscher, M.A.; Koop, J.I.; Klingelhutz, A.J.; McDougall, J.K. Telomere erosion and chromosomal instability in cells expressing the HPV oncogene 16E6. *Oncogene* **2004**, *23*, 3561–3571. [CrossRef] [PubMed]

40. Gabet, A.S.; Accardi, R.; Bellopede, A.; Popp, S.; Boukamp, P.; Sylla, B.S.; Londono-Vallejo, J.A.; Tommasino, M. Impairment of the telomere/telomerase system and genomic instability are associated with keratinocyte immortalization induced by the skin human papillomavirus type 38. *FASEB J.* **2008**, *22*, 622–632. [CrossRef] [PubMed]

41. Verdun, R.E.; Karlseder, J. Replication and protection of telomeres. *Nature* **2007**, *447*, 924–931. [CrossRef] [PubMed]

42. Janknecht, R. On the road to immortality: hTERT upregulation in cancer cells. *FEBS Lett.* **2004**, *564*, 9–13. [CrossRef]

43. Harley, C.B.; Futcher, A.B.; Greider, C.W. Telomeres shorten during ageing of human fibroblasts. *Nature* **1990**, *345*, 458–460. [CrossRef] [PubMed]

44. Kiyono, T.; Foster, S.A.; Koop, J.I.; McDougall, J.K.; Galloway, D.A.; Klingelhutz, A.J. Both Rb/p16INK4a inactivation and telomerase activity are required to immortalize human epithelial cells. *Nature* **1998**, *396*, 84–88. [PubMed]

45. Liu, X.; Roberts, J.; Dakic, A.; Zhang, Y.; Schlegel, R. HPV E7 contributes to the telomerase activity of immortalized and tumorigenic cells and augments E6-induced hTERT promoter function. *Virology* **2008**, *375*, 611–623. [CrossRef] [PubMed]

46. Jeong, S.E.; Jung, K.H.; Jae, L.C.; Tae, K.H.; Seong, H.E. The role of HPV oncoproteins and cellular factors in maintenance of hTERT expression in cervical carcinoma cells. *Gynecol. Oncol.* **2004**, *94*, 40–47. [CrossRef] [PubMed]

47. Van Doorslaer, K.; Burk, R.D. Association between hTERT activation by HPV E6 proteins and oncogenic risk. *Virology* **2012**, *433*, 216–219. [CrossRef] [PubMed]

48. Gewin, L.; Myers, H.; Kiyono, T.; Galloway, D.A. Identification of a novel telomerase repressor that interacts with the human papillomavirus type-16 E6/E6-AP complex. *Genes Dev.* **2004**, *18*, 2269–2282. [CrossRef] [PubMed]

49. Veldman, T.; Horikawa, I.; Barrett, J.C.; Schlegel, R. Transcriptional activation of the telomerase hTERT gene by human papillomavirus type 16 E6 oncoprotein. *J. Virol.* **2001**, *75*, 4467–4472. [CrossRef] [PubMed]

50. Oh, S.T.; Kyo, S.; Laimins, L.A. Telomerase activation by human papillomavirus type 16 E6 protein: Induction of human telomerase reverse transcriptase expression through Myc and GC-Rich Sp1 binding sites. *J. Virol.* **2001**, *75*, 5559–5566. [CrossRef] [PubMed]

51. Veldman, T.; Liu, X.; Yuan, H.; Schlegel, R. Human papillomavirus E6 and Myc proteins associate in vivo and bind to and cooperatively activate the telomerase reverse transcriptase promoter. *Proc. Natl. Acad. Sci. USA* **2003**, *100*, 8211–8216. [CrossRef] [PubMed]

52. Liu, X.; Dakic, A.; Zhang, Y.; Dai, Y.; Chen, R.; Schlegel, R. HPV E6 protein interacts physically and functionally with the cellular telomerase complex. *Proc. Natl. Acad. Sci. USA* **2009**, *106*, 18780–18785. [CrossRef] [PubMed]

53. James, M.A.; Lee, J.H.; Klingelhutz, A.J. HPV-16-E6 associated hTERT promoter acetylation is E6AP dependent, increased in later passage cells and enhanced by loss of p300. *Int. J. Cancer* **2006**, *119*, 1878–1885. [CrossRef] [PubMed]

54. Liu, X.; Yuan, H.; Fu, B.; Disbrow, G.L.; Apolinario, T.; Tomaic, V.; Kelley, M.L.; Baker, C.C.; Huibregtse, J.; Schlegel, R. The E6AP ubiquitin ligase is required for transactivation of the hTERT promoter by the human papillomavirus E6 oncoprotein. *J. Biol. Chem.* **2005**, *280*, 10807–10816. [CrossRef] [PubMed]

55. Kelley, M.L.; Keiger, K.E.; Lee, C.J.; Huibregtse, J.M. The global transcriptional effects of the human papillomavirus E6 protein in cervical carcinoma cell lines are mediated by the E6AP ubiquitin ligase. *J. Virol.* **2005**, *79*, 3737–3747. [CrossRef] [PubMed]

56. Handa, K.; Yugawa, T.; Narisawa-Saito, M.; Ohno, S.; Fujita, M.; Kiyono, T. E6AP-dependent degradation of DLG4/PSD95 by high-risk human papillomavirus type 18 E6 protein. *J. Virol.* **2007**, *81*, 1379–1389. [CrossRef] [PubMed]

57. Thomas, M.; Laura, R.; Hepner, K.; Guccione, E.; Sawyers, C.; Lasky, L.; Banks, L. Oncogenic human papillomavirus E6 proteins target the MAGI-2 and MAGI-3 proteins for degradation. *Oncogene* **2002**, *21*, 5088–5096. [CrossRef] [PubMed]

58. Jung, H.M.; Phillips, B.L.; Chan, E.K. miR-375 activates p21 and suppresses telomerase activity by coordinately regulating HPV E6/E7, E6AP, CIP2a, and 14–3-3ζ. *Mol. Cancer* **2014**, *13*, 80. [CrossRef] [PubMed]

59. Klingelhutz, A.J.; Foster, S.A.; McDougall, J.K. Telomerase activation by the E6 gene product of human papillomavirus type 16. *Nature* **1996**, *380*, 79–82. [CrossRef] [PubMed]

60. Cong, Y.S.; Wen, J.; Bacchetti, S. The human telomerase catalytic subunit hTERT: Organization of the gene and characterization of the promoter. *Hum. Mol. Genet.* **1999**, *8*, 137–142. [CrossRef] [PubMed]

61. Wick, M.; Zubov, D.; Hagen, G. Genomic organization and promoter characterization of the gene encoding the human telomerase reverse transcriptase (hTERT). *Gene* **1999**, *232*, 97–106. [CrossRef]

62. Takakura, M.; Kyo, S.; Kanaya, T.; Hirano, H.; Takeda, J.; Yutsudo, M.; Inoue, M. Cloning of human telomerase catalytic subunit (hTERT) gene promoter and identification of proximal core promoter sequences essential for transcriptional activation in immortalized and cancer cells. *Cancer Res.* **1999**, *59*, 551–557. [PubMed]

63. Fujimoto, K.; Kyo, S.; Takakura, M.; Kanaya, T.; Kitagawa, Y.; Itoh, H.; Takahashi, M.; Inoue, M. Identification and characterization of negative regulatory elements of the human telomerase catalytic subunit (hTERT) gene promoter: Possible role of MZF-2 in transcriptional repression of hTERT. *Nucleic Acids Res.* **2000**, *28*, 2557–2562. [CrossRef] [PubMed]

64. Oh, S.; Song, Y.H.; Yim, J.; Kim, T.K. Identification of Mad as a repressor of the human telomerase (hTERT) gene. *Oncogene* **2000**, *19*, 1485–1490. [CrossRef] [PubMed]

65. Gunes, C.; Lichtsteiner, S.; Vasserot, A.P.; Englert, C. Expression of the hTERT gene is regulated at the level of transcriptional initiation and repressed by Mad1. *Cancer Res.* **2000**, *60*, 2116–2121. [PubMed]

66. Horikawa, I.; Cable, P.L.; Mazur, S.J.; Appella, E.; Afshari, C.A.; Barrett, J.C. Downstream E-box-mediated regulation of the human telomerase reverse transcriptase (*hTERT*) gene transcription: Evidence for an endogenous mechanism of transcriptional repression. *Mol. Biol. Cell* **2002**, *13*, 2585–2597. [CrossRef] [PubMed]

67. Won, J.; Yim, J.; Kim, T.K. Sp1 and Sp3 recruit histone deacetylase to repress transcription of human telomerase reverse transcriptase (hTERT) promoter in normal human somatic cells. *J. Biol. Chem.* **2002**, *277*, 38230–38238. [CrossRef] [PubMed]

68. Renaud, S.; Loukinov, D.; Bosman, F.T.; Lobanenkov, V.; Benhattar, J. CTCF binds the proximal exonic region of hTERT and inhibits its transcription. *Nucleic Acids Res.* **2005**, *33*, 6850–6860. [CrossRef] [PubMed]

69. Racek, T.; Mise, N.; Li, Z.; Stoll, A.; Putzer, B.M. C-terminal p73 isoforms repress transcriptional activity of the human telomerase reverse transcriptase (hTERT) promoter. *J. Biol. Chem.* **2005**, *280*, 40402–40405. [CrossRef] [PubMed]

70. Lebel, R.; McDuff, F.O.; Lavigne, P.; Grandbois, M. Direct visualization of the binding of c-Myc/Max heterodimeric b-HLH-LZ to E-box sequences on the hTERT promoter. *Biochemistry* **2007**, *46*, 10279–10286. [CrossRef] [PubMed]

71. Xu, M.; Katzenellenbogen, R.A.; Grandori, C.; Galloway, D.A. An unbiased in vivo screen reveals multiple transcription factors that control HPV E6-regulated hTERT in keratinocytes. *Virology* **2013**, *446*, 17–24. [CrossRef] [PubMed]

72. Liu, X.; Dakic, A.; Chen, R.; Disbrow, G.L.; Zhang, Y.; Dai, Y.; Schlegel, R. Cell-restricted immortalization by human papillomavirus correlates with telomerase activation and engagement of the hTERT promoter by Myc. *J. Virol.* **2008**, *82*, 11568–11576. [CrossRef] [PubMed]

73. Chang, J.T.; Yang, H.T.; Wang, T.C.; Cheng, A.J. Upstream stimulatory factor (USF) as a transcriptional suppressor of human telomerase reverse transcriptase (hTERT) in oral cancer cells. *Mol. Carcinog.* **2005**, *44*, 183–192. [CrossRef] [PubMed]

74. McMurray, H.R.; McCance, D.J. Human papillomavirus type 16 E6 activates TERT gene transcription through induction of c-Myc and release of USF-mediated repression. *J. Virol.* **2003**, *77*, 9852–9861. [CrossRef] [PubMed]

75. Gewin, L.; Galloway, D.A. E box-dependent activation of telomerase by human papillomavirus type 16 E6 does not require induction of *c-Myc*. *J. Virol.* **2001**, *75*, 7198–7201. [CrossRef] [PubMed]

76. Sekaric, P.; Cherry, J.J.; Androphy, E.J. Binding of human papillomavirus type 16 E6 to E6AP is not required for activation of hTERT. *J. Virol.* **2008**, *82*, 71–76. [CrossRef] [PubMed]

77. Shai, A.; Pitot, H.C.; Lambert, P.F. E6-associated protein is required for human papillomavirus type 16 E6 to cause cervical cancer in mice. *Cancer Res.* **2010**, *70*, 5064–5073. [CrossRef] [PubMed]

78. Xu, M.; Luo, W.; Elzi, D.J.; Grandori, C.; Galloway, D.A. NFX1 interacts with mSin3a/histone deacetylase to repress hTERT transcription in keratinocytes. *Mol. Cell. Biol.* **2008**, *28*, 4819–4828. [CrossRef] [PubMed]

79. Schutze, D.M.; Kooter, J.M.; Wilting, S.M.; Meijer, C.J.; Quint, W.; Snijders, P.J.; Steenbergen, R.D. Longitudinal assessment of DNA methylation changes during HPVE6E7-induced immortalization of primary keratinocytes. *Epigenetics* **2015**, *10*, 73–81. [CrossRef] [PubMed]

80. De Wild, J.; Kooter, J.M.; Overmeer, R.M.; Claassen-Kramer, D.; Meijer, C.J.; Snijders, P.J.; Steenbergen, R.D. hTERT promoter activity and CpG methylation in HPV-induced carcinogenesis. *BMC Cancer* **2010**, *10*, 271. [CrossRef] [PubMed]

81. Jiang, J.; Zhao, L.J.; Zhao, C.; Zhang, G.; Zhao, Y.; Li, J.R.; Li, X.P.; Wei, L.H. Hypomethylated CpG around the transcription start site enables TERT expression and HPV16 E6 regulates TERT methylation in cervical cancer cells. *Gynecol. Oncol.* **2012**, *124*, 534–541. [CrossRef] [PubMed]

82. Zinn, R.L.; Pruitt, K.; Eguchi, S.; Baylin, S.B.; Herman, J.G. hTERT is expressed in cancer cell lines despite promoter DNA methylation by preservation of unmethylated DNA and active chromatin around the transcription start site. *Cancer Res.* **2007**, *67*, 194–201. [CrossRef] [PubMed]

83. Oikonomou, P.; Messinis, I.; Tsezou, A. DNA methylation is not likely to be responsible for hTERT expression in premalignant cervical lesions. *Exp. Biol. Med.* **2007**, *232*, 881–886.

84. Schutze, D.M.; Snijders, P.J.; Bosch, L.; Kramer, D.; Meijer, C.J.; Steenbergen, R.D. Differential in vitro immortalization capacity of eleven, probable high-risk human papillomavirus types. *J. Virol.* **2014**, *88*, 1714–1724. [CrossRef] [PubMed]

85. Kilian, A.; Bowtell, D.D.; Abud, H.E.; Hime, G.R.; Venter, D.J.; Keese, P.K.; Duncan, E.L.; Reddel, R.R.; Jefferson, R.A. Isolation of a candidate human telomerase catalytic subunit gene, which reveals complex splicing patterns in different cell types. *Hum. Mol. Genet.* **1997**, *6*, 2011–2019. [CrossRef] [PubMed]

86. Katzenellenbogen, R.A.; Vliet-Gregg, P.; Xu, M.; Galloway, D.A. Nfx1-123 increases hTERT expression and telomerase activity posttranscriptionally in human papillomavirus type 16 E6 keratinocytes. *J. Virol.* **2009**, *83*, 6446–6456. [CrossRef] [PubMed]

87. Cerezo, A.; Kalthoff, H.; Schuermann, M.; Schafer, B.; Boukamp, P. Dual regulation of telomerase activity through c-Myc-dependent inhibition and alternative splicing of hTERT. *J. Cell Sci.* **2002**, *115*, 1305–1312. [PubMed]

88. Mole, S.; McFarlane, M.; Chuen-Im, T.; Milligan, S.G.; Millan, D.; Graham, S.V. RNA splicing factors regulated by HPV16 during cervical tumour progression. *J. Pathol.* **2009**, *219*, 383–391. [CrossRef] [PubMed]

89. Katzenellenbogen, R.A.; Egelkrout, E.M.; Vliet-Gregg, P.; Gewin, L.C.; Gafken, P.R.; Galloway, D.A. Nfx1-123 and poly(A) binding proteins synergistically augment activation of telomerase in human papillomavirus type 16E6 expressing cells. *J. Virol.* **2007**, *81*, 3786–3796. [CrossRef] [PubMed]

90. Grishin, N.V. The R3H motif: A domain that binds single-stranded nucleic acids. *Trends Biochem. Sci.* **1998**, *23*, 329–330. [CrossRef]

91. Mangus, D.A.; Evans, M.C.; Jacobson, A. Poly(A)-binding proteins: Multifunctional scaffolds for the post-transcriptional control of gene expression. *Genome Biol.* **2003**, *4*, 223. [CrossRef] [PubMed]

92. Katzenellenbogen, R.A.; Vliet-Gregg, P.; Xu, M.; Galloway, D.A. Cytoplasmic poly(A) binding proteins regulate telomerase activity and cell growth in human papillomavirus type 16 E6-expressing keratinocytes. *J. Virol.* **2010**, *84*, 12934–12944. [CrossRef] [PubMed]

93. Bedard, K.M.; Underbrink, M.P.; Howie, H.L.; Galloway, D.A. The E6 oncoproteins from human betapapillomaviruses differentially activate telomerase through an E6AP-dependent mechanism and prolong the lifespan of primary keratinocytes. *J. Virol.* **2008**, *82*, 3894–3902. [CrossRef] [PubMed]

94. Mutirangura, A.; Sriuranpong, V.; Termrunggraunglert, W.; Tresukosol, D.; Lertsaguansinchai, P.; Voravud, N.; Niruthisard, S. Telomerase activity and human papillomavirus in malignant, premalignant and benign cervical lesions. *Br. J. Cancer* **1998**, *78*, 933–939. [CrossRef] [PubMed]

95. Ley, C.; Bauer, H.M.; Reingold, A.; Schiffman, M.H.; Chambers, J.C.; Tashiro, C.J.; Manos, M.M. Determinants of genital human papillomavirus infection in young women. *J. Natl. Cancer Inst.* **1991**, *83*, 997–1003. [CrossRef] [PubMed]

96. Branca, M.; Giorgi, C.; Ciotti, M.; Santini, D.; Di, B.L.; Costa, S.; Benedetto, A.; Bonifacio, D.; Di, B.P.; Paba, P.; et al. Upregulation of telomerase (hTERT) is related to the grade of cervical intraepithelial neoplasia, but is not an independent predictor of high-risk human papillomavirus, virus persistence, or disease outcome in cervical cancer. *Diagn. Cytopathol.* **2006**, *34*, 739–748. [CrossRef] [PubMed]

97. Nachajova, M.; Brany, D.; Dvorska, D. Telomerase and the process of cervical carcinogenesis. *Tumour Biol.* **2015**, *36*, 7335–7338. [CrossRef] [PubMed]

98. Snijders, P.J.; van Duin, M.; Walboomers, J.M.; Steenbergen, R.D.; Risse, E.K.; Helmerhorst, T.J.; Verheijen, R.H.; Meijer, C.J. Telomerase activity exclusively in cervical carcinomas and a subset of cervical intraepithelial neoplasia grade III lesions: Strong association with elevated messenger RNA levels of its catalytic subunit and high-risk human papillomavirus DNA. *Cancer Res.* **1998**, *58*, 3812–3818. [PubMed]

99. Miller, J.; Dakic, A.; Chen, R.; Palechor-Ceron, N.; Dai, Y.; Kallakury, B.; Schlegel, R.; Liu, X. HPV16 E7 protein and hTERT proteins defective for telomere maintenance cooperate to immortalize human keratinocytes. *PLoS. Pathog.* **2013**, *9*, e1003284. [CrossRef] [PubMed]

100. Jeon, S.; Lambert, P.F. Integration of human papillomavirus type 16 DNA into the human genome leads to increased stability of E6 and E7 mRNAs: Implications for cervical carcinogenesis. *Proc. Natl. Acad. Sci. USA* **1995**, *92*, 1654–1658. [CrossRef] [PubMed]

101. Reuschenbach, M.; Huebbers, C.U.; Prigge, E.S.; Bermejo, J.L.; Kalteis, M.S.; Preuss, S.F.; Seuthe, I.M.; Kolligs, J.; Speel, E.J.; Olthof, N.; et al. Methylation status of HPV16 E2-binding sites classifies subtypes of HPV-associated oropharyngeal cancers. *Cancer* **2015**, *121*, 1966–1976. [CrossRef] [PubMed]

102. Baege, A.C.; Berger, A.; Schlegel, R.; Veldman, T.; Schlegel, R. Cervical epithelial cells transduced with the papillomavirus E6/E7 oncogenes maintain stable levels of oncoprotein expression but exhibit progressive, major increases in hTERT gene expression and telomerase activity. *Am. J. Pathol.* **2002**, *160*, 1251–1257. [CrossRef]

MDPI

Review

Roles of APOBEC3A and APOBEC3B in Human Papillomavirus Infection and Disease Progression

Cody J. Warren [1,†], Joseph A. Westrich [1], Koenraad Van Doorslaer [2] and Dohun Pyeon [1,3,*]

[1]   Department of Immunology and Microbiology, University of Colorado School of Medicine, Aurora, CO 80045, USA; cody.warren@colorado.edu (C.J.W.); joseph.westrich@ucdenver.edu (J.A.W.)
[2]   BIO5 Institute, School of Animal and Comparative Biomedical Sciences, University of Arizona, Tucson, AZ 85721, USA; vandoorslaer@email.arizona.edu
[3]   Department of Medicine, University of Colorado School of Medicine, Aurora, CO 80045, USA
*   Correspondence: dohun.pyeon@ucdenver.edu; Tel.: +1-303-724-7279
†   Current address: BioFrontiers Institute, University of Colorado Boulder, Boulder, CO 80303, USA.

Academic Editors: Alison A. McBride and Karl Munger
Received: 26 July 2017; Accepted: 16 August 2017; Published: 21 August 2017

Abstract: The apolipoprotein B messenger RNA-editing, enzyme-catalytic, polypeptide-like 3 (APOBEC3) family of cytidine deaminases plays an important role in the innate immune response to viral infections by editing viral genomes. However, the cytidine deaminase activity of APOBEC3 enzymes also induces somatic mutations in host genomes, which may drive cancer progression. Recent studies of human papillomavirus (HPV) infection and disease outcome highlight this duality. HPV infection is potently inhibited by one family member, APOBEC3A. Expression of APOBEC3A and APOBEC3B is highly elevated by the HPV oncoproteins E6 and E7 during persistent virus infection and disease progression. Furthermore, there is a high prevalence of APOBEC3A and APOBEC3B mutation signatures in HPV-associated cancers. These findings suggest that induction of an APOBEC3-mediated antiviral response during HPV infection may inadvertently contribute to cancer mutagenesis and virus evolution. Here, we discuss current understanding of APOBEC3A and APOBEC3B biology in HPV restriction, evolution, and associated cancer mutagenesis.

Keywords: papillomavirus; human papillomavirus (HPV); apolipoprotein B messenger RNA-editing, enzyme-catalytic, polypeptide-like 3 (APOBEC3); innate immunity; virus restriction; cancer mutagenesis; cancer progression; somatic mutation; virus evolution

1. Introduction

The family members of apolipoprotein B messenger RNA-editing, enzyme-catalytic, polypeptide-like 3 (APOBEC3; A3) are DNA cytidine deaminases that remove the amino group from a cytosine, converting it to uracil. Cytosine deamination by A3 results in DNA degradation or mutations if not repaired (Figure 1A) [1–3]. For many years following their initial discovery, the A3 family members APOBEC3A (A3A) and APOBEC3B (A3B) were considered as viral restriction factors, important only for inhibiting the replication of endogenous retroviruses and retroelements [4–7]. However, several studies from our and other groups have revealed a broader range of viruses restricted by A3A and A3B: human immunodeficiency virus 1 (HIV-1) [8–10], parvovirus [11–13], herpesvirus [14,15], hepatitis B virus (HBV) [16,17], and human papillomavirus (HPV) [18–20] (Table 1). In addition, recent studies have identified additional important roles for these family members in diverse cellular processes, including (1) promoting catabolism of foreign DNA [21,22]; (2) editing of mRNA transcripts [23–25]; and (3) promoting host genome mutations and DNA damage that may contribute to cellular transformation [26–34].

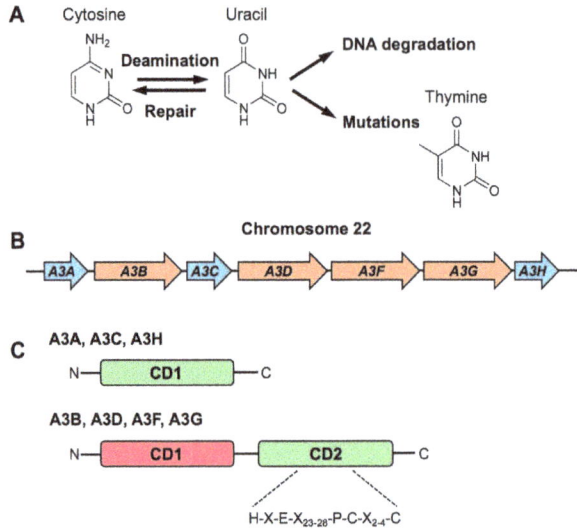

Figure 1. Structural features of the human apolipoprotein B messenger RNA-editing, enzyme-catalytic, polypeptide-like 3 (APOBEC3s). (**A**) APOBEC3 family members convert cytosine to uracil and induce DNA degradation or mutations if the APOBEC3-mediated conversion of cytosine to uracil is not repaired; (**B**) All *APOBEC3* family members are arrayed in tandem on chromosome 22; (**C**) Schematic of APOBEC3 family members containing one or two cytidine deaminase (CD) domains. The catalytically active (green) and inactive (red) cytidine deaminase domains are pictured. The conserved zinc-coordinating motif is pictured between dashed lines.

Table 1. Restriction of DNA viruses by APOBEC3 family members.

Virus	APOBEC3 Family Members	Functions on Viral Genome	References
Parvovirus	A3A	Unknown	[11–13,35]
Herpesvirus	A3A, A3G	DNA editing, unknown	[14,15]
Papillomavirus	A3A, A3C (?)	DNA editing, unknown	[18,20,36]
Hepadnavirus	A3A, A3B, A3C, A3F, A3G, A3H	DNA editing, deamination, and degradation	[16,17,37–40]

A3-mediated mutagenesis of cellular genomes is hypothesized to contribute to genetic aberrations that lead to cancer [30,41]. This A3 mutator hypothesis is supported by several lines of evidence: (1) A3A and A3B mutation signatures are distinguishable from those caused by other mutagens [29]; (2) A3A and A3B localize to the nucleus [42–44] and induce DNA damage [26,45]; and (3) mutation loads correlate with A3A and A3B mRNA expression levels [30,34]. It has been proposed that A3B might be responsible for up to half of all the mutations in breast cancer [46]. A3 mutation signatures are also prevalent in many other different types of cancers, including HPV-associated cervical (CxCa) and head/neck (HNC) cancers [30,31,41,46,47]. Studies from our group and others have shown that A3A and A3B are the only two A3s transcriptionally upregulated in HPV-positive keratinocytes and cancer cells [18,48]. In the context of HPV positive cells, A3A and A3B upregulation is mainly driven by the HPV oncoproteins E7 and E6, respectively. These findings suggest that high levels of E6 and E7 expression during HPV persistence may be a major trigger for A3A- and A3B-mediated mutations. A3A and A3B appear to be primary players in HPV infection and associated cancer mutagenesis, and therefore will be the main focus of this review article.

2. Biology of APOBEC3

2.1. Structural Features of APOBEC3s

There are seven human A3 genes (*A3A*, *A3B*, *A3C*, *A3D*, *A3F*, *A3G*, and *A3H*) that are arrayed in tandem on chromosome 22 (Figure 1B) [49]. Each of the seven A3 proteins has one (A3A, A3C, A3H) or two (A3B, A3D, A3F, A3G) cytidine deaminase (CD) domains that are characterized by a conserved zinc-coordinating motif (H-X-E-X_{23-28}-P-C-X_{2-4}-C) (Figure 1C) [50]. Previous research suggested that the C-terminal CD domain (CD2) of the double-domain A3s is catalytically active, while the N-terminal pseudocatalytic domain (CD1), lacking enzymatic activity, is involved in nucleic acid binding (reviewed in References [51,52]). This generalization is largely based on studies of A3F and A3G, and their roles in the inhibition of HIV-1 infection, and may not hold true for other A3s [53–56]. Bogerd et al. showed that point mutations in either domain significantly limited the ability of A3B to restrict HIV-1 infectivity. This suggests that both CD1 and CD2 of A3B are catalytically active [9]. Interestingly, C-to-T editing of HIV-1 reverse transcripts is still detected if one CD domain of A3B is left functional, but not when both are rendered inactive [9]. However, the ability of both CD1 and CD2 to edit DNA is likely context specific. For instance, mutations in CD1 of A3B had no effect on the overall ability to induce hypermutation of HBV and bacterial DNA, which is in contrast to effects on HIV-1 [9,57]. While both CD domains of A3B may be catalytically active, it is likely that cytidine deamination is either preferentially mediated by CD2 or context specific, in terms of the nature of the substrate recognized.

In contrast to A3B, A3A has a single CD domain that mediates both nucleic acid binding and cytidine deamination. The mechanism by which A3A coordinates both DNA binding and cytidine deaminase activities has recently been uncovered. A3A exists as both a monomer and dimer when in solution and bound to a substrate [58,59]. However, the formation of the homodimer is necessary for high affinity DNA binding [60]. Examination of the A3A crystal structure identified a positively charged groove that is formed upon A3A dimerization. The positively charged amino acids within this groove are positioned to bind to the negatively charged phosphate residues of single stranded DNA (ssDNA) [35,61,62]. These structural studies suggest that the catalytically active form of A3A exists as a homodimer when bound to a substrate.

The substrate specificity of single-domain A3s, like A3A, is likely dependent on relative protein abundance. Bohn et al. hypothesized that A3A, at low protein concentrations, is mostly found in monomeric form and has poor binding affinity for ssDNA. When protein levels are elevated, however, A3A molecules dimerize and deaminate target cytidines with high binding affinity and specificity to ssDNA [60]. On the other hand, the double-domain A3s may have evolved to separate DNA binding from cytidine deamination, resulting in proteins that are more refined to their target. Given that off-target activity of A3s is associated with cancer risk (discussed further in Section 4), targeting the activity of these proteins may be crucial for developing next-generation cancer therapies. Such developments will only be achieved by a thorough understanding of the physiological properties of A3s, which will be greatly aided by mechanistic insights from studies of A3 structure and function.

2.2. Evolution of APOBEC3s

The *A3* locus is specific to mammals, yet there is significant variation in the number and arrangement of individual *A3* family members across species (Figure 2) [50]. For instance, primates and rodents are relatives within the superorder of placental mammals, *Euarchontoglires* [63–65]. However, primate and rodent genomes contain dramatically different numbers of *A3* genes: seven *A3* genes are encoded in primates while only one is encoded in rodents [49,66–69]. Additionally, other mammals grouped together within the superorder *Laurasiatheria* [68] have varying numbers of A3s, including dog/pig (two) [70], sheep/cow (three) [71], cat (four) [70], and horse (six) [72]. These studies highlight the complex evolutionary trajectory of *A3s* during mammalian speciation, which arose following a series of gene duplications, fusions, and losses (Figure 2). In addition to having an expanded repertoire of A3s, primate A3 proteins also display signatures of rapid protein evolution. The rate of non-synonymous amino acid substitutions is significantly greater than synonymous substitutions for several primate A3s [73–76]. This feature, termed positive natural selection, indicates the existence of a strong selective pressure on the host protein to change and adapt (reviewed in Reference [77]). The relatively rapid expansion of the number of *A3* genes in primates may reflect the necessity, and likely non-redundant function, for specialization of the A3 family members against particular pathogens. Given that A3s restrict viral infections, and that mammalian viruses have evolved countermeasures to avoid A3 restriction, it is plausible that the diversification of primate A3s is a direct response to cope with a wide array of viruses.

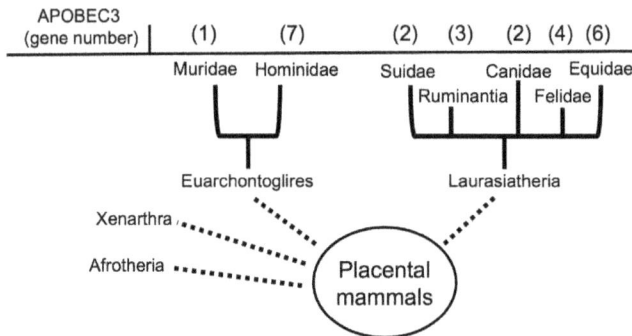

Figure 2. Copy number variation in mammalian *APOBEC3s*. Placental mammals are taxonomically split into four diverse groups (superorders), which include *Afrotheria*, *Xenarthra*, *Euarchontoglires*, and *Larasiatheria*. APOBEC3s have been described in the groups belonging to *Euarchontoglires* and *Laurasiatheria*. During mammalian speciation, A3s have evolved through a series of gene duplications, fusions, and losses. This has resulted in the copy number variability of mammalian *A3s* during speciation.

2.3. Target Specificity of APOBEC3s

While all A3s deaminate cytidine residues, specific dinucleotide motifs are preferably targeted by each A3. The dinucleotide motifs for A3 specificity are immediately 5′ to the target cytidine. For instance, 5′-CC dinucleotides (underline denotes the cytidine targeted for deamination) are the preferred target for A3G [1,78]. In contrast, A3A and A3B preferentially target 5′-TC dinucleotides [61,62,79,80]. By analyzing additional bases adjacent to target cytidine dinucleotides, a recent study further revealed tetra-nucleotide motifs, YTCA and RTCA, differentially targeted by A3A and A3B, respectively [29]. Target site specificity and relative expression levels in different cell and tissue types have been used as a proxy for determining the respective roles of individual A3s on cytidine deamination of both viral and host DNA [20,29,30,81,82].

All seven human A3s bind to ssDNA. A3G targets the minus strand DNA of HIV-1 that is generated during reverse transcription [1,2]. Additionally, A3A potently inhibits the infection of adeno-associated virus (AAV), whose genome consists of a linear ssDNA molecule [11]. However, given that these instances of ssDNA in the cell are relatively rare, additional mechanisms must be available for A3s to target DNA. In theory, the double-stranded genomes of most DNA viruses and their host should be protected from A3-mediated cytidine deamination. Nevertheless, A3-induced mutations have been frequently found in double-stranded DNA (dsDNA) genomes [20,36,82,83]. It has been speculated that transient ssDNA intermediates during gene transcription and genome replication may serve as substrates for A3 deamination. Using the yeast *Saccharomyces cerevisiae* as a model, Hoopes et al. recently demonstrated that A3A- and A3B-mediated mutations are mainly caused by the deamination of the lagging strand template during DNA replication [84]. Consistently, an independent group has shown that A3-associated mutations, identified in sequencing data from The Cancer Genome Atlas (TCGA), are highly enriched on the lagging strand during DNA replication [85].

In addition to ssDNA generated during replication and transcription, replication stress that leads to dsDNA breaks also generates ssDNA substrates that A3A and A3B may act upon [32,42]. These findings imply that cancer cells, given their high levels of cellular proliferation and replication stress, are prime targets for A3A- and A3B-mediated cytidine deamination. This topic is further explored in the sections that follow.

2.4. RNA Editing by APOBEC3A

Although ssDNA is the well-known substrate for A3s, recent studies have suggested that A3A can also mutate cellular RNAs [23–25]. By analyzing RNA sequencing data from monocytes and macrophages, Sharma et al. discovered widespread C-to-U editing of host cellular mRNAs under hypoxic conditions or after interferon (IFN) treatment [23]. Knockdown of A3A expression reduced RNA editing of succinate dehydrogenase B (SDHB), which was previously shown to be mutated under hypoxic conditions in monocytes [23,86]. Furthermore, editing of cellular RNAs was recapitulated in 293T cells by transient overexpression of A3A [23,24]. The discovery that A3A deaminates RNA, in addition to ssDNA, markedly expands the cellular roles of the A3 family.

Despite numerous antiviral roles for A3A, the precise mechanisms of A3A-mediated restriction are unknown. For instance, transgenic mice expressing human A3A are capable of restricting several murine retroviruses such as mouse mammary tumor virus and murine leukemia virus, yet very minimal DNA deamination was observed [87]. In the context of HPV infection, overexpression of A3A during HPV virion production markedly reduced infectivity [18]. Unexpectedly, despite their restriction being dependent on a functional A3A catalytic domain, no A3A-induced mutations were found in the HPV16 long control region or *E2* gene [18], which were previously identified as A3A mutation hotspots [20,36]. While it is still possible that other regions of the HPV genome may be edited by A3A, RNA editing may provide an alternative mechanism by which A3A restricts HPV infection in lieu of DNA editing. For example, deamination of the transcripts encoding L1 and L2 capsid proteins would likely have a dramatic effect on HPV virion infectivity. Supporting this idea, previous studies have revealed that virus restriction by A3A can occur in a deaminase-dependent mechanism without DNA sequence editing, or by a deaminase-independent mechanism [11,12]. A study from the Malim group further supports this concept by showing that cytidine deamination and DNA editing is not sufficient for antiviral activity during HIV-1 infection [88]. These results suggest alternative mechanisms by which A3A restricts virus infections beyond editing viral DNA sequences. Editing of viral transcripts may provide a novel mechanism by which A3A inhibits virus infection through cytidine deamination.

2.5. Transcriptional Regulation of APOBEC3A and APOBEC3B

The innate antiviral immune response is commonly initiated by cellular sensors that recognize foreign entities. These sensors relay intracellular signals to their effectors, which are responsible for clearing invading pathogens from the host cell. The A3 family members are important effectors of the innate antiviral immune defense (reviewed in Reference [89]). The detection of viral factors by cellular sensors leads to the activation of type 1 IFN signaling [90], which upregulates the expression of numerous antiviral genes, including the *A3* family members [36,38,91–93]. A3B transcription also appears to be activated through protein kinase C and nuclear factor kappa-light-chain-enhancer of B cells (NF-κB) signaling [94,95]. When expressed at high levels, A3s are capable of limiting infection of a diverse range of viruses, including retroviruses and DNA viruses. We and others have demonstrated that A3A and A3B transcription is also upregulated by the HPV oncoproteins E6 and E7 [18,48]. Interestingly, analyses of gene expression data from patient tissue specimens and cell lines confirm that A3A and A3B mRNA levels are elevated in HPV-positive cancers compared to normal tissues [18,30,48,96]. Together, these studies suggest that HPV infection induces the expression of A3A and A3B.

The *A3B* promoter is composed of a distal region for basal transactivation and a proximal region for transcriptional repression [97]. Interestingly, both the distal and proximal regions contain E6-responsive elements, which are essential for *A3B* promoter activity. DNA pull-down and chromatin immunoprecipitation assays further identified zinc finger protein 384 (ZNF384) as an important player for HPV-induced *A3B* transactivation [97]. Additionally, the same group has reported that HPV16 E6 upregulates *A3B* transcription by enhancing the expression of transcriptional enhancer factor (TEA) domain (TEAD) transcription factor that then binds to the *A3B* promoter [98]. These results suggest that *A3B* transcription may be directly activated by HPV16 E6. However, it is still possible that increased A3A and A3B mRNA levels may be, at least in part, due to an inadvertent consequence of a master transcription factor dysregulated by the HPV oncoproteins.

It has been speculated that HPV16 E6 expression increases *A3B* transcription through the functional inactivation of p53 [48]. Consistent with this notion, it was recently shown that wildtype p53 represses *A3B* transcription, and inactivating mutations in p53 protein leads to the upregulation of A3B expression [30,99,100]. In contrast to A3B, a study showed that A3A expression is upregulated by activated p53 [99]. A3A alters genome integrity by inducing DNA strand breaks and activating the DNA damage response [26,42], resulting in cell cycle arrest and apoptosis closely linked to p53 responses. As it is possible that the p53 regulation of A3 expression influences the course of a viral infection under conditions of cellular stress, understanding the interactions between p53 and A3s may provide novel avenues to treat persistent viral infections and neoplastic lesions.

3. Restriction of DNA Viruses by APOBEC3A

3.1. APOBEC3A Restriction of HPV Infection

A3A is localized throughout the cell in both cytoplasmic and nuclear compartments [26,42,43]. A growing body of evidence has implicated A3A as an important contributor to somatic mutations in human genomes [29], further emphasizing that A3A has access to nuclear DNA. Access to the nucleus may partially confer specificity to the types of viruses targeted by A3A (discussed further below). A3A is expressed in several cell types, including keratinocytes and myeloid cells [10,20,36,101,102]. Particularly, we have found that compared to cutaneous skin, A3A is expressed at high levels in mucosal tissue, which is vulnerable to the entry of foreign invaders like viruses [103].

Recent findings suggest that A3A is arguably the most important A3 family member targeting foreign DNA (Figure 3). A3A can localize to the nucleus [26,42,43], binds to ssDNA with high affinity [59,61], and deaminates cytidines in transient ssDNA undergoing transcription or replication [84,104]. The partially single-stranded genomes of HBV and AAV are highly susceptible to A3A restriction [11,17,37,81,105]. Several lines of evidence suggest that HPV genomes are also targeted by A3A [20,96,103,106,107]. Based on substrate specificity (TC dinucleotide targets) and high A3A expression levels in keratinocytes (the host cell for HPV infection), Vartanian et al. provided the first in vitro and in vivo evidence to suggest that A3A is a mutator of HPV genomes [20]. Since this seminal study, there have been multiple reports of HPV genome editing in patient tissue biopsies, including CxCa and oropharyngeal cancers [96,106,107]. These studies highlight that A3A may play an important role in restricting HPV infection.

Interferon-β (IFN-β) treatment significantly restricts HPV infection in keratinocytes as well as represses HPV DNA replication in infected keratinocytes [18,108–111]. Given that A3s are IFN-inducible proteins that target retroviruses and DNA viruses, Wang et al. sought to clarify whether A3s are also involved in the IFN-β-mediated response against HPV infections. This study revealed that IFN-β treatment upregulated A3A expression in cervical keratinocytes, and that knockdown of A3A expression reduced IFN-β-induced hypermutation of the viral *E2* gene [36]. However, A3A-induced hypermutation was detected only after enrichment by differential DNA denaturation PCR (3D-PCR), indicating that A3A-induced mutation events are rare. These interesting discoveries led us to question whether A3A affects HPV infectivity. Using our high-yield HPV production system [112], we have shown that virions packaged in cells overexpressing A3A is dramatically less infectious in keratinocytes. In contrast, the expression of other nuclear-localized A3s, A3B and A3C, had no effect on restricting viral infection [18]. HPV restriction by A3A is deaminase dependent, as a catalytically inactive mutant

A3A was unable to restrict HPV infection [18]. However, using highly sensitive next-generation sequencing, we were unable to detect A3A-induced mutations in genomic regions previously shown to be edited by A3s [18,20,36]. Further analysis of whole viral genome or RNA sequences may identify critical A3A mutation targets that disrupt HPV infectivity.

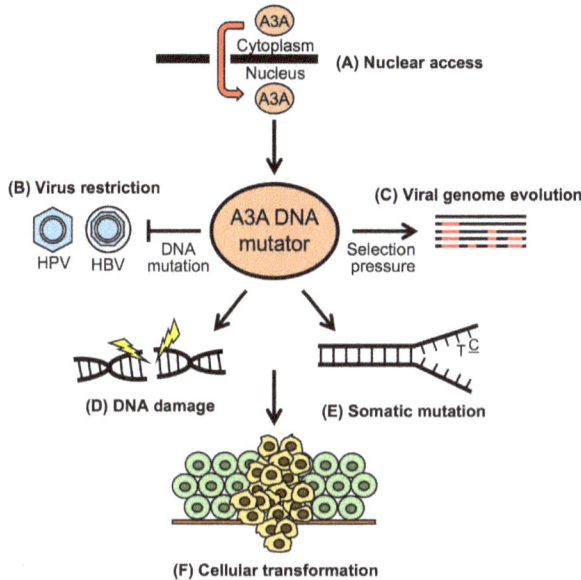

Figure 3. A3A is arguably the most important A3 family member in HPV restriction, evolution, and cancer mutagenesis. (**A**) A3A localizes to both the cytoplasm and nuclear compartments. Nuclear access broadens the substrates targeting by A3A; (**B**) In addition to restricting lentiviruses during reverse transcription in the cytoplasm, A3A also restricts DNA viruses that replicate in the nucleus, such as human papillomavirus (HPV) and hepatitis B virus (HBV); (**C**) Selection pressures imposed by A3A may lead to viral genome evolution (pictured in red) and partial escape from A3A restriction, as has been proposed for HPV [103]; (**D,E**) A3A activity enhanced by virus persistence and/or chronic inflammation may promote DNA damage and induce somatic mutations in host DNA; (**F**) Somatic mutagenesis and DNA damage further enables cancer cell evolution and drives disease progression.

3.2. APOBEC3A-Mediated Clearance of HPV DNA during Persistent Infection

A3A also plays an important role in mediating the clearance of foreign, circular DNA from cells [21,22]. Stenglein et al. have shown that A3A overexpression leads to the deamination and degradation of transfected foreign plasmid DNA [22]. Both HBV and HPV genomes are maintained as dsDNA episomes in the nucleus of persistently infected cells. Given that A3A is nuclear and restricts foreign circular DNA, it is possible that A3A may mediate the clearance of HBV and HPV DNA in persistently infected cells. Indeed, Lucifora et al. have shown that IFN-α-induced A3A triggers cytidine deamination and degradation of nuclear HBV DNA [16]. Confocal microscopy revealed that A3A colocalizes with the HBV core protein, which likely facilitates close contact with viral DNA in the nucleus. The degradation of HBV DNA prevents virus reactivation without hepatotoxicity [16], suggesting that A3A may be used as a tool for the treatment of persistent HBV infection, for which current therapies are limited. Accordingly, it would be of immense interest to the HPV field to determine whether A3A can similarly clear persistent HPV infection. In a series of elegant studies, the Coleman group has identified important roles for antiviral IFN responses in clearing persistent HPV infection, showing that IFN-β treatment of HPV-positive CxCa cells leads to a rapid loss of viral episomes [108,113]. In addition, loss of HPV DNA in CxCa cells during serial passaging is correlated with a surge in expression of IFN-inducible antiviral genes. This suggests that the antiviral IFN response is likely one of the key contributors in promoting spontaneous loss of extrachromosomal HPV DNA [113]. Along with other similar results, these findings collectively suggest that antiviral IFN responses likely play an important role in limiting the persistence of extrachromosomal HPV DNA [114–116]. Given that A3A is an IFN-inducible protein in keratinocytes [36], and that A3A can eliminate foreign DNA [22], it is plausible that A3A may contribute to the loss of HPV genomes in persistently infected cells, similar to clearance of HBV DNA [16]. However, the restriction pressure of IFN and A3A may also accelerate HPV-induced cancer progression by facilitating the integration of HPV DNA into the host chromosome. Kondo et al. have found that A3A expression is strongly linked to HPV integration in oropharyngeal cancers [96]. Similarly, either IFN-β or IFN-γ treatment significantly enhances HPV integration in persistently infected cervical keratinocytes [108,115]. These findings suggest that using IFN and/or A3A may not be feasible as antiviral agents to treat patients. Developing therapeutics to treat chronic HPV infections requires an in-depth understanding of the complicated interactions between A3A and HPV during persistent infection.

3.3. Viral Evasion of APOBEC3A-Mediated Restriction

Given the antiviral potency of A3A, it is likely that viruses have evolved countermeasures to combat or avoid A3A-mediated restriction. For instance, A3A significantly inhibits HIV-1 replication following infection of myeloid cells [10]. Interestingly, the viral accessory proteins Viral infectivity factor (Vif) and Viral protein X (Vpx) of HIV-1 and simian immunodeficiency virus of macaques (SIVmac), respectively, are capable of degrading A3A protein [10]. A3A degradation by viral proteins appears to be conserved in primate lentiviruses, further emphasizing the importance of antagonizing A3A during lentiviral infection. Similarly, as A3A restricts HPV infectivity, elevated A3A expression in HPV-infected cells is likely to be deleterious for viral fitness. Thus, one would predict that HPV has evolved strategies to evade A3A-mediated restriction. It is well known that the HIV-1 accessory protein Vif degrades another A3 family member, A3G, through a ubiquitin mediated, proteasome-dependent process that requires the cellular factors cullin 5, elongin B/C, and Ring-box protein 1 (RBX1) [117–119]. Interestingly, high-risk HPV E7 coordinates a similar process whereby interactions with the cullin 2 ubiquitin ligase complex, which also contains elongin B/C and RBX1, mediates the degradation of the tumor suppressor retinoblastoma protein (pRB) [120–122]. Given these striking similarities, it stands to reason that high-risk HPV E7 might facilitate the degradation of A3A protein in a process similar to HIV-1 Vif. Contrary to this hypothesis, we found that the expression of high-risk HPV E7 in keratinocytes significantly increased A3A protein levels [123]. While this relative increase in A3A protein may well explain the A3 mutation signatures in HPV-positive cancers, it implies that HPV likely employs other mechanisms to cope with elevated A3A levels during persistent infection.

As explained above, A3s deaminate cytidines within the context of preferred dinucleotides. A3A, for instance, prefers targeting cytidine residues that are preceded by thymidine (5'-T$\underline{C}$). An alternative mechanism to evade A3A deaminase activity would be to reduce the prevalence of A3A target sequences within the viral genome. Underrepresentation of CG dinucleotides in small DNA viruses, including polyomaviruses and papillomaviruses, has been suggested as a means to evade toll-like receptor 9 (TLR9) recognition and/or host DNA methylation of viral genomes [103,124–127]. As papillomaviruses have co-evolved with their host over millions of years, it is possible that A3A has exerted selective pressure on papillomavirus evolution, resulting in reduced TC dinucleotide contents in viral genomes (Figure 3). Analysis of 274 papillomavirus genomes has revealed that CG and TC dinucleotides are significantly depleted in all papillomavirus genomes [103]. Interestingly, the magnitude of TC depletion is greater in HPV genotypes from the *Alphapapillomavirus* genus (α-HPV) than β- or γ-HPV genotypes, while the degree of CG depletion is similar across all HPV genera. The significant difference in TC depletion between α- and β/γ-HPVs may be caused by either (1) the ancestral *Alphapapillomavirus* having low TC contents that subsequently radiated to all extant genotypes, or (2) a strong selective pressure that was exerted on the entire *Alphapapillomavirus* clade leading to extreme TC depletion. Phylogenetic reconstruction of the ancestral *Alphapapillomavirus* state revealed that the most recent common ancestor of all *Alphapapillomavirus* had TC contents significantly higher than extant members of this clade [103]. Thus, TC depletion likely occurred after *Alphapapillomavirus* began to diverge. This finding is suggestive of a possible role for A3 restriction that drove TC depletion within this clade. One major hallmark of most α-HPVs is tropism for mucosal tissues, while β- and γ-HPVs are typically found in association with cutaneous skin [128]. Analysis of publicly available RNA sequencing data revealed that A3A expression levels are significantly higher in mucosal skin compared to cutaneous skin [103]. Taken together, these findings suggest an evolutionary model in which HPV copes with elevated A3A expression by limiting the number of TC dinucleotides within their genomes [129]. In addition to the reduction of A3A target motifs, HPV may employ other means of escaping restriction by A3A. Further studies may provide additional clues about the complex interactions between HPV and A3s.

4. Cancer Mutagenesis by APOBEC3A

4.1. Sources of APOBEC3 Mutational Signatures in HPV-Positive Cancer

Although many studies have identified A3 mutational signatures in multiple human cancers, the molecular triggers resulting in off-target A3 activity on cellular DNA remain poorly understood. We and others have reported that the mRNA levels of A3A and A3B are significantly higher in HPV-positive keratinocytes and cancer tissues compared to uninfected keratinocytes and normal tissues, respectively [18,48]. A3A mRNA expression is increased by high-risk HPV E7, while A3B mRNA expression is upregulated by both the HPV oncoproteins E6 and E7 in cultured keratinocytes and human tonsillar epithelial cells. Interestingly, our results have shown that the mRNA expression levels of A3A and A3B are highly correlated, indicating that A3A and A3B may share common mechanisms for transcriptional regulation [18]. While mRNA expression of A3A and A3B has been well studied, little is known about their protein levels and posttranslational modification. Our unpublished study found that A3A protein levels are dramatically increased in human keratinocytes by high-risk HPV E7 mediated protein stabilization [123]. Consistent with these results, analysis of exome sequencing data from TCGA has revealed that HPV-positive HNC genomes contain high levels of A3-mediated driver mutations, while HPV-negative HNC displays a smoking-associated mutational signature [130]. Further, A3 deaminase activity is causally associated with helical domain hotspot mutations in the phosphatidylinositol-4,5-bisphosphate 3-kinase catalytic subunit alpha (*PIK3CA*) gene, which are more prevalent in HPV-positive cancers when compared to HPV-negative cancers [131]. Taken together, these findings strongly suggest that HPV oncoprotein expression during persistent viral infection may be the trigger for the increase of A3A and A3B expression, culminating in the accumulation of somatic mutations in HPV-positive cancers (Figure 3).

4.2. The Relative Contributions of APOBEC3A and APOBEC3B to Cancer Mutagenesis

The expression of the HPV oncoproteins E6 and E7 immediately inactivates numerous cellular proteins, including the tumor suppressors p53 and pRB [132,133]. While these mechanisms are important for promoting cell proliferation, they alone are not sufficient to drive cancer progression. The accumulation of additional somatic mutations over decades of persistent infection is necessary for cancer progression. Recent findings suggest that A3A and/or A3B potentially play important roles in cancer mutagenesis by mutating host genomic DNA [30,31,41,46,47,134]. Since both A3A and A3B are upregulated in HPV-positive epithelial cells and target the same TC dinucleotide motif, it is difficult to tease out the relative contributions of either in promoting somatic mutations in cancer cell genomes. Previous studies have correlated elevated A3B mRNA expression with A3-associated mutation loads, and proposed A3B as the source of A3 mutation signatures in cervical, bladder, lung, head and neck, ovarian, and breast cancers [30,47,135,136]. Burns et al. further showed that knockdown of A3B expression by short-hairpin RNA (shRNA) abrogates cytidine deaminase activity in the lysate of a breast cancer cell line [41]. Contrary to these findings, recent studies have shown that the *A3B* deletion polymorphism, highly prevalent in South East Asia, China, and Oceania, is associated with increased risk of breast and ovarian cancer [136–138]. Caval et al. revealed that *A3B* deletion frequently generates a chimeric *A3A–A3B* deletion allele in which the *A3A* gene is fused to the 3' untranslated region (UTR) of *A3B* [139]. The mRNA generated from the chimeric *A3A–A3B* deletion allele is more stable than wildtype A3A transcripts, and the resulting protein facilitates DNA damage. Using yeast models, Chan et al. recently found that A3A and A3B mutation signatures may be distinguishable by the preferred target motifs: YT<u>C</u>A favored by A3A and RT<u>C</u>A favored by A3B [29]. Further analysis of sequence data from yeast and human cancer genomes uncovered that A3A-like mutations are 10 times more abundant than A3B-like mutations. Based on these results, the authors propose that mutagenesis and DNA damage caused by A3A might be greater than A3B in HPV-associated cancers. Interestingly, A3A expression is tightly correlated with HPV DNA integration into host chromosomes, which is facilitated by dsDNA breaks [96,140]. Consistently, A3A protein dramatically accumulates in HPV-positive keratinocytes by E7-mediated protein stabilization [123]. Taken together, these findings suggest that both A3A and A3B may contribute to cancer mutagenesis, but likely have differential contributions in various cancers when triggered by different mechanisms.

4.3. Source of APOBEC3 Signature in Other Virus-Associated Cancers

Type I IFNs, commonly induced during various virus infections, highly upregulate A3A and A3B expression. This suggests that persistent inflammatory responses may generally facilitate somatic mutations by A3A and A3B [16,18,38,92]. A3B is also upregulated by several polyomaviruses (PyV), including JC PyV, Merkel cell PyV, and BK PyV through a mechanism dependent on large T antigen expression [141]. These results suggest that A3 mutation signatures from A3A and A3B may also be caused by other virus infections as well. However, it is not clear whether the increase of A3B expression observed in many HPV-negative cancers is mediated by type I IFN or by other factors such as viral proteins similar to the HPV oncoproteins. Further investigations are required to determine the mechanisms by which A3A and A3B are activated and contribute to cancer mutagenesis.

4.4. APOBEC3-Mediated Somatic Mutations and Clinical Outcomes of HPV-Positive Cancers

Somatic mutagenesis has been recognized as a key mechanism of carcinogenesis by generating driver mutations in numerous genes including *p53*, epidermal growth factor receptor (*EGFR*), *pRB*, and *PIK3CA* [142,143]. Given that the deaminase activity of A3A and A3B in epithelial cells mutates the T<u>C</u> motifs of host DNA as well as viral DNA, somatic mutagenesis by A3A and A3B is likely associated with cancer risk. Indeed, a high frequency of activating *PIK3CA* mutations was observed in HPV-positive HNCs compared to HPV-negative HNCs [131,144]. Our analysis also showed that all *PIK3CA* mutations in HPV-positive HNC, dominant with the E542K and E545K substitutions, are caused by GA-to-AA changes, while only about a half of *PIK3CA* mutations in HPV-negative HNC are from GA-to-AA changes (Figure 4). As *PIK3CA* is an oncogenic driver gene, A3A- and/or A3B-mediated somatic mutations may contribute to HPV-associated cancer progression through mutations in *PIK3CA*.

Contrary to the idea that A3-mediated somatic mutations may drive HPV-positive cancer progression, recent cancer immunology studies have shown that high levels of somatic mutations favor antitumor immune responses that also coincide with better prognosis after immunotherapies [145–147]. In these instances, tumor neoantigens are recognized as emerging targets for personalized cancer immunotherapies. This implies that cancers with a high degree of A3 mutation signatures may be beneficial for immunotherapies that induce robust antitumor T cell responses specific to neoantigens generated by A3-mediated mutations. A recent study has revealed that tumor infiltrating lymphocytes in CxCa are more reactive to neoantigens than to HPV viral epitopes [148]. This suggests that abundant neoantigens in HPV-positive cancers may be associated with the deaminase activity of upregulated A3A and A3B expression. If this is true, A3-mediated mutations could be utilized beneficially to identify

T cell epitopes and treat HPV-positive cancer patients. Thus, it would be interesting to investigate if A3 mutation loads in patients correlate to better outcome following current immunotherapies with immune checkpoint blockades.

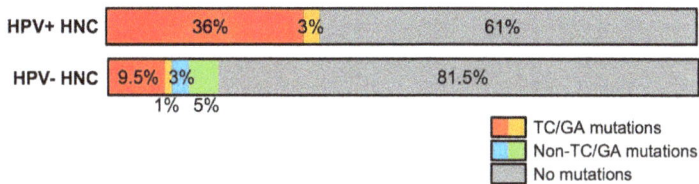

Figure 4. High levels of TC/GA mutations in the phosphatidylinositol-4,5-bisphosphate 3-kinase catalytic subunit alpha (*PIK3CA*) open reading frame in HPV-positive head and neck cancers (HNCs) compared to HPV-negative HNCs. Amino acid mutation data of PIK3CA in HNC patients was obtained from The Cancer Genome Atlas through cBioPortal (cbioportal.org) [149]: HPV-positive HNC (HPV+ HNC), *n* = 36; HPV-negative HNC (HPV- HNC), *n* = 243. Amino acid changes by TC/GA mutations are indicated as red (E542K and E545K) and orange (M1043V, R88Q, G1007R, G451R, R335G), and amino acid changes by non-TC/GA mutations are indicated as blue (H1047L, H1047R) and green (C604R, C901F, C971R, E110del, E365V, G363A, K111E, K111N, Q75E, R975S, V344G, W328S). Shown are the percentage of patients containing each mutation signature.

5. Conclusions and Perspectives

The inactivation of tumor suppressors by HPV oncoproteins is robust and quick. For example, p53 and pRB are degraded in host cells in which high-risk HPV E6 and E7 are expressed [132,133]. Nevertheless, HPV-associated cancer progression is a slow process, typically taking two to three decades. A growing number of studies have shown that the continuous expression of E6 and E7 is required through the full process of cancer progression and maintenance [150–156]. Our CxCa progression study has shown that many HPV-specific gene expression changes occur in a later stage or continuously throughout decades of cancer progression [157]. In this regard, the roles of A3A and A3B in HPV-associated cancer progression are particularly interesting. However, most of these new findings have generated more questions than answers, particularly due to the causal relations and the need for defining the mechanistic elements of these interactions. Now, most studies on A3-induced cancer mutagenesis have been limited to using highly biased sequencing approaches or are based on correlations between expression levels and preferred target sequence changes. Since A3-induced somatic mutations probably accumulate over decades, it would be technically challenging to recapitulate and confirm this process in experimental models. To overcome these barriers, developing transgenic animal models expressing human A3s along with HPV oncoproteins may provide useful tools to track cancer mutagenesis. Additionally, further work is needed to elucidate the mechanisms of A3A restriction of HPV infection, which are distinct from A3B and A3C. It would also be interesting to further investigate whether A3A and A3B restrict other small DNA tumor viruses and contribute to somatic mutagenesis of their associated cancers. Future studies may provide great insights into how virus-host interactions drive the evolution of viruses and host cells, and how these interactions may lead to unexpected consequences such as cancer development.

Acknowledgments: This work was supported in part by the National Institutes of Health (R01 AI091968 and R01 DE026125 to D.P., and T32 AI052066 to J.A.W.). K.V.D. is supported by the Technology Research Initiative Fund of Arizona and an award from the US Department of Agriculture—National Institute of Food and Agriculture (Hatch 1012632).

Author Contributions: C.J.W. and D.P. reviewed the literature, wrote the manuscript, and generated the figures. J.A.W. and K.V.D. provided edits and approved the final version.

Conflicts of Interest: The authors declare no conflict of interest.

References

1. Mangeat, B.; Turelli, P.; Caron, G.; Friedli, M.; Perrin, L.; Trono, D. Broad antiretroviral defence by human APOBEC3G through lethal editing of nascent reverse transcripts. *Nature* **2003**, *424*, 99–103. [CrossRef] [PubMed]
2. Zhang, H.; Yang, B.; Pomerantz, R.J.; Zhang, C.; Arunachalam, S.C.; Gao, L. The cytidine deaminase CEM15 induces hypermutation in newly synthesized HIV-1 DNA. *Nature* **2003**, *424*, 94–98. [CrossRef] [PubMed]
3. Yang, B.; Chen, K.; Zhang, C.; Huang, S.; Zhang, H. Virion-associated uracil DNA glycosylase-2 and apurinic/apyrimidinic endonuclease are involved in the degradation of APOBEC3G-edited nascent HIV-1 DNA. *J. Biol. Chem.* **2007**, *282*, 11667–11675. [CrossRef] [PubMed]
4. Bogerd, H.P.; Wiegand, H.L.; Doehle, B.P.; Lueders, K.K.; Cullen, B.R. APOBEC3A and APOBEC3B are potent inhibitors of LTR-retrotransposon function in human cells. *Nucleic Acids Res.* **2006**, *34*, 89–95. [CrossRef] [PubMed]
5. Bogerd, H.P.; Wiegand, H.L.; Hulme, A.E.; Garcia-Perez, J.L.; O'Shea, K.S.; Moran, J.V.; Cullen, B.R. Cellular inhibitors of long interspersed element 1 and Alu retrotransposition. *Proc. Natl. Acad. Sci. USA* **2006**, *103*, 8780–8785. [CrossRef] [PubMed]
6. Muckenfuss, H.; Hamdorf, M.; Held, U.; Perkovic, M.; Löwer, J.; Cichutek, K.; Flory, E.; Schumann, G.G.; Münk, C. APOBEC3 proteins inhibit human LINE-1 retrotransposition. *J. Biol. Chem.* **2006**, *281*, 22161–22172. [CrossRef] [PubMed]
7. Stenglein, M.D.; Harris, R.S. APOBEC3B and APOBEC3F inhibit L1 retrotransposition by a DNA deamination-independent mechanism. *J. Biol. Chem.* **2006**, *281*, 16837–16841. [CrossRef] [PubMed]
8. Doehle, B.P.; Sch fer, A.; Cullen, B.R. Human APOBEC3B is a potent inhibitor of HIV-1 infectivity and is resistant to HIV-1 Vif. *Virology* **2005**, *339*, 281–288. [CrossRef] [PubMed]
9. Bogerd, H.P.; Wiegand, H.L.; Doehle, B.P.; Cullen, B.R. The intrinsic antiretroviral factor APOBEC3B contains two enzymatically active cytidine deaminase domains. *Virology* **2007**, *364*, 486–493. [CrossRef] [PubMed]
10. Berger, G.; Durand, S.; Fargier, G.; Nguyen, X.-N.; Cordeil, S.; Bouaziz, S.; Muriaux, D.; Darlix, J.-L.; Cimarelli, A. APOBEC3A is a specific inhibitor of the early phases of HIV-1 infection in myeloid cells. *PLoS Pathog.* **2011**, *7*, e1002221. [CrossRef] [PubMed]
11. Chen, H.; Lilley, C.E.; Yu, Q.; Lee, D.V.; Chou, J.; Narvaiza, I.; Landau, N.R.; Weitzman, M.D. APOBEC3A is a potent inhibitor of adeno-associated virus and retrotransposons. *Curr. Biol.* **2006**, *16*, 480–485. [CrossRef] [PubMed]
12. Narvaiza, I.; Linfesty, D.C.; Greener, B.N.; Hakata, Y.; Pintel, D.J.; Logue, E.; Landau, N.R.; Weitzman, M.D. Deaminase-independent inhibition of parvoviruses by the APOBEC3A cytidine deaminase. *PLoS Pathog.* **2009**, *5*, e1000439. [CrossRef] [PubMed]
13. Nakaya, Y.; Stavrou, S.; Blouch, K.; Tattersall, P.; Ross, S.R. In vivo examination of mouse APOBEC3- and human APOBEC3A- and APOBEC3G-mediated restriction of parvovirus and herpesvirus infection in mouse models. *J. Virol.* **2016**, *90*, 8005–8012. [CrossRef] [PubMed]
14. Minkah, N.; Chavez, K.; Shah, P.; Maccarthy, T.; Chen, H.; Landau, N.; Krug, L.T. Host restriction of murine gammaherpesvirus 68 replication by human APOBEC3 cytidine deaminases but not murine APOBEC3. *Virology* **2014**, *454-455*, 215–226. [CrossRef] [PubMed]
15. Suspène, R.; Aynaud, M.-M.; Koch, S.; Pasdeloup, D.; Labetoulle, M.; Gaertner, B.; Vartanian, J.-P.; Meyerhans, A.; Wain-Hobson, S. Genetic editing of herpes simplex virus 1 and Epstein-Barr herpesvirus genomes by human APOBEC3 cytidine deaminases in culture and in vivo. *J. Virol.* **2011**, *85*, 7594–7602. [CrossRef] [PubMed]
16. Lucifora, J.; Xia, Y.; Reisinger, F.; Zhang, K.; Stadler, D.; Cheng, X.; Sprinzl, M.F.; Koppensteiner, H.; Makowska, Z.; Volz, T.; et al. Specific and nonhepatotoxic degradation of nuclear hepatitis B virus cccDNA. *Science* **2014**, *343*, 1221–1228. [CrossRef] [PubMed]
17. Henry, M.; Guétard, D.; Suspène, R.; Rusniok, C.; Wain-Hobson, S.; Vartanian, J.-P. Genetic editing of HBV DNA by monodomain human APOBEC3 cytidine deaminases and the recombinant nature of APOBEC3G. *PLoS ONE* **2009**, *4*, e4277. [CrossRef] [PubMed]
18. Warren, C.J.; Xu, T.; Guo, K.; Griffin, L.M.; Westrich, J.A.; Lee, D.; Lambert, P.F.; Santiago, M.L. Pyeon, APOBEC3A functions as a restriction factor of human papillomavir. *J. Virol.* **2015**, *89*, 688–702. [CrossRef] [PubMed]

19. Ahasan, M.M.; Wakae, K.; Wang, Z.; Kitamura, K.; Liu, G.; Koura, M.; Imayasu, M.; Sakamoto, N.; Hanaoka, K.; Nakamura, M.; et al. APOBEC3A and 3C decrease human papillomavirus 16 pseudovirion infectivity. *Biochem. Biophys. Res. Commun.* **2015**, *457*, 295–299. [CrossRef] [PubMed]
20. Vartanian, J.-P.; Guétard, D.; Henry, M.; Wain-Hobson, S. Evidence for editing of human papillomavirus DNA by APOBEC3 in benign and precancerous lesions. *Science* **2008**, *320*, 230–233. [CrossRef] [PubMed]
21. Kostrzak, A.; Henry, M.; Demoyen, P.L.; Wain-Hobson, S.; Vartanian, J.P. APOBEC3A catabolism of electroporated plasmid DNA in mouse muscle. *Gene Ther.* **2015**, *22*, 96–103. [CrossRef] [PubMed]
22. Stenglein, M.D.; Burns, M.B.; Li, M.; Lengyel, J.; Harris, R.S. APOBEC3 proteins mediate the clearance of foreign DNA from human cells. *Nat. Struct. Mol. Biol.* **2010**, *17*, 222–229. [CrossRef] [PubMed]
23. Sharma, S.; Patnaik, S.K.; Taggart, R.T.; Kannisto, E.D.; Enriquez, S.M.; Gollnick, P.; Baysal, B.E. APOBEC3A cytidine deaminase induces RNA editing in monocytes and macrophages. *Nat. Commun.* **2015**, *6*. [CrossRef] [PubMed]
24. Sharma, S.; Patnaik, S.K.; Kemer, Z.; Baysal, B.E. Transient overexpression of exogenous APOBEC3A causes C-to-U RNA editing of thousands of genes. *RNA Biol.* **2016**, *14*, 603–610. [CrossRef] [PubMed]
25. Niavarani, A.; Currie, E.; Reyal, Y.; Anjos-Afonso, F.; Horswell, S.; Griessinger, E.; Sardina, J.L.; Bonnet, D. APOBEC3A is implicated in a novel class of G-to-A mRNA editing in WT1 transcripts. *PLoS ONE* **2015**, *10*, e0120089. [CrossRef] [PubMed]
26. Landry, S.; Narvaiza, I.; Linfesty, D.C.; Weitzman, M.D. APOBEC3A can activate the DNA damage response and cause cell-cycle arrest. *EMBO Rep.* **2011**, *12*, 444–450. [CrossRef] [PubMed]
27. Nowarski, R.; Kotler, M. APOBEC3 Cytidine Deaminases in Double-Strand DNA Break Repair and Cancer Promotion. *Cancer Res.* **2013**, *73*, 3494–3498. [CrossRef] [PubMed]
28. Green, A.M.; Landry, S.; Budagyan, K.; Avgousti, D.C.; Shalhout, S.; Bhagwat, A.S.; Weitzman, M.D. APOBEC3A damages the cellular genome during DNA replication. *Cell Cycle* **2016**, *15*, 998–1008. [CrossRef] [PubMed]
29. Chan, K.; Roberts, S.A.; Klimczak, L.J.; Sterling, J.F.; Saini, N.; Malc, E.P.; Kim, J.; Kwiatkowski, D.J.; Fargo, D.C.; Mieczkowski, P.A.; et al. An APOBEC3A hypermutation signature is distinguishable from the signature of background mutagenesis by APOBEC3B in human cancers. *Nat. Genet.* **2015**, *47*, 1067–1072. [CrossRef] [PubMed]
30. Burns, M.B.; Temiz, N.A.; Harris, R.S. Evidence for APOBEC3B mutagenesis in multiple human cancers. *Nat. Genet.* **2013**, *45*, 977–983. [CrossRef] [PubMed]
31. Alexandrov, L.B.; Nik-Zainal, S.; Wedge, D.C.; Aparicio, S.A.J.R.; Behjati, S.; Biankin, A.V.; Bignell, G.R.; Bolli, N.; Borg, Å.; Børresen-Dale, A.-L.; et al. Signatures of mutational processes in human cancer. *Nature* **2013**, *500*, 415–421. [CrossRef] [PubMed]
32. Kanu, N.; Cerone, M.A.; Goh, G.; Zalmas, L.-P.; Bartkova, J.; Dietzen, M.; McGranahan, N.; Rogers, R.; Law, E.K.; Gromova, I.; et al. DNA replication stress mediates APOBEC3 family mutagenesis in breast cancer. *Genome Biol.* **2016**, *17*, 185. [CrossRef] [PubMed]
33. Long, J.; Delahanty, R.J.; Li, G.; Gao, Y.-T.; Lu, W.; Cai, Q.; Xiang, Y.-B.; Li, C.; Ji, B.-T.; Zheng, Y.; et al. A Common Deletion in the APOBEC3 Genes and Breast Cancer Risk. *JNCI J. Natl. Cancer Inst.* **2013**, *105*, 573–579. [CrossRef] [PubMed]
34. Nik-Zainal, S.; Wedge, D.C.; Alexandrov, L.B.; Petljak, M.; Butler, A.P.; Bolli, N.; Davies, H.R.; Knappskog, S.; Martin, S.; Papaemmanuil, E.; et al. Association of a germline copy number polymorphism of APOBEC3A and APOBEC3B with burden of putative APOBEC-dependent mutations in breast cancer. *Nat. Genet.* **2014**, *46*, 487–491. [CrossRef] [PubMed]
35. Bulliard, Y.; Narvaiza, I.; Bertero, A.; Peddi, S.; Röhrig, U.F.; Ortiz, M.; Zoete, V.; Castro-Díaz, N.; Turelli, P.; Telenti, A.; et al. Structure-function analyses point to a polynucleotide-accommodating groove essential for APOBEC3A restriction activities. *J. Virol.* **2011**, *85*, 1765–1776. [CrossRef] [PubMed]
36. Wang, Z.; Wakae, K.; Kitamura, K.; Aoyama, S.; Liu, G.; Koura, M.; Monjurul, A.M.; Kukimoto, I.; Muramatsu, M. APOBEC3 deaminases induce hypermutation in human papillomavirus 16 DNA upon beta interferon stimulation. *J. Virol.* **2014**, *88*, 1308–1317. [CrossRef] [PubMed]
37. Suspène, R.; Guétard, D.; Henry, M.; Sommer, P.; Wain-Hobson, S.; Vartanian, J.-P. Extensive editing of both hepatitis B virus DNA strands by APOBEC3 cytidine deaminases in vitro and in vivo. *Proc. Natl. Acad. Sci. USA* **2005**, *102*, 8321–8326. [CrossRef] [PubMed]

38. Bonvin, M.; Achermann, F.; Greeve, I.; Stroka, D.; Keogh, A.; Inderbitzin, D.; Candinas, D.; Sommer, P.; Wain-Hobson, S.; Vartanian, J.-P.; et al. Interferon-inducible expression of APOBEC3 editing enzymes in human hepatocytes and inhibition of hepatitis B virus replication. *Hepatology* **2006**, *43*, 1364–1374. [CrossRef] [PubMed]

39. Baumert, T.F.; Rösler, C.; Malim, M.H.; von Weizsäcker, F. Hepatitis B virus DNA is subject to extensive editing by the human deaminase APOBEC3C. *Hepatology* **2007**, *46*, 682–689. [CrossRef] [PubMed]

40. Nguyen, D.H.; Gummuluru, S.; Hu, J. Deamination-independent inhibition of hepatitis B virus reverse transcription by APOBEC3G. *J. Virol.* **2007**, *81*, 4465–4472. [CrossRef] [PubMed]

41. Burns, M.B.; Lackey, L.; Carpenter, M.A.; Rathore, A.; Land, A.M.; Leonard, B.; Refsland, E.W.; Kotandeniya, D.; Tretyakova, N.; Nikas, J.B.; et al. APOBEC3B is an enzymatic source of mutation in breast cancer. *Nature* **2013**, *494*, 366–370. [CrossRef] [PubMed]

42. Mussil, B.; Suspène, R.; Aynaud, M.-M.; Gauvrit, A.; Vartanian, J.-P.; Wain-Hobson, S. Human APOBEC3A Isoforms Translocate to the Nucleus and Induce DNA Double Strand Breaks Leading to Cell Stress and Death. *PLoS ONE* **2013**, *8*, e73641. [CrossRef] [PubMed]

43. Lackey, L.; Law, E.K.; Brown, W.L.; Harris, R.S. Subcellular localization of the APOBEC3 proteins during mitosis and implications for genomic DNA deamination. *Cell Cycle* **2013**, *12*, 762–772. [CrossRef] [PubMed]

44. Lackey, L.; Demorest, Z.L.; Land, A.M.; Hultquist, J.F.; Brown, W.L.; Harris, R.S. APOBEC3B and AID have similar nuclear import mechanisms. *J. Mol. Biol.* **2012**, *419*, 301–314. [CrossRef] [PubMed]

45. Shinohara, M.; Io, K.; Shindo, K.; Matsui, M.; Sakamoto, T.; Tada, K.; Kobayashi, M.; Kadowaki, N.; Takaori-Kondo, A. APOBEC3B can impair genomic stability by inducing base substitutions in genomic DNA in human cells. *Sci. Rep.* **2012**, *2*, 806. [CrossRef] [PubMed]

46. Nik-Zainal, S.; Van Loo, P.; Wedge, D.C.; Alexandrov, L.B.; Greenman, C.D.; Lau, K.W.; Raine, K.; Jones, D.; Marshall, J.; Ramakrishna, M.; et al. Breast Cancer Working Group of the International Cancer Genome Consortium The life history of 21 breast cancers. *Cell* **2012**, *149*, 994–1007. [CrossRef] [PubMed]

47. Kuong, K.J.; Loeb, L.A. APOBEC3B mutagenesis in cancer. *Nat. Genet.* **2013**. [CrossRef] [PubMed]

48. Vieira, V.C.; Leonard, B.; White, E.A.; Starrett, G.J.; Temiz, N.A.; Lorenz, L.D.; Lee, D.; Soares, M.A.; Lambert, P.F.; Howley, P.M.; et al. Human papillomavirus E6 triggers upregulation of the antiviral and cancer genomic DNA deaminase APOBEC3B. *MBio* **2014**, *5*, e02234-14. [CrossRef] [PubMed]

49. Jarmuz, A.; Chester, A.; Bayliss, J.; Gisbourne, J.; Dunham, I.; Scott, J.; Navaratnam, N. An anthropoid-specific locus of orphan C to U RNA-editing enzymes on chromosome 22. *Genomics* **2002**, *79*, 285–296. [CrossRef] [PubMed]

50. LaRue, R.S.; Andrésdóttir, V.; Blanchard, Y.; Conticello, S.G.; Derse, D.; Emerman, M.; Greene, W.C.; Jónsson, S.R.; Landau, N.R.; Löchelt, M.; et al. Guidelines for naming nonprimate APOBEC3 genes and proteins. *J. Virol.* **2009**, *83*, 494–497. [CrossRef] [PubMed]

51. Salter, J.D.; Bennett, R.P.; Smith, H.C. The APOBEC Protein Family: United by Structure, Divergent in Function. *Trends Biochem. Sci.* **2016**, *41*, 578–594. [CrossRef] [PubMed]

52. Bransteitter, R.; Prochnow, C.; Chen, X.S. The current structural and functional understanding of APOBEC deaminases. *Cell. Mol. Life Sci.* **2009**, *66*, 3137–3147. [CrossRef] [PubMed]

53. Navarro, F.; Bollman, B.; Chen, H.; König, R.; Yu, Q.; Chiles, K.; Landau, N.R. Complementary function of the two catalytic domains of APOBEC3G. *Virology* **2005**, *333*, 374–386. [CrossRef] [PubMed]

54. Newman, E.N.C.; Holmes, R.K.; Craig, H.M.; Klein, K.C.; Lingappa, J.R.; Malim, M.H.; Sheehy, A.M. Antiviral Function of APOBEC3G Can Be Dissociated from Cytidine Deaminase Activity. *Curr. Biol.* **2005**, *15*, 166–170. [CrossRef] [PubMed]

55. Jónsson, S.R.; Haché, G.; Stenglein, M.D.; Fahrenkrug, S.C.; Andrésdóttir, V.; Harris, R.S. Evolutionarily conserved and non-conserved retrovirus restriction activities of artiodactyl APOBEC3F proteins. *Nucleic Acids Res.* **2006**, *34*, 5683–5694. [CrossRef] [PubMed]

56. Holmes, R.K.; Koning, F.A.; Bishop, K.N.; Malim, M.H. APOBEC3F Can Inhibit the Accumulation of HIV-1 Reverse Transcription Products in the Absence of Hypermutation COMPARISONS WITH APOBEC3G. *J. Biol. Chem.* **2007**, *282*, 2587–2595. [CrossRef] [PubMed]

57. Bonvin, M.; Greeve, J. Effects of point mutations in the cytidine deaminase domains of APOBEC3B on replication and hypermutation of hepatitis B virus in vitro. *J. Gen. Virol.* **2007**, *88*, 3270–3274. [CrossRef] [PubMed]

58. Logue, E.C.; Bloch, N.; Dhuey, E.; Zhang, R.; Cao, P.; Herate, C.; Chauveau, L.; Hubbard, S.R.; Landau, N.R. A DNA Sequence Recognition Loop on APOBEC3A Controls Substrate Specificity. *PLoS ONE* **2014**, *9*, e97062. [CrossRef] [PubMed]

59. Shlyakhtenko, L.S.; Lushnikov, A.J.; Li, M.; Harris, R.S.; Lyubchenko, Y.L. Interaction of APOBEC3A with DNA assessed by atomic force microscopy. *PLoS ONE* **2014**, *9*, e99354. [CrossRef] [PubMed]

60. Bohn, M.-F.; Shandilya, S.M.D.; Silvas, T.V.; Nalivaika, E.A.; Kouno, T.; Kelch, B.A.; Ryder, S.P.; Kurt-Yilmaz, N.; Somasundaran, M.; Schiffer, C.A. The ssDNA mutator APOBEC3A is regulated by cooperative dimerization. *Structure* **2015**, *23*, 903–911. [CrossRef] [PubMed]

61. Byeon, I.-J.L.; Ahn, J.; Mitra, M.; Byeon, C.-H.; Hercík, K.; Hritz, J.; Charlton, L.M.; Levin, J.G.; Gronenborn, A.M. NMR structure of human restriction factor APOBEC3A reveals substrate binding and enzyme specificity. *Nat. Commun.* **2013**, *4*, 1890. [CrossRef] [PubMed]

62. Mitra, M.; Hercík, K.; Byeon, I.-J.L.; Ahn, J.; Hill, S.; Hinchee-Rodriguez, K.; Singer, D.; Byeon, C.-H.; Charlton, L.M.; Nam, G.; et al. Structural determinants of human APOBEC3A enzymatic and nucleic acid binding properties. *Nucleic Acids Res.* **2014**, *42*, 1095–1110. [CrossRef] [PubMed]

63. Amrine-Madsen, H.; Koepfli, K.-P.; Wayne, R.K.; Springer, M.S. A new phylogenetic marker, apolipoprotein B, provides compelling evidence for eutherian relationships. *Mol. Phylogenet. Evol.* **2003**, *28*, 225–240. [CrossRef]

64. Murphy, W.J.; Eizirik, E.; Johnson, W.E.; Zhang, Y.P.; Ryder, O.A.; O'Brien, S.J. Molecular phylogenetics and the origins of placental mammals. *Nature* **2001**, *409*, 614–618. [CrossRef] [PubMed]

65. Madsen, O.; Scally, M.; Douady, C.J.; Kao, D.J.; DeBry, R.W.; Adkins, R.; Amrine, H.M.; Stanhope, M.J.; de Jong, W.W.; Springer, M.S. Parallel adaptive radiations in two major clades of placental mammals. *Nature* **2001**, *409*, 610–614. [CrossRef] [PubMed]

66. Harris, R.S.; Liddament, M.T. Retroviral restriction by APOBEC proteins. *Nat. Rev. Immunol.* **2004**, *4*, 868–877. [CrossRef] [PubMed]

67. Mikl, M.C.; Watt, I.N.; Lu, M.; Reik, W.; Davies, S.L.; Neuberger, M.S.; Rada, C. Mice deficient in APOBEC2 and APOBEC3. *Mol. Cell Biol.* **2005**, *25*, 7270–7277. [CrossRef] [PubMed]

68. Münk, C.; Willemsen, A.; Bravo, I.G. An ancient history of gene duplications, fusions and losses in the evolution of APOBEC3 mutators in mammals. *BMC Evol. Biol.* **2012**, *12*, 71. [CrossRef] [PubMed]

69. Conticello, S.G. The AID/APOBEC family of nucleic acid mutators. *Genome Biol.* **2008**, *9*, 229. [CrossRef] [PubMed]

70. Münk, C.; Beck, T.; Zielonka, J.; Hotz-Wagenblatt, A.; Chareza, S.; Battenberg, M.; Thielebein, J.; Cichutek, K.; Bravo, I.G.; O'Brien, S.J.; et al. Functions, structure, and read-through alternative splicing of feline APOBEC3 genes. *Genome Biol.* **2008**, *9*, R48. [CrossRef] [PubMed]

71. LaRue, R.S.; Jónsson, S.R.; Silverstein, K.A.T.; Lajoie, M.; Bertrand, D.; El-Mabrouk, N.; Hötzel, I.; Andrésdóttir, V.; Smith, T.P.L.; Harris, R.S. The artiodactyl APOBEC3 innate immune repertoire shows evidence for a multi-functional domain organization that existed in the ancestor of placental mammals. *BMC Mol. Biol.* **2008**, *9*, 104. [CrossRef] [PubMed]

72. Bogerd, H.P.; Tallmadge, R.L.; Oaks, J.L.; Carpenter, S.; Cullen, B.R. Equine infectious anemia virus resists the antiretroviral activity of equine APOBEC3 proteins through a packaging-independent mechanism. *J. Virol.* **2008**, *82*, 11889–11901. [CrossRef] [PubMed]

73. Sawyer, S.L.; Emerman, M.; Malik, H.S. Ancient adaptive evolution of the primate antiviral DNA-editing enzyme APOBEC3G. *PLoS Biol.* **2004**, *2*, E275. [CrossRef] [PubMed]

74. Zhang, J.; Webb, D.M. Rapid evolution of primate antiviral enzyme APOBEC3G. *Hum. Mol. Genet.* **2004**, *13*, 1785–1791. [CrossRef] [PubMed]

75. McLaughlin, R.N.; Gable, J.T.; Wittkopp, C.J.; Emerman, M.; Malik, H.S. Conservation and Innovation of APOBEC3A Restriction Functions during Primate Evolution. *Mol. Biol. Evol.* **2016**, *33*, 1889–1901. [CrossRef] [PubMed]

76. Duggal, N.K.; Malik, H.S.; Emerman, M. The breadth of antiviral activity of Apobec3DE in chimpanzees has been driven by positive selection. *J. Virol.* **2011**, *85*, 11361–11371. [CrossRef] [PubMed]

77. Meyerson, N.R.; Sawyer, S.L. Two-stepping through time: mammals and viruses. *Trends Microbiol.* **2011**, *19*, 286–294. [CrossRef] [PubMed]

78. Harris, R.S.; Bishop, K.N.; Sheehy, A.M.; Craig, H.M.; Petersen-Mahrt, S.K.; Watt, I.N.; Neuberger, M.S.; Malim, M.H. DNA deamination mediates innate immunity to retroviral infection. *Cell* **2003**, *113*, 803–809. [CrossRef]

79. Shi, K.; Carpenter, M.A.; Banerjee, S.; Shaban, N.M.; Kurahashi, K.; Salamango, D.J.; McCann, J.L.; Starrett, G.J.; Duffy, J.V.; Demir, Ö.; et al. Structural basis for targeted DNA cytosine deamination and mutagenesis by APOBEC3A and APOBEC3B. *Nat. Struct. Mol. Biol.* **2017**, *24*, 131–139. [CrossRef] [PubMed]

80. Kouno, T.; Silvas, T.V.; Hilbert, B.J.; Shandilya, S.M.D.; Bohn, M.F.; Kelch, B.A.; Royer, W.E.; Somasundaran, M.; Kurt-Yilmaz, N.; Matsuo, H.; et al. Crystal structure of APOBEC3A bound to single-stranded DNA reveals structural basis for cytidine deamination and specificity. *Nat. Commun.* **2017**, *8*, 15024. [CrossRef] [PubMed]

81. Vartanian, J.-P.; Henry, M.; Marchio, A.; Suspène, R.; Aynaud, M.-M.; Guétard, D.; Cervantes-Gonzalez, M.; Battiston, C.; Mazzaferro, V.; Pineau, P.; et al. Massive APOBEC3 editing of hepatitis B viral DNA in cirrhosis. *PLoS Pathog.* **2010**, *6*, e1000928. [CrossRef] [PubMed]

82. Suspène, R.; Aynaud, M.-M.; Guétard, D.; Henry, M.; Eckhoff, G.; Marchio, A.; Pineau, P.; Dejean, A.; Vartanian, J.-P.; Wain-Hobson, S. Somatic hypermutation of human mitochondrial and nuclear DNA by APOBEC3 cytidine deaminases, a pathway for DNA catabolism. *Proc. Natl. Acad. Sci. USA* **2011**, *108*, 4858–4863. [CrossRef] [PubMed]

83. Roberts, S.A.; Lawrence, M.S.; Klimczak, L.J.; Grimm, S.A.; Fargo, D.; Stojanov, P.; Kiezun, A.; Kryukov, G.V.; Carter, S.L.; Saksena, G.; et al. An APOBEC cytidine deaminase mutagenesis pattern is widespread in human cancers. *Nat. Genet.* **2013**, *45*, 970–976. [CrossRef] [PubMed]

84. Hoopes, J.I.; Cortez, L.M.; Mertz, T.M.; Malc, E.P.; Mieczkowski, P.A.; Roberts, S.A. APOBEC3A and APOBEC3B Preferentially Deaminate the Lagging Strand Template during DNA Replication. *Cell Rep.* **2016**, *14*, 1273–1282. [CrossRef] [PubMed]

85. Seplyarskiy, V.B.; Soldatov, R.A.; Popadin, K.Y.; Antonarakis, S.E.; Bazykin, G.A.; Nikolaev, S.I. APOBEC-induced mutations in human cancers are strongly enriched on the lagging DNA strand during replication. *Genome Res.* **2016**, *26*, 174–182. [CrossRef] [PubMed]

86. Baysal, B.E.; De Jong, K.; Liu, B.; Wang, J.; Patnaik, S.K.; Wallace, P.K.; Taggart, R.T. Hypoxia-inducible C-to-U coding RNA editing downregulates SDHB in monocytes. *PeerJ* **2013**, *1*, e152. [CrossRef] [PubMed]

87. Stavrou, S.; Crawford, D.; Blouch, K.; Browne, E.P.; Kohli, R.M.; Ross, S.R. Different modes of retrovirus restriction by human APOBEC3A and APOBEC3G in vivo. *PLoS Pathog.* **2014**, *10*, e1004145. [CrossRef] [PubMed]

88. Bishop, K.N.; Holmes, R.K.; Malim, M.H. Antiviral potency of APOBEC proteins does not correlate with cytidine deamination. *J. Virol.* **2006**, *80*, 8450–8458. [CrossRef] [PubMed]

89. Malim, M.H. APOBEC proteins and intrinsic resistance to HIV-1 infection. *Philos. Trans. R. Soc. Lond. B Biol. Sci.* **2009**, *364*, 675–687. [CrossRef] [PubMed]

90. García-Sastre, A.; Biron, C.A. Type 1 Interferons and the Virus-Host Relationship: A Lesson in Détente. *Science* **2006**, *312*, 879–882. [CrossRef] [PubMed]

91. Peng, G.; Lei, K.J.; Jin, W.; Greenwell-Wild, T.; Wahl, S.M. Induction of APOBEC3 family proteins, a defensive maneuver underlying interferon-induced anti-HIV-1 activity. *J. Exp. Med.* **2006**, *203*, 41–46. [CrossRef] [PubMed]

92. Mohanram, V.; Sköld, A.E.; Bächle, S.M.; Pathak, S.K.; Spetz, A.-L. IFN-α induces APOBEC3G, F, and A in immature dendritic cells and limits HIV-1 spread to CD4+ T cells. *J. Immunol.* **2013**, *190*, 3346–3353. [CrossRef] [PubMed]

93. Wang, F.-X.; Huang, J.; Zhang, H.; Ma, X.; Zhang, H. APOBEC3G upregulation by alpha interferon restricts human immunodeficiency virus type 1 infection in human peripheral plasmacytoid dendritic cells. *J. Gen. Virol.* **2008**, *89*, 722–730. [CrossRef] [PubMed]

94. Leonard, B.; McCann, J.L.; Starrett, G.J.; Kosyakovsky, L.; Luengas, E.M.; Molan, A.M.; Burns, M.B.; McDougle, R.M.; Parker, P.J.; Brown, W.L.; et al. The PKC/NF-κB signaling pathway induces APOBEC3B expression in multiple human cancers. *Cancer Res.* **2015**, *75*, 4538–4547. [CrossRef] [PubMed]

95. Maruyama, W.; Shirakawa, K.; Matsui, H.; Matsumoto, T.; Yamazaki, H.; Sarca, A.D.; Kazuma, Y.; Kobayashi, M.; Shindo, K.; Takaori-Kondo, A. Classical NF-κB pathway is responsible for APOBEC3B expression in cancer cells. *Biochem. Biophys. Res. Commun.* **2016**, *478*, 1466–1471. [CrossRef] [PubMed]

96. Kondo, S.; Wakae, K.; Wakisaka, N.; Nakanishi, Y.; Ishikawa, K.; Komori, T.; Moriyama-Kita, M.; Endo, K.; Murono, S.; Wang, Z.; et al. APOBEC3A associates with human papillomavirus genome integration in oropharyngeal cancers. *Oncogene* **2017**, *36*, 1687–1697. [CrossRef] [PubMed]

97. Mori, S.; Takeuchi, T.; Ishii, Y.; Kukimoto, I. Identification of APOBEC3B promoter elements responsible for activation by human papillomavirus type 16 E6. *Biochem. Biophys. Res. Commun.* **2015**, *460*, 555–560. [CrossRef] [PubMed]

98. Mori, S.; Takeuchi, T.; Ishii, Y.; Yugawa, T.; Kiyono, T.; Nishina, H.; Kukimoto, I. Human papillomavirus 16 E6 upregulates APOBEC3B via the TEAD transcription factor. *J. Virol.* **2017**, *91*, e02413-16. [CrossRef] [PubMed]

99. Menendez, D.; Nguyen, T.-A.; Snipe, J.; Resnick, M.A. The cytidine deaminase APOBEC3 family is subject to transcriptional regulation by p53. *Mol. Cancer Res.* **2017**, *15*, 735–743. [CrossRef] [PubMed]

100. Gasco, M.; Shami, S.; Crook, T. The p53 pathway in breast cancer. *Breast Cancer Res.* **2002**, *4*, 70. [CrossRef] [PubMed]

101. Peng, G.; Greenwell-Wild, T.; Nares, S.; Jin, W.; Lei, K.J.; Rangel, Z.G.; Munson, P.J.; Wahl, S.M. Myeloid differentiation and susceptibility to HIV-1 are linked to APOBEC3 expression. *Blood* **2007**, *110*, 393–400. [CrossRef] [PubMed]

102. Koning, F.A.; Newman, E.N.C.; Kim, E.-Y.; Kunstman, K.J.; Wolinsky, S.M.; Malim, M.H. Defining APOBEC3 expression patterns in human tissues and hematopoietic cell subsets. *J. Virol.* **2009**, *83*, 9474–9485. [CrossRef] [PubMed]

103. Warren, C.J.; Van Doorslaer, K.; Pandey, A.; Espinosa, J.M.; Pyeon, D. Role of the host restriction factor APOBEC3 on papillomavirus evolution. *Virus Evol.* **2015**, *1*, vev015. [CrossRef] [PubMed]

104. Love, R.P.; Xu, H.; Chelico, L. Biochemical analysis of hypermutation by the deoxycytidine deaminase APOBEC3A. *J. Biol. Chem.* **2012**, *287*, 30812–30822. [CrossRef] [PubMed]

105. Beggel, B.; Münk, C.; Däumer, M.; Hauck, K.; Häussinger, D.; Lengauer, T.; Erhardt, A. Full genome ultra-deep pyrosequencing associates G-to-A hypermutation of the hepatitis B virus genome with the natural progression of hepatitis B. *J. Viral Hepat.* **2013**, *20*, 882–889. [CrossRef] [PubMed]

106. Wakae, K.; Aoyama, S.; Wang, Z.; Kitamura, K.; Liu, G.; Monjurul, A.M.; Koura, M.; Imayasu, M.; Sakamoto, N.; Nakamura, M.; et al. Detection of hypermutated human papillomavirus type 16 genome by Next-Generation Sequencing. *Virology* **2015**, *485*, 460–466. [CrossRef] [PubMed]

107. Kukimoto, I.; Mori, S.; Aoyama, S.; Wakae, K.; Muramatsu, M.; Kondo, K. Hypermutation in the E2 gene of human papillomavirus type 16 in cervical intraepithelial neoplasia. *J. Med. Virol.* **2015**, *87*, 1754–1760. [CrossRef] [PubMed]

108. Herdman, M.T.; Pett, M.R.; Roberts, I.; Alazawi, W.O.F.; Teschendorff, A.E.; Zhang, X.-Y.; Stanley, M.A.; Coleman, N. Interferon-β treatment of cervical keratinocytes naturally infected with human papillomavirus 16 episomes promotes rapid reduction in episome numbers and emergence of latent integrants. *Carcinogenesis* **2006**, *27*, 2341–2353. [CrossRef] [PubMed]

109. Chang, Y.E.; Pena, L.; Sen, G.C.; Park, J.K.; Laimins, L.A. Long-term effect of interferon on keratinocytes that maintain human papillomavirus type 31. *J. Virol.* **2002**, *76*, 8864–8874. [CrossRef] [PubMed]

110. Terenzi, F.; Saikia, P.; Sen, G.C. Interferon-inducible protein, P56, inhibits HPV DNA replication by binding to the viral protein E1. *EMBO* **2008**, *27*, 3311–3321. [CrossRef] [PubMed]

111. Warren, C.J.; Griffin, L.M.; Little, A.S.; Huang, I.-C.; Farzan, M.; Pyeon, D. The Antiviral Restriction Factors IFITM1, 2 and 3 Do Not Inhibit Infection of Human Papillomavirus, Cytomegalovirus and Adenovirus. *PLoS ONE* **2014**, *9*, e96579. [CrossRef] [PubMed]

112. Pyeon, D.; Lambert, P.F.; Ahlquist, P. Production of infectious human papillomavirus independently of viral replication and epithelial cell differentiation. *Proc. Natl. Acad. Sci. USA* **2005**, *102*, 9311–9316. [CrossRef] [PubMed]

113. Pett, M.R.; Herdman, M.T.; Palmer, R.D.; Yeo, G.S.H.; Shivji, M.K.; Stanley, M.A.; Coleman, N. Selection of cervical keratinocytes containing integrated HPV16 associates with episome loss and an endogenous antiviral response. *Proc. Natl. Acad. Sci. USA* **2006**, *103*, 3822–3827. [CrossRef] [PubMed]

114. Alazawi, W.; Pett, M.; Arch, B.; Scott, L.; Freeman, T.; Stanley, M.A.; Coleman, N. Changes in cervical keratinocyte gene expression associated with integration of human papillomavirus 16. *Cancer Res.* **2002**, *62*, 6959–6965. [PubMed]

115. Lace, M.J.; Anson, J.R.; Haugen, T.H.; Dierdorff, J.M.; Turek, L.P. Interferon treatment of human keratinocytes harboring extrachromosomal, persistent HPV-16 plasmid genomes induces de novo viral integration. *Carcinogenesis* **2015**, *36*, 151–159. [CrossRef] [PubMed]

116. Turek, L.P.; Byrne, J.C.; Lowy, D.R.; Dvoretzky, I.; Friedman, R.M.; Howley, P.M. Interferon induces morphologic reversion with elimination of extrachromosomal viral genomes in bovine papillomavirus-transformed mouse cells. *Proc. Natl. Acad. Sci. USA* **1982**, *79*, 7914–7918. [CrossRef] [PubMed]

117. Yu, X.; Yu, Y.; Liu, B.; Luo, K.; Kong, W.; Mao, P.; Yu, X.-F. Induction of APOBEC3G ubiquitination and degradation by an HIV-1 Vif-Cul5-SCF complex. *Science* **2003**, *302*, 1056–1060. [CrossRef] [PubMed]

118. Mehle, A.; Strack, B.; Ancuta, P.; Zhang, C.; McPike, M.; Gabuzda, D. Vif overcomes the innate antiviral activity of APOBEC3G by promoting its degradation in the ubiquitin-proteasome pathway. *J. Biol. Chem.* **2004**, *279*, 7792–7798. [CrossRef] [PubMed]

119. Yu, Y.; Xiao, Z.; Ehrlich, E.S.; Yu, X.; Yu, X.-F. Selective assembly of HIV-1 Vif-Cul5-ElonginB-ElonginC E3 ubiquitin ligase complex through a novel SOCS box and upstream cysteines. *Genes Dev.* **2004**, *18*, 2867–2872. [CrossRef] [PubMed]

120. Huh, K.; Zhou, X.; Hayakawa, H.; Cho, J.-Y.; Libermann, T.A.; Jin, J.; Harper, J.W.; Münger, K. Human papillomavirus type 16 E7 oncoprotein associates with the cullin 2 ubiquitin ligase complex, which contributes to degradation of the retinoblastoma tumor suppressor. *J. Virol.* **2007**, *81*, 9737–9747. [CrossRef] [PubMed]

121. Boyer, S.N.; Wazer, D.E.; Band, V. E7 protein of human papilloma virus-16 induces degradation of retinoblastoma protein through the ubiquitin-proteasome pathway. *Cancer Res.* **1996**, *56*, 4620–4624. [PubMed]

122. Jones, D.L.; Thompson, D.A.; Münger, K. Destabilization of the RB Tumor Suppressor Protein and Stabilization of p53 Contribute to HPV Type 16 E7-Induced Apoptosis. *Virology* **1997**, *239*, 97–107. [CrossRef] [PubMed]

123. Westrich, J.A.; Warren, C.J.; Klausner, M.J.; Vermeer, D.W.; Guo, K.; Lee, J.H.; Liu, C.; Santiago, M.L.; Pyeon, D. High-risk Human Papillomavirus E7 Stabilizes APOBEC3A Protein by Inhibiting Cullin 2-dependent Protein Degradation. *J. Virol.* **2017**, Under revision.

124. Shackelton, L.A.; Parrish, C.R.; Holmes, E.C. Evolutionary basis of codon usage and nucleotide composition bias in vertebrate DNA viruses. *J. Mol. Evol.* **2006**, *62*, 551–563. [CrossRef] [PubMed]

125. Hoelzer, K.; Shackelton, L.A.; Parrish, C.R. Presence and role of cytosine methylation in DNA viruses of animals. *Nucleic Acids Res.* **2008**, *36*, 2825–2837. [CrossRef] [PubMed]

126. Upadhyay, M.; Samal, J.; Kandpal, M.; Vasaikar, S.; Biswas, B.; Gomes, J.; Vivekanandan, P. CpG dinucleotide frequencies reveal the role of host methylation capabilities in parvovirus evolution. *J. Virol.* **2013**, *87*, 13816–13824. [CrossRef] [PubMed]

127. Upadhyay, M.; Vivekanandan, P. Depletion of CpG Dinucleotides in Papillomaviruses and Polyomaviruses: A Role for Divergent Evolutionary Pressures. *PLoS ONE* **2015**, *10*, e0142368. [CrossRef] [PubMed]

128. De Villiers, E.-M.; Fauquet, C.; Broker, T.R.; Bernard, H.-U.; zur Hausen, H. Classification of papillomaviruses. *Virology* **2004**, *324*, 17–27. [CrossRef] [PubMed]

129. Warren, C.J.; Pyeon, D. APOBEC3 in papillomavirus restriction, evolution and cancer progression. *Oncotarget* **2015**, *6*, 39385–39386. [CrossRef] [PubMed]

130. Henderson, S.; Chakravarthy, A.; Su, X.; Boshoff, C.; Fenton, T.R. APOBEC-mediated cytosine deamination links PIK3CA helical domain mutations to human papillomavirus-driven tumor development. *Cell Rep.* **2014**, *7*, 1833–1841. [CrossRef] [PubMed]

131. Zhang, Y.; Koneva, L.A.; Virani, S.; Arthur, A.E.; Virani, A.; Hall, P.B.; Warden, C.D.; Carey, T.E.; Chepeha, D.B.; Prince, M.E.; et al. Subtypes of HPV-Positive Head and Neck Cancers Are Associated with HPV Characteristics, Copy Number Alterations, PIK3CA Mutation, and Pathway Signatures. *Clin. Cancer Res.* **2016**, *22*, 4735–4745. [CrossRef] [PubMed]

132. Moody, C.A.; Laimins, L.A. Human papillomavirus oncoproteins: pathways to transformation. *Nat. Rev. Cancer* **2010**, *10*, 550–560. [CrossRef] [PubMed]

133. Münger, K.; Baldwin, A.; Edwards, K.M.; Hayakawa, H.; Nguyen, C.L.; Owens, M.; Grace, M.; Huh, K. Mechanisms of Human Papillomavirus-Induced Oncogenesis. *J. Virol.* **2004**, *78*, 11451–11460. [CrossRef] [PubMed]

134. Cescon, D.W.; Haibe-Kains, B.; Mak, T.W. APOBEC3B expression in breast cancer reflects cellular proliferation, while a deletion polymorphism is associated with immune activation. *Proc. Natl. Acad. Sci. USA* **2015**, *112*, 2841–2846. [CrossRef] [PubMed]
135. Leonard, B.; Hart, S.N.; Burns, M.B.; Carpenter, M.A.; Temiz, N.A.; Rathore, A.; Vogel, R.I.; Nikas, J.B.; Law, E.K.; Brown, W.L.; et al. APOBEC3B upregulation and genomic mutation patterns in serous ovarian carcinoma. *Cancer Res.* **2013**, *73*, 7222–7231. [CrossRef] [PubMed]
136. Middlebrooks, C.D.; Banday, A.R.; Matsuda, K.; Udquim, K.-I.; Onabajo, O.O.; Paquin, A.; Figueroa, J.D.; Zhu, B.; Koutros, S.; Kubo, M.; et al. Association of germline variants in the APOBEC3 region with cancer risk and enrichment with APOBEC-signature mutations in tumors. *Nat. Genet.* **2016**, *48*, 1330–1338. [CrossRef] [PubMed]
137. Marouf, C.; Göhler, S.; Filho, M.I.D.S.; Hajji, O.; Hemminki, K.; Nadifi, S.; Försti, A. Analysis of functional germline variants in APOBEC3 and driver genes on breast cancer risk in Moroccan study population. *BMC Cancer* **2016**, *16*, 165. [CrossRef] [PubMed]
138. Revathidevi, S.; Manikandan, M.; Rao, A.K.D.M.; Vinothkumar, V.; Arunkumar, G.; Rajkumar, K.S.; Ramani, R.; Rajaraman, R.; Ajay, C.; Munirajan, A.K. Analysis of APOBEC3A/3B germline deletion polymorphism in breast, cervical and oral cancers from South India and its impact on miRNA regulation. *Tumour Biol.* **2016**, *37*, 11983–11990. [CrossRef] [PubMed]
139. Caval, V.; Suspène, R.; Shapira, M.; Vartanian, J.-P.; Wain-Hobson, S. A prevalent cancer susceptibility APOBEC3A hybrid allele bearing APOBEC3B 3'UTR enhances chromosomal DNA damage. *Nat. Commun.* **2014**, *5*, 5129. [CrossRef] [PubMed]
140. Winder, D.M.; Pett, M.R.; Foster, N.; Shivji, M.K.K.; Herdman, M.T.; Stanley, M.A.; Venkitaraman, A.R.; Coleman, N. An increase in DNA double-strand breaks, induced by Ku70 depletion, is associated with human papillomavirus 16 episome loss and de novo viral integration events. *J. Pathol.* **2007**, *213*, 27–34. [CrossRef] [PubMed]
141. Verhalen, B.; Starrett, G.J.; Harris, R.S.; Jiang, M. Functional upregulation of the DNA cytosine deaminase APOBEC3B by polyomaviruses. *J. Virol.* **2016**, *90*, 6379–6386. [CrossRef] [PubMed]
142. Ding, L.; Getz, G.; Wheeler, D.A.; Mardis, E.R.; McLellan, M.D.; Cibulskis, K.; Sougnez, C.; Greulich, H.; Muzny, D.M.; Morgan, M.B.; et al. Somatic mutations affect key pathways in lung adenocarcinoma. *Nature* **2008**, *455*, 1069–1075. [CrossRef] [PubMed]
143. Martincorena, I.; Campbell, P.J. Somatic mutation in cancer and normal cells. *Science* **2015**, *349*, 1483–1489. [CrossRef] [PubMed]
144. Nichols, A.C.; Palma, D.A.; Chow, W.; Tan, S.; Rajakumar, C.; Rizzo, G.; Fung, K.; Kwan, K.; Wehrli, B.; Winquist, E.; et al. High frequency of activating PIK3CA mutations in human papillomavirus-positive oropharyngeal cancer. *JAMA Otolaryngol. Head Neck Surg.* **2013**, *139*, 617–622. [CrossRef] [PubMed]
145. Schumacher, T.N.; Schreiber, R.D. Neoantigens in cancer immunotherapy. *Science* **2015**, *348*, 69–74. [CrossRef] [PubMed]
146. McGranahan, N.; Furness, A.J.S.; Rosenthal, R.; Ramskov, S.; Lyngaa, R.; Saini, S.K.; Jamal-Hanjani, M.; Wilson, G.A.; Birkbak, N.J.; Hiley, C.T.; et al. Clonal neoantigens elicit T cell immunoreactivity and sensitivity to immune checkpoint blockade. *Science* **2016**, *351*, 1463–1469. [CrossRef] [PubMed]
147. Anagnostou, V.; Smith, K.N.; Forde, P.M.; Niknafs, N.; Bhattacharya, R.; White, J.; Zhang, T.; Adleff, V.; Phallen, J.; Wali, N.; et al. Evolution of Neoantigen Landscape during Immune Checkpoint Blockade in Non-Small Cell Lung Cancer. *Cancer Discov.* **2017**, *7*, 264–276. [CrossRef] [PubMed]
148. Stevanović, S.; Pasetto, A.; Helman, S.R.; Gartner, J.J.; Prickett, T.D.; Howie, B.; Robins, H.S.; Robbins, P.F.; Klebanoff, C.A.; Rosenberg, S.A.; et al. Landscape of immunogenic tumor antigens in successful immunotherapy of virally induced epithelial cancer. *Science* **2017**, *356*, 200–205. [CrossRef] [PubMed]
149. Cancer Genome Atlas Network. Comprehensive genomic characterization of head and neck squamous cell carcinomas. *Nature* **2015**, *517*, 576–582. [CrossRef]
150. Jabbar, S.F.; Park, S.; Schweizer, J.; Berard-Bergery, M.; Pitot, H.C.; Lee, D.; Lambert, P.F. Cervical cancers require the continuous expression of the human papillomavirus type 16 E7 oncoprotein even in the presence of the viral E6 oncoprotein. *Cancer Res.* **2012**, *72*, 4008–4016. [CrossRef] [PubMed]
151. Kennedy, E.M.; Kornepati, A.V.R.; Goldstein, M.; Bogerd, H.P.; Poling, B.C.; Whisnant, A.W.; Kastan, M.B.; Cullen, B.R. Inactivation of the human papillomavirus E6 or E7 gene in cervical carcinoma cells by using a bacterial CRISPR/Cas RNA-guided endonuclease. *J. Virol.* **2014**, *88*, 11965–11972. [CrossRef] [PubMed]

152. Hanning, J.E.; Saini, H.K.; Murray, M.J.; Caffarel, M.M.; van Dongen, S.; Ward, D.; Barker, E.M.; Scarpini, C.G.; Groves, I.J.; Stanley, M.A.; et al. Depletion of HPV16 early genes induces autophagy and senescence in a cervical carcinogenesis model, regardless of viral physical state. *J. Pathol.* **2013**, *231*, 354–366. [CrossRef] [PubMed]

153. Sima, N.; Wang, S.; Wang, W.; Kong, D.; Xu, Q.; Tian, X.; Luo, A.; Zhou, J.; Xu, G.; Meng, L.; et al. Antisense targeting human papillomavirus type 16 E6 and E7 genes contributes to apoptosis and senescence in SiHa cervical carcinoma cells. *Gynecol. Oncol.* **2007**, *106*, 299–304. [CrossRef] [PubMed]

154. Johung, K.; Goodwin, E.C.; DiMaio, D. Human papillomavirus E7 repression in cervical carcinoma cells initiates a transcriptional cascade driven by the retinoblastoma family, resulting in senescence. *J. Virol.* **2007**, *81*, 2102–2116. [CrossRef] [PubMed]

155. Gu, W.; Putral, L.; Hengst, K.; Minto, K.; Saunders, N.A.; Leggatt, G.; McMillan, N.A.J. Inhibition of cervical cancer cell growth in vitro and in vivo with lentiviral-vector delivered short hairpin RNA targeting human papillomavirus E6 and E7 oncogenes. *Cancer Gene Ther.* **2006**, *13*, 1023–1032. [CrossRef] [PubMed]

156. Horner, S.M.; DeFilippis, R.A.; Manuelidis, L.; DiMaio, D. Repression of the human papillomavirus E6 gene initiates p53-dependent, telomerase-independent senescence and apoptosis in HeLa cervical carcinoma cells. *J. Virol.* **2004**, *78*, 4063–4073. [CrossRef] [PubMed]

157. Boonden, J.A.; Pyeon, D.; Wang, S.S.; Horswill, M.; Schiffman, M.; Sherman, M.; Zuna, R.E.; Wang, Z.; Hewitt, S.M.; Pearson, R.; et al. Molecular transitions from papillomavirus infection to cervical precancer and cancer: Role of stromal estrogen receptor signaling. *Proc. Natl. Acad. Sci. USA* **2015**. [CrossRef]

viruses

MDPI

Review

Exosomes and Other Extracellular Vesicles in HPV Transmission and Carcinogenesis

David Guenat [1,2,3], François Hermetet [4], Jean-Luc Prétet [1,2] and Christiane Mougin [1,2,*]

[1] EA3181, University Bourgogne Franche-Comté, LabEx LipSTIC ANR-11-LABX-0021, Rue Ambroise Paré, 25000 Besançon, France; david.guenat@univ-fcomte.fr (D.G.); jean_luc.pretet@univ-fcomte.fr (J.-L.P.)
[2] CNR Papillomavirus, CHRU, Boulevard Alexandre Fleming, 25000 Besançon, France
[3] Department of Medicine, Division of Oncology, Stanford Cancer Institute, Stanford University, Stanford, CA 94305, USA
[4] INSERM LNC-UMR1231, University Bourgogne Franche-Comté, LabEx LipSTIC ANR-11-LABX-0021, Fondation de Coopération Scientifique Bourgogne Franche-Comté, 21000 Dijon, France; francois.hermetet@u-bourgogne.fr
*   Correspondence: christiane.mougin@univ-fcomte.fr; Tel.: +33-3-70-63-20-53

Academic Editors: Alison A. McBride and Karl Munger
Received: 7 July 2017; Accepted: 31 July 2017; Published: 7 August 2017

Abstract: Extracellular vesicles (EVs), including exosomes (Exos), microvesicles (MVs) and apoptotic bodies (ABs) are released in biofluids by virtually all living cells. Tumor-derived Exos and MVs are garnering increasing attention because of their ability to participate in cellular communication or transfer of bioactive molecules (mRNAs, microRNAs, DNA and proteins) between neighboring cancerous or normal cells, and to contribute to human cancer progression. Malignant traits can also be transferred from apoptotic cancer cells to phagocytizing cells, either professional or non-professional. In this review, we focus on Exos and ABs and their relationship with human papillomavirus (HPV)-associated tumor development. The potential implication of EVs as theranostic biomarkers is also addressed.

Keywords: exosomes; microvesicles; apoptotic bodies; papillomavirus; horizontal gene transfer; cell-to-cell communication; carcinogenesis

---

1. Introduction

Extracellular vesicles (EVs) including exosomes (Exos), microvesicles (MVs), and apoptotic bodies (ABs) have recently attracted great attention in cancer research. EVs take part in cell-to-cell communication between cancer cells and neighboring cells, including fibroblasts and endothelial and immune cells [1]. A growing body of evidence indicates that EVs can modulate tumor progression. Their pro-tumorigenic properties are fully described and reviewed elsewhere [2–7]. These properties are linked to direct effects through horizontal transfer of bioactive molecules (mRNAs, microRNAs (miRNAs), DNA or proteins) in recipient cells, and to indirect effects by remodeling the cancer microenvironment.

One of the most exciting applications of EV research in cancer is their potential use as biomarkers to monitor cancer progression by means of liquid biopsy, a non-invasive procedure [8]. In recent years, an exponential number of methods for minimally invasive early detection, diagnosis and follow-up of cancers has been developed using either circulating tumor cells (CTCs) or circulating tumor DNA (ctDNA). Circulating EVs from liquid biopsy may advantageously be used instead of CTCs or ctDNA. Indeed, they bear crucial protein and genetic material in their cargo and studying them may potentially provide information on the cellular origin of cancer. Circulating EVs are released virtually by all tumor cells as well as by the tumor microenvironment. Thus, the study of circulating EVs gives a snapshot of the heterogeneity of both the tumor and the microenvironment. The current limitation of liquid biopsy is the fragmentation of circulating DNA and the poor conservation of circulating RNA. Nucleic acids associated with EVs seem to be well conserved, perhaps shielded from degradation by nucleases inside the lipid bilayer membranes of the vesicles. This offers an opportunity for minimally invasive whole genome and/or whole transcriptome analysis from the blood of any cancer patient. Sequencing of both exosomal DNA and RNA may detect all driver and passenger mutations, translocations and amplifications which can be molecular targets for new drugs [3,9].

Few studies related to the role of EVs in virus transmission or carcinogenesis are available in the literature. Published reports describe how EVs modulate viral cell life, activate proliferation pathways, or modify the microenvironment to promote viral-associated carcinogenesis [10].

Although the role of EVs derived from human papillomavirus (HPV)-induced cancer cells is not yet completely understood, we review here recent data demonstrating their importance in tumorigenesis.

2. Extracellular Vesicles and Cancer

2.1. Biogenesis and Characterization of Extracellular Vesicles

"Extracellular vesicle" is a term given to a heterogeneous group of particles released by cells. A range of different names, referring to several types of EVs, can be found in the literature, such as exosomes, ectosomes, oncosomes, prostasomes, apoptotic bodies, shedding vesicles, microparticles or microvesicles amongst others, but no consensus on their classification has emerged thus far. However, based on the mechanisms of biogenesis, the scientific community classically distinguishes three main subgroups of EVs, namely exosomes, microvesicles and apoptotic bodies. Exos were first described in 1983 by Harding et al. [11]. These small vesicles, initially considered as a way to eliminate altered and/or unnecessary cellular constituents, originate from the invagination of endosome-forming multivesicular endosomes (MVEs). MVEs then fuse with the cell surface membrane to liberate Exos in the extracellular compartment. MVs refer to a subgroup of EVs produced by direct budding at the cell membrane (Figure 1). ABs are vesicles shed by the fragmentation of cells following the induction of apoptosis. The diameter of vesicles precludes specifically distinguishing different types of EVs from each other, but it is usually accepted that the size of Exos ranges mostly from 30 to 100 nm (approximately overlapping the size of virus particles), the size of MVs from 100 nm to 1 µm, and the size of ABs from 1 to 5 µm [12].

EVs derive from a large proportion of normal and cancer cell lines in vitro. In vivo, EVs can be isolated in most (if not all) biofluids, including plasma, urine, cerebrospinal fluid, saliva or breast milk and are produced in physiological and pathological conditions, especially by cancer cells. EV isolation protocols based on differential ultracentrifugation are probably the gold standard [13,14] and allow for the recovery of high amounts of EVs with intact physicochemical and functional properties. Highly pure EV preparations can be obtained with an additional step of density gradient ultracentrifugation [15]. Nevertheless, these protocols are time-consuming. In order to reduce the laboratory turnaround time, alternative EV isolation methods including commercial kits based on immunocapture, chromatography, microfluidics or polyethylene glycol precipitation have been developed by different research groups and companies in a very competitive market [16–19].

To date, international guidelines regarding the standardization of pre-analytical methods (specimen handling, isolation protocols) and analytical techniques to characterize EVs (electron microscopy, nanoparticle tracking analysis, proteomics) are not clearly defined and would be useful to choose the most relevant procedure for specific applications [20]. This is why the International Society for Extracellular Vesicles published recommendations defining the minimal requirements (e.g., biochemical, biophysical and functional standards of EVs) to conduct research on EVs [21]. Recently, a worldwide survey confirmed the great diversity of methods used by scientists working on EVs; it further pointed out that the EV field is rapidly evolving and that the standardization of practices remains challenging [22].

Of note, given the absence of specific markers of Exos or MVs, it remains difficult to clearly distinguish the origins of EVs after isolation. Thus, most of the published works on Exos probably relies on mixes of Exos and MVs and it should be kept in mind that the term "exosomes" generally refers to EVs or exosomes-enriched EVs. In the different sections below we will use the proper term EVs, except when referring to articles in which the authors used the term "Exos" instead of EVs.

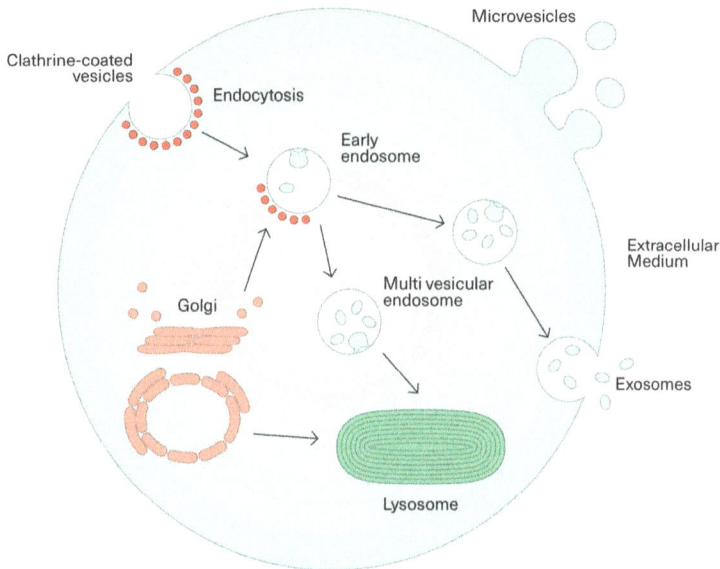

Figure 1. Exosome and microvesicle biogenesis. Exosomes originate from the inward budding of endosomes to form multivesicular endosomes (MVEs). MVEs then fuse with the cell surface membrane to release exosomes into the extracellular medium. Some MVEs fuse with lysosomes. Microvesicles are produced by direct budding from the plasma membrane.

2.2. Content and Functions of Extracellular Vesicles

EVs are packed with proteins, lipids and nucleic acids that mirror the parental cell-of-origin [23]. Studies have shown that Exos are specifically enriched in some biologically active molecules via a non-random process whose molecular mechanism remains unknown [24]. In particular, cytoskeletal proteins (actin, cofilin), adhesion molecules (tetraspanins, integrins), peptide-binding proteins such as the heat shock protein family proteins, major histocompatibility complex (MHC) class I and II, and membrane fusion proteins (*Ras* genes from rat brain (RAB) family proteins, annexins) are enriched in Exos and may serve as biomarkers. Interestingly, the exosomal nucleic acid content, mainly mRNA and miRNA, can constitute a specific molecular signature to characterize Exos. Some authors have reported that Exos contain high molecular weight double-stranded DNA. Using comparative genomic hybridization array [25] or, more recently, whole genome sequencing [26], it has been shown that exosomal DNA represents the entire genome of the parental cells. Moreover, in cancer patients, the exosomal DNA is enriched with tumor DNA [26]. From a structural point of view, one unsolved question concerns the association of nucleic acids with Exos. Indeed, to the best of our knowledge, no study has clearly demonstrated whether exosomal nucleic acids are located within vesicles, adsorbed onto the vesicle surface, or simply co-purified with vesicles as nucleic acid–protein complexes [12]. The methods used to isolate Exos are likely to influence their nucleic acids cargo.

Thus, the physiological and pathological functions of nucleic acids associated with Exos remain a matter of debate.

ABs represent the ultimate stage of apoptosis. Indeed, apoptosis is presented as a three-stage process: an initiation phase followed by an integration phase and finally an irreversible execution phase (reviewed in [27]). Specific ultrastructural, biochemical and mechanical modifications lead to AB formation [28,29]. Caspases activated by apoptotic signals constitute the main agents of cellular demolition through the cleavage of numerous cytoplasmic (e.g., constituents of the cell cytoskeleton) and nuclear (e.g., nuclear envelope) substrates. From a dynamic point of view, apoptotic cells undergo plasma membrane blebbing followed by karyorrhexis and formation of ABs which are eliminated by phagocytes. ABs present characteristic morphological features. They are delineated by a plasma membrane whose phosphatidylserines are located at the outer leaflet. The plasma membrane of ABs also exhibits intrinsic proteins, some of which serve as "eat me" signals recognized by receptors expressed on phagocytes. Furthermore, ABs contain altered mitochondria, fragmented organelles (Golgi apparatus, endoplasmic reticulum) and fragments of nucleus with typical internucleosomal DNA fragmentation [28,29].

EVs may be considered as vehicles for local and systemic cell-to-cell communication. Indeed, EVs are small messengers that interact with target cells, directly via cell surface molecules, or after endocytosis and internalization of bioactive molecules (Figure 2).

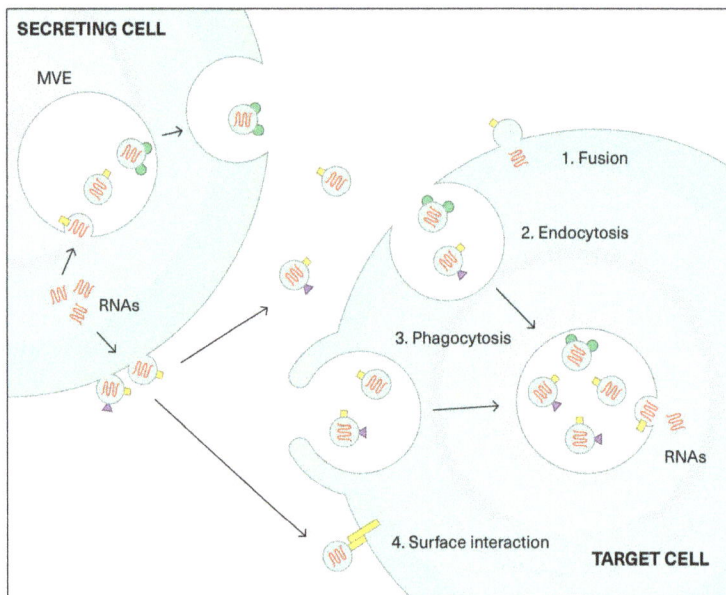

Figure 2. The four different modes of communication by exosomes and microvesicles. Extracellular vesicles serve as vehicles for cell-to-cell communication through horizontal transfer of bioactive molecules (proteins, lipids and nucleic acids). Extracellular vesicles (microvesicles (MVs) or exosomes (Exos)) produced from a secreting cell may be internalized by fusion (1), endocytosis (2), phagocytosis (3) or may interact with the membrane proteins of the target cell (4). Squares, triangles and circles represent membrane-associated proteins.

The function of EVs has long been studied, especially by immunologists because of their important role in antigen presentation. For instance, EVs have the potential to present antigens via MHC molecules and to activate cellular immune responses. However, efficient priming of naïve T cells seems to be restricted to Exos released from mature dendritic cells and requires the expression of intercellular adhesion molecule 1 [30]. Exos can also activate the host innate immunity. Indeed, Exos derived from cells infected by intracellular pathogens such as mycobacteria [31] or directly derived from pathogens such as *Leishmania* [32] can spread molecular patterns of infection leading to the modulation of the host immune responses.

Cell-to-cell communication also involves the transfer of biologically active molecules to target cells following the internalization of Exos. Valadi et al. showed that mRNA purified from mast cell Exos were functional as they could be translated in vitro [33]. More interestingly, when exosomal mRNAs isolated from mouse cells were delivered to human cells, the corresponding mouse proteins were efficiently expressed and detected in the recipient cells. Thus, it appears consistent that Exos are readily able to reprogram the phenotype of recipient cells via the transfer of RNA [12]. Nevertheless, numerous questions regarding processing, delivery and functions of EV-associated RNA remain unsolved and important considerations have to be properly addressed to refine understanding of RNA components [34].

Horizontal transfer, also called lateral transfer, of malignant traits from one mammalian cell to another involves cancer pathogenic effectors like DNA, mRNA, miRNA, lipids, proteins, transcription factors and cell-surface receptors circulating as free molecules or via tumor-derived MVs or Exos [35–39]. It can also occur through the clearance of apoptotic cancer cells, termed efferocytosis [40–43]. During cancer progression, many cells are lost through apoptosis [44]. Growth/survival factor depletion, hypoxia, loss of cell-matrix interactions and DNA damage are implicated in cell death [45]. Thereafter, professional phagocytic cells largely contribute to dead cell clearance by inactivating and degrading cellular components [46]. Over the last few decades, it has become clear that amateur phagocytes like epithelial cells [47–49] or fibroblasts [48,50] are able to engulf apoptotic corpses. Through this endocytic process, apoptotic cells can act as a vector of biomolecules which may confer a selective advantage on the recipient cell.

2.3. Pro-Tumorigenic Properties of Exosomes

Cancer-derived Exos can activate transforming signaling pathways in recipients cells by transferring oncogenic proteins such as mutated Kirsten rat sarcoma viral oncogene homolog (KRAS) proteins [51] or the epidermal growth factor receptor variant III [52]. This generally leads to enhanced growth of the target cells, sometimes in an anchorage-independent manner, and in this regard Exos participate in the horizontal transfer of oncogenes. Exos also have the ability to remodel the tumor microenvironment by fostering angiogenesis. Umezu et al. confirmed that multiple myeloma-derived Exos promote the angiogenic potential of endothelial cells. This occurs via the transfer of miR-135b that targets factor-inhibiting hypoxia-inducible factor 1 (HIF-1). Consequently, this releases HIF-1 which in turn can execute its transcriptional activity to maintain oxygen homeostasis [53]. Tumor-derived Exos were also shown to favor the expression of numerous angiogenic factors such as vascular endothelial growth factor (VEGF), VEGF receptor 2 or von Willebrand factor following their engulfment by endothelial cells. Furthermore, they are important mediators that induce the differentiation of fibroblasts into a "wound-healing" myofibroblastic phenotype that is associated with tumor growth [54]. Recently, Peinado et al. reported that Exos have a major role in the dissemination of tumor cells at distant sites. In particular, they showed that tumor-derived Exos educate bone marrow cells to prepare a pre-metastatic niche that will favor the recruitment and growth of metastasizing cancer cells [55,56]. Last but not least, cancer-derived Exos can modulate anti-tumor immune responses. Since they express tumor-specific antigens at their surface, Exos derived from cancer cells can elicit anti-tumor immune responses. Specifically they can transfer tumor antigens to dendritic cells which in turn induce efficient CD8+ T-cell-dependent antitumor effects [57]. Based on these observations, the concept of exosome-based anti-tumor vaccines has been proposed, but the translation of this into the clinic is tempered by the fact that cancer-derived Exos mostly present immunosuppressive effects. Indeed, cancer-derived EVs mediated both direct and indirect immunosuppressive effects notably through the modulation of inflammatory mediators. First, they harbor immunosuppressive molecules such as Fas-Ligand or tumor-necrosis-factor related apoptosis inducing ligand (TRAIL), check point receptor ligands (PD-L1) or inhibitory cytokines (interleukin-10, transforming growth factor-β1, and prostaglandin E2) [6]. They can induce the cell death of CD95+ lymphocytes [58], impair the activity of natural killer cells, and decrease the proliferative response of CD8+ T cells to interleukin-2 due to the proliferation of forkhead box P3 (Foxp-3) regulatory T cells [59]. Second EVs promote the proliferation of myeloid-derived suppressor cells, favor the polarization of macrophages toward a M2 phenotype, or inhibit the differentiation of lymphoid or myeloid progenitors (for review see [60]). Finally, tumor antigens at the surface of Exos may represent decoy targets for antibodies, thereby impairing antibody-dependent cell-mediated cytotoxicity [61] or limiting the bioavailability of therapeutic antibodies such as trastuzumab [62].

2.4. Horizontal Oncogene Transfer by Apoptotic Bodies

In 1999, de la Taille et al. revealed for the first time that tumor cells derived from a prostate adenocarcinoma (LNCaP) are able to exchange genetic information [63]. LNCaP$^{bcl-2/neo-r}$ cells highly resistant to in vitro apoptotic signals were cocultured with LNCaP^{hygr-r} cells sensitive to serum starvation apoptosis. Cocultures gave rise to progeny cells containing both genetic markers when they were exposed to an apoptotic stimulus that preferentially kills LNCaP^{hygr-r} cells. This provides evidence that genetic information can be transferred from apoptotic cells to viable partners through the phagocytosis process. The authors termed this latter event as an apoptotic conversion and put forward the hypothesis that this process has the potential to mediate the passage of genetic defects between cancer cells in a tumor, leading to the development of more aggressive cancer cell types after therapy of some cancer patients. In the same year, Holmgren et al. demonstrated that irradiation-induced ABs of Burkitt-derived lymphocytes harboring integrated Epstein-Barr virus (EBV) DNA can transfer viral sequences to professional and non-professional phagocytes such as endothelial cells and fibroblasts during co-cultivation [64]. Recipient human macrophages and bovine aortic endothelial cells expressed viral sequences (Epstein-Barr nuclear antigen 1, Epstein-Barr virus-encoded RNA 1 and 2) at a higher frequency than human fetal fibroblasts, which is likely related to their higher phagocytic activity. Interestingly, viral DNA fragments are preferentially transferred compared to human DNA. However, horizontal EBV DNA transfer was not observed when lymphocyte cell lines harboring episomal copies of EBV were cocultured with recipient cells. This suggests that viral episomes were more prone to degradation by apoptosis-activated nucleases than integrated DNA copies.

Two years later, Holmgren's team reported that DNA could be horizontally transferred from apoptotic c-myc and H-ras^{V12}-transfected rat embryonic fibroblasts (REF) to p53$^{-/-}$ mouse embryonic fibroblasts (MEF) resulting in loss of contact inhibition and anchorage-independence in vitro as well as a tumorigenic phenotype in severe combined immunodeficiency (SCID) mice [65]. When ABs derived from normal REF were exposed to recipient cells, none of the above effects was observed. Similarly, ABs were not capable of conferring the transformed characteristics of cancer cells onto recipient cells with intact p53. Thus, in phagocytic recipient cells, genomic instability results from the horizontal transfer of malignant traits and the inactivation of the *tp53* tumor suppressor gene [65]. Afterwards, Bergsmedh et al. showed that MEF cells lacking the p21$^{Cip1/Waf1}$ cyclin-kinase inhibitor are able to propagate DNA after the uptake of ABs derived from a rat fibrosarcoma, resulting in foci formation in vitro and tumor growth in SCID mice [66]. Taken together, the data from Holmgren's team indicate that p53 via the activation of its target p21 protects recipient cells from the propagation of potentially pathological exogenous DNA acquired from apoptotic cells [66]. This protective mechanism is dependent on DNA fragmentation and *DNase II* together with the *Chk2* DNA damage pathway, to form a protective barrier against horizontally transferred DNA [67].

More recently, Ehnfors et al. demonstrated that DNA from apoptotic rat fibrosarcoma cells expressing H-ras^{V12}, c-myc and simian virus SV40 large T (SV40LT) can be transferred to, and transform fibroblasts and endothelial cells in vitro [68]. Interestingly, tumor DNA can be spontaneously transferred to stromal murine host cells in vivo. A sub-population of endothelial cells expressing SV40LT and injected together with Matrigel into SCID mice form new functional blood vessels. Inactivation of the p53 pathway through the transfer of SV40LT might overcome the barrier to tumor DNA propagation [68].

3. Exosomes in Viral Transmission and Carcinogenesis Associated with Viruses Other than HPV

Studies agreed that EVs may enhance viral transmission and carcinogenesis in different models and this has been recently reviewed [69]. Ramakrishnaiah et al. showed that Exos derived from hepatitis C virus (HCV)-infected cells contain full-length viral RNA. Furthermore, these Exos were capable of delivering viral RNA into non-infected cells and establishing a productive infection [70]. This mode of transmission, which does not depend on the production of infectious virions, may be an efficient mechanism for the virus to escape the immune system since Exos derived from HCV-infected cells are not neutralized by immunoglobulin (Ig) Gs purified from HCV-infected patients.

Since Exos are constitutively produced by cells, it may be considered that their composition is altered following viral infection. To address this question, Meckes et al. investigated the protein composition of Exos derived from uninfected B cells or Epstein–Barr virus (EBV) or Kaposi sarcoma-associated herpesvirus (KSHV)-infected B cells [71]. Two hundred and thirty proteins were specifically expressed in virus-infected cells, 93 of them being specific to EBV-infected Exos and 22 to KSHV-infected Exos. In EBV-infected Exos, the main oncogenic viral protein latent membrane protein 1 (LMP1) was systematically detected, confirming a previous study conducted from a lymphoblastoid cell line infected by EBV [72]. LMP1-enriched Exos also exhibit high levels of other proteins that are associated with cancer and supposed to affect in particular survival/proliferation/death pathways, cell movement, cell-to-cell signaling as well as protein synthesis pathways and immune response [71,73–75]. Interestingly, Exos derived from both EBV and KSHV-infected cell lines also harbor the Ezrin protein, which is absent from the parental non-infected B cell line. This protein, a plasma membrane-microfilament linker involved in cellular architecture, was shown to activate the protein kinase B survival pathway in epithelial cells [76].

The involvement of LMP1-enriched Exos (also called LMP1$^+$Exos) in oncogenesis can be considered in several ways. First, LMP1$^+$Exos can remodel the tumor microenvironment by facilitating the secretion of the angiogenic fibroblast growth factor 2 [77]. Second, it has been shown, at least in vitro, that LMP1$^+$Exos are readily engulfed through the endocytic pathway [75] leading to the activation of several intracellular pathways in recipient cells. For example, the engulfment of LMP1$^+$Exos by B cells activates their proliferation and their differentiation in plasmablast-like cells [74]. As for nasopharyngeal carcinoma, EBV-derived Exos are readily able to induce an epithelial–mesenchymal transition to recipient cells. They also favor the metastatic potential of the recipient cells through HIF-1 α [73]. Moreover, EBV-modified Exos can play a role in immune evasion by inducing T cell anergy and protecting neighboring tumor cells from immune effector cells [72,78]. Furthermore, galectin-9-containing Exos derived from nasopharyngeal cancer cells induce the apoptosis of EBV-specific CD4+ T cells. The binding of galectin-9 to the death-inducing receptor anti-T-cell immunoglobulin and mucin-domain containing-3 (Tim-3) expressed by mature T helper (Th)1 cells appeared necessary given that anti-Tim-3 and antigalectin-9-blocking antibodies tempered the suppressive activity of Exos [79].

4. Extracellular Vesicles and HPV-Associated Carcinogenesis

Persistent infection with high risk (hr) human papillomaviruses (HPV) is the necessary cause of cervical cancer [80]. Among hrHPV, HPV16 and HPV18 are responsible for more than 70% of invasive cervical cancers in the world [81]. Expression of the viral E6 and E7 proteins is crucial for cellular immortalization, transformation and maintenance of the transformed phenotype [82]. E6 and E7 oncoproteins target many cellular growth regulatory circuits, among them the guardian of the genome protein p53 [83] and the growth-suppressive form of the retinoblastoma (Rb) tumor suppressor Rb

Protein (pRb) [84], respectively. Interestingly, silencing of E6/E7 in HeLa cervical carcinoma cells reactivates tumor suppressor pathways leading to cellular senescence [85,86].

4.1. E6/E7 and Anti-Apoptotic Exosomal Proteins

It has been previously shown that various cancer cell lines, among which HeLa, an HPV18-infected cell line derived from cervix adenocarcinoma, actively release Exos harboring survivin [87,88], a member of the inhibitory of apoptosis protein (IAP) family. Other members of the IAP protein family (cellular inhibitor of apoptosis protein (c-IAP), X-linked inhibitor of apoptosis protein1/2 (XIAP1/2)) were also identified in exosomal fractions released from HeLa cells, but only mRNA encoding survivin was detected [89]. In a model of HeLa cells stably transfected to overexpress survivin, it has been shown that a sublethal dose of proton irradiation significantly increases the survivin content of Exos, but not the secretory rate of Exos [88].

Since tumor virus infection modulates the composition of Exos, Honegger et al. determined whether E6/E7 could alter the amounts and/or contents of Exos released from HPV-infected cancer cells [90,91]. To address this question, the authors used a small interfering RNA (siRNA) approach to knock-down E6/E7, bearing in mind that a genome-wide transcriptome analysis previously showed that survivin mRNA was downregulated following E6/E7 repression [92]. In the experiments of Honegger et al., E6/E7 expression was effectively downregulated in HeLa cells by siRNA for up to 96 h and the authors confirmed that survivin expression was reduced. As for Exos, no major modification in their morphology was noted following E6/E7 downregulation but their number was largely increased when compared to mock-treated cells. Regarding the content of IAP family members, the relative level of survivin was lower in exosomal fractions derived from E6/E7 siRNA treated cells than from control cells. Nevertheless, the relative level of survivin remained higher in exosomal fractions than in whole cell extracts, indicating that survivin is specifically enriched in Exos, even if E6/E7 is downregulated. The authors also documented the presence of XIAP, c-IAP1 and livin alpha in Exos, raising the question of whether these vesicles could play a role in modulation of apoptosis.

In HPV-associated cancer models, it has been shown that conditioned media obtained from cells transfected with survivin construct present anti-apoptotic properties and protect HeLa cells from ultra-violet (UV)-induced apoptosis [87]. Thus, survivin liberated in the microenvironment via Exos may act as a protumorigenic factor [87–91]. Indeed, these Exos are assumed to be engulfed by neighboring cancers cells that in turn become more proliferative and resistant to therapy.

4.2. E6/E7 and miRNA Content of Exosomes

Honegger et al. investigated the exosomal miRNA content using a small RNA deep sequencing approach and compared it to the miRNA pattern expressed from Exo-producing HeLa cells whose E6/E7 expression was silenced [91]. They first observed an absence of 18S and 28S RNA in exosomal fractions confirming previous publications, but clearly identified transfer RNAs and miRNAs. Then, they documented that the relative levels of miRNAs among small RNAs were 50% less abundant in Exos than in exosome-producing cells. While E6/E7 silencing did not affect the intracellular content of miRNA in whole cells, it increased two-fold the relative miRNA content in Exos, suggesting that E6/E7 regulate exosomal miRNA sorting. Among the 47 most expressed exosomal miRNAs, 21 were upregulated and 4 were downregulated following E6/E7 silencing. The modulation of miRNA expression following E6/E7 siRNA treatment was further confirmed by real-time quantitative PCR (RT-qPCR) for approximately three-quarters of the targets and among them, let-7d-5p, miR-20a-5p, miR-378a-3p, miR-423-3p, miR-7-5p, miR-92a-3p were downregulated and miR-21-5p was upregulated. Interestingly, this signature of seven miRNAs was altered in a similar manner in Exos derived from E6/E7 siRNA-treated SiHa cells, a HPV16-infected cell line derived from cervix squamous cell carcinoma, suggesting that exosomal miRNA content is neither dependent on the type of tumor tissue nor on HPV genotype, but most likely on E6/E7 expression.

Very recently, Harden and Münger assessed whether HPV16 E6/E7 expression altered miR expression in Exos derived from human foreskin keratinocytes (HFKs). After E6/E7 retroviral transduction, both intracellular and Exo miRNA contents were examined and compared to non-transduced HFKs [93]. Forty-eight miRNAs were deregulated in E6/E7 HFKs and 31 were differentially expressed in Exos showing that E6/E7 oncoproteins alter miRNA composition of Exos. HPV16 E6/E7 expression modifies the expression of many miRNAs in a similar manner intracellularly and in Exos. However, seven miRNAs were either upregulated or downregulated in Exos when compared to whole cells: miR-16-5p, miR-21-5p, miR-200b-3p, miR-205-5p, miR-222-3p, miR-320a, miR-378-3p. Interestingly, the biological functions of miRNAs selectively packaged in Exos are predicted to inhibit apoptosis and necrosis [93]. These results are consistent with those published by Honegger et al. in their E6/E7 extinction model [91]. They also confirm the data published by Chiantore et al. who demonstrated that specific miRNAs were packaged in Exos derived from HFKs transduced with HPV16 E6/E7 [94].

Thus, tumor-promoting activity of Exos derived from hrHPV-positive cells might rely on the transfer of oncogenic miRNAs to recipient cells. Indeed, most of the miRNAs altered by E6/E7 display oncogenic activities. Among these miRNAs, miR-222 appears to be of particular interest since it was shown to play a role in cervical carcinogenesis, notably through the downregulation of p27 and phosphatase and tensin homolog deleted on chromosome 10 (PTEN) [95]. Moreover, the miR-7-5p favors cell proliferation [96], the miR-20a-5p blocks oncogene-induced senescence by targeting p21 [97], and miR-92a-3p possesses anti-apoptotic properties [98].

4.3. Exosome Release and Senescence Induction

E6/E7 siRNA-treated HeLa cells release higher amounts of Exos than control cells and become senescent but not apoptotic. Interestingly, the relationship between Exo release and senescence has also been reported for irradiated prostate cancer cells [99]. This secretory phenotype was clearly dependent on the reactivation of p53, which in turn activates the transcription of the tumor suppressor-activated pathway 6 (*TSAP6*) and the charge multivesicular body protein 4C (*CHMP4C*) encoding proteins involved in Exo production/sorting [100–102]. Honegger et al. investigated *TSAP6* and *CHMP4C* transcript levels in E6/E7 siRNA-treated HeLa cells [90]. Silencing endogenous HPV18 E6/E7 oncogene expression led to reconstitution of p53 linked to increased expression of *TSAP6* and *CHMP4C* contributing to senescence induction. While such a senescence should reduce tumor growth, it favors the release of a broad spectrum of molecules associated with inflammation and malignancy [103]. This is probably why senescence is associated with cancer progression in certain circumstances [104], and secretion of IAP family protein-enriched Exos (especially when E6/E7 are targeted in HPV-associated cancer) may participate in tumor development/progression.

4.4. Exosomes as Potential Biomarkers

Numerous studies have reported biological effects mediated by EVs. However, we should keep in mind that experimental procedures do not reflect physiological conditions. Therefore, the in vitro observations have to be cautiously interpreted. Nevertheless, given that Exos released from HPV-associated cancer cells present key features, they may serve as very specific cancer biomarkers [105]. For this purpose, analysis of Exos purified from cervico-vaginal lavage samples of patients with cervical cancer made it possible to demonstrate high levels of miR-21 and miR-146a [106] or long non-coding RNAs (especially metastasis associated lung adenocarcinoma transcript 1 (MALAT1) which is known to be activated by E6) [107]. Thus, like liquid biopsies taken from blood sampling [108,109], analysis of Exos from cervico-vaginal lavages may serve as a non-invasive procedure for early diagnosis and follow-up of patients with HPV-associated cancers.

4.5. HPV-Positive Apoptotic Bodies and Cell Transformation

In our laboratory, we investigated whether HPV16 and HPV18 might be capable of conferring transformed characteristics of apoptotic cancer cells onto normal recipient human fibroblasts, and we hypothesized that the horizontal transfer of HPV oncogenes could be an alternative mechanism of carcinogenesis [40]. We provided evidence that apoptotic wild-type p53 HeLa cells (harboring integrated HPV18) and Ca Ski cells (derived from a metastatic cervical epidermoid carcinoma harboring integrated HPV16) but not p53-mutated and HPV-negative C-33 A cells (derived from a rare HPV-negative cervical carcinoma) were able to transform human primary fibroblasts (HPFs) [40,42]. HPFs engulf late apoptotic cells more efficiently than early apoptotic cells, but their phagocytic activity remains limited compared to professional phagocytes [41,43]. The uptake of ABs requires phosphatidylserine recognition, which is mainly mediated by brain-specific angiogenesis inhibitor 1 (BAI-1). Microtubule-associated protein1-light chain 3 (LC3)-associated phagocytosis (LAP) is required for the clearance of dying cells, as demonstrated by confocal microscopy analysis with organelle-specific markers [41,43] and is well described in the review of Green et al. [110]. To the best of our knowledge, this is the first time that LAP has been observed in amateur phagocytes. However, its significance in the clearance of apoptotic corpses and possible altered intracellular processing, especially in the context of horizontal gene transfer (HGT), remains to be clarified. While there is accumulating data for a role of cancer-associated fibroblasts in cancer progression [111,112], studies are necessary to clarify the role of efferocytosis during HPV-induced tumor development, metastasis and chemotherapy-mediated tumor clearance.

HPFs with phagocytosed apoptotic HeLa and Ca Ski cells were able to form colonies when grown under anchorage-independent conditions [40–42], which is a major hallmark of cellular transformation [113]. This process was further tested by limit-dilution assays. Horizontal transfer of viral oncogenes in fibroblast recipients was confirmed using in situ hybridization with a probe hybridizing specifically with hrHPV DNA [40,42]. Moreover, the expression of E6 transcripts studied by RT-qPCR were detected in the transformed recipient cells with, however, lower levels than in the parental HeLa and Ca Ski cells. In addition, the levels of p53 and p21 in the host cells decreased substantially. Chromosomal rearrangements (loss and gain of alleles) were confirmed by studying microsatellite amplifications. As described by Holmgren, small and large DNA fragments were transferred [64]. Chromosomal instability and aneuploidy observed in our model are common characteristics of cancer cells that increase the acquisition of mutations conferring aggressive or drug-resistant phenotypes during cancer progression [114,115]. Our findings suggest that horizontal transfer of viral oncogenes may confer the transformed characteristics of cancer cells on fibroblasts and epithelial cells, which are major constituents of the tumor microenvironment [40–43]. Even though we did not highlight the whole complexity of cell transformation following engulfment of ABs, the inactivation of p53 through the transfer of HPV oncogenes likely disrupts the protective barrier against tumor development.

In 2012, Trejo-Becerril et al. [116] hypothesized that circulating oncogenic DNA, also called virtosome [117], spontaneously released by living cells, was able to foster tumor progression. Indeed, the authors confirmed that cancer progression is mediated by HGT when animals were injected in the dorsum with human SW480 colon carcinoma cells as a source of circulating oncogenic DNA. However, results were less meaningful when immunocompetent rats were intraperitoneally injected daily for 30 days with supernatant of HeLa cells or when intravenously injected every day for 60 days with ABs from HeLa cells, which were obtained after exposure to 75 µM cisplatin for 24 h. In this last condition, no clinical alteration was observed and no dysplastic or neoplastic cells were found in different organs. To provide evidence of nucleic acid transfer, PCR and RT-PCR targeting E6 and E7 of HPV18 were performed. Viral DNA was detected in the liver, spleen, colon, uterus, lung and kidney while viral mRNAs were expressed only in the liver. Ongoing research in our laboratory shows that apoptotic HeLa cells subcutaneously injected in the flank of Swiss nu/nu mice led to tumor formation at the injection site. Major organs (kidney, liver, lung, spleen) are currently under investigation.

Our findings shed important light on unconventional and poorly understood mechanisms of cell-to-cell communication and how that communication may have significant consequences in human HPV-associated cancer progression. Transfer of oncogenic HPV DNA sequences from cancer epithelial cells to fibroblasts makes it possible to explain the detection of HPV DNA in stromal cells of unusual undifferentiated cervix carcinomas [118].

5. Conclusions

There is increasing evidence in support of the pro-tumorigenic properties of Exos derived from cancer cells. These properties rely either on a direct effect of Exos by transferring oncogenes to recipient cells or on an indirect effect by remodeling the microenvironment. In virus-infected cancer cells, it is noteworthy that Exos composition is modified compared to non-infected cells. These EVs appear to be enriched with viral or cellular oncogenenic factors, such as LMP1, survivin or certain oncomiRs. Thus, viral-modified Exos might be involved in intercellular communication by delivering oncogenic molecules to the neighboring cells. On the other hand, they are armed to manipulate the tumor environment by favoring angiogenesis or exerting immunosuppressive effects, for example. ABs can also serve as efficient vectors for carrying oncogenic signals to normal cells. Indeed, they can transfer viral and cellular oncogenes to amateur phagocytes cells through HGT and induce their immortalization/transformation. Further investigations are necessary to better understand these alternative modes of oncogenesis mediated by EVs. Together, this will lead to an improved understanding of viral carcinogenesis and will offer new opportunities in cancer diagnostics and therapeutics [119].

Acknowledgments: The authors would like to thank Emilie Gaiffe for her excellent contribution in the execution of the Horizontal Gene Transfer study. The technical assistance of Sophie Launay is also much appreciated. The HGT work was supported by grant from the "Région de Franche-Comté" and the "Ligue Contre le Cancer (CCIR-GE)". The authors also acknowledge the "Comités Départementaux de la Ligue Contre le Cancer" (Besançon and Montbéliard—Doubs, Jura and Haute-Saône) and the "Cancéropôle Grand-Est" (CGE) for the financial support of the post-doctoral fellowship program of David Guenat. The funders had no role in study design, data collection and analysis, decision to publish or preparation of the manuscript. We would like also to thank Fiona Ecarnot for English language editing and Collectif MBC for designing Figures 1 and 2.

Conflicts of Interest: The authors declare no conflict of interest.

References

1. Lopatina, T.; Gai, C.; Deregibus, M.C.; Kholia, S.; Camussi, G. Cross Talk between Cancer and Mesenchymal Stem Cells through Extracellular Vesicles Carrying Nucleic Acids. *Front. Oncol.* **2016**, *6*, 125. [CrossRef] [PubMed]
2. Kaiser, J. Malignant messengers. *Science* **2016**, *352*, 164–166. [CrossRef] [PubMed]
3. Kalluri, R. The biology and function of exosomes in cancer. *J. Clin. Investig.* **2016**, *126*, 1208–1215. [CrossRef] [PubMed]
4. Verma, M.; Lam, T.K.; Hebert, E.; Divi, R.L. Extracellular vesicles: Potential applications in cancer diagnosis, prognosis, and epidemiology. *BMC Clin. Pathol.* **2015**, *15*, 6. [CrossRef] [PubMed]
5. Webber, J.; Yeung, V.; Clayton, A. Extracellular vesicles as modulators of the cancer microenvironment. *Semin. Cell Dev. Biol.* **2015**, *40*, 27–34. [CrossRef] [PubMed]
6. Whiteside, T.L. Tumor-Derived Exosomes and Their Role in Cancer Progression. *Adv. Clin. Chem.* **2016**, *74*, 103–141. [PubMed]
7. Zoller, M. Exosomes in Cancer Disease. *Methods Mol. Biol.* **2016**, *1381*, 111–149.
8. Becker, A.; Thakur, B.K.; Weiss, J.M.; Kim, H.S.; Peinado, H.; Lyden, D. Extracellular Vesicles in Cancer: Cell-to-Cell Mediators of Metastasis. *Cancer Cell* **2016**, *30*, 836–848. [CrossRef] [PubMed]
9. Whiteside, T.L. The potential of tumor-derived exosomes for noninvasive cancer monitoring. *Expert Rev. Mol. Diagn.* **2015**, *15*, 1293–1310. [CrossRef] [PubMed]
10. Meckes, D.G., Jr. Exosomal communication goes viral. *J. Virol.* **2015**, *89*, 5200–5203. [CrossRef] [PubMed]

11. Harding, C.V.; Heuser, J.E.; Stahl, P.D. Exosomes: Looking back three decades and into the future. *J. Cell Biol.* **2013**, *200*, 367–371. [CrossRef] [PubMed]
12. Raposo, G.; Stoorvogel, W. Extracellular vesicles: Exosomes, microvesicles, and friends. *J. Cell Biol.* **2013**, *200*, 373–383. [CrossRef] [PubMed]
13. Thery, C.; Zitvogel, L.; Amigorena, S. Exosomes: Composition, biogenesis and function. *Nat. Rev. Immunol.* **2002**, *2*, 569–579. [PubMed]
14. Thery, C.; Amigorena, S.; Raposo, G.; Clayton, A. Isolation and characterization of exosomes from cell culture supernatants and biological fluids. *Curr. Protoc. Cell Biol.* **2006**, *30*, 3.22.1–3.22.29.
15. Van Deun, J.; Mestdagh, P.; Sormunen, R.; Cocquyt, V.; Vermaelen, K.; Vandesompele, J.; Bracke, M.; De Wever, O.; Hendrix, A. The impact of disparate isolation methods for extracellular vesicles on downstream RNA profiling. *J. Extracell. Vesicles* **2014**, *3*, 24858. [CrossRef] [PubMed]
16. Baranyai, T.; Herczeg, K.; Onodi, Z.; Voszka, I.; Modos, K.; Marton, N.; Nagy, G.; Mager, I.; Wood, M.J.; El Andaloussi, S.; et al. Isolation of Exosomes from Blood Plasma: Qualitative and Quantitative Comparison of Ultracentrifugation and Size Exclusion Chromatography Methods. *PLoS ONE* **2015**, *10*, e0145686. [CrossRef] [PubMed]
17. Lane, R.E.; Korbie, D.; Anderson, W.; Vaidyanathan, R.; Trau, M. Analysis of exosome purification methods using a model liposome system and tunable-resistive pulse sensing. *Sci. Rep.* **2015**, *5*, 7639. [CrossRef] [PubMed]
18. Lobb, R.J.; Becker, M.; Wen, S.W.; Wong, C.S.; Wiegmans, A.P.; Leimgruber, A.; Moller, A. Optimized exosome isolation protocol for cell culture supernatant and human plasma. *J. Extracell. Vesicles* **2015**, *4*, 27031. [CrossRef] [PubMed]
19. Nordin, J.Z.; Lee, Y.; Vader, P.; Mager, I.; Johansson, H.J.; Heusermann, W.; Wiklander, O.P.; Hallbrink, M.; Seow, Y.; Bultema, J.J.; et al. Ultrafiltration with size-exclusion liquid chromatography for high yield isolation of extracellular vesicles preserving intact biophysical and functional properties. *Nanomedicine* **2015**, *11*, 879–883. [CrossRef] [PubMed]
20. Witwer, K.W.; Buzas, E.I.; Bemis, L.T.; Bora, A.; Lasser, C.; Lotvall, J.; Nolte-'t Hoen, E.N.; Piper, M.G.; Sivaraman, S.; Skog, J.; et al. Standardization of sample collection, isolation and analysis methods in extracellular vesicle research. *J. Extracell. Vesicles* **2013**, *2*, 20360. [CrossRef] [PubMed]
21. Lotvall, J.; Hill, A.F.; Hochberg, F.; Buzas, E.I.; Di Vizio, D.; Gardiner, C.; Gho, Y.S.; Kurochkin, I.V.; Mathivanan, S.; Quesenberry, P.; et al. Minimal experimental requirements for definition of extracellular vesicles and their functions: A position statement from the International Society for Extracellular Vesicles. *J. Extracell. Vesicles* **2014**, *3*, 26913. [CrossRef] [PubMed]
22. Gardiner, C.; Di Vizio, D.; Sahoo, S.; Thery, C.; Witwer, K.W.; Wauben, M.; Hill, A.F. Techniques used for the isolation and characterization of extracellular vesicles: Results of a worldwide survey. *J. Extracell. Vesicles* **2016**, *5*, 32945. [CrossRef] [PubMed]
23. Park, J.O.; Choi, D.Y.; Choi, D.S.; Kim, H.J.; Kang, J.W.; Jung, J.H.; Lee, J.H.; Kim, J.; Freeman, M.R.; Lee, K.Y.; et al. Identification and characterization of proteins isolated from microvesicles derived from human lung cancer pleural effusions. *Proteomics* **2013**, *13*, 2125–2134. [CrossRef] [PubMed]
24. Kowal, J.; Arras, G.; Colombo, M.; Jouve, M.; Morath, J.P.; Primdal-Bengtson, B.; Dingli, F.; Loew, D.; Tkach, M.; Thery, C. Proteomic comparison defines novel markers to characterize heterogeneous populations of extracellular vesicle subtypes. *Proc. Natl. Acad. Sci. USA* **2016**, *113*, E968–E977. [CrossRef] [PubMed]
25. Thakur, B.K.; Zhang, H.; Becker, A.; Matei, I.; Huang, Y.; Costa-Silva, B.; Zheng, Y.; Hoshino, A.; Brazier, H.; Xiang, J.; et al. Double-stranded DNA in exosomes: A novel biomarker in cancer detection. *Cell Res.* **2014**, *24*, 766–769. [CrossRef] [PubMed]
26. San Lucas, F.A.; Allenson, K.; Bernard, V.; Castillo, J.; Kim, D.U.; Ellis, K.; Ehli, E.A.; Davies, G.E.; Petersen, J.L.; Li, D.; et al. Minimally invasive genomic and transcriptomic profiling of visceral cancers by next-generation sequencing of circulating exosomes. *Ann. Oncol.* **2016**, *27*, 635–641. [CrossRef] [PubMed]
27. Eastman, A.; Rigas, J.R. Modulation of apoptosis signaling pathways and cell cycle regulation. *Semin. Oncol.* **1999**, *26*, 7–16. [PubMed]
28. Taylor, R.C.; Cullen, S.P.; Martin, S.J. Apoptosis: Controlled demolition at the cellular level. *Nat. Rev. Mol. Cell. Biol.* **2008**, *9*, 231–241. [CrossRef] [PubMed]

29. Zhang, B.; Li, L.; Li, Z.; Liu, Y.; Zhang, H.; Wang, J. Carbon Ion-Irradiated Hepatoma Cells Exhibit Coupling Interplay between Apoptotic Signaling and Morphological and Mechanical Remodeling. *Sci. Rep.* **2016**, *6*, 35131. [CrossRef] [PubMed]

30. Segura, E.; Nicco, C.; Lombard, B.; Veron, P.; Raposo, G.; Batteux, F.; Amigorena, S.; Thery, C. ICAM-1 on exosomes from mature dendritic cells is critical for efficient naive T-cell priming. *Blood* **2005**, *106*, 216–223. [CrossRef] [PubMed]

31. Bhatnagar, S.; Schorey, J.S. Exosomes released from infected macrophages contain *Mycobacterium avium* glycopeptidolipids and are proinflammatory. *J. Biol. Chem.* **2007**, *282*, 25779–25789. [CrossRef] [PubMed]

32. Silverman, J.M.; Clos, J.; Horakova, E.; Wang, A.Y.; Wiesgigl, M.; Kelly, I.; Lynn, M.A.; McMaster, W.R.; Foster, L.J.; Levings, M.K.; et al. Leishmania exosomes modulate innate and adaptive immune responses through effects on monocytes and dendritic cells. *J. Immunol.* **2010**, *185*, 5011–5022. [CrossRef] [PubMed]

33. Valadi, H.; Ekstrom, K.; Bossios, A.; Sjostrand, M.; Lee, J.J.; Lotvall, J.O. Exosome-mediated transfer of mRNAs and microRNAs is a novel mechanism of genetic exchange between cells. *Nat. Cell Biol.* **2007**, *9*, 654–659. [CrossRef] [PubMed]

34. Mateescu, B.; Kowal, E.J.; van Balkom, B.W.; Bartel, S.; Bhattacharyya, S.N.; Buzas, E.I.; Buck, A.H.; de Candia, P.; Chow, F.W.; Das, S.; et al. Obstacles and opportunities in the functional analysis of extracellular vesicle RNA—An ISEV position paper. *J. Extracell. Vesicles* **2017**, *6*, 1286095. [CrossRef] [PubMed]

35. Balaj, L.; Lessard, R.; Dai, L.; Cho, Y.J.; Pomeroy, S.L.; Breakefield, X.O.; Skog, J. Tumour microvesicles contain retrotransposon elements and amplified oncogene sequences. *Nat. Commun.* **2011**, *2*, 180. [CrossRef] [PubMed]

36. Fleischhacker, M.; Schmidt, B. Circulating nucleic acids (CNAs) and cancer—A survey. *Biochim. Biophys. Acta* **2007**, *1775*, 181–232. [CrossRef] [PubMed]

37. Grange, C.; Tapparo, M.; Collino, F.; Vitillo, L.; Damasco, C.; Deregibus, M.C.; Tetta, C.; Bussolati, B.; Camussi, G. Microvesicles released from human renal cancer stem cells stimulate angiogenesis and formation of lung premetastatic niche. *Cancer Res.* **2011**, *71*, 5346–5356. [CrossRef] [PubMed]

38. Hood, J.L.; San, R.S.; Wickline, S.A. Exosomes released by melanoma cells prepare sentinel lymph nodes for tumor metastasis. *Cancer Res.* **2011**, *71*, 3792–3801. [CrossRef] [PubMed]

39. Runz, S.; Keller, S.; Rupp, C.; Stoeck, A.; Issa, Y.; Koensgen, D.; Mustea, A.; Sehouli, J.; Kristiansen, G.; Altevogt, P. Malignant ascites-derived exosomes of ovarian carcinoma patients contain CD24 and EpCAM. *Gynecol. Oncol.* **2007**, *107*, 563–571. [CrossRef] [PubMed]

40. Gaiffe, E.; Pretet, J.L.; Launay, S.; Jacquin, E.; Saunier, M.; Hetzel, G.; Oudet, P.; Mougin, C. Apoptotic HPV positive cancer cells exhibit transforming properties. *PLoS ONE* **2012**, *7*, e36766. [CrossRef] [PubMed]

41. Hermetet, F.; Jacquin, E.; Launay, S.; Gaiffe, E.; Couturier, M.; Hirchaud, F.; Sandoz, P.; Pretet, J.L.; Mougin, C. Efferocytosis of apoptotic human papillomavirus-positive cervical cancer cells by human primary fibroblasts. *Biol. Cell* **2016**, *108*, 189–204. [CrossRef] [PubMed]

42. Gaiffe, E. Apoptotic Cells as Vectors of Viral Oncogenes: An Alternative Way of HPV-Associated Carcinogenesis. Ph.D. Thesis, University of Franche-Comté, Besançon, France, 2011.

43. Hermetet, F. Duality of Apoptosis of Cervical Cancer Cells or Hidden Face of Janus: A Therapeutic Objective and An Implication in Horizontal Viral Oncogene Transfer. Ph.D. Thesis, University of Franche-Comté, Besançon, France, 2015.

44. Kerr, J.F.; Searle, J. A mode of cell loss in malignant neoplasms. *J. Pathol.* **1972**, *106*, Pxi.

45. Lowe, S.W.; Lin, A.W. Apoptosis in cancer. *Carcinogenesis* **2000**, *21*, 485–495. [CrossRef] [PubMed]

46. Kroemer, G.; Petit, P.; Zamzami, N.; Vayssiere, J.L.; Mignotte, B. The biochemistry of programmed cell death. *FASEB J.* **1995**, *9*, 1277–1287. [PubMed]

47. Monks, J.; Rosner, D.; Geske, F.J.; Lehman, L.; Hanson, L.; Neville, M.C.; Fadok, V.A. Epithelial cells as phagocytes: Apoptotic epithelial cells are engulfed by mammary alveolar epithelial cells and repress inflammatory mediator release. *Cell Death Differ.* **2005**, *12*, 107–114. [CrossRef] [PubMed]

48. Parnaik, R.; Raff, M.C.; Scholes, J. Differences between the clearance of apoptotic cells by professional and non-professional phagocytes. *Curr. Biol.* **2000**, *10*, 857–860. [CrossRef]

49. Patel, V.A.; Lee, D.J.; Feng, L.; Antoni, A.; Lieberthal, W.; Schwartz, J.H.; Rauch, J.; Ucker, D.S.; Levine, J.S. Recognition of apoptotic cells by epithelial cells: Conserved versus tissue-specific signaling responses. *J. Biol. Chem.* **2010**, *285*, 1829–1840. [CrossRef] [PubMed]

50. Hall, S.E.; Savill, J.S.; Henson, P.M.; Haslett, C. Apoptotic neutrophils are phagocytosed by fibroblasts with participation of the fibroblast vitronectin receptor and involvement of a mannose/fucose-specific lectin. *J. Immunol.* **1994**, *153*, 3218–3227. [PubMed]

51. Demory Beckler, M.; Higginbotham, J.N.; Franklin, J.L.; Ham, A.J.; Halvey, P.J.; Imasuen, I.E.; Whitwell, C.; Li, M.; Liebler, D.C.; Coffey, R.J. Proteomic analysis of exosomes from mutant KRAS colon cancer cells identifies intercellular transfer of mutant KRAS. *Mol. Cell. Proteom.* **2013**, *12*, 343–355. [CrossRef] [PubMed]

52. Al-Nedawi, K.; Meehan, B.; Micallef, J.; Lhotak, V.; May, L.; Guha, A.; Rak, J. Intercellular transfer of the oncogenic receptor EGFRvIII by microvesicles derived from tumour cells. *Nat. Cell Biol.* **2008**, *10*, 619–624. [CrossRef] [PubMed]

53. Umezu, T.; Tadokoro, H.; Azuma, K.; Yoshizawa, S.; Ohyashiki, K.; Ohyashiki, J.H. Exosomal miR-135b shed from hypoxic multiple myeloma cells enhances angiogenesis by targeting factor-inhibiting HIF-1. *Blood* **2014**, *124*, 3748–3757. [CrossRef] [PubMed]

54. Webber, J.; Steadman, R.; Mason, M.D.; Tabi, Z.; Clayton, A. Cancer exosomes trigger fibroblast to myofibroblast differentiation. *Cancer Res.* **2010**, *70*, 9621–9630. [CrossRef] [PubMed]

55. Peinado, H.; Aleckovic, M.; Lavotshkin, S.; Matei, I.; Costa-Silva, B.; Moreno-Bueno, G.; Hergueta-Redondo, M.; Williams, C.; Garcia-Santos, G.; Ghajar, C.; et al. Melanoma exosomes educate bone marrow progenitor cells toward a pro-metastatic phenotype through MET. *Nat. Med.* **2012**, *18*, 883–891. [CrossRef] [PubMed]

56. Peinado, H.; Zhang, H.; Matei, I.R.; Costa-Silva, B.; Hoshino, A.; Rodrigues, G.; Psaila, B.; Kaplan, R.N.; Bromberg, J.F.; Kang, Y.; et al. Pre-metastatic niches: Organ-specific homes for metastases. *Nat. Rev. Cancer* **2017**, *17*, 302–317. [CrossRef] [PubMed]

57. Wolfers, J.; Lozier, A.; Raposo, G.; Regnault, A.; Thery, C.; Masurier, C.; Flament, C.; Pouzieux, S.; Faure, F.; Tursz, T.; et al. Tumor-derived exosomes are a source of shared tumor rejection antigens for CTL cross-priming. *Nat. Med.* **2001**, *7*, 297–303. [CrossRef] [PubMed]

58. Andreola, G.; Rivoltini, L.; Castelli, C.; Huber, V.; Perego, P.; Deho, P.; Squarcina, P.; Accornero, P.; Lozupone, F.; Lugini, L.; et al. Induction of lymphocyte apoptosis by tumor cell secretion of FasL-bearing microvesicles. *J. Exp. Med.* **2002**, *195*, 1303–1316. [CrossRef] [PubMed]

59. Clayton, A.; Mitchell, J.P.; Court, J.; Mason, M.D.; Tabi, Z. Human tumor-derived exosomes selectively impair lymphocyte responses to interleukin-2. *Cancer Res.* **2007**, *67*, 7458–7466. [CrossRef] [PubMed]

60. Whiteside, T.L. Exosomes and tumor-mediated immune suppression. *J. Clin. Investig.* **2016**, *126*, 1216–1223. [CrossRef] [PubMed]

61. Battke, C.; Ruiss, R.; Welsch, U.; Wimberger, P.; Lang, S.; Jochum, S.; Zeidler, R. Tumour exosomes inhibit binding of tumour-reactive antibodies to tumour cells and reduce ADCC. *Cancer Immunol. Immunother.* **2011**, *60*, 639–648. [CrossRef] [PubMed]

62. Ciravolo, V.; Huber, V.; Ghedini, G.C.; Venturelli, E.; Bianchi, F.; Campiglio, M.; Morelli, D.; Villa, A.; Della Mina, P.; Menard, S.; et al. Potential role of HER2-overexpressing exosomes in countering trastuzumab-based therapy. *J. Cell. Physiol.* **2012**, *227*, 658–667. [CrossRef] [PubMed]

63. De la Taille, A.; Chen, M.W.; Burchardt, M.; Chopin, D.K.; Buttyan, R. Apoptotic conversion: Evidence for exchange of genetic information between prostate cancer cells mediated by apoptosis. *Cancer Res.* **1999**, *59*, 5461–5463. [PubMed]

64. Holmgren, L.; Szeles, A.; Rajnavolgyi, E.; Folkman, J.; Klein, G.; Ernberg, I.; Falk, K.I. Horizontal transfer of DNA by the uptake of apoptotic bodies. *Blood* **1999**, *93*, 3956–3963. [CrossRef] [PubMed]

65. Bergsmedh, A.; Szeles, A.; Henriksson, M.; Bratt, A.; Folkman, M.J.; Spetz, A.L.; Holmgren, L. Horizontal transfer of oncogenes by uptake of apoptotic bodies. *Proc. Natl. Acad. Sci. USA* **2001**, *98*, 6407–6411. [CrossRef] [PubMed]

66. Bergsmedh, A.; Szeles, A.; Spetz, A.L.; Holmgren, L. Loss of the p21(Cip1/Waf1) cyclin kinase inhibitor results in propagation of horizontally transferred DNA. *Cancer Res.* **2002**, *62*, 575–579. [PubMed]

67. Bergsmedh, A.; Ehnfors, J.; Kawane, K.; Motoyama, N.; Nagata, S.; Holmgren, L. DNase II and the Chk2 DNA damage pathway form a genetic barrier blocking replication of horizontally transferred DNA. *Mol. Cancer Res.* **2006**, *4*, 187–195. [CrossRef] [PubMed]

68. Ehnfors, J.; Kost-Alimova, M.; Persson, N.L.; Bergsmedh, A.; Castro, J.; Levchenko-Tegnebratt, T.; Yang, L.; Panaretakis, T.; Holmgren, L. Horizontal transfer of tumor DNA to endothelial cells in vivo. *Cell Death Differ.* **2009**, *16*, 749–757. [CrossRef] [PubMed]

69. Alenquer, M.; Amorim, M.J. Exosome Biogenesis, Regulation, and Function in Viral Infection. *Viruses* **2015**, *7*, 5066–5083. [CrossRef] [PubMed]

70. Ramakrishnaiah, V.; Thumann, C.; Fofana, I.; Habersetzer, F.; Pan, Q.; de Ruiter, P.E.; Willemsen, R.; Demmers, J.A.; Stalin Raj, V.; Jenster, G.; et al. Exosome-mediated transmission of hepatitis C virus between human hepatoma Huh7.5 cells. *Proc. Natl. Acad. Sci. USA* **2013**, *110*, 13109–13113. [CrossRef] [PubMed]

71. Meckes, D.G., Jr.; Gunawardena, H.P.; Dekroon, R.M.; Heaton, P.R.; Edwards, R.H.; Ozgur, S.; Griffith, J.D.; Damania, B.; Raab-Traub, N. Modulation of B-cell exosome proteins by gamma herpesvirus infection. *Proc. Natl. Acad. Sci. USA* **2013**, *110*, E2925–E2933. [CrossRef] [PubMed]

72. Flanagan, J.; Middeldorp, J.; Sculley, T. Localization of the Epstein-Barr virus protein LMP 1 to exosomes. *J. Gen. Virol.* **2003**, *84*, 1871–1879. [CrossRef] [PubMed]

73. Aga, M.; Bentz, G.L.; Raffa, S.; Torrisi, M.R.; Kondo, S.; Wakisaka, N.; Yoshizaki, T.; Pagano, J.S.; Shackelford, J. Exosomal HIF1α supports invasive potential of nasopharyngeal carcinoma-associated LMP1-positive exosomes. *Oncogene* **2014**, *33*, 4613–4622. [CrossRef] [PubMed]

74. Gutzeit, C.; Nagy, N.; Gentile, M.; Lyberg, K.; Gumz, J.; Vallhov, H.; Puga, I.; Klein, E.; Gabrielsson, S.; Cerutti, A.; et al. Exosomes derived from Burkitt's lymphoma cell lines induce proliferation, differentiation, and class-switch recombination in B cells. *J. Immunol.* **2014**, *192*, 5852–5862. [CrossRef] [PubMed]

75. Nanbo, A.; Kawanishi, E.; Yoshida, R.; Yoshiyama, H. Exosomes derived from Epstein-Barr virus-infected cells are internalized via caveola-dependent endocytosis and promote phenotypic modulation in target cells. *J. Virol.* **2013**, *87*, 10334–10347. [CrossRef] [PubMed]

76. Gautreau, A.; Poullet, P.; Louvard, D.; Arpin, M. Ezrin, a plasma membrane-microfilament linker, signals cell survival through the phosphatidylinositol 3-kinase/Akt pathway. *Proc. Natl. Acad. Sci. USA* **1999**, *96*, 7300–7305. [CrossRef] [PubMed]

77. Ceccarelli, S.; Visco, V.; Raffa, S.; Wakisaka, N.; Pagano, J.S.; Torrisi, M.R. Epstein-Barr virus latent membrane protein 1 promotes concentration in multivesicular bodies of fibroblast growth factor 2 and its release through exosomes. *Int. J. Cancer* **2007**, *121*, 1494–1506. [CrossRef] [PubMed]

78. Keryer-Bibens, C.; Pioche-Durieu, C.; Villemant, C.; Souquere, S.; Nishi, N.; Hirashima, M.; Middeldorp, J.; Busson, P. Exosomes released by EBV-infected nasopharyngeal carcinoma cells convey the viral latent membrane protein 1 and the immunomodulatory protein galectin 9. *BMC Cancer* **2006**, *6*, 283. [CrossRef] [PubMed]

79. Klibi, J.; Niki, T.; Riedel, A.; Pioche-Durieu, C.; Souquere, S.; Rubinstein, E.; Le Moulec, S.; Guigay, J.; Hirashima, M.; Guemira, F.; et al. Blood diffusion and Th1-suppressive effects of galectin-9-containing exosomes released by Epstein-Barr virus-infected nasopharyngeal carcinoma cells. *Blood* **2009**, *113*, 1957–1966. [CrossRef] [PubMed]

80. Walboomers, J.M.; Jacobs, M.V.; Manos, M.M.; Bosch, F.X.; Kummer, J.A.; Shah, K.V.; Snijders, P.J.; Peto, J.; Meijer, C.J.; Munoz, N. Human papillomavirus is a necessary cause of invasive cervical cancer worldwide. *J. Pathol.* **1999**, *189*, 12–19. [CrossRef]

81. Li, N.; Franceschi, S.; Howell-Jones, R.; Snijders, P.J.; Clifford, G.M. Human papillomavirus type distribution in 30,848 invasive cervical cancers worldwide: Variation by geographical region, histological type and year of publication. *Int. J. Cancer* **2011**, *128*, 927–935. [CrossRef] [PubMed]

82. Zur Hausen, H. Papillomaviruses and cancer: From basic studies to clinical application. *Nat. Rev. Cancer* **2002**, *2*, 342–350. [CrossRef] [PubMed]

83. Scheffner, M.; Werness, B.A.; Huibregtse, J.M.; Levine, A.J.; Howley, P.M. The E6 oncoprotein encoded by human papillomavirus types 16 and 18 promotes the degradation of p53. *Cell* **1990**, *63*, 1129–1136. [CrossRef]

84. Heck, D.V.; Yee, C.L.; Howley, P.M.; Munger, K. Efficiency of binding the retinoblastoma protein correlates with the transforming capacity of the E7 oncoproteins of the human papillomaviruses. *Proc. Natl. Acad. Sci. USA* **1992**, *89*, 4442–4446. [CrossRef] [PubMed]

85. Goodwin, E.C.; DiMaio, D. Repression of human papillomavirus oncogenes in HeLa cervical carcinoma cells causes the orderly reactivation of dormant tumor suppressor pathways. *Proc. Natl. Acad. Sci. USA* **2000**, *97*, 12513–12518. [CrossRef] [PubMed]

86. Goodwin, E.C.; Yang, E.; Lee, C.J.; Lee, H.W.; DiMaio, D.; Hwang, E.S. Rapid induction of senescence in human cervical carcinoma cells. *Proc. Natl. Acad. Sci. USA* **2000**, *97*, 10978–10983. [CrossRef] [PubMed]

87. Khan, S.; Aspe, J.R.; Asumen, M.G.; Almaguel, F.; Odumosu, O.; Acevedo-Martinez, S.; De Leon, M.; Langridge, W.H.; Wall, N.R. Extracellular, cell-permeable survivin inhibits apoptosis while promoting proliferative and metastatic potential. *Br. J. Cancer* **2009**, *100*, 1073–1086. [CrossRef] [PubMed]

88. Khan, S.; Jutzy, J.M.; Aspe, J.R.; McGregor, D.W.; Neidigh, J.W.; Wall, N.R. Survivin is released from cancer cells via exosomes. *Apoptosis* **2011**, *16*, 1–12. [CrossRef] [PubMed]

89. Valenzuela, M.M.; Ferguson Bennit, H.R.; Gonda, A.; Diaz Osterman, C.J.; Hibma, A.; Khan, S.; Wall, N.R. Exosomes Secreted from Human Cancer Cell Lines Contain Inhibitors of Apoptosis (IAP). *Cancer Microenviron.* **2015**, *8*, 65–73. [CrossRef] [PubMed]

90. Honegger, A.; Leitz, J.; Bulkescher, J.; Hoppe-Seyler, K.; Hoppe-Seyler, F. Silencing of human papillomavirus (HPV) E6/E7 oncogene expression affects both the contents and the amounts of extracellular microvesicles released from HPV-positive cancer cells. *Int. J. Cancer* **2013**, *133*, 1631–1642. [CrossRef] [PubMed]

91. Honegger, A.; Schilling, D.; Bastian, S.; Sponagel, J.; Kuryshev, V.; Sultmann, H.; Scheffner, M.; Hoppe-Seyler, K.; Hoppe-Seyler, F. Dependence of intracellular and exosomal microRNAs on viral E6/E7 oncogene expression in HPV-positive tumor cells. *PLoS Pathog.* **2015**, *11*, e1004712. [CrossRef] [PubMed]

92. Kuner, R.; Vogt, M.; Sultmann, H.; Buness, A.; Dymalla, S.; Bulkescher, J.; Fellmann, M.; Butz, K.; Poustka, A.; Hoppe-Seyler, F. Identification of cellular targets for the human papillomavirus E6 and E7 oncogenes by RNA interference and transcriptome analyses. *J. Mol. Med.* **2007**, *85*, 1253–1262. [CrossRef] [PubMed]

93. Harden, M.E.; Munger, K. Human papillomavirus 16 E6 and E7 oncoprotein expression alters microRNA expression in extracellular vesicles. *Virology* **2017**, *508*, 63–69. [CrossRef] [PubMed]

94. Chiantore, M.V.; Mangino, G.; Iuliano, M.; Zangrillo, M.S.; De Lillis, I.; Vaccari, G.; Accardi, R.; Tommasino, M.; Columba Cabezas, S.; Federico, M.; et al. Human papillomavirus E6 and E7 oncoproteins affect the expression of cancer-related microRNAs: Additional evidence in HPV-induced tumorigenesis. *J. Cancer Res. Clin. Oncol.* **2016**, *142*, 1751–1763. [CrossRef] [PubMed]

95. Sun, Y.; Zhang, B.; Cheng, J.; Wu, Y.; Xing, F.; Wang, Y.; Wang, Q.; Qiu, J. MicroRNA-222 promotes the proliferation and migration of cervical cancer cells. *Clin. Investig. Med.* **2014**, *37*, E131. [CrossRef]

96. Chou, Y.T.; Lin, H.H.; Lien, Y.C.; Wang, Y.H.; Hong, C.F.; Kao, Y.R.; Lin, S.C.; Chang, Y.C.; Lin, S.Y.; Chen, S.J.; et al. EGFR promotes lung tumorigenesis by activating miR-7 through a Ras/ERK/Myc pathway that targets the Ets2 transcriptional repressor ERF. *Cancer Res.* **2010**, *70*, 8822–8831. [CrossRef] [PubMed]

97. Hong, L.; Lai, M.; Chen, M.; Xie, C.; Liao, R.; Kang, Y.J.; Xiao, C.; Hu, W.Y.; Han, J.; Sun, P. The miR-17-92 cluster of microRNAs confers tumorigenicity by inhibiting oncogene-induced senescence. *Cancer Res.* **2010**, *70*, 8547–8557. [CrossRef] [PubMed]

98. Li, M.; Guan, X.; Sun, Y.; Mi, J.; Shu, X.; Liu, F.; Li, C. miR-92a family and their target genes in tumorigenesis and metastasis. *Exp. Cell Res.* **2014**, *323*, 1–6. [CrossRef] [PubMed]

99. Lehmann, B.D.; Paine, M.S.; Brooks, A.M.; McCubrey, J.A.; Renegar, R.H.; Wang, R.; Terrian, D.M. Senescence-associated exosome release from human prostate cancer cells. *Cancer Res.* **2008**, *68*, 7864–7871. [CrossRef] [PubMed]

100. Lespagnol, A.; Duflaut, D.; Beekman, C.; Blanc, L.; Fiucci, G.; Marine, J.C.; Vidal, M.; Amson, R.; Telerman, A. Exosome secretion, including the DNA damage-induced p53-dependent secretory pathway, is severely compromised in TSAP6/Steap3-null mice. *Cell Death Differ.* **2008**, *15*, 1723–1733. [CrossRef] [PubMed]

101. Yu, X.; Harris, S.L.; Levine, A.J. The regulation of exosome secretion: A novel function of the p53 protein. *Cancer Res.* **2006**, *66*, 4795–4801. [CrossRef] [PubMed]

102. Yu, X.; Riley, T.; Levine, A.J. The regulation of the endosomal compartment by p53 the tumor suppressor gene. *FEBS J.* **2009**, *276*, 2201–2212. [CrossRef] [PubMed]

103. Coppe, J.P.; Patil, C.K.; Rodier, F.; Sun, Y.; Munoz, D.P.; Goldstein, J.; Nelson, P.S.; Desprez, P.Y.; Campisi, J. Senescence-associated secretory phenotypes reveal cell-nonautonomous functions of oncogenic RAS and the p53 tumor suppressor. *PLoS Biol.* **2008**, *6*, 2853–2868. [CrossRef] [PubMed]

104. Perez-Mancera, P.A.; Young, A.R.; Narita, M. Inside and out: The activities of senescence in cancer. *Nat. Rev. Cancer* **2014**, *14*, 547–558. [CrossRef] [PubMed]

105. Khan, S.; Jutzy, J.M.; Valenzuela, M.M.; Turay, D.; Aspe, J.R.; Ashok, A.; Mirshahidi, S.; Mercola, D.; Lilly, M.B.; Wall, N.R. Plasma-derived exosomal survivin, a plausible biomarker for early detection of prostate cancer. *PLoS ONE* **2012**, *7*, e46737. [CrossRef] [PubMed]

106. Liu, J.; Sun, H.; Wang, X.; Yu, Q.; Li, S.; Yu, X.; Gong, W. Increased exosomal microRNA-21 and microRNA-146a levels in the cervicovaginal lavage specimens of patients with cervical cancer. *Int. J. Mol. Sci.* **2014**, *15*, 758–773. [CrossRef] [PubMed]

107. Zhang, J.; Liu, S.C.; Luo, X.H.; Tao, G.X.; Guan, M.; Yuan, H.; Hu, D.K. Exosomal Long Noncoding RNAs are Differentially Expressed in the Cervicovaginal Lavage Samples of Cervical Cancer Patients. *J. Clin. Lab. Anal.* **2016**, *30*, 1116–1121. [CrossRef] [PubMed]

108. Campitelli, M.; Jeannot, E.; Peter, M.; Lappartient, E.; Saada, S.; de la Rochefordiere, A.; Fourchotte, V.; Alran, S.; Petrow, P.; Cottu, P.; et al. Human papillomavirus mutational insertion: Specific marker of circulating tumor DNA in cervical cancer patients. *PLoS ONE* **2012**, *7*, e43393. [CrossRef] [PubMed]

109. Jeannot, E.; Becette, V.; Campitelli, M.; Calmejane, M.A.; Lappartient, E.; Ruff, E.; Saada, S.; Holmes, A.; Bellet, D.; Sastre-Garau, X. Circulating human papillomavirus DNA detected using droplet digital PCR in the serum of patients diagnosed with early stage human papillomavirus-associated invasive carcinoma. *J. Pathol. Clin. Res.* **2016**, *2*, 201–209. [CrossRef] [PubMed]

110. Green, D.R.; Oguin, T.H.; Martinez, J. The clearance of dying cells: Table for two. *Cell Death Differ.* **2016**, *23*, 915–926. [CrossRef] [PubMed]

111. Harper, J.; Sainson, R.C. Regulation of the anti-tumour immune response by cancer-associated fibroblasts. *Semin. Cancer Biol.* **2014**, *25*, 69–77. [CrossRef] [PubMed]

112. Karagiannis, G.S.; Poutahidis, T.; Erdman, S.E.; Kirsch, R.; Riddell, R.H.; Diamandis, E.P. Cancer-associated fibroblasts drive the progression of metastasis through both paracrine and mechanical pressure on cancer tissue. *Mol. Cancer Res.* **2012**, *10*, 1403–1418. [CrossRef] [PubMed]

113. Colburn, N.H.; Bruegge, W.F.; Bates, J.R.; Gray, R.H.; Rossen, J.D.; Kelsey, W.H.; Shimada, T. Correlation of anchorage-independent growth with tumorigenicity of chemically transformed mouse epidermal cells. *Cancer Res.* **1978**, *38*, 624–634. [PubMed]

114. Rajagopalan, H.; Lengauer, C. Aneuploidy and cancer. *Nature* **2004**, *432*, 338–341. [CrossRef] [PubMed]

115. Giam, M.; Rancati, G. Aneuploidy and chromosomal instability in cancer: A jackpot to chaos. *Cell Div.* **2015**, *10*, 3. [CrossRef] [PubMed]

116. Trejo-Becerril, C.; Perez-Cardenas, E.; Taja-Chayeb, L.; Anker, P.; Herrera-Goepfert, R.; Medina-Velazquez, L.A.; Hidalgo-Miranda, A.; Perez-Montiel, D.; Chavez-Blanco, A.; Cruz-Velazquez, J.; et al. Cancer progression mediated by horizontal gene transfer in an in vivo model. *PLoS ONE* **2012**, *7*, e52754. [CrossRef] [PubMed]

117. Gahan, P.B.; Stroun, M. The virtosome-a novel cytosolic informative entity and intercellular messenger. *Cell Biochem. Funct.* **2010**, *28*, 529–538. [CrossRef] [PubMed]

118. Unger, E.R.; Vernon, S.D.; Hewan-Lowe, K.O.; Lee, D.R.; Thoms, W.W.; Reeves, W.C. An unusual cervical carcinoma showing exception to epitheliotropism of human papillomavirus. *Hum. Pathol.* **1999**, *30*, 483–485. [CrossRef]

119. Wendler, F.; Favicchio, R.; Simon, T.; Alifrangis, C.; Stebbing, J.; Giamas, G. Extracellular vesicles swarm the cancer microenvironment: From tumor-stroma communication to drug intervention. *Oncogene* **2017**, *36*, 877–884. [CrossRef] [PubMed]

Review

Risk of Human Papillomavirus Infection in Cancer-Prone Individuals: What We Know

Ruby Khoury [1], Sharon Sauter [2], Melinda Butsch Kovacic [2], Adam S. Nelson [1],
Kasiani C. Myers [1], Parinda A. Mehta [1], Stella M. Davies [1] and Susanne I. Wells [3],*

[1]   Divisions of Bone Marrow Transplantation and Immune Deficiency, Cincinnati Children's Hospital Medical
      Center, Cincinnati, OH 45229, USA; Ruby.khoury@cchmc.org (R.K.); adam.nelson@cchmc.org (A.S.N.);
      kasiani.myers@cchmc.org (K.C.M.); parinda.mehta@cchmc.org (P.A.M.); stella.davies@cchmc.org (S.M.D.)
[2]   Divisions of Asthma Research, Cincinnati Children's Hospital Medical Center, Cincinnati, OH 45229, USA;
      sharon.sauter@cchmc.org (S.S.); Melinda.Butsch.Kovacic@cchmc.org (M.B.K.)
[3]   Divisions of Oncology, Cincinnati Children's Hospital Medical Center, Cincinnati, OH 45229, USA
*   Correspondence: susanne.wells@cchmc.org

Received: 10 December 2017; Accepted: 16 January 2018; Published: 20 January 2018

Abstract: Human papillomavirus (HPV) infections cause a significant proportion of cancers worldwide, predominantly squamous cell carcinomas (SCC) of the mucosas and skin. High-risk HPV types are associated with SCCs of the anogenital and oropharyngeal tract. HPV oncogene activities and the biology of SCCs have been intensely studied in laboratory models and humans. What remains largely unknown are host tissue and immune-related factors that determine an individual's susceptibility to infection and/or carcinogenesis. Such susceptibility factors could serve to identify those at greatest risk and spark individually tailored HPV and SCC prevention efforts. Fanconi anemia (FA) is an inherited DNA repair disorder that is in part characterized by extreme susceptibility to SCCs. An increased prevalence of HPV has been reported in affected individuals, and molecular and functional connections between FA, SCC, and HPV were established in laboratory models. However, the presence of HPV in some human FA tumors is controversial, and the extent of the etiological connections remains to be established. Herein, we discuss cellular, immunological, and phenotypic features of FA, placed into the context of HPV pathogenesis. The goal is to highlight this orphan disease as a unique model system to uncover host genetic and molecular HPV features, as well as SCC susceptibility factors.

Keywords: human papillomavirus; Fanconi anemia; orphan disease; squamous cell carcinoma; inherited cancer susceptibility

---

1. HPV Infection, a Significant Threat to Public Health

Human Papillomaviruses (HPVs) are a group of more than 150 related viruses which are known to cause 5% of all human cancers by infecting keratinocytes in the skin and mucosas [1–3]. Cutaneous HPV types are widely present in the normal human skin and may contribute to skin cancer in cooperation with UV radiation [4]. Mucosal HPV types are the most common sexually transmitted viruses in the US [5]. There are two major HPV subtypes: high-risk types (e.g., HPV16 and 18) which cause anogenital and oropharyngeal squamous cell carcinomas (SCCs), and low-risk types (eg, HPV6 and 11) which cause genital warts and recurrent respiratory papilloma. Although most individuals infected with HPV clear the virus, a minority harbor persistent high-risk HPV which puts them at risk of HPV-associated diseases, including cancer. HPV-related health-care costs are exorbitant and, despite three Food and Drug Administration (FDA)-approved HPV vaccines, are likely to remain high for decades for a number of reasons: (a) individuals who are already infected are likely not protected by the vaccine for that particular subtype, (b) vaccine coverage is poor—in part due to racial, ethnic, and income disparities, and (c) a long latency separates infection from tumorigenesis [1,2,6–8]. Furthermore, effective cures for the infection are not available. While HPV-specific antivirals are greatly needed, their design has been hindered by a paucity of information about the viral life cycle—the molecular interactions between HPV and the heterogeneous keratinocytes that permit infection and induce and sustain virus amplification in the human epidermis [9–12].

The human epidermis is composed of four distinct layers. The deepest, basal layer contains epidermal stem and progenitor cells (ESPCs), keratinocytes whose proliferative activity is tightly regulated to regenerate the cells that are shed from the surface. Keratinocyte connectivity to the basement membrane and the underlying dermis occurs via hemidesmosomes. Despite that connectivity, the ESPC progeny exit the basal layer continuously and migrate outward through the suprabasal spinous, granular, and cornified layers, where they reside as terminally differentiated keratinocytes. Granular and cornified keratinocytes form an impermeable barrier to resist physical, chemical, and infectious insults. For example, the disruption of tight junctions by two Human Immunodeficiency Virus (HIV) proteins promotes HPV pseudovirus penetration, thus linking cellular adhesion defects to HPV infection [13]. Overall cell–cell connectivity is crucial for adherence and communication in the epidermis and is provided by intercellular junctions: desmosomes, adherence, gap, and tight junctions. Desmosomes are abundant throughout the epidermis and anchor to the intermediate filament cytoskeleton to ensure proper architecture and mechanical resistance. Desmosomes also function as signaling centers, regulating fundamental processes such as proliferation, migration, and morphogenesis [14–16]. In the cornified layer, desmosomes mature into corneodesmosomes, key components of the epidermal barrier function together with the tight junctions in the granular layer [17].

HPV virions contain circular double-stranded DNA genomes of approximately 8 kbp within an icosahedral capsid. For many high risk HPV types, the viral genome encodes six early (E) nonstructural, and two late capsid proteins (L1, L2). HPV gene expression is complex, involving a synchronization of transcription, mRNA stability, splicing, and polyadenylation with keratinocyte differentiation and distinct phases of the viral life cycle [18,19]. In the initial phase of the HPV life cycle, basal keratinocytes are infected by the virus that has permeated the above epidermal barrier through microwounds. The second phase is characterized by a short burst of viral genome amplification, followed by maintenance at 50–200 copies per cell. Upon upward migration and differentiation, the third phase, i.e., genome amplification, produces thousands of HPV copies in a fraction of keratinocytes in the granular and cornified layers. L1 and L2 expression and virus assembly are limited to these terminally differentiated layers. The switch from phase 2 (viral maintenance) to phase 3 (productive replication) is poorly understood, but is key to HPV disease and transmission, as it leads to the production of viral progeny that will infect the same or a different host. Therefore, the identification of regulators of HPV infection and amplification is a critical step towards new approaches to attenuate viral production. This review highlights how a DNA repair pathway, the Fanconi anemia pathway, interacts with HPV at a molecular, cellular, and epidemiological level and can help further our understanding of HPV pathogenesis and host response to identify potential targets for novel therapeutic interventions. As most of the studies cited below tested high-risk HPV subtypes, from this point on "HPV" will be used to refer to high-risk HPV, unless otherwise indicated.

2. Fanconi Anemia, an Inherited DNA Repair Syndrome

Fanconi anemia (FA) is an inherited autosomal recessive, rarely X-linked or dominant negative, disorder characterized by congenital abnormalities, progressive bone marrow failure, and a predisposition to malignancies. FA arises from a germline loss of function of any one of 22 known FA genes involved in DNA repair. This pathway specializes in the repair of interstrand cross-links (ICL) (Figure 1), but is also implicated in stabilizing stalled replication forks, ensuring proper cytokinesis, and suppressing nonhomologous end-joining [20–24]. ICLs occur endogenously during the S phase or upon exposure to exogenous crosslinkers, such as chemotherapeutics, aldehydes, or other breakdown products from alcohol, tobacco, and dietary fats [25]. ICLs block transcription and replication near the lesion, and therefore their continuous removal is required for sustained cellular survival. In response to ICL formation, the protein products of eight FA genes assemble to form the nuclear "FA core complex" at the lesion, trigger the monoubiquitination of Fanconi Anemia Complementation Group (FANC) D2 (FANCD2) and FANCI, and then activate the FANCD2/FANCI dimer that orchestrates the coordinated recruitment of downstream proteins to repair the lesion. These include nucleolytic processing proteins, translesion polymerases, and homologous recombination proteins [26]. Two DNA damage sensor kinases, ataxia-telangectasia mutated (ATM) and Rad-3 related (ATR), play intricate roles in this process by phosphorylating and activating a number of FA proteins. Once the repair is completed, the deubiquitination of the FANCD2/FANCI complex by ubiquitin specific peptidase 1 USP1 and USP1- associated factor (USP1-UAF1) leads to complex release from chromatin and to pathway reset. Loss of the FA pathway results in the accumulation of ICLs and in global genome instability. The intact HPV genome is an episome [27]. However, HPV integration into the host genome and sustained expression of the two oncogenes E6 and E7 can also lead to genome instability, but additionally inactivate members of the p53 and Retinoblastoma (RB) tumor suppressor families, playing an important role in inducing carcinogenesis [28–30]. Genomic instability and accumulation of mutations in the context of the FA DNA-repair pathway dysfunction are likely contributors to FA patients' susceptibility to HPV-independent and HPV-associated SCCs. Below, we summarize the clinical features of FA, the epidemiology of HPV in FA, and cellular mechanisms underlying altered HPV biology and perhaps HPV-associated SCC in FA.

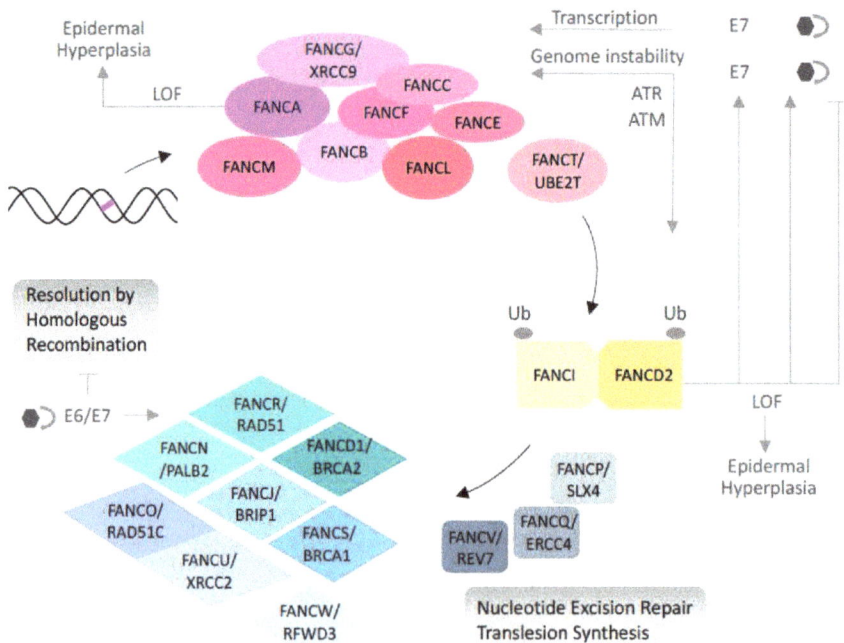

Figure 1. Only the classical role of the Fanconi anemia (FA) pathway in the repair of interstrand cross-links (ICLs) is depicted, which involves highly coordinated protein–DNA and protein–protein interactions in the nucleus. Other reported functions of the FA pathway in replication stabilization, origin firing, and cytokinesis are not shown, but are reviewed in detail in [22]. In response to ICL formation, the FA core complex (pink circles), including Fanconi Anemia Complementation Group (FANC) A (FANCA), assembles near the lesion, and this activation drives the ubiquitination, by FANCL,

of two central pathway components FANCD2 and FANCI (yellow squares), and their recruitment. These initial steps require the activity of the ataxia-telangectasia mutated (ATM) and ataxia-telangectasia and Rad-3 related (ATR) DNA damage sensor kinases. The formation of the central "ID" complex is followed by the activity of endonucleases and translesion polymerases (grey squares) for ICL unhooking and bypass. Finally, the recruitment of homologous recombination proteins (blue diamonds) completes the repair process. Some examples of HPV interactions with the FA pathway are depicted in grey. Pink bar: interstrand cross link in the DNA double helix. Arrow: activation, hatched line: inhibition, hexagon: HPV, hexagon with circular arrow: HPV replication. LOF: loss of function. BRCA: Breast cancer susceptibility protein; PALB: Partner and localizer of BRCA2; BRIP: BRCA interacting protein; RFWD: ring finger and WD repeat domain protein; XRCC: X-ray repair cross complimenting protein; UBE2T: Ubiquitin conjugating enzyme E2 T; REV7: DNA polymerase zeta processivity subunit also known as MAD2L2 (mitotic arrest deficient 2 like 2); SLX: structure specific endonuclease subunit; ERCC: excision repair 1 protein.

3. The Many Phenotypes of Fanconi Anemia

FA is characterized by congenital abnormalities, progressive bone marrow failure, and a predisposition to leukemia and solid tumors. Congenital abnormalities occur in approximately 75% of patients with FA and commonly include short stature, abnormal skin pigmentation, skeletal malformations, and genitourinary tract abnormalities [31]. The age of the onset of bone marrow failure varies significantly, with a median age of 7.6 years [32]. Individuals with FA carry a higher risk of developing myelodysplastic syndrome (MDS), acute myelogenous leukemia (AML), or both. The risk of developing AML is increased 500-fold as compared to the general population, and most individuals are typically diagnosed between the ages of 15 and 35 [33]. The relative risk of developing MDS is even higher, 5000-fold greater than that of the general population [34]. The only curative therapeutic option for the hematologic manifestations of FA is a hematopoietic stem cell transplant (HSCT), which should ideally be performed prior to the onset of MDS and AML. HSCT outcomes have dramatically improved over the last 20 years because of the attenuation of the preparative regimen intensity overall, with the addition of fludarabine and the elimination of total body irradiation. The five-year survival after a matched sibling transplant is now close to 90%, and excellent outcomes have also been achieved with alternate donor transplants, especially in patients under the age of 10 years [35–37].

While a successful stem cell transplant can cure the hematologic manifestations of FA, including bone marrow failure, AML, and MDS, post-HSCT individuals remain at a significantly increased risk for solid tumors. The most common tumors seen in FA are head and neck squamous cell carcinomas (HNSCC), particularly in the oral cavity, and vaginal squamous cell carcinomas, which occur at a younger age and are very challenging to treat because of the inherent chemotherapy and radiation therapy sensitivity of individuals with FA. A cohort study of 35 individuals with FA and HNSCC from the International Fanconi Anemia Registry reported a low overall 5-year survival of 39%, with 49% of individuals with HNSCC experiencing recurrence after treatment [38]. In this study, 15 out of 20 tumors that were tested for HPV were positive (75%) [39]. One hypothesis for the predisposition of individuals with FA to SCCs in the oral cavity and genital region is an increased susceptibility to exogenous carcinogens, including oncogenic viruses such as HPV. A subpopulation of tumors harbor HPV sequences, but the role of HPV in the development of SCC in patients with FA has been controversial [40]. Some studies have failed to detect HPV in HNSCCs derived from patients with FA, others have found a high prevalence of HPV positivity in both FA-related HNSCC tumors and oral samples from individuals with FA [39,41–44]. While the evidence implicating HPV in HNSCC is controversial, there is a clearer association with a subset of anogenital SCCs. More studies are needed to establish whether HPV plays a role in the development of many or most SCC in patients with FA, and to determine whether FA imparts a greater likelihood of viral infection, replication and persistence, as might be indicated by experimental data.

4. FA Loss Stimulates HPV Genome Amplification, Integration, and Oncogenicity

Early studies of HNSCC, anogenital warts, and other SCCs in individuals with FA suggested a possible role for HPV in FA-associated cancers [39]. This report was consistent with another US-based study that found HPV DNA prevalence to be significantly greater in 25 FA-related HNSCC tumors compared to non-FA controls [41]. However, a Dutch study failed to detect HPV DNA in any of the FA-related tumors tested, and HPV was not detected in five HNSCC or four anogenital squamous cell carcinoma samples from subjects with FA in a more recent report [41,42]. While the detection of HPV in FA HNSCC remains controversial, there is growing evidence from cell lines, animal studies, expression analysis, and computational biology approaches that supports a virus–cellular crosstalk between the FA pathway and HPV infection and oncogenicity [45]. HPV interacts with a number of DNA damage pathways. For instance, the E1 and E2 replication proteins activate a DNA damage response which includes ATM and ATR signaling. The E6 and E7 oncogenes promote viral replication by both activating and deactivating DNA damage repair pathway components and by uncoupling inappropriate cell cycle progression from apoptosis. Detailed HPV interactions with DNA damage repair pathways are described in [46]. Our focus below remains on the FA pathway.

First, microarray analysis of vulvar tissue infected with both noncarcinogenic and carcinogenic HPV types indicated that FANCA, FANCD2, and other DNA damage markers were significantly induced following high-risk HPV infection, along with increased DNA damage in the tissues [47]. Similarly, the expression of the high-risk HPV16 E7 oncogene, individually or together with E6, led to a coordinated upregulation of FA genes by E7 expression via Rb/E2F (retinoblastoma/E2F transcription factor 1) regulation [48]. Second, a link between HPV16 and the FA DNA repair pathway was identified using a computational biology approach and then validated experimentally by evaluating the impact of overexpressing the E6 or E7 proteins in primary fibroblasts and keratinocytes, using global gene expression analysis [45]. Third, the FA pathway loss in human HPV16+ or HPV31+ organotypic epithelial rafts increased DNA damage as expected from the loss of a key DNA repair pathway. However, quite unexpectedly, cell proliferation, expansion of the basal ESPC compartment, and tissue hyperplasia were observed. FA correction rescued the abnormal phenotype [49,50]. Increased cellular and viral replication were correlated with elevated levels of the E7 oncoprotein, but relevant host–viral networks whereby the FA pathway limits HPV genome amplification and thus, perhaps, infectious virus yields are unclear [46]. In subsequent experiments with these models, FANCD2 loss stimulated HPV16 and HPV31 genome amplification on the basis of qPCR and in situ hybridization—including in the basal and spinous cell layers where genome amplification does not normally occur. It is likely that the stabilization of the viral E7 protein contributes to both viral and cellular hyper-replication [49]. Since increased viral and cellular proliferation in 3D epidermis was an immediate response to FA loss, these phenotypes are not likely a consequence of adaptation by selection. Fourth, in the context of an intact FA pathway, the central FANCD2 protein was recently shown to localize to HPV viral replication centers, and FANCD2 knockdown promoted integration at the expense of viral replication [51]. Thus, depending upon the experimental circumstances, FA loss can lead to either amplified viral replication or increased integration with reduced replication. Regardless of the outcome of these processes, increased viral load or integration, would be expected to stimulate neoplastic transformation. Fifth, the HPV E6 and E7 oncoproteins repress homologous replication [52]. Sixth, *Fancd2* knockout mice do not develop SCC spontaneously and, therefore, are not a model for human SCC susceptibility in the absence of other gene modifications or environmental carcinogens. However, *Fancd2* knockout mice, bred to mice with transgenic expression of the HPV16 E7 oncoprotein targeted to basal epithelial cells, harbor increased DNA damage in mutagen-treated epidermis and are more likely to develop head and neck SCCs [53], cervical, and vaginal SCCs, compared to E7-transgenic control animals [54]. These effects of E7 are due to the inactivation of the Rb family of tumor suppressors that normally limit DNA damage [55]. Altogether, a multitude of physical, molecular, and functional connections between the FA pathway and HPV oncogenes in epidermal models may support a clinically important relationship in humans.

Together, these diverse data point to a common theme. It appears that HPV infection results in elevated DNA damage that then triggers the FA pathway to repair this DNA damage [56] and reprograms the FA pathway to participate in viral genome processing. In individuals where this pathway is defective, it is likely that the DNA damage will not be repaired in HPV E7-expressing, highly proliferative cells, compounding the likelihood of tumor development over time. For these reasons, it is now critical to reconsider these studies in the context of the conflicting human data. Even if HPV is suppressed or cleared to levels undetectable by PCR assays, one might speculate that the resulting DNA damage is the trigger for increased HNSCCs and anogenital carcinomas clinically evident years later. This may have relevance for sporadic tumors where the FA pathway is frequently inactivated, either mutationally or through transcriptional silencing. Exome sequencing data and whole genome sequencing data demonstrated that 11% and 18%, respectively, of both HPV+ and HPV- HNSCCs in the general population harbor nonsynonymous mutations in FA genes [57,58], suggesting selective pressure for FA pathway loss during tumor development or progression. The depletion of components of the FA pathway in sporadic HPV-positive and -negative HNSCC cell lines induced epithelial to mesenchymal transition (EMT)-like phenotypes and invasion, features of advanced tumors, by mechanisms that involve the activation of the DNA-PK (DNA-protein kinase) DNA damage sensor kinase and downstream signaling through the Rac1 GTPase (Rac Family Small GTPase1) [58]. Collectively, there is impressive evidence pointing to a role for HPV in FA SCC, and a role for HPV-independent phenotypes, including DNA damage induction and cellular tumor phenotypes. Despite this, etiological associations remain unproven, and studies of the natural history of tumor development in the HPV-positive (and -negative) hosts are now needed to identify the underlying mechanisms of infection by HPV and to explore the role of other viruses or pathogens as possible contributors to cancer risk. Intriguingly, recent in vivo data from the Lambert laboratory may even hint at a possible hit-and-run mechanism for SCC development following high-risk HPV infection. This 2016 study again used HPV16 E7 transgenic mice, wherein the transgene expression is conditionally controlled [59]. Following the conventional paradigm in *Fancd2*-proficient mice, the persistence of cervical neoplasia was highly dependent upon the continued expression of HPV16 E7. In *Fancd2*-knockout mice, however, cervical cancers persisted after HPV16 E7 expression was turned off, suggesting that FA loss relieves tumors from sustained E7 dependency. Since HPV oncogenes cause an accumulation of DNA damage, the authors hypothesized that HPV-induced DNA damage leads to an accumulation of mutations in FA-deficient mice, which might allow HPV-driven cancers to acquire a rapid independence from the viral oncogenes. If true for individuals with FA, scenarios of hit-and-run HPV infection would arise for situations where the FA pathway is inactivated. This paradigm would be consistent with the ongoing debate about HPV detection in SCC specimens from the FA population and would explain past difficulties to establish or rule out HPV causality in FA SCC. Immune dysfunction in patients with FA may also play a role in cancer development in addition to the underlying genome instability and the inability of the cells to repair the DNA damage caused by HPV16 infection.

5. Immune dysfunction in FA and HPV susceptibility

To date, only limited studies have been published investigating the immune function in children and adults with FA. Froom et al. reported a low natural killer (NK) cell activity in two FA patients and their family members, suggesting a potential intrinsic defect in NK cell activity associated with FA [60]. Three additional studies subsequently supported these findings [61–63]. High levels of tumor necrosis factor alpha (TNF-α) [64,65] and low production of interleukins (IL), including IL1, IL2, and IL6, as well as interferon gamma (IFN-γ) and granulocyte-macrophage colony-stimulating factor (GM-CSF) have also been reported, while immunoglobulin (Ig) levels were shown to be within the normal range [66]. Another study evaluated 11 individuals with FA and found their peripheral blood mononuclear cells (PBMCs) to have lower lymphoproliferative responses to phytohemagglutinin (PHA) and pokeweed mitogens compared to controls, implying that they may have a general immune defect [67]. Further, this study showed that PBMCs of individuals with FA responded poorly compared to controls to a stimulation with Tetanus toxoid and a purified protein derivative of mycobacterium, but not cytomegalovirus (CMV), antigen, suggesting that individuals with FA may have lower activation and proliferation capabilities, and therefore increased susceptibility to some, but not all, infections. Using flow cytometry to study PBMCs, Castello et al., reported the potential to grade immunologic defects in individuals with FA, as well as in their asymptomatic parents (obligate heterozygotes), based on the differential expression of cell surface markers in lymphocytes, including cluster of differentiation (CD)20, CD4, CD8 (reflected in the CD4/CD8 ratio), CD25, and HLA-DR (Human leukocyte antigen-DR) [68]. Most compelling in this study was the activated status of T cells in patients and their parents, on the basis of CD25 and HLA-DR expression [60].

Similarly, Petridou et al. studied the parents of individuals with FA and found low levels of NK cell subsets as well as reduced mitogen-induced proliferation of PBMCs [62]. Another group observed a decrease in NK CD56dimCD16+ and CD8+ T cells in FA *(n = 42)* compared to non-FA controls [69]. These data further imply that the impaired differentiation of the NK cells subsets may be directly related to the impairment of the immune surveillance of viruses. Our own retrospective, cross-sectional analysis of a small group of children with FA ($n = 10$) showed a heterogeneous immune dysfunction compared to non-FA children [70]. Overall, we found that children with FA had decreased numbers of natural killer (NK) cells with impaired function (decreased NK lytic units and perforin and granzyme levels), fewer CD19+ B cells and tetanus responses, and diminished cytotoxic T lymphocyte (CTL) function [70]. A more recent study ($n = 31$) found that FA adults, but not children, had significantly lower IgG, IgA, IgM, total lymphocytes, and CD4 T cells than their relatives or compared to reference values ($p < 0.001$) [71]. Consistent with our study, this study found that both children and adults with FA had fewer B- and NK cells compared to controls.

More recently, we characterized the immune competence of a larger cohort of individuals with FA and assessed lymphocyte populations and functional status in the blood of 29 pre-BMT (bone marrow transplant) individuals with FA (2–47 years), with no history of cancer or severe bone marrow failure [72]. Strikingly, few individuals with FA were normal in all parameters tested. Many individuals with FA had reduced total NK cells, confirming the previously observed decreases in absolute CD16+ NK cells in FA. In those with normal NK levels, NK function was significantly impaired. Similar to previous results, decreased absolute CD19+ B cells were observed, but further investigation revealed also fewer CD19+CD27+ memory B cells and lower levels of immunoglobulins. In fifteen individuals with corresponding clinical bone marrows, we also compared blood immune parameters to bone marrow parameters. Not unexpectedly, we observed a strong positive correlation ($r = 0.96$) between mature B cell percentages in the bone marrow and CD19+ B cells in the blood, suggesting that the peripheral blood represents the bone marrow well. Further, we observed a moderate negative correlation ($r = -0.54$) between mature B cell % in the bone marrow and CD19+CD27+ memory B cells in the blood, suggesting that memory B cells may remain in the bone marrow longer, unlike what is often observed in individuals with typical immune deficiencies. While total CD3+ T cells and CD8+ T cells were not significantly different, we observed fewer CD4+T cells and reduced CTL function in individuals with FA. Since other laboratories had observed aberrant CD8+ T cells in FA, we took a closer look at the regression of the CD8+ *z*-scores by time and stratified by age. Indeed, the subanalysis uncovered more CD8+ T cells in older FA subjects ($p = 0.0002$), who also had fewer CTL lytic units ($p = 0.03$).

Taken together, these findings point to potential alterations of the immune function in individuals with FA and in their parents and provide support for an immunologic basis of the viral susceptibility, or susceptibility to other pathogens observed in FA. Importantly, a reduced NK cell activity may critically alter immune surveillance with regard to neoplastic cells in a population with known predisposition to DNA damage and malignancy. The differences in the study populations (median cohort age, status of bone marrow function, percentage having had a BMT, etc.) likely contribute to the variable findings between studies. Longitudinal studies are needed to determine whether the observed immune defects are present at birth or develop with age.

6. Epidemiological Studies in Fanconi Anemia Demonstrate Increased Oral HPV Prevalence

Currently there are no known published longitudinal studies of HPV in people living with Fanconi anemia. However, epidemiological studies have provided some evidence of risk. One study collected oral swabs from 67 participants with Fanconi anemia and tested for 27 HPV genotypes using polymerase chain reaction-based methods [73]. The study reported that the prevalence of oral HPV infection was 7.5%, and the prevalence of high-risk HPV infection was 6.0%. The prevalence was higher in adult males than in adult females (25.0% versus 9.1%, respectively). Another related study collected oral rinse samples from 126 individuals with FA and 162 unaffected first-degree relatives, testing for 37 HPV types [44]. The study found that 11.1% of individuals with FA tested HPV+, a percentage that was significantly higher ($p = 0.003$) than that corresponding to the primary relatives (2.5%). The HPV prevalence was even higher for sexually active individuals with FA (17.7% versus 2.4% for the family members; $p = 0.003$). HPV positivity also tended to be higher in the sexually inactive (8.7% for FA versus 2.9% for their siblings). Indeed, having FA increased HPV positivity 4.9-fold (95% confidence interval (CI): 1.6–15.4), considering age and sexual experience, but did not affect other potential risk factors. Among the 14 individuals with Fanconi anemia who tested positive for oral HPV, 8 individuals had corresponding immune data [44]. Only three of the eight oral HPV+ subjects had been vaccinated for HPV prior to blood sampling and all had positive titers of all four HPV types included in the Gardasil vaccine. Two of the oral HPV+ individuals who indicated that they had not yet been vaccinated for HPV were also serologically HPV+ for types other than those identified in their oral sample, suggesting prior natural infections with these types. Three of the oral HPV+ participants had no detectable HPV titers. Two individuals were HPV6+ and HPV16+. Interestingly, these individuals were deficient in either their absolute B cell count or their memory B cells count. While there were no unifying characteristics that were shared by oral HPV or seropositive individuals with FA, these data support that individuals with FA have heterogeneous and frequently reduced immune responses [70,72]. Considering these oral results, even young children with FA are commonly infected with HPV, likely from nonsexual routes. A recent study found that HPV can survive outside of its host to potentially infect people by nonsexual means [74]. Testing for HPV in multiple anatomical sites at all ages and particularly before and after vaccination could inform vaccine guidelines for FA and for other cancer-prone and immunodeficient populations. A continued study of both immunity and HPV infection is needed to better understand their contributions to SCC.

7. Response to HPV Vaccine in Individuals with FA

Three HPV vaccines are currently licensed: two vaccines containing L1 Virus-Like Particles (VLPs) for HPV types 6, 11, 16, and 18 (Gardasil and Gardasil-9, which additionally covers HPV types 31, 33, 45, 52, and 58 (Merck and Co., USA), and a third vaccine containing L1 VLPs for HPV types 16 and 18 (Cervarix, GlaxoSmithKline Biologicals, Belgium, which has since been withdrawn by GSK from the US market) [75]. The impaired B cell function in FA raises concern about the efficacy of vaccination in FA. Our earlier study of the immune function in FA tested the type-specific seropositivity using a bead Luminex assay. Seropositivity ranged from 7 to 21% for skin HPV types and from 7 to 38% for mucosal HPV types in unvaccinated individuals with FA; in self-reported vaccine recipients, the serological positivity to HPV vaccine types was 75 to 96% [76]. Our more recently published study of the immune response in FA reported antibody responses to natural HPV infection and HPV vaccination in 39 unvaccinated and 24 vaccinated patients for HPV [77]. Similar to the earlier study, 30% of reportedly unvaccinated individuals were seropositive. As expected, seropositivity was significantly associated with having had sex regardless of age ($p = 0.03$). Importantly, among the 23 unvaccinated children younger than 13 years old, 5 were positive to more than one HPV types by multiplex ELISA (M4ELISA—HPV16, 18, 6, 11). Seropositivity among individuals vaccinated for HPV16, 18, 6, and 11 was also high. However, HPV titers for all four subtypes were significantly lower in the post-HSCT participants compared to those who had also received the vaccine but had not undergone HSCT. Currently, there are few other published US studies that have reported HPV titers for individuals with

FA. One study of 34 unvaccinated and 12 mainly Gardasil vaccinated individuals with FA suggests that individuals with FA are, to some level, protected by the vaccines up to 5 years, similar to the general population [78]. While none of the immune function measures predicted whether the participants' responses would be in the expected range for their age or whether they would have lower HPV titers following vaccination, 23% of participants (n = 3) with lower HPV titers had a low total memory B cell count, while none of the participants with titers in the expected range had a low memory B cell count. Indeed, this group tended to have more participants with a higher memory B cell count, suggesting a possible association between preserved memory B cells and HPV titer. Further study of host immune responses is needed to determine whether inherent virus-specific immune defects contribute to persistent HPV infection.

8. Epidermal abnormalities in Fanconi Anemia

A prominent feature of FA is an abnormal pigmentation of the skin and mucosas, and, as mentioned previously, individuals harbor an extreme disposition to keratinocyte-based SCCs early in life, primarily oral, esophageal, and anogenital tumors, as well as skin tumors [39,79–82]. First, individuals with FA have a significantly increased risk of HPV positivity (compared to unaffected family members) [44]. Second, the viral presence or the seropositivity for cutaneous and mucosal HPV types is detectable in FA individuals [43,73,76]. Anogenital SCCs in FA are linked to a history of low-risk HPV infection-associated genital warts; a causal role for the high-risk HPV types 16 and 18 in FA head and neck SCCs remains controversial [38,39,41–43,73,76,78,83–87]. Several laboratories have published clinical data that suggest that FA genes limit mucosal and cutaneous HPV infection and amplification. It is unclear whether frequent pigmentation defects are associated with any subclinical skin vulnerabilities.

Interestingly, an unrelated SCC predisposition syndrome, epidermolysis bullosa (EB), is known for defective collagen production, or defects in cell–cell or cell–substratum adhesion complexes which result in skin fragility, skin blistering and scarring, and increased risk of SCC formation [88,89]. Our published and preliminary data suggest the presence of adhesion defects in SCCs and perhaps the skin of individuals with FA [90]. However, unlike individuals with EB, those with FA do not exhibit any clinically evident skin blistering at baseline. Defects in the structure or barrier function of the skin may allow exogenous stressor permeation into the deeper layers of the skin, further stimulating the existing genetic instability, progression to dysplasia, and cancer development. Similarly, in these individuals, the skin is a target organ for viral infection (including but not limited to HPV), as seen in other skin fragility conditions with increased SCC development, where reduced T cell numbers allow HPV to overcome host defenses and contribute to malignancy [91,92]. In the FA-deficient HPV immortalized epidermis, an increased epithelial hyperplasia was observed suggesting a role for the FA pathway in the maintenance of alternative epithelial properties, such as its structure and cell cycle control for the prevention of HPV-associated SCCs [50]. There may be other nonstructural abnormalities of the epithelium that predispose to the development of SCCs in individuals with FA. Naturally occurring aldehydes are known to damage the DNA by ICLs, and relevant repair mechanisms are deficient in individuals with FA. Acetaldehyde-mediated carcinogenesis has been reported in the squamous epithelium, and aldehyde dehydrogenase-deficient keratinocytes demonstrate increased aldehyde-derived DNA damage [93,94].

Another important question is whether keratinocyte abnormalities in individuals with FA contribute to growth factor abnormalities that could be involved in the growth of tumors or in SCC development. Individuals with FA commonly have pigmentation abnormalities of the skin, with café-au-lait or hypopigmented areas noted on clinical examination. Neurofibromatosis (NF) Type 1 (NF-1) is associated with café-au-lait lesions of the skin. Elevated levels of hepatocyte growth factor and stem cell factor were demonstrated in cultured fibroblasts of individuals with NF compared to controls and in keratinocytes from non-café-au-lait lesions in individuals with neurofibromatosis [95]. Keratinocytes from café-au-lait lesions in non-FA individuals were shown to produce elevated levels of endothelin-1 [96]. Deregulated levels of expression of potential growth-related cytokines may promote tumor development in individuals with the inability to repair DNA damage. However, further studies on the secretion of growth factors from FA-keratinocytes are needed, as are expanded epidemiological studies on the potential skin SCC susceptibility in FA.

Taken together, there may be an integral role for epidermal structural and functional abnormalities in the development of SCCs in individuals with FA. If true, this could significantly compound the effects of inherent genome instability by increasing the risk of mechanical stress or of carcinogen exposure. A multifactorial network involving potentially increased skin fragility, antigen entry, growth factor secretion, and the inability to effectively repair DNA damage may coalesce to foster SCC development in these individuals. Further work exploring the mechanisms behind these skin defects and the possibilities of therapeutic prevention or intervention is crucial to prevent the life-limiting complication of SCC in FA.

9. Summary

Individuals with Fanconi anemia have a dramatically increased risk of developing SCC when compared to the general population, especially as more FA individuals are surviving into adulthood because of improved outcomes due to HSCT. High-risk HPVs have been found to be causative agents of SCCs in the general population, with subtype 16 alone found to be present in 90% of HPV-related HNSCCSs. The association between HPV and SCCs in individuals with FA remains controversial. While there has been conflicting clinical evidence linking HPV positivity to HNSCCs, HPV seems to be more common in anogenital SCCs in this patient population.

While there are a few studies with a limited sample size assessing the immune function in individuals with FA, these studies have shown that individuals with FA do have some defects in cellular and humoral immunities. There are also studies showing that the FA pathway limits cellular proliferation and the amplification and integration of HPV. Taken together, multiple factors might predispose individuals with FA to HPV infection and oncogenicity. This might not have been reliably reproduced in the literature for various reasons. One reason is that, according to studies from the Kovacic laboratory, there are more types of HPV that might be implicated in HNSCC but that were not tested in human tumors. Another reason is that, perhaps, HPV infection is less implicated in the development of SCCs, particularly in the head and neck, in individuals with FA as compared to the general population. In this case, other factors ranging from other oncogenic pathogens to DNA damaging exposures might be at play. Yet another hypothesis is that, perhaps, the initial HPV infection supports the development of SCCs by a hit-and-run mechanism, in the absence of HPV persistence. We believe that more epidemiological and basic research studies are needed to study the prevalence of HPV in individuals with FA and the possible predisposing factors, such as defects in immunity and the loss of proliferative control and epidermal integrity that might be at play.

Acknowledgments: This work was supported by NCI R01 CA102357 and by the Fanconi Anemia Research Fund to Susanne I. Wells, and by NIH R01 HL108102 to Melinda Butsch Kovacic.

Conflicts of Interest: The authors declare no conflict of interest.

References

1. De Martel, C.; Ferlay, J.; Franceschi, S.; Vignat, J.; Bray, F.; Forman, D.; Plummer, M. Global burden of cancers attributable to infections in 2008: A review and synthetic analysis. *Lancet Oncol.* **2012**, *13*, 607–615. [CrossRef]
2. Forman, D.; de Martel, C.; Lacey, C.J.; Soerjomataram, I.; Lortet-Tieulent, J.; Bruni, L.; Vignat, J.; Ferlay, J.; Bray, F.; Plummer, M.; et al. Global burden of human papillomavirus and related diseases. *Vaccine* **2012**, *30*, F12–F23. [CrossRef] [PubMed]
3. Zur Hausen, H. The search for infectious causes of human cancers: Where and why (Nobel lecture). *Angew. Chem. Int. Ed.* **2009**, *48*, 5798–5808. [CrossRef] [PubMed]

4. McLaughlin-Drubin, M.E. Human papillomaviruses and non-melanoma skin cancer. *Semin. Oncol.* **2015**, *42*, 284–290. [CrossRef] [PubMed]

5. Prigge, E.S.; von Knebel Doeberitz, M.; Reuschenbach, M. Clinical relevance and implications of HPV-induced neoplasia in different anatomical locations. *Mutat. Res. Rev. Mutat. Res.* **2017**, *772*, 51–66. [CrossRef] [PubMed]

6. Frazer, I.H. Prevention of cervical cancer through papillomavirus vaccination. *Nat. Rev. Immunol.* **2004**, *4*, 46–54. [CrossRef] [PubMed]

7. Schiller, J.T.; Muller, M. Next generation prophylactic human papillomavirus vaccines. *Lancet Oncol.* **2015**, *16*, e217–e225. [CrossRef]

8. Hofstetter, A.M.; Rosenthal, S.L. Factors impacting HPV vaccination: Lessons for health care professionals. *Expert Rev. Vaccines* **2014**, *13*, 1013–1026. [CrossRef] [PubMed]

9. Bodily, J.; Laimins, L.A. Persistence of human papillomavirus infection: Keys to malignant progression. *Trends Microbiol.* **2011**, *19*, 33–39. [CrossRef] [PubMed]

10. Chow, L.T. Model systems to study the life cycle of human papillomaviruses and HPV-associated cancers. *Virol. Sin.* **2015**, *30*, 92–100. [CrossRef] [PubMed]

11. Donati, G.; Watt, F.M. Stem cell heterogeneity and plasticity in epithelia. *Cell Stem Cell* **2015**, *16*, 465–476. [CrossRef] [PubMed]

12. Doorbar, J. Model systems of human papillomavirus-associated disease. *J. Pathol.* **2016**, *238*, 166–179. [CrossRef] [PubMed]

13. Tugizov, S.M.; Herrera, R.; Chin-Hong, P.; Veluppillai, P.; Greenspan, D.; Michael Berry, J.; Pilcher, C.D.; Shiboski, C.H.; Jay, N.; Rubin, M.; et al. HIV-associated disruption of mucosal epithelium facilitates paracellular penetration by human papillomavirus. *Virology* **2013**, *446*, 378–388. [CrossRef] [PubMed]

14. Brooke, M.A.; Nitoiu, D.; Kelsell, D.P. Cell-cell connectivity: Desmosomes and disease. *J. Pathol.* **2012**, *226*, 158–171. [CrossRef] [PubMed]

15. Broussard, J.A.; Getsios, S.; Green, K.J. Desmosome regulation and signaling in disease. *Cell Tissue Res.* **2015**, *360*, 501–512. [CrossRef] [PubMed]

16. Ishida-Yamamoto, A.; Igawa, S. Genetic skin diseases related to desmosomes and corneodesmosomes. *J. Dermatol. Sci.* **2014**, *74*, 99–105. [CrossRef] [PubMed]

17. Brandner, J.M.; Zorn-Kruppa, M.; Yoshida, T.; Moll, I.; Beck, L.A.; de Benedetto, A. Epidermal tight junctions in health and disease. *Tissue Barriers* **2015**, *3*, e974451. [CrossRef] [PubMed]

18. Ajiro, M.; Zheng, Z.M. Oncogenes and RNA splicing of human tumor viruses. *Emerg. Microbes Infect.* **2014**, *3*, e63. [CrossRef] [PubMed]

19. Graham, S.V. Human papillomavirus: Gene expression, regulation and prospects for novel diagnostic methods and antiviral therapies. *Future Microbiol.* **2010**, *5*, 1493–1506. [CrossRef] [PubMed]

20. Pace, P.; Mosedale, G.; Hodskinson, M.R.; Rosado, I.V.; Sivasubramaniam, M.; Patel, K.J. Ku70 corrupts DNA repair in the absence of the Fanconi anemia pathway. *Science* **2010**, *329*, 219–223. [CrossRef] [PubMed]

21. Schlacher, K.; Wu, H.; Jasin, M. A distinct replication fork protection pathway connects Fanconi anemia tumor suppressors to RAD51-BRCA1/2. *Cancer Cell* **2012**, *22*, 106–116. [CrossRef] [PubMed]

22. Ceccaldi, R.; Sarangi, P.; D'Andrea, A.D. The Fanconi anaemia pathway: New players and new functions. *Nat. Rev. Mo.l Cell. Biol.* **2016**, *17*, 337–349. [CrossRef] [PubMed]

23. Naim, V.; Rosselli, F. The FANC pathway and BLM collaborate during mitosis to prevent micro-nucleation and chromosome abnormalities. *Nat. Cell Biol.* **2009**, *11*, 761–768. [CrossRef] [PubMed]

24. Adamo, A.; Collis, S.J.; Adelman, C.A.; Silva, N.; Horejsi, Z.; Ward, J.D.; Martinez-Perez, E.; Boulton, S.J.; La Volpe, A. Preventing nonhomologous end joining suppresses DNA repair defects of Fanconi anemia. *Mol. Cell* **2010**, *39*, 25–35. [CrossRef] [PubMed]

25. Stone, M.P.; Cho, Y.J.; Huang, H.; Kim, H.Y.; Kozekov, I.D.; Kozekova, A.; Wang, H.; Minko, I.G.; Lloyd, R.S.; Harris, T.M.; et al. Interstrand DNA cross-links induced by α, β-unsaturated aldehydes derived from lipid peroxidation and environmental sources. *Acc. Chem. Res.* **2008**, *41*, 793–804. [CrossRef] [PubMed]

26. Kottemann, M.C.; Smogorzewska, A. Fanconi anaemia and the repair of Watson and Crick DNA crosslinks. *Nature* **2013**, *493*, 356–363. [CrossRef] [PubMed]

27. Zur Hausen, H. Papillomaviruses and cancer: From basic studies to clinical application. *Nat. Rev. Cancer* **2002**, *2*, 342–350. [CrossRef] [PubMed]

28. Ghittoni, R.; Accardi, R.; Hasan, U.; Gheit, T.; Sylla, B.; Tommasino, M. The biological properties of E6 and E7 oncoproteins from human papillomaviruses. *Virus Genes* **2010**, *40*, 1–13. [CrossRef] [PubMed]
29. Duensing, S.; Munger, K. Mechanisms of genomic instability in human cancer: Insights from studies with human papillomavirus oncoproteins. *Int. J. Cancer* **2004**, *109*, 157–162. [CrossRef] [PubMed]
30. Negrini, S.; Gorgoulis, V.G.; Halazonetis, T.D. Genomic instability—An evolving hallmark of cancer. *Nat. Rev. Mol. Cell Biol.* **2010**, *11*, 220–228. [CrossRef] [PubMed]
31. Mehta, P.A.; Tolar, J. Fanconi Anemia. In *Genereviews(r)*; Pagon, R.A., Adam, M.P., Ardinger, H.H., Wallace, S.E., Amemiya, A., Bean, L.J.H., Bird, T.D., Ledbetter, N., Mefford, H.C., Smith, R.J.H., et al., Eds.; University of Washington: Seattle, WA, USA, 1993.
32. Shimamura, A.; Alter, B.P. Pathophysiology and management of inherited bone marrow failure syndromes. *Blood Rev.* **2010**, *24*, 101–122. [CrossRef] [PubMed]
33. Rosenberg, P.S.; Huang, Y.; Alter, B.P. Individualized risks of first adverse events in patients with Fanconi anemia. *Blood* **2004**, *104*, 350–355. [CrossRef] [PubMed]
34. Alter, B.P. Fanconi anemia and the development of leukemia. *Best Pract. Res. Clin. Haematol.* **2014**, *27*, 214–221. [CrossRef] [PubMed]
35. Mehta, P.A.; Davies, S.M.; Leemhuis, T.; Myers, K.; Kernan, N.A.; Prockop, S.E.; Scaradavou, A.; O'Reilly, R.J.; Williams, D.A.; Lehmann, L.; et al. Radiation-free, alternative-donor HCT for Fanconi anemia patients: Results from a prospective multi-institutional study. *Blood* **2017**, *129*, 2308–2315. [CrossRef] [PubMed]
36. Boulad, F.; Gillio, A.; Small, T.N.; George, D.; Prasad, V.; Torok-Castanza, J.; Regan, A.D.; Collins, N.; Auerbach, A.D.; Kernan, N.A.; et al. Stem cell transplantation for the treatment of Fanconi anaemia using a fludarabine-based cytoreductive regimen and T-cell-depleted related HLA-mismatched peripheral blood stem cell grafts. *Br. J. Haematol.* **2000**, *111*, 1153–1157. [CrossRef] [PubMed]
37. Locatelli, F.; Zecca, M.; Pession, A.; Morreale, G.; Longoni, D.; Di Bartolomeo, P.; Porta, F.; Fagioli, F.; Nobili, B.; Bernardo, M.E.; et al. The outcome of children with Fanconi anemia given hematopoietic stem cell transplantation and the influence of fludarabine in the conditioning regimen: A report from the Italian pediatric group. *Haematologica* **2007**, *92*, 1381–1388. [CrossRef] [PubMed]
38. Kutler, D.I.; Patel, K.R.; Auerbach, A.D.; Kennedy, J.; Lach, F.P.; Sanborn, E.; Cohen, M.A.; Kuhel, W.I.; Smogorzewska, A. Natural history and management of Fanconi anemia patients with head and neck cancer: A 10-year follow-up. *Laryngoscope* **2016**, *126*, 870–879. [CrossRef] [PubMed]
39. Kutler, D.I.; Singh, B.; Satagopan, J.; Batish, S.D.; Berwick, M.; Giampietro, P.F.; Hanenberg, H.; Auerbach, A.D. A 20-year perspective on the international Fanconi anemia registry (IFAR). *Blood* **2003**, *101*, 1249–1256. [CrossRef] [PubMed]
40. Mehanna, H.; Beech, T.; Nicholson, T.; El-Hariry, I.; McConkey, C.; Paleri, V.; Roberts, S. Prevalence of human papillomavirus in oropharyngeal and nonoropharyngeal head and neck cancer—Systematic review and meta-analysis of trends by time and region. *Head Neck* **2013**, *35*, 747–755. [CrossRef] [PubMed]
41. Van Zeeburg, H.J.; Snijders, P.J.; Wu, T.; Gluckman, E.; Soulier, J.; Surralles, J.; Castella, M.; van der Wal, J.E.; Wennerberg, J.; Califano, J.; et al. Clinical and molecular characteristics of squamous cell carcinomas from Fanconi anemia patients. *J. Natl. Cancer Inst.* **2008**, *100*, 1649–1653. [CrossRef] [PubMed]
42. Alter, B.P.; Giri, N.; Savage, S.A.; Quint, W.G.; de Koning, M.N.; Schiffman, M. Squamous cell carcinomas in patients with Fanconi anemia and dyskeratosis congenita: A search for human papillomavirus. *Int. J. Cancer* **2013**, *133*, 1513–1515. [CrossRef] [PubMed]
43. De Araujo, M.R.; Rubira-Bullen, I.R.; Santos, C.F.; Dionisio, T.J.; Bonfim, C.M.; De Marco, L.; Gillio-Tos, A.; Merletti, F. High prevalence of oral human papillomavirus infection in Fanconi's anemia patients. *Oral Dis.* **2011**, *17*, 572–576. [CrossRef] [PubMed]
44. Sauter, S.L.; Wells, S.I.; Zhang, X.; Hoskins, E.E.; Davies, S.M.; Myers, K.C.; Mueller, R.; Panicker, G.; Unger, E.R.; Sivaprasad, U.; et al. Oral human papillomavirus is common in individuals with Fanconi anemia. *Cancer Epidemiol. Prev. Biomark.* **2015**, *24*, 864–872. [CrossRef] [PubMed]
45. Gulbahce, N.; Yan, H.; Dricot, A.; Padi, M.; Byrdsong, D.; Franchi, R.; Lee, D.S.; Rozenblatt-Rosen, O.; Mar, J.C.; Calderwood, M.A.; et al. Viral perturbations of host networks reflect disease etiology. *PLoS Comput. Biol.* **2012**, *8*, e1002531. [CrossRef] [PubMed]
46. Wallace, N.A.; Galloway, D.A. Manipulation of cellular DNA damage repair machinery facilitates propagation of human papillomaviruses. *Semin. Cancer Biol.* **2014**, *26*, 30–42. [CrossRef] [PubMed]

47. Santegoets, L.A.; van Baars, R.; Terlou, A.; Heijmans-Antonissen, C.; Swagemakers, S.M.; van der Spek, P.J.; Ewing, P.C.; van Beurden, M.; van der Meijden, W.I.; Helmerhorst, T.J.; et al. Different DNA damage and cell cycle checkpoint control in low- and high-risk human papillomavirus infections of the vulva. *Int. J. Cancer* **2012**, *130*, 2874–2885. [CrossRef] [PubMed]

48. Hoskins, E.E.; Gunawardena, R.W.; Habash, K.B.; Wise-Draper, T.M.; Jansen, M.; Knudsen, E.S.; Wells, S.I. Coordinate regulation of Fanconi anemia gene expression occurs through the RB/E2F pathway. *Oncogene* **2008**, *27*, 4798–4808. [CrossRef] [PubMed]

49. Hoskins, E.E.; Morreale, R.J.; Werner, S.P.; Higginbotham, J.M.; Laimins, L.A.; Lambert, P.F.; Brown, D.R.; Gillison, M.L.; Nuovo, G.J.; Witte, D.P.; et al. The Fanconi anemia pathway limits human papillomavirus replication. *J. Virol.* **2012**, *86*, 8131–8138. [CrossRef] [PubMed]

50. Hoskins, E.E.; Morris, T.A.; Higginbotham, J.M.; Spardy, N.; Cha, E.; Kelly, P.; Williams, D.A.; Wikenheiser-Brokamp, K.A.; Duensing, S.; Wells, S.I. Fanconi anemia deficiency stimulates HPV-associated hyperplastic growth in organotypic epithelial raft culture. *Oncogene* **2009**, *28*, 674–685. [CrossRef] [PubMed]

51. Spriggs, C.C.; Laimins, L.A. FANCD2 binds human papillomavirus genomes and associates with a distinct set of DNA repair proteins to regulate viral replication. *mBio* **2017**, *8*, e02340-16. [CrossRef] [PubMed]

52. Wallace, N.A.; Khanal, S.; Robinson, K.L.; Wendel, S.O.; Messer, J.J.; Galloway, D.A. High-risk alphapapillomavirus oncogenes impair the homologous recombination pathway. *J. Virol.* **2017**, *91*, e01084-17. [CrossRef] [PubMed]

53. Park, J.W.; Pitot, H.C.; Strati, K.; Spardy, N.; Duensing, S.; Grompe, M.; Lambert, P.F. Deficiencies in the Fanconi anemia DNA damage response pathway increase sensitivity to HPV-associated head and neck cancer. *Cancer Res.* **2010**, *70*, 9959–9968. [CrossRef] [PubMed]

54. Park, J.W.; Shin, M.K.; Lambert, P.F. High incidence of female reproductive tract cancers in FA-deficient HPV16-transgenic mice correlates with E7's induction of DNA damage response, an activity mediated by E7's inactivation of pocket proteins. *Oncogene* **2014**, *33*, 3383–3391. [CrossRef] [PubMed]

55. Park, J.W.; Shin, M.K.; Pitot, H.C.; Lambert, P.F. High incidence of HPV-associated head and neck cancers in FA deficient mice is associated with E7's induction of DNA damage through its inactivation of pocket proteins. *PLoS ONE* **2013**, *8*, e75056. [CrossRef] [PubMed]

56. Moody, C.A.; Laimins, L.A. Human papillomavirus oncoproteins: Pathways to transformation. *Nat. Rev. Cancer* **2010**, *10*, 550–560. [CrossRef] [PubMed]

57. Romick-Rosendale, L.E.; Lui, V.W.; Grandis, J.R.; Wells, S.I. The Fanconi anemia pathway: Repairing the link between DNA damage and squamous cell carcinoma. *Mutat. Res.* **2013**, *743*, 78–88. [CrossRef] [PubMed]

58. Romick-Rosendale, L.E.; Hoskins, E.E.; Vinnedge, L.M.P.; Foglesong, G.D.; Brusadelli, M.G.; Potter, S.S.; Komurov, K.; Brugmann, S.A.; Lambert, P.F.; Kimple, R.J.; et al. Defects in the Fanconi anemia pathway in head and neck cancer cells stimulate tumor cell invasion through DNA-PK and RAC1 signaling. *Clin. Cancer Res.* **2016**, *22*, 2062–2073. [CrossRef] [PubMed]

59. Park, S.; Park, J.W.; Pitot, H.C.; Lambert, P.F. Loss of dependence on continued expression of the human papillomavirus 16 E7 oncogene in cervical cancers and precancerous lesions arising in Fanconi anemia pathway-deficient mice. *mBio* **2016**, *7*, e00628-16. [CrossRef] [PubMed]

60. Froom, P.; Aghai, E.; Dobinsky, J.B.; Quitt, M.; Lahat, N. Reduced natural killer activity in patients with Fanconi's anemia and in family members. *Leuk. Res.* **1987**, *11*, 197–199. [CrossRef]

61. Hersey, P.; Edwards, A.M.; Lewis, R.; Kemp, A.H.; McInnes, J. Deficient natural killer cell activity in a patient with Fanconi's anaemia and squamous cell carcinoma. *Clin. Exp. Immunol.* **1982**, *48*, 205–212. [PubMed]

62. Petridou, M.; Barrett, A. Physical and laboratory characteristics of heterozygote carriers of the Fanconi aplasia gene. *Acta Paediatr.* **1990**, *79*, 1069–1074. [CrossRef]

63. Lebbe, C.; Pinquier, L.; Rybojad, M.; Chomienne, C.; Ochonisky, S.; Miclea, J.M.; Gluckman, E.; Morel, P. Fanconi's anaemia associated with multicentric bowen's disease and decreased NK cytotoxicity. *Br. J. Dermatol.* **1993**, *129*, 615–618. [CrossRef] [PubMed]

64. Schultz, J.; Shahidi, N. Tumor necrosis factor-alpha overproduction in Fanconi's anemia. *Am. J. Hematol.* **1993**, *42*, 196–201. [CrossRef] [PubMed]

65. Rosselli, F.; Sanceau, J.; Gluckman, E.; Wietzerbin, J.; Moustacchi, E. Abnormal lymphokine production: A novel feature of the genetic disease Fanconi anemia. II. In vitro and in vivo spontaneous overproduction of tumor necrosis factor alpha. *Blood* **1994**, *83*, 1216–1225. [PubMed]

66. Roxo, P., Jr.; Arruda, L.K.; Nagao, A.T.; Carneiro-Sampaio, M.M.; Ferriani, V.P. Allergic and immunologic parameters in patients with Fanconi's anemia. *Int. Arch. Allergy Immunol.* **2001**, *125*, 349–355. [CrossRef] [PubMed]
67. Suzergoz, F.; Gurol, A.; Evcimik, F.; Yalman, N. Lymphoproliferative response of Fanconi anemia patients to mitogens, bacterial and viral antigens in vitro. *Harran Üniv. Tıp Fak. Derg.* **2008**, *5*, 19–23.
68. Castello, G.; Gallo, C.; Napolitano, M.; Ascierto, P.A. Immunological phenotype analysis of patients with Fanconi's anaemia and their family members. *Acta Haematol.* **1998**, *100*, 39–43. [CrossRef] [PubMed]
69. Justo, G.A.; Bitencourt, M.A.; Pasquini, R.; Castelo-Branco, M.T.; Almeida-Oliveira, A.; Diamond, H.R.; Rumjanek, V.M. Immune status of Fanconi anemia patients: Decrease in T CD8 and CD56dim CD16+ NK lymphocytes. *Ann. Hematol.* **2014**, *93*, 761–767. [CrossRef] [PubMed]
70. Myers, K.C.; Bleesing, J.J.; Davies, S.M.; Zhang, X.; Martin, L.J.; Mueller, R.; Harris, R.E.; Filipovich, A.H.; Kovacic, M.B.; Wells, S.I.; et al. Impaired immune function in children with Fanconi anaemia. *Br. J. Haematol.* **2011**, *154*, 234–240. [CrossRef] [PubMed]
71. Giri, N.; Alter, B.P.; Penrose, K.; Falk, R.T.; Pan, Y.; Savage, S.A.; Williams, M.; Kemp, T.J.; Pinto, L.A. Immune status of patients with inherited bone marrow failure syndromes. *Am. J. Hematol.* **2015**, *90*, 702–708. [CrossRef] [PubMed]
72. Myers, K.C.; Sauter, S.; Zhang, X.; Bleesing, J.J.; Davies, S.M.; Wells, S.I.; Mehta, P.A.; Kumar, A.; Marmer, D.; Marsh, R.; et al. Impaired immune function in children and adults with Fanconi anemia. *Pediatr. Blood Cancer* **2017**, *64*, e26599. [CrossRef] [PubMed]
73. Winer, R.L.; Huang, C.E.; Cherne, S.; Stern, J.E.; Butsch Kovacic, M.S.; Mehta, P.A.; Sauter, S.L.; Galloway, D.A.; Katzenellenbogen, R.A. Detection of human papillomavirus in the oral cavities of persons with Fanconi anemia. *Oral Dis.* **2015**, *21*, 349–354. [CrossRef] [PubMed]
74. Ryndock, E.J.; Meyers, C. A risk for non-sexual transmission of human papillomavirus? *Expert Rev. Anti-Infect. Ther.* **2014**, *12*, 1165–1170. [CrossRef] [PubMed]
75. Dinshaw, J.E.; Frazer, I.H.; Garcia, P.J.; Kahn, J.; Markowitz, L.E.; Munoz, N.; Ndumbe, P.M.; Pitisuttithum, P.; Beutels, P.; Chirenje, M.; et al. *Human Papillomavirus (HPV) Vaccine Background Paper*; World Health Organization: Geneva, Switzerland, 2008.
76. Katzenellenbogen, R.A.; Carter, J.J.; Stern, J.E.; Butsch Kovacic, M.S.; Mehta, P.A.; Sauter, S.L.; Galloway, D.A.; Winer, R.L. Skin and mucosal human papillomavirus seroprevalence in persons with Fanconi anemia. *Clin. Vaccine Immunol.* **2015**, *22*, 413–420. [CrossRef] [PubMed]
77. Mehta, P.A.; Sauter, S.; Zhang, X.; Davies, S.M.; Wells, S.I.; Myers, K.C.; Panicker, G.; Unger, E.R.; Butsch Kovacic, M. Antibody response to human papillomavirus vaccination and natural exposure in individuals with Fanconi anemia. *Vaccine* **2017**, *35*, 6712–6719. [CrossRef] [PubMed]
78. Alter, B.P.; Giri, N.; Pan, Y.; Savage, S.A.; Pinto, L.A. Antibody response to human papillomavirus vaccine in subjects with inherited bone marrow failure syndromes. *Vaccine* **2014**, *32*, 1169–1173. [CrossRef] [PubMed]
79. De Araujo, M.R.; de Oliveira Ribas, M.; Koubik, A.C.; Mattioli, T.; de Lima, A.A.; Franca, B.H. Fanconi's anemia: Clinical and radiographic oral manifestations. *Oral Dis.* **2007**, *13*, 291–295. [CrossRef] [PubMed]
80. De Kerviler, E.; Guermazi, A.; Zagdanski, A.M.; Gluckman, E.; Frija, J. The clinical and radiological features of Fanconi's anaemia. *Clin. Radiol.* **2000**, *55*, 340–345. [CrossRef] [PubMed]
81. Karalis, A.; Tischkowitz, M.; Millington, G.W. Dermatological manifestations of inherited cancer syndromes in children. *Br. J. Dermatol.* **2011**, *164*, 245–256. [CrossRef] [PubMed]
82. Tischkowitz, M.; Dokal, I. Fanconi anaemia and leukaemia—Clinical and molecular aspects. *Br. J. Haematol.* **2004**, *126*, 176–191. [CrossRef] [PubMed]
83. Han, T.J.; Lee, C.H.; Yoo, C.W.; Shin, H.J.; Park, H.J.; Cho, K.H.; Park, J.Y.; Choi, S.W.; Kim, J.Y. Synchronous multifocal HPV-related neoplasm involving both the genital tract and the head-and-neck area: A case report of Fanconi anemia. *Radiother. Oncol.* **2009**, *92*, 138–141. [CrossRef] [PubMed]
84. Kutler, D.I.; Wreesmann, V.B.; Goberdhan, A.; Ben-Porat, L.; Satagopan, J.; Ngai, I.; Huvos, A.G.; Giampietro, P.; Levran, O.; Pujara, K.; et al. Human papillomavirus DNA and p53 polymorphisms in squamous cell carcinomas from Fanconi anemia patients. *J. Natl. Cancer Inst.* **2003**, *95*, 1718–1721. [CrossRef] [PubMed]
85. Alter, B.P. Fanconi's anemia and malignancies. *Am. J. Hematol.* **1996**, *53*, 99–110. [CrossRef]
86. Rosenberg, P.S.; Greene, M.H.; Alter, B.P. Cancer incidence in persons with Fanconi anemia. *Blood* **2003**, *101*, 822–826. [CrossRef] [PubMed]

87. Rosenberg, P.S.; Alter, B.P.; Ebell, W. Cancer risks in Fanconi anemia: Findings from the German Fanconi anemia registry. *Haematologica* **2008**, *93*, 511–517. [CrossRef] [PubMed]

88. Gorell, E.S.; Nguyen, N.; Siprashvili, Z.; Marinkovich, M.P.; Lane, A.T. Characterization of patients with dystrophic epidermolysis bullosa for collagen VII therapy. *Br. J. Dermatol.* **2015**, *173*, 821–823. [CrossRef] [PubMed]

89. Pourreyron, C.; Cox, G.; Mao, X.; Volz, A.; Baksh, N.; Wong, T.; Fassihi, H.; Arita, K.; O'Toole, E.A.; Ocampo-Candiani, J.; et al. Patients with recessive dystrophic epidermolysis bullosa develop squamous-cell carcinoma regardless of type VII collagen expression. *J. Investig. Dermatol.* **2007**, *127*, 2438–2444. [CrossRef] [PubMed]

90. Romick-Rosendale, L.E. Cincinnati Children's Hospital Medical Center, USA, Unpublished work. 2016.

91. Li, S.L.; Duo, L.N.; Wang, H.J.; Dai, W.; Zhou, E.H.; Xu, Y.N.; Zhao, T.; Xiao, Y.Y.; Xia, L.; Yang, Z.H.; et al. Identification of LCK mutation in a family with atypical epidermodysplasia verruciformis with T-cell defects and virus-induced squamous cell carcinoma. *Br. J. Dermatol.* **2016**, *175*, 1204–1209. [CrossRef] [PubMed]

92. Vuillier, F.; Gaud, G.; Guillemot, D.; Commere, P.H.; Pons, C.; Favre, M. Loss of the HPV-infection resistance EVER2 protein impairs NF-kappaB signaling pathways in keratinocytes. *PLoS ONE* **2014**, *9*, e89479. [CrossRef] [PubMed]

93. Amanuma, Y.; Ohashi, S.; Itatani, Y.; Tsurumaki, M.; Matsuda, S.; Kikuchi, O.; Nakai, Y.; Miyamoto, S.; Oyama, T.; Kawamoto, T.; et al. Protective role of ALDH2 against acetaldehyde-derived DNA damage in oesophageal squamous epithelium. *Sci. Rep.* **2015**, *5*, 14142. [CrossRef] [PubMed]

94. Mizumoto, A.; Ohashi, S.; Hirohashi, K.; Amanuma, Y.; Matsuda, T.; Muto, M. Molecular mechanisms of acetaldehyde-mediated carcinogenesis in squamous epithelium. *Int. J. Mol. Sci.* **2017**, *18*, 1943. [CrossRef] [PubMed]

95. Okazaki, M.; Yoshimura, K.; Suzuki, Y.; Uchida, G.; Kitano, Y.; Harii, K.; Imokawa, G. The mechanism of epidermal hyperpigmentation in cafe-au-lait macules of neurofibromatosis type 1 (von Recklinghausen's disease) may be associated with dermal fibroblast-derived stem cell factor and hepatocyte growth factor. *Br. J. Dermatol.* **2003**, *148*, 689–697. [CrossRef] [PubMed]

96. Okazaki, M.; Yoshimura, K.; Uchida, G.; Suzuki, Y.; Kitano, Y.; Harii, K. Epidermal hyperpigmentation in non-syndromic solitary cafe-au-lait macules may be associated with increased secretion of endothelin-1 by lesional keratinocytes. *Scand. J. Plast. Reconstr. Surg. Hand Surg.* **2005**, *39*, 213–217. [CrossRef] [PubMed]

Meeting Report

The Intersection of HPV Epidemiology, Genomics and Mechanistic Studies of HPV-Mediated Carcinogenesis

Lisa Mirabello [1,*], Megan A. Clarke [1], Chase W. Nelson [1,2], Michael Dean [1],
Nicolas Wentzensen [1], Meredith Yeager [1,3], Michael Cullen [1,3], Joseph F. Boland [1,3],
NCI HPV Workshop [4,†], Mark Schiffman [1] and Robert D. Burk [5,*]

[1] Division of Cancer Epidemiology and Genetics (DCEG), National Cancer Institute, National Institutes of Health, Rockville, MD 20850, USA; megan.clarke@nih.gov (M.A.C.); cnelson@amnh.org (C.W.N.); deanm@mail.nih.gov (M.D.); wentzenn@mail.nih.gov (N.W.); yeagerm@mail.nih.gov (M.Y.); michael.cullen@nih.gov (M.C.); bolandj2@mail.nih.gov (J.F.B.); schiffmm@mail.nih.gov (M.S.)
[2] Sackler Institute for Comparative Genomics, American Museum of Natural History, New York, NY 10024, USA
[3] Cancer Genomics Research Laboratory, Leidos Biomedical Research, Inc., Frederick, MD 21701, USA
[4] NCI HPV Workshop, DCEG, Rockville, MD 20850, USA; sara.bass2@nih.gov
[5] Departments of Pediatrics, Microbiology and Immunology, Epidemiology and Population Health, and Obstetrics & Gynecology and Women's Health, Albert Einstein College of Medicine, Bronx, NY 10461, USA
* Correspondence: mirabellol@mail.nih.gov (L.M.); robert.burk@einstein.yu.edu (R.D.B.)
† Attendees (not already listed as co-authors, alphabetical order): Laia Alemany, Lawrence Banks, Sara Bass, Chris Buck, Laurie Burdett, Anil Chaturvedi, Gary Clifford, Daniel DiMaio, John Doorbar, Allan Hildesheim, Lou Laimins, Paul Lambert, Hong Lou, Alison McBride, Karl Munger, Miguel Angel Pavón Ribas, Ligia Pinto, John Schiller, Richard Schlegel, Stefan Schwartz, Mia Steinberg, Sarah Wagner, Yanzi Xiao, Qi Yang, Kai Yu, Bin Zhu (Supplementary Materials Table S1 lists attendee affiliations).

Received: 9 December 2017; Accepted: 12 February 2018; Published: 13 February 2018

Abstract: Of the ~60 human papillomavirus (HPV) genotypes that infect the cervicovaginal epithelium, only 12–13 "high-risk" types are well-established as causing cervical cancer, with HPV16 accounting for over half of all cases worldwide. While HPV16 is the most important carcinogenic type, variants of HPV16 can differ in their carcinogenicity by 10-fold or more in epidemiologic studies. Strong genotype-phenotype associations embedded in the small 8-kb HPV16 genome motivate molecular studies to understand the underlying molecular mechanisms. Understanding the mechanisms of HPV genomic findings is complicated by the linkage of HPV genome variants. A panel of experts in various disciplines gathered on 21 November 2016 to discuss the interdisciplinary science of HPV oncogenesis. Here, we summarize the discussion of the complexity of the viral–host interaction and highlight important next steps for selected applied basic laboratory studies guided by epidemiological genomic findings.

Keywords: HPV carcinogenesis; HPV epidemiology; HPV genomics; viral–host interactions; HPV16

1. Background

High-risk human papillomaviruses (HR-HPVs) cause a heavy burden of cancer with more than 600,000 cancers attributed to HR-HPVs worldwide in 2008 [1–3]. Of these patients, 236,000 deaths are estimated to be caused by cervical cancer alone each year [4]. Yet, the genetic basis of HPV oncogenicity, firmly established for one clade of the genus *Alphapapillomavirus* [5], has not been solved. This research question is now tractable, given the strong genotype-phenotype associations, the availability of fundamental molecular biological knowledge regarding the small HPV genome, and recent advances in high-throughput sequencing of HPV genomes.

Papillomaviruses (PVs) are circular, double-stranded DNA viruses that are believed to have co-diverged with animal and human host populations for millions of years, and are ubiquitous throughout the world [6–9]. HPV genomes are approximately 8000 base pairs in length and contain eight to nine open reading frames (ORFs) that encode highly conserved core proteins involved in viral genome replication (E1 and E2/E4) and assembly (L1 and L2), as well as accessory proteins (E5, E6, and E7). While E1 and E2 are involved in HPV replication and in regulating viral transcription, the primary oncogenes, E6 and E7, are thought to be largely responsible for niche adaptation, viral amplification, and inadvertently driving carcinogenesis [10]. Niche adaptation refers to the virus adapting to a specific anatomic/cellular region. All HPV genomes contain E7, and a few lack E6 and E5 [11]. The upstream regulatory region (URR) is a non-coding region containing cis-responsive elements that regulate replication and transcription of viral proteins [12].

HPV types are classified based on pairwise nucleotide sequence identity within the highly conserved *L1* gene, and distinct types (e.g., HPV16 vs. HPV31) are defined by differences of at least 10% at the nucleotide level [13]. An HPV species group (e.g., *Alphapapillomaviruses-9*) comprises HPV types sharing ≥70% of their L1 sequences [13]. To date, more than 200 HPV types have been identified and characterized [13]. Some HPV types within the alpha-HPVs infect mucosal epithelia and are associated with a variety of outcomes, ranging from benign asymptomatic infections to genital warts and cervical cancer. Of the ~60 alpha-HPV types, 13 from a single clade (i.e., branch of the phylogenetic tree) have been classified as definitely or probably carcinogenic (high-risk) and account for >95% of all cervical cancers worldwide. These include HPV16, HPV31, HPV33, HPV35, HPV52 and HPV58 (*Alphapapillomaviruses-9* species group); HPV18, HPV39, HPV45, HPV59, and HPV68 (*Alphapapillomaviruses-7* species group); HPV51 (*Alphapapillomaviruses-5* species group); and HPV56 (*Alphapapillomaviruses-6* species group) [14]. On a finer scale, within each of these HPV types there are variant lineages and sublineages with intratypic genome sequence differences of 1.0–10% and 0.5–1.0%, respectively [15].

HPV16 is by far the most carcinogenic HPV type, associated with approximately 50% of all cervical cancers, the majority of other HPV-related anogenital cancers, and more than 80% of HPV-positive head and neck cancers [16–20]. HPV16 variant lineages have been extensively studied, with four major variant lineages and up to 16 sublineages identified to date (Figure 1), including: sublineages A1-3 (traditionally classified as European), A4 (Asian), B1-4 (African-1), C1-4 (African-2), D1 (North American), D2 and D3 (Asian-American), and D4 [15,21]. Several studies have demonstrated that HPV16 variant lineages and sublineages confer differential risks of persistence, and progression to cervical precancer and cancer [21–33] (reviewed in Burk et al. [15]). The epidemiologically-defined co-factors, smoking and use of hormonal contraceptives, do not modify HPV16 variant lineage risk substantially [28], however, HPV16 lineages have not been evaluated considering host genetic factors and this could modify risk slightly. Some authors further showed that specific HPV16 lineages are associated with glandular versus squamous histology [21,26,32,34–36].

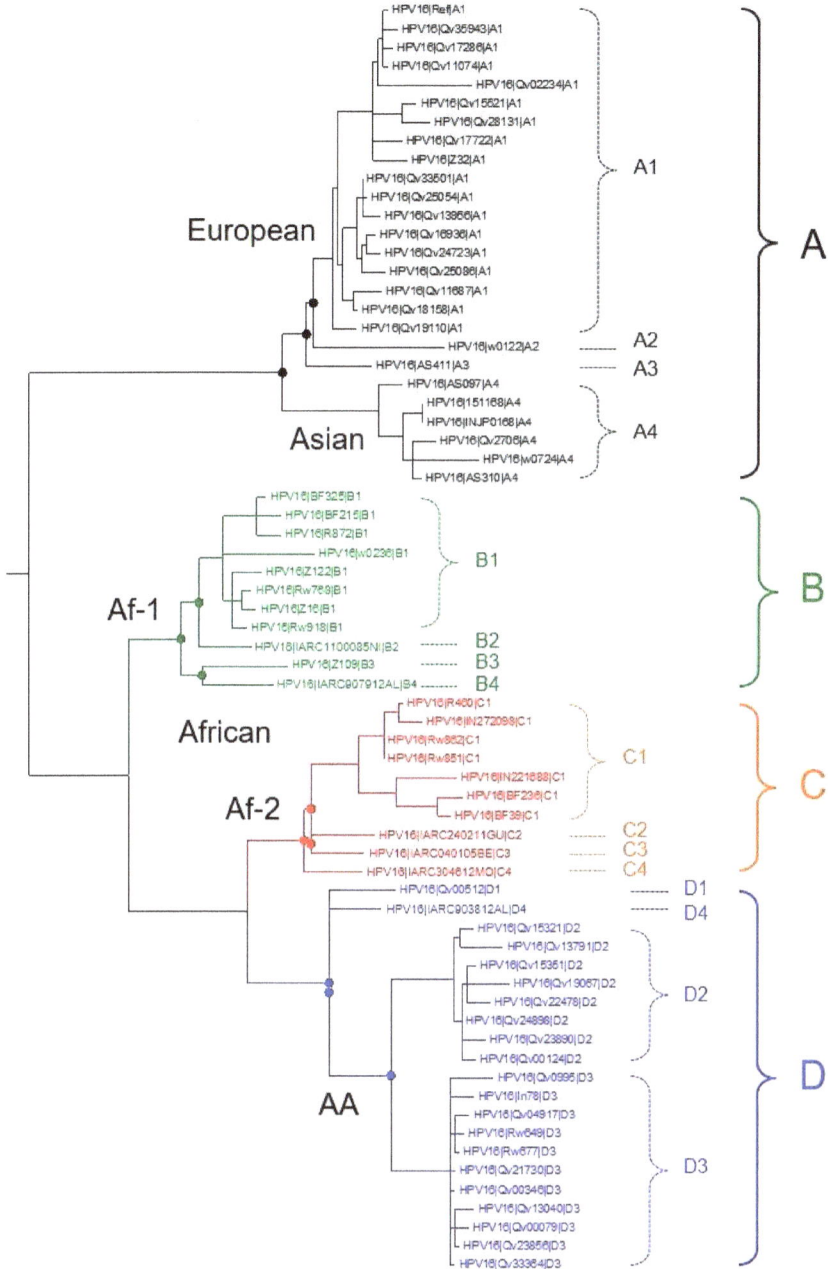

Figure 1. Phylogenetic tree illustrating human papillomavirus 16 (HPV16) lineages (A–D) and sublineage (A1–4, B1–4, C1–4, D1–4) relationships. Colors indicate main lineage branches. A maximum likelihood tree is shown inferred from 66 HPV16 whole-genome sequences, modified from Burk et al., 2013 [15], including additional reference sequence data from Mirabello et al. 2016 [21], 2017 [37]. Af-1, African-1; Af-2, African-2; AA, Asian-American.

Recent advances in high-throughput next-generation sequencing [38] have enabled the large-scale study of HPV genome variability and led to new discoveries in HPV genomic research. These findings based on empirical population-based studies provide opportunities for further investigation at the intersection of molecular biology and epidemiology that could enhance our molecular understanding of HPV-related carcinogenesis. In recognition of this exciting and unique opportunity, the National Cancer Institute's (NCI) Division of Cancer Epidemiology and Genetics (DCEG) sponsored a workshop entitled "Mechanistic Understanding of Cervical Carcinogenesis" on 21 November 2016. Planning and organization of the workshop was led by Lisa Mirabello and Robert D. Burk of the NCI-DCEG HPV Genomics Group. The primary goal of this workshop was to promote interdisciplinary discussions on the potential mechanisms underlying differences in carcinogenicity at the HPV type, lineage, and nucleotide levels and the potential next steps. The workshop brought together an expert panel spanning biochemistry, molecular biology, evolution, pathology, epidemiology, bioinformatics and statistics (Supplementary Materials Table S1). Some of the members of the NCI-DCEG HPV Genomics Group (Robert D. Burk, Mark Schiffman, Nicolas Wentzensen, and Lisa Mirabello) briefly presented the latest HPV epidemiologic and genomic data, and the majority of the workshop was focused on panel discussions addressing specific questions about the molecular mechanisms of HPV carcinogenesis defined by differences in HPV genomes that have remained unresolved (Table 1). This report summarizes the main highlights from this Workshop, with the goal of stimulating further research to understand the specific mechanisms underlying HPV carcinogenesis.

Table 1. Topics and specific questions addressed at the National Cancer Institute (NCI) HPV Workshop.

Topic Area	Questions
HPV studies at the species and/or type level and risk of cancer	What features of the biology and/or biochemistry of HPV16 make it so uniquely carcinogenic?
	What features of HPV16 biology, or interaction with the host cells, enable it to have a wider tissue tropism and disease association?
	What are the experimental approaches to investigate this? • This should include comparisons of closely related HPV types (i.e., HPV16, 31, and 35) with large differences in carcinogenesis • Comparison of appropriate type(s) to HPV16/18 biology/biochemistry (HPV6/11 are probably not a good choice because they are evolutionarily distant)
Studies of HPV variant lineages within a type to elucidate differences and risk of cancer	Are there functional differences between HPV16 A1 vs. D sublineage viruses that could help explain their pathological differences in cancer risk? • How do we mechanistically explain the genetic variants associated with glandular lesions for the HPV16 A4, D2, D3 sublineages compared to A1 and A2?
	Why are specific HPV16 sublineages (i.e., A4, D2, D3), and HPV18 and HPV45 prone to adenocarcinoma?
	What are the next steps after the SNP/gene-based epidemiological approach using case-control datasets? What cell based and/or biochemical experiments could be used to identify the mechanisms of different genetic associations?
Synthesizing current knowledge and moving forward in the era of NGS, systems biology and large data sets	How do we define the characteristics of HPV fitness? How do we annotate HPV genomes to be able to capture common functional motifs with disparate genomes in large datasets? (e.g., it's hard to align a large number of genome sequences of HPV16 with HPV31, HPV52, etc) • How do we go beyond the annotations in NCBI and PaVE?
	How do we incorporate information on viral suppression/invisibility to the host immune system?
	Where does epigenetics of the viral genome fit into the discussion of dissecting viral genome differences?
	How to best approach viral—host interactions?

2. Recent Discoveries in HPV Genomics

Through an international collaborative effort, the NCI HPV Genomics Project has sequenced many thousands of HPV genomes from well-characterized populations. With an initial focus on HPV16, recent data have confirmed and expanded earlier reports relating precancer/cancer risk to particular HPV16 lineages and uncovered several remarkably strong associations between HPV16 genetic variation and cervical carcinogenicity, as well as providing new insights into HPV diversity in the population. We have applied lineage-based and agnostic, gene and single nucleotide polymorphism (SNP)-based approaches to studying HPV genetic variation [21,38]. At the lineage-level, HPV16 sublineages confer differing risks of precancer and cancer, and most strikingly, differing risks of glandular lesions [21,26,32,34–36] which are of rising public health importance—they are more difficult to detect with cytology, have a poorer prognosis than squamous cell carcinoma (SCC), and their proportion among all cancers have been shown to be increasing in many developed regions [39–44]. The HPV16 sublineages, A4, D2, and D3, have significantly increased risks of adenocarcinoma in situ (AIS) and adenocarcinoma (ADC) compared to the most common A1/A2 sublineages; D3 and D2 have the strongest risks of ADC with relative risks of 59 and 137, respectively [21]. This indicates that only a small number of genetic differences (e.g., D and A HPV16 sublineages differ by <2.0% of 7906 nt) lead to large differences in risk of ADC.

Next-generation sequencing (NGS) HPV genome data with deep sequence coverage allows the sensitive identification of within HPV16 variant lineage co-infections. Using these data, within type co-infections are suspected in women with multiple "heterozygous" allele calls (HPV is a monoploid genome). The co-infections can be confirmed by the identification and visualization of multiple lineage-specific sequence variants occurring in shared sequence reads, representing two separate HPV16 variant lineage isolates. In the case of multiple lineages present in a specimen, a predominant lineage can usually be assigned based on presence in at least 60% of the sequence reads, and nucleotide variants only included from the predominant lineage; and, if a predominant lineage cannot be assigned (~50/50 of each lineage), that sample excluded (for more detail, see Cullen et al., [38]).

To address whether even finer genetic variation (i.e., SNPs) is associated with HPV16 carcinogenicity, we evaluated non-lineage-specific SNPs and the distribution of rare variants occurring within HPV16 lineages in a collection of 5,570 HPV16-infected case-control samples [37]. Thousands of unique HPV16 viral isolates were identified among women, suggesting that each may represent a distinct viral genome sequence possibly differing in carcinogenic potential. The controls had higher levels of rare sequence variants (particularly nonsynonymous and nonsense variants, i.e., amino acid changing) compared with cases across the genome and in specific regions. Interestingly, focusing on non-silent variation, *E7* was more variable in the controls compared to the cases, and we confirmed that *E7* showed extremely low variability in ~1700 cervical cancers from around the world. The *E7* gene was significantly less variable than all other gene regions in the cancers, including *E6*. This highlights that genetic conservation of *E7* (but not *E6*) is critical for HPV16 carcinogenesis. These rare nucleotide variants in *E7* frequently occurred in DNA motifs that are associated with the antiviral activity of human APOBEC3 (apolipoprotein B mRNA-editing, enzyme-catalytic polypeptide-like 3) family of cytidine deaminases. Interestingly, the majority of cervical cancer cases have been shown to be enriched for somatic APOBEC mutation signatures, suggesting that APOBEC antiviral activity is also a major source of somatic mutations in cervical cancers [45,46].

As discussed by the panel, these findings suggest that genetic variation within specific regions of the genome may differentially allow the virus to be cleared by the host; alternatively, the phenotype of cancer may require a fixed set of nucleotide variants at the viral genome level.

3. Summary/Next Steps: Molecular Mechanisms Underlying HPV Carcinogenesis

The Workshop agenda outlined many key questions related to three broad topic areas for discussion (Table 1). The first two sessions addressed molecular mechanisms underlying HPV carcinogenesis at the type and variant lineage/sublineage level, respectively. Each attendee was asked to contribute to these discussions by providing their unique perspective based on their expertise, for example, in basic science, epidemiology and/or computational biology. The following sections present an overview of the major themes that emerged.

3.1. Viral Molecules and Their Interactions with Host Cellular Machinery

The HPV life cycle is tightly linked to the differentiation state of infected epithelial cells. HPV infects basal keratinocytes that are exposed as the result of micro-abrasions in the epithelial surface [47,48]. Viral genome replication occurs at low levels in the basal layer and, as infected cells undergo terminal differentiation in the upper layers of the epithelium, E6 and E7 drive genome amplification by promoting cell-cycle re-entry and proliferation of HPV-infected cells [49,50]. Functional differences in E6 and E7 are thought to determine differences in carcinogenicity between high-risk and low-risk HPV types. Further, differences in the regulation of host protein interactions have been observed across high-risk types, which may contribute to known differences in carcinogenicity [51,52]. In addition, most viral proteins are expressed from spliced RNA molecules that have a complex regulation. Correlated changes across the viral genome could account for large changes in infection outcome (clearance, persistence and progression), as described in epidemiologic studies.

3.2. Tissue Tropism and Site of Infection

HPV infections can occur within specific sites of transitional epithelial cells (e.g., the squamocolumnar junction) with complex patterns of regulation that may render them more prone to viral transformation [48,53,54]. At glandular sites such as the endocervix and tonsils, the diagnosis and/or presence of precursor lesions compared to invasive cancers is much lower than the diagnosis of cervical squamous intraepithelial lesions to squamous cell cancers [55]. Whether this is due to the position of the lesions and impact on screening efficacy, morphologic features of the precursors, or is related to actual differences in HPV natural history and oncogenesis is not fully understood [56]. Since E6 and E7 may induce a stem cell-like state, the type of cell that becomes infected may contribute to disease outcome. More research is needed to understand whether patterns of viral gene expression and protein function are site-specific, and how they vary across different high-risk HPV types. Specific observations that need explanation include the relatively high prevalence of precancers (i.e., CIN3) that do not translate into similar rates of cancer, suggesting that precancer is a distinct endpoint that does not serve as a perfect surrogate for cancer risk. For example, the ratio of HPV31 to HPV16 is much higher for CIN3 than for cancer, suggesting that a higher proportion of CIN3s caused by HPV31 do not lead to cancer.

3.3. Regulation of HPV Transcription

High-risk HPV types have evolved regulatory strategies to tightly control viral gene expression during productive and quiescent infections. Because of its critical role in regulating gene expression at different stages of the HPV life cycle, mRNA splicing efficiency may contribute to carcinogenic potential [57]. Recent findings from The Cancer Genome Atlas (TCGA) suggest that in HPV16 there is a lower ratio of spliced E6 transcripts (coding for E7) to unspliced transcripts (coding for E6) compared with HPV18 [45]. Thus providing some of the evidence that mechanisms of carcinogenesis differ between these *Alphapapillomaviruses-9* and *Alphapapillomaviruses-7* genomes and probably relates to their genetic differences. Efficiency of splicing may also differ [57,58], since HPV mRNA splicing and polyadenylation are regulated by cis-acting HPV RNA elements and cellular RNA-binding proteins. Synonymous, nonsynonymous and non-coding sequence differences in binding motifs or RNA structures may induce subtle changes in splicing and/or polyadenylation efficiencies that could have significant effects on viral gene expression and thus, carcinogenicity. However, to date, over 1000 host RNA binding proteins have been identified [59] and their ability to recognize and bind to multiple sequence motifs makes it particularly challenging to predict differences in splicing and/or polyadenylation across types or lineages by sequencing alone. In addition to primary sequence variation, secondary structures may influence the efficiency of RNA binding and can be difficult to predict. One approach discussed at the meeting involves transfecting various HPV16 isolates differing in oncogenicity and measuring viral RNA molecules through RNA sequencing approaches (RNA-Seq or whole transcriptome sequencing).

In vitro studies have recently shown that HPV16 and related types express a fusion protein encoded by subregions of the *E1* and *E2* ORFs (termed E8^E2), which limits viral transcription and replication through the virus life cycle in undifferentiated keratinocytes [60]. This may play an important role in keeping the expression of early viral proteins at low levels so as to evade immune detection. Whether this protein occurs in natural infections remains untested, but it is an additional region that should be evaluated for any genome variation that might influence the viral life cycle and pathogenesis.

3.4. HPV Integration into Host Genomes

Integration of HPV DNA into the host genome occurs in the majority of cervical cancers, but not all [61,62]. Mechanisms by which HPV integrates into the host cell genome and promotes carcinogenesis are not well understood. Sites of integration tend to occur in regions of genomic instability [63–65], and have also been reported to occur in short regions of HPV and host genome sequence homology (i.e., "micro-homologies") [66–68], suggesting a potential role for DNA repair processes in integrating the HPV and host cell genomes based on nucleotide sequence similarities [69]. The prevalence of HPV integration in cervical cancers has been shown to vary by type, with lower frequencies observed for HPV types 31 and 33, compared with HPV types 16, 18, and 45 [61,69]. As a finer distinction, not all HPV16-associated cancers have integrated HPV DNA, whereas HPV18 integration is present in almost all HPV18-associated cancers. Viral-cellular fusion transcripts have been detected in all HPV18-positive cancers, some occurring in previously identified hotspots, such as 8q24 [45]. Interestingly, integration events associated with HPV18 appear to be more common at 8q24.21 near the *MYC* oncogene compared with HPV16-associated cancers [69,70]. At the HPV variant level, a recent study characterizing integration events by the HPV16 D and A variant lineages suggested differences in variant-specific integration potential, potentially mediated by E6 [71]. More studies are needed to confirm these findings and determine if and how viral genetic variation might relate to integration.

3.5. Viral–Host Interactions

In response to infection with HPV, humans can mount an adaptive immune response including the development of specific antibodies to the virion L1 coat protein. Antibody and/or human cytotoxic T-lymphocyte (CTL) epitopes have been predicted within the peptides encoded by all HPV16 ORFs: 100% of *E5*, *E6*, and *E7* residues; 65–83% of *E2*, *E4*, and *L1* residues; and only 7% of *E1* residues (Immune Epitope Database). *E6* and *E2* epitopes appear to be the most important for a CTL response [72].

Alternative approaches for identifying potentially important HPV epitopes are based on evolutionary methods to identify positive selection that might indicate a pressure for immune escape, and these have mainly detected codons in the *E5*, *E6*, *L1*, and *L2* ORFs [11,73]. Future research in this area will take advantage of increasingly available sequence data to detect regions undergoing positive selection within sublineages, helping to elucidate sublineage- and case/control-specific immune responses.

Both genetic and environmental host factors play key roles in determining viral oncogenicity. Epidemiologically defined co-factors, such as smoking and use of hormonal contraceptives, also play a role—for example, smoking has been associated with an approximate two-fold increased risk of precancer and cancer [74]. One goal of future research should be to link both host factors/genetics and viral genetics to infection outcomes. The host human leukocyte antigen (HLA) allele repertoire in particular, which is crucial for cell-mediated immune responses, may be a critical factor in determining which HPV variants will clear, and which will persist and potentially evade the immune system.

In fact, these host immune alleles show signals for an inherited risk of cervix precancer/cancer [75–77]. Furthermore, specific HLA class I alleles have been associated with the oncogenicity of specific HPV16 variants [78,79], which highlights the importance of the HLA type combined with the HPV16 variants for immune surveillance in cervical carcinogenesis. The development of cancer may include such steps as an HPV variant infecting a host, who has an insufficient HLA repertoire for clearing that particular variant.

4. Synthesizing Current Knowledge and Moving Forward in the Era of NGS, Systems Biology, and Big Data

The final session covered a range of topics related to characterizing and defining HPV fitness, annotation of HPV genomes, and host-viral interactions. These topics have important implications for HPV genomics research and could serve as a model for other genetic systems.

4.1. Defining HPV "Fitness"

Evolutionary fitness in biology is usually defined as reproductive success. The definition of how to define viral fitness in general, and HPV fitness in particular, remains unresolved and was not agreed upon by the workshop panel. An increase in viral replicative success may have conflicting proximal and ultimate outcomes for the host. For example, a particular viral genotype may replicate to high viral load in a particular cell, but the outcome may be to increase the likelihood of an immune response, thereby drastically decreasing the actual fitness of the virus. Surprisingly, much of the feedback from the panel was that a consensus definition for fitness might not be useful for describing carcinogenic features of different HPV types and variants. The intellectual divide resided in whether viral evolution, niche adaptation, and fitness represent the key drivers of carcinogenesis, although carcinogenesis does not support viral replication. Some attendees suggested a more direct paradigm using viral outcomes, such as causing cancer or not, as the "viral" phenotype (Table 2). Others suggested defining fitness as viral prevalence in the infected population, i.e., the outcome of incidence and persistence. In addition, the steps to cancer can be considered either as independent outcomes (e.g., persistence, precancer), or as a sequential set of steps that could be studied using functional assays.

Table 2. Distinct steps in the pathway from HPV infection to carcinogenesis.

Functional Step	Relevant Features
1. Infection	Cell receptor(s) for entry Tissue tropism
2. Persistence	Continued productive infection Persistence without productive infection Cellular immunity Latency Early, inapparent transformation
3. Transformation	Increased E6/E7 expression Chromosomal instability Somatic mutations Viral integration
4. Invasion	Increasing somatic mutations Integration, disruption, and partial loss of viral genome Epithelial-stromal interactions

Given that the most prevalent anogenital HPV type (HPV16) is also the most carcinogenic, future research should consider the relationship between viral prevalence and oncogenicity. It appears that viral traits that improve reproductive success also tend to initiate processes that predispose host cells to cancer. In either case, the ability to induce cancer is neither necessarily, nor inextricably, linked to HPV's ability to successfully propagate in populations, such that oncogenicity may be considered an unfortunate byproduct that is not itself under selection. Adaptation to a specific cellular environment may define features of the HPV genome that induce cancer as "collateral damage" rather than a selective trait, since cancer does not support the production of infectious virus.

4.2. HPV Genome Annotation and Other New Emerging Data Concepts

The importance of genome annotation is critical for evaluating the impact of sequence variations across viral variants and between viral types. In fact, since the variation between viral types is so large (approximately 30%), unequivocal alignment and position assignments are not currently feasible. Therefore, annotation of functions could serve as a common database to connect different features of disparate genomes. The way forward was not defined, but a bioinformatic approach is a promising area that could build upon work done on the annotation of various mammalian genomes that face similar challenges.

5. Conclusions

The field of HPV genomics is undergoing a major paradigm shift from thinking of an HPV type infection as an evolutionarily static entity to thinking of thousands of unique viral genomes with differences in carcinogenic potential. Findings from recent large epidemiologic studies defining the association of HPV variant lineages/sublineages/SNPs with cervical cancer risk have led to new discoveries that call for HPV natural history and carcinogenesis to be re-visited. These findings also merit additional experimental studies using tools developed in the "omics" era. These novel discoveries underscore the importance of designing relevant comparisons to help sort out the differences in viral genetic features of carcinogenesis at the biochemical and mechanistic level. For example, across HPV genotypes, a large number of nucleotide differences may reveal more broad associations between HPV type and processes such as viral–host interactions, tissue tropism at the cellular level, splicing, and protein translation. In contrast, variant lineage/sublineage studies within a particular HPV type will allow for the identification of individual variants, or small groups of variant sites (haplotypes), related to differences in carcinogenicity. Integrating epidemiologic findings with functional studies may transform our basic understanding of HPV-associated carcinogenesis and may eventually elucidate the genetic basis defining what makes some HPVs, especially HPV16, such powerful carcinogens.

HPV carcinogenesis is a multifactorial complex process that involves a confluence of viral and host factors. However, compared with the complexities associated with studying the human genome, the genetic basis of HPV carcinogenicity in an 8000 bp genome is a more tractable problem that deserves immediate attention.

Supplementary Materials: The following are available online at www.mdpi.com/1999-4915/10/2/80/s1, Table S1: Full list of workshop attendees.

Acknowledgments: We thank Zigui Chen (The Chinese University of Hong Kong) for help with the HPV16 phylogenetic tree. This study was funded by the intramural research program of the Division of Cancer Epidemiology and Genetics, National Cancer Institute, National Institutes of Health. The content of this publication does not necessarily reflect the views or policies of the Department of Health and Human Services, nor does mention of trade names, commercial products, or organizations imply endorsement by the U.S. Government. This work was supported in part by the National Cancer Institute (CA78527) and the Einstein Cancer Research Center (P30CA013330) from the National Cancer Institute (to Robert D. Burk). Chase W. Nelson was supported by a Gerstner Scholars Fellowship from the Gerstner Family Foundation at the American Museum of Natural History.

Author Contributions: Lisa Mirabello and Robert D. Burk organized and led the workshop; all authors and attendees contributed to the workshop; Lisa Mirabello, Robert D. Burk, Megan A. Clarke and Chase W. Nelson drafted the manuscript; all authors participated in the editing of the manuscript.

Conflicts of Interest: The authors declare no conflict of interest.

References

1. Forman, D.; de Martel, C.; Lacey, C.J.; Soerjomataram, I.; Lortet-Tieulent, J.; Bruni, L.; Vignat, J.; Ferlay, J.; Bray, F.; Plummer, M.; et al. Global Burden of Human Papillomavirus and Related Diseases. *Vaccine* **2012**, *30* (Suppl. 5), F12–F23. [CrossRef] [PubMed]
2. De Martel, C.; Ferlay, J.; Franceschi, S.; Vignat, J.; Bray, F.; Forman, D.; Plummer, M. Global burden of cancers attributable to infections in 2008: A review and synthetic analysis. *Lancet Oncol.* **2012**, *13*, 607–615. [CrossRef]
3. De Martel, C.; Plummer, M.; Vignat, J.; Franceschi, S. Worldwide burden of cancer attributable to HPV by site, country and HPV type. *Int. J. Cancer* **2017**, *141*, 664–670. [CrossRef] [PubMed]
4. Global Burden of Disease Cancer Collaboration. The global burden of cancer 2013. *JAMA Oncol.* **2015**, *1*, 505–527.
5. Schiffman, M.; Herrero, R.; Desalle, R.; Hildesheim, A.; Wacholder, S.; Rodriguez, A.C.; Bratti, M.C.; Sherman, M.E.; Morales, J.; Guillen, D.; et al. The carcinogenicity of human papillomavirus types reflects viral evolution. *Virology* **2005**, *337*, 76–84. [CrossRef] [PubMed]
6. Chan, S.Y.; Bernard, H.U.; Ong, C.K.; Chan, S.P.; Hofmann, B.; Delius, H. Phylogenetic analysis of 48 papillomavirus types and 28 subtypes and variants: A showcase for the molecular evolution of DNA viruses. *J. Virol.* **1992**, *66*, 5714–5725. [PubMed]
7. Yamada, T.; Manos, M.M.; Peto, J.; Greer, C.E.; Munoz, N.; Bosch, F.X.; Wheeler, C.M. Human papillomavirus type 16 sequence variation in cervical cancers: A worldwide perspective. *J. Virol.* **1997**, *71*, 2463–2472. [PubMed]
8. Pimenoff, V.N.; de Oliveira, C.M.; Bravo, I.G. Transmission between Archaic and Modern Human Ancestors during the Evolution of the Oncogenic Human Papillomavirus 16. *Mol. Biol. Evol.* **2017**, *34*, 4–19. [CrossRef] [PubMed]
9. Chen, Z.; van Doorslaer, K.; DeSalle, R.; Wood, C.E.; Kaplan, J.R.; Wagner, J.D.; Burk, R.D. Genomic diversity and interspecies host infection of alpha12 Macaca fascicularis papillomaviruses (MfPVs). *Virology* **2009**, *393*, 304–310. [CrossRef] [PubMed]
10. Schiffman, M.; Doorbar, J.; Wentzensen, N.; de Sanjosé, S.; Fakhry, C.; Monk, B.J.; Stanley, M.A.; Franceschi, S. Carcinogenic human papillomavirus infection. *Nat. Rev. Dis. Primers* **2016**, *2*, 16086. [CrossRef] [PubMed]
11. Van Doorslaer, K. Evolution of the papillomaviridae. *Virology* **2013**, *445*, 11–20. [CrossRef] [PubMed]
12. Bernard, H.U. Regulatory elements in the viral genome. *Virology* **2013**, *445*, 197–204. [CrossRef] [PubMed]
13. Bernard, H.U.; Burk, R.D.; Chen, Z.; van Doorslaer, K.; zur Hausen, H.; de Villiers, E.M. Classification of papillomaviruses (PVs) based on 189 PV types and proposal of taxonomic amendments. *Virology* **2010**, *401*, 70–79. [CrossRef] [PubMed]
14. Bouvard, V.; Baan, R.; Straif, K.; Grosse, Y.; Secretan, B.; El Ghissassi, F.; Benbrahim-Tallaa, L.; Guha, N.; Freeman, C.; Galichet, L.; et al. A review of human carcinogens—Part B: Biological agents. *Lancet Oncol.* **2009**, *10*, 321–322. [CrossRef]
15. Burk, R.D.; Harari, A.; Chen, Z. Human papillomavirus genome variants. *Virology* **2013**, *445*, 232–243. [CrossRef] [PubMed]
16. Guan, P.; Howell-Jones, R.; Li, N.; Bruni, L.; de Sanjosé, S.; Franceschi, S.; Clifford, G.M. Human papillomavirus types in 115,789 HPV-positive women: A meta-analysis from cervical infection to cancer. *Int. J. Cancer* **2012**, *131*, 2349–2359. [CrossRef] [PubMed]
17. De Sanjose, S.; Quint, W.G.; Alemany, L.; Geraets, D.T.; Klaustermeier, J.E.; Lloveras, B.; Tous, S.; Felix, A.; Bravo, L.E.; Shin, H.R.; et al. Human papillomavirus genotype attribution in invasive cervical cancer: A retrospective cross-sectional worldwide study. *Lancet Oncol.* **2010**, *11*, 1048–1056. [CrossRef]
18. Serrano, B.; de Sanjosé, S.; Tous, S.; Quiros, B.; Muñoz, N.; Bosch, X.; Alemany, L. Human papillomavirus genotype attribution for HPVs 6, 11, 16, 18, 31, 33, 45, 52 and 58 in female anogenital lesions. *Eur. J. Cancer* **2015**, *51*, 1732–1741. [CrossRef] [PubMed]
19. Taylor, S.; Bunge, E.; Bakker, M.; Castellsagué, X. The incidence, clearance and persistence of non-cervical human papillomavirus infections: A systematic review of the literature. *BMC Infect. Dis.* **2016**, *16*, 293. [CrossRef] [PubMed]
20. Ndiaye, C.; Mena, M.; Alemany, L.; Arbyn, M.; Castellsagué, X.; Laporte, L.; Bosch, F.X.; de Sanjosé, S.; Trottier, H. HPV DNA, E6/E7 mRNA, and p16INK4a detection in head and neck cancers: A systematic review and meta-analysis. *Lancet Oncol.* **2014**, *15*, 1319–1331. [CrossRef]

21. Mirabello, L.; Yeager, M.; Cullen, M.; Boland, J.F.; Chen, Z.; Wentzensen, N.; Zhang, X.; Yu, K.; Yang, Q.; Mitchell, J.; et al. HPV16 Sublineage Associations with Histology-Specific Cancer Risk Using HPV Whole-Genome Sequences in 3200 Women. *J. Natl. Cancer Inst.* **2016**, *108*, djw100. [CrossRef] [PubMed]

22. Cornet, I.; Gheit, T.; Iannacone, M.R.; Vignat, J.; Sylla, B.S.; Del Mistro, A.; Franceschi, S.; Tommasino, M.; Clifford, G.M. HPV16 genetic variation and the development of cervical cancer worldwide. *Br. J. Cancer* **2013**, *108*, 240–244. [CrossRef] [PubMed]

23. Gheit, T.; Cornet, I.; Clifford, G.M.; Iftner, T.; Munk, C.; Tommasino, M.; Kjaer, S.K. Risks for persistence and progression by human papillomavirus type 16 variant lineages among a population-based sample of Danish women. *Cancer Epidemiol. Biomark. Prev.* **2011**, *20*, 1315–1321. [CrossRef] [PubMed]

24. Hildesheim, A.; Schiffman, M.; Bromley, C.; Wacholder, S.; Herrero, R.; Rodriguez, A.C.; Bratti, M.C.; Sherman, M.E.; Scarpidis, U.; Lin, Q.-Q.; et al. Human papillomavirus type 16 variants and risk of cervical cancer. *J. Natl. Cancer Inst.* **2001**, *93*, 315–318. [CrossRef] [PubMed]

25. Pientong, C.; Wongwarissara, P.; EkalaksananEmail, T.; Swangphon, P.; Kleebkaow, P.; Kongyingyoes, B.; Siriaunkgul, S.; Tungsinmunkong, K.; Suthipintawong, C. Association of human papillomavirus type 16 long control region mutation and cervical cancer. *Virol. J.* **2013**, *10*, 30. [CrossRef] [PubMed]

26. Rabelo-Santos, S.H.; Villa, L.L.; Derchain, S.F.; Ferreira, S.; Sarian, L.O.; Angelo-Andrade, L.A.; do Amaral Westin, M.C.; Zeferino, L.C. Variants of human papillomavirus types 16 and 18: Histological findings in women referred for atypical glandular cells or adenocarcinoma in situ in cervical smear. *Int. J. Gynecol. Pathol.* **2006**, *25*, 393–397. [CrossRef] [PubMed]

27. Schiffman, M.; Rodriguez, A.C.; Chen, Z.; Wacholder, S.; Herrero, R.; Hildesheim, A.; Desalle, R.; Befano, B.; Yu, K.; Safaeian, M.; et al. A population-based prospective study of carcinogenic human papillomavirus variant lineages, viral persistence, and cervical neoplasia. *Cancer Res.* **2010**, *70*, 3159–3169. [CrossRef] [PubMed]

28. Xi, L.F.; Koutsky, L.A.; Hildesheim, A.; Galloway, D.A.; Wheeler, C.M.; Winer, R.L.; Ho, J.; Kiviat, N.B. Risk for high-grade cervical intraepithelial neoplasia associated with variants of human papillomavirus types 16 and 18. *Cancer Epidemiol. Biomark. Prev.* **2007**, *16*, 4–10. [CrossRef] [PubMed]

29. Zehbe, I.; Voglino, G.; Delius, H.; Wilander, E.; Tommasino, M. Risk of cervical cancer and geographical variations of human papillomavirus 16 E6 polymorphisms. *Lancet* **1998**, *352*, 1441–1442. [CrossRef]

30. Zuna, R.E.; Moore, W.E.; Shanesmith, R.P.; Dunn, S.T.; Wang, S.S.; Schiffman, M.; Blakey, G.L.; Teel, T. Association of HPV16 E6 variants with diagnostic severity in cervical cytology samples of 354 women in a US population. *Int. J. Cancer* **2009**, *125*, 2609–2613. [CrossRef] [PubMed]

31. Sichero, L.; Ferreira, S.; Trottier, H.; Duarte-Franco, E.; Ferenczy, A.; Franco, E.L.; Villa, L.L. High grade cervical lesions are caused preferentially by non-European variants of HPVs 16 and 18. *International J. Cancer* **2007**, *120*, 1763–1768. [CrossRef] [PubMed]

32. Berumen, J.; Ordoñez, R.M.; Lazcano, E.; Salmeron, J.; Galvan, S.C.; Estrada, R.A.; Yunes, E.; Garcia-Carranca, A.; Gonzalez-Lira, G.; Madrigal-de la Campa, A.; et al. Asian-American Variants of Human Papillomavirus 16 and Risk for Cervical Cancer: A Case–Control Study. *J. Natl. Cancer Inst.* **2001**, *93*, 1325–1330. [CrossRef] [PubMed]

33. Freitas, L.B.; Chen, Z.; Muqui, E.F.; Boldrini, N.A.T.; Miranda, A.E.; Spano, L.C.; Burk, R.D. Human Papillomavirus 16 Non-European Variants Are Preferentially Associated with High-Grade Cervical Lesions. *PLoS ONE* **2014**, *9*, e100746. [CrossRef] [PubMed]

34. Burk, R.D.; Terai, M.; Gravitt, P.E.; Brinton, L.A.; Kurman, R.J.; Barnes, W.A.; Greenberg, M.D.; Hadjimichael, O.C.; Fu, L.; McGowan, L.; et al. Distribution of Human Papillomavirus Types 16 and 18 Variants in Squamous Cell Carcinomas and Adenocarcinomas of the Cervix. *Cancer Res.* **2003**, *63*, 7215–7220. [PubMed]

35. Quint, K.D.; de Koning, M.N.; van Doorn, L.J.; Quint, W.G.; Pirog, E.C. HPV genotyping and HPV16 variant analysis in glandular and squamous neoplastic lesions of the uterine cervix. *Gynecol. Oncol.* **2010**, *117*, 297–301. [CrossRef] [PubMed]

36. Nicolás-Párraga, S.; Alemany, L.; de Sanjosé, S.; Bosch, F.X.; Bravo, I.G.; RIS HPV TT and HPV VVAP Study Groups. Differential HPV16 variant distribution in squamous cell carcinoma, adenocarcinoma and adenosquamous cell carcinoma. *Int. J. Cancer* **2017**, *140*, 2092–2100.

37. Mirabello, L.; Yeager, M.; Yu, K.; Clifford, G.M.; Xiao, Y.; Zhu, B.; Cullen, M.; Boland, J.F.; Wentzensen, N.; Nelson, C.W.; et al. HPV16 E7 genetic conservation is critical to carcinogenesis. *Cell* **2017**, *170*, 1164–1174. [CrossRef] [PubMed]

38. Cullen, M.; Boland, J.F.; Schiffman, M.; Zhang, X.; Wenzensen, N.; Yang, Q.; Chen, Z.; Yu, K.; Mitchell, J.; Roberson, D.; et al. Deep sequencing of HPV16 genomes: A new high-throughput tool for exploring the carcinogenicity and natural history of HPV16 infection. *Papillomavirus Res.* **2015**, *1*, 3–11. [CrossRef] [PubMed]

39. Schiffman, M.; Kinney, W.K.; Cheung, L.C.; Gage, J.C.; Fetterman, B.; Poitras, N.E.; Lorey, T.S.; Wentzensen, N.; Befano, B.; Schussler, J.; et al. Relative Performance of HPV and Cytology Components of Cotesting in Cervical Screening. *JNCI J. Natl. Cancer Inst.* **2017**. [CrossRef] [PubMed]

40. Bray, F.; Carstensen, B.; Møller, H.; Zappa, M.; Zakelj, M.P.; Lawrence, G.; Hakama, M.; Weiderpass, E. Incidence Trends of Adenocarcinoma of the Cervix in 13 European Countries. *Cancer Epidemiol. Biomark. Prev.* **2005**, *14*, 2191–2199. [CrossRef] [PubMed]

41. Gien, L.T.; Beauchemin, M.-C.; Thomas, G. Adenocarcinoma: A unique cervical cancer. *Gynecol. Oncol.* **2010**, *116*, 140–146. [CrossRef] [PubMed]

42. Di Bonito, L.; Bergeron, C. Cytological screening of endocervical adenocarcinoma. *Ann. Pathol.* **2012**, *32*, e8–e14. [CrossRef] [PubMed]

43. Davy, M.L.J.; Dodd, T.J.; Luke, C.G.; Roder, D.M. Cervical cancer: Effect of glandular cell type on prognosis, treatment, and survival. *Obstet. Gynecol.* **2003**, *101*, 38–45. [CrossRef] [PubMed]

44. Ault, K.A.; Joura, E.A.; Kjaer, S.K.; Iversen, O.E.; Wheeler, C.M.; Perez, G.; Brown, D.R.; Koutsky, L.A.; Garland, S.M.; Olsson, S.E.; et al. Adenocarcinoma in situ and associated human papillomavirus type distribution observed in two clinical trials of a quadrivalent human papillomavirus vaccine. *Int. J. Cancer* **2011**, *128*, 1344–1353. [CrossRef] [PubMed]

45. The Cancer Genome Atlas Research Network. Integrated genomic and molecular characterization of cervical cancer. *Nature* **2017**, *543*, 378–384.

46. Litwin, T.; Clarke, M.A.; Dean, M.; Wentzensen, N. Somatic Host Cell Alterations in HPV Carcinogenesis. *Viruses* **2017**, *9*, 206. [CrossRef] [PubMed]

47. Doorbar, J.; Quint, W.; Banks, L.; Bravo, I.G.; Stoler, M.; Broker, T.R.; Stanley, M.A. The biology and life-cycle of human papillomaviruses. *Vaccine* **2012**, *30* (Suppl. 5), F55–F70. [CrossRef] [PubMed]

48. Herfs, M.; Yomamoto, Y.; Laury, A.; Wang, X.; Nucci, M.R.; McLaughlin-Drubin, M.E.; Münger, K.; Feldman, S.; McKeon, F.D.; Xian, W.; et al. A discrete population of squamocolumnar junction cells implicated in the pathogenesis of cervical cancer. *Proc. Natl. Acad. Sci. USA* **2012**, *109*, 10516–10521. [CrossRef] [PubMed]

49. Vande Pol, S.B.; Klingelhutz, A.J. Papillomavirus E6 oncoproteins. *Virology* **2013**, *445*, 115–137. [CrossRef] [PubMed]

50. Roman, A.; Munger, K. The papillomavirus E7 proteins. *Virology* **2013**, *445*, 138–168. [CrossRef] [PubMed]

51. Boon, S.S.; Banks, L. High-risk human papillomavirus E6 oncoproteins interact with 14-3-3zeta in a PDZ binding motif-dependent manner. *J. Virol.* **2013**, *87*, 1586–1595. [CrossRef] [PubMed]

52. Boon, S.S.; Tomaića, V.; Thomasa, M.; Robertsb, S.; Banksa, L. Cancer-causing human papillomavirus E6 proteins display major differences in the phospho-regulation of their PDZ interactions. *J. Virol.* **2015**, *89*, 1579–1586. [CrossRef] [PubMed]

53. Egawa, N.; Egawa, K.; Griffin, H.; Doorbar, J. Human Papillomaviruses; Epithelial Tropisms, and the Development of Neoplasia. *Viruses* **2015**, *7*, 3863–3890. [CrossRef] [PubMed]

54. Kranjec, C.; Doorbar, J. Human papillomavirus infection and induction of neoplasia: A matter of fitness. *Curr. Opin. Virol.* **2016**, *20*, 129–136. [CrossRef] [PubMed]

55. Palmer, E.; Newcombe, R.G.; Green, A.C.; Kelly, C.; Noel Gill, O.; Hall, G.; Fiander, A.N.; Pirotte, E.; Hibbitts, S.J.; Homer, J.; et al. Human papillomavirus infection is rare in nonmalignant tonsil tissue in the UK: Implications for tonsil cancer precursor lesions. *Int. J. Cancer* **2014**, *135*, 2437–2443. [CrossRef] [PubMed]

56. Schiffman, M.; Wentzensen, N. Human papillomavirus infection and the multistage carcinogenesis of cervical cancer. *Cancer Epidemiol. Biomark. Prev.* **2013**, *22*, 553–560. [CrossRef] [PubMed]

57. Schwartz, S. Papillomavirus transcripts and posttranscriptional regulation. *Virology* **2013**, *445*, 187–196. [CrossRef] [PubMed]

58. Johansson, C.; Schwartz, S. Regulation of human papillomavirus gene expression by splicing and polyadenylation. *Nat. Rev. Microbiol.* **2013**, *11*, 239–251. [CrossRef] [PubMed]

59. Kajitani, N.; Schwartz, S. RNA Binding Proteins that Control Human Papillomavirus Gene Expression. *Biomolecules* **2015**, *5*, 758–774. [CrossRef] [PubMed]

60. Straub, E.; Dreer, M.; Fertey, J.; Iftner, T.; Stubenrauch, F. The viral E8^E2C repressor limits productive replication of human papillomavirus 16. *J. Virol.* **2014**, *88*, 937–947. [CrossRef] [PubMed]

61. Vinokurova, S.; Wentzensen, N.; Kraus, I.; Klaes, R.; Driesch, C.; Melsheimer, P.; Kisseljov, F.; Dürst, M.; Schneider, A.; von Knebel Doeberitz, M.; et al. Type-dependent integration frequency of human papillomavirus genomes in cervical lesions. *Cancer Res.* **2008**, *68*, 307–313. [CrossRef] [PubMed]

62. Wentzensen, N.; Ridder, R.; Klaes, R.; Vinokurova, S.; Schaefer, U.; von Knebel Doeberitz, M. Characterization of viral-cellular fusion transcripts in a large series of HPV16 and 18 positive anogenital lesions. *Oncogene* **2002**, *21*, 419–426. [CrossRef] [PubMed]

63. Akagi, K.; Li, J.; Broutian, T.R.; Padilla-Nash, H.; Xiao, W.; Jiang, B.; Rocco, J.W.; Teknos, T.N.; Kumar, B.; Wangsa, D.; et al. Genome-wide analysis of HPV integration in human cancers reveals recurrent, focal genomic instability. *Genome Res.* **2014**, *24*, 185–199. [CrossRef] [PubMed]

64. Wentzensen, N.; Vinokurova, S.; von Knebel Doeberitz, M. Systematic Review of Genomic Integration Sites of Human Papillomavirus Genomes in Epithelial Dysplasia and Invasive Cancer of the Female Lower Genital Tract. *Cancer Res.* **2004**, *64*, 3878–3884. [CrossRef] [PubMed]

65. Bodelon, C.; Vinokurova, S.; Sampson, J.N.; den Boon, J.A.; Walker, J.L.; Horswill, M.A.; Korthauer, K.; Schiffman, M.; Sherman, M.E.; Zuna, R.E.; et al. Chromosomal copy number alterations and HPV integration in cervical precancer and invasive cancer. *Carcinogenesis* **2016**, *37*, 188–196. [CrossRef] [PubMed]

66. Hu, Z.; Zhu, D.; Wang, W.; Li, W.; Jia, W.; Zeng, X.; Ding, W.; Yu, L.; Wang, X.; Wang, L.; et al. Genome-wide profiling of HPV integration in cervical cancer identifies clustered genomic hot spots and a potential microhomology-mediated integration mechanism. *Nat Genet.* **2015**, *47*, 158–163. [CrossRef] [PubMed]

67. Schmitz, M.; Driesch, C.; Jansen, L.; Runnebaum, I.B.; Dürst, M. Non-random integration of the HPV genome in cervical cancer. *PLoS ONE* **2012**, *7*, e39632. [CrossRef] [PubMed]

68. El Awady, M.K.; Kaplan, J.B.; O'Brien, S.J.; Burk, R.D. Molecular analysis of integrated human papillomavirus 16 sequences in the cervical cancer cell line SiHa. *Virology* **1987**, *159*, 389–398. [CrossRef]

69. Bodelon, C.; Untereiner, M.E.; Machiela, M.J.; Vinokurova, S.; Wentzensen, N. Genomic characterization of viral integration sites in HPV-related cancers. *Int. J. Cancer* **2016**, *139*, 2001–2011. [CrossRef] [PubMed]

70. Couturier, J.; Sastre-Garau, X.; Schneider-Maunoury, S.; Labib, A.; Orth, G. Integration of papillomavirus DNA near myc genes in genital carcinomas and its consequences for proto-oncogene expression. *J. Virol.* **1991**, *65*, 4534–4538. [PubMed]

71. Jackson, R.; Rosa, B.A.; Lameiras, S.; Cuninghame, S.; Bernard, J.; Floriano, W.B.; Lambert, P.F.; Nicolas, A.; Zehbe, I. Functional variants of human papillomavirus type 16 demonstrate host genome integration and transcriptional alterations corresponding to their unique cancer epidemiology. *BMC Genom.* **2016**, *17*, 851. [CrossRef] [PubMed]

72. De Jong, A.; van Poelgeest, M.I.; van der Hulst, J.M.; Drijfhout, J.W.; Fleuren, G.J.; Melief, C.J.; Kenter, G.; Offringa, R.; van der Burg, S.H. Human Papillomavirus Type 16-Positive Cervical Cancer Is Associated with Impaired CD4+ T-Cell Immunity against Early Antigens E2 and E6. *Cancer Res.* **2004**, *64*, 5449–5455. [CrossRef] [PubMed]

73. Chen, Z.; Terai, M.; Fu, L.; Herrero, R.; DeSalle, R.; Burk, R.D. Diversifying Selection in Human Papillomavirus Type 16 Lineages Based on Complete Genome Analyses. *J. Virol.* **2005**, *79*, 7014–7023. [CrossRef] [PubMed]

74. Roura, E.; Castellsagué, X.; Pawlita, M.; Travier, N.; Waterboer, T.; Margall, N.; Bosch, F.X.; de Sanjosé, S.; Dillner, J.; Gram, I.T.; et al. Smoking as a major risk factor for cervical cancer and pre-cancer: Results from the EPIC cohort. *Int. J. Cancer* **2014**, *135*, 453–466. [CrossRef] [PubMed]

75. Chen, D.; Juko-Pecirep, I.; Hammer, J.; Ivansson, E.; Enroth, S.; Gustavsson, I.; Feuk, L.; Magnusson, P.K.; McKay, J.D.; Wilander, E.; et al. Genome-wide Association Study of Susceptibility Loci for Cervical Cancer. *J. Natl. Cancer Inst.* **2013**, *105*, 624–633. [CrossRef] [PubMed]

76. Shi, Y.; Li, L.; Hu, Z.; Li, S.; Wang, S.; Liu, J.; Wu, C.; He, L.; Zhou, J.; Li, Z.; et al. A genome-wide association study identifies two new cervical cancer susceptibility loci at 4q12 and 17q12. *Nat Genet.* **2013**, *45*, 918–922. [CrossRef] [PubMed]

77. Ivansson, E.L.; Juko-Pecirep, I.; Erlich, H.A.; Gyllensten, U.B. Pathway-based analysis of genetic susceptibility to cervical cancer in situ: HLA-DPB1 affects risk in Swedish women. *Genes Immun.* **2011**, *12*, 605–614. [CrossRef] [PubMed]

78. Zehbe, I.; Mytilineos, J.; Wikström, I.; Henriksen, R.; Edler, L.; Tommasino, M. Association between human papillomavirus 16 E6 variants and human leukocyte antigen class I polymorphism in cervical cancer of Swedish women. *Hum. Immunol.* **2003**, *64*, 538–542. [CrossRef]

79. Zehbe, I.; Tachezy, R.; Mytilineos, J.; Voglino, G.; Mikyskova, I.; Delius, H.; Marongiu, A.; Gissmann, L.; Wilander, E.; Tommasino, M. Human papillomavirus 16 E6 polymorphisms in cervical lesions from different European populations and their correlation with human leukocyte antigen class II haplotypes. *Int. J. Cancer* **2001**, *94*, 711–716. [CrossRef] [PubMed]

viruses

MDPI

Communication

Molecular Mechanisms of Human Papillomavirus Induced Skin Carcinogenesis

Martin Hufbauer and Baki Akgül * (ID)

Institute of Virology, University of Cologne, Fürst-Pückler-Str. 56, 50935 Cologne, Germany;
martin.hufbauer@uk-koeln.de
* Correspondence: baki.akguel@uk-koeln.de; Tel.: +49-221-478-85820; Fax: +49-221-478-85802

Academic Editors: Alison A. McBride and Karl Munger
Received: 27 June 2017; Accepted: 7 July 2017; Published: 14 July 2017

Abstract: Infection of the cutaneous skin with human papillomaviruses (HPV) of genus *betapapillomavirus* (βHPV) is associated with the development of premalignant actinic keratoses and squamous cell carcinoma. Due to the higher viral loads of βHPVs in actinic keratoses than in cancerous lesions, it is currently discussed that these viruses play a carcinogenic role in cancer initiation. In vitro assays performed to characterize the cell transforming activities of high-risk HPV types of genus *alphapapillomavirus* have markedly contributed to the present knowledge on their oncogenic functions. However, these assays failed to detect oncogenic functions of βHPV early proteins. They were not suitable for investigations aiming to study the interactive role of βHPV positive epidermis with mesenchymal cells and the extracellular matrix. This review focuses on βHPV gene functions with special focus on oncogenic mechanisms that may be relevant for skin cancer development.

Keywords: *betapapillomavirus*, extracellular matrix; invasion; cancer initiating cells; wound healing; squamous cell carcinoma

1. Introduction

Keratinocyte derived squamous cell carcinoma (SCC) is the most common metastatic skin cancer, and its incidence is increasing worldwide [1]. Exposure to ultraviolet (UV) radiation is accepted to be the main risk factor for skin carcinogenesis. Most skin SCC arise in association with a distinct precancerous lesion, the actinic keratosis. Epidemiological data indicate that human papillomaviruses (HPV) of genus *betapapillomavirus* (βHPV) may have a co-factorial role in this process [2–5]. The oncogenic potential of βHPV in skin carcinogenesis was originally identified in patients suffering from the rare inherited disease Epidermodysplasia verruciformis (EV), who have an increased susceptibility to βHPV infections. These viruses can also be found in skin cancers of non-EV patients. Also, immunosuppressed organ-transplant-recipients (OTR) have a higher susceptibility to βHPV infection in the skin as well as an increased risk of developing SCC compared with healthy individuals [6–8]. However, the association of βHPV infection and skin SCC development in the normal population remains controversial. Here, higher βHPV prevalence rates and viral loads can only be found in precancerous actinic keratoses but are missing in SCC [9]. These observations, however, are compatible with a carcinogenic role of these viruses in skin cancer initiation, probably through a hit-and-run mechanism where the virus acts as a co-factor along with UV to promote driver mutations in stem cells.

In contrast to high-risk *alphapapillomaviruses* (αHPV), that are causative factors for the development of cervical, anal, and oropharyngeal cancers [10], the molecular mechanisms that underlie the role of βHPV types in SCC development are less well understood. The specific tropism of cutaneous and mucosal HPV strains suggests that the life cycle is tied to particular epithelial cell types, which implies that cancer models for cutaneous HPVs cannot be unrestrictedly adopted from those for mucosal types. In vitro studies performed with conventional submerged keratinocyte cultures have markedly contributed to the present knowledge on the keratinocyte transforming activities by αHPV. However, these in vitro assays were not suitable for investigations aiming to study the interactive role of βHPV positive keratinocytes with mesenchymal cells and the extracellular matrix (ECM), as the spatial tissue organization in monolayer cultures is missing. This review highlights the recent progress in the field and outlines some unresolved questions related to oncogenic functions of βHPV during skin cancer initiation.

2. Invasion is Regulated Both at the Level of the Tumor Cell and the Extracellular Matrix

Alterations in the microenvironment of keratinocytes are required for initiation and progression over the course of cancer development. Difficulties in generating differentiating epithelia in vitro have hampered molecular studies involving cutaneous types and the extracellular matrix. Three dimensional organotypic skin cultures are useful systems for the in vitro analysis of skin biology because they mimic keratinocyte differentiation far better than monolayer cultures [11,12]. These experimental setups employ human cutaneous keratinocytes and a mesenchymal matrix that is repopulated with dermal fibroblasts. Cellular functions requiring epithelial differentiation, cell and extracellular matrix interactions, or keratinocyte-fibroblast paracrine communications can be observed and analyzed to a much higher degree. Boxman et al. [13] were the first to analyze the potential cell transforming activities of cutaneous HPV types in such an organotypic skin culture model. To study the effect of cutaneous HPV gene expression on keratinocytes, organotypic cultures based on a collagen type I matrix were repopulated with 3T3 mouse fibroblasts and primary keratinocytes expressing the *E6/E7* genes of βHPV types HPV5, 12, 15, 17, 20, and 38 as well as the high-risk αHPV type HPV16. These cultures showed varying degrees of dysregulated keratinocyte differentiation but lacked the features of cancer progression including the key step of basement membrane invasion for any of the HPV types analyzed. However, collagen-based organotypic cultures seem to mimic an in vivo environment in which invasion-specific βHPV oncoprotein functions cannot be carried out. In light of this observation, it needs to be noted that such exclusively collagen-comprised organotypic cultures lack the epidermal basement membrane and other ECM proteins. Over the last several years, it has become increasingly evident that tumor development requires cross-talk between different cell types within the tumor and its surrounding stroma. The latter provides a connective tissue, which is believed to possess a defined composition of structural and cellular components [14]. The knowledge of the most important constituents of the extracellular matrix, their metabolism and degradation provides insight into the pathophysiology of malignant growth [15]. With the published data of Boxman and colleagues in mind [13], we set out to investigate the effects of E6 and E7 of the βHPV type 8 (HPV8) in a more physiological skin-equivalent three dimensional (3D) model which sets itself apart from other skin culture models by using a de-epidermalized human dermis repopulated with either *E6* or *E7* positive cutaneous keratinocytes. Under these conditions, HPV8-E6 positive regenerated skin showed less epithelial layers compared to matched controls, which indicated at that time that E6 inhibits the normal differentiation program in keratinocytes [16,17]. In particular HPV8-E7 positive keratinocytes displayed significantly altered proliferation and differentiation. These skin cultures displayed enhanced terminal cell differentiation and a hyperproliferative phenotype, as evidenced by increased cornification and the presence of dividing cells in suprabasal layers. In addition and most strikingly, the cells lost their normal polarity and gained the ability to invade the dermal matrix. Migration of keratinocytes downward into the dermis was facilitated by degradation of components of the basement membrane and the extracellular matrix through the induction of the expression of matrix metalloproteinases [18,19]. As a result, basement membrane integrity was compromised in a time-dependent manner as evidenced by the degradation of collagen VII, collagen IV, and laminin-V. These observations revealed the hitherto unknown activity of βHPV—the ability to promote cell invasion. In addition, the accumulated data suggested that understanding of the composition of the tumor stroma and the interaction of βHPV positive keratinocytes with the ECM are important in discerning mechanisms regulating cell invasion. However, the molecular basis regulating the invasion of βHPV positive keratinocytes was still not known.

In the context of carcinoma pathogenesis, the conversion of normal keratinocytes to cancer cells induces an epithelial-mesenchymal-transition (EMT), which is associated with changes in intercellular adhesion molecules [20]. At the molecular level, this involves a reorganization of cell-cell adhesion complexes and modifications in cell-matrix interactions. Related to cell-cell connections, HPV8-E7 possesses the capacity to deregulate cell-cell junctions through the upregulation of β-catenin, zona occuldens protein 1 (ZO-1) [21], and the AKT serine/threonine kinase 2 (AKT2) [17]. Studies based on cell-matrix interactions revealed that the ECM proteins collagen IV, laminin V, and fibronectin all trigger HPV8-E7-mediated invasion. However, invasion associated with an EMT phenotype was only observed on fibronectin matrices. This indicates that ECM proteins can exert different modes of invasion. Most interestingly, fibronectin, but not collagen variants nor laminin-V were found to be deposited in peritumoral areas in HPV8 positive skin SCC that could have been produced and secreted by infected keratinocytes and by keratinocyte-stimulated fibroblasts. Again, only fibronectin led to a shift in the cell surface expression profiles of integrin, which are glycoprotein receptors mediating cell-matrix adhesion [15,22]. Invasive keratinocytes showed enhanced cell surface localization of integrin α3β1. Silencing of the α3 chain or using the 8E7 mutant L23A which is incapable of inducing invasion, prevented keratinocyte invasion, thus providing evidence for a role of the α3β1 integrin and fibronectin interplay in the invasion of HPV8 expressing keratinocytes [23]. Sonnenberg's group identified that epidermal-specific deletion of α3β1 integrin leads to a reduction of skin tumorigenesis in mice. However, tumors that did form progressed more rapidly to invasive carcinoma, implying a requirement for elevated α3β1 cell surface presence during tumor initiation and early growth [24]. These findings and our observations provide some evidence that α3β1 integrin and fibronectin may represent mediators of invasion during the initiation of the invasion cascade by βHPV.

Unfortunately, the answer how fibronectin and α3β1 integrin regulate keratinocyte invasion is not clear as yet because α3β1 is mainly considered to be a laminin-V receptor. It also needs to be noted that increased invasion often results from reduced adhesion to the ECM [25,26]. Furthermore, previous investigative efforts to ascertain a potential fibronectin binding by α3β1 have been contradictory. However, studies showing minimal binding of α3β1 to fibronectin were performed on cells cultured on plastic, whereas several other studies confirming this interaction utilized ECM interaction assays [27]. Yet many of these studies employed cell-based attachment assays, in which the binding affinity of α3β1 to ECM ligands was influenced in the presence of additional integrins [28,29]. Studies on the cross-talk between α3β1 with other fibronectin binding integrins may shed light on the underlying mechanism leading to keratinocyte invasion.

3. Mouse Models

Further in vivo evidence for the cell transforming ability of HPV8 early proteins was provided with the generation of HPV8 transgenic mice. Nearly all transgenics containing the complete early genome region of HPV8 under the control of the human keratin-14 promoter (K14-HPV8-CER) developed papillomas, partially with moderate or severe dysplasia. In 6% of the animals, SCC developed without previous exposure to physical or chemical carcinogens [30,31]. A mechanistic link between HPV8 protein expression and UV exposure in the development of skin tumors could be established by induction of skin papillomas in UVA/B treated animals in about three weeks after treatment [32,33]. Proteolytic activity identified in the peritumoral stroma and epidermal sheets hinted at a cross-talk between HPV8 positive keratinocytes and stromal cells [31]. Cross-breeding K14-HPV8-CER animals with *Stat3* heterozygous animals or mice with epidermal knock-out of *Rac1* inhibited papilloma formation. This pointed to an important role of Stat3- and Rac1-dependent pathways in HPV8 induced tumorigenesis [34,35]. By expressing individual HPV8 proteins under the control of the keratin14 promoter, the tumorigenic potential of the individual viral proteins could be analyzed in greater detail. HPV8-E2 mice presented mild to severe skin dysplasia mostly in their second year of life [36]. The kinetics of tumor development in K14-HPV8-E6 mice were comparable to K14-HPV8-CER mice [37]. This and the observation that E6 expression is necessary and sufficient for induction of papilloma formation [32] pointed to E6 as the major oncogene of HPV8 in the murine epidermis. While K14-HPV8-E6wt mice developed papillomas within three weeks post UV irradiation, skin tumor formation was significantly inhibited in mice, in which the E6 mutant K136N was expressed (K14-HPV8-E6K136N) which, in contrast to E6wt, is unable to impair DNA damage repair. These results provided further evidence for an interplay of UV-light and βHPV gene expression [38]. K14-HPV8-E7 animals showed no papilloma formation despite very low transgene expression. However, they exhibited carcinoma in situ formation early after chronic UVA/B treatment [23]. In conclusion, studies performed on HPV8 as well as HPV38 transgenic mice [39,40] uncovered several oncogenic functions of βHPV early proteins and represent suitable models to mimic high viral early gene expression as seen in EV or in OTR patients. However, to establish a possible hit-and-run mechanism of tumorigenesis it would be a required prerequisite to establish suitable animal models where the kinetics of βHPV gene expression and disease onset can be studied with translational applicability to the normal human population. Animals with βHPV genes under the control of inducible promoters may represent appropriate models for mechanistic evaluations on HPV oncogene functions in early phases of skin cancer development. In addition, the multimammate rat Mastomys coucha, whose skin is naturally infected—similar to βHPV in humans—with the Mastomys natalensis papillomavirus (MnPV) [41,42] may provide an animal model in which tumorigenic processes can be studied with a background of the complete viral life cycle.

4. Do βHPV Affect Self-Renewal of Infected Keratinocytes?

An apparent prerequisite for cancer initiation is the property to generate cancer stem cells [43]. Epidermal stem cells are located either in the basal epidermis or in the hair follicles. Like all stem cells, they are able to undergo asymmetric cell division, giving rise to one true stem cell which ensures that their population is not depleted, and a second, transiently amplifying cell that maintains and further amplifies along its differentiation axis before undergoing terminal differentiation [44]. Since βHPV were consistently detected on plucked hairs, it is postulated that the natural reservoir for βHPV latent infection resides within the hair follicular stem cells [45–47]. Considering that in skin SCC only a minority of cells harbor βHPV and viral functions are not required for cancer maintenance, a "hit-and-run" mechanism would be an explanation for the contribution of βHPV to skin cancer development. This implies that the expansion of the tumor is driven by virally-infected cancer stem cells, in which UV-induced driver mutations accumulated are caused by the interference of the viral E6 protein with the DNA repair mechanisms [38,48,49]. The marked decrease in the number of apoptotic cells in βHPV positive SCC compared to HPV negative cancers, and the continued expression of proliferation markers [50,51], indicates that the balance between apoptosis and proliferation is modified in βHPV-containing lesions. This may be the result of a cellular cross-talk between the βHPV positive stem cells and normal neighboring keratinocytes. The identification of soluble mediators or still unknown mechanisms may aid in our efforts to understand how βHPV positive stem cells communicate with and modulate their cellular environment.

In a very recent study, Marisa Gariglio and her team addressed the role of different hair follicle stem cells in HPV8-induced skin cancer development utilizing K14-HPV8-CER mice [52]. They identified the leucine rich repeats and immunoglobulin like domains 1 (Lrig1) positive stem cell population, residing in the hair follicle junctional zone [53], to be expanded in HPV8 transgenic animals. Proliferation in these cells is induced by the overexpression of the p63 protein lacking the N-terminal domain (ΔNp63) that is accepted as an epidermal stemness marker [52]. At least for HPV8-E6, it is known to trigger ΔNp63 levels [54,55]. It still needs to be determined whether other HPV8 early proteins also target ΔNp63 and which viral protein is responsible for the expansion of the Lrig1 positive stem cell population. Our group has previously shown that HPV8 early proteins can increase the number of keratinocytes with stem cell-like properties in monolayer culture. In these assays, HPV8-E7 in particular increased the clonogenicity of transduced keratinocytes and led to the formation of tumor spheres, which are generally regarded as cancer stem cells with self-renewal and tumorigenic capacities. Stem cell-like characteristics were associated with an increase in the epidermal stemness markers cluster of differentiation (CD)44 and epithelial cell adhesion molecule (EpCAM) [56]. Interestingly, the α3 integrin chain, crucial for the regulation of cell invasion, is known to be co-expressed with CD44 and EpCAM on tumor cell lines [57] and was indeed expressed in higher levels on *E7* positive cells with a CD44high/EpCAMhigh immunophenotype. Accumulating evidence indicates that EMT processes contribute to the progression of several carcinoma types (reviewed in reference [20]) and result in the generation of epithelial cells with stem-like properties [58]. It remains to be seen whether the growth of βHPV oncoprotein expressing keratinocytes on a fibronectin matrix may trigger cells to enter faster into this state. In line with the established fact that βHPV types (including HPV8) inhibit keratinocyte differentiation, apoptosis and DNA damage repair processes [38,55,59,60], these observations indicate that βHPV may increase the pool of cells in which DNA damage can persist and lead to the generation of stem cells with malignant properties. Additional experimental evidence is required to address and clarify more precisely the events occurring in βHPV positive stem cells. The identification and molecular characterization of these βHPV positive keratinocytes may aid in understanding the hit-and-run mechanism of βHPV-initiated skin cancer development (Figure 1). A precise understanding of the molecular program involved in regulating self-renewal and long-term survival in these cells may lead to potential drug targets and is therefore also of clinical relevance.

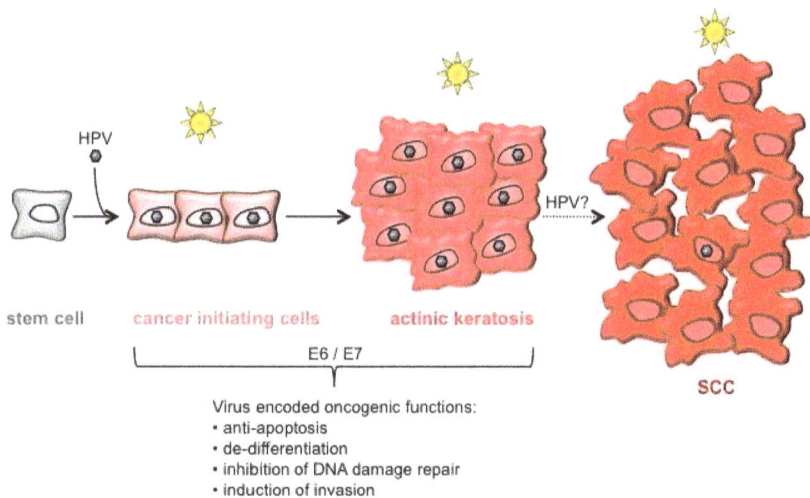

Figure 1. Model of keratinocyte transformation stages that may represent initial phases of *betapapillomavirus* (βHPV)-induced squamous cell carcinoma (SCC) development. The accepted risk factor for SCC development is ultra-violet (UV)-irradiation. Infection with βHPV may play a carcinogenic role in the early phases of SCC development. The human papillomavirus (HPV)-mediated expansion of the stem cell pool may allow the generation of cancer-initiating stem cells, which can give rise to actinic keratoses. Yet unknown is the exact role of HPV in the progression of actinic keratoses to SCC, in which only few cells harbor viral genomes.

5. Do Wound-Healing Processes Play a Role in βHPV-Mediated Cancer Initiation?

Another open question is the issue of whether signals that result in basal keratinocyte activation, spreading, and migration during cutaneous wound re-epithelialization may also contribute to βHPV-mediated cancer initiation. It is well established that tissue repair relies on stem cells and that chronic wounds predispose for tumor formation [61]. Such associations, together with similarities in the histology of wounds and tumors, led Dvorak to the often-cited conclusion that "tumors are wounds that do not heal" [62]. Several lines of evidence indicate that wound-healing processes may contribute to βHPV-induced cancer initiation as well:

(1) K14-HPV8-CER mice spontaneously developed papillomas and skin SCC without any treatment with physical or chemical carcinogens. Interestingly, these spontaneous skin lesions arose mostly in places where the animals had also scratched themselves. These results led to the hypothesis that activated keratinocytes that drive re-epithelialization fail to switch off the activated status upon finalized wound-healing and do not return to their normal differentiation pathway.

(2) While full-thickness wounding induced papilloma growth at the wound sites in 100% of K14-HPV8-E6wt animals, in marked contrast, none of the *FVB/n* control nor of K14-HPV8-E6K136N mice developed lesions following wounding (Figure 2A). It has been demonstrated that wound-healing-induced oxidative stress and the resulting reactive oxygen species (ROS) can affect the transcription of DNA damage repair enzymes. Yet, it may also directly inactivate DNA repair by oxidation. ROS can induce a number of modifications to DNA including single strand breaks, double strand breaks (DSBs), strand cross-links, along with protein-DNA cross-links and therefore increase the rate of mutations which may contribute to tumorigenesis [63]. We found that phosphorylation of the Ser-139 residue of the histone variant H2AX, forming γH2AX, a surrogate marker for DSB, were present in the skin tumors of K14-HPV8-E6wt mice taken 24 days after full-thickness wounding but not in healed skin of *FVB/n* and K14-HPV8-E6K136N animals (Figure 2B). These results support the hypothesis that wound-healing processes together with βHPV oncogene expression may contribute to the persistence of DSB and consequently to skin tumorigenesis.

(3) Tumor progression in HPV8 transgenic mice was paralleled by a strong inflammatory response within the tumor stroma, dominated by macrophages promoting tumor initiation and progression. In addition, wounding-associated tumor initiation was prevented after epidermal-specific depletion of vascular endothelial growth factor (VEGF) in HPV8 transgenic mice [64].

(4) The HPV8 target proteins fibronectin and α3β1 integrin are critical factors regulating epithelial repair during wound-healing [65–67].

The understanding of the similarities among wounding and UV-induced skin tumors and how βHPV-induced oxidative stress interferes with normal DNA damage repair may yield further insights into the mechanisms behind the impaired DNA damage response that may drive cancer initiation in infected cells.

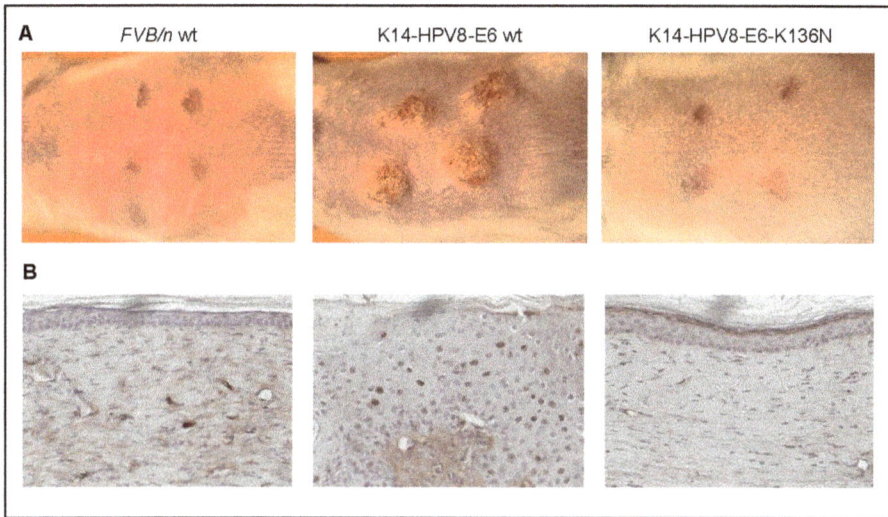

Figure 2. Persistence of DNA damage in full-thickness wounded K14-HPV8-E6 wt mice. Five weeks old *FVB/n* wt, K14-HPV8-E6 wt and K14-HPV8-E6K136N mice (*n* = 8 for each mouse line) were anaesthetized and wounded dorsal-caudal. Four punch biopsies were taken from the skin of the back creating four circular 4-mm wounds according to previously published procedures [37]. (**A**) Representative macroscopical images of animals taken 24 days after full-thickness wounding; (**B**) Paraffin embedded sections of from treated skin areas were stained for γH2AX (clone EP854(2)Y, Millipore (Darmstadt, Germany), 1:750 dilution used overnight). Sections were developed using the Vectastain Elite ABC kit (Linaris, Dossenheim, Germany; brown staining) and counterstained with haematoxylin (violet staining) (magnification: 400×). These experiments were approved by the governmental animal care office North-Rhine-Westphalia (approval no. 8.87-50.10.35.08.163).

6. Conclusions

While much of the underlying general mechanisms of βHPV-initiated keratinocyte transformation are unknown, recent discoveries have aided in directing future work towards the molecular basis of βHPV oncogenesis. We still need to understand the molecular alterations of how βHPV-mediated clonal expansion of cells with enriched UV signature mutations can result in the formation of pre-cancerous lesions and drive the progression to SCC in the absence of the virus. Since several βHPV types are potentially involved in the development of skin SCC, future research may hopefully identify more general oncogenic mechanisms of βHPV—in addition to the known genus-specific targeting of Mastermind-like protein-1 (MAML1) by E6 [68–70] and pRb, UBR4, KCMF1, and PTPN14 by E7 [69,71,72]—that underline the proposed hit-and-run mechanism of tumorigenesis.

Acknowledgments: This work was supported financially by support from the German Cancer Aid (Deutsche Krebshilfe, 111087) and the Wilhelm-Sander Stiftung (2012.105.3).

Author Contributions: B.A. reviewed the literature and wrote the paper. M.H. performed the mouse experiments, generated the figures and approved the final version of the manuscript.

Conflicts of Interest: The authors declare no conflict of interest.

References

1. Rogers, H.W.; Weinstock, M.A.; Harris, A.R.; Hinckley, M.R.; Feldman, S.R.; Fleischer, A.B.; Coldiron, B.M. Incidence estimate of nonmelanoma skin cancer in the United States, 2006. *Arch. Dermatol.* **2010**, *146*, 283–287. [CrossRef] [PubMed]
2. Iannacone, M.R.; Gheit, T.; Waterboer, T.; Giuliano, A.R.; Messina, J.L.; Fenske, N.A.; Cherpelis, B.S.; Sondak, V.K.; Roetzheim, R.G.; Michael, K.M.; Tommasino, M.; Pawlita, M.; Rollison, D.E. Case-control study of cutaneous human papillomaviruses in squamous cell carcinoma of the skin. *Cancer Epidemiol. Biomark. Prev.* **2012**, *21*, 1303–1313. [CrossRef] [PubMed]
3. Neale, R.E.; Weissenborn, S.; Abeni, D.; Bavinck, J.N.; Euvrard, S.; Feltkamp, M.C.; Green, A.C.; Harwood, C.; de Koning, M.; Naldi, L.; Nindl, I.; Pawlita, M.; Proby, C.; Quint, W.G.; Waterboer, T.; Wieland, U.; Pfister, H. Human papillomavirus load in eyebrow hair follicles and risk of cutaneous squamous cell carcinoma. *Cancer Epidemiol. Biomark. Prev.* **2013**, *22*, 719–727. [CrossRef] [PubMed]
4. Iannacone, M.R.; Gheit, T.; Pfister, H.; Giuliano, A.R.; Messina, J.L.; Fenske, N.A.; Cherpelis, B.S.; Sondak, V.K.; Roetzheim, R.G.; Silling, S.; Pawlita, M.; Tommasino, M.; Rollison, D.E. Case-control study of genus-β *human papillomaviruses* in plucked eyebrow hairs and cutaneous squamous cell carcinoma. *Int. J. Cancer* **2014**, *134*, 2231–2244. [CrossRef] [PubMed]
5. Genders, R.E.; Mazlom, H.; Michel, A.; Plasmeijer, E.I.; Quint, K.D.; Pawlita, M.; van der Meijden, E.; Waterboer, T.; de Fijter, H.; Claas, F.H.; et al. The presence of βpapillomavirus antibodies around transplantation predicts the development of keratinocyte carcinoma in organ transplant recipients: A cohort study. *J. Investig. Dermatol.* **2015**, *135*, 1275–1282. [CrossRef] [PubMed]
6. Howley, P.M.; Pfister, H.J. Beta genus papillomaviruses and skin cancer. *Virology* **2015**, *479–480*, 290–296. [CrossRef] [PubMed]
7. Smola, S. Human papillomaviruses and skin cancer. *Adv. Exp. Med. Biol.* **2014**, *810*, 192–207. [PubMed]
8. Tommasino, M. The biology of β human papillomaviruses. *Virus Res.* **2017**, *231*, 128–138. [CrossRef] [PubMed]
9. Weissenborn, S.J.; Nindl, I.; Purdie, K.; Harwood, C.; Proby, C.; Breuer, J.; Majewski, S.; Pfister, H.; Wieland, U. Human papillomavirus-DNA loads in actinic keratoses exceed those in non-melanoma skin cancers. *J. Investig. Dermatol.* **2005**, *125*, 93–97. [CrossRef] [PubMed]
10. Hübbers, C.U.; Akgül, B. HPV and cancer of the oral cavity. *Virulence* **2015**, *6*, 244–248. [CrossRef] [PubMed]
11. Banerjee, N.S.; Chow, L.T.; Broker, T.R. Retrovirus-mediated gene transfer to analyze HPV gene regulation and protein functions in organotypic "raft" cultures. *Methods Mol. Med.* **2005**, *119*, 187–202. [PubMed]
12. Lee, D.; Norby, K.; Hayes, M.; Chiu, Y.F.; Sugden, B.; Lambert, P.F. Using Organotypic Epithelial Tissue Culture to Study the Human Papillomavirus Life Cycle. *Curr. Protoc. Microbiol.* **2016**, *41*, 14B-8-1–14B-8-19. [PubMed]
13. Boxman, I.L.; Mulder, L.H.; Noya, F.; de Waard, V.; Gibbs, S.; Broker, T.R.; ten Kate, F.; Chow, L.T.; ter Schegget, J. Transduction of the *E6* and *E7* genes of epidermodysplasia-verruciformis-associated human papillomaviruses alters human keratinocyte growth and differentiation in organotypic cultures. *J. Investig. Dermatol.* **2001**, *117*, 1397–1404. [CrossRef] [PubMed]
14. Lim, Y.Z.; South, A.P. Tumour-stroma crosstalk in the development of squamous cell carcinoma. *Int. J. Biochem. Cell Biol.* **2014**, *53*, 450–458. [CrossRef] [PubMed]
15. Watt, F.M.; Fujiwara, H. Cell-extracellular matrix interactions in normal and diseased skin. *Cold Spring Harb. Perspect. Biol.* **2011**, *3*. [CrossRef] [PubMed]
16. Leverrier, S.; Bergamaschi, D.; Ghali, L.; Ola, A.; Warnes, G.; Akgül, B.; Blight, K.; Garcia-Escudero, R.; Penna, A.; Eddaoudi, A.; et al. Role of HPV E6 proteins in preventing UVB-induced release of pro-apoptotic factors from the mitochondria. *Apoptosis* **2007**, *12*, 549–560. [CrossRef] [PubMed]
17. O'Shaughnessy, R.F.; Akgül, B.; Storey, A.; Pfister, H.; Harwood, C.A.; Byrne, C. Cutaneous human papillomaviruses down-regulate AKT1, whereas AKT2 up-regulation and activation associates with tumors. *Cancer Res.* **2007**, *67*, 8207–8215. [CrossRef] [PubMed]
18. Akgül, B.; Garcia-Escudero, R.; Ghali, L.; Pfister, H.J.; Fuchs, P.G.; Navsaria, H.; Storey, A. The E7 protein of cutaneous human papillomavirus type 8 causes invasion of human keratinocytes into the dermis in organotypic cultures of skin. *Cancer Res.* **2005**, *65*, 2216–2223. [CrossRef] [PubMed]

19. Smola-Hess, S.; Pahne, J.; Mauch, C.; Zigrino, P.; Smola, H.; Pfister, H.J. Expression of membrane type 1 matrix metalloproteinase in papillomavirus-positive cells: Role of the human papillomavirus (HPV) 16 and HPV8 E7 gene products. *J. Gen. Virol.* **2005**, *86*, 1291–1296. [CrossRef] [PubMed]
20. Pattabiraman, D.R.; Weinberg, R.A. Tackling the cancer stem cells - what challenges do they pose? *Nat. Rev. Drug Discov.* **2014**, *13*, 497–512. [CrossRef] [PubMed]
21. Heuser, S.; Hufbauer, M.; Marx, B.; Tok, A.; Majewski, S.; Pfister, H.; Akgul, B. The levels of epithelial anchor proteins β-catenin and zona occludens-1 are altered by E7 of human papillomaviruses 5 and 8. *J. Gen. Virol.* **2016**, *97*, 463–472. [CrossRef] [PubMed]
22. De Franceschi, N.; Hamidi, H.; Alanko, J.; Sahgal, P.; Ivaska, J. Integrin traffic—the update. *J. Cell Sci.* **2015**, *128*, 839–852. [CrossRef] [PubMed]
23. Heuser, S.; Hufbauer, M.; Steiger, J.; Marshall, J.; Sterner-Kock, A.; Mauch, C.; Zigrino, P.; Akgul, B. The fibronectin/α3β1 integrin axis serves as molecular basis for keratinocyte invasion induced by βHPV. *Oncogene* **2016**, *35*, 4529–4539. [CrossRef] [PubMed]
24. Sachs, N.; Secades, P.; van Hulst, L.; Kreft, M.; Song, J.Y.; Sonnenberg, A. Loss of integrin α3 prevents skin tumor formation by promoting epidermal turnover and depletion of slow-cycling cells. *Proc. Natl. Acad. Sci. USA* **2012**, *109*, 21468–21473. [CrossRef] [PubMed]
25. Thiery, J.P.; Sleeman, J.P. Complex networks orchestrate epithelial-mesenchymal transitions. *Nat. Rev. Mol. Cell Biol.* **2006**, *7*, 131–142. [CrossRef] [PubMed]
26. Thiery, J.P.; Acloque, H.; Huang, R.Y.; Nieto, M.A. Epithelial-mesenchymal transitions in development and disease. *Cell* **2009**, *139*, 871–890. [CrossRef] [PubMed]
27. Brown, A.C.; Dysart, M.M.; Clarke, K.C.; Stabenfeldt, S.E.; Barker, T.H. Integrin α3β1 Binding to Fibronectin Is Dependent on the Ninth Type III Repeat. *J. Biol. Chem.* **2015**, *290*, 25534–25547. [CrossRef] [PubMed]
28. Hodivala-Dilke, K.M.; DiPersio, C.M.; Kreidberg, J.A.; Hynes, R.O. Novel roles for α3β1 integrin as a regulator of cytoskeletal assembly and as a trans-dominant inhibitor of integrin receptor function in mouse keratinocytes. *J. Cell Biol.* **1998**, *142*, 1357–1369. [CrossRef] [PubMed]
29. Longmate, W.M.; Lyons, S.P.; Chittur, S.V.; Pumiglia, K.M.; Van De Water, L.; DiPersio, C.M. Suppression of integrin α3β1 by α9β1 in the epidermis controls the paracrine resolution of wound angiogenesis. *J. Cell Biol.* **2017**, *216*, 1473–1488. [CrossRef] [PubMed]
30. Schaper, I.D.; Marcuzzi, G.P.; Weissenborn, S.J.; Kasper, H.U.; Dries, V.; Smyth, N.; Fuchs, P.; Pfister, H. Development of skin tumors in mice transgenic for early genes of human papillomavirus type 8. *Cancer Res.* **2005**, *65*, 1394–1400. [CrossRef] [PubMed]
31. Akgül, B.; Pfefferle, R.; Marcuzzi, G.P.; Zigrino, P.; Krieg, T.; Pfister, H.; Mauch, C. Expression of matrix metalloproteinase (MMP)-2, MMP-9, MMP-13, and MT1-MMP in skin tumors of human papillomavirus type 8 transgenic mice. *Exp. Dermatol.* **2006**, *15*, 35–42. [CrossRef] [PubMed]
32. Hufbauer, M.; Lazic, D.; Akgül, B.; Brandsma, J.L.; Pfister, H.; Weissenborn, S.J. Enhanced human papillomavirus type 8 oncogene expression levels are crucial for skin tumorigenesis in transgenic mice. *Virology* **2010**, *403*, 128–136. [CrossRef] [PubMed]
33. Hufbauer, M.; Lazic, D.; Reinartz, M.; Akgül, B.; Pfister, H.; Weissenborn, S.J. Skin tumor formation in human papillomavirus 8 transgenic mice is associated with a deregulation of oncogenic miRNAs and their tumor suppressive targets. *J. Dermatol. Sci.* **2011**, *64*, 7–15. [CrossRef] [PubMed]
34. De Andrea, M.; Ritta, M.; Landini, M.M.; Borgogna, C.; Mondini, M.; Kern, F.; Ehrenreiter, K.; Baccarini, M.; Marcuzzi, G.P.; Smola, S.; et al. Keratinocyte-specific Stat3 heterozygosity impairs development of skin tumors in human papillomavirus 8 transgenic mice. *Cancer Res.* **2010**, *70*, 7938–7948. [CrossRef] [PubMed]
35. Deshmukh, J.; Pofahl, R.; Pfister, H.; Haase, I. Deletion of epidermal Rac1 inhibits HPV-8 induced skin papilloma formation and facilitates HPV-8- and UV-light induced skin carcinogenesis. *Oncotarget* **2016**, *7*, 57841–57850. [CrossRef] [PubMed]
36. Pfefferle, R.; Marcuzzi, G.P.; Akgül, B.; Kasper, H.U.; Schulze, F.; Haase, I.; Wickenhauser, C.; Pfister, H. The human papillomavirus type 8 E2 protein induces skin tumors in transgenic mice. *J. Investig. Dermatol.* **2008**, *128*, 2310–2315. [CrossRef] [PubMed]
37. Marcuzzi, G.P.; Hufbauer, M.; Kasper, H.U.; Weissenborn, S.J.; Smola, S.; Pfister, H. Spontaneous tumour development in human papillomavirus type 8 E6 transgenic mice and rapid induction by UV-light exposure and wounding. *J. Gen. Virol.* **2009**, *90*, 2855–2864. [CrossRef] [PubMed]

38. Hufbauer, M.; Cooke, J.; van der Horst, G.T.; Pfister, H.; Storey, A.; Akgül, B. Human papillomavirus mediated inhibition of DNA damage sensing and repair drives skin carcinogenesis. *Mol. Cancer* **2015**, *14*, 183. [CrossRef] [PubMed]

39. Viarisio, D.; Mueller-Decker, K.; Kloz, U.; Aengeneyndt, B.; Kopp-Schneider, A.; Grone, H.J.; Gheit, T.; Flechtenmacher, C.; Gissmann, L.; Tommasino, M. E6 and E7 from beta HPV38 cooperate with ultraviolet light in the development of actinic keratosis-like lesions and squamous cell carcinoma in mice. *PLoS Pathog.* **2011**, *7*, e1002125. [CrossRef] [PubMed]

40. Viarisio, D.; Muller-Decker, K.; Hassel, J.C.; Alvarez, J.C.; Flechtenmacher, C.; Pawlita, M.; Gissmann, L.; Tommasino, M. The BRAF Inhibitor Vemurafenib Enhances UV-Induced Skin Carcinogenesis in β HPV38 E6 and E7 Transgenic Mice. *J. Investig. Dermatol.* **2017**, *137*, 261–264. [CrossRef] [PubMed]

41. Vinzon, S.E.; Braspenning-Wesch, I.; Muller, M.; Geissler, E.K.; Nindl, I.; Grone, H.J.; Schafer, K.; Rosl, F. Protective vaccination against papillomavirus-induced skin tumors under immunocompetent and immunosuppressive conditions: A preclinical study using a natural outbred animal model. *PLoS Pathog.* **2014**, *10*, e1003924. [CrossRef] [PubMed]

42. Vinzon, S.E.; Rosl, F. HPV vaccination for prevention of skin cancer. *Hum. Vaccines Immunother.* **2015**, *11*, 353–357. [CrossRef] [PubMed]

43. Lambert, A.W.; Pattabiraman, D.R.; Weinberg, R.A. Emerging Biological Principles of Metastasis. *Cell* **2017**, *168*, 670–691. [CrossRef] [PubMed]

44. Fuchs, E. Epithelial Skin Biology: Three Decades of Developmental Biology, a Hundred Questions Answered and a Thousand New Ones to Address. *Curr. Top. Dev. Biol.* **2016**, *116*, 357–374. [PubMed]

45. Boxman, I.L.; Berkhout, R.J.; Mulder, L.H.; Wolkers, M.C.; Bouwes Bavinck, J.N.; Vermeer, B.J.; ter Schegget, J. Detection of human papillomavirus DNA in plucked hairs from renal transplant recipients and healthy volunteers. *J. Investig. Dermatol.* **1997**, *108*, 712–715. [CrossRef] [PubMed]

46. De Koning, M.N.; Struijk, L.; Bavinck, J.N.; Kleter, B.; ter Schegget, J.; Quint, W.G.; Feltkamp, M.C. Betapapillomaviruses frequently persist in the skin of healthy individuals. *J. Gen. Virol.* **2007**, *88*, 1489–1495. [CrossRef] [PubMed]

47. Bouwes Bavinck, J.N.; Plasmeijer, E.I.; Feltkamp, M.C. β-papillomavirus infection and skin cancer. *J. Investig. Dermatol.* **2008**, *128*, 1355–1358. [CrossRef] [PubMed]

48. Wallace, N.A.; Robinson, K.; Howie, H.L.; Galloway, D.A. HPV 5 and 8 E6 abrogate ATR activity resulting in increased persistence of UVB induced DNA damage. *PLoS Pathog.* **2012**, *8*, e1002807. [CrossRef] [PubMed]

49. Wallace, N.A.; Robinson, K.; Howie, H.L.; Galloway, D.A. beta-HPV 5 and 8 E6 disrupt homology dependent double strand break repair by attenuating BRCA1 and BRCA2 expression and foci formation. *PLoS Pathog.* **2015**, *11*, e1004687. [CrossRef] [PubMed]

50. Jackson, S.; Harwood, C.; Thomas, M.; Banks, L.; Storey, A. Role of Bak in UV-induced apoptosis in skin cancer and abrogation by HPV E6 proteins. *Genes Dev.* **2000**, *14*, 3065–3073. [CrossRef] [PubMed]

51. Jackson, S.; Ghali, L.; Harwood, C.; Storey, A. Reduced apoptotic levels in squamous but not basal cell carcinomas correlates with detection of cutaneous human papillomavirus. *Br. J. Cancer* **2002**, *87*, 319–323. [CrossRef] [PubMed]

52. Lanfredini, S.; Olivero, C.; Borgogna, C.; Calati, F.; Powell, K.; Davies, K.J.; De Andrea, M.; Harries, S.; Tang, H.K.C.; Pfister, H.; et al. HPV8 Field Cancerization in a Transgenic Mouse Model is due to Lrig1+ Keratinocyte Stem Cell Expansion. *J. Investig. Dermatol.* **2017**. [CrossRef] [PubMed]

53. Kretzschmar, K.; Watt, F.M. Markers of epidermal stem cell subpopulations in adult mammalian skin. *Cold Spring Harb. Perspect. Med.* **2014**, *4*. [CrossRef] [PubMed]

54. Marthaler, A.M.; Podgorska, M.; Feld, P.; Fingerle, A.; Knerr-Rupp, K.; Grasser, F.; Smola, H.; Roemer, K.; Ebert, E.; Kim, Y.J.; et al. Identification of C/EBPα as a novel target of the HPV8 E6 protein regulating miR-203 in human keratinocytes. *PLoS Pathog.* **2017**, *13*, e1006406. [CrossRef] [PubMed]

55. Meyers, J.M.; Spangle, J.M.; Munger, K. The human papillomavirus type 8 E6 protein interferes with NOTCH activation during keratinocyte differentiation. *J. Virol.* **2013**, *87*, 4762–4767. [CrossRef] [PubMed]

56. Hufbauer, M.; Biddle, A.; Borgogna, C.; Gariglio, M.; Doorbar, J.; Storey, A.; Pfister, H.; Mackenzie, I.; Akgül, B. Expression of betapapillomavirus oncogenes increases the number of keratinocytes with stem cell-like properties. *J. Virol.* **2013**, *87*, 12158–12165. [CrossRef] [PubMed]

57. Schmidt, D.S.; Klingbeil, P.; Schnolzer, M.; Zoller, M. CD44 variant isoforms associate with tetraspanins and EpCAM. *Exp. Cell. Res.* **2004**, *297*, 329–347. [CrossRef] [PubMed]

58. Mani, S.A.; Guo, W.; Liao, M.J.; Eaton, E.N.; Ayyanan, A.; Zhou, A.Y.; Brooks, M.; Reinhard, F.; Zhang, C.C.; Shipitsin, M.; et al. The epithelial-mesenchymal transition generates cells with properties of stem cells. *Cell* **2008**, *133*, 704–715. [CrossRef] [PubMed]

59. Holloway, A.; Simmonds, M.; Azad, A.; Fox, J.L.; Storey, A. Resistance to UV-induced apoptosis by β-HPV5 E6 involves targeting of activated BAK for proteolysis by recruitment of the HERC1 ubiquitin ligase. *Int. J. Cancer* **2015**, *136*, 2831–2843. [CrossRef] [PubMed]

60. Wallace, N.A.; Galloway, D.A. Manipulation of cellular DNA damage repair machinery facilitates propagation of human papillomaviruses. *Semi. Cancer Biol.* **2014**, *26*, 30–42. [CrossRef] [PubMed]

61. Arwert, E.N.; Hoste, E.; Watt, F.M. Epithelial stem cells, wound healing and cancer. *Nat. Rev. Cancer* **2012**, *12*, 170–180. [CrossRef] [PubMed]

62. Dvorak, H.F. Tumors: Wounds that do not heal. Similarities between tumor stroma generation and wound healing. *N. Engl. J. Med.* **1986**, *315*, 1650–1659. [PubMed]

63. Melis, J.P.; van Steeg, H.; Luijten, M. Oxidative DNA damage and nucleotide excision repair. *Antioxid Redox Signal.* **2013**, *18*, 2409–2419. [CrossRef] [PubMed]

64. Ding, X.; Lucas, T.; Marcuzzi, G.P.; Pfister, H.; Eming, S.A. Distinct functions of epidermal and myeloid-derived VEGF-A in skin tumorigenesis mediated by HPV8. *Cancer Res.* **2015**, *75*, 330–343. [CrossRef] [PubMed]

65. Singh, P.; Carraher, C.; Schwarzbauer, J.E. Assembly of fibronectin extracellular matrix. *Annu. Rev. Cell Dev. Biol.* **2010**, *26*, 397–419. [CrossRef] [PubMed]

66. Sun, X.; Fa, P.; Cui, Z.; Xia, Y.; Sun, L.; Li, Z.; Tang, A.; Gui, Y.; Cai, Z. The EDA-containing cellular fibronectin induces epithelial-mesenchymal transition in lung cancer cells through integrin $\alpha 9 \beta 1$-mediated activation of PI3-K/AKT and ERK1/2. *Carcinogenesis* **2014**, *35*, 184–191. [PubMed]

67. Wen, T.; Zhang, Z.; Yu, Y.; Qu, H.; Koch, M.; Aumailley, M. Integrin $\alpha 3$ subunit regulates events linked to epithelial repair, including keratinocyte migration and protein expression. *Wound Repair Regen.* **2010**, *18*, 325–334. [CrossRef] [PubMed]

68. White, E.A.; Kramer, R.E.; Tan, M.J.; Hayes, S.D.; Harper, J.W.; Howley, P.M. Comprehensive analysis of host cellular interactions with human papillomavirus E6 proteins identifies new E6 binding partners and reflects viral diversity. *J. Virol.* **2012**, *86*, 13174–13186. [CrossRef] [PubMed]

69. White, E.A.; Howley, P.M. Proteomic approaches to the study of papillomavirus-host interactions. *Virology* **2013**, *435*, 57–69. [CrossRef] [PubMed]

70. Meyers, J.M.; Uberoi, A.; Grace, M.; Lambert, P.F.; Munger, K. Cutaneous HPV8 and MmuPV1 E6 Proteins Target the NOTCH and TGF-β Tumor Suppressors to Inhibit Differentiation and Sustain Keratinocyte Proliferation. *PLoS Pathog.* **2017**, *13*, e1006171. [CrossRef] [PubMed]

71. Akgül, B.; Ghali, L.; Davies, D.; Pfister, H.; Leigh, I.M.; Storey, A. HPV8 early genes modulate differentiation and cell cycle of primary human adult keratinocytes. *Exp. Dermatol.* **2007**, *16*, 590–599. [CrossRef] [PubMed]

72. Buitrago-Perez, A.; Hachimi, M.; Duenas, M.; Lloveras, B.; Santos, A.; Holguin, A.; Duarte, B.; Santiago, J.L.; Akgül, B.; Rodriguez-Peralto, J.L.; Storey, A.; Ribas, C.; Larcher, F.; del Rio, M.; Paramio, J.M.; Garcia-Escudero, R. A humanized mouse model of HPV-associated pathology driven by E7 expression. *PLoS ONE* **2012**, *7*, e41743. [CrossRef] [PubMed]

viruses

MDPI

Review

Integration of Human Papillomavirus Genomes in Head and Neck Cancer: Is It Time to Consider a Paradigm Shift?

Iain M. Morgan [1,2,*], Laurence J. DiNardo [2,3] and Brad Windle [1,2]

1   Philips Institute for Oral Health Research, Virginia Commonwealth University (VCU) School of Dentistry, Department of Oral and Craniofacial Molecular Biology, Richmond, VA 23298, USA; bwindle@vcu.edu
2   VCU Massey Cancer Center, Richmond, VA 23298, USA; laurence.dinardo@vcuhealth.org
3   VCU Department of Otolaryngology, Richmond, VA 23298, USA
*   Correspondence: immorgan@vcu.edu, Tel.: +1-804-828-0149

Academic Editors: Alison A. McBride and Karl Munger
Received: 30 June 2017; Accepted: 31 July 2017; Published: 3 August 2017

Abstract: Human papillomaviruses (HPV) are detected in 70–80% of oropharyngeal cancers in the developed world, the incidence of which has reached epidemic proportions. The current paradigm regarding the status of the viral genome in these cancers is that there are three situations: one where the viral genome remains episomal, one where the viral genome integrates into the host genome and a third where there is a mixture of both integrated and episomal HPV genomes. Our recent work suggests that this third category has been mischaracterized as having integrated HPV genomes; evidence indicates that this category consists of virus–human hybrid episomes. Most of these hybrid episomes are consistent with being maintained by replication from HPV origin. We discuss our evidence to support this new paradigm, how such genomes can arise, and more importantly the implications for the clinical management of HPV positive head and neck cancers following accurate determination of the viral genome status.

Keywords: head and neck cancer; human papillomavirus; human papillomavirus 16; HPV16; The Cancer Genome Atlas; oral keratinocytes; integration; episomal; mixed

1. Introduction

Recently we published an analysis of head and neck cancer data from The Cancer Genome Atlas (TCGA) providing evidence that the three types of genomic status for human papillomavirus (HPV) in head and neck cancer (HNC) are episomal, integrated and virus–human hybrid episomes replicating autonomously; there was very little evidence for the presence of so-called mixed tumors that have both episomal and integrated viral genomes [1]. In this review, we will describe HPV replication and the mechanisms that could promote viral genome breakage and therefore association with host DNA. We will then discuss the current status of understanding of HPV integration in head and neck cancer and explain why our results are not in conflict with those of others. The implications of our model for management of HPV-positive HNC will be described, particularly for individuals who are receiving de-escalation therapy in clinical trials. Finally, we will propose a set of diagnostic assays for predicting the genomic status of HPV, and the clinical importance of this status.

2. Human Papillomavirus Replication

The HPV life cycle is inextricably linked to the differentiation of the host epithelium [2,3]. Infection begins with targeting of basal epithelial cells, thought to be stem cells, followed by entry of the viral genome into the cell nucleus that requires mitosis of the infected cell [4,5]. Cellular factors then interact with the long control region (LCR) of papillomaviruses to activate transcription from the viral genome [6] and this results in expression of the viral genes and proteins. In high-risk HPV (HR-HPV, that causes cancer), E6 and E7 are viral oncoproteins that target the tumor suppressor proteins p53 and pRb [7] and this promotes replication of the infected cell, which then migrates through the differentiating epithelium. The initial phase of replication in the infected cells is called establishment, where the viral genome copy number increases to 20–50 copies per cell. Therefore, during this process the viral replication process is not under a strict once and once only per cell cycle control. By definition, the HPV has to overcome this limitation. The second replication phase of the viral life cycle is called the maintenance phase where, during the differentiation and proliferation of the infected cell, the viral genome copy number is controlled at 20–50 copies. Finally, in the differentiated epithelium when cell proliferation has been arrested the viral genome amplifies to around 1000 copies per cell before being encapsulated by L1 and L2 prior to viral particle egress from the cell. Given the number of genomes generated during the establishment and amplification replication phases, viral replication is very different from that of the host. This difference includes evasion of the tight control over host cell replication timing and quality.

There are two viral proteins that coordinate viral replication for all papillomaviruses in association with host proteins; E1 and E2. The E2 protein forms homodimers and binds to 12-bp palindromic DNA sequences surrounding the viral origin of replication in the LCR and recruits the viral helicase E1 to the origin of replication [8,9]. E1 interacts with host polymerases and other factors involved in DNA replication including single-stranded binding proteins [10–14], while E2 also interacts with cellular proteins to promote viral replication including TopBP1 [15–18], Brd4 [17,19] and ChlR1 [20].

3. Human Papillomavirus Replication and the DNA Damage Response: Primed for Integration

High-risk HPV activate a DNA damage response (DDR) that promotes the viral life cycle and several DDR proteins are involved directly with HPV replication [15,16,21–27]. The E7 protein elevates the levels of proteins involved in the DDR [28] while the E1 helicase activates a DDR when expressed in cells [29–31]. It also seems likely there is a combination of activation of the DDR by E1 combined with E7-mediated elevation of factors involved in this process. Therefore, HR-HPV cells are unusual in that they replicate and go through a cell cycle while a DDR is turned on in the cells; ordinarily the DDR that is activated in response to genotoxic stress arrests the cell cycle to promote repair of DNA damage. One proposed benefit to the virus for activation of the DDR is to recruit factors that promote homologous recombination to the viral genome in order to promote the amplification stage of the viral life cycle [32,33]. Significantly, DNA damaging agents do not arrest E1–E2-mediated DNA replication, even if the host cell replication is arrested, perhaps due to E1 not being a substrate for ATR/ATM [29,34]. This is not surprising as the virus activates the DDR to promote its own life cycle; inhibitors of the DDR block the viral life cycle [22] and therefore the virus must be able to replicate in the presence of DNA damage. However, the E1–E2-mediated DNA replication that occurs in the presence of DNA damaging agents is extremely mutagenic [29] even though the replication levels are not affected. The precise reason behind the requirement of an activated DDR to promote the HPV life cycle remains unclear but it does build in a degree of viral genomic instability into the viral life cycle. If the infected cell is genotoxically stressed, then the viral replication will continue and this replication in the presence of DNA-damaging agents could promote double-strand breaks in the viral genome. Such breakage would provide a substrate for integration into the host genome via non-homologous end joining mechanisms. Indeed, integration is observed in HR-HPV positive cancers.

4. Human Papillomavirus and Integration in Cervical Cancer

A large majority of cervical cancers are caused by HR-HPV [35]. There are a number of HR-HPV genomes that are causative in cervical cancer; HPV16 is present in around 50% while HPV18 is present in around 20%. The first demonstration that the HPV genome was adjoined with human DNA was made using cervical cancer DNA samples and also cervical cancer cell lines [36–52]. Originally this was primarily confirmed using Southern blotting of DNA digested with enzymes that cut the HPV16 genome in a single position or not at all. In many cases the disruption in cancer samples was in the E1 and E2 genes and this could also be observed in cell lines with HPV16 genomes; lack of E2 is associated with more aggressive tumor and cell line growth [53–59]. Overexpression of E2 can repress HPV LCRs in transient transfection experiments [60–62], therefore it was originally proposed that the loss of E2 is required to increase E6 and E7 expression and subsequent progression of the transformed cell to a more malignant phenotype. E2 has no effect on transcription from episomal HPV16 DNA in W12 cells (a cell line containing HPV16 derived from a cervical lesion) [63], although it can repress transcription from integrated genomes in cervical lines [63–67]. However, given the toxicity of the E1 protein in cells due to binding to and unwinding of host DNA resulting in a DDR [30,31], it is equally possible that the expression of E1 must be disrupted in order for the integrated cells to survive. E1 and E2 expression from integrated DNA would initiate DNA replication from the integrated viral LCR and this would create replication stress within the host genome throughout S phase and beyond as E1 and E2 replication is not under the control of host DNA rules with regards to a once and once only per cell cycle replication firing. However, there is a mechanism that allows for the presence of the E2 and E1 genes in tandemly integrated HPV genomes and that is methylation of the viral DNA [68–74]. In derivatives of W12 cells with integrated DNA there are two types that occur; Type 1 has only one copy of the viral genome and that is non-methylated while Type 2 has tandem integrants that are methylated to silence the viral genome, only one copy produces E6/E7 transcripts and there is no E1/E2 expressed. There are two cell lines that model this, SiHa has one copy of an integrated genome that is hypomethylated, while CaSki has multiple genomes integrated that are hypermethylated [75]. When methylation is reversed in CaSki cells using 5-aza-2'-deoxycytidine the cells die, perhaps due to overexpression of other viral genes such as E1 and E2.

5. Human Papillomavirus and Head and Neck Cancer

Head and neck cancer is the sixth most common worldwide, with around 600,000 cases diagnosed annually. Cancer incidence in the developed world is decreasing in general, due to smoking cessation; the one exception to this is the increasing cases of head and neck cancer. This increase is due to an increase in the incidence of HPV-positive head and neck cancer in the oropharyngeal region (HPV + OPC, oropharyngeal cancer) over the past several decades; HPV16 is causative in 80–90% of these tumors [76–81]. The incidence of HPV + OPC continues to increase [81]. The 5-year overall survival of patients with HPV + OPC is increased over two-fold when compared with HPV-negative head and neck cancers, and this is attributed to a good response to chemo–radiation therapy (CRT) [82–84]; the precise reason for the improved response of HPV + OPC to CRT is not clear. There are no diagnostics (such as Pap smears for cervical cancer) available for assisting with the management of HPV + OPC and no therapeutics available that directly target the viral life cycle.

6. Human Papillomavirus and Integration in Head and Neck Cancer

As described above, integration of the HPV genome into the host is a common feature of cervical cancer, and integration is an indicator of a poorer clinical outcome. The situation with HPV integration in head and neck cancer with relation to clinical outcome is much less clear. Early studies on HPV and integration in head and neck cancer employed fluorescence in situ hybridization (FISH) applied to interphase nuclei to visualize the viral genomes and concluded that the appearance of defined "dots" in the nuclei were indicative of viral genome integration [85,86]. This FISH approach was based around a report that suggested dispersed FISH staining indicates episomal genomes while punctate FISH staining suggests integrated viral genomes [87]. Other studies assumed the loss of the E2 gene during integration, as happens in cervical cancer, and measured the ratio between the E2 and E6 genes; an E2/E6 ratio of less than 1 is presumed to contain integrated DNA [88,89]. Using these approaches, investigators concluded that the majority of HPV-positive head and neck cancers had integrated viral genomes, similarly to cervical cancer. An additional approach for investigating viral genome integration is to carry out Amplification of Papillomavirus Oncogene Transcripts PCR (APOT-PCR) and also Detection of Integrated Papillomavirus Sequences PCR (DIPS-PCR); both of these techniques search for the association of viral with human DNA. These techniques were used on HPV-positive head and neck cancers and combined with analysis for the expression of the E6 and E2 viral genes [90]. This study identified 39% of the tumors with integrated DNA, while the remaining 61% had episomal viral genomes. In addition, even though they could see integration of the viral genome with loss of E2, they could still detect E2 transcripts in some of these tumors which they postulate is due to the presence of episomal viral genomes alongside the integrated DNA. Our own analysis indicated that ca. 66% of HPV-positive head and neck cancer samples had HPV joined to human DNA, however, our results lead us to a different interpretation of what these HPV–human DNA junctions represent.

Full RNA and DNA genomic sequences for HPV-positive head and neck cancers became available following publication of data from The Cancer Genome Atlas Network [91]. This original report from TCGA concentrated on the gene changes and mutations of the tumors and was not focused on the HPV status of the tumors; however, they did conclude that gene expression levels of genes associated with HPV integration were higher than those detected in corresponding non-integrated HPV-positive head and neck cancers. A subsequent publication from TCGA Network did focus on the status of the HPV genome in these tumors [92], carrying out the analysis of 279 tumors, 35 of which had evidence of HR-HPV types 16, 33 or 35. From this analysis they determined that 25 of these cases had evidence for integration of the viral genome into that of the host in regions that were more likely to be associated with genes. However, there was no strong association with any particular host gene although the limited sample size may be an explanation of the failure to find such an association. They also concluded that integration was not associated with clinical outcome, unlike for cervical cancer where integration does statistically result in a worse clinical outcome. Several other studies looking at HPV integration in head and neck cancer also determined that viral genome integration did not result in a worse clinical outcome [93,94].

Recently we published a report detailing our genomic analysis of all of the head and neck cancer samples from The Cancer Genome Atlas [1]. Here we will summarize these findings as it is pertinent to the main thrust of this review, that there is an alternative interpretation of the genomic structure of HPV in head and neck cancer. In this manuscript, we analyzed data from all 520 head and neck cancer samples and determined that 72 were HPV-positive as determined by viral gene expression and of those, 83% had HPV16 which we focused on for this report. After considering the HPV16 status and the availability of RNA-seq and DNA-seq data we analyzed 30 tumors in depth. At the genomic level, there were three types of tumors: (1) those that had deletions in the E1–E2 region of the viral genome suggesting an integration event; (2) those that had a constant number of reads across the entire HPV16 genome suggesting an intact genome and therefore an episomal genome; and (3) those that contained DNA covering the entire viral genome but that had a portion of the genome represented at around half the reads as the rest of the genome. This latter category could be arrived at if there was a mixture of integrated and episomal viral genomes present in the tumors. However, the striking feature of these tumors was the copy number for supposed integrated HPV and the copy number for supposed episomal HPV was ordinarily equivalent (there was as many predicted integrated genomes as there were predicted episomal). This was true for a broad spectrum of copy numbers, from 5 to 130. This struck us as unlikely to represent truly mixed tumors that contained episomal and integrated versions of the viral genome whether in the same cell or separate cells; one might expect the episomal genomes to be in large excess over the integrated numbers, but certainly not conspicuously equal at a statistically significant level. Another striking feature of our analysis of the three types of genomic HPV16 tumors was that in the purely integrated samples there was a low number of viral genomes per cell. This is to be expected for an integrated tumor; the median copy number was 1.7, compared to greater than 14 for episomal HPV samples. The RNA-seq data from the tumors confirmed that Category 1 tumors were truly integrated as there was no viral RNA transcript that went past the proposed integration site. There was a truncation of transcription within the E1 gene in all Category 1 samples, thus no functional E1 was produced and there was no expression of E2, E4, or E5. For the other two categories of tumors, the HPV genome was expressed intact throughout the early region with no deletions or differences in the levels of the particular viral genes between the two categories.

Analysis of the DNA-seq data supported our conclusion that Category 1 tumors were integrated tumors as hybrid viral–human reads that describe the integration point could be detected. Category 2 tumors were confirmed as predominantly episomal as there was no consistent detection of viral–human reads. The conclusion about our Category 1 and 2 tumors is no different from that arrived at by others. The difference in our interpretation is with the Category 3 tumors. Others would have called this a "mixed" tumor that contains both episomal and integrated viral DNA. This concept is difficult to reconcile with the observation that E1 can bind to and amplify DNA when it is integrated into that of the host [30,31,95]. Therefore, if there was a viral genome containing an origin of replication integrated into the host, and E1 and E2 expressed from an episome in the same cell, there would be nothing to prevent the viral replication complex recognizing the origin in the host genome and initiating DNA replication. This would be toxic; during initial infection E1–E2 replication is not controlled by a once and once only rule as is host replication, therefore repeated initiation of replication from host associated DNA would result in genome breakage and activated DNA damage response and repair machinery. In the long term, the presence of both integrated and episomal viral genomes in the same cell seems incompatible for this reason. All samples we assessed as integrated in our study had no expression of intact E1 and E2, consistent with this scenario. It is not clear from cervical cancer studies, where the presence of "mixed" tumors has also been proposed, whether the integrated and episomal genomes exist in the same cell in the tumor or whether they exist in separate cells within the same tumor. This is also a possibility for the head and neck cancer samples. However, the maintenance of the same number of both integrated and episomal genomes in separate cells in most samples is unlikely. Therefore, we considered alternative interpretations of the status of the HPV genome Category 3 tumors.

Our conclusion from analyzing TCGA Whole Genome Sequencing (WGS) data for Category 3 tumors is that the virus genome is not integrated into that of the host genome. Rather, the HPV genome replicates as an independent viral–human hybrid episome. This suggests that at some point the original episomal viral genome broke and ends joined to that of the host but subsequently the host DNA associated with the HPV DNA broke and a ligation event occurred resulting in excision of DNA that forms a viral–human hybrid circular episome. Potential mechanisms to explain this are shown in Figure 1. In both cases shown in this figure, E1 and E2 would still be expressed and would initiate replication from the viral origin of replication. This initiation is not controlled in cell cycle manner therefore multiple initiation events could occur forming an "onion skin" replication structure that could be resolved by double-strand DNA breaks. Following breakage of the host DNA the linear viral–human DNA could be ligated to form a circular viral–human hybrid episome. There are multiple pieces of evidence that support the presence of these viral–human hybrid episomes. Firstly, genomic DNA for HPV and joined human DNA maps as a contiguous circular structure, not a linear structure. Secondly, viral-associated human DNA is equivalent in amplification to that of viral DNA, which would be expected if within an episome. Thirdly, there were novel human–human DNA junctions mapped that represent the excision and joining of ends to form the viral–human hybrid episomal DNA elements (the green arrow in Figure 1B). The copy number for these hybrid human–human junctions are the same as of the viral genome structures and for the amplified associated human DNA. Fourthly, the viral–human DNA junctions had the same prevalence as the viral genomes; if there were multimers of the viral genome integrated into the host then this number should be much lower. Overall the evidence overwhelmingly suggests that in samples people have considered "mixed" tumors, the virus in fact is episomal and replicates joined to a segment of human DNA.

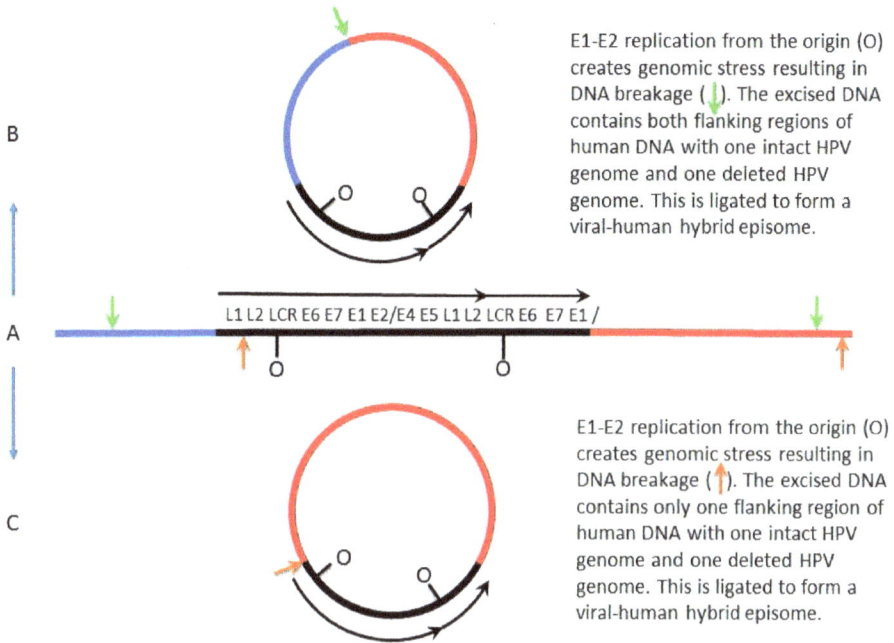

B

E1-E2 replication from the origin (O) creates genomic stress resulting in DNA breakage (↓). The excised DNA contains both flanking regions of human DNA with one intact HPV genome and one deleted HPV genome. This is ligated to form a viral–human hybrid episome.

A

L1 L2 LCR E6 E7 E1 E2/E4 E5 L1 L2 LCR E6 E7 E1 /

C

E1-E2 replication from the origin (O) creates genomic stress resulting in DNA breakage (↑). The excised DNA contains only one flanking region of human DNA with one intact HPV genome and one deleted HPV genome. This is ligated to form a viral–human hybrid episome.

Figure 1. Mechanisms for formation of viral–human hybrid episomes. In (**A**), a human papillomavirus 16 (HPV16) dimer has broken and integrated into the host genome, losing a part of one of the viral genomes as often happens during integration. E1 and E2 can still be expressed from the intact HPV genome and therefore the potential to initiate replication from the viral origins is retained; this initiation is not restricted to once per cell cycle therefore repeated initiation would form an "onion skin" replication bubble. This would create stress on the host genome resulting in double strand breaks promoting excision from the host genome. This viral–human hybrid DNA could then be ligated to form an episome. In (**B**), the breaks occur in the host genome flanking the integrated viral genome (the green arrows) and ligation occurs to form an episome that consists of the viral genome with two flanking regions of human DNA. In (**C**) the breaks in the DNA occur in the viral genome and in flanking human DNA (orange arrows) and ligation occurs to form a viral–human hybrid episome that consists of the viral genome with one flanking region of human DNA. LCR = long control region.

The conclusion from our analysis is that the viral genome is maintained episomally either as an intact HPV genome structure or an HPV–human DNA hybrid episome. If the HPV genome co-exists as both an integrated and episomal structure in tumors, it is not common. In Section 7 we discuss why our conclusions are not incompatible with the work of others, but that the differences are in interpretation of the data. TCGA data has been crucial to the development of our model as it provides much more in-depth information than simple diagnostic assays that have been used in the past to characterize the viral genome in head and neck cancers.

7. A Model for Integration and Excision of HPV DNA

The proposed initial driving force for integration is DNA breakage in which HPV is linearized in the E1 region and some DNA is removed from the ends to result in deletion. Breakage of the human DNA then leads to recombination of the free ends of the HPV DNA with the free ends of the human DNA, resulting in integration. For a monomeric HPV episome, the integrated HPV DNA will have lost the ability to express an intact E1 and any E2. The integrated HPV DNA will no longer initiate replication from its own origin within the LCR and will be replicated from human origins as part of the human genome. However, if the same breakage and integration events occurred for a dimeric HPV episomal genome, the result could be quite different. An integrated dimer of HPV DNA will retain the ability to express E1 and E2 from the unaltered copy of the HPV genome. This would allow the initiation of replication from the HPV origin while being within the human genome. Since HPV-origin firing can evade the one initiation per cell cycle control used for human origins, the HPV genome could go through multiple rounds of replication in a single cell cycle, which is predicted to form the proposed onion skin structure causing harm to the cell. This problem can be averted or solved by the excision of the HPV DNA to form a new episome. While we propose the excision encompasses the HPV DNA, there are possible options as to what DNA is excised. The HPV episome could be formed from excision of human DNA on both sides of the integrated HPV DNA encompassing all HPV DNA. The excised DNA could encompass human DNA on only one side of the integrated HPV DNA and most of the HPV DNA. Figure 1 summarizes these mechanisms graphically.

8. Conflicting Interpretations Rather Than Conflicting Results

The question that arises from our work [1] is why have others not observed this before, or why this interpretation has not been made before in head and neck cancer? One of the answers lies in the fact that there have not been a large number of studies in this area with clinical samples. Sequencing of head and neck cancer lines have been done and revealed a host of genomic changes associated with integrated HPV genomes [96]. However, in head and neck cancer cell lines we suggest that episomal HPV, whether by itself or as a viral–human episome, integrates into the host genome during culture; such an observation is common in cervical cell lines containing episomal HPV16 genomes [58]. Therefore, cell lines may not be an accurate model for studying the physical status of the HPV genome in actual cancers. Some studies have used APOT-PCR [90] and detected viral–human transcripts suggesting integration. However, such transcripts would also be detected in hybrid viral–human episomes, therefore they do not prove definitively that the viral genome is permanently integrated into that of the host. Indeed, in this particular study they observed deletion in the E2 region but the presence of E2 transcripts and concluded the presence of mixed tumors. However, their data is not incompatible with the existence of viral–human hybrid episomes that have lost one of two copies of the E2 gene. A standard approach for the characterization of the HPV genome status in both cervical and head and neck cancer is to measure the ratio of E2 DNA to E6 DNA and declare that if that ratio is less than one then there must be an integration event [88,89]. If the ratio is zero (i.e., there is no E2 DNA) then it would be consistent with an integrated tumor. However, in these types of studies it is common to observe ratios around 0.5 similar to our Category 3 tumors. If the virus replicates as a dimer and loses a copy of the E2 gene, then the ratio of E2 to E6 would be 0.5. Therefore, using an E2 to E6 ratio from PCR is not a good indicator of HPV genome integration. One of the earlier ways that the integration was predicted was to use FISH analysis where it was suggested that punctate staining of the viral genome indicated an integrated tumor while more dispersed staining indicated an episomal genome [87]. However, this is not a reliable mechanism for truly detecting integration. For example, Parvenov and colleagues [92] suggested that there may be viral–human structures in HPV-positive head and neck cancers and to investigate this used a FISH approach for the human DNA and observed punctate staining. It is of course possible that in this clinical sample the viral–human hybrid DNA is indeed integrated, but integration would have to have occurred in multiple points on the human genome as there were multiple FISH signals detected. In differentiating cervical cell lines that retain HPV episomes, the viral genome is amplified in large replication foci [32] and when this happens they do look punctate and would be interpreted as integrated using this classification. They are not. In addition, we also see in oral keratinocytes containing episomal HPV16 genomes the presence of detectable discreet viral foci by FISH even in non-differentiated epithelial cells [97]. These would be interpreted as integrated using this criterion, but they are not. Therefore, the punctate/dispersed staining criterion is not a reliable mechanism for truly predicting whether the viral genome is integrated or episomal.

We are not claiming that the work of others is wrong in any way. What we are proposing is that, in light of data from The Cancer Genome Atlas, it may be worthwhile re-interpreting older results. Our results are not in conflict with those of others. Figure 2 summarizes these thoughts.

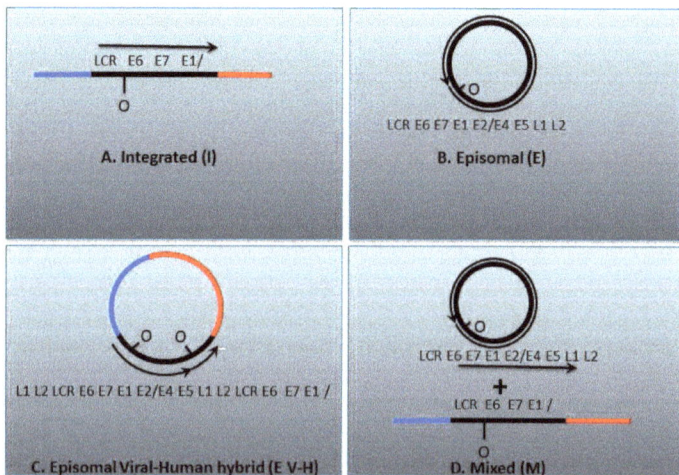

	I	E	E V-H	M
E2/E6 DNA ratio is less than 1	Y	N	Y	Y
Viral–human DNA hybrids	Y	N	Y	Y
Viral–human RNA hybrids	Y	N	Y	Y
Full length E2 RNA expression	N	Y	Y	Y
E5 RNA expression	N	Y	Y	Y
8-kbp band on Southern blot following digestion with single cutter	N	Y	Y	Y

Figure 2. The genomic status of HPV16 in head and neck cancers. (**A**) The HPV genome has broken and become integrated permanently into the host; (**B**) The HPV genome remains as an episome; (**C**) An HPV dimer (or multimer) has integrated and been excised along with human DNA and ligated together to form a viral–human hybrid DNA episome (see Figure 1); (**D**) Previously the viral–human hybrid episome tumors would have been assigned as a mixed tumor that contains both integrated and episomal viral genomes. The box below describes various tests that have been used to characterize the status of the HPV genome in cancers; Y = yes, N = no. It is notable that all of these tests carried out together can differentiate between integrated and episomal tumors, but they do not allow differentiation between whether the viral genome exists as an episomal viral–human hybrid or as a mixed tumor. We propose that these tumors have viral–human hybrid episomes following our analysis of The Cancer Genome Atlas data; so-called mixed tumors, if they exist at all in head and neck cancers, are rare. See the text for details.

9. Why This Matters: De-Escalation Therapy for Human Papillomavirus-Positive Head and Neck Cancer

HPV-positive head and neck cancer patients have a much improved overall survival when compared with HPV-negative patients [82]. For this reason, it has been proposed to treat HPV positive patients with a de-escalated therapy [98] with the aim of reducing the cytotoxic effects of CRT on these patients; there are ongoing clinical trials in this area. However, although HPV-positive head and neck cancers do in general respond better to therapy, a significant percentage do not, varying from 10 to 20% of patients [99]. Therefore, in planned de-escalation therapy it is important to be able to identify those HPV-positive patients who are predicted to not respond well to therapy; at the moment, there is no way to do this. p16 staining in HPV-positive head and neck cancers is the best prognostic marker for predicting clinical outcome; traditional staging parameters based around primary tumor extension, lymph node involvement and distant metastasis historically had no prognostic value for HPV-positive head and neck cancers [100]. More recent attempts at staging HPV-positive oropharyngeal cancers have met with some success in being able to predict worse clinical outcomes, but none of these systems offers a guaranteed path for identifying patients who will definitely fare worse clinically although there are detectable trends [101–104]. Almost all HPV-positive tumors exhibit p16 overexpression, therefore this marker is not able to identify those HPV-positive patients who will not respond well to therapy. For these reasons, even though there are ongoing de-escalation clinical trials, the recommended standard treatment for HPV-positive head and neck is no different from that used for HPV-negative patients.

One indicator for worse outcome in HPV-positive cervical cancer is integration of the viral genome into that of the host. However, several reports have claimed that this is not the case in HPV-positive head and neck cancer [93,94]. We propose that mischaracterization of integrated tumors is the reason for this and that in fact, those tumors that do have truly integrated viral genomes are those that do worse clinically. Our preliminary analysis of clinical outcomes for TCGA tumors based on our categorization of those that are truly integrated (Category 1) demonstrates that integration does predict a worse outcome (in preparation). Those tumors that we describe as Category 3, where there are viral–human hybrid genomes replicating as an episome, do as well clinically as Category 2 tumors, which have virus-only episomes. Previous studies in this area are confused by the definition of integration being defined by E2 to E6 ratios and/or the conclusion that "mixed" characterized tumors with integrated and episomal genomes co-existing exist. These studies would include Category 3 tumors (virus–human episomes) as having integrated viral genomes ("mixed") along with the truly integrated Category 1 tumors. It would be interesting for the authors of these reports to go back and reanalyze the data by including their "mixed" tumors as being predominantly episomal, or reanalyzing the samples using different techniques.

10. Future Approaches to the Diagnostic Management of Human Papillomavirus-Positive Head and Neck Cancer

What is the best approach for characterizing tumors that have truly integrated viral DNA? We can easily eliminate some tests. Measuring an E2 to E6 DNA ratio as an indicator of integration is invalid for several reasons. Firstly, there are several head and neck tumors where the integration of the viral genome is in the E1 gene and the E2 gene is actually retained intact, therefore a 1:1 ratio would falsely suggest these tumors were episomal. Secondly, in our model we show that the viral genome exists as dimers/multimers in episomal forms and that in some occasions the E2 gene from one of the viral genomes can be deleted. This would give an E2 to E6 ratio of less than 1 suggesting integration but this would not be the case. Thirdly, in our Category 3 tumors where we propose the virus is replicating as a viral–human episome it is a viral dimer/multimer that fuses to the host genome and again E2 can be lost in one of the copies of the viral genome. Therefore, in these tumors the E2 to E6 ratio would be less than 1 but the viral genome is not permanently integrated into that of the host. Another technique that has been used to confirm the presence of integrated and/or mixed tumors is Southern blotting. But again, in our model the results from such analysis are open to misinterpretation. For example, if a viral–human hybrid episome was digested with *Bam*H1 (a single cutter for HPV16) and bands are seen at 8 kbp and additional positions on the gel it is presumed that this represents a mixed tumor. The proposed viral–human hybrid genomes could give the same signal. In addition, Southern blotting of tumor DNA that does not cut the viral genome has been used to suggest integration but again, if the human DNA replicating with the virus in a hybrid episome is cut by the non-cutter enzyme confusing results can occur in addition to the size of the uncut virus–human hybrid also confusing the results. Two other assays that are used, APOT-PCR and DIPS-PCR, are used to monitor the presence of viral-human RNA and DNA products, respectively, and have been used to demonstrate integration. However, these techniques would give positive signals in Category 3 tumors where the viral genome is replicating as an episome as a viral–human hybrid. FISH has also been used to indicate the presence of integrated HPV DNA but this technique can also pick up episomal viral DNA and does not precisely define integration. Therefore, none of these techniques are appropriate for categorizing the genomic status of HPV in head and neck cancer.

From our preliminary analysis of outcome data from TCGA HPV-positive head and neck cancers the data demonstrates that our Category 1 patients (those that we predict only have integrated DNA) do worse clinically than those with tumors containing the virus as an episome, whether replicating with human DNA as a hybrid or not. This makes our challenge easier in many ways as, for clinical outcome concerns, a simple identification of HPV-positive tumors with truly integrated DNA is required. A simple assay for identifying these patients who are truly integrated is the absence of E2 through to E5 RNA. Indeed, a recent report looking at just this facet of HPV positive head and neck cancers demonstrated that, using the lack of E2 RNA as an indicator of integration, patients who lacked E2 expression had poorer clinical outcomes [105]. Therefore, a simple RNA in situ hybridization probe of formalin-fixed, paraffin-embedded clinical samples would allow stratification of HPV-positive tumors into integrated and non-integrated types; the presence of E6 RNA and the absence of E2/E5 RNA would predict integration. Such a simple characterization could be done with existing tumor samples allowing for outcome data analysis.

Therefore, we propose that as part of de-escalation clinical trials, the status of HPV-positive head and neck tumors with regards to E2/E5 expression should be considered. Patients who lack E2/E5 expression should be followed very carefully in such trials.

11. Conclusions

Looking at the head and neck cancer data from TCGA resulted in our model of the three different types of HPV genome status. Considering all possible explanations of what is present in the data, we do think it reasonable that "mixed" tumors are a category where the viral and human sequences replicate together as an episome. This matters as including this category as having integrated viral genomes has muddied the interpretation of outcome for HPV-positive patients. During de-escalation trials, it is important to identify patients who are at increased risk and we predict that these patients will be those who have truly integrated tumors. Much work remains to be done. We are currently working on establishing cell lines from head and neck cancers where we will look at the status of the viral genome at very early passage to ensure that the viral genome does not become integrated following medium-term cell culture. We are in the process of using E6/E2/E5 RNA in situ hybridization with tumor samples to determine if those that lack E2/E5 expression do have worse clinical outcomes. We conclude by asking the HPV and clinical communities to have an open mind with respect to our proposed model and consider it for testing.

Acknowledgments: This work was supported by the grant 1R03DE026230-01 sponsored by the National Institute for Dental and Craniofacial Research and by P30 CA016059 from the National Cancer Institute.

Author Contributions: I.M.M., L.J.D. and B.W. drafted and wrote the text.

Conflicts of Interest: The authors declare no conflict of interest.

References

1. Nulton, T.J.; Olex, A.L.; Dozmorov, M.; Morgan, I.M.; Windle, B. Analysis of the Cancer Genome Atlas Sequencing Data Reveals Novel Properties of the Human Papillomavirus 16 Genome in Head and Neck Squamous Cell Carcinoma. *Oncotarget* **2017**, *8*, 17684–17699. [CrossRef] [PubMed]
2. Hebner, C.M.; Laimins, L.A. Human Papillomaviruses: Basic Mechanisms of Pathogenesis and Oncogenicity. *Rev. Med. Virol.* **2006**, *16*, 83–97. [CrossRef] [PubMed]

3.	Doorbar, J.; Quint, W.; Banks, L.; Bravo, I.G.; Stoler, M.; Broker, T.R.; Stanley, M.A. The Biology and Life-Cycle of Human Papillomaviruses. *Vaccine* **2012**, *30* (Suppl. 5), F55–F70. [CrossRef] [PubMed]

4.	Calton, C.M.; Bronnimann, M.P.; Manson, A.R.; Li, S.; Chapman, J.A.; Suarez-Berumen, M.; Williamson, T.R.; Molugu, S.K.; Bernal, R.A.; Campos, S.K. Translocation of the Papillomavirus L2/vDNA Complex Across the Limiting Membrane Requires the Onset of Mitosis. *PLoS Pathog.* **2017**, *13*, e1006200. [CrossRef] [PubMed]

5.	Aydin, I.; Weber, S.; Snijder, B.; Samperio Ventayol, P.; Kuhbacher, A.; Becker, M.; Day, P.M.; Schiller, J.T.; Kann, M.; Pelkmans, L.; et al. Large Scale RNAi Reveals the Requirement of Nuclear Envelope Breakdown for Nuclear Import of Human Papillomaviruses. *PLoS Pathog.* **2014**, *10*, e1004162. [CrossRef] [PubMed]

6.	Thierry, F. Transcriptional Regulation of the Papillomavirus Oncogenes by Cellular and Viral Transcription Factors in Cervical Carcinoma. *Virology* **2009**, *384*, 375–379. [CrossRef] [PubMed]

7.	Moody, C.A.; Laimins, L.A. Human Papillomavirus Oncoproteins: Pathways to Transformation. *Nat. Rev. Cancer* **2010**, *10*, 550–560. [CrossRef] [PubMed]

8.	McBride, A.A. The Papillomavirus E2 Proteins. *Virology* **2013**, *445*, 57–79. [CrossRef] [PubMed]

9.	Bergvall, M.; Melendy, T.; Archambault, J. The E1 Proteins. *Virology* **2013**, *445*, 35–56. [CrossRef] [PubMed]

10.	Masterson, P.J.; Stanley, M.A.; Lewis, A.P.; Romanos, M.A. A C-Terminal Helicase Domain of the Human Papillomavirus E1 Protein Binds E2 and the DNA Polymerase Alpha-Primase p68 Subunit. *J. Virol.* **1998**, *72*, 7407–7419. [PubMed]

11.	Melendy, T.; Sedman, J.; Stenlund, A. Cellular Factors Required for Papillomavirus DNA Replication. *J. Virol.* **1995**, *69*, 7857–7867. [PubMed]

12.	Hu, Y.; Clower, R.V.; Melendy, T. Cellular Topoisomerase I Modulates Origin Binding by Bovine Papillomavirus Type 1 E1. *J. Virol.* **2006**, *80*, 4363–4371. [CrossRef] [PubMed]

13.	Clower, R.V.; Fisk, J.C.; Melendy, T. Papillomavirus E1 Protein Binds to and Stimulates Human Topoisomerase I. *J. Virol.* **2006**, *80*, 1584–1587. [CrossRef] [PubMed]

14.	Loo, Y.M.; Melendy, T. Recruitment of Replication Protein A by the Papillomavirus E1 Protein and Modulation by Single-Stranded DNA. *J. Virol.* **2004**, *78*, 1605–1615. [CrossRef] [PubMed]

15.	Boner, W.; Taylor, E.R.; Tsirimonaki, E.; Yamane, K.; Campo, M.S.; Morgan, I.M. A Functional Interaction between the Human Papillomavirus 16 Transcription/Replication Factor E2 and the DNA Damage Response Protein TopBP1. *J. Biol. Chem.* **2002**, *277*, 22297–22303. [CrossRef] [PubMed]

16.	Donaldson, M.M.; Mackintosh, L.J.; Bodily, J.M.; Dornan, E.S.; Laimins, L.A.; Morgan, I.M. An Interaction between Human Papillomavirus 16 E2 and TopBP1 is Required for Optimum Viral DNA Replication and Episomal Genome Establishment. *J. Virol.* **2012**, *86*, 12806–12815. [CrossRef] [PubMed]

17.	Gauson, E.J.; Donaldson, M.M.; Dornan, E.S.; Wang, X.; Bristol, M.; Bodily, J.M.; Morgan, I.M. Evidence Supporting a Role for TopBP1 and Brd4 in the Initiation but Not Continuation of Human Papillomavirus 16 E1/E2-Mediated DNA Replication. *J. Virol.* **2015**, *89*, 17684–17699. [CrossRef] [PubMed]

18.	Donaldson, M.M.; Boner, W.; Morgan, I.M. TopBP1 Regulates Human Papillomavirus Type 16 E2 Interaction with Chromatin. *J. Virol.* **2007**, *81*, 4338–4342. [CrossRef] [PubMed]

19.	You, J.; Croyle, J.L.; Nishimura, A.; Ozato, K.; Howley, P.M. Interaction of the Bovine Papillomavirus E2 Protein with Brd4 Tethers the Viral DNA to Host Mitotic Chromosomes. *Cell* **2004**, *117*, 349–360. [CrossRef]

20.	Parish, J.L.; Bean, A.M.; Park, R.B.; Androphy, E.J. ChlR1 is Required for Loading Papillomavirus E2 Onto Mitotic Chromosomes and Viral Genome Maintenance. *Mol. Cell* **2006**, *24*, 867–876. [CrossRef] [PubMed]

21.	Chappell, W.H.; Gautam, D.; Ok, S.T.; Johnson, B.A.; Anacker, D.C.; Moody, C.A. Homologous Recombination Repair Factors, Rad51 and BRCA1, are Necessary for Productive Replication of Human Papillomavirus 31. *J. Virol.* **2015**, *90*, 2639–2652. [CrossRef] [PubMed]

22.	Moody, C.A.; Laimins, L.A. Human Papillomaviruses Activate the ATM DNA Damage Pathway for Viral Genome Amplification upon Differentiation. *PLoS Pathog.* **2009**, *5*, e1000605. [CrossRef] [PubMed]

23.	Gillespie, K.A.; Mehta, K.P.; Laimins, L.A.; Moody, C.A. Human Papillomaviruses Recruit Cellular DNA Repair and Homologous Recombination Factors to Viral Replication Centers. *J. Virol.* **2012**, *86*, 9520–9526. [CrossRef] [PubMed]

24.	Gautam, D.; Moody, C.A.; Weitzman, M.D.; Weitzman, J.B.; Moody, C.A.; Laimins, L.A.; Hong, S.; Dutta, A.; Laimins, L.A.; Gillespie, K.A.; et al. Impact of the DNA Damage Response on Human Papillomavirus Chromatin. *PLoS Pathog.* **2016**, *12*, e1005613. [CrossRef] [PubMed]

25. Fradet-Turcotte, A.; Bergeron-Labrecque, F.; Moody, C.A.; Lehoux, M.; Laimins, L.A.; Archambault, J. Nuclear Accumulation of the Papillomavirus E1 Helicase Blocks S-Phase Progression and Triggers an ATM-Dependent DNA Damage Response. *J. Virol.* **2011**, *85*, 8996–9012. [CrossRef] [PubMed]
26. Anacker, D.C.; Gautam, D.; Gillespie, K.A.; Chappell, W.H.; Moody, C.A. Productive Replication of Human Papillomavirus 31 Requires DNA Repair Factor Nbs1. *J. Virol.* **2014**, *88*, 8528–8544. [CrossRef] [PubMed]
27. Anacker, D.C.; Aloor, H.L.; Shepard, C.N.; Lenzi, G.M.; Johnson, B.A.; Kim, B.; Moody, C.A. HPV31 Utilizes the ATR-Chk1 Pathway to Maintain Elevated RRM2 Levels and a Replication-Competent Environment in Differentiating Keratinocytes. *Virology* **2016**, *499*, 383–396. [CrossRef] [PubMed]
28. Johnson, B.A.; Aloor, H.L.; Moody, C.A. The Rb Binding Domain of HPV31 E7 is Required to Maintain High Levels of DNA Repair Factors in Infected Cells. *Virology* **2017**, *500*, 22–34. [CrossRef] [PubMed]
29. Bristol, M.L.; Wang, X.; Smith, N.W.; Son, M.P.; Evans, M.R.; Morgan, I.M. DNA Damage Reduces the Quality, but Not the Quantity of Human Papillomavirus 16 E1 and E2 DNA Replication. *Viruses* **2016**, *8*, 175. [CrossRef] [PubMed]
30. Reinson, T.; Toots, M.; Kadaja, M.; Pipitch, R.; Allik, M.; Ustav, E.; Ustav, M. Engagement of the ATR-Dependent DNA Damage Response at the HPV18 Replication Centers during the Initial Amplification. *J. Virol.* **2013**, *87*, 951–964. [CrossRef] [PubMed]
31. Sakakibara, N.; Mitra, R.; McBride, A.A. The Papillomavirus E1 Helicase Activates a Cellular DNA Damage Response in Viral Replication Foci. *J. Virol.* **2011**, *85*, 8981–8995. [CrossRef] [PubMed]
32. Sakakibara, N.; Chen, D.; Jang, M.K.; Kang, D.W.; Luecke, H.F.; Wu, S.Y.; Chiang, C.M.; McBride, A.A. Brd4 is Displaced from HPV Replication Factories as they Expand and Amplify Viral DNA. *PLoS Pathog.* **2013**, *9*, e1003777. [CrossRef] [PubMed]
33. Sakakibara, N.; Chen, D.; McBride, A.A. Papillomaviruses use Recombination-Dependent Replication to Vegetatively Amplify their Genomes in Differentiated Cells. *PLoS Pathog.* **2013**, *9*, e1003321. [CrossRef] [PubMed]
34. King, L.E.; Fisk, J.C.; Dornan, E.S.; Donaldson, M.M.; Melendy, T.; Morgan, I.M. Human Papillomavirus E1 and E2 Mediated DNA Replication is Not Arrested by DNA Damage Signalling. *Virology* **2010**, *406*, 95–102. [CrossRef] [PubMed]
35. Zur Hausen, H. Papillomaviruses in the Causation of Human Cancers—A Brief Historical Account. *Virology* **2009**, *384*, 260–265. [CrossRef] [PubMed]
36. Cullen, A.P.; Reid, R.; Campion, M.; Lorincz, A.T. Analysis of the Physical State of Different Human Papillomavirus DNAs in Intraepithelial and Invasive Cervical Neoplasm. *J. Virol.* **1991**, *65*, 606–612. [PubMed]
37. Durst, M.; Kleinheinz, A.; Hotz, M.; Gissmann, L. The Physical State of Human Papillomavirus Type 16 DNA in Benign and Malignant Genital Tumours. *J. Gen. Virol.* **1985**, *66*, 1515–1522. [CrossRef] [PubMed]
38. Kristiansen, E.; Jenkins, A.; Holm, R. Coexistence of Episomal and Integrated HPV16 DNA in Squamous Cell Carcinoma of the Cervix. *J. Clin. Pathol.* **1994**, *47*, 253–256. [CrossRef] [PubMed]
39. Jeon, S.; Allen-Hoffmann, B.L.; Lambert, P.F. Integration of Human Papillomavirus Type 16 into the Human Genome Correlates with a Selective Growth Advantage of Cells. *J. Virol.* **1995**, *69*, 2989–2997. [PubMed]
40. Cooper, K.; Herrington, C.S.; Lo, E.S.; Evans, M.F.; McGee, J.O. Integration of Human Papillomavirus Types 16 and 18 in Cervical Adenocarcinoma. *J. Clin. Pathol.* **1992**, *45*, 382–384. [CrossRef] [PubMed]
41. Couturier, J.; Sastre-Garau, X.; Schneider-Maunoury, S.; Labib, A.; Orth, G. Integration of Papillomavirus DNA near Myc Genes in Genital Carcinomas and its Consequences for Proto-Oncogene Expression. *J. Virol.* **1991**, *65*, 4534–4538. [PubMed]
42. Cooper, K.; Herrington, C.S.; Graham, A.K.; Evans, M.F.; McGee, J.O. In Situ Human Papillomavirus (HPV) Genotyping of Cervical Intraepithelial Neoplasia in South African and British Patients: Evidence for Putative HPV Integration in Vivo. *J. Clin. Pathol.* **1991**, *44*, 400–405. [CrossRef] [PubMed]
43. Fukushima, M.; Yamakawa, Y.; Shimano, S.; Hashimoto, M.; Sawada, Y.; Fujinaga, K. The Physical State of Human Papillomavirus 16 DNA in Cervical Carcinoma and Cervical Intraepithelial Neoplasia. *Cancer* **1990**, *66*, 2155–2161. [CrossRef]
44. Wagatsuma, M.; Hashimoto, K.; Matsukura, T. Analysis of Integrated Human Papillomavirus Type 16 DNA in Cervical Cancers: Amplification of Viral Sequences Together with Cellular Flanking Sequences. *J. Virol.* **1990**, *64*, 813–821. [PubMed]

45. Woodworth, C.D.; Bowden, P.E.; Doniger, J.; Pirisi, L.; Barnes, W.; Lancaster, W.D.; DiPaolo, J.A. Characterization of Normal Human Exocervical Epithelial Cells Immortalized in Vitro by Papillomavirus Types 16 and 18 DNA. *Cancer Res.* **1988**, *48*, 4620–4628. [PubMed]

46. Lehn, H.; Villa, L.L.; Marziona, F.; Hilgarth, M.; Hillemans, H.G.; Sauer, G. Physical State and Biological Activity of Human Papillomavirus Genomes in Precancerous Lesions of the Female Genital Tract. *J. Gen. Virol.* **1988**, *69*, 187–196. [CrossRef] [PubMed]

47. Choo, K.B.; Pan, C.C.; Han, S.H. Integration of Human Papillomavirus Type 16 into Cellular DNA of Cervical Carcinoma: Preferential Deletion of the E2 Gene and Invariable Retention of the Long Control Region and the E6/E7 Open Reading Frames. *Virology* **1987**, *161*, 259–261. [CrossRef]

48. Baker, C.C.; Phelps, W.C.; Lindgren, V.; Braun, M.J.; Gonda, M.A.; Howley, P.M. Structural and Transcriptional Analysis of Human Papillomavirus Type 16 Sequences in Cervical Carcinoma Cell Lines. *J. Virol.* **1987**, *61*, 962–971. [PubMed]

49. Durst, M.; Croce, C.M.; Gissmann, L.; Schwarz, E.; Huebner, K. Papillomavirus Sequences Integrate Near Cellular Oncogenes in some Cervical Carcinomas. *Proc. Natl. Acad. Sci. USA* **1987**, *84*, 1070–1074. [CrossRef] [PubMed]

50. Choo, K.B.; Pan, C.C.; Liu, M.S.; Ng, H.T.; Chen, C.P.; Lee, Y.N.; Chao, C.F.; Meng, C.L.; Yeh, M.Y.; Han, S.H. Presence of Episomal and Integrated Human Papillomavirus DNA Sequences in Cervical Carcinoma. *J. Med. Virol.* **1987**, *21*, 101–107. [CrossRef] [PubMed]

51. Shirasawa, H.; Tomita, Y.; Sekiya, S.; Takamizawa, H.; Simizu, B. Integration and Transcription of Human Papillomavirus Type 16 and 18 Sequences in Cell Lines Derived from Cervical Carcinomas. *J. Gen. Virol.* **1987**, *68*, 583–591. [CrossRef] [PubMed]

52. Mincheva, A.; Gissmann, L.; zur Hausen, H. Chromosomal Integration Sites of Human Papillomavirus DNA in Three Cervical Cancer Cell Lines Mapped by in Situ Hybridization. *Med. Microbiol. Immunol.* **1987**, *176*, 245–256. [CrossRef] [PubMed]

53. Romanczuk, H.; Howley, P.M. Disruption of either the E1 or the E2 Regulatory Gene of Human Papillomavirus Type 16 Increases Viral Immortalization Capacity. *Proc. Natl. Acad. Sci. USA* **1992**, *89*, 3159–3163. [CrossRef] [PubMed]

54. Daniel, B.; Mukherjee, G.; Seshadri, L.; Vallikad, E.; Krishna, S. Changes in the Physical State and Expression of Human Papillomavirus Type 16 in the Progression of Cervical Intraepithelial Neoplasia Lesions Analysed by PCR. *J. Gen. Virol.* **1995**, *76 Pt 10*, 2589–2593. [CrossRef] [PubMed]

55. Vernon, S.D.; Unger, E.R.; Miller, D.L.; Lee, D.R.; Reeves, W.C. Association of Human Papillomavirus Type 16 Integration in the E2 Gene with Poor Disease-Free Survival from Cervical Cancer. *Int. J. Cancer* **1997**, *74*, 50–56. [CrossRef]

56. Park, J.S.; Hwang, E.S.; Park, S.N.; Ahn, H.K.; Um, S.J.; Kim, C.J.; Kim, S.J.; Namkoong, S.E. Physical Status and Expression of HPV Genes in Cervical Cancers. *Gynecol. Oncol.* **1997**, *65*, 121–129. [CrossRef] [PubMed]

57. Yoshinouchi, M.; Hongo, A.; Nakamura, K.; Kodama, J.; Itoh, S.; Sakai, H.; Kudo, T. Analysis by Multiplex PCR of the Physical Status of Human Papillomavirus Type 16 DNA in Cervical Cancers. *J. Clin. Microbiol.* **1999**, *37*, 3514–3517. [PubMed]

58. Gray, E.; Pett, M.R.; Ward, D.; Winder, D.M.; Stanley, M.A.; Roberts, I.; Scarpini, C.G.; Coleman, N. In Vitro Progression of Human Papillomavirus 16 Episome-Associated Cervical Neoplasia Displays Fundamental Similarities to Integrant-Associated Carcinogenesis. *Cancer Res.* **2010**, *70*, 4081–4091. [CrossRef] [PubMed]

59. Herdman, M.T.; Pett, M.R.; Roberts, I.; Alazawi, W.O.F.; Teschendorff, A.E.; Zhang, X.; Stanley, M.A.; Coleman, N. Interferon-Beta Treatment of Cervical Keratinocytes Naturally Infected with Human Papillomavirus 16 Episomes Promotes Rapid Reduction in Episome Numbers and Emergence of Latent Integrants. *Carcinogenesis* **2006**, *27*, 2341–2353. [CrossRef] [PubMed]

60. Bouvard, V.; Storey, A.; Pim, D.; Banks, L. Characterization of the Human Papillomavirus E2 Protein: Evidence of Trans-Activation and Trans-Repression in Cervical Keratinocytes. *EMBO J.* **1994**, *13*, 5451–5459. [PubMed]

61. Demeret, C.; Desaintes, C.; Yaniv, M.; Thierry, F. Different Mechanisms Contribute to the E2-Mediated Transcriptional Repression of Human Papillomavirus Type 18 Viral Oncogenes. *J. Virol.* **1997**, *71*, 9343–9349. [PubMed]

62. Romanczuk, H.; Thierry, F.; Howley, P.M. Mutational Analysis of *cis* Elements Involved in E2 Modulation of Human Papillomavirus Type 16 P$_{97}$ and Type 18 P$_{105}$ Promoters. *J. Virol.* **1990**, *64*, 2849–2859. [PubMed]

63. Bechtold, V.; Beard, P.; Raj, K. Human Papillomavirus Type 16 E2 Protein has no Effect on Transcription from Episomal Viral DNA. *J. Virol.* **2003**, *77*, 2021–2028. [CrossRef] [PubMed]
64. Hwang, E.S.; Naeger, L.K.; DiMaio, D. Activation of the Endogenous p53 Growth Inhibitory Pathway in HeLa Cervical Carcinoma Cells by Expression of the Bovine Papillomavirus E2 Gene. *Oncogene* **1996**, *12*, 795–803. [PubMed]
65. DeFilippis, R.A.; Goodwin, E.C.; Wu, L.; DiMaio, D. Endogenous Human Papillomavirus E6 and E7 Proteins Differentially Regulate Proliferation, Senescence, and Apoptosis in HeLa Cervical Carcinoma Cells. *J. Virol.* **2003**, *77*, 1551–1563. [CrossRef] [PubMed]
66. Goodwin, E.C.; DiMaio, D. Induced Senescence in HeLa Cervical Carcinoma Cells Containing Elevated Telomerase Activity and Extended Telomeres. *Cell Growth Differ.* **2001**, *12*, 525–534. [PubMed]
67. Parish, J.L.; Kowalczyk, A.; Chen, H.; Roeder, G.E.; Sessions, R.; Buckle, M.; Gaston, K. E2 Proteins from High- and Low-Risk Human Papillomavirus Types Differ in their Ability to Bind p53 and Induce Apoptotic Cell Death E2 Proteins from High- and Low-Risk Human Papillomavirus Types Differ in their Ability to Bind p53 and Induce Apoptotic Cell. *J. Virol.* **2006**, *80*, 4580–4590. [CrossRef] [PubMed]
68. Kalantari, M.; Bernard, H.U. Assessment of the HPV DNA Methylation Status in Cervical Lesions. *Methods Mol. Biol.* **2015**, *1249*, 267–280.
69. Kalantari, M.; Calleja-Macias, I.E.; Tewari, D.; Hagmar, B.; Lie, K.; Barrera-Saldana, H.A.; Wiley, D.J.; Bernard, H.U. Conserved Methylation Patterns of Human Papillomavirus Type 16 DNA in Asymptomatic Infection and Cervical Neoplasia. *J. Virol.* **2004**, *78*, 12762–12772. [CrossRef] [PubMed]
70. Das, D.; Bhattacharjee, B.; Sen, S.; Mukhopadhyay, I.; Sengupta, S. Association of Viral Load with HPV16 Positive Cervical Cancer Pathogenesis: Causal Relevance in Isolates Harboring Intact Viral E2 Gene. *Virology* **2010**, *402*, 197–202. [CrossRef] [PubMed]
71. Bhattacharjee, B.; Sengupta, S. CpG Methylation of HPV 16 LCR at E2 Binding Site Proximal to P97 is Associated with Cervical Cancer in Presence of Intact E2. *Virology* **2006**, *354*, 280–285. [CrossRef] [PubMed]
72. Fernandez, A.F.; Rosales, C.; Lopez-Nieva, P.; Grana, O.; Ballestar, E.; Ropero, S.; Espada, J.; Melo, S.A.; Lujambio, A.; Fraga, M.F.; et al. The Dynamic DNA Methylomes of Double-Stranded DNA Viruses Associated with Human Cancer. *Genome Res.* **2009**, *19*, 438–451. [CrossRef] [PubMed]
73. Wentzensen, N.; Sun, C.; Ghosh, A.; Kinney, W.; Mirabello, L.; Wacholder, S.; Shaber, R.; LaMere, B.; Clarke, M.; Lorincz, A.T.; et al. Methylation of HPV18, HPV31, and HPV45 Genomes and Cervical Intraepithelial Neoplasia Grade 3. *J. Natl. Cancer Inst.* **2012**, *104*, 1738–1749. [CrossRef] [PubMed]
74. Badal, V.; Chuang, L.S.; Tan, E.H.; Badal, S.; Villa, L.L.; Wheeler, C.M.; Li, B.F.; Bernard, H.U. CpG Methylation of Human Papillomavirus Type 16 DNA in Cervical Cancer Cell Lines and in Clinical Specimens: Genomic Hypomethylation Correlates with Carcinogenic Progression. *J. Virol.* **2003**, *77*, 6227–6234. [CrossRef] [PubMed]
75. Kalantari, M.; Lee, D.; Calleja-Macias, I.E.; Lambert, P.F.; Bernard, H.U. Effects of Cellular Differentiation, Chromosomal Integration and 5-aza-2′-Deoxycytidine Treatment on Human Papillomavirus-16 DNA Methylation in Cultured Cell Lines. *Virology* **2008**, *374*, 292–303. [CrossRef] [PubMed]
76. Gillison, M.L.; Koch, W.M.; Capone, R.B.; Spafford, M.; Westra, W.H.; Wu, L.; Zahurak, M.L.; Daniel, R.W.; Viglione, M.; Symer, D.E.; et al. Evidence for a Causal Association between Human Papillomavirus and a Subset of Head and Neck Cancers. *J. Natl. Cancer Inst.* **2000**, *92*, 709–720. [CrossRef] [PubMed]
77. Gillison, M.L. Human Papillomavirus-Associated Head and Neck Cancer is a Distinct Epidemiologic, Clinical, and Molecular Entity. *Semin. Oncol.* **2004**, *31*, 744–754. [CrossRef] [PubMed]
78. Gillison, M.L.; Shah, K.V. Human Papillomavirus-Associated Head and Neck Squamous Cell Carcinoma: Mounting Evidence for an Etiologic Role for Human Papillomavirus in a Subset of Head and Neck Cancers. *Curr. Opin. Oncol.* **2001**, *13*, 183–188. [CrossRef] [PubMed]
79. Garnaes, E.; Kiss, K.; Andersen, L.; Therkildsen, M.H.; Franzmann, M.B.; Filtenborg-Barnkob, B.; Hoegdall, E.; Lajer, C.B.; Andersen, E.; Specht, L.; et al. Increasing Incidence of Base of Tongue Cancers from 2000 to 2010 due to HPV: The Largest Demographic Study of 210 Danish Patients. *Br. J. Cancer* **2015**, *113*, 131–134. [CrossRef] [PubMed]
80. Garnaes, E.; Kiss, K.; Andersen, L.; Therkildsen, M.H.; Franzmann, M.B.; Filtenborg-Barnkob, B.; Hoegdall, E.; Krenk, L.; Josiassen, M.; Lajer, C.B.; et al. A High and Increasing HPV Prevalence in Tonsillar Cancers in Eastern Denmark, 2000–2010: The Largest Registry-Based Study to Date. *Int. J. Cancer* **2015**, *136*, 2196–2203. [CrossRef] [PubMed]

81. Carlander, A.F.; Gronhoj Larsen, C.; Jensen, D.H.; Garnaes, E.; Kiss, K.; Andersen, L.; Olsen, C.H.; Franzmann, M.; Hogdall, E.; Kjaer, S.K.; et al. Continuing Rise in Oropharyngeal Cancer in a High HPV Prevalence Area: A Danish Population-Based Study from 2011 to 2014. *Eur. J. Cancer* **2017**, *70*, 75–82. [CrossRef] [PubMed]

82. Ang, K.K.; Harris, J.; Wheeler, R.; Weber, R.; Rosenthal, D.I.; Nguyen-Tân, P.F.; Westra, W.H.; Chung, C.H.; Jordan, R.C.; Lu, C.; et al. Human Papillomavirus and Survival of Patients with Oropharyngeal Cancer. *N. Engl. J. Med.* **2010**, *363*, 24–35. [CrossRef] [PubMed]

83. Garnaes, E.; Frederiksen, K.; Kiss, K.; Andersen, L.; Therkildsen, M.H.; Franzmann, M.B.; Specht, L.; Andersen, E.; Norrild, B.; Kjaer, S.K.; et al. Double Positivity for HPV DNA/p16 in Tonsillar and Base of Tongue Cancer Improves Prognostication: Insights from a Large Population-Based Study. *Int. J. Cancer* **2016**, *139*, 2598–2605. [CrossRef] [PubMed]

84. Larsen, C.G.; Jensen, D.H.; Carlander, A.F.; Kiss, K.; Andersen, L.; Olsen, C.H.; Andersen, E.; Garnaes, E.; Cilius, F.; Specht, L.; et al. Novel Nomograms for Survival and Progression in HPV+ and HPV− Oropharyngeal Cancer: A Population-Based Study of 1542 Consecutive Patients. *Oncotarget* **2016**, *7*, 71761–71772. [PubMed]

85. Hafkamp, H.C.; Speel, E.J.; Haesevoets, A.; Bot, F.J.; Dinjens, W.N.; Ramaekers, F.C.; Hopman, A.H.; Manni, J.J. A Subset of Head and Neck Squamous Cell Carcinomas Exhibits Integration of HPV 16/18 DNA and Overexpression of p16INK4A and p53 in the Absence of Mutations in p53 Exons 5–8. *Int. J. Cancer* **2003**, *107*, 394–400. [CrossRef] [PubMed]

86. Begum, S.; Cao, D.; Gillison, M.; Zahurak, M.; Westra, W.H. Tissue Distribution of Human Papillomavirus 16 DNA Integration in Patients with Tonsillar Carcinoma. *Clin. Cancer Res.* **2005**, *11*, 5694–5699. [CrossRef] [PubMed]

87. Samama, B.; Plas-Roser, S.; Schaeffer, C.; Chateau, D.; Fabre, M.; Boehm, N. HPV DNA Detection by in Situ Hybridization with Catalyzed Signal Amplification on Thin-Layer Cervical Smears. *J. Histochem. Cytochem.* **2002**, *50*, 1417–1420. [CrossRef] [PubMed]

88. Kim, S.H.; Koo, B.S.; Kang, S.; Park, K.; Kim, H.; Lee, K.R.; Lee, M.J.; Kim, J.M.; Choi, E.C.; Cho, N.H. HPV Integration Begins in the Tonsillar Crypt and Leads to the Alteration of p16, EGFR and c-Myc during Tumor Formation. *Int. J. Cancer* **2007**, *120*, 1418–1425. [CrossRef] [PubMed]

89. Deng, Z.; Hasegawa, M.; Kiyuna, A.; Matayoshi, S.; Uehara, T.; Agena, S.; Yamashita, Y.; Ogawa, K.; Maeda, H.; Suzuki, M. Viral Load, Physical Status, and E6/E7 mRNA Expression of Human Papillomavirus in Head and Neck Squamous Cell Carcinoma. *Head Neck* **2013**, *35*, 800–808. [CrossRef] [PubMed]

90. Olthof, N.C.; Speel, E.M.; Kolligs, J.; Haesevoets, A.; Henfling, M.; Ramaekers, F.C.S.; Preuss, S.F.; Drebber, U.; Wieland, U.; Silling, S.; et al. Comprehensive Analysis of HPV16 Integration in OSCC Reveals no Significant Impact of Physical Status on Viral Oncogene and Virally Disrupted Human Gene Expression. *PLoS ONE* **2014**, *9*, e88718. [CrossRef] [PubMed]

91. Cancer Genome Atlas Network. Comprehensive Genomic Characterization of Head and Neck Squamous Cell Carcinomas. *Nature* **2015**, *517*, 576–582.

92. Parfenov, M.; Pedamallu, C.S.; Gehlenborg, N.; Freeman, S.S.; Danilova, L.; Bristow, C.A.; Lee, S.; Hadjipanayis, A.G.; Ivanova, E.V.; Wilkerson, M.D.; et al. Characterization of HPV and Host Genome Interactions in Primary Head and Neck Cancers. *Proc. Natl. Acad. Sci. USA* **2014**, *111*, 15544–15549. [CrossRef] [PubMed]

93. Vojtechova, Z.; Sabol, I.; Salakova, M.; Turek, L.; Grega, M.; Smahelova, J.; Vencalek, O.; Lukesova, E.; Klozar, J.; Tachezy, R. Analysis of the Integration of Human Papillomaviruses in Head and Neck Tumours in Relation to Patients' Prognosis. *Int. J. Cancer* **2016**, *138*, 386–395. [CrossRef] [PubMed]

94. Lim, M.Y.; Dahlstrom, K.R.; Sturgis, E.M.; Li, G. Human Papillomavirus Integration Pattern and Demographic, Clinical, and Survival Characteristics of Patients with Oropharyngeal Squamous Cell Carcinoma. *Head Neck* **2016**, *38*, 1139–1144. [CrossRef] [PubMed]

95. Kadaja, M.; Sumerina, A.; Verst, T.; Ojarand, M.; Ustav, E.; Ustav, M. Genomic Instability of the Host Cell Induced by the Human Papillomavirus Replication Machinery. *EMBO J.* **2007**, *26*, 2180–2191. [CrossRef] [PubMed]

96. Akagi, K.; Li, J.; Broutian, T.R.; Padilla-Nash, H.; Xiao, W.; Jiang, B.; Rocco, J.W.; Teknos, T.N.; Kumar, B.; Wangsa, D.; et al. Genome-Wide Analysis of HPV Integration in Human Cancers Reveals Recurrent, Focal Genomic Instability. *Genome Res.* **2014**, *24*, 185–199. [CrossRef] [PubMed]

97. Evans, M.R.; James, C.D.; Loughran, O.; Nulton, T.J.; Wang, X.; Bristol, M.L.; Windle, B.; Morgan, I.M. An Oral Keratinocyte Life Cycle Model Identifies Novel Host Genome Regulation by Human Papillomavirus 16 Relevant to HPV Positive Head and Neck Cancer. *Oncotarget* **2017**. [CrossRef] [PubMed]

98. Kelly, J.R.; Husain, Z.A.; Burtness, B. Treatment De-Intensification Strategies for Head and Neck Cancer. *Eur. J. Cancer* **2016**, *68*, 125–133. [CrossRef] [PubMed]

99. Fakhry, C.; Westra, W.H.; Li, S.; Cmelak, A.; Ridge, J.A.; Pinto, H.; Forastiere, A.; Gillison, M.L. Improved Survival of Patients with Human Papillomavirus-Positive Head and Neck Squamous Cell Carcinoma in a Prospective Clinical Trial. *J. Natl. Cancer Inst.* **2008**, *100*, 261–269. [CrossRef] [PubMed]

100. Fischer, C.A.; Kampmann, M.; Zlobec, I.; Green, E.; Tornillo, L.; Lugli, A.; Wolfensberger, M.; Terracciano, L.M. P16 Expression in Oropharyngeal Cancer: Its Impact on Staging and Prognosis Compared with the Conventional Clinical Staging Parameters. *Ann. Oncol.* **2010**, *21*, 1961–1966. [CrossRef] [PubMed]

101. Haughey, B.H.; Sinha, P.; Kallogjeri, D.; Goldberg, R.L.; Lewis, J.S., Jr.; Piccirillo, J.F.; Jackson, R.S.; Moore, E.J.; Brandwein-Gensler, M.; Magnuson, S.J.; et al. Pathology-Based Staging for HPV-Positive Squamous Carcinoma of the Oropharynx. *Oral Oncol.* **2016**, *62*, 11–19. [CrossRef] [PubMed]

102. Lydiatt, W.M.; Patel, S.G.; O'Sullivan, B.; Brandwein, M.S.; Ridge, J.A.; Migliacci, J.C.; Loomis, A.M.; Shah, J.P. Head and Neck Cancers-Major Changes in the American Joint Committee on Cancer Eighth Edition Cancer Staging Manual. *CA Cancer. J. Clin.* **2017**, *67*, 122–137. [CrossRef] [PubMed]

103. Malm, I.J.; Fan, C.J.; Yin, L.X.; Li, D.X.; Koch, W.M.; Gourin, C.G.; Pitman, K.T.; Richmon, J.D.; Westra, W.H.; Kang, H.; et al. Evaluation of Proposed Staging Systems for Human Papillomavirus-Related Oropharyngeal Squamous Cell Carcinoma. *Cancer* **2017**, *123*, 1768–1777. [CrossRef] [PubMed]

104. Porceddu, S.V.; Milne, R.; Brown, E.; Bernard, A.; Rahbari, R.; Cartmill, B.; Foote, M.; McGrath, M.; Coward, J.; Panizza, B. Validation of the ICON-S Staging for HPV-Associated Oropharyngeal Carcinoma using a Pre-Defined Treatment Policy. *Oral Oncol.* **2017**, *66*, 81–86. [CrossRef] [PubMed]

105. Ramqvist, T.; Mints, M.; Tertipis, N.; Nasman, A.; Romanitan, M.; Dalianis, T. Studies on Human Papillomavirus (HPV) 16 E2, E5 and E7 mRNA in HPV-Positive Tonsillar and Base of Tongue Cancer in Relation to Clinical Outcome and Immunological Parameters. *Oral Oncol.* **2015**, *51*, 1126–1131. [CrossRef] [PubMed]

![viruses logo] *viruses*

MDPI

Review

Somatic Host Cell Alterations in HPV Carcinogenesis

Tamara R. Litwin [1,2], Megan A. Clarke [1,2], Michael Dean [3] and Nicolas Wentzensen [2,*]

1   Cancer Prevention Fellowship Program, Division of Cancer Prevention, National Cancer Institute, Rockville, MD 20850, USA; tamara.litwin@nih.gov (T.R.L.); megan.clarke@nih.gov (M.A.C.)
2   Clinical Genetics Branch, Division of Cancer Epidemiology and Genetics, National Cancer Institute, Rockville, MD 20850, USA
3   Laboratory of Translational Genomics, Division of Cancer Epidemiology and Genetics, National Cancer Institute, Gaithersburg, MD 20850, USA; deanm@mail.nih.gov
*   Correspondence: wentzenn@mail.nih.gov; Tel.: +1-240-276-7303

Academic Editor: Karl Munger
Received: 11 July 2017; Accepted: 25 July 2017; Published: 3 August 2017

Abstract: High-risk human papilloma virus (HPV) infections cause cancers in different organ sites, most commonly cervical and head and neck cancers. While carcinogenesis is initiated by two viral oncoproteins, E6 and E7, increasing evidence shows the importance of specific somatic events in host cells for malignant transformation. HPV-driven cancers share characteristic somatic changes, including apolipoprotein B mRNA editing catalytic polypeptide-like (APOBEC)-driven mutations and genomic instability leading to copy number variations and large chromosomal rearrangements. HPV-associated cancers have recurrent somatic mutations in phosphatidylinositol-4,5-bisphosphate 3-kinase catalytic subunit alpha (*PIK3CA*) and phosphatase and tensin homolog (*PTEN*), human leukocyte antigen A and B (*HLA-A* and *HLA-B*)-*A/B*, and the transforming growth factor beta (TGFβ) pathway, and rarely have mutations in the tumor protein p53 (*TP53*) and RB transcriptional corepressor 1 (*RB1*) tumor suppressor genes. There are some variations by tumor site, such as *NOTCH1* mutations which are primarily found in head and neck cancers. Understanding the somatic events following HPV infection and persistence can aid the development of early detection biomarkers, particularly when mutations in precancers are characterized. Somatic mutations may also influence prognosis and treatment decisions.

Keywords: HPV; somatic mutation; cervical cancer; APOBEC; significantly mutated gene; copy number variation; chromosomal instability; head and neck cancer; integration

1. Introduction

High-risk human papilloma virus (HPV) infections cause cancers at many sites. It is estimated that almost all cervical cancers [1], 20–70% of oropharyngeal cancers and 5–30% of other head and neck cancers [2–4], 88% of anal cancers [5], 48% of penile cancers [6,7], 19% of vulvar cancers [5], and 71% of vaginal cancers [5] are caused by HPV, with some geographic variation observed for the non-cervical cancers. Together, these cancers resulted in approximately 610,000, or 5%, of all cancer diagnoses worldwide in 2008 [8,9].

HPV infection alone is an insufficient cause of carcinogenesis. Most HPV infections become undetectable after a few months and never result in malignancies, with 91% becoming undetectable after two years, although it has been proposed that there may be some level of persistent latent infection that is undetectable by PCR [10,11]. High-risk HPV types persist longer on average than low-risk types [12]. A failure to clear the virus results in viral persistence, but many persistent infections never develop into precancerous lesions [13]. Finally, even advanced precancerous cervical intraepithelial neoplasias grade 3 (CIN3) only progress to invasive cancer in 30% of cases over 30 years [14]. When infections persist over time, somatic mutations may accumulate and contribute to the development of precancerous lesions, and then finally to malignant cancers. Understanding the complete carcinogenic pathways is important for developing new strategies to prevent HPV-associated cancer mortality, both through early detection and through targeted therapies [15,16].

HPV-derived cancers share many carcinogenic features across cancer sites, suggesting that the viral oncoproteins E6 and E7 work similarly at different sites. A previous review on this topic [17] predates recent publications of large genomic data from HPV-driven cervical and head and neck cancers in The Cancer Genome Atlas (TCGA) [2,18]. Here, we review common somatic mutations, copy number alterations, and related pathways identified by TCGA and other recent efforts. While the focus of this review is on somatic changes, genome-wide association (GWAS) studies of cervical and HPV-related head and neck cancers have shown that there is also a heritable component. At both cancer sites, human leukocyte antigen (HLA) variants are among the few consistent, independently replicated findings from GWAS studies [19–21].

2. Mechanisms of HPV-Mediated Mutagenesis

There is a great diversity of HPV genotypes, but only a small subset is carcinogenic; among these, HPV16 alone accounts for 50–90% of HPV-driven cancers depending on the site, with some regional variations [22,23]. Most cancers evaluated in studies included in this review are caused by HPV16, and there may be variations in somatic mutation load and type by HPV genotype that are currently not adequately captured. Two of the eight proteins encoded by the HPV genome, E6 and E7, account for most carcinogenic effects of high-risk HPV types [15]. They promote carcinogenesis in several ways, including creating genomic instability and inhibiting tumor suppressor genes. E6 and E7 directly promote genomic instability, which can result in large chromosomal rearrangements and copy number variations, by interfering with centromere duplication during mitosis [24,25]. Both oncoproteins interfere with important cellular tumor suppressor pathways: E6 inhibits the p53 tumor suppressor by promoting its proteasomal degradation [26,27], while E7 disrupts the retinoblastoma (Rb) pathway resulting in uncontrolled activation of the cell cycle and induction of p16^{INK4A}, a cyclin-dependent kinase inhibitor, through a disrupted feedback loop (Figure 1) [28–30]. Theoretically, since HPV oncoproteins are important carcinogenic drivers interfering with several cellular pathways, it could be expected that fewer somatic alterations are required for malignant transformation in HPV-associated compared to non-HPV associated cancers. There is some evidence of lower mutation load in HPV-positive compared to HPV-negative penile cancers [31]. However, the evidence is inconclusive for head and neck cancers, with one study showing evidence of a reduced somatic mutation load in HPV-positive compared to HPV-negative cancers [3] while the TCGA head and neck study did not find evidence of a difference [2].

In addition to direct viral effects, specific mutation signatures may be overrepresented in HPV-positive cancers due to host–viral interactions. The apolipoprotein B mRNA editing catalytic polypeptide-like (APOBEC) mutation signature in particular is very common in HPV-positive cancers, likely triggered by the host response to HPV infection [32].

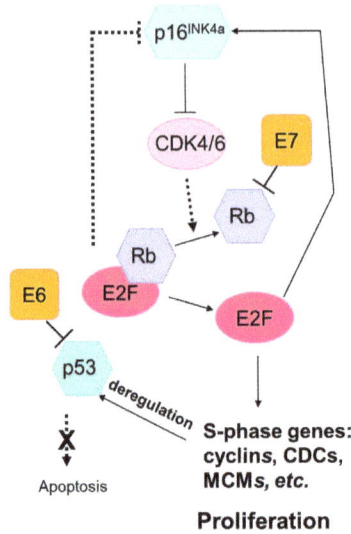

Figure 1. The Rb and p53 pathways are disrupted by the human papilloma virus (HPV) oncoproteins E7 and E6, respectively. The HPV E7 protein binds to Rb with high affinity, disrupting its interaction with the transcription factor E2F. This results in the release and activation of E2F, driving expression of S-phase genes and cell cycle progression. P16^{INK4A} is a cyclin-dependent kinase inhibitor that regulates the cell cycle by inactivating cyclin-dependent kinases involved in Rb phosphorylation. Upregulation of p16^{INK4A} is induced by HPV-mediated disruption of E7, leading to the accumulation of p16^{INK4A} in HPV-transformed cells. The HPV E6 protein inhibits apoptosis by targeting the tumor suppressor protein, p53, for degradation. HPV E6 inhibition of p53 promotes cell proliferation and can lead to genomic instability and the accumulation of somatic mutations. Abbreviations: Rb, retinoblastoma protein; p16^{INK4A}, cyclin-dependent kinase inhibitor 2A; CDK, cyclin-dependent kinases; E2F, E2F transcription factor; CDC, cell-division-cycle genes; MCM, minichromosome maintenance family.

2.1. Genomic Instability

Rates of copy number alterations vary across cancer sites. Cervical cancers average 88 copy number alterations in the TCGA dataset, including 26 amplifications and 37 losses [18]. Focal amplifications of loci containing genes discussed elsewhere in this review in order of frequency include 3q28 (tumor protein p63 (*TP63*), altered in 77% of samples), 3q24.1 (transforming growth factor beta receptor 2 (*TGFBR2*), 36%), 10q23.31 (phosphatase and tensin homolog (*PTEN*), 31%), 18q21.2 (SMAD family member 4 (*SMAD4*), 28%), and 7p11.2 (epidermal growth factor receptor (*EGFR*), 17%) [18]. Greater numbers of copy number variations were reported in cervical squamous cell carcinomas than in cervical adenocarcinomas [18]. A review of cervical squamous cell carcinomas from other datasets as well as limited information on HPV-positive vulvar squamous cell carcinomas also showed gains at 3q (55%), losses at 3p (36%), and losses at 11q (33%) [33]. A study of CIN3 lesions and invasive cancers reported an average of 36.3 copy number alterations in cancers, with the most frequent amplification at 3q (50% of cancers and 25% of CIN3) [34]. Notably, this region contains the phosphatidylinositol-4,5-bisphosphate 3-kinase catalytic subunit alpha (*PIK3CA*) gene, which is the most commonly mutated gene in HPV-driven cancers across sites (see below). Losses were most common in 3p (40% of cancers and 10% of CIN3) [34]. A summary of copy number alterations reported in HPV-driven cancers can be found in Table 1. Figure 2 shows the frequency of chromosomal amplifications and deletions across the whole genome in cervical cancers from TCGA [18].

Table 1. Common copy number alterations in HPV-driven cancers.

Arm or Location [1]	Gene	Pathway	Alteration	Cancer Site [2]	Fraction Altered in HPV+ Cancers (%)	References
1p	–	–	Gain	Cervix squamous	33	[35]
1q	–	–	Gain	Cervix squamous	29–36	[33,35]
	–	–	Gain	Cervix adeno	22–35	[33,35]
2q	–	–	Loss	Cervix squamous	22	[33]
3p	–	–	Loss	Cervix squamous	36–51	[33–35]
	–	–	Loss	Vulva	45	[33]
3p24.1	TGFBR2	TGFβ	Gain	Cervix	36	[18]
3p14.1	FOXP1	Transcription	Loss	Cervix squamous	42	[34]
3q	–	–	Gain	Cervix	66	[18,33,35]
	–	–	Gain	Cervix squamous	44–62	[33–35]
	–	–	Gain	Cervix adeno	29–39	[33,35]
	–	–	Gain	Vulva	58	[33]
3q25.32	MLF1	Phenotypic determination	Gain	Cervix squamous	60	[34]
3q26.32	PIK3CA	PI3K/AKT	Gain	Head and neck	30–56	[2,36]
3q26.33	SOX2	Transcription-Sox2	Gain	Head and neck	11–28	[2,36]
3q27.1	KLHL6	Immune signaling	Gain	Head and neck	1–25	[2,36]
3q27.3	BCL6	RTK–JAK–STAT	Gain	Head and neck	1–25	[2,36]
3q28	TP63	p53	Gain	Cervix	77	[18]
3q28	LPP	Cell-cell adhesion	Gain	Cervix squamous	60	[34]
4p	–	–	Loss	Cervix squamous	24–47	[33,35]
	–	–	Loss	Cervix adeno	10–46	[33,35]
	–	–	Loss	Vulva	27	[33]
4q	–	–	Loss	Cervix squamous	21–34	[33,35]
	–	–	Loss	Cervix adeno	17–42	[33,35]
4q31.3	FBXW7	Notch	Loss	Head and neck	3–12	[2,36]
5p	–	–	Gain	Cervix squamous	27–28	[33,35]
	–	–	Gain	Vulva	15	[33]
5p13.1	RICTOR	PI3K/AKT	Gain	Head and neck	4–6	[2,36]
6p	–	–	Loss	Cervix squamous	24	[35]
6q	–	–	Loss	Cervix squamous	20–29	[33,35]
7p	EGFR	RAS/EGFR/ERK	Gain	Cervix	17	[18]
8p	–	–	Loss	Cervix squamous	27	[35]
8q	–	–	Gain	Cervix squamous	25	[34,35]
8q24.21	MYC	TGFβ	Gain	Head and neck	3–6	[2,36]
9p	CD274	Immune response	Gain	Cervix	8	[18]
10q23.31	PTEN	PI3K/AKT	Loss	Cervix	31	[2,18,36]
			Loss	Head and neck	3–15	[2,36]
11p	–	–	Loss	Cervix squamous	32	[35]
	–	–	Loss	Cervix adeno	35	[35]
11q	–	–	Loss	Cervix squamous	32–33	[33,35]
	–	–	Loss	Cervix adeno	9–35	[33,35]
	–	–	Loss	Vulva	30	[33]
11q13.3	FGF19	RAS/EGFR/ERK	Gain	Head and neck	4–6	[2,36]
11q13.3	FGF3	RAS/EGFR/ERK	Gain	Head and neck	4–6	[2,36]
11q13.3	FGF4	RAS/EGFR/ERK	Gain	Head and neck	4–6	[2,36]
11q22.1	YAP1	Hippo	Gain	Cervix	16	[18]
13q	–	–	Loss	Cervix squamous	24–41	[33,35]
	–	–	Loss	Cervix adeno	21	[33]
	–	–	Loss	Vulva	12	[33]
13q14.2	RB1	Rb	Loss	Head and neck	6–24	[2,36]
14q	–	–	Gain	Cervix squamous	26	[35]
14q32.32	TRAF3	NF-κB	Loss	Head and neck	14	[2]
14q32.33	AKT1	PI3K/AKT	Gain	Head and neck	5	[2]
16p	–	–	Loss	Cervix adeno	33	[35]
16p13.13	BCAR4	Hedgehog	Gain	Cervix	7	[18]
16q	–	–	Loss	Cervix adeno	45	[35]
17p	–	–	Loss	Cervix squamous	34	[35]
18q	–	–	Loss	Cervix adeno	54	[35]
18q21.2	SMAD4	TGFβ	Gain	Cervix	28	[18]

Table 1. *Cont.*

Arm or Location [1]	Gene	Pathway	Alteration	Cancer Site [2]	Fraction Altered in HPV+ Cancers (%)	References
19p	–	–	Loss	Cervix adeno	30	[35]
19q	–	–	Gain	Cervix squamous	23	[35]
	–	–	Gain	Cervix adeno	32	[35]
20p	–	–	Gain	Cervix squamous	33	[35]
	–	–	Gain	Cervix adeno	26	[35]
20q	–	–	Gain	Cervix squamous	31	[35]
Xp11.3	KDM6A	Chromatin organization	Loss	Head and neck	3–7	[2,36]

[1] Ordered by chromosome arm or location. [2] Cervix is listed when references did not differentiate between squamous cell carcinoma and adenocarcinoma. Sites other than cervix always refer to squamous cell carcinoma. Abbreviations: Adeno, adenocarcinoma; TGFBR2, transforming growth factor beta receptor 2; TGFβ, transforming growth factor beta; FOXP1, forkhead box P1; MLF1, myeloid leukemia factor 1; PIK3CA, phosphatidylinositol-4,5-bisphosphate 3-kinase catalytic subunit alpha; PI3K/AKT, phosphatidylinositol 3-kinase/protein kinase B; SOX2, SRY-box 2; KLHL6, kelch like family member 6; BCL6, B-cell CLL/lymphoma 6; RTK-JAK–STAT, RTK, receptor tyrosine kinase-Janus kinase-signal transducer and activator of transcription; TP63, tumor protein p63; LPP, LIM domain containing preferred translocation partner in lipoma; FBXW7, F-box and WD repeat domain containing 7; RICTOR, RPTOR independent companion of MTOR complex 2; EGFR, epidermal growth factor receptor; RAS, retrovirus-associated DNA sequences; ERK, extracellular signal-regulated kinases; MYC, MYC proto-oncogene, bHLH transcription factor; PTEN, phosphatase and tensin homolog; FGF19, fibroblast growth factor 19; FGF3, fibroblast growth factor 3; FGF4, fibroblast growth factor 4; YAP1, Yes associated protein 1; RB1, RB transcriptional corepressor 1; Rb, retinoblastoma; TRAF3, TNF receptor associated factor 3; NF-κB, nuclear factor kappa-light-chain-enhancer of activated B cells; AKT1, AKT serine/threonine kinase 1; BCAR4, breast cancer anti-estrogen resistance 4; SMAD4, SMAD family member 4; KDM6A, lysine demethylase 6A.

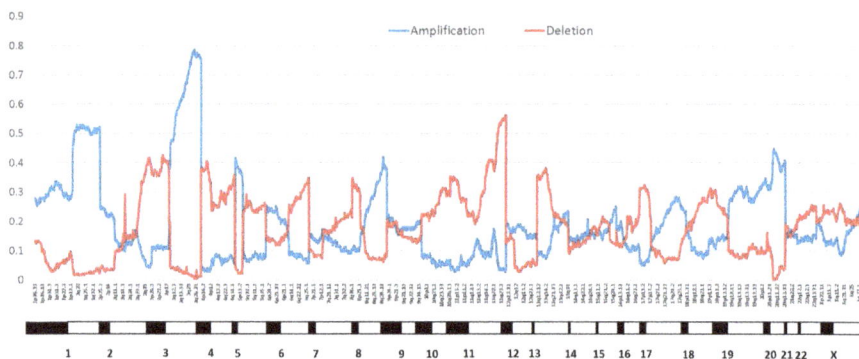

Figure 2. Proportion of cervical cancers with copy number variation by chromosome position from The Cancer Genome Atlas (TCGA) cervical cancer data [18]. Amplifications are in blue and deletions are in red.

In HPV-positive head and neck cancers, significant copy number losses have been reported in 22 genes and gains in 65 genes, including RB transcriptional corepressor 1 (*RB1*) and *PIK3CA* [37]. The 3q26-28 region is amplified in both HPV-positive and HPV-negative cancers, while 3p deletions are primarily found in HPV-negative head and neck cancers [37].

In penile cancers, greater copy number gains in 15 regions and losses in four regions are seen in HPV-positive compared to HPV-negative cancers [38]. Autosomal copy number variations are most frequently observed on chromosomes 3 and 8, including losses in 3p and gains in 3q, and are also associated with worse prognosis [38]. A small study of HPV-positive anal cancers reported recurrent gains in 17q, 3q, 19p, and 19q [39].

In HPV-driven cancers of the cervix and head and neck, copy number variations often co-localize with sites of viral integration [2,18], a phenomenon that occurs in many HPV-associated cancers, and has been shown to vary by HPV type [40–42]. Though the mechanisms by which HPV integrates into the host cell genome are not well understood, these events tend to occur at regions of genomic instability [34,42–45]. It has been proposed that copy number alterations commonly occur in regions of genomic instability, which in turn may promote viral integration in those locations, explaining why viral integration is more common at sites with copy number alterations than expected by chance [34]. Viral integration has also been observed in short regions of HPV and host genome sequence homology (i.e., "micro-homologies"), suggesting a potential role for DNA repair processes to integrate HPV and host cell genomes based on nucleotide sequence similarities [45,46].

Recurrent large chromosomal rearrangements have been reported in 23 locations in cervical cancers in TCGA [18]. One notable recurrent rearrangement is the 16p13 zinc finger CCCH-type containing 7Abreast cancer anti-estrogen resistance 4 (*ZC3H7A-BCAR4*) fusion, which together with copy number gain of the locus containing *BCAR4* (16p13.13, found in 20% of tumors) and duplication detected by whole genome sequencing suggest a potential role of this gene in cervical carcinogenesis [18].

HPV-driven cancers of the cervix, head and neck, and penis share copy number alteration sites, most notably copy number gains in 3q, which in addition to *PIK3CA* contains the telomerase RNA component (*TERC*), MDS1 and EVI1 complex locus (*MECOM*), SRY-box 2 (*SOX2*), and *TP63* genes [18,34,37,38]. It is worth noting that both HPV-positive and HPV-negative cancers display recurrent focal amplifications of this region [2]. Together with the extremely high somatic mutation rate of *PIK3CA* (see Section 3.2), this supports an important role for *PIK3CA* in HPV-mediated carcinogenesis.

2.2. Mutational Signatures

2.2.1. APOBEC

The APOBEC family of cytosine deaminases causes cytosine to thymine or guanine mutations [47–49]. APOBEC3B, a subclass of these proteins, causes characteristic mutations that are enriched in many cervical and head and neck cancers [18,35,50–52]. During DNA repair, APOBEC-mediated cytosine deamination can result in characteristic mutational signatures that occur at motifs involving a thymine immediately 5′ to the target cytosine, collectively referred to as "TCW" mutations, where W corresponds to an A or T [52]. APOBEC-mediated mutagenesis is also enriched in HPV-positive subsets of many head and neck cancers [53] as well as in penile cancers [54] suggesting the activation of APOBEC enzymes in HPV-driven cancers across sites.

APOBEC-associated mutations are responsible for many mutations of genes in the HPV-associated carcinogenesis pathways discussed below, including common *PIK3CA* point mutations [53]. APOBEC signature enrichment was reported in 150 of 192 exomes in TCGA cervical cancer data, with the fraction of ABOPEC signature mutations by gene reproduced in Figure 3 [18].

The APOBEC pathway drives mutations in many cancer sites including cervix, head and neck, bladder, lung, and breast [51,52]. However, APOBEC mutations are likely enriched in HPV-positive cancers due to its role in the host response to the viral infection. The APOBEC3A protein may inhibit HPV infectivity, so upregulation assists in viral clearance and reduces persistence [32], although it has also been suggested that APOBEC3B is likely to be the primary APOBEC involved HPV-related carcinogenesis because unlike APOBEC3A it is expressed in the nucleus [51]. The APOBEC mutagenesis pathway has also been reported to be upregulated by the HPV oncoprotein E6 [55]. Upregulation of APOBEC proteins in response to viral infection can cause "collateral damage" to the host DNA [56]. However, the exact mechanism of induction of the APOBEC pathway and its contribution to carcinogenesis once activated remain unclear, since it is also found in many cancer types not associated with infectious agents, including breast cancer and ovarian serous carcinoma [57–59]. Due to insufficient data from cancer precursors, it is currently not clear at what stage in the carcinogenic process APOBEC mutations start to accumulate and whether APOBEC mutations occur before non-APOBEC mutations.

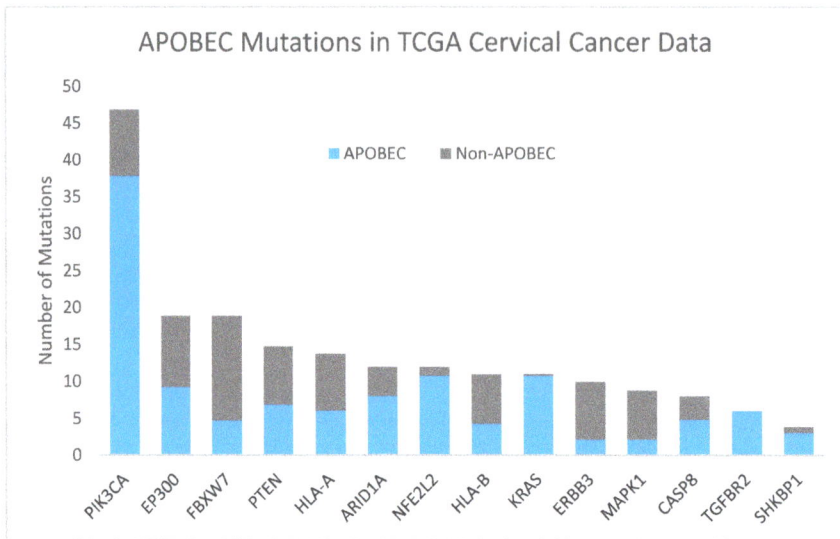

Figure 3. Apolipoprotein B mRNA editing catalytic polypeptide-like (APOBEC, blue) and non-APOBEC (gray) mutations in significantly mutated genes in TCGA cervical cancer data [18]. Abbreviations: PIK3CA, phosphatidylinositol-4,5-bisphosphate 3-kinase catalytic subunit alpha; EP300, E1A binding protein p300; FBXW7, F-box and WD repeat domain containing 7; PTEN, phosphatase and tensin homolog; HLA-A, human leukocyte antigen A; NFE2L2, nuclear factor, erythroid 2 like 2; ARID1A, AT-rich interaction domain 1A; HLA-B, human leukocyte antigen B; KRAS, KRAS proto-oncogene, GTPase; ERBB3, erb-b2 receptor tyrosine kinase 2; MAPK1, mitogen-activated protein kinase 1; CASP8, caspase 8; TGFBR2, transforming growth factor beta receptor 2; SHKBP1, SH3KBP1 binding protein 1.

2.2.2. Other Mutational Signatures

Cervical cancer, which has an attributable risk for HPV of close to 100% [1], has two primary mutational signatures, classified as signature 1B and signature 2 by Alexandrov et al. [50]. Signature 2 is the above-discussed APOBEC signature. Signature 1B is a common pattern across many cancer sites that is characterized by cytosine to thymine mutations at methylated cytosine-guanine (CpG) sites along the DNA and is associated with age [50]. Other cancers associated with signature 1B include head and neck, the only other HPV-associated cancer characterized by this study, as well as ovarian and endometrial, the other major gynecological cancers [50].

3. Genes and Pathways

Many somatic mutations overlap across HPV-associated cancer sites. Frequently somatically mutated genes are summarized in Table 2. In the following sections, common mutations are discussed in the context of their respective pathways.

Table 2. Common Somatic Mutations in HPV-Driven Cancers.

Gene [1]	Pathway	Mutation	Cancer Site [2]	Fraction Mutated in HPV+ Cancers (%)	References
PIK3CA[3]	PI3K/AKT	Activating	Cervix squamous	6–42	[18,35,60–69]
			Cervix adeno	10–42	[18,35,60–69]
			Head and neck	22–56	[2,36,37]
EGFR	RAS/EGFR/ERK	Activating	Cervix squamous	3–33	[18,62,70,71]
			Cervix adeno	6	[18,62,70]
SMAD4	TGFβ	Inactivating	Cervix	28	[18]
ERBB2	PI3K/AKT	Activating	Cervix squamous	4	[18]
			Cervix adeno	26	[18]
TP53	DNA repair	Inactivating	Cervix squamous	5	[35]
			Head and neck	0–25	[2,3,36,37,72]
			Vulvar	8–16	[73,74]
RB1	Rb	Inactivating	Head and neck	6–24	[2,36,37]
FGFR2 & FGFR3	RAS/EGFR/ERK	Activating	Head and neck	1–24	[2,36,37]
KRAS	RAS/EGFR/ERK	Activating	Cervix squamous	4	[18,35,62]
			Cervix adeno	8–23	[18,35,62,75]
			Head and neck	6	[37]
MLL2	Chromatin organization	Activating	Head and neck	10–20	[2,36]
ASXL1		Inactivating	Head and neck	5–19	[36]
NOTCH1	Notch	Activating, inactivating	Head and neck	6–18	[2,36,37,76,77]
EP300	TGFβ	Inactivating	Cervix squamous	13–16	[18]
			Cervix adeno	10	[18]
			Head and neck	10–14	[36]
ERBB3	PI3K/AKT	Activating	Cervix squamous	4	[18]
			Cervix adeno	16	[18]
ATM	DNA repair	Inactivating	Head and neck	1–16	[36]
FBXW7	Notch	Inactivating	Cervix	11–15	[18,35]
			Head and neck	3–12	[36]
PTEN	PI3K/AKT	Inactivating	Cervix squamous	6–13	[18,35]
			Cervix adeno	13	[18,35]
			Head and neck	3–15	[2,36]
BRCA1	DNA repair	Inactivating	Head and neck	2–14	[36]
NF1	RAS/EGFR/ERK	Inactivating	Head and neck	0–14	[2,36]
ELF3	PI3K/AKT	Inactivating	Cervix adeno	13	[35]
FLG			Head and neck	12	[2]
BRCA2	DNA repair	Inactivating	Head and neck	3–12	[36]
LRP1B	RTK		Head and neck	2–12	[36]
HRAS	RAS/EGFR/ERK	Activating	Head and neck	1–12	[36]
HLA-A/B	MHC	Inactivating	Cervix	6–9	[18,35,78]
			Head and neck	11	[2,37]
MLL3	Chromatin organization	Activating	Head and neck	10	[2]
TGFBR2	TGFβ	Inactivating	Cervix squamous	8	[18]
CREBBP	TGFβ	Inactivating	Cervix squamous	8	[18]
			Cervix adeno	6	[18]
TRAF3	NF-κB	Truncating mutations	Head and neck	8	[2]
MAPK1	MEK/ERK	Activating	Cervix squamous	8	[18,35]
CBFB	RUNX1/RUNX2	Inactivating	Cervix adeno	8	[35]
DDX3X		Inactivating	Head and neck	8	[2]
ARID1A	Chromatin organization	Inactivating	Cervix	7	[18]
NFE2L2		Inactivating	Cervix squamous	4–7	[18,35]
TPRX1			Head and neck	6	[2]
CYLD	NF-κB	Inactivating	Head and neck	6	[2]
RIPK4	NF-κB, Notch		Head and neck	6	[2]
UBR5	Proteolysis		Head and neck	6	[2]
CASP8	Fas apoptosis	Inactivating	Cervix	4	[18]
STK11	PI3K/AKT	Inactivating	Cervix squamous	4	[35]
SHKBP1	RAS/EGFR/ERK		Cervix	2	[18]

[1] Ordered by higher reported fraction mutated. [2] Cervix is listed when references did not differentiate between squamous cell carcinoma and adenocarcinoma. Sites other than cervix always refer to squamous cell carcinoma. [3] Significantly mutated genes in TCGA cervical cancer data are bolded [18]. Abbreviations: ERBB2, erb-b2 receptor tyrosine kinase 2; KRAS, KRAS proto-oncogene, GTPase; MLL2, lysine methyltransferase 2D; ASXL1, additional sex combs like 1, transcriptional regulator; EP300, E1A binding protein p300; ERBB3, erb-b2 receptor tyrosine kinase 2; ATM, ATM serine/threonine kinase; BRCA1, BRCA1, DNA repair associated; NF1, neurofibromin 1; ELF3, E74 like ETS transcription factor 3; FLG, filaggrin; BRCA2, BRCA2, DNA repair associated; LRP1B, LDL receptor related protein 1B; HRAS, HRas proto-oncogene, GTPase; HLA-A/B, human leukocyte antigen A/B; MHC, major histocompatibility complex; MLL3, lysine methyltransferase 2C; CREBBP, CREB binding protein; MAPK1, mitogen-activated protein kinase 1; CBFB, core-binding factor beta subunit; RUNX1/RUNX2, runt related transcription factor 1/2; DDX3X, DEAD-box helicase 3, X-linked; ARID1A, AT-rich interaction domain 1A; NFE2L2, nuclear factor, erythroid 2 like 2; TPRX1, tetrapeptide repeat homeobox 1; CYLD, CYLD lysine 63 deubiquitinase; RIPK4, receptor interacting serine/threonine kinase 4; UBR5, ubiquitin protein ligase E3 component n-recognin 5; CASP8, caspase 8; STK11, serine/threonine kinase 11; SHKBP1, SH3KBP1 binding protein 1.

3.1. Lack of Mutations in TP53 and RB1

The HPV oncogenic proteins E6 and E7 target the tumor suppressor proteins p53 and pRB, respectively, for degradation [79]. They therefore obviate the need for somatic deactivation of the *TP53* and *RB1* genes during the carcinogenesis process, and mutations in these genes infrequently occur in HPV-positive cancers compared to corresponding HPV-negative cancers at the same sites (Figure 1).

In cervical squamous cell carcinoma, *TP53* mutations have been reported with a frequency of 5% [35]. Although fewer than 1% of cervical squamous cell carcinomas are HPV-negative, one study reported a difference in *TP53* mutation status by classifying tumors in the TCGA-CESC data set as "HPV active" (expressing HPV transcripts; 4% TP53 mutation rate) versus "HPV inactive" (not expressing HPV transcripts; 47% TP53 mutation rate and 8% of the total number of HPV-positive samples) [80]. This is consistent with the idea that TP53 inactivation is exceedingly common, and that the TP53 mutation rates are negatively correlated with HPV activity. Vulvar squamous cell carcinoma has an 8–16% TP53 mutation prevalence in HPV-positive tumors versus 30–76% prevalence in HPV-negative tumors, and vulvar intraepithelial neoplasia (VIN) precancerous lesions have a 3% *TP53* mutation prevalence in HPV-positive and 21% prevalence in HPV-negative lesions [73,74]. Likewise, TP53 mutations appear to be more prevalent in HPV-negative than in HPV-positive penile squamous cell carcinomas [31].

Numerous studies have reported significantly higher *TP53* mutation rates in HPV-negative (52–86%) compared to HPV-positive (0–25%) head and neck tumors [2,3,36,37,72]. A complete absence of *TP53* mutations in tumors with high-risk HPV types present has also been found in laryngeal [81] and esophageal [82] cancers. It has been suggested that *TP53* inactivation, either through HPV infection or somatic mutation, is nearly ubiquitous in head and neck squamous cell carcinomas, even those that are HPV-negative and therefore must achieve this inactivation via other pathways [3].

Head and neck cancers with wild-type *TP53* have a better prognosis than those with *TP53* mutations [2,83]. HPV positivity and p16 [INK4A] expression, which are both related to retention of wild type *TP53*, are also positively correlated with overall 3-year survival in anal cancers [84]. Evidence in penile cancers is mixed [85–88].

The Rb pathway controls the cell cycle and regulates growth and proliferation [89]. *RB1* mutations are very rare in cervical cancers because HPV E7 activity inactivates Rb tumor suppression activity by disrupting its interaction with the transcription factor E2F, making mutations in this gene unnecessary in HPV-positive cancers [90,91]. *RB1* is mutated in 6–24% of HPV-positive head and neck cancers, a similar fraction to HPV-negative head and neck cancers (4%) [2,36,37]. Cyclin dependent kinase inhibitor 2A (*CDKN2A*) encodes p16[INK4A], an Rb pathway gene which as described above is nearly ubiquitously expressed in HPV-positive cancers due to activation of a negative feedback loop triggered by E2F release [92,93]. Overexpression of p16[INK4A] is also common in HPV-related precancers, which has led to development of p16[INK4A]-based biomarkers for cervical cancer screening and triage [94,95]. *CDKN2A* is rarely altered in HPV-positive (0%) compared with HPV-negative head and neck cancers (25% mutation rate, frequent alterations in 9p21.3 chromosomal region containing the *CDKN2A* gene) [2]. An absence of *CDKN2A* alterations in HPV-positive penile squamous cell carcinomas has also been reported, compared with 16% mutation prevalence and 24% copy number reduction in HPV-negative tumors [31].

3.2. PI3K/AKT Pathway

PIK3CA is a part of the phosphatidylinositol 3-kinase (PI3K)/protein kinase B (AKT)/mammalian target of rapamycin (mTOR) pathway, a very commonly disrupted pathway observed across several cancer sites that is involved in the regulation of cell growth, proliferation, differentiation, glucose metabolism, protein synthesis, and apoptosis [96–99] (Figure 4).

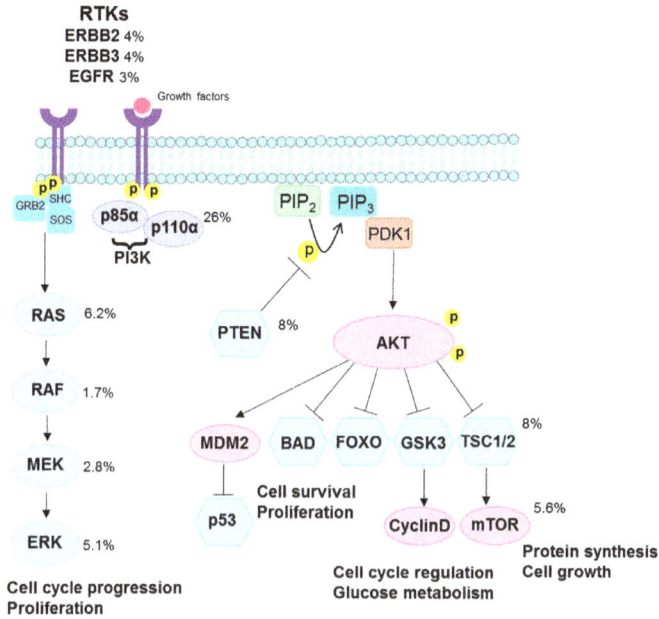

Figure 4. PI3K/AKT and RAS/EGFR/ERK pathways. Class IA PI3K are heterodimers consisting of a p85α regulatory subunit and a p110α catalytic subunit (encoded by PIK3CA). The p85α regulatory subunit normally stabilizes p110α and inhibits its catalytic activity. Activation of the PI3K pathway via ligand binding to transmembrane RTKs such as EGFR, ERBB2, and ERBB3, results in phosphorylation of p85α and activation of the p110α catalytic subunit. Once activated, PI3K phosphorylates PIP$_2$ at the plasma membrane to produce the lipid second messenger, PIP$_3$. This step is inhibited by PTEN, which dephosphorylates PIP$_3$ to PIP$_2$. PIP$_3$ binds to PDK1 which phosphorylates and activates AKT. Activated AKT phosphorylates TSC1/2, leading to mTOR activation and increased protein synthesis and cell growth. AKT increases cell proliferation by phosphorylating GSK3 which normally regulates the degradation of cyclin D. In addition, activation of AKT promotes cell survival by inhibiting proapoptotic factors such as BAD and FOXO transcription factors, and by phosphorylating MDM2 which antagonizes p53-mediated apoptosis. Other PI3K activation pathways depend on adaptor proteins such as GRB2, which binds to and activates SOS, stimulating RAS and independent activation of p110α. A Ras-binding domain in p110α also mediates activation by RAS. RAS-mediated recruitment to the plasma membrane activates RAF, which in turn activates MEK and ERK, respectively. ERK phosphorylates several proteins that control cell proliferation and cell cycle progression. Somatic mutation frequencies in cervical squamous cell carcinomas are shown next to each gene [18]. Abbreviations: RTKs, receptor tyrosine kinases; ERBB2, erb-b2 receptor tyrosine kinase 2; ERBB3, erb-b2 receptor tyrosine kinase 3; EGFR, epidermal growth factor receptor; PI3K, phosphatidylinositol 3-kinases; AKT, protein kinase B; mTOR, mammalian target of rapamycin; PTEN, phosphatase and tensin homolog; PIK3CA, phosphatidylinositol 3-kinase catalytic subunit alpha; PDK1, phosphoinositide-dependent kinase; PIP$_2$, phosphatidylinositol 4,5-bisphosphate; PIP$_3$, phosphatidylinositol 3,4,5-trisphosphate; TSCL1/2, T-cell leukemia 1 and 2; GSK, glycogen synthase kinase; BAD, Bcl-2-associated death promoter; FOXO, forkhead box, O subclass; MDM2, mouse double minute 2 homolog; SHC, Src homology 2 domain-containing; GRB2, growth factor receptor-bound protein 2; SOS, son of sevenless; RAS, retrovirus-associated DNA sequences; RAF, rapidly accelerated fibrosarcoma; MEK, mitogen-activated protein kinase; ERK, extracellular signal–regulated kinases.

PIK3CA encodes p110α, the catalytic subunit of PI3K, and is considered an oncogene; mutations and copy number variations of *PIK3CA* and other related genes in this pathway can contribute to unchecked growth, invasion, and metastasis [100]. *PIK3CA* is the most frequently mutated gene in HPV-positive cancers, with frequencies ranging from 6 to 42% in cervical squamous cell carcinoma, 10–42% in cervical adenocarcinoma, and 22–56% in HPV-positive head and neck cancers [2,18,35–37,60–69].

The most common *PIK3CA* mutations occur in "hotspots" E542K and E545K in the helical domain (exon 9) of p110α. Mutations in these sites have been shown to increase phosphatidylinositol 3,4,5-trisphosphate (PIP$_3$) levels, activate downstream effectors such as phosphoinositide-dependent kinase (PDK1) and AKT, and promote cellular transformation. Although the mechanisms by which these mutations activate PI3K signaling are not fully understood, current data suggests these mutations block the inhibitory effect of the p85α regulatory subunit on p110α activity [101]. In HPV-positive head and neck and cervical squamous cancers, mutations in *PIK3CA* are almost exclusively found in E542K (c.1624G > A) and E545K (1633G > A) corresponding to a C to T single base change at a TCW motif, indicative of APOBEC-induced mutagenesis [35,53,102–105]. In contrast, these mutations are less common in HPV-negative head and neck cancers, suggesting that APOBEC activity is the major source of *PIK3CA* mutations in HPV-driven carcinogenesis. Evidence from a limited number of studies suggests that these mutations may represent a late event in cervical carcinogenesis [63,67,105]; however, a comprehensive deep-sequencing study of cervical precancers has not been conducted.

PTEN is a cell cycle regulator that inhibits rapid cell growth and functions as a tumor suppressor [106]. Signaling of the PI3K pathway is regulated by PTEN through dephosphorylation of PIP$_3$ (Figure 4) [107]. *PTEN* mutations are less frequent than *PIK3CA* mutations but are found in 6–13% of cervical carcinomas and 6–10% of HPV-positive head and neck cancers [2,18,35,36]. High rates of concurrent *PIK3CA* mutations with *PTEN* loss have been documented in HPV-positive tumors, ranging from 24 to 56% in head and neck cancers to over 80% in anal cancers [99,108]. In the context of *PTEN* loss or deficiency, helical mutations in *PIK3CA* have been shown to induce tumorigenesis through AKT-dependent signaling; whereas in tumors with intact *PTEN*, helical mutations in *PIK3CA* have been shown to promote cell growth and transformation through AKT-independent pathways involving *PDK1* and its substrate serine/threonine protein kinase family member 3 (*SGK3*) [109].

Overall, more than 50% of cancers of the cervix and anus have at least one mutation in the PI3K/AKT pathway [110]. Similarly, mutations in this pathway have been reported in 61% of HPV-positive head and neck cancers (and a similar number of HPV-negative head and neck cancers) [2]. The average across all solid tumors was 38%, suggesting that compared with the known driver mutations in other cancers, PI3K pathway alterations are uniquely high in HPV-driven cancers [110]. It is interesting to note that *PIK3CA* is also commonly mutated in endometrial and some ovarian cancers [111,112], which could make it a hallmark of gynecological cancers as well as of HPV-driven cancers.

3.3. Human Leukocyte Antigen

Human leukocyte antigen (HLA) alleles are important components of host cell-mediated immune responses to viral infections and are essential to the major histocompatibility complex (MHC) immune response pathway. *HLA-A* and *HLA-B* are MHC class I molecules that present viral antigens on the cell surface to alert the immune system to infection [113] (Figure 5). Germline HLA variants have been associated with cervical cancer and with HPV-positive oropharyngeal cancer susceptibility [19–21]. Somatic mutations are found in *HLA-A* in 8% and *HLA-B* in 6–9% of cervical squamous cell carcinomas [18,35]. In a small study evaluating cervical cancer cell suspensions, 90% of tumors showed some *HLA* gene alterations including gene mutations, loss of heterozygosity, and other genetic changes [78]. *HLA* alterations are found frequently in cervical precancers as well, suggesting that it is an early event in cervical carcinogenesis [114]. Rates of *HLA-A/B* mutations are somewhat more common in HPV-positive (11%) than HPV-negative (7%) head and neck cancers [2,37]. Loss of *HLA-A* or *HLA-B* could lead to loss of presentation of tumor antigens and immune cell recognition. One small study reported frequent mutations in the HLA pathway-associated transporter associated with antigen processing (*TAP*) gene (52%) in cervical carcinomas [115]. However, another candidate gene study failed to replicate this finding [116] and the large cervical cancer studies did not identify recurrent mutations in this gene [18,35]. Given the observed associations of both germline and somatic changes with the antigen presentation pathway, it is clear that it plays an important role in the host response to viral invasion that can alter the probability of persistence and potentially subsequent steps in the carcinogenesis process.

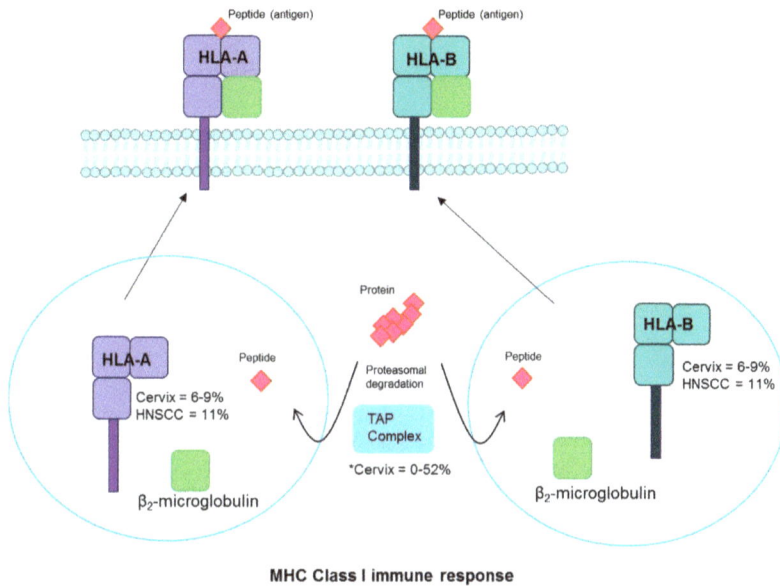

Figure 5. HLA pathway. Proteins undergo proteasomal degradation and the resulting peptides are transported to the endoplasmic reticulum by the TAP complex. There they are bound with MHC Class I into HLA-A or HLA-B and bound to β_2-microglobulin. The complex is transported to the plasma membrane, where the peptide antigen is displayed for cytotoxic T-cell recognition. Fraction of cervical and head and neck cancers with each gene mutated are noted [2,18,115,116]. * There are conflicting reports of TAP mutation prevalence [115,116]. Abbreviations: MHC, major histocompatibility complex; HLA, human leukocyte antigen; TAP, transporter associated with antigen processing; HNSCC, head and neck squamous cell carcinoma.

3.4. Transforming Growth Factor Beta Pathway

The transforming growth factor beta (TGFβ) pathway inhibits DNA synthesis and plays a tumor suppressor role, although it can also promote cancer progression once carcinogenesis has been initiated [117–119]. Inhibition of this pathway by the HPV oncoprotein E7 contributes to early tumor development in HPV-positive cervical and head and neck cancers [120–123] (Figure 6). Commonly mutated genes in the TGFβ pathway include *TGFBR2* (a receptor), CREB binding protein (*CREBBP*) and E1A binding protein p300 (*EP300*) (activators), and *SMAD4* (a transcription factor and tumor suppressor), and mutations in at least one of these genes have been reported in 30% of cervical squamous cell carcinomas [18]. In contrast, among TGFβ genes, only *EP300* was in the top 30 mutated genes in head and neck squamous cell carcinomas [36]. This implies that somatic alterations in *TGFBR2*, *CREBBP*, and *SMAD4* may be cervical squamous cell carcinoma-specific, although E7-driven expression effects in the TGFβ pathway may still play a role in carcinogenesis in other HPV-positive cancers. *SMAD4* downregulation is also associated with HPV-negative head and neck cancers [124], and SMAD signaling pathway alterations have been found in both HPV-positive and HPV-negative tumors [37].

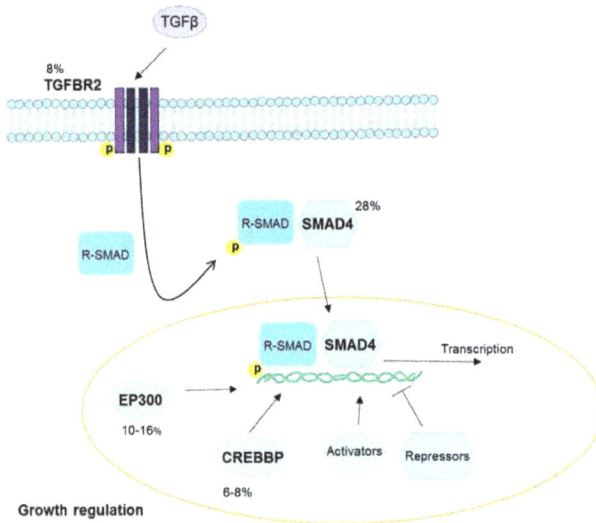

Figure 6. TGFβ pathway. TGFβ binds to TGFBR2 and other receptors to form a complex which becomes phosphorylated. This triggers the phosphorylation of R-SMADs. The phosphorylated R-SMADs form a complex with SMAD4 and are transported into the nucleus, where they promote transcription by binding to promotor regions of the DNA. EP300 and CREBBP are two activators commonly mutated in HPV-driven cancers, and many other activators and repressors also act to regulate this pathway. Fractions of cervical cancers with each gene mutated are noted [18]. Abbreviations: TGFβ, transforming growth factor beta; TGFBR2, TGFβ receptor 2; R-SMAD, receptor-regulated SMAD; EP300, E1A binding protein p300; CREBBP, CREB binding protein.

3.5. Notch Pathway

The Notch signaling pathway is responsible for cellular differentiation. Mutations in the *NOTCH1* receptor are found in both HPV-negative (12–26%) and in HPV-positive (6–17%) head and neck cancers, albeit somewhat more frequently in HPV-negative tumors, and are not commonly reported in cervix or other HPV-driven cancer sites [2,36,37,76,77]. This mutation may, therefore, be specific to head and neck carcinogenesis rather than to HPV infection, and *NOTCH1* has indeed been reported as a driver gene in oral tumorigenesis independent of HPV status [125]. F-box and WD repeat domain containing 7 (*FBXW7*) is involved in angiogenesis through regulation of the Notch pathway [126] and is mutated at higher rates in cervix (11–15%) and HPV-positive head and neck (12%) squamous cell carcinomas than in combined head and neck squamous cell carcinomas (HPV status not specified) (5%) [18,35,36].

3.6. RAS/EGFR/ERK Pathway

The RAS/EGFR/ERK (retrovirus-associated DNA sequences/ epidermal growth factor receptor/ extracellular signal–regulated kinases) pathway is involved in cellular proliferation and survival (Figure 3). It consists of a signaling cascade that regulates transcription of genes affecting many functions including differentiation, growth, and senescence, which can contribute to carcinogenesis [127]. KRAS proto-oncogene, GTPase (*KRAS*) is an oncogene in which mutations are found in 8–23% in cervical adenocarcinomas but rarely in cervical squamous cell carcinomas [18,35,62,75]. The mutation rate of *KRAS* in head and neck cancers is 6% [37]. In contrast, *EGFR* is a tumor suppressor in the same pathway in which mutations are found in 3–33% of cervical squamous cell carcinomas but rarely in cervical adenocarcinomas [18,62,70,71]. Other genes in this pathway are mutated in fewer than 10% of HPV-positive tumors except for *FGFR2* and *FGFR3*, which have combined mutation rates of 10–17% in HPV-positive head and neck cancers [2,18,35–37]. This is notable because, as kinases, the *FGFR* genes may potentially be therapeutic targets [37].

3.7. Other Genes

The tumor necrosis factor (TNF) receptor associated factor 3 (*TRAF3*) is involved in viral immune responses [128] and was recently reported to have truncating mutations (8%) or deletions (14%) in HPV-positive head and neck cancers [2]. It is not commonly mutated in cervical cancers [18], and it remains to be investigated whether this gene is mutated in HPV-positive cancers at other sites. Other genes differentially mutated in HPV-positive versus HPV-negative head and neck squamous cell carcinomas include *E2F1*, a cell cycle related gene more commonly mutated in HPV-positive cancers (19% versus 2%), and FAT atypical cadherin 1 (*FAT1*) and ajuba LIM protein (*AJUBA*), two genes involved in differentiation that are more commonly mutated in HPV-negative cancers (32% versus 3% and 7% versus 0%, respectively) [2].

4. Discussion

While HPV infection is a necessary cause of many cancers, the interplay between the virus and the host cell is what ultimately causes cancers to develop. There are many similarities across sites in the mechanisms and mutations found in HPV-driven cancers, suggesting that mechanisms are likely to be similar in rarer cancers such as penile and vaginal carcinomas in which it is difficult to complete large genomic studies. For example, one recent candidate gene study found no statistically significant differences in gene mutations in any of 48 candidate genes including *PIK3CA*, *EGFR*, *NOTCH1*, and *KRAS* or copy number alterations in any of six candidate genes across HPV-positive squamous cell carcinomas at four anatomical sites [99]. While HPV-positive cancers share many characteristic mutagenesis mechanisms and somatic mutations, there are also site-specific aspects. The other major gynecological cancers, endometrial and ovarian cancer, share with cervical cancer high rates of *PIK3CA* mutations and APOBEC and signature 1B mutational signatures. HPV-positive and HPV-negative tumors arising in the head and neck also share properties such as recurrent focal amplifications of the 3q26-28 chromosomal region. Recent data have shown that HPV genetic variation is very common and that HPV variant sublineages influence the risk of different histologic types of cervical precancer and cancer. It will be important to study the interplay between viral genetics and host genomic changes to better understand HPV-driven carcinogenesis [129,130].

Characterizing somatic mutations in HPV-related carcinogenesis could be highly relevant for early detection, prognosis, and treatment. To date, very few studies have attempted to characterize the somatic landscape of precancerous lesions, none comprehensively [63,74,131]. Several important steps are required to develop early detection assays based on somatic mutations. First, the sequence of somatic mutation events in the transition from precancers to cancers needs to be established. Next, a promising panel of mutations needs to be selected and evaluated in cervical cytology samples. Similar efforts have been evaluated for other gynecological cancers [132].

In addition to early detection, somatic characterization can be important for prognosis and targeted treatment strategies. For example, *PIK3CA*-mutated cervical cancers have worse prognosis than cancer with wild-type *PIK3CA* [61]. Site-specific mutations in *PIK3CA* have been shown to have varying responses to treatment, with evidence suggesting a greater response to PI3K/AKT/mTOR pathway inhibitors for tumors with mutations in the H1047R kinase domain (which are not commonly found in cervical cancers) compared with mutations at other sites [133]. Another prospective therapeutic target is *BCAR4*, in which amplifications and gene fusions have been found in cervical cancer and which is targeted by lapatinib [18,134]. *CD274* and *PDCD1LG2* are immunotherapy targets with amplifications reported in cervical cancer [18]. Erb-b2 receptor tyrosine kinase 2 (*ERBB2*;

HER2) and erb-b2 receptor tyrosine kinase 3 (*ERBB3*; *HER3*) are mutated in a subset of cervical adenocarcinomas and these tumors may be susceptible to targeted therapies, and *PTEN* and AT-rich interaction domain 1A (*ARID1A*) alterations are also potential targets [18]. The PI3K/AKT and TGFβ signaling pathways, at least one of which is altered in over 70% of cervical cancers, are very promising in that targeted therapies may be broadly applicable due to their high prevalence [18]. The development of somatic marker panels for HPV-driven cancers will enable oncologists to more precisely tailor treatments.

Acknowledgments: This work was supported by the Intramural Research Program of the National Cancer Institute, Z01 CP010124-21.

Conflicts of Interest: The authors declare no conflict of interest.

References

1. Walboomers, J.M.; Jacobs, M.V.; Manos, M.M.; Bosch, F.X.; Kummer, J.A.; Shah, K.V.; Snijders, P.J.; Peto, J.; Meijer, C.J.; Munoz, N. Human papillomavirus is a necessary cause of invasive cervical cancer worldwide. *J. Pathol.* **1999**, *189*, 12–19. [CrossRef]
2. Cancer Genome Atlas Network. Comprehensive genomic characterization of head and neck squamous cell carcinomas. *Nature* **2015**, *517*, 576–582.
3. Stransky, N.; Egloff, A.M.; Tward, A.D.; Kostic, A.D.; Cibulskis, K.; Sivachenko, A.; Kryukov, G.V.; Lawrence, M.S.; Sougnez, C.; McKenna, A.; et al. The mutational landscape of head and neck squamous cell carcinoma. *Science* **2011**, *333*, 1157–1160. [CrossRef] [PubMed]
4. Ndiaye, C.; Mena, M.; Alemany, L.; Arbyn, M.; Castellsague, X.; Laporte, L.; Bosch, F.X.; de Sanjose, S.; Trottier, H. HPV DNA, E6/E7 mRNA, and p16^{INK4A} detection in head and neck cancers: A systematic review and meta-analysis. *Lancet Oncol.* **2014**, *15*, 1319–1331. [CrossRef]
5. Hartwig, S.; Baldauf, J.-J.; Dominiak-Felden, G.; Simondon, F.; Alemany, L.; de Sanjosé, S.; Castellsagué, X. Estimation of the epidemiological burden of HPV-related anogenital cancers, precancerous lesions, and genital warts in women and men in europe: Potential additional benefit of a nine-valent second generation HPV vaccine compared to first generation HPV vaccines. *Papillomavirus Res.* **2015**, *1*, 90–100.
6. Backes, D.M.; Kurman, R.J.; Pimenta, J.M.; Smith, J.S. Systematic review of human papillomavirus prevalence in invasive penile cancer. *Cancer Causes Control* **2009**, *20*, 449–457. [CrossRef] [PubMed]
7. Miralles-Guri, C.; Bruni, L.; Cubilla, A.L.; Castellsague, X.; Bosch, F.X.; de Sanjose, S. Human papillomavirus prevalence and type distribution in penile carcinoma. *J. Clin. Pathol.* **2009**, *62*, 870–878. [CrossRef] [PubMed]
8. Forman, D.; de Martel, C.; Lacey, C.J.; Soerjomataram, I.; Lortet-Tieulent, J.; Bruni, L.; Vignat, J.; Ferlay, J.; Bray, F.; Plummer, M.; et al. Global burden of human papillomavirus and related diseases. *Vaccine* **2012**, *30*, F12–F23. [CrossRef] [PubMed]
9. De Martel, C.; Ferlay, J.; Franceschi, S.; Vignat, J.; Bray, F.; Forman, D.; Plummer, M. Global burden of cancers attributable to infections in 2008: A review and synthetic analysis. *Lancet Oncol.* **2012**, *13*, 607–615. [CrossRef]
10. Ho, G.Y.; Bierman, R.; Beardsley, L.; Chang, C.J.; Burk, R.D. Natural history of cervicovaginal papillomavirus infection in young women. *N. Engl. J. Med.* **1998**, *338*, 423–428. [CrossRef] [PubMed]
11. Schiffman, M.; Kjaer, S.K. Chapter 2: Natural history of anogenital human papillomavirus infection and neoplasia. *J. Natl. Cancer Inst. Monogr.* **2003**, *2003*, 14–19. [CrossRef]
12. Rositch, A.F.; Koshiol, J.; Hudgens, M.G.; Razzaghi, H.; Backes, D.M.; Pimenta, J.M.; Franco, E.L.; Poole, C.; Smith, J.S. Patterns of persistent genital human papillomavirus infection among women worldwide: A literature review and meta-analysis. *Int. J. Cancer* **2013**, *133*, 1271–1285. [CrossRef] [PubMed]
13. Moscicki, A.B.; Schiffman, M.; Kjaer, S.; Villa, L.L. Chapter 5: Updating the natural history of HPV and anogenital cancer. *Vaccine* **2006**, *2*, S42–S51. [CrossRef] [PubMed]
14. McCredie, M.R.; Sharples, K.J.; Paul, C.; Baranyai, J.; Medley, G.; Jones, R.W.; Skegg, D.C. Natural history of cervical neoplasia and risk of invasive cancer in women with cervical intraepithelial neoplasia 3: A retrospective cohort study. *Lancet Oncol.* **2008**, *9*, 425–434. [CrossRef]
15. Schiffman, M.; Doorbar, J.; Wentzensen, N.; de Sanjose, S.; Fakhry, C.; Monk, B.J.; Stanley, M.A.; Franceschi, S. Carcinogenic human papillomavirus infection. *Nat. Rev. Dis. Primers* **2016**, *2*, 16086. [CrossRef] [PubMed]

16. Wentzensen, N.; Arbyn, M.; Berkhof, J.; Bower, M.; Canfell, K.; Einstein, M.; Farley, C.; Monsonego, J.; Franceschi, S. Eurogin 2016 roadmap: How HPV knowledge is changing screening practice. *Int. J. Cancer* **2017**, *140*, 2192–2200. [CrossRef] [PubMed]

17. Rusan, M.; Li, Y.Y.; Hammerman, P.S. Genomic landscape of human papillomavirus-associated cancers. *Clin. Cancer Res.* **2015**, *21*, 2009–2019. [CrossRef] [PubMed]

18. The Cancer Genome Atlas Research Network. Integrated genomic and molecular characterization of cervical cancer. *Nature* **2017**, *543*, 378–384.

19. Chen, D.; Gyllensten, U. Lessons and implications from association studies and post-gwas analyses of cervical cancer. *Trends Genet.* **2015**, *31*, 41–54. [CrossRef] [PubMed]

20. Martínez-Nava, G.A.; Fernández-Niño, J.A.; Madrid-Marina, V.; Torres-Poveda, K. Cervical cancer genetic susceptibility: A systematic review and meta-analyses of recent evidence. *PLoS ONE* **2016**, *11*, e0157344. [CrossRef] [PubMed]

21. Lesseur, C.; Diergaarde, B.; Olshan, A.F.; Wunsch-Filho, V.; Ness, A.R.; Liu, G.; Lacko, M.; Eluf-Neto, J.; Franceschi, S.; Lagiou, P.; et al. Genome-wide association analyses identify new susceptibility loci for oral cavity and pharyngeal cancer. *Nat. Genet.* **2016**, *48*, 1544–1550. [CrossRef] [PubMed]

22. Li, N.; Franceschi, S.; Howell-Jones, R.; Snijders, P.J.; Clifford, G.M. Human papillomavirus type distribution in 30,848 invasive cervical cancers worldwide: Variation by geographical region, histological type and year of publication. *Int. J. Cancer* **2011**, *128*, 927–935. [CrossRef] [PubMed]

23. Prigge, E.S.; von Knebel Doeberitz, M.; Reuschenbach, M. Clinical relevance and implications of HPV-induced neoplasia in different anatomical locations. *Mutat. Res.* **2017**, *772*, 51–66. [CrossRef] [PubMed]

24. Duensing, S.; Munger, K. Centrosome abnormalities, genomic instability and carcinogenic progression. *Biochim. Biophys. Acta* **2001**, *1471*, M81–M88. [CrossRef]

25. Duensing, S.; Lee, L.Y.; Duensing, A.; Basile, J.; Piboonniyom, S.; Gonzalez, S.; Crum, C.P.; Munger, K. The human papillomavirus type 16 E6 and E7 oncoproteins cooperate to induce mitotic defects and genomic instability by uncoupling centrosome duplication from the cell division cycle. *Proc. Natl. Acad. Sci. USA* **2000**, *97*, 10002–10007. [CrossRef] [PubMed]

26. Werness, B.A.; Levine, A.J.; Howley, P.M. Association of human papillomavirus types 16 and 18 E6 proteins with p53. *Science* **1990**, *248*, 76–79. [CrossRef] [PubMed]

27. Scheffner, M.; Werness, B.A.; Huibregtse, J.M.; Levine, A.J.; Howley, P.M. The E6 oncoprotein encoded by human papillomavirus types 16 and 18 promotes the degradation of p53. *Cell* **1990**, *63*, 1129–1136. [CrossRef]

28. Dyson, N.; Howley, P.M.; Munger, K.; Harlow, E. The human papilloma virus-16 E7 oncoprotein is able to bind to the retinoblastoma gene product. *Science* **1989**, *243*, 934–937. [CrossRef] [PubMed]

29. Munger, K.; Werness, B.A.; Dyson, N.; Phelps, W.C.; Harlow, E.; Howley, P.M. Complex formation of human papillomavirus E7 proteins with the retinoblastoma tumor suppressor gene product. *EMBO J.* **1989**, *8*, 4099–4105. [PubMed]

30. Boyer, S.N.; Wazer, D.E.; Band, V. E7 protein of human papilloma virus-16 induces degradation of retinoblastoma protein through the ubiquitin-proteasome pathway. *Cancer Res.* **1996**, *56*, 4620–4624. [PubMed]

31. McDaniel, A.S.; Hovelson, D.H.; Cani, A.K.; Liu, C.J.; Zhai, Y.; Zhang, Y.; Weizer, A.Z.; Mehra, R.; Feng, F.Y.; Alva, A.S.; et al. Genomic profiling of penile squamous cell carcinoma reveals new opportunities for targeted therapy. *Cancer Res.* **2015**, *75*, 5219–5227. [CrossRef] [PubMed]

32. Warren, C.J.; Xu, T.; Guo, K.; Griffin, L.M.; Westrich, J.A.; Lee, D.; Lambert, P.F.; Santiago, M.L.; Pyeon, D. APOBEC3A functions as a restriction factor of human papillomavirus. *J. Virol.* **2015**, *89*, 688–702. [CrossRef] [PubMed]

33. Thomas, L.K.; Bermejo, J.L.; Vinokurova, S.; Jensen, K.; Bierkens, M.; Steenbergen, R.; Bergmann, M.; von Knebel Doeberitz, M.; Reuschenbach, M. Chromosomal gains and losses in human papillomavirus-associated neoplasia of the lower genital tract—A systematic review and meta-analysis. *Eur. J. Cancer* **2014**, *50*, 85–98. [CrossRef] [PubMed]

34. Bodelon, C.; Vinokurova, S.; Sampson, J.N.; den Boon, J.A.; Walker, J.L.; Horswill, M.A.; Korthauer, K.; Schiffman, M.; Sherman, M.E.; Zuna, R.E.; et al. Chromosomal copy number alterations and HPV integration in cervical precancer and invasive cancer. *Carcinogenesis* **2016**, *37*, 188–196. [CrossRef] [PubMed]

35. Ojesina, A.I.; Lichtenstein, L.; Freeman, S.S.; Pedamallu, C.S.; Imaz-Rosshandler, I.; Pugh, T.J.; Cherniack, A.D.; Ambrogio, L.; Cibulskis, K.; Bertelsen, B.; et al. Landscape of genomic alterations in cervical carcinomas. *Nature* **2014**, *506*, 371–375. [CrossRef] [PubMed]
36. Chung, C.H.; Guthrie, V.B.; Masica, D.L.; Tokheim, C.; Kang, H.; Richmon, J.; Agrawal, N.; Fakhry, C.; Quon, H.; Subramaniam, R.M.; et al. Genomic alterations in head and neck squamous cell carcinoma determined by cancer gene-targeted sequencing. *Ann. Oncol.* **2015**, *26*, 1216–1223. [CrossRef] [PubMed]
37. Seiwert, T.Y.; Zuo, Z.; Keck, M.K.; Khattri, A.; Pedamallu, C.S.; Stricker, T.; Brown, C.; Pugh, T.J.; Stojanov, P.; Cho, J.; et al. Integrative and comparative genomic analysis of HPV-positive and HPV-negative head and neck squamous cell carcinomas. *Clin. Cancer Res.* **2015**, *21*, 632–641. [CrossRef] [PubMed]
38. Busso-Lopes, A.F.; Marchi, F.A.; Kuasne, H.; Scapulatempo-Neto, C.; Trindade-Filho, J.C.; de Jesus, C.M.; Lopes, A.; Guimaraes, G.C.; Rogatto, S.R. Genomic profiling of human penile carcinoma predicts worse prognosis and survival. *Cancer Prev. Res.* **2015**, *8*, 149–156. [CrossRef] [PubMed]
39. Heselmeyer, K.; du Manoir, S.; Blegen, H.; Friberg, B.; Svensson, C.; Schrock, E.; Veldman, T.; Shah, K.; Auer, G.; Ried, T. A recurrent pattern of chromosomal aberrations and immunophenotypic appearance defines anal squamous cell carcinomas. *Br. J. Cancer* **1997**, *76*, 1271–1278. [CrossRef] [PubMed]
40. Wentzensen, N.; Ridder, R.; Klaes, R.; Vinokurova, S.; Schaefer, U.; Doeberitz, M. Characterization of viral-cellular fusion transcripts in a large series of HPV16 and 18 positive anogenital lesions. *Oncogene* **2002**, *21*, 419–426. [CrossRef] [PubMed]
41. Vinokurova, S.; Wentzensen, N.; Kraus, I.; Klaes, R.; Driesch, C.; Melsheimer, P.; Kisseljov, F.; Durst, M.; Schneider, A.; von Knebel Doeberitz, M. Type-dependent integration frequency of human papillomavirus genomes in cervical lesions. *Cancer Res.* **2008**, *68*, 307–313. [CrossRef] [PubMed]
42. Bodelon, C.; Untereiner, M.E.; Machiela, M.J.; Vinokurova, S.; Wentzensen, N. Genomic characterization of viral integration sites in HPV-related cancers. *Int. J. Cancer* **2016**, *139*, 2001–2011. [CrossRef] [PubMed]
43. Akagi, K.; Li, J.; Broutian, T.R.; Padilla-Nash, H.; Xiao, W.; Jiang, B.; Rocco, J.W.; Teknos, T.N.; Kumar, B.; Wangsa, D.; et al. Genome-wide analysis of HPV integration in human cancers reveals recurrent, focal genomic instability. *Genome Res.* **2014**, *24*, 185–199. [CrossRef] [PubMed]
44. Schmitz, M.; Driesch, C.; Jansen, L.; Runnebaum, I.B.; Durst, M. Non-random integration of the HPV genome in cervical cancer. *PLoS ONE* **2012**, *7*, e39632. [CrossRef] [PubMed]
45. Wentzensen, N.; Vinokurova, S.; von Knebel Doeberitz, M. Systematic review of genomic integration sites of human papillomavirus genomes in epithelial dysplasia and invasive cancer of the female lower genital tract. *Cancer Res.* **2004**, *64*, 3878–3884. [CrossRef] [PubMed]
46. Hu, Z.; Zhu, D.; Wang, W.; Li, W.; Jia, W.; Zeng, X.; Ding, W.; Yu, L.; Wang, X.; Wang, L.; et al. Genome-wide profiling of HPV integration in cervical cancer identifies clustered genomic hot spots and a potential microhomology-mediated integration mechanism. *Nat. Genet.* **2015**, *47*, 158–163. [CrossRef] [PubMed]
47. Conticello, S.G. The AID/APOBEC family of nucleic acid mutators. *Genome Biol.* **2008**, *9*, 229. [CrossRef] [PubMed]
48. Roberts, S.A.; Sterling, J.; Thompson, C.; Harris, S.; Mav, D.; Shah, R.; Klimczak, L.J.; Kryukov, G.V.; Malc, E.; Mieczkowski, P.A.; et al. Clustered mutations in yeast and in human cancers can arise from damaged long single-strand DNA regions. *Mol. Cell* **2012**, *46*, 424–435. [CrossRef] [PubMed]
49. Simonelli, V.; Narciso, L.; Dogliotti, E.; Fortini, P. Base excision repair intermediates are mutagenic in mammalian cells. *Nucleic Acids Res.* **2005**, *33*, 4404–4411. [CrossRef] [PubMed]
50. Alexandrov, L.B.; Nik-Zainal, S.; Wedge, D.C.; Aparicio, S.A.; Behjati, S.; Biankin, A.V.; Bignell, G.R.; Bolli, N.; Borg, A.; Borresen-Dale, A.L.; et al. Signatures of mutational processes in human cancer. *Nature* **2013**, *500*, 415–421. [CrossRef] [PubMed]
51. Burns, M.B.; Temiz, N.A.; Harris, R.S. Evidence for APOBEC3b mutagenesis in multiple human cancers. *Nat. Genet.* **2013**, *45*, 977–983. [CrossRef] [PubMed]
52. Roberts, S.A.; Lawrence, M.S.; Klimczak, L.J.; Grimm, S.A.; Fargo, D.; Stojanov, P.; Kiezun, A.; Kryukov, G.V.; Carter, S.L.; Saksena, G.; et al. An APOBEC cytidine deaminase mutagenesis pattern is widespread in human cancers. *Nat. Genet.* **2013**, *45*, 970–976. [CrossRef] [PubMed]
53. Henderson, S.; Chakravarthy, A.; Su, X.; Boshoff, C.; Fenton, T.R. APOBEC-mediated cytosine deamination links PIK3CA helical domain mutations to human papillomavirus-driven tumor development. *Cell Rep.* **2014**, *7*, 1833–1841. [CrossRef] [PubMed]

54. Feber, A.; Worth, D.C.; Chakravarthy, A.; de Winter, P.; Shah, K.; Arya, M.; Saqib, M.; Nigam, R.; Malone, P.R.; Tan, W.S.; et al. CSN1 somatic mutations in penile squamous cell carcinoma. *Cancer Res.* **2016**, *76*, 4720–4727. [CrossRef] [PubMed]

55. Vieira, V.C.; Leonard, B.; White, E.A.; Starrett, G.J.; Temiz, N.A.; Lorenz, L.D.; Lee, D.; Soares, M.A.; Lambert, P.F.; Howley, P.M.; et al. Human papillomavirus E6 triggers upregulation of the antiviral and cancer genomic DNA deaminase APOBEC3B. *MBio* **2014**, *5*. [CrossRef] [PubMed]

56. Rebhandl, S.; Huemer, M.; Greil, R.; Geisberger, R. AID/APOBEC deaminases and cancer. *Oncoscience* **2015**, *2*, 320–333. [CrossRef] [PubMed]

57. Nik-Zainal, S.; Alexandrov, L.B.; Wedge, D.C.; Van Loo, P.; Greenman, C.D.; Raine, K.; Jones, D.; Hinton, J.; Marshall, J.; Stebbings, L.A.; et al. Mutational processes molding the genomes of 21 breast cancers. *Cell* **2012**, *149*, 979–993. [CrossRef] [PubMed]

58. Taylor, B.J.; Nik-Zainal, S.; Wu, Y.L.; Stebbings, L.A.; Raine, K.; Campbell, P.J.; Rada, C.; Stratton, M.R.; Neuberger, M.S. DNA deaminases induce break-associated mutation showers with implication of APOBEC3B and 3A in breast cancer kataegis. *Elife* **2013**, *2*, e00534. [CrossRef] [PubMed]

59. Leonard, B.; Hart, S.N.; Burns, M.B.; Carpenter, M.A.; Temiz, N.A.; Rathore, A.; Vogel, R.I.; Nikas, J.B.; Law, E.K.; Brown, W.L.; et al. APOBEC3B upregulation and genomic mutation patterns in serous ovarian carcinoma. *Cancer Res.* **2013**, *73*, 7222–7231. [CrossRef] [PubMed]

60. McIntyre, J.B.; Wu, J.S.; Craighead, P.S.; Phan, T.; Kobel, M.; Lees-Miller, S.P.; Ghatage, P.; Magliocco, A.M.; Doll, C.M. PIK3CA mutational status and overall survival in patients with cervical cancer treated with radical chemoradiotherapy. *Gynecol. Oncol.* **2013**, *128*, 409–414. [CrossRef] [PubMed]

61. Lou, H.; Villagran, G.; Boland, J.F.; Im, K.M.; Polo, S.; Zhou, W.; Odey, U.; Juarez-Torres, E.; Medina-Martinez, I.; Roman-Basaure, E.; et al. Genome analysis of latin american cervical cancer: Frequent activation of the PIK3CA pathway. *Clin. Cancer Res.* **2015**, *21*, 5360–5370. [CrossRef] [PubMed]

62. Wright, A.A.; Howitt, B.E.; Myers, A.P.; Dahlberg, S.E.; Palescandolo, E.; Van Hummelen, P.; MacConaill, L.E.; Shoni, M.; Wagle, N.; Jones, R.T.; et al. Oncogenic mutations in cervical cancer: Genomic differences between adenocarcinomas and squamous cell carcinomas of the cervix. *Cancer* **2013**, *119*, 3776–3783. [CrossRef] [PubMed]

63. Cui, B.; Zheng, B.; Zhang, X.; Stendahl, U.; Andersson, S.; Wallin, K.L. Mutation of PIK3CA: Possible risk factor for cervical carcinogenesis in older women. *Int. J. Oncol.* **2009**, *34*, 409–416. [PubMed]

64. Rashmi, R.; DeSelm, C.; Helms, C.; Bowcock, A.; Rogers, B.E.; Rader, J.L.; Grigsby, P.W.; Schwarz, J.K. Akt inhibitors promote cell death in cervical cancer through disruption of mtor signaling and glucose uptake. *PLoS ONE* **2014**, *9*, e92948. [CrossRef] [PubMed]

65. Spaans, V.M.; Trietsch, M.D.; Crobach, S.; Stelloo, E.; Kremer, D.; Osse, E.M.; Haar, N.T.; van Eijk, R.; Muller, S.; van Wezel, T.; et al. Designing a high-throughput somatic mutation profiling panel specifically for gynaecological cancers. *PLoS ONE* **2014**, *9*, e93451. [CrossRef] [PubMed]

66. Spaans, V.M.; Trietsch, M.D.; Peters, A.A.; Osse, M.; Ter Haar, N.; Fleuren, G.J.; Jordanova, E.S. Precise classification of cervical carcinomas combined with somatic mutation profiling contributes to predicting disease outcome. *PLoS ONE* **2015**, *10*, e0133670. [CrossRef] [PubMed]

67. Tornesello, M.L.; Annunziata, C.; Buonaguro, L.; Losito, S.; Greggi, S.; Buonaguro, F.M. TP53 and PIK3CA gene mutations in adenocarcinoma, squamous cell carcinoma and high-grade intraepithelial neoplasia of the cervix. *J. Transl. Med.* **2014**, *12*, 255. [CrossRef] [PubMed]

68. Hou, M.M.; Liu, X.; Wheler, J.; Naing, A.; Hong, D.; Coleman, R.L.; Tsimberidou, A.; Janku, F.; Zinner, R.; Lu, K.; et al. Targeted PI3K/AKT/mTOR therapy for metastatic carcinomas of the cervix: A phase I clinical experience. *Oncotarget* **2014**, *5*, 11168–11179. [CrossRef] [PubMed]

69. Chung, T.K.H.; Cheung, T.H.; Yim, S.F.; Yu, M.Y.; Chiu, R.W.K.; Lo, K.W.K.; Lee, I.P.C.; Wong, R.R.Y.; Lau, K.K.M.; Wang, V.W.; et al. Liquid biopsy of PIK3CA mutations in cervical cancer in Hong Kong Chinese women. *Gynecol. Oncol.* **2017**, *146*, 334–339. [CrossRef] [PubMed]

70. Qureshi, R.; Arora, H.; Biswas, S.; Perwez, A.; Naseem, A.; Wajid, S.; Gandhi, G.; Rizvi, M.A. Mutation analysis of EGFR and its correlation with the HPV in indian cervical cancer patients. *Tumour Biol.* **2016**, *37*, 9089–9098. [CrossRef] [PubMed]

71. Iida, K.; Nakayama, K.; Rahman, M.T.; Rahman, M.; Ishikawa, M.; Katagiri, A.; Yeasmin, S.; Otsuki, Y.; Kobayashi, H.; Nakayama, S.; et al. EGFR gene amplification is related to adverse clinical outcomes in cervical squamous cell carcinoma, making the EGFR pathway a novel therapeutic target. *Br. J. Cancer* **2011**, *105*, 420–427. [CrossRef] [PubMed]
72. Westra, W.H.; Taube, J.M.; Poeta, M.L.; Begum, S.; Sidransky, D.; Koch, W.M. Inverse relationship between human papillomavirus-16 infection and disruptive p53 gene mutations in squamous cell carcinoma of the head and neck. *Clin. Cancer Res.* **2008**, *14*, 366–369. [CrossRef] [PubMed]
73. Kashofer, K.; Regauer, S. Analysis of full coding sequence of the TP53 gene in invasive vulvar cancers: Implications for therapy. *Gynecol. Oncol.* **2017**, *146*, 314–318. [CrossRef] [PubMed]
74. Trietsch, M.D.; Nooij, L.S.; Gaarenstroom, K.N.; van Poelgeest, M.I. Genetic and epigenetic changes in vulvar squamous cell carcinoma and its precursor lesions: A review of the current literature. *Gynecol. Oncol.* **2015**, *136*, 143–157. [CrossRef] [PubMed]
75. Kang, S.; Kim, H.S.; Seo, S.S.; Park, S.Y.; Sidransky, D.; Dong, S.M. Inverse correlation between RASSF1A hypermethylation, kras and braf mutations in cervical adenocarcinoma. *Gynecol. Oncol.* **2007**, *105*, 662–666. [CrossRef] [PubMed]
76. Tinhofer, I.; Stenzinger, A.; Eder, T.; Konschak, R.; Niehr, F.; Endris, V.; Distel, L.; Hautmann, M.G.; Mandic, R.; Stromberger, C.; et al. Targeted next-generation sequencing identifies molecular subgroups in squamous cell carcinoma of the head and neck with distinct outcome after concurrent chemoradiation. *Ann. Oncol.* **2016**, *27*, 2262–2268. [CrossRef] [PubMed]
77. Agrawal, N.; Frederick, M.J.; Pickering, C.R.; Bettegowda, C.; Chang, K.; Li, R.J.; Fakhry, C.; Xie, T.X.; Zhang, J.; Wang, J.; et al. Exome sequencing of head and neck squamous cell carcinoma reveals inactivating mutations in NOTCH1. *Science* **2011**, *333*, 1154–1157. [CrossRef] [PubMed]
78. Koopman, L.A.; Corver, W.E.; van der Slik, A.R.; Giphart, M.J.; Fleuren, G.J. Multiple genetic alterations cause frequent and heterogeneous human histocompatibility leukocyte antigen class I loss in cervical cancer. *J. Exp. Med.* **2000**, *191*, 961–976. [CrossRef] [PubMed]
79. Moody, C.A.; Laimins, L.A. Human papillomavirus oncoproteins: Pathways to transformation. *Nat. Rev. Cancer* **2010**, *10*, 550–560. [CrossRef] [PubMed]
80. Banister, C.E.; Liu, C.; Pirisi, L.; Creek, K.E.; Buckhaults, P.J. Identification and characterization of HPV-independent cervical cancers. *Oncotarget* **2017**, *8*, 13375–13386. [CrossRef] [PubMed]
81. Salam, M.; Rockett, J.; Morris, A. The prevalence of different human papillomavirus types and p53 mutations in laryngeal carcinomas: Is there a reciprocal relationship? *Eur. J. Surg. Oncol.* **1995**, *21*, 290–296. [CrossRef]
82. Rajendra, S.; Wang, B.; Merrett, N.; Sharma, P.; Humphris, J.; Lee, H.C.; Wu, J. Genomic analysis of HPV-positive versus HPV-negative oesophageal adenocarcinoma identifies a differential mutational landscape. *J. Med. Genet.* **2016**, *53*, 227–231. [CrossRef] [PubMed]
83. Zhang, W.; Edwards, A.; Fang, Z.; Flemington, E.K.; Zhang, K. Integrative genomics and transcriptomics analysis reveals potential mechanisms for favorable prognosis of patients with HPV-positive head and neck carcinomas. *Sci. Rep.* **2016**, *6*, 24927. [CrossRef] [PubMed]
84. Serup-Hansen, E.; Linnemann, D.; Skovrider-Ruminski, W.; Hogdall, E.; Geertsen, P.F.; Havsteen, H. Human papillomavirus genotyping and p16 expression as prognostic factors for patients with american joint committee on cancer stages i to iii carcinoma of the anal canal. *J. Clin. Oncol.* **2014**, *32*, 1812–1817. [CrossRef] [PubMed]
85. Lont, A.P.; Kroon, B.K.; Horenblas, S.; Gallee, M.P.; Berkhof, J.; Meijer, C.J.; Snijders, P.J. Presence of high-risk human papillomavirus DNA in penile carcinoma predicts favorable outcome in survival. *Int. J. Cancer* **2006**, *119*, 1078–1081. [CrossRef] [PubMed]
86. Djajadiningrat, R.S.; Jordanova, E.S.; Kroon, B.K.; van Werkhoven, E.; de Jong, J.; Pronk, D.T.; Snijders, P.J.; Horenblas, S.; Heideman, D.A. Human papillomavirus prevalence in invasive penile cancer and association with clinical outcome. *J. Urol.* **2015**, *193*, 526–531. [CrossRef] [PubMed]
87. Bezerra, A.L.; Lopes, A.; Santiago, G.H.; Ribeiro, K.C.; Latorre, M.R.; Villa, L.L. Human papillomavirus as a prognostic factor in carcinoma of the penis: Analysis of 82 patients treated with amputation and bilateral lymphadenectomy. *Cancer* **2001**, *91*, 2315–2321. [CrossRef]

88. Bezerra, S.M.; Chaux, A.; Ball, M.W.; Faraj, S.F.; Munari, E.; Gonzalez-Roibon, N.; Sharma, R.; Bivalacqua, T.J.; Burnett, A.L.; Netto, G.J. Human papillomavirus infection and immunohistochemical p16(INK4A) expression as predictors of outcome in penile squamous cell carcinomas. *Hum. Pathol.* **2015**, *46*, 532–540. [CrossRef] [PubMed]

89. Schaal, C.; Pillai, S.; Chellappan, S.P. The RB-E2F transcriptional regulatory pathway in tumor angiogenesis and metastasis. *Adv. Cancer Res.* **2014**, *121*, 147–182. [PubMed]

90. Doorbar, J.; Egawa, N.; Griffin, H.; Kranjec, C.; Murakami, I. Human papillomavirus molecular biology and disease association. *Rev. Med. Virol.* **2015**, *25*, 2–23. [CrossRef] [PubMed]

91. Felsani, A.; Mileo, A.M.; Paggi, M.G. Retinoblastoma family proteins as key targets of the small DNA virus oncoproteins. *Oncogene* **2006**, *25*, 5277–5285. [CrossRef] [PubMed]

92. El-Naggar, A.K.; Westra, W.H. p16 expression as a surrogate marker for HPV-related oropharyngeal carcinoma: A guide for interpretative relevance and consistency. *Head Neck* **2012**, *34*, 459–461. [CrossRef] [PubMed]

93. Klaes, R.; Friedrich, T.; Spitkovsky, D.; Ridder, R.; Rudy, W.; Petry, U.; Dallenbach-Hellweg, G.; Schmidt, D.; von Knebel Doeberitz, M. Overexpression of p16^{INK4A} as a specific marker for dysplastic and neoplastic epithelial cells of the cervix uteri. *Int. J. Cancer* **2001**, *92*, 276–284. [CrossRef] [PubMed]

94. Wentzensen, N.; Fetterman, B.; Castle, P.E.; Schiffman, M.; Wood, S.N.; Stiemerling, E.; Tokugawa, D.; Bodelon, C.; Poitras, N.; Lorey, T.; et al. p16/KI-67 dual stain cytology for detection of cervical precancer in HPV-positive women. *J. Natl. Cancer Inst.* **2015**, *107*, djv257. [CrossRef] [PubMed]

95. Cuschieri, K.; Wentzensen, N. Human papillomavirus mRNA and p16 detection as biomarkers for the improved diagnosis of cervical neoplasia. *Cancer Epidemiol. Biomark. Prev.* **2008**, *17*, 2536–2545. [CrossRef] [PubMed]

96. Kennedy, S.G.; Wagner, A.J.; Conzen, S.D.; Jordan, J.; Bellacosa, A.; Tsichlis, P.N.; Hay, N. The PI 3-kinase/AKT signaling pathway delivers an anti-apoptotic signal. *Genes Dev.* **1997**, *11*, 701–713. [CrossRef] [PubMed]

97. Klippel, A.; Escobedo, M.A.; Wachowicz, M.S.; Apell, G.; Brown, T.W.; Giedlin, M.A.; Kavanaugh, W.M.; Williams, L.T. Activation of phosphatidylinositol 3-kinase is sufficient for cell cycle entry and promotes cellular changes characteristic of oncogenic transformation. *Mol. Cell Biol.* **1998**, *18*, 5699–5711. [CrossRef] [PubMed]

98. Ma, Y.Y.; Wei, S.J.; Lin, Y.C.; Lung, J.C.; Chang, T.C.; Whang-Peng, J.; Liu, J.M.; Yang, D.M.; Yang, W.K.; Shen, C.Y. PIK3CA as an oncogene in cervical cancer. *Oncogene* **2000**, *19*, 2739–2744. [CrossRef] [PubMed]

99. Koncar, R.F.; Feldman, R.; Bahassi, E.M.; Hashemi Sadraei, N. Comparative molecular profiling of HPV-induced squamous cell carcinomas. *Cancer Med.* **2017**, *6*, 1673–1685. [CrossRef] [PubMed]

100. Samuels, Y.; Ericson, K. Oncogenic PI3K and its role in cancer. *Curr. Opin. Oncol.* **2006**, *18*, 77–82. [CrossRef] [PubMed]

101. Liu, P.; Cheng, H.; Roberts, T.M.; Zhao, J.J. Targeting the phosphoinositide 3-kinase pathway in cancer. *Nat. Rev. Drug Discov.* **2009**, *8*, 627–644. [CrossRef] [PubMed]

102. Vogt, P.K.; Kang, S.; Elsliger, M.A.; Gymnopoulos, M. Cancer-specific mutations in phosphatidylinositol 3-kinase. *Trends Biochem. Sci.* **2007**, *32*, 342–349. [CrossRef] [PubMed]

103. Lui, V.W.; Hedberg, M.L.; Li, H.; Vangara, B.S.; Pendleton, K.; Zeng, Y.; Lu, Y.; Zhang, Q.; Du, Y.; Gilbert, B.R.; et al. Frequent mutation of the PI3K pathway in head and neck cancer defines predictive biomarkers. *Cancer Discov.* **2013**, *3*, 761–769. [CrossRef] [PubMed]

104. Lechner, M.; Frampton, G.M.; Fenton, T.; Feber, A.; Palmer, G.; Jay, A.; Pillay, N.; Forster, M.; Cronin, M.T.; Lipson, D.; et al. Targeted next-generation sequencing of head and neck squamous cell carcinoma identifies novel genetic alterations in HPV+ and HPV− tumors. *Genome Med.* **2013**, *5*, 49. [CrossRef] [PubMed]

105. Verlaat, W.; Snijders, P.J.; van Moorsel, M.I.; Bleeker, M.; Rozendaal, L.; Sie, D.; Ylstra, B.; Meijer, C.J.; Steenbergen, R.D.; Heideman, D.A. Somatic mutation in PIK3CA is a late event in cervical carcinogenesis. *J. Pathol. Clin. Res.* **2015**, *1*, 207–211. [CrossRef] [PubMed]

106. Husain, R.S.; Ramakrishnan, V. Global variation of human papillomavirus genotypes and selected genes involved in cervical malignancies. *Ann. Glob. Health* **2015**, *81*, 675–683. [CrossRef] [PubMed]

107. Carracedo, A.; Pandolfi, P.P. The PTEN-PI3K pathway: Of feedbacks and cross-talks. *Oncogene* **2008**, *27*, 5527–5541. [CrossRef] [PubMed]

108. Feldman, R.; Gatalica, Z.; Knezetic, J.; Reddy, S.; Nathan, C.A.; Javadi, N.; Teknos, T. Molecular profiling of head and neck squamous cell carcinoma. *Head Neck* **2016**, *38*, E1625–1638. [CrossRef] [PubMed]

109. Vasudevan, K.M.; Barbie, D.A.; Davies, M.A.; Rabinovsky, R.; McNear, C.J.; Kim, J.J.; Hennessy, B.T.; Tseng, H.; Pochanard, P.; Kim, S.Y.; et al. AKT-independent signaling downstream of oncogenic PIK3CA mutations in human cancer. *Cancer Cell* **2009**, *16*, 21–32. [CrossRef] [PubMed]

110. Millis, S.Z.; Ikeda, S.; Reddy, S.; Gatalica, Z.; Kurzrock, R. Landscape of phosphatidylinositol-3-kinase pathway alterations across 19,784 diverse solid tumors. *JAMA Oncol.* **2016**, *2*, 1565–1573. [CrossRef] [PubMed]

111. Shayesteh, L.; Lu, Y.; Kuo, W.L.; Baldocchi, R.; Godfrey, T.; Collins, C.; Pinkel, D.; Powell, B.; Mills, G.B.; Gray, J.W. PIK3CA is implicated as an oncogene in ovarian cancer. *Nat. Genet.* **1999**, *21*, 99–102. [CrossRef] [PubMed]

112. Markowska, A.; Pawalowska, M.; Lubin, J.; Markowska, J. Signalling pathways in endometrial cancer. *Contemp. Oncol.* **2014**, *18*, 143–148.

113. Hewitt, E.W. The MHC class I antigen presentation pathway: Strategies for viral immune evasion. *Immunology* **2003**, *110*, 163–169. [CrossRef] [PubMed]

114. Vermeulen, C.F.; Jordanova, E.S.; Zomerdijk-Nooijen, Y.A.; ter Haar, N.T.; Peters, A.A.; Fleuren, G.J. Frequent HLA class I loss is an early event in cervical carcinogenesis. *Hum. Immunol.* **2005**, *66*, 1167–1173. [CrossRef] [PubMed]

115. Fowler, N.L.; Frazer, I.H. Mutations in TAP genes are common in cervical carcinomas. *Gynecol. Oncol.* **2004**, *92*, 914–921. [CrossRef] [PubMed]

116. Vermeulen, C.F.; Jordanova, E.S.; ter Haar, N.T.; Kolkman-Uljee, S.M.; de Miranda, N.F.; Ferrone, S.; Peters, A.A.; Fleuren, G.J. Expression and genetic analysis of transporter associated with antigen processing in cervical carcinoma. *Gynecol. Oncol.* **2007**, *105*, 593–599. [CrossRef] [PubMed]

117. Huang, J.J.; Blobe, G.C. Dichotomous roles of TGF-beta in human cancer. *Biochem. Soc. Trans.* **2016**, *44*, 1441–1454. [CrossRef] [PubMed]

118. Deng, W.; Tsao, S.W.; Kwok, Y.K.; Wong, E.; Huang, X.R.; Liu, S.; Tsang, C.M.; Ngan, H.Y.; Cheung, A.N.; Lan, H.Y.; et al. Transforming growth factor beta1 promotes chromosomal instability in human papillomavirus 16 E6E7-infected cervical epithelial cells. *Cancer Res.* **2008**, *68*, 7200–7209. [CrossRef] [PubMed]

119. Zhu, H.; Luo, H.; Shen, Z.; Hu, X.; Sun, L.; Zhu, X. Transforming growth factor-beta1 in carcinogenesis, progression, and therapy in cervical cancer. *Tumour Biol.* **2016**, *37*, 7075–7083. [CrossRef] [PubMed]

120. Chang, H.S.; Lin, C.H.; Yang, C.H.; Liang, Y.J.; Yu, W.C. The human papillomavirus-16 (HPV-16) oncoprotein E7 conjugates with and mediates the role of the transforming growth factor-beta inducible early gene 1 (TIEG1) in apoptosis. *Int. J. Biochem. Cell Biol.* **2010**, *42*, 1831–1839. [CrossRef] [PubMed]

121. Habig, M.; Smola, H.; Dole, V.S.; Derynck, R.; Pfister, H.; Smola-Hess, S. E7 proteins from high- and low-risk human papillomaviruses bind to TGF-beta-regulated smad proteins and inhibit their transcriptional activity. *Arch. Virol.* **2006**, *151*, 1961–1972. [CrossRef] [PubMed]

122. Murvai, M.; Borbely, A.A.; Konya, J.; Gergely, L.; Veress, G. Effect of human papillomavirus type 16 E6 and E7 oncogenes on the activity of the transforming growth factor-beta2 (TGF-beta2) promoter. *Arch. Virol.* **2004**, *149*, 2379–2392. [CrossRef] [PubMed]

123. Lee, D.K.; Kim, B.C.; Kim, I.Y.; Cho, E.A.; Satterwhite, D.J.; Kim, S.J. The human papilloma virus E7 oncoprotein inhibits transforming growth factor-beta signaling by blocking binding of the smad complex to its target sequence. *J. Biol. Chem.* **2002**, *277*, 38557–38564. [CrossRef] [PubMed]

124. Cheng, H.; Fertig, E.J.; Ozawa, H.; Hatakeyama, H.; Howard, J.D.; Perez, J.; Considine, M.; Thakar, M.; Ranaweera, R.; Krigsfeld, G.; et al. Decreased SMAD4 expression is associated with induction of epithelial-to-mesenchymal transition and cetuximab resistance in head and neck squamous cell carcinoma. *Cancer Biol. Ther.* **2015**, *16*, 1252–1258. [CrossRef] [PubMed]

125. Izumchenko, E.; Sun, K.; Jones, S.; Brait, M.; Agrawal, N.; Koch, W.; McCord, C.L.; Riley, D.R.; Angiuoli, S.V.; Velculescu, V.E.; et al. NOTCH1 mutations are drivers of oral tumorigenesis. *Cancer Prev. Res.* **2015**, *8*, 277–286. [CrossRef] [PubMed]

126. Izumi, N.; Helker, C.; Ehling, M.; Behrens, A.; Herzog, W.; Adams, R.H. FBXW7 controls angiogenesis by regulating endothelial notch activity. *PLoS ONE* **2012**, *7*, e41116. [CrossRef] [PubMed]

127. Santarpia, L.; Lippman, S.M.; El-Naggar, A.K. Targeting the MAPK-RAS-RAF signaling pathway in cancer therapy. *Expert Opin. Ther. Targets* **2012**, *16*, 103–119. [CrossRef] [PubMed]

128. Oganesyan, G.; Saha, S.K.; Guo, B.; He, J.Q.; Shahangian, A.; Zarnegar, B.; Perry, A.; Cheng, G. Critical role of TRAF3 in the toll-like receptor-dependent and -independent antiviral response. *Nature* **2006**, *439*, 208–211. [CrossRef] [PubMed]
129. Mirabello, L.; Yeager, M.; Cullen, M.; Boland, J.F.; Chen, Z.; Wentzensen, N.; Zhang, X.; Yu, K.; Yang, Q.; Mitchell, J.; et al. HPV16 sublineage associations with histology-specific cancer risk using HPV whole-genome sequences in 3200 women. *J. Natl. Cancer Inst.* **2016**, *108*, djw100. [CrossRef] [PubMed]
130. Mirabello, L.; Yeager, M.; Yu, K.; Clifford, G.; Xiao, Y.; Zhu, B.; Cullen, M.; Boland, J.F.; Wentzensen, N.; Nelson, C.W.; et al. HPV16 E7 genetic conservation is critical to carcinogenesis. *Cell* **2017**, in press.
131. Pinto, A.P.; Miron, A.; Yassin, Y.; Monte, N.; Woo, T.Y.; Mehra, K.K.; Medeiros, F.; Crum, C.P. Differentiated vulvar intraepithelial neoplasia contains TP53 mutations and is genetically linked to vulvar squamous cell carcinoma. *Mod. Pathol.* **2010**, *23*, 404–412. [CrossRef] [PubMed]
132. Kinde, I.; Bettegowda, C.; Wang, Y.; Wu, J.; Agrawal, N.; Shih Ie, M.; Kurman, R.; Dao, F.; Levine, D.A.; Giuntoli, R.; et al. Evaluation of DNA from the Papanicolaou test to detect ovarian and endometrial cancers. *Sci. Transl. Med.* **2013**, *5*, 167ra164. [CrossRef] [PubMed]
133. Janku, F.; Wheler, J.J.; Naing, A.; Falchook, G.S.; Hong, D.S.; Stepanek, V.M.; Fu, S.; Piha-Paul, S.A.; Lee, J.J.; Luthra, R.; et al. PIK3CA mutation H1047R is associated with response to PI3K/AKT/mTOR signaling pathway inhibitors in early-phase clinical trials. *Cancer Res.* **2013**, *73*, 276–284. [CrossRef] [PubMed]
134. Godinho, M.F.; Wulfkuhle, J.D.; Look, M.P.; Sieuwerts, A.M.; Sleijfer, S.; Foekens, J.A.; Petricoin, E.F., 3rd; Dorssers, L.C.; van Agthoven, T. BCAR4 induces antioestrogen resistance but sensitises breast cancer to lapatinib. *Br. J. Cancer* **2012**, *107*, 947–955. [CrossRef] [PubMed]

viruses

MDPI

Review

Epigenetic Alterations in Human Papillomavirus-Associated Cancers

David Soto [ID], Christine Song and Margaret E. McLaughlin-Drubin *

Division of Infectious Disease, Department of Medicine, Brigham & Women's Hospital, Harvard Medical School, 181 Longwood Avenue, Boston, MA 02115, USA; dsoto7@bwh.harvard.edu (D.S.); cksong@partners.org (C.S.)
* Correspondence: mdrubin@rics.bwh.harvard.edu; Tel.: +1-617-525-4262

Academic Editors: Alison A. McBride and Karl Munger
Received: 14 August 2017; Accepted: 25 August 2017; Published: 1 September 2017

Abstract: Approximately 15–20% of human cancers are caused by viruses, including human papillomaviruses (HPVs). Viruses are obligatory intracellular parasites and encode proteins that reprogram the regulatory networks governing host cellular signaling pathways that control recognition by the immune system, proliferation, differentiation, genomic integrity, and cell death. Given that key proteins in these regulatory networks are also subject to mutation in non-virally associated diseases and cancers, the study of oncogenic viruses has also been instrumental to the discovery and analysis of many fundamental cellular processes, including messenger RNA (mRNA) splicing, transcriptional enhancers, oncogenes and tumor suppressors, signal transduction, immune regulation, and cell cycle control. More recently, tumor viruses, in particular HPV, have proven themselves invaluable in the study of the cancer epigenome. Epigenetic silencing or de-silencing of genes can have cellular consequences that are akin to genetic mutations, i.e., the loss and gain of expression of genes that are not usually expressed in a certain cell type and/or genes that have tumor suppressive or oncogenic activities, respectively. Unlike genetic mutations, the reversible nature of epigenetic modifications affords an opportunity of epigenetic therapy for cancer. This review summarizes the current knowledge on epigenetic regulation in HPV-infected cells with a focus on those elements with relevance to carcinogenesis.

Keywords: Human papillomavirus; HPV; cervical cancer; epigenetics; histone

1. Introduction

Approximately 15–20% of the 12.7 million incident cancer cases per year have a viral etiology [1,2]. Carcinogenesis is a complex, multi-step process, and oncogenic viruses, including high-risk human papillomaviruses (HPVs), the Epstein-Barr virus (EBV), the hepatitis B virus (HBV), the hepatitis C virus (HCV), human T cell lymphotrophic virus-1 (HTLV-1), Kaposi's sarcoma herpesvirus (KSHV), and the Merkel cell polyoma virus (MCV), contribute to different steps of this process (reviewed in [3]). Viruses are obligatory intracellular parasites and encode proteins that reprogram the regulatory networks governing host cellular signaling pathways that control recognition by the immune system, proliferation, differentiation, genomic integrity, and cell death. The study of oncogenic viruses, as well as the manner in which they target regulatory nodes, has been key to the understanding of the etiology of several human cancers. It has led to the development of prophylactic vaccines for HBV along with the most abundant low- and high-risk HPVs. Given that key proteins in these regulatory networks are also subject to mutation in non-virally associated diseases and cancers, the study of oncogenic viruses has also been instrumental to the discovery and analysis of many fundamental cellular processes, including mRNA splicing, transcriptional enhancers, oncogenes and tumor suppressors, signal transduction, immune regulation, and cell cycle control (reviewed in [3,4]).

More recently, tumor viruses, in particular HPV, have proven themselves invaluable in the study of the cancer epigenome.

The concept that cancer is equally an epigenetic and a genetic disease has been increasingly validated during the past decade, particularly since the advent of whole-genome approaches. Epigenetic abnormalities in cancer involve aberrations in virtually every aspect of chromatin biology, including post-translational modifications of histone proteins, DNA methylation, chromatin remodeling, and non-coding RNAs (ncRNAs). The cancer epigenome harbors numerous abnormalities that distinguish it from its normal counterpart (reviewed in [5]). Aberrations in virtually every aspect of chromatin biology have been identified in cancer-harboring epigenetic abnormalities, including post-translational modifications of histone proteins, DNA methylation, chromatin remodeling, and ncRNAs. For example, aberrant methylation patterns and histone modifications are found in both virus-associated and non-viral cancers [6–8]. Indeed, viral oncoproteins can induce the expression of, as well as interact with, DNA methyltransferases (DNMTs) and histone-modifying enzymes, including histone deacetylases (HDACs), histone acetyltransferases (HATs), histone methyltransferases (HMTs), and histone demethylases (reviewed in [9,10]). Moreover, viral oncoproteins can also alter the activity of chromatin-remodeling, complex-associated proteins and miRNA processing-associated proteins (reviewed in [9,10]). In addition to causing alterations in the host epigenome, the tumor virus genome itself also undergoes epigenetic modification (reviewed in [11]).

A full understanding of the cancer epigenome has numerous translational implications. It is evident that cancer cells have global epigenome changes involving entire pathways. These epigenetic alterations are often found early in tumorigenesis and are likely to be key initiating events in certain cancers (reviewed in [12]). In addition to tumor initiation, epigenetic events also contribute to tumor progression (reviewed in [5]). Virus-associated cancers present unique experimental systems to determine what role epigenetic modifications play in carcinogenesis. Similar to hematological tumors that are often driven by a single initiating mutation, virus-associated cancers are initiated by a uniform oncogenic hit of viral oncogene expression. This has been quite impressively demonstrated for HPV-associated cervical cancers where cancer initiation and progression are driven by the expression of the E6 and E7 oncogenes (reviewed in [13]). Much has already been learned from detailed molecular analyses of epigenetic mechanisms in virus-induced tumorigenesis (reviewed in [9]). The study of tumor viruses such as HPV should continue to provide answers regarding the importance of epigenetic alterations in viral cancers as well as, hopefully, non-viral-associated cancers. It is predicted that epigenetic factors, including readers, writers, and erasers, that are targeted by HPV, are important drug targets for both viral and non-viral associated cancers. This includes factors that are induced by HPV or triggered in response to viral infection/viral protein expression. The reversibility of epigenetic modifications makes such epigenetic factors ideal therapeutic targets. Several drugs targeting chromatin modifiers are already in use in the clinic (reviewed in [5]).

Human Papillomaviruses

HPVs are small, double-stranded DNA virus members of the *Papillomaviridae*, a large family with a tropism for squamous epithelium. To date, more than 200 HPV types have been described, which are divided into cutaneous and mucosal HPVs based on the tissue they infect. The mucosal HPVs are clinically classified as "high-risk" and "low-risk" based on the propensity for malignant progression. Low-risk HPVs, such as HPVs 6 and 11, cause benign genital warts, while high-risk HPVs cause intraepithelial lesions that are at risk for malignant progression. Infection with high-risk HPVs are associated with approximately 5% of all human cancers, in particular with cervical carcinoma, the third most common cancer in women worldwide [1,14]. HPV infections are also frequently associated with other anogenital cancers, including anal, vulvar, vaginal, and penile cancers, as well as oropharyngeal cancers [15,16]. While prophylactic vaccination prevents infections with HPV types represented in the vaccine, no therapeutic efficacy is associated with these vaccines. In addition to the fact that HPV-associated cervical cancers arise years after initial infection, vaccination rates are low in many countries. Therefore, it will be decades before the current vaccination efforts have a measurable impact on the incidence of HPV-associated tumors [17].

The viral E6 and E7 proteins are consistently expressed in HPV-associated lesions and cancers, and are the major drivers of cell transformation (reviewed in [4,13]). The HPV E6 and E7 proteins lack enzymatic activities and instead function by associating with host cellular proteins. These proteins reprogram cellular signal transduction pathways (reviewed in [18]), causing alterations in the "hallmarks of cancer" [4,19]. Notably, high-risk mucosal HPV E6 and E7 proteins, respectively, target p53 and retinoblastoma (pRB) tumor suppressors; these tumor suppressor pathways are also rendered dysfunctional by mutation in almost all human solid tumors [20,21]. High-risk HPV E6 and E7 also interact with a number of other proteins, such as transcription factors, thus altering cellular gene expression. In addition to targeting specific transcriptional programs, the HPV E6 and E7 oncoproteins can globally alter the transcriptional competence of the infected cells by affecting epigenetic control mechanisms. Indeed, epigenetic alterations such as changes in the DNA methylation pattern of the viral and host genomes, as well as changes in histone modifications, are often found associated with HPV infection and cervical carcinogenesis. This article focuses on HPV-induced changes in these epigenetic control mechanisms, including DNA methylation, histone modifications, chromatin remodeling proteins, and ncRNAs.

2. DNA Methylation

DNA methyltransferases (DNMTs) methylate the carbon-5 position of cytosine nucleotides; this covalent modification occurs predominantly on cytosines preceding guanine nucleotides (CpG dinucleotides). In normal cells, methylation of DNA is involved in the regulation of gene expression, including the organization of active and inactive chromatin, tissue-specific gene expression, and genomic imprinting (reviewed in [22]). In contrast, global DNA hypomethylation in repetitive regions and hypermethylation in CpG islands of tumor suppressor gene promoters are frequently observed in tumors [23,24], and the activity of DNMT1, which is the maintenance methyltransferase, is often increased (reviewed in [6,7]). These alterations are also observed in HPV-induced carcinogenesis. HPV E7 binds to DNMT1 and stimulates its DNA methyltransferase activity [25], and may be able to activate transcription of DNMT1 through the pRB/E2F pathway [26], while HPV E6 upregulates DNMT1 by suppression of p53 [27]. As a consequence of the association of HPV E7 with DNMT1, E-cadherin expression is suppressed and adhesion between squamous epithelial cells is reduced [28,29]. Similarly, increased expression of DNMT3A and 3B has also been observed in HPV-positive cells [28,30,31]. The effects of HPV on the DNA methylation machinery have the ability to alter both the host and the viral genome.

2.1. HPV Genome Methylation

While methylation of CpG islands in human gene promotors generally represses gene transcription, the methylation of viral DNA both negatively and positively regulates viral gene transcription. Although it is unclear if viral DNA methylation provides a growth advantage to the infected cell, it has been suggested that viral DNA methylation is due to a host defense response to silence viral replication and transcription [32–35]. HPV gene methylation, particularly in the L1 and L2 genes, varies during the viral life cycle as well as with the disease stage [36–41]. Methylation of the upstream regulatory region (URR) appears to be associated with latent infection [42], although results from different studies are inconsistent, possibly due to the integrated or episomal state of the viral genome and/or the stage of the lesion examined. When comparing URR methylation in cervical intraepithelial neoplasia (CIN) and cancer samples compared to normal samples, some studies described decreased methylation [43–45], while others showed an increase in URR methylation [33–35,41,46]. URR methylation also differs based on type 1 versus type 2 HPV integration [36,37,45–48]. These differences highlight the need to take into account not only the methodology used to analyze methylation but also the HPV genome and disease status when comparing across studies.

Methylation of the E2 binding sites (E2BSs) in the URR reduces E2 binding, thus deregulating E6 and E7 expression [49], and methylation of E2BSs in reporter plasmids inhibits the transcriptional transactivation activity of E2 in transfected cells [36]. E2 also functions in the initiation of viral DNA replication and in partitioning the viral DNA to the daughter cells during cell division; both of these activities also rely on its ability to bind E2BSs and are thus thought to be affected by the methylation status of the E2BSs. E2BSs in the immortalized HPV16-positive W12 cells are hypomethylated upon differentiation in vivo, providing evidence that the methylation of the E2BSs varies during epithelial differentiation and thus during the viral life cycle [36,42,47]. An analysis of the methylation status of the HPV16 URR in distinct stages of the viral life cycle from patient-derived tissues confirmed a decrease in the methylation of the transcriptional enhancer region of the URR, but also indicated hypermethylation of the E2BSs [42]. Additional studies indicate that methylation of HPV DNA may differentiate between an acute HPV infection and CIN2+ (reviewed in [40]). Indeed, it has been proposed that CpG methylation status is a potential biomarker for cervical cancer [50].

2.2. Cellular Gene Methylation

Aberrant methylation occurs frequently in cervical cancer, leading to inappropriate gene expression, the activation of oncogenes and transposable elements, loss of imprinting, and the inactivation of tumor suppressor genes (reviewed in [51]). Of note, a number of tumor suppressor genes are hypermethylated in HPV-associated lesions and carcinomas, including *CCNA1* and *hTERT* [8,52–59]. The most frequently methylated genes in cervical cancer are cell adhesion molecule 1 (*CADM1*), cadherin 1 (*CDH1*), death-associated protein kinase 1 (*DAPK1*), *EPB41L3*, *FAM1A4*, myelin and lymphocyte (*MAL*), paired box 1 (*PAX1*), PR domain containing 14 (*PRDM14*,) and telomerase reverse transcriptase (*hTERT*) [52,58,59], however a single gene target has not proven amenable as a biomarker [52] indicating that a panel of methylated genes may be more useful.

3. Regulation of Histone Modifications

In addition to DNA methylation, the epigenetic regulation of gene expression is also impacted by histone modifications and the remodeling of nucleosomes. Post-translational modifications of histone tails, including acetylation, methylation, phosphorylation, sumoylation, and ubiquitination, impact the physical state and the transcriptional competence of chromatin. These modifications play a crucial role in the regulation of cellular processes such as stem cell maintenance, cell fate determination and maintenance, cell cycle control, and epigenetic heritability of transcriptional programs (reviewed in [60,61]). Distinct posttranslational modifications on histones, or combinations thereof, characterize transcriptionally active and silent chromatin. In general, transcriptionally active genes are characterized by promoters with unmethylated CpG dinucleotides and nucleosomes. These active genes are arranged such that transcription and regulatory factors are allowed access. Transcriptionally active genes usually have extensive H3 and H4 acetylation and are marked by trimethylation of lysine 4 on histone H3 (H3K4me3), trimethylation of lysine 79 on histone H3 (H3K79me3), ubiquitylation of H2B (H2Bub), and trimethylation of lysine 36 on histone H3 (H3K36me3), while transcriptionally inactive genes are characterized by low levels of acetylation and high levels of trimethylation of lysine 9 on histone H3 (H3K9me3), trimethylation of lysine 27 on histone H3 (H3K27me3), trimethylation of lysine 20 on histone H4 (H4K20me3), and ubiquitylation of lysine 119 on histone H2A (H2AK199ub) (reviewed in [62]). The different patterns of histone modifications associated with distinct transcriptional states are established via interplay between histone readers, writers, and erasers. Enzymes that modify histones and other chromatin components are designated writer proteins, and include HATs, histone methyltransferases (KMTs), and histone ubiquitin ligases; these modifications are reversible and are removed by erasers such as HDACs, histone demethylases, and histone deubiquitinases. The modifications are recognized by reader proteins, which bind to the modified histones and recruit additional proteins [62,63], and ultimately realize the functional translation of the epigenetic mark.

3.1. Histone Modification of the Human Papillomavirus Genome

Human papillomavirus genomes are bound by nucleosomes around the viral promoters [64–66]. Chromatin immunoprecipitation (ChIP) analysis of the histones bound to the HPV genome throughout the differentiation-dependent viral life cycle demonstrated the presence of acetylated H3 and H4 histones and the dimethylation of lysine 4 on histone H3 (H3K4me2) at the HPV early and late promoters, indicating that they are in an active conformation throughout the viral life cycle [67]. The levels of acetylation and the demethylation of the histones at the early and late promoter regions increase upon differentiation, and the binding of a number of transcription factors was increased upon differentiation [67]. In summary, this study indicated that both the early and late HPV promoter regions are in an active chromatin state throughout the viral life cycle. In a study on the HPV18-positive HeLa cervical cancer cell line, localized distinctions in the status of histone modifications of the chromatin on the HPV18 genome were observed; these correlated with the occupancy of the host transcriptional machinery [41]. The viral E6 and E7 oncoproteins modulate the host epigenetic machinery and histone modification enzymes, which has implications for the epigenetic regulation of both the viral and host genomes, and has implications in both the viral life cycle and the carcinogenic process.

3.2. Acetylation

One mechanism by which the HPV E6 and E7 oncoproteins alter the transcriptional competence of infected cells is by associating with and/or modulating the expression, as well as the activities, of histone-modifying and chromatin-remodeling enzymes (Figure 1) [68–77]. For example, acetylation of lysine residues of histones 3 and 4 (H3 and H4) by HATs leads to transcriptionally active chromatin, while the removal of these marks by HDACs results in transcriptionally repressed chromatin. HPV E6 and E7 can associate with and modulate the activity of the HATs p300 and CBP [70,71,73,78–80]; p300/CBP regulates a number of genes [73,81–87]. HPV E6 inhibits p300/CBP-mediated acetylation of p53 [88], while HPV E7 forms a complex with p300/CBP and pRB, acetylating pRB and decreasing p300/CBP levels [80]. HPV E7 also associates with p300/CBP-Associated Factor (pCAF), reducing its ability to acetylate histones [70] and the steroid-receptor coactivator (SRC1), and abrogating SRC1-associated HAT activity [72]. Moreover, the HPV E7 oncoprotein interacts with class I HDACs [68,69], which function as transcriptional co-repressors by inducing chromatin remodeling via the reversal of acetyl modifications on histone lysine residues. The association of E7 and HDAC1/2 occurs in an RB-independent manner through the intermediary Mi2β, a member of the nucleosome remodeling and histone deacetylation (NuRD) complex; the NuRD complex remodels chromatin structure through the deactylation of histones and ATP-dependent nucleosome repositioning [89,90]. The association of E7 and HDAC1/2 does not result in the inhibition of HDAC activity [68], but does play a role in HPV E7-associated transcriptional regulation. For example, this association results in increased levels of E2F2-mediated transcription in differentiating cells, which may affect S-phase progression [91]. Furthermore, HPV E7 can interact with interferon response factor 1 (IRF1) and recruit HDACs to suppress IRF1 transcriptional activity [92,93].

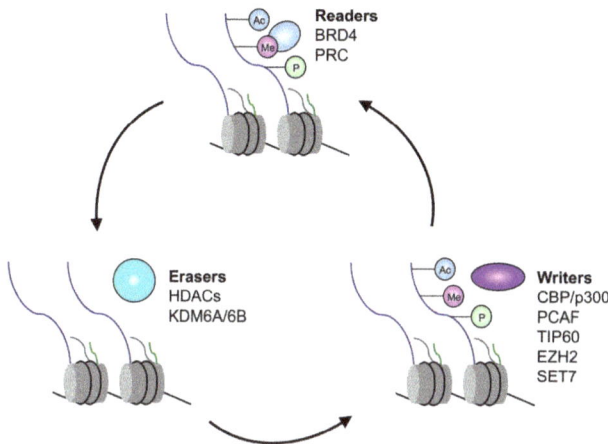

Figure 1. Summary of histone readers, writers, and erasers targeted by human papillomavirus (HPV). Abbreviations: BRD4: bromodomain-containing protein 4; PRC: polycomb repressive complex; CBP/p300: CREB-binding protein/p300; pCAF: p300/CBP-Associated Factor; TIP60: Tat interactive protein, 60 kDa; EZH2: enhancer of zeste homolog 2; SET7: SET domain containing lysine methyltransferase 7; HDACs: histone deacetylases; KDM6A/6B: histone lysine demethylases 6A/6B.

3.3. Polycomb Group Proteins and Histone Lysine Modifications

Global levels of the polycomb-regulated H3K27me3 repressive mark are dramatically decreased in HPV16 E7-expressing primary human foreskin keratinocytes and in HPV16-positive cervical lesions and cancers [75,77]. The function of the H3K27me3 mark is exerted by the formation of two polycomb repressive complex (PRC) species, PRC1 and PRC2. PRC2 contains the histone methyltransferase (KMT) EZH2 (KMT6), which places the H3K27me3 mark. The H3K27me3 marked chromatin is occupied by PRC1, and the chromatin is further silenced by mono-ubiquitination of lysine 119 on histone H2A (H2AK119Ub). Gene expression can also be silenced by certain PRC1 complexes in the absence of H3K27me3, as H2AK119Ub is a binding site for L3MBTL2, which establishes repressive structures [94] that play an important role in pluripotent stem cells [95].

PcG proteins regulate both epithelial cell differentiation and the expansion of basal cell pools during the wound healing process [96–98], two processes that HPVs may target during the viral life cycle. Thus, it is not surprising that HPVs target components of the PRC machinery (reviewed in [99]). Indeed, HPV16 E7 associates with, as well as potentially modifies, activities of E2F6-containing PRCs and causes a reduction in the number of nuclear E2F6-containing polycomb bodies [76]. Moreover, PcG proteins are likely best known for their role in maintaining stable transcriptional repression of Homeobox (*HOX*) genes during development [100,101], and HOX family members are frequently dysregulated during carcinogenesis, including cervical carcinogenesis and in HPV16 E7-expressing cells [75,102–105].

While the decrease in H3K27me3 observed in HPV16 E7-expressing cells offered a potential explanation for the decrease in polycomb body number and the dysregulation of *HOX* genes, this decrease is observed despite the fact that the enhancer of the zeste homolog 2 (EZH2) component of the polycomb repressive complex 2 (PRC2) is highly overexpressed in cervical lesions and tumors in an E2F-dependent manner [106]. A number of possible mechanisms have been proposed to explain the seemingly paradoxical finding of decreased H3K27me3 in the presence of increased EZH2. AKT-mediated phosphorylation of EZH2 negatively regulates the enzymatic activity of EZH2 [107], and both HPV16 E6 and E7 activate AKT [108,109]. Thus, it is possible that PRC2-associated EZH2 enzymatic activity is low despite high EZH2 levels in HPV-expressing cells. EZH2 overexpression has also been shown to enhance PRC4 formation [110]. PRC4 causes histone H1K26 deacetylation and methylation [110], which then serves as a binding site for L3MBTL1. Hence, increased EZH2 expression in E7-expressing cells may be predicted to cause enhanced H1K26 methylation. Additionally, another mechanistic explanation for the decrease in H3K27me3 was provided by the finding that the histone lysine demethylases (KDMs) KDM6A (UTX) and KDM6B (JMJD3) are expressed at markedly higher levels in these cells [74,75,77]. Interestingly, cervical cancer cells are dependent on the expression of KDM6A and KDM6B [74,75]. Although KDM6A and KDM6B appear identical with regards to catalytic activities and histone substrate specificities, they have a number of unique biological targets. KDM6B, but not KDM6A, regulate RAS/RAF and HPV E7-induced oncogene-induced senescence (OIS) [74,111,112]. OIS is a cell-intrinsic tumor-suppressive mechanism that protects cells from unrestrained proliferation following an oncogenic insult (reviewed in [113]). In order for a lesion to progress, OIS must be evaded or bypassed, as evidenced by the fact the OIS is observed in premalignant lesions much more than in frank lesions [114]. OIS is signaled through transcriptional upregulation of the p16^{INK4A} tumor suppressor [115]. The p16^{INK4A} tumor suppressor is a biomarker for high-risk, HPV-associated lesions and cancers, and is induced by HPV E7 [116,117]. These high levels of p16INK4A expression are a readout of HPV E7-induced OIS [74]. Interestingly, HPV E7-expressing and some cervical cancer cells are "addicted" to the expression of p16^{INK4A}, suggesting that the biological activity of p16^{INK4A} in HPV-associated cancers is more like that of an oncogene, as opposed to its well-established role as a tumor suppressor in most other human cancer types [74].

3.4. Histone Arginine Modifications

Histone methylation also takes place on arginine residues, and HPV modulates the activity of two coactivator histone arginine methyltransferases, CARM1 and PRMT1 [118]. HPV E6 downregulates their expression, and these HMTs are needed for HPV E6 to attenuate p53 transactivation [118]. E6 hinders CARM1- and PRMT1-mediated histone methylation at p53-responsive promoters and suppresses p53 binding to DNA [118]. E6 also inhibits SET7, which, in addition to catalyzing H3K4 monomethylation, methylates non-histone proteins, including p53 [119]. HPV E6 downregulates p53K372 mono-methylation, thereby reducing p53 stability [118]. Together, modulation of CARM1, PRMT1, and SET7 provides another mechanism by which HPV alters p53 function.

3.5. Epigenetic Readers

Bromodomain-containing protein 4 (Brd4) is a member of the bromodomain and extra-terminal domain (BET) family of chromatin-binding proteins [120] and plays a crucial role in transcription. The bromodomains of Brd4 interact with methylated histones H3 and H4 [121] and mark genes that are expressed shortly after exit from mitosis [122,123]. Brd4 recruits transcription initiation and elongation factors to these genes [124], including the transcriptional elongation factor, p-TEFb [125,126]. Brd4 plays a key role in the transcriptional regulation and replication of papillomaviruses (reviewed in [127]).

The papillomavirus E2 protein interacts with Brd4, stabilizing its association with chromatin [128–133]. E2 interacts with the C-terminal domain (CTD) of Brd4, blocking the formation of Brd4-pTEFb [134], and thus acting as an E2-dependent transcriptional repressor of E6 and E7. Brd4 also represses the HPV early promoter, and the binding of Brd4 to the HPV early promoter is dependent on histone H4 acetylation by TIP60 [135]. HPV E6 destabilizes TIP60 in a proteasome-dependent manner, derepressing the early promoter, resulting in HPV oncoprotein expression [135,136].

4. Non-Coding RNAs

It has recently become evident that the non-coding portion of the human genome plays an important role in the regulation of the expression of activities of cellular proteins. ncRNAs are classified according to their length and include microRNAs (miRNAs) and long non-coding RNAs (lncRNAs) (reviewed in [137]).

4.1. MicroRNAs

MicroRNAs (miRNAs) are small (~22 nucleotides), ncRNAs that regulate their target mRNAs at the post-transcriptional level. miRNAs bind to the 3′-untranslated regions (UTRs) of target mRNAs, mediating translational repression or mRNA destruction [138,139]. A single miRNA can affect the expression of hundreds of targets [140], and multiple miRNAs can affect the same target. miRNAs play a key role in the development of human cancer with tumor suppressor miRNAs and oncogenic miRNAs (onco-miRs). To date, no HPV-encoded miRNAs have been discovered [141]. However, host miRNA expression is altered in the presence of HPV in cervical cancer tissue and precursor lesions, as well as in cervical cancer cell lines and keratinocytes expressing the HPV oncoproteins [142–148]. Moreover, a number of microRNAs, including miR-9, miR-21, miR-143, miR-203, and miR-372, among others, have been implicated in different aspects of cervical carcinogenesis, with the expression of some microRNAs increased (miR-21, miR-143, miR-9) and others decreased (miR-34a, miR-203, miR-372) [55,149–154]. Bioinformatic analyses of microRNA expression, coupled with changes in RNA expression as a result of HPV16E6/E7 in human keratinocytes, identified a number of canonical pathways targeted by miR-modulated mRNAs, including cyclins, cell cycle regulation, estrogen-mediated S-phase entry, and aryl hydrocarbon reception signaling [155]. Experiments to dissect the molecular mechanisms underlying the mode of action of particular microRNAs in cervical carcinogenesis revealed that miR-21 targets chemokine (C-C) motif ligand 20 (*CCL20*), and its overexpression regulates proliferation, apoptosis, and migration of HPV16-positive cervical cancer cells [156]. Increased levels of miR-203 inhibit HPV amplification, and HPV E7 suppresses miR-203 to allow for productive replication to occur [157]. mir372 is downregulated and targets CDK2 and Cyclin A1 in cervical cancer [152]. When comparing studies such as these, which focus on a single microRNA and the modulation of a single target mRNA, with studies that investigate the modulation of cellular microRNAs by HPV gene expression, one must take into consideration the global landscape of microRNA expression, the cell type studied (differentiating versus undifferentiated epithelial cells), the HPV type studied, and the whole HPV genome versus just HPV E6 and/or E7. In fact, these considerations should be kept in mind when comparing all of the studies mentioned in this review.

A number of miRNAs are epigenetically regulated, suggesting that aberrant methylation of miRNA promoters is one of the possible mechanisms for deregulation of miRNAs in cervical cancer [149,158,159]. The miRNA biogenesis machinery is often dysregulated in human cancers, including cervical carcinoma (reviewed in [160,161]). Chromosome 5p amplifications are found in some cervical carcinomas, and *DROSHA* is the most significantly overexpressed transcript in cervical tumors with 5p gain [162,163]. Expression of high-risk HPV E6 and E7 in HPV-negative C33A cervical carcinoma cells and primary human epithelial cells causes increased expression of *DROSHA* and *DICER* [164], and many *DROSHA*-regulated miRNAs are dysregulated in high-risk HPV16 E6/E7 expressing cells [155,164].

4.2. Long Non-Coding RNAs

Long non-coding RNAs (lncRNAs) are non-coding RNA transcripts with a length greater than 200 nucleotides; to date, 27,919 lncRNA have been discovered in humans [165]. Although their function is not fully elucidated, they do contribute to many biological processes including cellular development, differentiation, and transformation. However, it is known that lncRNAs bind to PRC1 and PRC2, function as antisense molecules, and organize enhancer activity (reviewed in [166,167]). A number of lncRNAs are differentially expressed in cancer, including HOX transcript antisense intergenic RNA (HOTAIR) [168–173]. HOTAIR regulates gene expression through association with chromatin remodeling complexes [174]; it bridges PRC2 with the lysine-specific histone demethylase1A complex (LSD1), resulting in gene silencing [168,174,175]. Down-regulation of HOTAIR, with corresponding upregulation of the HOTAIR target HOXD10, has been observed in cervical cancer [176].

5. Concluding Remarks

High-risk HPVs are associated with approximately five percent of human cancers, including virtually all cervical cancers, as well as anal, vaginal, vulvar, penile, and oropharyngeal cancers. Although highly efficacious prophylactic vaccines appear promising for preventing a large fraction of HPV-associated cancers, they do not protect from pre-existing infections or prevent malignant progression, and are not expected to impact the frequency of these cancers for decades. In the meantime, millions will develop HPV-associated cancers, and many will die of these cancers worldwide. It is imperative that we identify novel therapeutic targets to control and, ideally, eradicate HPV-associated cancers. A number of epigenetic alterations have been identified that occur in both the HPV and the cellular genome, including DNA hypomethylation, hypermethylation of tumor suppressor genes, histone modifications, and alterations in ncRNAs. These alterations have the potential to be used as biomarkers for early detection. In addition, epigenetic alterations, unlike genetic mutations, may be reversed by inhibiting the associated enzymes, and as such should be evaluated as therapeutic modalities for HPV-associated lesions and cancers. Moreover, we can apply the findings of these studies to other, non-HPV associated cancers.

Acknowledgments: The authors would like to thank Catherine Xie for her helpful discussion. This work was supported by a grant from the American Cancer Society (ACS) (126540-RSG-12-203-01-MPC).

Author Contributions: David Soto, Christine Song and Margaret E. McLaughlin-Drubin wrote the paper.

Conflicts of Interest: The authors declare no conflict of interest. The founding sponsors had no role in the design of the study; in the collection, analyses, or interpretation of data; in the writing of the manuscript, and in the decision to publish the results.

References

1. Ferlay, J.; Shin, H.R.; Bray, F.; Forman, D.; Mathers, C.; Parkin, D.M. Estimates of worldwide burden of cancer in 2008: Globocan 2008. *Int. J. Cancer* **2010**, *127*, 2893–2917. [CrossRef] [PubMed]
2. De Martel, C.; Ferlay, J.; Franceschi, S.; Vignat, J.; Bray, F.; Forman, D.; Plummer, M. Global burden of cancers attributable to infections in 2008: A review and synthetic analysis. *Lancet Oncol.* **2012**, *13*, 607–615. [CrossRef]
3. McLaughlin-Drubin, M.E.; Munger, K. Viruses associated with human cancer. *Biochim. Biophys. Acta* **2008**, *1782*, 127–150. [CrossRef] [PubMed]
4. Mesri, E.A.; Feitelson, M.A.; Munger, K. Human viral oncogenesis: A cancer hallmarks analysis. *Cell Host Microbe* **2014**, *15*, 266–282. [CrossRef] [PubMed]
5. Baylin, S.B.; Jones, P.A. A decade of exploring the cancer epigenome—Biological and translational implications. *Nat. Rev. Cancer* **2011**, *11*, 726–734. [CrossRef] [PubMed]
6. Jones, P.A.; Baylin, S.B. The fundamental role of epigenetic events in cancer. *Nat. Rev. Genet.* **2002**, *3*, 415–428. [PubMed]
7. Robertson, K.D. DNA methylation, methyltransferases, and cancer. *Oncogene* **2001**, *20*, 3139–3155. [CrossRef] [PubMed]
8. Szalmas, A.; Konya, J. Epigenetic alterations in cervical carcinogenesis. *Semin. Cancer Biol.* **2009**, *19*, 144–152. [CrossRef] [PubMed]
9. Poreba, E.; Broniarczyk, J.K.; Gozdzicka-Jozefiak, A. Epigenetic mechanisms in virus-induced tumorigenesis. *Clin. Epigenet.* **2011**, *2*, 233–247. [CrossRef] [PubMed]
10. Minarovits, J.; Niller, H. Patho-epigenetics. *Med. Epigenet.* **2013**, *1*, 37–45. [CrossRef]
11. Fernandez, A.F.; Esteller, M. Viral epigenomes in human tumorigenesis. *Oncogene* **2010**, *29*, 1405–1420. [CrossRef] [PubMed]
12. Sharma, S.; Kelly, T.K.; Jones, P.A. Epigenetics in cancer. *Carcinogenesis* **2010**, *31*, 27–36. [CrossRef] [PubMed]
13. McLaughlin-Drubin, M.E.; Munger, K. Oncogenic activities of human papillomaviruses. *Virus Res.* **2009**, *143*, 195–208. [CrossRef] [PubMed]
14. Parkin, D.M. The global health burden of infection-associated cancers in the year 2002. *Int. J. Cancer* **2006**, *118*, 3030–3044. [CrossRef] [PubMed]

15. Schiffman, M.; Castle, P.E.; Jeronimo, J.; Rodríguez, A.C.; Wacholder, S. Human papillomavirus and cervical cancer. *Lancet* **2007**, *370*, 890–907. [CrossRef]
16. Gillison, M.L.; Lowy, D.R. A causal role for human papillomavirus in head and neck cancer. *Lancet* **2004**, *363*, 1488–1489. [CrossRef]
17. Frazer, I.H. Prevention of cervical cancer through papillomavirus vaccination. *Nat. Rev. Immunol.* **2004**, *4*, 46–54. [CrossRef] [PubMed]
18. McLaughlin-Drubin, M.E.; Meyers, J.; Munger, K. Cancer associated human papillomaviruses. *Curr. Opin. Virol.* **2012**, *2*, 459–466. [CrossRef] [PubMed]
19. Hanahan, D.; Weinberg, R.A. Hallmarks of cancer: The next generation. *Cell* **2011**, *144*, 646–674. [CrossRef] [PubMed]
20. Dyson, N.; Howley, P.M.; Münger, K.; Harlow, E. The human papillomavirus-16 E7 oncoprotein is able to bind to the retinoblastoma gene product. *Science* **1989**, *243*, 934–937. [CrossRef] [PubMed]
21. Werness, B.A.; Levine, A.J.; Howley, P.M. Association of human papillomavirus types 16 and 18 E6 proteins with p53. *Science* **1990**, *248*, 76–79. [CrossRef] [PubMed]
22. Robertson, K.D. DNA methylation and human disease. *Nat. Rev. Genet.* **2005**, *6*, 597–610. [CrossRef] [PubMed]
23. Yang, H.J. Aberrant DNA methylation in cervical carcinogenesis. *Chin. J. Cancer* **2013**, *32*, 42–48. [CrossRef] [PubMed]
24. Ehrlich, M. DNA hypomethylation in cancer cells. *Epigenomics* **2009**, *1*, 239–259. [CrossRef] [PubMed]
25. Burgers, W.A.; Blanchon, L.; Pradhan, S.; de Launoit, Y.; Kouzarides, T.; Fuks, F. Viral oncoproteins target the DNA methyltransferases. *Oncogene* **2007**, *26*, 1650–1655. [CrossRef] [PubMed]
26. McCabe, M.T.; Davis, J.N.; Day, M.L. Regulation of DNA methyltransferase 1 by the pRb/E2F1 pathway. *Cancer Res.* **2005**, *65*, 3624–3632. [CrossRef] [PubMed]
27. Yeung, C.L.; Tsang, T.Y.; Yau, P.L.; Kwok, T.T. Human papillomavirus type 16 E6 suppresses microRNA-23b expression in human cervical cancer cells through DNA methylation of the host gene C9orf3. *Oncotarget* **2017**, *8*, 12158–12173. [PubMed]
28. Laurson, J.; Khan, S.; Chung, R.; Cross, K.; Raj, K. Epigenetic repression of E-cadherin by human papillomavirus 16 E7 protein. *Carcinogenesis* **2010**, *31*, 918–926. [CrossRef] [PubMed]
29. D'Costa, Z.J.; Jolly, C.; Androphy, E.J.; Mercer, A.; Matthews, C.M.; Hibma, M.H. Transcriptional repression of E-cadherin by human papillomavirus type 16 E6. *PLoS ONE* **2012**, *7*, e48954. [CrossRef] [PubMed]
30. Sartor, M.A.; Dolinoy, D.C.; Jones, T.R.; Colacino, J.A.; Prince, M.E.; Carey, T.E.; Rozek, L.S. Genome-wide methylation and expression differences in HPV(+) and HPV(−) squamous cell carcinoma cell lines are consistent with divergent mechanisms of carcinogenesis. *Epigenetics* **2011**, *6*, 777–787. [CrossRef] [PubMed]
31. Leonard, S.M.; Wei, W.; Collins, S.I.; Pereira, M.; Diyaf, A.; Constandinou-Williams, C.; Young, L.S.; Roberts, S.; Woodman, C.B. Oncogenic human papillomavirus imposes an instructive pattern of DNA methylation changes which parallel the natural history of cervical HPV infection in young women. *Carcinogenesis* **2012**, *33*, 1286–1293. [CrossRef] [PubMed]
32. Doerfler, W.; Remus, R.; Muller, K.; Heller, H.; Hohlweg, U.; Schubbert, R. The fate of foreign DNA in mammalian cells and organisms. *Dev. Biol.* **2001**, *106*, 89–97.
33. Fernandez, A.F.; Rosales, C.; Lopez-Nieva, P.; Grana, O.; Ballestar, E.; Ropero, S.; Espada, J.; Melo, S.A.; Lujambio, A.; Fraga, M.F.; et al. The dynamic DNA methylomes of double-stranded DNA viruses associated with human cancer. *Genome Res.* **2009**, *19*, 438–451. [CrossRef] [PubMed]
34. Ding, D.C.; Chiang, M.H.; Lai, H.C.; Hsiung, C.A.; Hsieh, C.Y.; Chu, T.Y. Methylation of the long control region of HPV16 is related to the severity of cervical neoplasia. *Eur. J. Obstet. Gynecol. Reprod. Biol.* **2009**, *147*, 215–220. [CrossRef] [PubMed]
35. Hong, D.; Ye, F.; Lu, W.; Hu, Y.; Wan, X.; Chen, Y.; Xie, X. Methylation status of the long control region of HPV 16 in clinical cervical specimens. *Mol. Med. Rep.* **2008**, *1*, 555–560. [CrossRef] [PubMed]
36. Kim, K.; Garner-Hamrick, P.A.; Fisher, C.; Lee, D.; Lambert, P.F. Methylation patterns of papillomavirus DNA, its influence on E2 function, and implications in viral infection. *J. Virol.* **2003**, *77*, 12450–12459. [CrossRef] [PubMed]
37. Badal, S.; Badal, V.; Calleja-Macias, I.E.; Kalantari, M.; Chuang, L.S.; Li, B.F.; Bernard, H.U. The human papillomavirus-18 genome is efficiently targeted by cellular DNA methylation. *Virology* **2004**, *324*, 483–492. [CrossRef] [PubMed]

38. Kalantari, M.; Calleja-Macias, I.E.; Tewari, D.; Hagmar, B.; Lie, K.; Barrera-Saldana, H.A.; Wiley, D.J.; Bernard, H.U. Conserved methylation patterns of human papillomavirus type 16 DNA in asymptomatic infection and cervical neoplasia. *J. Virol.* **2004**, *78*, 12762–12772. [CrossRef] [PubMed]
39. Turan, T.; Kalantari, M.; Calleja-Macias, I.E.; Cubie, H.A.; Cuschieri, K.; Villa, L.L.; Skomedal, H.; Barrera-Saldana, H.A.; Bernard, H.U. Methylation of the human papillomavirus-18 l1 gene: A biomarker of neoplastic progression? *Virology* **2006**, *349*, 175–183. [CrossRef] [PubMed]
40. Clarke, M.A.; Wentzensen, N.; Mirabello, L.; Ghosh, A.; Wacholder, S.; Harari, A.; Lorincz, A.; Schiffman, M.; Burk, R.D. Human papillomavirus DNA methylation as a potential biomarker for cervical cancer. *Cancer Epidemiol. Biomark. Prev.* **2012**, *21*, 2125–2137. [CrossRef] [PubMed]
41. Johannsen, E.; Lambert, P.F. Epigenetics of human papillomaviruses. *Virology* **2013**, *445*, 205–212. [CrossRef] [PubMed]
42. Vinokurova, S.; von Knebel Doeberitz, M. Differential methylation of the HPV 16 upstream regulatory region during epithelial differentiation and neoplastic transformation. *PLoS ONE* **2011**, *6*, e24451. [CrossRef] [PubMed]
43. Mazumder Indra, D.; Singh, R.K.; Mitra, S.; Dutta, S.; Chakraborty, C.; Basu, P.S.; Mondal, R.K.; Roychoudhury, S.; Panda, C.K. Genetic and epigenetic changes of HPV16 in cervical cancer differentially regulate E6/E7 expression and associate with disease progression. *Gynecol. Oncol.* **2011**, *123*, 597–604. [CrossRef] [PubMed]
44. Hublarova, P.; Hrstka, R.; Rotterova, P.; Rotter, L.; Coupkova, M.; Badal, V.; Nenutil, R.; Vojtesek, B. Prediction of human papillomavirus 16 E6 gene expression and cervical intraepithelial neoplasia progression by methylation status. *Int. J. Gynecol. Cancer* **2009**, *19*, 321–325. [CrossRef] [PubMed]
45. Badal, V.; Chuang, L.S.; Tan, E.H.; Badal, S.; Villa, L.L.; Wheeler, C.M.; Li, B.F.; Bernard, H.U. Cpg methylation of human papillomavirus type 16 DNA in cervical cancer cell lines and in clinical specimens: Genomic hypomethylation correlates with carcinogenic progression. *J. Virol.* **2003**, *77*, 6227–6234. [CrossRef] [PubMed]
46. Snellenberg, S.; Schutze, D.M.; Claassen-Kramer, D.; Meijer, C.J.; Snijders, P.J.; Steenbergen, R.D. Methylation status of the E2 binding sites of HPV16 in cervical lesions determined with the Luminex xMAP system. *Virology* **2012**, *422*, 357–365. [CrossRef] [PubMed]
47. Kalantari, M.; Lee, D.; Calleja-Macias, I.E.; Lambert, P.F.; Bernard, H.U. Effects of cellular differentiation, chromosomal integration and 5-aza-2′-deoxycytidine treatment on human papillomavirus-16 DNA methylation in cultured cell lines. *Virology* **2008**, *374*, 292–303. [CrossRef] [PubMed]
48. Chaiwongkot, A.; Vinokurova, S.; Pientong, C.; Ekalaksananan, T.; Kongyingyoes, B.; Kleebkaow, P.; Chumworathayi, B.; Patarapadungkit, N.; Reuschenbach, M.; von Knebel Doeberitz, M. Differential methylation of E2 binding sites in episomal and integrated HPV 16 genomes in preinvasive and invasive cervical lesions. *Int. J. Cancer* **2013**, *132*, 2087–2094. [CrossRef] [PubMed]
49. Thain, A.; Jenkins, O.; Clarke, A.R.; Gaston, K. CpG methylation directly inhibits binding of the human papillomavirus type 16 E2 protein to specific DNA sequences. *J. Virol.* **1996**, *70*, 7233–7235. [PubMed]
50. Marongiu, L.; Godi, A.; Parry, J.V.; Beddows, S. Human papillomavirus 16, 18, 31 and 45 viral load, integration and methylation status stratified by cervical disease stage. *BMC Cancer* **2014**, *14*, 384. [CrossRef] [PubMed]
51. Duenas-Gonzalez, A.; Lizano, M.; Candelaria, M.; Cetina, L.; Arce, C.; Cervera, E. Epigenetics of cervical cancer. An overview and therapeutic perspectives. *Mol. Cancer* **2005**, *4*, 38. [CrossRef] [PubMed]
52. Wentzensen, N.; Sherman, M.E.; Schiffman, M.; Wang, S.S. Utility of methylation markers in cervical cancer early detection: Appraisal of the state-of-the-science. *Gynecol. Oncol.* **2009**, *112*, 293–299. [CrossRef] [PubMed]
53. Brebi, P.; Maldonado, L.; Noordhuis, M.G.; Ili, C.; Leal, P.; Garcia, P.; Brait, M.; Ribas, J.; Michailidi, C.; Perez, J.; et al. Genome-wide methylation profiling reveals Zinc finger protein 516 (ZNF516) and FK-506-binding protein 6 (FKBP6) promoters frequently methylated in cervical neoplasia, associated with HPV status and ethnicity in a chilean population. *Epigenetics* **2014**, *9*, 308–317. [CrossRef] [PubMed]
54. Huang, R.L.; Chang, C.C.; Su, P.H.; Chen, Y.C.; Liao, Y.P.; Wang, H.C.; Yo, Y.T.; Chao, T.K.; Huang, H.C.; Lin, C.Y.; et al. Methylomic analysis identifies frequent DNA methylation of zinc finger protein 582 (ZNF582) in cervical neoplasms. *PLoS ONE* **2012**, *7*, e41060. [CrossRef] [PubMed]
55. Saavedra, K.P.; Brebi, P.M.; Roa, J.C. Epigenetic alterations in preneoplastic and neoplastic lesions of the cervix. *Clin. Epigenet.* **2012**, *4*, 13. [CrossRef] [PubMed]

56. Hansel, A.; Steinbach, D.; Greinke, C.; Schmitz, M.; Eiselt, J.; Scheungraber, C.; Gajda, M.; Hoyer, H.; Runnebaum, I.B.; Durst, M. A promising DNA methylation signature for the triage of high-risk human papillomavirus DNA-positive women. *PLoS ONE* **2014**, *9*, e91905. [CrossRef] [PubMed]

57. Siegel, E.M.; Riggs, B.M.; Delmas, A.L.; Koch, A.; Hakam, A.; Brown, K.D. Quantitative DNA methylation analysis of candidate genes in cervical cancer. *PLoS ONE* **2015**, *10*, e0122495. [CrossRef] [PubMed]

58. Kitkumthorn, N.; Yanatatsanajit, P.; Kiatpongsan, S.; Phokaew, C.; Triratanachat, S.; Trivijitsilp, P.; Termrungruanglert, W.; Tresukosol, D.; Niruthisard, S.; Mutirangura, A. Cyclin A1 promoter hypermethylation in human papillomavirus-associated cervical cancer. *BMC Cancer* **2006**, *6*, 55. [CrossRef] [PubMed]

59. De Wilde, J.; Kooter, J.M.; Overmeer, R.M.; Claassen-Kramer, D.; Meijer, C.J.; Snijders, P.J.; Steenbergen, R.D. Htert promoter activity and CpG methylation in HPV-induced carcinogenesis. *BMC Cancer* **2010**, *10*, 271. [CrossRef] [PubMed]

60. Schwartz, Y.B.; Pirrotta, V. Polycomb silencing mechanisms and the management of genomic programmes. *Nat. Rev. Genet.* **2007**, *8*, 9–22. [CrossRef] [PubMed]

61. Tolhuis, B.; de Wit, E.; Muijrers, I.; Teunissen, H.; Talhout, W.; van Steensel, B.; van Lohuizen, M. Genome-wide profiling of PRC1 and PRC2 polycomb chromatin binding in drosophila melanogaster. *Nat. Genet.* **2006**, *38*, 694–699. [CrossRef] [PubMed]

62. Li, B.; Carey, M.; Workman, J.L. The role of chromatin during transcription. *Cell* **2007**, *128*, 707–719. [CrossRef] [PubMed]

63. Zhang, T.; Cooper, S.; Brockdorff, N. The interplay of histone modifications—Writers that read. *EMBO Rep.* **2015**, *16*, 1467–1481. [CrossRef] [PubMed]

64. Del Mar Pena, L.M.; Laimins, L.A. Differentiation-dependent chromatin rearrangement coincides with activation of human papillomavirus type 31 late gene expression. *J. Virol.* **2001**, *75*, 10005–10013. [CrossRef] [PubMed]

65. Stunkel, W.; Bernard, H.U. The chromatin structure of the long control region of human papillomavirus type 16 represses viral oncoprotein expression. *J. Virol.* **1999**, *73*, 1918–1930. [PubMed]

66. Swindle, C.S.; Engler, J.A. Association of the human papillomavirus type 11 E1 protein with histone H1. *J. Virol.* **1998**, *72*, 1994–2001. [PubMed]

67. Wooldridge, T.R.; Laimins, L.A. Regulation of human papillomavirus type 31 gene expression during the differentiation-dependent life cycle through histone modifications and transcription factor binding. *Virology* **2008**, *374*, 371–380. [CrossRef] [PubMed]

68. Brehm, A.; Nielsen, S.J.; Miska, E.A.; McCance, D.J.; Reid, J.L.; Bannister, A.J.; Kouzarides, T. The E7 oncoprotein associates with Mi2 and histone deacetylase activity to promote cell growth. *EMBO J.* **1999**, *18*, 2449–2458. [CrossRef] [PubMed]

69. Longworth, M.S.; Laimins, L.A. The binding of histone deacetylases and the integrity of Zinc finger-like motifs of the E7 protein are essential for the life cycle of human papillomavirus type 31. *J. Virol.* **2004**, *78*, 3533–3541. [CrossRef] [PubMed]

70. Avvakumov, N.; Torchia, J.; Mymryk, J.S. Interaction of the hpv E7 proteins with the pCAF acetyltransferase. *Oncogene* **2003**, *22*, 3833–3841. [CrossRef] [PubMed]

71. Bernat, A.; Avvakumov, N.; Mymryk, J.S.; Banks, L. Interaction between the HPV E7 oncoprotein and the transcriptional coactivator p300. *Oncogene* **2003**, *22*, 7871–7881. [CrossRef] [PubMed]

72. Baldwin, A.; Huh, K.W.; Munger, K. Human papillomavirus E7 oncoprotein dysregulates steroid receptor coactivator 1 localization and function. *J. Virol.* **2006**, *80*, 6669–6677. [CrossRef] [PubMed]

73. Huang, S.M.; McCance, D.J. Down regulation of the interleukin-8 promoter by human papillomavirus type 16 E6 and E7 through effects on creb binding protein/p300 and P/CAF. *J. Virol.* **2002**, *76*, 8710–8721. [CrossRef] [PubMed]

74. McLaughlin-Drubin, M.E.; Park, D.; Munger, K. Tumor suppressor p16INK4A is necessary for survival of cervical carcinoma cell lines. *Proc. Natl. Acad. Sci. USA* **2013**, *110*, 16175–16180. [CrossRef] [PubMed]

75. McLaughlin-Drubin, M.E.; Crum, C.P.; Munger, K. Human papillomavirus E7 oncoprotein induces KDM6A and KDM6B histone demethylase expression and causes epigenetic reprogramming. *Proc. Natl. Acad. Sci. USA* **2011**, *108*, 2130–2135. [CrossRef] [PubMed]

76. McLaughlin-Drubin, M.E.; Huh, K.W.; Munger, K. Human papillomavirus type 16 E7 oncoprotein associates with E2F6. *J. Virol.* **2008**, *82*, 8695–8705. [CrossRef] [PubMed]

77. Hyland, P.L.; McDade, S.S.; McCloskey, R.; Dickson, G.J.; Arthur, K.; McCance, D.J.; Patel, D. Evidence for alteration of EZH2, BMI1, and KDM6A and epigenetic reprogramming in human papillomavirus type 16 E6/E7-expressing keratinocytes. *J. Virol.* **2011**, *85*, 10999–11006. [CrossRef] [PubMed]

78. Patel, D.; Huang, S.M.; Baglia, L.A.; McCance, D.J. The E6 protein of human papillomavirus type 16 binds to and inhibits co- activation by CBP and p300. *EMBO J.* **1999**, *18*, 5061–5072. [CrossRef] [PubMed]

79. Zimmermann, H.; Degenkolbe, R.; Bernard, H.U.; O'Connor, M.J. The human papillomavirus type 16 E6 oncoprotein can down-regulate p53 activity by targeting the transcriptional coactivator CBP/p300. *J. Virol.* **1999**, *73*, 6209–6219. [PubMed]

80. Jansma, A.L.; Martinez-Yamout, M.A.; Liao, R.; Sun, P.; Dyson, H.J.; Wright, P.E. The high-risk HPV16 E7 oncoprotein mediates interaction between the transcriptional coactivator CBP and the retinoblastoma protein prb. *J. Mol. Biol.* **2014**, *426*, 4030–4048. [CrossRef] [PubMed]

81. Avantaggiati, M.L.; Ogryzko, V.; Gardner, K.; Giordano, A.; Levine, A.S.; Kelly, K. Recruitment of p300/CBP in p53-dependent signal pathways. *Cell* **1997**, *89*, 1175–1184. [CrossRef]

82. Gu, W.; Shi, X.L.; Roeder, R.G. Synergistic activation of transcription by CBP and p53. *Nature* **1997**, *387*, 819–823. [CrossRef] [PubMed]

83. Lill, N.L.; Grossman, S.R.; Ginsberg, D.; DeCaprio, J.; Livingston, D.M. Binding and modulation of p53 by p300/CBP coactivators. *Nature* **1997**, *387*, 823–827. [CrossRef] [PubMed]

84. Perkins, N.D.; Felzien, L.K.; Betts, J.C.; Leung, K.; Beach, D.H.; Nabel, G.J. Regulation of NF-κB by cyclin-dependent kinases associated with the p300 coactivator. *Science* **1997**, *275*, 523–527. [CrossRef] [PubMed]

85. Snowden, A.W.; Perkins, N.D. Cell cycle regulation of the transcriptional coactivators p300 and creb binding protein. *Biochem. Pharmacol.* **1998**, *55*, 1947–1954. [CrossRef]

86. Ito, A.; Lai, C.H.; Zhao, X.; Saito, S.; Hamilton, M.H.; Appella, E.; Yao, T.P. P300/CBP-mediated p53 acetylation is commonly induced by p53-activating agents and inhibited by MDM2. *EMBO J.* **2001**, *20*, 1331–1340. [CrossRef] [PubMed]

87. Gray, M.J.; Zhang, J.; Ellis, L.M.; Semenza, G.L.; Evans, D.B.; Watowich, S.S.; Gallick, G.E. HIF-1alpha, STAT3, CBP/p300 and Ref-1/APE are components of a transcriptional complex that regulates Src-dependent hypoxia-induced expression of VEGF in pancreatic and prostate carcinomas. *Oncogene* **2005**, *24*, 3110–3120. [CrossRef] [PubMed]

88. Thomas, M.C.; Chiang, C.M. E6 oncoprotein represses p53-dependent gene activation via inhibition of protein acetylation independently of inducing p53 degradation. *Mol. Cell* **2005**, *17*, 251–264. [CrossRef] [PubMed]

89. Xue, Y.; Wong, J.; Moreno, G.T.; Young, M.K.; Cote, J.; Wang, W. Nurd, a novel complex with both ATP-dependent chromatin-remodeling and histone deacetylase activities. *Mol. Cell* **1998**, *2*, 851–861. [CrossRef]

90. Zhang, Y.; LeRoy, G.; Seelig, H.P.; Lane, W.S.; Reinberg, D. The dermatomyositis-specific autoantigen MI2 is a component of a complex containing histone deacetylase and nucleosome remodeling activities. *Cell* **1998**, *95*, 279–289. [CrossRef]

91. Longworth, M.S.; Wilson, R.; Laimins, L.A. HPV31 E7 facilitates replication by activating E2F2 transcription through its interaction with hdacs. *EMBO J.* **2005**, *24*, 1821–1830. [CrossRef] [PubMed]

92. Park, J.S.; Kim, E.J.; Kwon, H.J.; Hwang, E.S.; Namkoong, S.E.; Um, S.J. Inactivation of interferon regulatory factor-1 tumor suppressor protein by HPV E7 oncoprotein. Implication for the E7-mediated immune evasion mechanism in cervical carcinogenesis. *J. Biol. Chem.* **2000**, *275*, 6764–6769. [CrossRef] [PubMed]

93. Um, S.J.; Rhyu, J.W.; Kim, E.J.; Jeon, K.C.; Hwang, E.S.; Park, J.S. Abrogation of IRF-1 response by high-risk HPV E7 protein in vivo. *Cancer Lett.* **2002**, *179*, 205–212. [CrossRef]

94. Trojer, P.; Cao, A.R.; Gao, Z.; Li, Y.; Zhang, J.; Xu, X.; Li, G.; Losson, R.; Erdjument-Bromage, H.; Tempst, P.; et al. L3MBTL2 protein acts in concert with PcG protein-mediated monoubiquitination of H2A to establish a repressive chromatin structure. *Mol. Cell* **2011**, *42*, 438–450. [CrossRef] [PubMed]

95. Qin, J.; Whyte, W.A.; Anderssen, E.; Apostolou, E.; Chen, H.H.; Akbarian, S.; Bronson, R.T.; Hochedlinger, K.; Ramaswamy, S.; Young, R.A.; et al. The polycomb group protein L3MBTL2 assembles an atypical PRC1-family complex that is essential in pluripotent stem cells and early development. *Cell Stem Cell* **2012**, *11*, 319–332. [CrossRef] [PubMed]

96. Perdigoto, C.N.; Valdes, V.J.; Bardot, E.S.; Ezhkova, E. Epigenetic regulation of skin: Focus on the polycomb complex. *Cell. Mol. Life Sci.* **2012**, *69*, 2161–2172. [PubMed]

97. Eckert, R.L.; Adhikary, G.; Rorke, E.A.; Chew, Y.C.; Balasubramanian, S. Polycomb group proteins are key regulators of keratinocyte function. *J. Investig. Dermatol.* **2011**, *131*, 295–301. [CrossRef] [PubMed]

98. Shaw, T.; Martin, P. Epigenetic reprogramming during wound healing: Loss of polycomb-mediated silencing may enable upregulation of repair genes. *EMBO Rep.* **2009**, *10*, 881–886. [CrossRef] [PubMed]

99. McLaughlin-Drubin, M.E.; Munger, K. Biochemical and functional interactions of human papillomavirus proteins with polycomb group proteins. *Viruses* **2013**, *5*, 1231–1249. [CrossRef] [PubMed]

100. Schumacher, A.; Magnuson, T. Murine Polycomb- and trithorax-group genes regulate homeotic pathways and beyond. *Trends Genet.* **1997**, *13*, 167–170. [CrossRef]

101. Gould, A. Functions of mammalian Polycomb group and trithorax group related genes. *Curr. Opin. Genet. Dev.* **1997**, *7*, 488–494. [CrossRef]

102. Hung, Y.C.; Ueda, M.; Terai, Y.; Kumagai, K.; Ueki, K.; Kanda, K.; Yamaguchi, H.; Akise, D.; Ueki, M. Homeobox gene expression and mutation in cervical carcinoma cells. *Cancer Sci.* **2003**, *94*, 437–441. [CrossRef] [PubMed]

103. Alami, Y.; Castronovo, V.; Belotti, D.; Flagiello, D.; Clausse, N. HOXC5 and HOXC8 expression are selectively turned on in human cervical cancer cells compared to normal keratinocytes. *Biochem. Biophys. Res. Commun.* **1999**, *257*, 738–745. [CrossRef] [PubMed]

104. Barba-de la Rosa, A.P.; Briones-Cerecero, E.; Lugo-Melchor, O.; De Leon-Rodriguez, A.; Santos, L.; Castelo-Ruelas, J.; Valdivia, A.; Pina, P.; Chagolla-Lopez, A.; Hernandez-Cueto, D.; et al. HOX B4 as potential marker of non-differentiated cells in human cervical cancer cells. *J. Cancer Res. Clin. Oncol.* **2012**, *138*, 293–300. [CrossRef] [PubMed]

105. Zhai, Y.; Kuick, R.; Nan, B.; Ota, I.; Weiss, S.J.; Trimble, C.L.; Fearon, E.R.; Cho, K.R. Gene expression analysis of preinvasive and invasive cervical squamous cell carcinomas identifies HOXC10 as a key mediator of invasion. *Cancer Res.* **2007**, *67*, 10163–10172. [CrossRef] [PubMed]

106. Holland, D.; Hoppe-Seyler, K.; Schuller, B.; Lohrey, C.; Maroldt, J.; Durst, M.; Hoppe-Seyler, F. Activation of the enhancer of zeste homologue 2 gene by the human papillomavirus E7 oncoprotein. *Cancer Res.* **2008**, *68*, 9964–9972. [CrossRef] [PubMed]

107. Cha, T.L.; Zhou, B.P.; Xia, W.; Wu, Y.; Yang, C.C.; Chen, C.T.; Ping, B.; Otte, A.P.; Hung, M.C. AKT-mediated phosphorylation of EZH2 suppresses methylation of lysine 27 in histone H3. *Science* **2005**, *310*, 306–310. [CrossRef] [PubMed]

108. Menges, C.W.; Baglia, L.A.; Lapoint, R.; McCance, D.J. Human papillomavirus type 16 E7 up-regulates AKT activity through the retinoblastoma protein. *Cancer Res.* **2006**, *66*, 5555–5559. [CrossRef] [PubMed]

109. Spangle, J.M.; Munger, K. The human papillomavirus type 16 E6 oncoprotein activates mTORC1 signaling and increases protein synthesis. *J. Virol.* **2010**, *84*, 9398–9407. [CrossRef] [PubMed]

110. Kuzmichev, A.; Margueron, R.; Vaquero, A.; Preissner, T.S.; Scher, M.; Kirmizis, A.; Ouyang, X.; Brockdorff, N.; Abate-Shen, C.; Farnham, P.; et al. Composition and histone substrates of polycomb repressive group complexes change during cellular differentiation. *Proc. Natl. Acad. Sci. USA* **2005**, *102*, 1859–1864. [CrossRef] [PubMed]

111. Agger, K.; Cloos, P.A.; Rudkjaer, L.; Williams, K.; Andersen, G.; Christensen, J.; Helin, K. The H3K27ME3 demethylase JMJD3 contributes to the activation of the *INK4A-ARF* locus in response to oncogene- and stress-induced senescence. *Genes Dev.* **2009**, *23*, 1171–1176. [CrossRef] [PubMed]

112. Barradas, M.; Anderton, E.; Acosta, J.C.; Li, S.; Banito, A.; Rodriguez-Niedenfuhr, M.; Maertens, G.; Banck, M.; Zhou, M.M.; Walsh, M.J.; et al. Histone demethylase JMJD3 contributes to epigenetic control of INK4A/ARF by oncogenic RAS. *Genes Dev.* **2009**, *23*, 1177–1182. [CrossRef] [PubMed]

113. Campisi, J.; d'Adda di Fagagna, F. Cellular senescence: When bad things happen to good cells. *Nat. Rev. Mol. Cell Biol.* **2007**, *8*, 729–740. [CrossRef] [PubMed]

114. Collado, M.; Gil, J.; Efeyan, A.; Guerra, C.; Schuhmacher, A.J.; Barradas, M.; Benguria, A.; Zaballos, A.; Flores, J.M.; Barbacid, M.; et al. Tumour biology: Senescence in premalignant tumours. *Nature* **2005**, *436*, 642. [CrossRef] [PubMed]

115. Serrano, M.; Lin, A.W.; McCurrach, M.E.; Beach, D.; Lowe, S.W. Oncogenic ras provokes premature cell senescence associated with accumulation of p53 and P16INK4A. *Cell* **1997**, *88*, 593–602. [CrossRef]

116. Sano, T.; Oyama, T.; Kashiwabara, K.; Fukuda, T.; Nakajima, T. Expression status of p16 protein is associated with human papillomavirus oncogenic potential in cervical and genital lesions. *Am. J. Pathol.* **1998**, *153*, 1741–1748. [CrossRef]

117. Klaes, R.; Friedrich, T.; Spitkovsky, D.; Ridder, R.; Rudy, W.; Petry, U.; Dallenbach-Hellweg, G.; Schmidt, D.; von Knebel Doeberitz, M. Overexpression of p16(INK4A) as a specific marker for dysplastic and neoplastic epithelial cells of the cervix uteri. *Int. J. Cancer* **2001**, *92*, 276–284. [CrossRef] [PubMed]

118. Hsu, C.H.; Peng, K.L.; Jhang, H.C.; Lin, C.H.; Wu, S.Y.; Chiang, C.M.; Lee, S.C.; Yu, W.C.; Juan, L.J. The HPV E6 oncoprotein targets histone methyltransferases for modulating specific gene transcription. *Oncogene* **2012**, *31*, 2335–2349. [CrossRef] [PubMed]

119. Pradhan, S.; Chin, H.G.; Esteve, P.O.; Jacobsen, S.E. SET7/9 mediated methylation of non-histone proteins in mammalian cells. *Epigenetics* **2009**, *4*, 383–387. [CrossRef] [PubMed]

120. Florence, B.; Faller, D.V. You BET-Cha: A novel family of transcriptional regulators. *Front. Biosci.* **2001**, *6*, D1008–D1018. [PubMed]

121. Dey, A.; Chitsaz, F.; Abbasi, A.; Misteli, T.; Ozato, K. The double bromodomain protein Brd4 binds to acetylated chromatin during interphase and mitosis. *Proc. Natl. Acad. Sci. USA* **2003**, *100*, 8758–8763. [CrossRef] [PubMed]

122. Dey, A.; Nishiyama, A.; Karpova, T.; McNally, J.; Ozato, K. Brd4 marks select genes on mitotic chromatin and directs postmitotic transcription. *Mol. Biol. Cell* **2009**, *20*, 4899–4909. [CrossRef] [PubMed]

123. Mochizuki, K.; Nishiyama, A.; Jang, M.K.; Dey, A.; Ghosh, A.; Tamura, T.; Natsume, H.; Yao, H.; Ozato, K. The bromodomain protein Brd4 stimulates G1 gene transcription and promotes progression to S phase. *J. Biol. Chem.* **2008**, *283*, 9040–9048. [CrossRef] [PubMed]

124. Zhao, R.; Nakamura, T.; Fu, Y.; Lazar, Z.; Spector, D.L. Gene bookmarking accelerates the kinetics of post-mitotic transcriptional re-activation. *Nat. Cell Biol.* **2011**, *13*, 1295–1304. [CrossRef] [PubMed]

125. Jang, M.K.; Mochizuki, K.; Zhou, M.; Jeong, H.S.; Brady, J.N.; Ozato, K. The bromodomain protein Brd4 is a positive regulatory component of P-TEFB and stimulates RNA polymerase II-dependent transcription. *Mol. Cell* **2005**, *19*, 523–534. [CrossRef] [PubMed]

126. Yang, Z.; Yik, J.H.; Chen, R.; He, N.; Jang, M.K.; Ozato, K.; Zhou, Q. Recruitment of P-TEFb for stimulation of transcriptional elongation by the bromodomain protein Brd4. *Mol. Cell* **2005**, *19*, 535–545. [CrossRef] [PubMed]

127. McBride, A.A.; Jang, M.K. Current understanding of the role of the Brd4 protein in the papillomavirus lifecycle. *Viruses* **2013**, *5*, 1374–1394. [CrossRef] [PubMed]

128. Wu, S.Y.; Lee, A.Y.; Hou, S.Y.; Kemper, J.K.; Erdjument-Bromage, H.; Tempst, P.; Chiang, C.M. Brd4 links chromatin targeting to hpv transcriptional silencing. *Genes Dev.* **2006**, *20*, 2383–2396. [CrossRef] [PubMed]

129. You, J.; Croyle, J.L.; Nishimura, A.; Ozato, K.; Howley, P.M. Interaction of the bovine papillomavirus E2 protein with Brd4 tethers the viral DNA to host mitotic chromosomes. *Cell* **2004**, *117*, 349–360. [CrossRef]

130. Olejnik-Schmidt, A.K.; Schmidt, M.T.; Kedzia, W.; Gozdzicka-Jozefiak, A. Search for cellular partners of human papillomavirus type 16 E2 protein. *Arch. Virol.* **2008**, *153*, 983–990. [CrossRef] [PubMed]

131. McPhillips, M.G.; Ozato, K.; McBride, A.A. Interaction of bovine papillomavirus E2 protein with Brd4 stabilizes its association with chromatin. *J. Virol.* **2005**, *79*, 8920–8932. [CrossRef] [PubMed]

132. Baxter, M.K.; McPhillips, M.G.; Ozato, K.; McBride, A.A. The mitotic chromosome binding activity of the papillomavirus E2 protein correlates with interaction with the cellular chromosomal protein, Brd4. *J. Virol.* **2005**, *79*, 4806–4818. [CrossRef] [PubMed]

133. Jang, M.K.; Anderson, D.E.; van Doorslaer, K.; McBride, A.A. A proteomic approach to discover and compare interacting partners of papillomavirus E2 proteins from diverse phylogenetic groups. *Proteomics* **2015**, *15*, 2038–2050. [CrossRef] [PubMed]

134. Yan, J.; Li, Q.; Lievens, S.; Tavernier, J.; You, J. Abrogation of the Brd4-positive transcription elongation factor b complex by papillomavirus E2 protein contributes to viral oncogene repression. *J. Virol.* **2010**, *84*, 76–87. [CrossRef] [PubMed]

135. Jha, S.; Vande Pol, S.; Banerjee, N.S.; Dutta, A.B.; Chow, L.T.; Dutta, A. Destabilization of TIP60 by human papillomavirus E6 results in attenuation of TIP60-dependent transcriptional regulation and apoptotic pathway. *Mol. Cell* **2010**, *38*, 700–711. [CrossRef] [PubMed]

136. Subbaiah, V.K.; Zhang, Y.; Rajagopalan, D.; Abdullah, L.N.; Yeo-Teh, N.S.; Tomaic, V.; Banks, L.; Myers, M.P.; Chow, E.K.; Jha, S. E3 ligase EDD1/UBR5 is utilized by the HPV E6 oncogene to destabilize tumor suppressor TIP60. *Oncogene* **2016**, *35*, 2062–2074. [CrossRef] [PubMed]

137. Cech, T.R.; Steitz, J.A. The noncoding RNA revolution-trashing old rules to forge new ones. *Cell* **2014**, *157*, 77–94. [CrossRef] [PubMed]

138. Carthew, R.W.; Sontheimer, E.J. Origins and mechanisms of miRNAs and siRNAs. *Cell* **2009**, *136*, 642–655. [CrossRef] [PubMed]

139. Singh, S.K.; Pal Bhadra, M.; Girschick, H.J.; Bhadra, U. MicroRNAs—Micro in size but macro in function. *FEBS J.* **2008**, *275*, 4929–4944. [CrossRef] [PubMed]

140. Lim, L.P.; Lau, N.C.; Garrett-Engele, P.; Grimson, A.; Schelter, J.M.; Castle, J.; Bartel, D.P.; Linsley, P.S.; Johnson, J.M. Microarray analysis shows that some microRNAs downregulate large numbers of target mRNAs. *Nature* **2005**, *433*, 769–773. [CrossRef] [PubMed]

141. Cai, X.; Li, G.; Laimins, L.A.; Cullen, B.R. Human papillomavirus genotype 31 does not express detectable microRNA levels during latent or productive virus replication. *J. Virol.* **2006**, *80*, 10890–10893. [CrossRef] [PubMed]

142. Wang, X.; Wang, H.K.; Li, Y.; Hafner, M.; Banerjee, N.S.; Tang, S.; Briskin, D.; Meyers, C.; Chow, L.T.; Xie, X.; et al. MicroRNAs are biomarkers of oncogenic human papillomavirus infections. *Proc. Natl. Acad. Sci. USA* **2014**, *111*, 4262–4267. [CrossRef] [PubMed]

143. Gunasekharan, V.; Laimins, L.A. Human papillomaviruses modulate microRNA 145 expression to directly control genome amplification. *J. Virol.* **2013**, *87*, 6037–6043. [CrossRef] [PubMed]

144. Harden, M.E.; Prasad, N.; Griffiths, A.; Munger, K. Modulation of microRNA-mRNA target pairs by human papillomavirus 16 oncoproteins. *MBio* **2017**, *8*. [CrossRef] [PubMed]

145. Martinez, I.; Gardiner, A.S.; Board, K.F.; Monzon, F.A.; Edwards, R.P.; Khan, S.A. Human papillomavirus type 16 reduces the expression of microRNA-218 in cervical carcinoma cells. *Oncogene* **2008**, *27*, 2575–2582. [CrossRef] [PubMed]

146. Pereira, P.M.; Marques, J.P.; Soares, A.R.; Carreto, L.; Santos, M.A. MicroRNA expression variability in human cervical tissues. *PLoS ONE* **2010**, *5*, e11780. [CrossRef] [PubMed]

147. Au Yeung, C.L.; Tsang, T.Y.; Yau, P.L.; Kwok, T.T. Human papillomavirus type 16 E6 induces cervical cancer cell migration through the p53/microRNA-23b/urokinase-type plasminogen activator pathway. *Oncogene* **2011**, *30*, 2401–2410. [CrossRef] [PubMed]

148. Hu, X.; Schwarz, J.K.; Lewis, J.S., Jr.; Huettner, P.C.; Rader, J.S.; Deasy, J.O.; Grigsby, P.W.; Wang, X. A microRNA expression signature for cervical cancer prognosis. *Cancer Res.* **2010**, *70*, 1441–1448. [CrossRef] [PubMed]

149. Wilting, S.M.; van Boerdonk, R.A.; Henken, F.E.; Meijer, C.J.; Diosdado, B.; Meijer, G.A.; le Sage, C.; Agami, R.; Snijders, P.J.; Steenbergen, R.D. Methylation-mediated silencing and tumour suppressive function of HSA-MIR-124 in cervical cancer. *Mol. Cancer* **2010**, *9*, 167. [CrossRef] [PubMed]

150. Lui, W.O.; Pourmand, N.; Patterson, B.K.; Fire, A. Patterns of known and novel small RNAs in human cervical cancer. *Cancer Res.* **2007**, *67*, 6031–6043. [CrossRef] [PubMed]

151. Patron, J.P.; Fendler, A.; Bild, M.; Jung, U.; Muller, H.; Arntzen, M.O.; Piso, C.; Stephan, C.; Thiede, B.; Mollenkopf, H.J.; et al. MIR-133b targets antiapoptotic genes and enhances death receptor-induced apoptosis. *PLoS ONE* **2012**, *7*, e35345. [CrossRef] [PubMed]

152. Tian, R.Q.; Wang, X.H.; Hou, L.J.; Jia, W.H.; Yang, Q.; Li, Y.X.; Liu, M.; Li, X.; Tang, H. MicroRNA-372 is down-regulated and targets cyclin-dependent kinase 2 (CDK2) and cyclin a1 in human cervical cancer, which may contribute to tumorigenesis. *J. Biol. Chem.* **2011**, *286*, 25556–25563. [CrossRef] [PubMed]

153. Xu, X.M.; Wang, X.B.; Chen, M.M.; Liu, T.; Li, Y.X.; Jia, W.H.; Liu, M.; Li, X.; Tang, H. MicroRNA-19a and -19b regulate cervical carcinoma cell proliferation and invasion by targeting CUL5. *Cancer Lett.* **2012**, *322*, 148–158. [CrossRef] [PubMed]

154. Miller, D.L.; Davis, J.W.; Taylor, K.H.; Johnson, J.; Shi, Z.; Williams, R.; Atasoy, U.; Lewis, J.S., Jr.; Stack, M.S. Identification of a human papillomavirus-associated oncogenic miRNA panel in human oropharyngeal squamous cell carcinoma validated by bioinformatics analysis of the cancer genome atlas. *Am. J. Pathol.* **2015**, *185*, 679–692. [CrossRef] [PubMed]

155. Harden, M.E.; Munger, K. Human papillomavirus 16 E6 and E7 oncoprotein expression alters microRNA expression in extracellular vesicles. *Virology* **2017**, *508*, 63–69. [CrossRef] [PubMed]

156. Yao, T.; Lin, Z. MIR-21 is involved in cervical squamous cell tumorigenesis and regulates CCL20. *Biochim. Biophys. Acta* **2012**, *1822*, 248–260. [CrossRef] [PubMed]

157. Melar-New, M.; Laimins, L.A. Human papillomaviruses modulate expression of microRNA 203 upon epithelial differentiation to control levels of p63 proteins. *J. Virol.* **2010**, *84*, 5212–5221. [CrossRef] [PubMed]

158. Zheng, Z.M.; Wang, X. Regulation of cellular miRNA expression by human papillomaviruses. *Biochim. Biophys. Acta* **2011**, *1809*, 668–677. [CrossRef] [PubMed]

159. Sato, F.; Tsuchiya, S.; Meltzer, S.J.; Shimizu, K. MicroRNAs and epigenetics. *FEBS J.* **2011**, *278*, 1598–1609. [CrossRef] [PubMed]

160. Adams, B.D.; Kasinski, A.L.; Slack, F.J. Aberrant regulation and function of microRNAs in cancer. *Curr. Biol.* **2014**, *24*, R762–R776. [CrossRef] [PubMed]

161. Hata, A.; Lieberman, J. Dysregulation of microRNA biogenesis and gene silencing in cancer. *Sci. Signal.* **2015**, *8*, re3. [CrossRef] [PubMed]

162. Muralidhar, B.; Goldstein, L.D.; Ng, G.; Winder, D.M.; Palmer, R.D.; Gooding, E.L.; Barbosa-Morais, N.L.; Mukherjee, G.; Thorne, N.P.; Roberts, I.; et al. Global microRNA profiles in cervical squamous cell carcinoma depend on Drosha expression levels. *J. Pathol.* **2007**, *212*, 368–377. [CrossRef] [PubMed]

163. Scotto, L.; Narayan, G.; Nandula, S.V.; Subramaniyam, S.; Kaufmann, A.M.; Wright, J.D.; Pothuri, B.; Mansukhani, M.; Schneider, A.; Arias-Pulido, H.; et al. Integrative genomics analysis of chromosome 5p gain in cervical cancer reveals target over-expressed genes, including Drosha. *Mol. Cancer* **2008**, *7*, 58. [CrossRef] [PubMed]

164. Harden, M.E.; Munger, K. Perturbation of drosha and dicer expression by human papillomavirus 16 oncoproteins. *Virology* **2017**, *507*, 192–198. [CrossRef] [PubMed]

165. Hon, C.C.; Ramilowski, J.A.; Harshbarger, J.; Bertin, N.; Rackham, O.J.; Gough, J.; Denisenko, E.; Schmeier, S.; Poulsen, T.M.; Severin, J.; et al. An atlas of human long non-coding RNAs with accurate 5′ ends. *Nature* **2017**, *543*, 199–204. [CrossRef] [PubMed]

166. Fatica, A.; Bozzoni, I. Long non-coding RNAs: New players in cell differentiation and development. *Nat. Rev. Genet.* **2014**, *15*, 7–21. [CrossRef] [PubMed]

167. Ponting, C.P.; Oliver, P.L.; Reik, W. Evolution and functions of long noncoding RNAs. *Cell* **2009**, *136*, 629–641. [CrossRef] [PubMed]

168. Gupta, R.A.; Shah, N.; Wang, K.C.; Kim, J.; Horlings, H.M.; Wong, D.J.; Tsai, M.C.; Hung, T.; Argani, P.; Rinn, J.L.; et al. Long non-coding RNA HOTAIR reprograms chromatin state to promote cancer metastasis. *Nature* **2010**, *464*, 1071–1076. [CrossRef] [PubMed]

169. Li, X.; Wu, Z.; Mei, Q.; Li, X.; Guo, M.; Fu, X.; Han, W. Long non-coding RNA HOTAIR, a driver of malignancy, predicts negative prognosis and exhibits oncogenic activity in oesophageal squamous cell carcinoma. *Br. J. Cancer* **2013**, *109*, 2266–2278. [CrossRef] [PubMed]

170. Huarte, M.; Rinn, J.L. Large non-coding RNAs: Missing links in cancer? *Hum. Mol. Genet.* **2010**, *19*, R152–R161. [CrossRef] [PubMed]

171. Wu, Z.H.; Wang, X.L.; Tang, H.M.; Jiang, T.; Chen, J.; Lu, S.; Qiu, G.Q.; Peng, Z.H.; Yan, D.W. Long non-coding RNA HOTAIR is a powerful predictor of metastasis and poor prognosis and is associated with epithelial-mesenchymal transition in colon cancer. *Oncol. Rep.* **2014**, *32*, 395–402. [CrossRef] [PubMed]

172. Ishibashi, M.; Kogo, R.; Shibata, K.; Sawada, G.; Takahashi, Y.; Kurashige, J.; Akiyoshi, S.; Sasaki, S.; Iwaya, T.; Sudo, T.; et al. Clinical significance of the expression of long non-coding RNA HOTAIR in primary hepatocellular carcinoma. *Oncol. Rep.* **2013**, *29*, 946–950. [CrossRef] [PubMed]

173. Kogo, R.; Shimamura, T.; Mimori, K.; Kawahara, K.; Imoto, S.; Sudo, T.; Tanaka, F.; Shibata, K.; Suzuki, A.; Komune, S.; et al. Long noncoding RNA HOTAIR regulates polycomb-dependent chromatin modification and is associated with poor prognosis in colorectal cancers. *Cancer Res.* **2011**, *71*, 6320–6326. [CrossRef] [PubMed]

174. Tsai, M.C.; Manor, O.; Wan, Y.; Mosammaparast, N.; Wang, J.K.; Lan, F.; Shi, Y.; Segal, E.; Chang, H.Y. Long noncoding RNA as modular scaffold of histone modification complexes. *Science* **2010**, *329*, 689–693. [CrossRef] [PubMed]

175. Rinn, J.L.; Kertesz, M.; Wang, J.K.; Squazzo, S.L.; Xu, X.; Brugmann, S.A.; Goodnough, L.H.; Helms, J.A.; Farnham, P.J.; Segal, E.; et al. Functional demarcation of active and silent chromatin domains in human HOX loci by noncoding RNAs. *Cell* **2007**, *129*, 1311–1323. [CrossRef] [PubMed]
176. Sharma, S.; Mandal, P.; Sadhukhan, T.; Roy Chowdhury, R.; Ranjan Mondal, N.; Chakravarty, B.; Chatterjee, T.; Roy, S.; Sengupta, S. Bridging links between long noncoding RNA HOTAIR and HPV oncoprotein E7 in cervical cancer pathogenesis. *Sci. Rep.* **2015**, *5*, 11724. [CrossRef] [PubMed]

Review

Virus/Host Cell Crosstalk in Hypoxic HPV-Positive Cancer Cells

Karin Hoppe-Seyler [1], Julia Mändl [1,2], Svenja Adrian [1], Bianca J. Kuhn [1] and Felix Hoppe-Seyler [1,*]

[1] Molecular Therapy of Virus-Associated Cancers (F065), German Cancer Research Center (DKFZ), D-69120 Heidelberg, Germany; k.hoppe-seyler@dkfz.de (K.H.-S.); j.maendl@dkfz.de (J.M.); s.adrian@dkfz.de (S.A.); bianca.kuhn@dkfz.de (B.J.K.)

[2] Viral Transformation Mechanisms (F030), German Cancer Research Center (DKFZ), D-69120 Heidelberg, Germany

* Correspondence: hoppe-seyler@dkfz.de; Tel.: +49-6221-424872

Academic Editors: Alison A. McBride and Karl Munger
Received: 22 June 2017; Accepted: 29 June 2017; Published: 5 July 2017

Abstract: Oncogenic types of human papillomaviruses (HPVs) are major human carcinogens. The expression of the viral *E6/E7* oncogenes plays a key role for HPV-linked oncogenesis. It recently has been found that low oxygen concentrations ("hypoxia"), as present in sub-regions of HPV-positive cancers, strongly affect the interplay between the HPV oncogenes and their transformed host cell. As a result, a state of dormancy is induced in hypoxic HPV-positive cancer cells, which is characterized by a shutdown of viral oncogene expression and a proliferative arrest that can be reversed by reoxygenation. In this review, these findings are put into the context of the current concepts of both HPV-linked carcinogenesis and of the effects of hypoxia on tumor biology. Moreover, we discuss the consequences for the phenotype of HPV-positive cancer cells as well as for their clinical behavior and response towards established and prospective therapeutic strategies.

Keywords: human papillomavirus; hypoxia; cervical cancer; head and neck cancer; senescence; metabolism; mTOR; therapy

---

1. Introduction

Approximately 50–60% of solid tumors exhibit pronounced hypoxic regions (commonly defined as oxygen concentrations below 1.5–2%) [1–3]. Hypoxia is of clinical importance because tumors with higher proportions of hypoxic cells usually have a poor prognosis, including human papillomavirus (HPV)-linked cancers, such as cervical cancers or head and neck squamous cell carcinomas (HNSCCs) [1,4,5]. In the process of HPV-linked carcinogenesis, the viral *E6* and *E7* oncogenes play a central role both for the induction of malignant cell transformation and for the maintenance of the oncogenic phenotype of HPV-positive cancer cells. Recent data indicate that hypoxia has profound effects on the crosstalk between the HPV oncogenes and their host cell, with implications for the malignant phenotype of HPV-positive cancer cells and for their clinical behavior.

2. Hypoxia and Cancer

In principle, two major forms of hypoxia can be differentiated in solid cancers. Firstly, chronic hypoxia occurs in a time frame of hours to days or weeks. This form of hypoxia is primarily caused by diffusion limitations, e.g., due to enlarged distances (>70 μm [1]) between tumor blood vessels and remote tumor cells (Figure 1A). Secondly, cycling (alternatively called acute, intermittent, or fluctuating) hypoxia primarily results from perfusion limitations, e.g., following the temporary partial or total occlusion of tumor microvessels (which are often structurally and functionally abnormal) through blood cell aggregates (Figure 1B). This latter form of hypoxia exposes tumor cells to repeated cycles of hypoxia and reoxygenation, which can be highly variable in their duration and frequency, for example occurring several times within one hour [1,6,7].

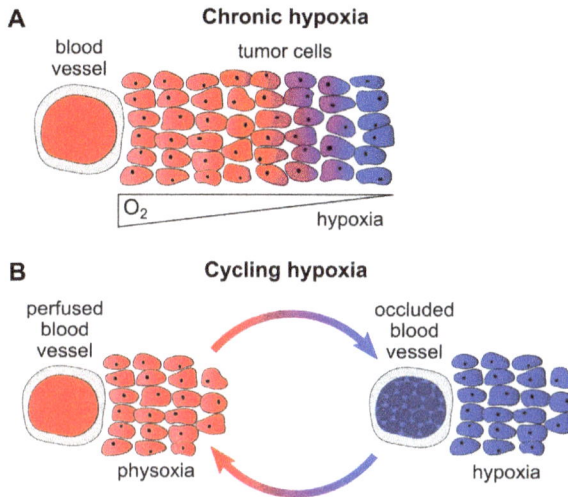

Figure 1. Chronic and cycling hypoxia. (**A**) Diffusion-limited chronic hypoxia due to enlarged distances between tumor blood vessels and tumor cells. Remote tumor cells (>70 μm away from the blood vessel [1]) are inadequately supplied with O_2 and become hypoxic. Red: oyxgenated tumor cells, blue: hypoxic tumor cells; (**B**) Perfusion-limited cycling hypoxia. Tumor vessels are often abnormally structured and can be temporarily occluded, e.g., through blood cell aggregates. Surrounding tumor cells will be exposed to fluctuating cycles of physoxia (left) or hypoxia (right). Red: oyxgenated tumor cells, blue: hypoxic tumor cells.

The poor clinical prognosis of hypoxic tumors could, on the one hand, mechanistically be due to hypoxia-linked processes that support the clonal evolution of more malignant tumor cells. These processes include (i) the induction of genetic instability through the downregulation of DNA repair mechanisms [8] and increased production of reactive oxygen species (ROS) [9]; (ii) the metabolic reprogramming of cancer cells, e.g., leading to a higher rate of glycolysis and enhanced lactate production and excretion [10,11]; (iii) the inhibition of cellular tumor suppressor pathways, such as apoptosis [12] or senescence [13]; (iv) the activation of autophagy, thereby supporting tumor cell survival [14]; (v) the induction of angiogenesis [15], and (vi) the promotion of invasion and metastasis [16]. On the other hand, hypoxia is also linked to an enhanced resistance towards anticancer treatments in clinical use, such as radiotherapy (RT) or chemotherapy (CT) [4,17], as further discussed below.

Notably, although the pronounced effects of hypoxia on cancer cell biology are well established, they are not taken into account in standard cell culture experiments, where cells are usually incubated at approximately 20% O_2 (corresponding to 95% air composed of 79% N_2 and 21% O_2, plus 5% CO_2). Although called "normoxia", these conditions do not reflect the median O_2 content of normal tissues ("physoxia"), which lies around 5–6% O_2 in most organs, with some exceptions [1,5]. Importantly, due to the diffusion and perfusion limitations discussed above, tumors often are substantially less oxygenated than the corresponding normal tissue, and many cancer entities exhibit median O_2 concentrations of less than 2% [1,5]. These considerations raise the question whether our conceptions of carcinogenesis that are derived from standard cell culture conditions are reflected under the low oxygen concentrations as they are present in the hypoxic sub-regions of many cancers.

3. Human Papillomaviruses and Cancer

At least 20% of the total cancer incidence in humans is attributable to infections [18]. In this context, oncogenic HPV types play a prominent role, since approximately every fourth of the infection-linked cancers worldwide is caused by this group of viruses [19]. HPV-induced cancers include cervical carcinomas, which alone account for more than 500,000 new cancer cases and over 250,000 cancer deaths per year [20]. Nearly 100% of cervical cancers are HPV-positive, with HPV16 (ca. 55–60%) and HPV18 (ca. 10–15%) being the most frequent types [21]. A substantial number of additional anogenital malignancies are HPV-positive as well, such as carcinomas of the anus (88%), vagina (78%), vulva (25%), and penis (50%) [19]. Moreover, a significant fraction of HNSCCs are linked to HPV infections, in particular oropharyngeal cancers (OPCs), which, in highly industrialized countries (e.g., Western Europe and the U.S.A.), are HPV-positive in 70–80% of cases. Differing from the virus type distribution in cervical cancers, HPV-positive OPCs contain HPV16 sequences in approximately 95% of cases [19,22].

In the process of HPV-induced carcinogenesis, the viral E6 and E7 oncoproteins target crucial tumor suppressor pathways for functional inactivation. For example, E6 forms a trimeric complex with the cellular ubiquitin ligase E6AP (E6-associated protein) and the p53 tumor suppressor protein [23], eventually leading to the proteolytic degradation of p53 [24]. The E7 oncoprotein binds to and inactivates the retinoblastoma tumor suppressor protein pRb [25]. Notably, the p53 and pRb pathways are also commonly impaired in HPV-negative human cancers by other routes, including somatic mutations of the *TP53* and *RB1* genes [26,27]. Thus, targeting the p53 and pRb proteins by the HPV oncoproteins represents an alternative strategy to block these key tumor suppressor pathways. Importantly, however, and in contrast to—for example—somatic mutations of the *TP53* and *RB1* genes, the inactivation of the p53 and pRb pathways by the HPV oncoproteins is reversible. In particular, the inhibition of E6/E7 expression in HPV-positive cancer cells leads to the reconstitution of p53 and pRb signaling and to the induction of cellular senescence [28–32], which is classically defined as an essentially irreversible growth arrest [33].

Consequently, it is widely assumed that HPV-positive cancer cells are "oncogene addicted" [34] in that they must continuously express E6/E7 in order to avoid the re-induction of dormant tumor suppressor pathways, such as p53 and pRb signaling. Notably, however, cervical cancers are often characterized by a heterogeneous distribution of hypoxic and better oxygenated sub-areas, and exhibit a median O_2 concentration of only 1.2%. HNSCCs appear to be only slightly better ventilated, with median O_2 concentrations between 1.3% and 1.9% [1,5,35]. The evidence that hypoxia can profoundly alter tumor cell biology raises the question whether the hypoxic conditions present in sub-regions of HPV-positive cancers may affect the crosstalk between the viral oncogenes and their host cell.

4. Regulation of Senescence in Normoxic and Hypoxic HPV-Positive Cancer Cells

A recent study investigated the effects of chronic hypoxia (1% O_2) on HPV-positive cancer cells [36]. It was observed that viral E6/E7 expression is strongly repressed. Furthermore, in contrast to normoxia where E6/E7 repression results in a strong p53 and p21 upregulation, the downregulation of E6/E7 under hypoxia is not linked to an increase of p53 or p21 levels. Moreover, unlike under normoxic conditions, E6/E7 repression under hypoxia does not result in the induction of senescence in HPV-positive cancer cells. Instead, the cells react with a reversible proliferation stop that can be overcome by reoxygenation. These results indicate that hypoxic HPV-positive cancer cells have the potential to evade the presumed selection pressure to sustain viral oncogene expression, without undergoing senescence.

What is the mechanism underlying the discrepant phenotypic responses following E6/E7 repression in normoxic and hypoxic HPV-positive cancer cells? In normoxic HPV-positive cancer cells, the induction of senescence upon E6/E7 silencing is linked to the reconstitution of the anti-proliferative p53 (and subsequently the p53-mediated stimulation of the negative cell cycle regulator p21) and pRb pathways [30,37]. This is in line with the well-documented pro-senescent potential of p53 and pRb signaling [33]. More recently, it has been found that the activity of the mechanistic target of rapamycin (mTOR) signaling cascade plays a critical role for the induction of senescence in many tumor cell models [38,39]. This is also the case for normoxic HPV-positive cancer cells, since treatment with chemical inhibitors of mTOR signaling, such as rapamycin or KU-0063794, allows the cells to evade senescence under conditions of efficient *E6/E7* silencing by RNA interference (RNAi) [36]. Collectively, these observations could be accommodated into a model [39] according to which the induction of senescence basically requires two major events: (i) a proliferative arrest (e.g., through the activation of the p53/p21 and/or pRb pathways) in the presence of conflicting growth promoting stimuli (e.g., through active mTOR signaling); and (ii) the conversion ("geroconversion") of the reversible proliferation arrest to senescence, a process which can be driven by mTOR signaling.

The anti-senescent effects of the mTOR inhibitors in normoxic HPV-positive cancer cells also indicate that the mTOR pathway remains active in these cells despite strong E6/E7 downregulation. In line with this, and indicative for sustained mTOR signaling, the amounts of phosphorylated mTOR downstream targets are not (phospho-S6, phospho-p70-S6 kinase) or are only partially (phospho-4E-binding protein 1) reduced when *E6/E7* is silenced in HPV-positive cancer cells. This observation is also remarkable in the light of previous studies, which show that the HPV E6 protein stimulates mTOR signaling upon ectopic expression [40,41], including in primary human keratinocytes, the natural target cells for HPV infections [41]. It thus will be interesting to determine how mTOR signaling is maintained in normoxic cervical cancer cell lines when the endogenous E6/E7 expression is silenced.

Importantly, the mTOR pathway is strongly repressed under hypoxia in many cell systems [42], including cervical cancer cells [36]. This impairment of mTOR signaling is associated with the hypoxia-induced stimulation of REDD1 (regulated in development and DNA damage response 1) expression, which activates the TSC2 (tuberous sclerosis complex 2) protein, a negative regulator of mTOR signaling [43]. Blocking *REDD1* or *TSC2* expression by RNAi leads to the stimulation of mTOR signaling and to the emergence of senescent HPV-positive cancer cells under hypoxia [36]. Collectively, these data argue that the hypoxic impairment of mTOR signaling, which occurs at least in part via the inhibitory REDD1/TSC2 axis, enables HPV-positive cancer cells to evade senescence under conditions of E6/E7 repression.

5. HPV-Positive Cancer Cells under Hypoxia: Clinical Implications

Besides surgical excision, RT and radiochemotherapy (RCT) play central roles for the clinical management of HPV-positive cancers, such as cervical carcinomas and HNSCCs. CT alone usually has only limited therapeutic efficacy, and often is applied as a palliative treatment when surgery or RT are not possible, e.g., in patients with recurrent or metastatic disease [44–46]. Notably, both the general effects of tumor hypoxia as well as the specific hypoxic response of HPV-positive cancer cells bear the potential to increase resistance towards all of these treatment strategies.

In particular, hypoxia is considered to be a major obstacle for the therapeutic efficacy of RT, mainly due to the fact that O_2 is required to manifest the DNA lesions that are induced by ionizing radiation [47,48]. As a consequence, hypoxic cells exhibit an approximately three-fold increase in their resistance towards RT compared with well-oxygenated cells [47,48]. The presence of hypoxic sub-regions in tumors can also cause resistance against CT, for example (i) as a consequence of the limited perfusion-dependent delivery of chemotherapeutic drugs in these areas; (ii) through the induction of genes that can protect tumor cells against CT, such as the *MDR1* (multidrug resistance 1) gene [49]; and (iii) through a hypoxia-linked inhibition of cancer cell proliferation, since many chemotherapeutic agents are preferentially active against dividing cells [50]. Thus, the proliferative stop of HPV-positive cancer cells that is observed under hypoxia could contribute to their resistance towards CT. It further should be noted that CT exerts its anti-tumorigenic properties not only by eliminating cancer cells via apoptosis, which could be attenuated by the anti-apoptotic effects of hypoxia [12], but also by inducing cellular senescence [51–53]. However, this latter activity could be counteracted by the hypoxia-induced impairment of mTOR signaling, which not only protects hypoxic HPV-positive cancer cells from the pro-senescent effects of E6/E7 inhibition, but also from pro-senescent CT [36].

Besides these classical treatment regimens, there are intense efforts to develop novel therapeutic strategies targeting E6/E7 in HPV-positive cancers. The theoretical basis for most of these approaches is the conception that the HPV oncoproteins are regularly expressed in HPV-positive cancer cells and that they play a crucial role in the maintenance of their malignant phenotype. Accordingly, antigens derived from the viral oncoproteins are believed to represent attractive targets for immunotherapy of cervical cancer or HPV-positive HNSCCs, since HPV-positive cancer cells would be unable to downregulate E6/E7 expression as an immune evasion mechanism [54,55]. However, this could be an oversimplified view, since hypoxic HPV-positive cancer cells can efficiently shut down E6/E7 expression [36], and consequently repress viral antigen production. Moreover, the hypoxic microenvironment itself is known to exert immunosuppressive effects by affecting the activities of various types of immune cells. These include myeloid-derived suppressor cells (MDSCs), regulatory T cells (Tregs), and tumor-associated macrophages (TAMs) that have been linked to the suppression of an effective anti-tumor immune response under hypoxia [56–58]. Thus, both viral oncogene repression and general immunosuppression in hypoxic tumor sub-regions could be major obstacles for the efficacy of immunotherapeutic approaches targeting E6- or E7-derived antigens, which, in most instances, have thus far been rather limited with regard to their therapeutic benefit [54,59].

Furthermore, the rapid and efficient senescence response that is observed upon E6/E7 repression in normoxic HPV-positive cancer cells raises the possibility that inhibitors of E6/E7 expression or function possess therapeutic potential for the treatment of HPV-positive tumors [28–32,60]. Indeed, there is evidence that the induction of senescence in tumor cells ("pro-senescence therapy") holds promise as a therapeutic anticancer strategy [61–63]. However, there are also potential pitfalls associated with this approach. Firstly, some observations suggest that the senescent phenotype may not be completely stable under all conditions. For example, the cellular senescence and concomitant growth arrest induced by RT or CT may not be as irreversible as it is in its classical definition, and rare tumor cells evading from this regulation could grow even more aggressively and exhibit an increased therapeutic resistance [53,64,65]. Secondly, whereas the induction of stable senescence in cancer cells would be therapeutically desirable, cells in the tumor microenvironment (such as stromal fibroblasts) that also senesce in response to CT or RT can acquire a "senescence-associated secretory phenotype (SASP)" [33], leading to the secretion of both anti-tumorigenic and pro-tumorigenic factors [33,66]. Thus, as a result of the pro-tumorigenic potential of the SASP, senescent cells in the tumor microenvironment may actually augment the oncogenicity of cancer cells that have escaped from the pro-senescent or pro-apoptotic effects of CT or RT, by supporting their proliferation and survival and by increasing their metastatic potential [66]. In line with this possibility, mTOR inhibitors that can interfere with the secretion of major components of the SASP have been reported to sensitize tumors towards CT [67,68].

In view of the potentially pro-tumorigenic effects of the SASP of surrounding stromal cells, it thus might be beneficial to induce senescence selectively in tumor cells. In principle, this specificity could be achieved by interfering with E6/E7 expression or function, since these therapeutic targets are not present in normal (HPV-negative) cells. Unfortunately, however, hypoxic HPV-positive cancer cells would be expected to resist a pro-senescence therapy that is based on E6/E7 inhibition, since the expression of the therapeutic targets is blocked and pro-senescent mTOR signaling is impaired [36].

Yet, it should be emphasized that these considerations do not preclude a therapeutic use of prospective E6/E7 inhibitors, since they would be expected to act in a pro-senescent way in non-hypoxic HPV-positive cancer cells where the therapeutic targets are expressed and mTOR signaling is active. The inhibition of E6/E7 could thus be combined with treatment strategies currently under development that aim to attack hypoxic cancer cells [4]. These include agents which interfere with the unfolded protein response (UPR), a mechanism that contributes to tumor cell survival under hypoxia, or substances which block HIF-1α and HIF-2α (hypoxia-induced factors 1α and 2α)-linked signaling pathways that adapt tumor cells to hypoxia, e.g., by inducing metabolic reprogramming, enhancing tumor cell survival, and supporting angiogenesis and metastasis [4,69]. In addition, although their therapeutic efficacy in the clinic has been mostly disappointing thus far, much hope is also put on the development of improved hypoxia-activated prodrugs (HAPs) that are metabolized to cytotoxic agents under hypoxic conditions [70]. Other approaches to target the hypoxic sub-regions of cancers include the application of certain bacteria, such as the oxygen gradient-sensing *Magnetococcus marinus* strain MC-1 that can transport drug-containing nanoliposomes into hypoxic tumor regions [71], or *Clostridium novyi*-NT that spreads to and destructs hypoxic cancer areas upon intratumoral injection [72].

Notably, the phenotype of hypoxic HPV-positive tumor cells may not only provide therapeutic resistance, but could also be a risk factor for tumor recurrence. In particular, by their ability to reinduce cellular proliferation upon reoxygenation, hypoxic HPV-positive cancer cells may serve as a reservoir for tumor regrowth when their oxygen supply is increased. This can occur, for example, through neoangiogenesis [73], or following the therapeutic shrinkage of tumors [74].

6. Conclusions and Perspectives

Hypoxia has the potential to strongly affect the biology of HPV-positive cancers, both through its general effects in tumors as well as through the modulation of the interplay between oncogenic HPVs and their host cell (Figure 2). In the latter context, hypoxic HPV-positive cancer cells can induce a state of dormancy, which is characterized by a shutdown of viral oncogene expression, an impairment of mTOR signaling, and a reversible growth arrest. Hypoxic HPV-positive cancer cells thus can escape from key regulatory principles that have been observed under normoxia, with consequences for both their cellular phenotype and therapeutic susceptibility. However, important questions remain to be solved.

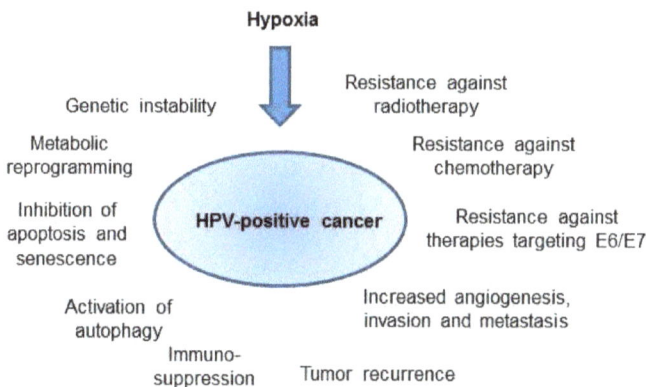

Figure 2. Potential effects of hypoxia on the biology and clinical behavior of human papillomavirus (HPV)-positive cancers. For further details please refer to the text.

Firstly, which molecular mechanisms are responsible for the strong inhibition of *E6/E7* oncogene expression under hypoxia? Secondly, why are the p53 levels not restored in hypoxic HPV-positive cancer cells, although E6 is repressed? Thirdly, how is mTOR signaling maintained in normoxic HPV-positive cancer cells under conditions of *E6/E7* silencing and cellular growth inhibition? Fourthly, since the impairment of mTOR signaling is crucial for the ability of hypoxic HPV-positive cancer cells to evade senescence, will it be possible to identify agents which override this regulatory principle and thereby induce senescence in an mTOR-independent manner? Could they be of therapeutic value in combination with RT, CT, RCT, or E6/E7 inhibition, which preferentially act on non-hypoxic cells? Fifthly, in view of the fact that standard cell culture conditions (20% O_2) do not reflect the O_2 levels in normal cervical tissue (median concentration of 5.5% O_2) [1,5], are there differences in the phenotypic responses listed above between HPV-positive cancer cells that are grown under physoxia instead of normoxia?

Moreover, it should be noted that analyses of the crosstalk between the viral *E6/E7* oncogenes and their host cell thus far have focused on the effects of chronic hypoxia (1% O_2 over days) [36]. However, there is evidence that cycling hypoxia (Figure 1B) may also be highly relevant in terms of increased tumor aggressiveness and enhanced therapeutic resistance [6,7,75]. Cycling hypoxia leads to a huge rise of ROS production, and paradoxically to a strong activation of HIF-1 signaling during the reoxygenation phases [76]. The latter observation is explained by experimental evidence that HIF-1-regulated transcripts are kept in cellular stress granules under hypoxia, which disaggregate upon reoxygenation and allow for the translation of the HIF-1-regulated RNAs [76]. Enhanced HIF-1 signaling and/or increased ROS production affect a broad range of different cancer-linked processes, such as the control of cell proliferation, apoptosis, senescence, cellular metabolism, genetic stability, angiogenesis, and metastasis. This can culminate in more aggressive cancer growth and higher therapeutic resistance (reviewed in [69,77]). Interestingly, in a transgenic mouse model of HPV-induced cervical carcinogenesis, enhanced HIF-1α expression resulted in an increase of tumor cell proliferation and invasion [78]. This finding indicates a possible cooperation between the HPV oncogenes and HIF-1α during cervical cancer progression. It thus should be informative to investigate the effects of cycling hypoxia on the expression of HIF-1α and the HPV oncogenes in HPV-positive cancer cells, and the resulting consequences for their cellular phenotype and therapeutic sensitivity. The further elucidation of the crosstalk between oncogenic HPVs and their host cell under hypoxia may therefore not only increase our current understanding of the molecular mechanisms of HPV-induced carcinogenesis, but also be revealing in regards to the clinical behavior of HPV-positive cancers and their therapeutic resistance.

Acknowledgments: The authors wish to thank the Deutsche Krebshilfe (Grant 112132) and the Wilhelm Sander-Stiftung (Grant 2015.137.1) for funding.

Conflicts of Interest: The authors declare no conflict of interest.

References

1. Vaupel, P.; Mayer, A. Hypoxia in cancer: Significance and impact on clinical outcome. *Cancer Metastasis Rev.* **2007**, *26*, 225–239. [CrossRef] [PubMed]
2. Bertout, J.A.; Patel, S.A.; Simon, M.C. The impact of O_2 availability on human cancer. *Nat. Rev. Cancer* **2008**, *8*, 967–975. [CrossRef] [PubMed]
3. Vaupel, P. Pathophysiology of solid tumors. In *The Impact of Tumor Biology on Cancer Treatment and Multidisciplinary Strategies*; Molls, M., Vaupel, P., Nieder, C., Anscher, M.S., Eds.; Springer: Berlin/Heidelberg, Germany, 2009; pp. 51–92.
4. Wilson, W.R.; Hay, M.P. Targeting hypoxia in cancer therapy. *Nat. Rev. Cancer* **2011**, *11*, 393–410. [CrossRef] [PubMed]
5. McKeown, S.R. Defining normoxia, physoxia and hypoxia in tumours-implications for treatment response. *Br. J. Radiol.* **2014**, *87*, 20130676. [CrossRef] [PubMed]
6. Vaupel, P.; Mayer, A. Hypoxia in tumors: Pathogenesis-related classification, characterization of hypoxia subtypes, and associated biological and clinical implications. *Adv. Exp. Med. Biol.* **2014**, *812*, 19–24. [PubMed]
7. Michiels, C.; Tellier, C.; Feron, O. Cycling hypoxia: A key feature of the tumor microenvironment. *Biochim. Biophys. Acta* **2016**, *1866*, 76–86. [CrossRef] [PubMed]
8. Bristow, R.G.; Hill, R.P. Hypoxia and metabolism. Hypoxia, DNA repair and genetic instability. *Nat. Rev. Cancer* **2008**, *8*, 180–192. [CrossRef] [PubMed]
9. Guzy, R.D.; Hoyos, B.; Robin, E.; Chen, H.; Liu, L.; Mansfield, K.D.; Simon, M.C.; Hammerling, U.; Schumacker, P.T. Mitochondrial complex III is required for hypoxia-induced ROS production and cellular oxygen sensing. *Cell Metab.* **2005**, *1*, 401–408. [CrossRef] [PubMed]

10. Eales, K.L.; Hollinshead, K.E.; Tennant, D.A. Hypoxia and metabolic adaptation of cancer cells. *Oncogenesis* **2016**, *5*, e190. [CrossRef] [PubMed]

11. Nakazawa, M.S.; Keith, B.; Simon, M.C. Oxygen availability and metabolic adaptations. *Nat. Rev. Cancer* **2016**, *16*, 663–673. [CrossRef] [PubMed]

12. Erler, J.T.; Cawthorne, C.J.; Williams, K.J.; Koritzinsky, M.; Wouters, B.G.; Wilson, C.; Miller, C.; Demonacos, C.; Stratford, I.J.; Dive, C. Hypoxia-Mediated down-regulation of Bid and Bax in tumors occurs via hypoxia-inducible factor 1-dependent and -independent mechanisms and contributes to drug resistance. *Mol. Cell. Biol.* **2004**, *24*, 2875–2889. [CrossRef] [PubMed]

13. Leontieva, O.V.; Natarajan, V.; Demidenko, Z.N.; Burdelya, L.G.; Gudkov, A.V.; Blagosklonny, M.V. Hypoxia suppresses conversion from proliferative arrest to cellular senescence. *Proc. Natl. Acad. Sci. USA* **2012**, *109*, 13314–13318. [CrossRef] [PubMed]

14. Rouschop, K.M.; van den Beucken, T.; Dubois, L.; Niessen, H.; Bussink, J.; Savelkouls, K.; Keulers, T.; Mujcic, H.; Landuyt, W.; Voncken, J.W.; et al. The unfolded protein response protects human tumor cells during hypoxia through regulation of the autophagy genes *MAP1LC3B* and *ATG5*. *J. Clin. Investig.* **2010**, *120*, 127–141. [CrossRef] [PubMed]

15. Krock, B.L.; Skuli, N.; Simon, M.C. Hypoxia-induced angiogenesis: Good and evil. *Genes Cancer* **2011**, *2*, 1117–1133. [CrossRef] [PubMed]

16. Rankin, E.B.; Giaccia, A.J. Hypoxic control of metastasis. *Science* **2016**, *352*, 175–180. [CrossRef] [PubMed]

17. Höckel, M.; Vaupel, P. Tumor hypoxia: Definitions and current clinical, biologic, and molecular aspects. *J. Natl. Cancer Inst.* **2001**, *93*, 266–276. [CrossRef] [PubMed]

18. Zur Hausen, H. The search for infectious causes of human cancers: Where and why (Nobel lecture). *Angew. Chem. Int. Ed. Engl.* **2009**, *48*, 5798–5808. [CrossRef] [PubMed]

19. De Martel, C.; Plummer, M.; Vignat, J.; Franceschi, S. Worldwide burden of cancer attributable to HPV by site, country and HPV type. *Int. J. Cancer* **2017**. [CrossRef] [PubMed]

20. American Cancer Society. *Cancer Facts & Figures 2015*; American Cancer Society: Atlanta, GA, USA, 2015.

21. Saslow, D.; Solomon, D.; Lawson, H.W.; Killackey, M.; Kulasingam, S.L.; Cain, J.; Garcia, F.A.; Moriarty, A.T.; Waxman, A.G.; Wilbur, D.C.; et al. American Cancer Society, American Society for Colposcopy and Cervical Pathology, and American Society for Clinical Pathology screening guidelines for the prevention and early detection of cervical cancer. *CA Cancer J. Clin.* **2012**, *62*, 147–172. [CrossRef] [PubMed]

22. Berman, T.A.; Schiller, J.T. Human papillomavirus in cervical cancer and oropharyngeal cancer: One cause, two diseases. *Cancer* **2017**, *123*, 2219–2229. [CrossRef] [PubMed]

23. Martinez-Zapien, D.; Ruiz, F.X.; Poirson, J.; Mitschler, A.; Ramirez, J.; Forster, A.; Cousido-Siah, A.; Masson, M.; Vande Pol, S.; Podjarny, A.; et al. Structure of the E6/E6AP/p53 complex required for HPV-mediated degradation of p53. *Nature* **2016**, *529*, 541–545. [CrossRef] [PubMed]

24. Scheffner, M.; Werness, B.A.; Huibregtse, J.M.; Levine, A.J.; Howley, P.M. The E6 oncoprotein encoded by human papillomavirus types 16 and 18 promotes the degradation of p53. *Cell* **1990**, *63*, 1129–1136. [CrossRef]

25. Roman, A.; Munger, K. The papillomavirus E7 proteins. *Virology* **2013**, *445*, 138–168. [CrossRef] [PubMed]

26. Rivlin, N.; Brosh, R.; Oren, M.; Rotter, V. Mutations in the p53 Tumor Suppressor Gene: Important Milestones at the Various Steps of Tumorigenesis. *Genes Cancer* **2011**, *2*, 466–474. [CrossRef] [PubMed]

27. Dyson, N.J. RB1: A prototype tumor suppressor and an enigma. *Genes Dev.* **2016**, *30*, 1492–1502. [CrossRef] [PubMed]

28. Goodwin, E.C.; DiMaio, D. Repression of human papillomavirus oncogenes in HeLa cervical carcinoma cells causes the orderly reactivation of dormant tumor suppressor pathways. *Proc. Natl. Acad. Sci. USA* **2000**, *97*, 12513–12518. [CrossRef] [PubMed]

29. Goodwin, E.C.; Yang, E.; Lee, C.J.; Lee, H.W.; DiMaio, D.; Hwang, E.S. Rapid induction of senescence in human cervical carcinoma cells. *Proc. Natl. Acad. Sci. USA* **2000**, *97*, 10978–10983. [CrossRef] [PubMed]

30. Wells, S.I.; Francis, D.A.; Karpova, A.Y.; Dowhanick, J.J.; Benson, J.D.; Howley, P.M. Papillomavirus E2 induces senescence in HPV-positive cells via pRB- and p21(CIP)-dependent pathways. *EMBO J.* **2000**, *19*, 5762–5771. [CrossRef] [PubMed]

31. Hall, A.H.; Alexander, K.A. RNA interference of human papillomavirus type 18 E6 and E7 induces senescence in HeLa cells. *J. Virol.* **2003**, *77*, 6066–6069. [CrossRef] [PubMed]

32. Magaldi, T.G.; Almstead, L.L.; Bellone, S.; Prevatt, E.G.; Santin, A.D.; DiMaio, D. Primary human cervical carcinoma cells require human papillomavirus E6 and E7 expression for ongoing proliferation. *Virology* **2012**, *422*, 114–124. [CrossRef] [PubMed]

33. Campisi, J. Aging, cellular senescence, and cancer. *Annu. Rev. Physiol.* **2013**, *75*, 685–705. [CrossRef] [PubMed]

34. Weinstein, I.B.; Joe, A. Oncogene addiction. *Cancer Res.* **2008**, *68*, 3077–3080. [CrossRef] [PubMed]

35. Vaupel, P.; Höckel, M.; Mayer, A. Detection and characterization of tumor hypoxia using pO_2 histography. *Antioxid. Redox Signal.* **2007**, *9*, 1221–1235. [CrossRef] [PubMed]

36. Hoppe-Seyler, K.; Bossler, F.; Lohrey, C.; Bulkescher, J.; Rösl, F.; Jansen, L.; Mayer, A.; Vaupel, P.; Dürst, M.; Hoppe-Seyler, F. Induction of dormancy in hypoxic human papillomavirus-positive cancer cells. *Proc. Natl. Acad. Sci. USA* **2017**, *114*, E990–E998. [CrossRef] [PubMed]

37. DeFilippis, R.A.; Goodwin, E.C.; Wu, L.; DiMaio, D. Endogenous human papillomavirus E6 and E7 proteins differentially regulate proliferation, senescence, and apoptosis in HeLa cervical carcinoma cells. *J. Virol.* **2003**, *77*, 1551–1563. [CrossRef] [PubMed]

38. Xu, S.; Cai, Y.; Wie, Y. mTOR Signaling from Cellular Senescence to Organismal Aging. *Aging Dis.* **2013**, *5*, 263–273.

39. Blagosklonny, M.V. Geroconversion: Irreversible step to cellular senescence. *Cell Cycle* **2014**, *13*, 3628–3635. [CrossRef] [PubMed]

40. Lu, Z.; Hu, X.; Li, Y.; Zheng, L.; Zhou, Y.; Jiang, H.; Ning, T.; Basang, Z.; Zhang, C.; Ke, Y. Human papillomavirus 16 E6 oncoprotein interferences with insulin signaling pathway by binding to tuberin. *J. Biol. Chem.* **2004**, *279*, 35664–35670. [CrossRef] [PubMed]

41. Spangle, J.M.; Münger, K. The human papillomavirus type 16 E6 oncoprotein activates mTORC1 signaling and increases protein synthesis. *J. Virol.* **2010**, *84*, 9398–9407. [CrossRef] [PubMed]

42. Leontieva, O.V.; Blagosklonny, M.V. Hypoxia and gerosuppression: The mTOR saga continues. *Cell Cycle* **2012**, *11*, 3926–3931. [CrossRef] [PubMed]

43. Brugarolas, J.; Lei, K.; Hurley, R.L.; Manning, B.D.; Reiling, J.H.; Hafen, E.; Witters, L.A.; Ellisen, L.W.; Kaelin, W.G., Jr. Regulation of mTOR function in response to hypoxia by REDD1 and the TSC1/TSC2 tumor suppressor complex. *Genes Dev.* **2004**, *18*, 2893–2904. [CrossRef] [PubMed]

44. Colombo, N.; Carinelli, S.; Colombo, A.; Marini, C.; Rollo, D.; Sessa, C.; ESMO Guidelines Working Group. Cervical cancer: ESMO Clinical Practice Guidelines for diagnosis, treatment and follow-up. *Ann. Oncol.* **2012**, *23* (Suppl. 7), vii27–vii32. [PubMed]

45. Sacco, A.G.; Cohen, E.E. Current Treatment Options for Recurrent or Metastatic Head and Neck Squamous Cell Carcinoma. *J. Clin. Oncol.* **2015**, *33*, 3305–3313. [CrossRef] [PubMed]

46. Boussios, S.; Seraj, E.; Zarkavelis, G.; Petrakis, D.; Kollas, A.; Kafantari, A.; Assi, A.; Tatsi, K.; Pavlidis, N.; Pentheroudakis, G. Management of patients with recurrent/advanced cervical cancer beyond first line platinum regimens: Where do we stand? A literature review. *Crit. Rev. Oncol. Hematol.* **2016**, *108*, 164–174. [CrossRef] [PubMed]

47. Hill, R.P.; Bristow, R.G.; Fyles, A.; Koritzinsky, M.; Milosevic, M.; Wouters, B.G. Hypoxia and Predicting Radiation Response. *Semin. Radiat. Oncol.* **2015**, *25*, 260–272. [CrossRef] [PubMed]

48. Overgaard, J. Hypoxic radiosensitization: adored and ignored. *J. Clin. Oncol.* **2007**, *25*, 4066–4074. [CrossRef] [PubMed]

49. Comerford, K.M.; Wallace, T.J.; Karhausen, J.; Louis, N.A.; Montalto, M.C.; Colgan, S.P. Hypoxia-inducible factor-1-dependent regulation of the multidrug resistance (*MDR1*) gene. *Cancer Res.* **2002**, *62*, 3387–3394. [PubMed]

50. Trédan, O.; Galmarini, C.M.; Patel, K.; Tannock, I.F. Drug resistance and the solid tumor microenvironment. *J. Natl. Cancer Inst.* **2007**, *99*, 1441–1454. [CrossRef] [PubMed]

51. Roninson, I.B. Tumor cell senescence in cancer treatment. *Cancer Res.* **2003**, *63*, 2705–2715. [PubMed]

52. Ewald, J.A.; Desotelle, J.A.; Wilding, G.; Jarrard, D.F. Therapy-induced senescence in cancer. *J. Natl. Cancer Inst.* **2010**, *102*, 1536–1546. [CrossRef] [PubMed]

53. Wu, P.C.; Wang, Q.; Grobman, L.; Chu, E.; Wu, D.Y. Accelerated cellular senescence in solid tumor therapy. *Exp. Oncol.* **2012**, *34*, 298–305. [PubMed]

54. Stern, P.L.; van der Burg, S.H.; Hampson, I.N.; Broker, T.R.; Fiander, A.; Lacey, C.J.; Kitchener, H.C.; Einstein, M.H. Therapy of human papillomavirus-related disease. *Vaccine* **2012**, *30*, 71–82. [CrossRef] [PubMed]

55. Nizard, M.; Sandoval, F.; Badoual, C.; Pere, H.; Terme, M.; Hans, S.; Benhamouda, N.; Granier, C.; Brasnu, D.; Tartour, E. Immunotherapy of HPV-associated head and neck cancer: Critical parameters. *Oncoimmunology* **2013**, *2*, e24534. [CrossRef] [PubMed]

56. Kumar, V.; Gabrilovich, D.I. Hypoxia-inducible factors in regulation of immune responses in tumour microenvironment. *Immunology* **2014**, *143*, C512–C519. [CrossRef] [PubMed]

57. Noman, M.Z.; Hasmim, M.; Messai, Y.; Terry, S.; Kieda, C.; Janji, B.; Chouaib, S. Hypoxia: A key player in antitumor immune response. A Review in the Theme: Cellular Responses to Hypoxia. *Am. J. Physiol. Cell. Physiol.* **2015**, *309*, 569–579. [CrossRef] [PubMed]

58. McDonald, P.C.; Chafe, S.C.; Dedhar, S. Overcoming Hypoxia-Mediated Tumor Progression: Combinatorial Approaches Targeting pH Regulation, Angiogenesis and Immune Dysfunction. *Front. Cell. Dev. Biol.* **2016**, *4*, 27. [CrossRef] [PubMed]

59. Skeate, J.G.; Woodham, A.W.; Einstein, M.H.; Da Silva, D.M.; Kast, W.M. Current therapeutic vaccination and immunotherapy strategies for HPV-related diseases. *Hum. Vaccines Immunother.* **2016**, *12*, 1418–1429. [CrossRef] [PubMed]

60. Honegger, A.; Schilling, D.; Bastian, S.; Sponagel, J.; Kuryshev, V.; Sültmann, H.; Scheffner, M.; Hoppe-Seyler, K.; Hoppe-Seyler, F. Dependence of intracellular and exosomal microRNAs on viral E6/E7 oncogene expression in HPV-positive tumor cells. *PLoS Pathog.* **2015**, *11*, e1004712. [CrossRef] [PubMed]

61. Nardella, C.; Clohessy, J.G.; Alimonti, A.; Pandolfi, P.P. Pro-senescence therapy for cancer treatment. *Nat. Rev. Cancer* **2011**, *11*, 503–511. [CrossRef] [PubMed]

62. Acosta, J.C.; Gil, J. Senescence: A new weapon for cancer therapy. *Trends Cell Biol.* **2012**, *22*, 211–219. [CrossRef] [PubMed]

63. Calcinotto, A.; Alimonti, A. Aging tumour cells to cure cancer: "pro-senescence" therapy for cancer. *Swiss Med. Wkly.* **2017**, *147*, w14367. [CrossRef] [PubMed]

64. Kahlem, P.; Dörken, B.; Schmitt, C.A. Cellular senescence in cancer treatment: Friend or foe? *J. Clin. Investig.* **2004**, *113*, 169–174. [CrossRef] [PubMed]

65. Gordon, R.R.; Nelson, P.S. Cellular senescence and cancer chemotherapy resistance. *Drug Resist. Updates* **2012**, *15*, 123–131. [CrossRef] [PubMed]

66. Rao, S.G.; Jackson, J.G. SASP: Tumor Suppressor or Promoter? Yes! *Trends Cancer* **2016**, *2*, 676–687. [CrossRef]

67. Herranz, N.; Gallage, S.; Mellone, M.; Wuestefeld, T.; Klotz, S.; Hanley, C.J.; Raguz, S.; Acosta, J.C.; Innes, A.J.; Banito, A.; et al. mTOR regulates MAPKAPK2 translation to control the senescence-associated secretory phenotype. *Nat. Cell Biol.* **2015**, *17*, 1205–1217. [CrossRef] [PubMed]

68. Laberge, R.M.; Sun, Y.; Orjalo, A.V.; Patil, C.K.; Freund, A.; Zhou, L.; Curran, S.C.; Davalos, A.R.; Wilson-Edell, K.A.; Liu, S.; et al. mTOR regulates the pro-tumorigenic senescence-associated secretory phenotype by promoting IL1α translation. *Nat. Cell Biol.* **2015**, *17*, 1049–1061. [CrossRef] [PubMed]

69. Semenza, G.L. Oxygen sensing, hypoxia-inducible factors, and disease pathophysiology. *Annu. Rev. Pathol.* **2014**, *9*, 47–71. [CrossRef] [PubMed]

70. Baran, N.; Konopleva, M. Molecular Pathways: Hypoxia-Activated Prodrugs in Cancer Therapy. *Clin. Cancer Res.* **2017**. [CrossRef] [PubMed]

71. Felfoul, O.; Mohammadi, M.; Taherkhani, S.; de Lanauze, D.; Xu, Y.Z.; Loghin, D.; Essa, S.; Jancik, S.; Houle, D.; Lafleur, M.; et al. Magneto-aerotactic bacteria deliver drug-containing nanoliposomes to tumour hypoxic regions. *Nat. Nanotechnol.* **2016**, *11*, 941–947. [CrossRef] [PubMed]

72. Roberts, N.J.; Zhang, L.; Janku, F.; Collins, A.; Bai, R.Y.; Staedtke, V.; Rusk, A.W.; Tung, D.; Miller, M.; Roix, J.; et al. Intratumoral injection of *Clostridium novyi*-NT spores induces antitumor responses. *Sci. Transl. Med.* **2014**, *6*, 249ra111. [CrossRef] [PubMed]

73. Carmeliet, P.; Jain, R.K. Angiogenesis in cancer and other diseases. *Nature* **2000**, *407*, 249–257. [CrossRef] [PubMed]

74. Seiwert, T.Y.; Salama, J.K.; Vokes, E.E. The concurrent chemoradiation paradigm–general principles. *Nat. Clin. Pract. Oncol.* **2007**, *4*, 86–100. [CrossRef] [PubMed]

75. Bayer, C.; Vaupel, P. Acute versus chronic hypoxia in tumors: Controversial data concerning time frames and biological consequences. *Strahlenther. Onkol.* **2012**, *188*, 616–627. [CrossRef] [PubMed]

76. Moeller, B.J.; Cao, Y.; Li, C.Y.; Dewhirst, M.W. Radiation activates HIF-1 to regulate vascular radiosensitivity in tumors: Role of reoxygenation, free radicals, and stress granules. *Cancer Cell* **2004**, *5*, 429–441. [CrossRef]
77. Reczek, C.R.; Chandel, N.S. The two faces of reactive oxygen species in cancer. *Ann. Rev. Cancer Biol.* **2017**, *1*, 79–98. [CrossRef]
78. Lu, Z.H.; Wright, J.D.; Belt, B.; Cardiff, R.D.; Arbeit, J.M. Hypoxia-inducible factor-1 facilitates cervical cancer progression in human papillomavirus type 16 transgenic mice. *Am. J. Pathol.* **2007**, *171*, 667–681. [CrossRef] [PubMed]

Review

Targeting Persistent Human Papillomavirus Infection

Srinidhi Shanmugasundaram and Jianxin You * [ORCID]

Department of Microbiology, Perelman School of Medicine, University of Pennsylvania, Philadelphia, PA 19104, USA; sshanmu@sas.upenn.edu
* Correspondence: jianyou@mail.med.upenn.edu; Tel.: +1-215-573-6781

Academic Editor: Alison A. McBride
Received: 30 July 2017; Accepted: 15 August 2017; Published: 18 August 2017

Abstract: While the majority of Human papillomavirus (HPV) infections are transient and cleared within a couple of years following exposure, 10–20% of infections persist latently, leading to disease progression and, ultimately, various forms of invasive cancer. Despite the clinical efficiency of recently developed multivalent prophylactic HPV vaccines, these preventive measures are not effective against pre-existing infection. Additionally, considering that the burden associated with HPV is greatest in regions with limited access to preventative vaccination, the development of effective therapies targeting persistent infection remains imperative. This review discusses not only the mechanisms underlying persistent HPV infection, but also the promise of immunomodulatory therapeutic vaccines and small-molecular inhibitors, which aim to augment the host immune response against the viral infection as well as obstruct critical viral–host interactions.

Keywords: HPV; persistent infection; cervical cancer; therapeutics; vaccines; episome maintenance; E2 protein

1. Persistent HPV Infection

Human papillomavirus (HPV) is a small, double-stranded DNA virus with a genome consisting of approximately 8000 base pairs. The HPV genome encodes six early genes (*E1, E2, E4, E5, E6,* and *E7*), two late genes (*L1* and *L2*), along with a non-coding region (Figure 1). Among the early genes, *E6* and *E7* are of particular significance due to their roles in inactivation of host tumor-suppressor genes and oncogenic progression [1]. The other early genes play critical roles in viral replication, transcriptional regulation, and viral genome maintenance—all necessary processes for sustaining persistent HPV infection [2].

It has been established that persistent infection with HPV is associated with cervical, anogenital, as well as head and neck cancers [3,4]. In the majority of infected individuals, HPV infection is cleared by the immune system within a couple years of onset; however, the viral infection can continue to persist latently in a subset of the population (Figure 2). These patients with persistent HPV infection have an increased chance of acquiring epithelial cell abnormalities and subsequently developing cancers at the site of infection [5,6]. Though such progression to cancer is relatively rare, the prevalence of the virus among the general population makes HPV-associated persistent infection a statistically significant affliction.

Risk factors that may prevent the natural clearance of HPV persistent infection in certain populations have been a major source of interest. Several studies have discovered that genetic and lifestyle factors can significantly increase the probability of developing persistent infection [7,8]. For instance, multiple studies have found both smoking and alcohol use to be significant risk factors of persistent oral and genital HPV infection [7,9,10]. It has been proposed that the carcinogens in cigarette smoke increase viral load as well as the likelihood of cancerous transformation of the epithelial cells infected with HPV [11,12].

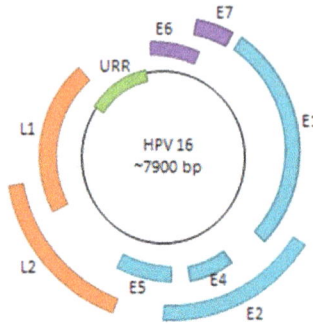

Figure 1. HPV Genome. The HPV genome consists of six early genes (*E1*, *E2*, *E4*, *E5*, *E6*, and *E7*) and two late genes (*L1* and *L2*). Many of the early genes are implicated in viral replication, transcriptional regulation, genome maintenance, along with immune system evasion. *E6* and *E7* are of particular interest as they are viral oncogenes that bind to and inactivate p53 and pRB, respectively. The URR (upstream regulatory region) consists of various promoter and enhancer elements as well as the viral origin of replication (ori).

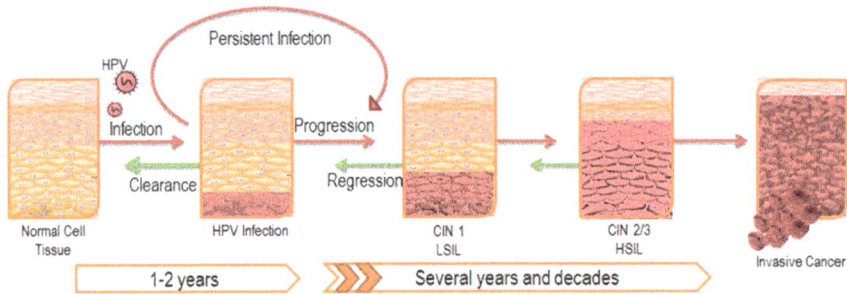

Figure 2. Progression of HPV Infection and Associated Disease. HPV typically establishes infection in the basal epithelial layer. A majority of these infections are transient and are cleared by the immune system within a couple of years. However, 10–20% of infections persist latently, leading to disease progression as illustrated by the red arrows. The lesion that develops as a result is also known as a central intraepithelial neoplasia (CIN) and is classified according to its severity. Eventually, low-grade squamous intraepithelial lesions (LSIL) advance to high-grade squamous intraepithelial lesions (HSIL), ultimately leading to invasive carcinoma. Despite tumor regression in response to initial treatment as illustrated by the green arrows, most cases of latent infection prevent complete clearance of the viral infection, and eventually results in lesion reoccurrence.

Interestingly, several genetic risk factors that predispose an individual to persistent HPV infection have also been identified, although the association is not particularly strong. The human leukocyte antigen (HLA) is one such genetic marker, of which certain alleles seem to have a more prominent association with an inability to clear HPV infection and the subsequent development of cervical cancer [13–15]. Given the variation in immunogenic profiles and associated risks among distinct ethnic groups, one study suggests that further investigation into these genetic markers for each population should be done in order to identify patients at an elevated risk for HPV persistence and to provide comprehensive preventative care accordingly [16].

Given the prevalence of infection with more than one type of HPV among patients, co-infection with multiple HPV types was investigated as a potential predictor of subsequent persistent infection. Results suggest that previous infection with HPV increases the chances of acquiring another

HPV infection [17,18]. However, it is not definitive whether HPV persistence is dependent on co-infection [18,19]. Additionally, it has been found that variants within the specific HPV type may predispose an individual to persistent infection as well. For instance, one recent study discovered that three of the six HPV 16 E6 variants were associated with persistent infection; furthermore, of the nine HPV 16 E2 variants, two were linked to persistent infection [20]. While it is unclear at the time how these mutations mechanistically affect HPV persistence, it has been proposed that they may be linked to the virus' ability to evade the immune system. In summary, studies seem to indicate that various risk factors may have contributing roles in HPV persistence among a small subset of infected patients [6,8,21]. However, additional investigation of the impact of these risk factors on the host immune system may paint a clearer picture of the development of persistent infection overall.

2. The Impact of Persistent Infection on Cancer

Approximately 95% of cervical cancer biopsies contain HPV viral genomes [1,22,23]. With cervical cancers being the second most common cancer among women globally, HPV is a significant infectious carcinogen that necessitates further investigation.

HPVs are classified into two major subcategories depending on the site of primary infection. Alpha HPVs generally infect genital epithelia and are further designated as high-risk and low-risk depending on their ability to induce cancer [24]. High-risk HPVs that are most frequently associated with malignant genital cancers include HPV 16, 18, 31, 33, and 45; on the contrary, low-risk HPVs, such as HPV 6 and 11, are mostly associated with benign papillomas at the site of infection [22,25]. Although more than one hundred types of HPV have been identified, two high-risk types, HPV 16 and HPV 18, are responsible for roughly 70% of cervical cancer cases [26].

In addition to being associated with cervical and anogenital cancers, persistent HPV infection has also been linked to head and neck cancers [4,27,28]. Several studies have additionally observed that women with cervical cancer had a greater risk of subsequently developing oral cancer [29]. These findings along with several similar studies established the presence of persistent HPV infection as a notable precursor to many genital as well as oropharyngeal cancers [3,30,31].

In cases of persistent HPV infection, an environment of genomic instability increases the likelihood of viral genome integration into the host genome [32]. In approximately 72% of cell samples taken from cervical carcinoma biopsies, HPV16 was integrated into the host genome [33,34]. However, the presence of carcinomas with solely episomal HPV and no detectable HPV integration implies that the progression to carcinoma does not necessitate integration of the papillomavirus genome [33]. Nonetheless, it is interesting to note that gene expression and DNA methylation patterns differ between cancers with integrated and non-integrated HPV genomes, suggesting that distinct oncogenic mechanisms may play a role in each setting [35].

A proposed consequence of viral genome integration that can result in oncogenic progression is the disruption of the *E2* open reading frame (ORF) [34]. Normally, the HPV E2 protein plays a critical role in regulating the activation and repression of viral promoters [36]. It was initially proposed that E2, by binding to sites proximal to the *E6/E7* promoter, is able to displace other transcriptional factors and thus prevent the formation of a transcription initiation complex [36,37]. The loss of *E2* expression is associated with the oncogenic progression of HPV as it consequently deregulates and increases the expression of viral oncogenic proteins, E6 and E7, which are known to disrupt tumor-suppressor genes *p53* and *pRB*, respectively [38,39]. It has also been shown that E6/E7 mRNA transcripts derived from integrated HPV 16 DNA display increased levels of stability, which in part contributes to the elevated steady-state levels of E6/E7 mRNA found in cervical cancers [40]. Integration of HPV may also contribute to cancer progression by interrupting the expression and function of key host cellular genes, thus promoting genomic instability [32,34]. Ultimately, it is clear that the progression to cancer from persistent infection is rather varied and can be affected by both the occurrence and location of integration.

3. Molecular Mechanisms Underlying Persistent HPV Infection

3.1. Viral Life Cycle and Immune System Evasion

After escaping the initial immune response, viruses must maintain their genomes within the host nucleus in order to achieve persistent infection. While integration into the host genome is an option favored by many chronic viruses, papillomaviruses like HPV maintain their genomes as extrachromosomal episomes that tether to host DNA [41]. By continuously replicating at low levels in a differentiating tissue, such as the basal epithelium, the papillomavirus is able to maintain a reserve in the host while simultaneously avoiding detection by the immune system [5,42]. During this stage of the infectious cycle, known as maintenance replication, viral genomes are able to partition themselves into the newly formed daughter cells by coordinating their replication with that of the host cell. Later in the infectious cycle, the virus enters a stage of vegetative amplification in which it replicates high levels of genomic products that are fated to be assembled into complete viral particles. This last stage tends to occur in terminally differentiating cell tissues, such as the upper epidermal layers that are destined to be sloughed off and thus are not strictly monitored by the host immune system. Higher levels of viral replication and assembly are observed in these layers because they tend not to trigger an immune response [42].

3.2. Viral Episome Hitchhiking on Host Mitotic Chromosomes

In order to ensure the viral genome is not lost during cell division, there is a tethering mechanism in place that attaches the viral genome to host mitotic chromosomes through protein intermediates [43]. Tethering viral genomes to mitotic chromosomes in dividing host cells is a common strategy used by many persistent DNA viruses, such as Epstein–Barr virus and Kaposi's sarcoma-associated herpesvirus to name a few [44–46]. In the early stages of infection, the HPV E2 protein plays the major role in establishing persistence by tethering viral episomes to host mitotic chromosomes [47]. E2 is a multifunctional protein that is critical for supporting HPV infection. Its roles in viral transcription and replication have been extensively studied and reviewed [48,49]. In order to tether viral episomes to mitotic chromosomes, E2 binds to specific sites in the HPV episome using its C-terminal DNA-binding domain while interacting with chromosome-associated proteins through its N-terminal domain [50]. Unlike many cellular proteins which transiently bind chromosomes during mitosis, bovine papillomavirus (BPV) E2 was found to be bound to the chromosomes throughout all stages of mitosis [41]. Such a mechanism ensures that the viral genome is maintained in the nucleus of the daughter cell following cell division.

Interestingly, one study demonstrated that E2 protein encoded by HPV 11, 16 and 18 has the property to directly interact with the mitotic spindles, thereby maintaining HPV ori-containing DNA as "mini-chromosomes" in dividing cells. The likely region of HPV E2 involved in this mitotic spindle interaction is thought to be a 14-amino acid sequence that is highly divergent from its analogous sequence in BPV 1 E2 [51]. It is likewise important to note that, as indicated by the varying levels of viral genome copies per cell, segregation of the viral episomes is not an extremely specific process [43,47]. This suggests that viral genomes most likely associate randomly with host chromosomes and/or mitotic spindles during mitosis as tethered passengers. Since the levels of E2 also vary to a similar extent between cells, it is predicted that the E2 protein levels may correlate with the number of viral genome replicates in different cells [43]. Nevertheless, while it is well established that E2 is critical for episomal maintenance, it is not the sole player in episomal tethering. Interactions with host cellular proteins, such as those described below, are integral for maintaining persistent infection.

3.3. E2 Interaction with Host Receptors

Because of the critical role E2 plays in maintaining persistent infection, its interactions with host proteins are of particular interest. Though nearly all papillomaviruses express E2 and maintain their genomes through episomal tethering to mitotic chromatin, the host proteins targeted by the viral E2 protein are largely dependent upon the specific papillomavirus type [47].

An early proteomic study identified Bromodomain-containing protein 4 (BRD4), also known as mitotic chromosome-associated protein (MCAP), as a critical binding partner for E2 [52]. BRD4 is a member of a large family of proteins known as Bromo- and Extra-Terminal (BET) proteins which interact with acetylated histones in chromatin and function as "readers" of the histone code [53,54]. It is thought that BRD4 may have a "post-it note" function by associating with and marking specific segments of the mitotic chromatin in order to pass on epigenetic information to daughter cells. You et al. observed co-localization of E2 and BRD4 on condensed mitotic chromosomes and determined that the E2–BRD4 complex plays a role in BPV E2-mediated viral episome segregation [52]. X-ray crystallography experiments have clarified the specific binding between the N-terminal transactivation domain of HPV E2 and a highly conserved region of the C-terminal domain of BRD4 [55]. Furthermore, mutagenesis studies showed that deletions and mutations in the E2 N-terminal domain negatively impact E2–BRD4 binding as well as the chromosomal localization of E2 [56,57]. Additional studies have demonstrated that treating cells maintaining PV episomes with merely the C-terminal domain of BRD4 interferes with E2's ability to tether viral episomes to mitotic chromosomes by competitively inhibiting its interaction with the functional full-length, chromosome-associated BRD4 [52,55]. Many, but not all, papillomaviruses seem to rely on BRD4 in order to maintain episomal tethering throughout mitosis [58].

In addition to episome maintenance, it has been demonstrated that BRD4 is vital for E2's ability to activate transcription in all papillomaviruses [58–60]. Additional studies examining the E2–BRD4 interaction have shown that BRD4 may play a role in viral DNA replication as well [61–63]. In summary, the E2–BRD4 interaction plays a critical role in multiple stages of the HPV life cycle and therefore represents an excellent target for developing anti-viral therapeutics to terminate persistent infection.

The diversity within the papillomavirus clade supports the ability of its viruses to interact with various host cellular partners. Researchers have shown that α and β HPV E2 proteins likely interact with cellular targets in addition to BRD4 and furthermore, bind to distinct regions of the host chromosome [47,64]. In a study by Oliveira et al., E2 proteins from four α papillomaviruses have been shown to interact with host chromosomes in a temporally differential manner [47]. Possible factors that may contribute to the interaction between E2 and host chromatin include DNA helicase, TopBP1, and ChIR1 [64,65]. It was shown that HPV16 E2 co-localizes with TopBP1 possibly implying that TopBP1 is the cellular chromatin-associated receptor for E2 [66]. TopBP1 is a cellular protein closely regulated by the cell cycle, playing a role in initiation of cellular DNA replication, mitotic progression, as well as viral DNA replication. Experimentally, it has been demonstrated that the interaction between HPV16 E2 and TopBP1 contributes to viral DNA replication and episomal genome maintenance [64,67].

In addition to TopBP1, ChIR1 has been shown to be a key cellular partner of E2. Parish et al. demonstrated through intracellular co-localization experiments that HPV 11 and BPV 1 E2s co-localize with ChIR1 on the chromosome during prophase [65,68]. While ChIR1 migrates to the spindle poles following the transition to metaphase however, the E2 proteins from both viruses remain bound to the chromosome throughout mitosis. It has therefore been suggested that ChIR1 is essential for loading E2 onto mitotic chromosomes [65]. Since it has been shown that ChIR1 dissociates from the E2 mitotic foci at the conclusion of prophase, it is still likely that E2 relies on other host proteins such as TopBP1 or BRD4 to maintain its attachment to the chromosome; once again, the specific protein interactions may vary to an extent between papillomavirus types [47]. In summary, the E2–host protein interactions that support mechanisms of episome maintenance have revealed novel therapeutic targets for eliminating persistent infection.

4. Current Therapeutic Strategies for Clearing Persistent HPV Infection

4.1. HPV Prophylactic Vaccines

The development of prophylactic vaccines, such as Gardasil which targets HPV 6, 11, 16 and 18, marked a turning point in the HPV field as it provided an efficient preventative public health measure against HPV [21]. However, since immunological protection is limited to specific HPV types, Gardasil does not offer universal protection against HPV infection nor is it effective as a treatment for existing HPV infection. Nevertheless, because Gardasil proves highly effective against the most prevalent types of HPV, it has significantly reduced the overall statistical toll of the associated disease [69].

National two-dose vaccine programs remain both cost-effective and successful in preventing persistent HPV infection [70]. In recent years, a nine-valent HPV vaccine (HPV 6/11/16/18/31/33/45/52/58) that provides protection against approximately 90% of HPV-associated cancers and other diseases has been licensed by the FDA for use in young adults [71]. Despite the improving range and efficacy of these vaccines, HPV screening will remain a vital tool especially for at-risk women who have not been vaccinated or women in which the vaccine was infective [69,70,72]. It is also important to note that the HPV vaccines currently in use do not provide protection against all types of HPVs that are associated with cervical and other human cancers, thereby justifying the need for continual screening and development of additional therapeutic options to resolve cases post-infection.

4.2. The Promise of Therapeutic Vaccines

Currently, clinical studies are also being conducted on potential therapeutic vaccines, which, unlike prophylactic vaccines, fight HPV infection via immunotherapy [73]. Most cases of HPV infection tend to be cleared by the immune system without intervention 1–2 years post-exposure; it is thought that persistent infection is most likely due to a lack of HPV-specific T-cell immunity [74]. Studies show that HPV-induced diseases indeed correlate with a weak HPV-specific CD4+ and CD8+ T-cell response [75]. Unlike prophylactic vaccines which rely on inducing specific antibodies and memory B-cells [76], therapeutic vaccines attempt to bolster HPV T-cell adaptive immunity. This is achieved through priming naïve T-cells to produce cytotoxic T lymphocytes (CTLs) that target HPV-infected cells, generating CD4+ T-cells to produce the necessary cytokines, and strengthening antigen-presenting cells (APCs). Dendritic cells are an important subset of APCs, which are involved in capturing and presenting antigens to T-cells and have been a central focus of many therapeutic vaccines [77,78] (Figure 3).

Numerous DNA-based vaccines have been developed to target persistent HPV infection and are currently in various stages of clinical study. These vaccines function by introducing a significant amount of viral DNA intradermally or intramuscularly to myocytes, which then express the antigen encoded by the DNA. The expressed, secreted antigen is then recognized and engulfed by dendritic cells and is subsequently expressed on MHC complexes, which are presented to CD8+ T-cells [77,78] (Figure 3A). GX-188E is one such recently developed therapeutic DNA vaccine that is designed to express E6 and E7 fusion proteins in order to increase the presentation of these HPV antigens by dendritic cells. The arrangement of the *E6* and *E7* genes within the DNA vaccine is shuffled to render the recombinant proteins incapable of degrading p53 and pRb [74]. GX-188E seems to be a promising therapeutic due to its demonstrated ability to induce an E6/E7-specific T-cell immune response in patients with high-grade lesions. Specifically, a polyfunctional CD8+ T-cell response was associated with clinical clearance of HPV as well as complete regression of lesions in seven of the nine patients who participated in the study [74]. However, despite the effectiveness demonstrated in this study, a greater sample size might be necessary to more thoroughly evaluate GX-188E's response rate.

Electroporation, alongside DNA vaccination, has been shown to augment vaccine efficiency by modifying the cell membrane to increase DNA uptake and promoting inflammation and recruitment of APC's at the site of vaccination [77]. In vitro studies have demonstrated that one method to stimulate a more effective HPV E6/E7 specific cytotoxic T lymphocyte response is to introduce HPV DNA through an oligomannose liposome (OML) as opposed to a standard liposome [79]. Further in vivo studies will be needed to evaluate the potential of OML-HPV to function as an effective therapeutic vaccine. Nevertheless, among the various classes of therapeutic vaccines, DNA- and protein-based HPV vaccines seem to be the less efficient, likely because they fail to produce a sufficient initial immune response [75,80] (Figure 3C). Adjuvants, such as imiquimod or cidofavir, which serve as agonists to various toll-like receptors, are necessary alongside these vaccines to augment the initial immune response to the vaccine and imbue long-lasting protection [73].

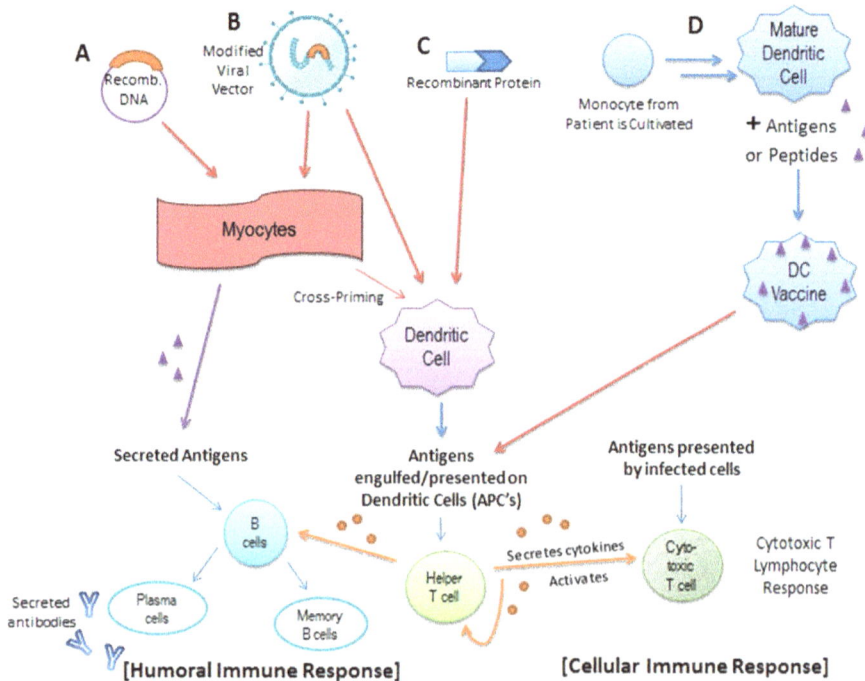

Figure 3. Activation of the Humoral and Cellular Immune Response by Various Forms of Therapeutic Vaccination. Therapeutic vaccines aim to strengthen and broaden the immune response to HPV by introducing specific antigens to a subset of antigen-presenting cells known as dendritic cells. (**A**) Recombinant DNA in the form of a plasmid encoding various antigens may be introduced intramuscularly to myocytes where they are transcribed and translated into antigen proteins, which are subsequently engulfed, processed, and presented by dendritic cells to downstream T-cells. The secreted antigens may also interact with B-cells to initiate a humoral immune response; (**B**) Recombinant DNA can be introduced in the form of a modified viral vector as well (i.e., Influenza A, Modified Vaccinia Ankara, etc.); (**C**) Additionally, the protein products can be introduced directly to the dendritic cells; in the case of HPV, recombinant protein vaccines usually consist of an E6/E7 fusion protein; (**D**) Dendritic whole-cell based vaccines are constructed from cultivated monocytes derived from the patients in combination with a particular antigen or peptide. These modified dendritic cells are then introduced to the patient's immune system and similar humoral and cellular immune responses are incited through the activation of helper T-cells, B-cells, and cytotoxic T-cells.

To counter this problem of low vaccine efficacy, new models of therapeutic vaccines have been proposed. For instance, a recent study utilizing live attenuated influenza A virus as the vaccine vector for the expression of HPV E6–E7 fusion transgenes demonstrated its ability to elicit a broader immune response relative to vaccination with the recombinant protein alone [81]. In this case, the influenza vector itself functions as an adjuvant (Figure 3B). This study reveals the potential of this recombinant vaccine not only to induce a stronger cellular immune response but also to promote lesion regression in mice. Additionally, because no DNA intermediates are created during the influenza life cycle, any theoretical integration of E6 or E7 into the host genome is elegantly prevented, providing an additional measure of safety [81].

Clinical studies have demonstrated the potential of utilizing Modified Vaccinia Ankara (MVA) as an alternative attenuated viral vector for designing therapeutic vaccines as well. Recombinant MVA engineered to express BPV E2 (MVA E2) has demonstrated its potential to serve as an alternative intralesional treatment for HPV-induced lesions [82,83]. Following a phase II clinical trial involving 34 patients with high grade lesions, it was observed that MVA E2 promoted a specific cytotoxic response in all patients, as demonstrated by the generation of antibodies specific to MVA E2 [82–84]. Additionally, treatment with MVA E2 resulted in elimination of lesion in 58.9% of patients and significant reduction (up to 60%) of lesion size in 41.2% of patients [84]. Unlike control patients who were treated with conization, patients receiving MVA E2 did not show signs of lesion recurrence [84]. These results illustrate that physical removal of the lesion, though temporarily effective, is not a permanent method to eliminate persistent basal infection. The efficient elimination of lesions, prevention of recurrence, along with the absence of adverse side effects, make MVA E2 and similar therapeutic vaccines potentially strong candidates in the future of HPV therapeutics.

Dendritic whole-cell based (DC) vaccines have also been proposed as a therapeutic strategy for patients with early stages of cervical cancer (Figure 3D). These vaccines are developed by culturing patient-derived monocytes into mature dendritic cells followed by pulsing with a particular antigen or peptide. When these modified dendritic cells are then introduced to the patient, a similar humoral and cellular immune responses is incited through the activation of downstream immune processes [85]. Clinical studies of full-length HPV E7-pulsed dendritic cell vaccines have demonstrated the tolerability of DC vaccines as well as an E7-specific antibody response in all immunized patients, although the observed T-cell response was somewhat variable among the patients tested [86].

Despite the advances made, there are significant challenges in the development of therapeutic vaccines [87]. Many forms of vaccines, though promising in vitro and in vivo, have shown to be ineffective clinically, possibly because the induced CD4+ and CD8+ T cell responses were neither strong enough nor broad enough [74,75]. Several of these challenges stem from HPV persistent infection triggering a series of events that downregulate the immune system [69,75,80]. Currently, studies concerned with immune-suppressive mechanisms involving T regulatory cells are being conducted, which may lead to suitable therapeutics that promote a more favorable balance of immune reactions [69,75,80]. Given these findings, the most pragmatic approach to achieve complete clearance may be through a combination of antiviral and immunomodulatory treatments.

4.3. Chemopreventive Strategies

An interesting pilot study has found that an intra-vaginal infusion of CIZAR®, a zinc-citrate compound, was effective in eliminating several types of cervical high-risk HPV infection [88]. Although the mechanism by which zinc operates to bring about this effect has not yet been thoroughly investigated, it is thought that the zinc component activates a cellular immune response by inducing T-cells [88]. Further studies will need to be conducted to conclusively characterize the mechanisms by which this zinc-citrate compound functions and determine its efficacy towards clearance of HPV infection and prevention of lesion recurrence.

4.4. Small Molecular Inhibitors

As may be inferred from previous discussion, protein–protein interactions provide more specific therapeutic windows to clearing persistent HPV replication. By interrupting processes critical to the HPV life cycle such as DNA replication, episome maintenance, or viral transcription, latent infection may be terminated because the virus is unable to maintain its genome in dividing cells. Small molecular inhibitors and nucleic acid-based treatments that target these interactions have shown to be promising non-conventional therapeutics and are currently undergoing clinical trial [89]. However, targeting specific protein–protein interactions will also be heavily dependent on the stage of cancer progression as well as the papillomavirus type. Therefore, in addition to advancing therapeutic options, specific diagnostic and classification systems of HPV disease progression also need to be developed.

The interaction between E2 and its host binding partners is a strong prospective target for therapeutic development because of its role in episomal maintenance, as well as viral transcription and replication. For example, blocking the BPV E2–BRD4 interaction prevents mitotic chromosomal localization of E2 and viral genome, thereby preventing episomal maintenance [52]. In later studies, a similar result was also found to occur as a result of inhibiting HPV 16 E2–BRD4 interaction [57]. Moreover, recent studies have demonstrated that interruption of E2–BRD4 function can additionally inhibit viral gene expression and replication [58,60,61]. Additional studies illustrate that targeting the interaction between E2 and other identified host cellular proteins could be an efficient therapeutic strategy to eliminate persistent infection [47,48,64–66], providing a framework with which to approach developing HPV-specific therapeutics. Highly specific visualization/detection techniques such as bimolecular fluorescence complementation and Mammalian Protein–Protein Interaction Trap (MAPPIT) will likewise prove integral to identifying these synthetic and naturally occurring small molecule inhibitors [57,90–92].

Nevertheless, it is crucial to consider that therapeutics targeting E2 and host interactions may prove only to be relevant in the stages of HPV infection in which integration has yet to occur in order to eliminate episomal HPV. Although proposed E1–E2 inhibitors have the same theoretical effect in terms of inhibiting persistent infection by preventing viral replication, the nuanced variations in E1–E2 structural binding between papillomavirus genera make it a difficult therapeutic target [49]. Due to the divergence of amino acid sequences among HPV proteins of different types, realistically, various classes of drugs will likely be necessary to interfere with multiple specific viral–host interactions.

5. Perspectives and Future Direction

While the advancement of Gardasil and other multivalent prophylactic vaccines provide preventative public health measures against HPV infection, the burden associated with HPV remains high in regions where access to regular screening and vaccination are unavailable. Therefore, the development of antiviral agents along with therapeutic vaccines to treat HPV post-infection remains imperative.

Interrupting virus-host interactions critical to HPV persistence demonstrates great potential to terminate persistent HPV infection, which may otherwise proceed to malignant cancer. Notably, these protein–protein interactions are relatively divergent between PV types and miniscule changes in relevant target protein amino acid sequences can drastically affect drug efficiency. Therefore, the likelihood of developing a pan-HPV antiviral currently remains low. Targeting the better-conserved cellular proteins with which HPV proteins interact provides an alternative option for developing a pan-HPV therapeutic; however, the deleterious effects caused by interrupting these proteins during normal cellular processes may outweigh the potential benefits derived from interfering with viral function. Specific drug delivery measures and research into the comprehensive cellular pathways associated with HPV pathogenesis may minimize these unfavorable outcomes. Practically speaking, multiple classes of antiviral drugs may be necessary in order to efficiently treat persistent HPV infection based on HPV types, stages of infection, and sites of infection. Despite these reservations, insights into the molecular mechanisms of HPV pathogenesis revealed in the last decade hold promise for the development of highly specific and effective antiviral agents and immunomodulatory vaccines for treating HPV persistent infection.

Acknowledgments: The authors would like to thank the members of our laboratory for helpful discussion. This work was supported in part by the National Institutes of Health (NIH) Grant R01CA187718 and the NCI Cancer Center Support Grant (NCI P30 CA016520).

Author Contributions: Srinidhi Shanmugasundaram and Jianxin You wrote the paper.

Conflicts of Interest: The authors declare no conflict of interest.

References

1. McLaughlin-Drubin, M.E.; Münger, K. Oncogenic activities of human papillomaviruses. *Virus Res.* **2009**, *143*, 195–208. [CrossRef] [PubMed]
2. Galloway, D.A.; Laimins, L.A. Human papillomaviruses: Shared and distinct pathways for pathogenesis. *Curr. Opin. Virol.* **2015**, *14*, 87–92. [CrossRef] [PubMed]
3. Radley, D.; Saah, A.; Stanley, M. Persistent infection with human papillomavirus 16 or 18 is strongly linked with high-grade cervical disease. *Hum. Vaccines Immunother.* **2015**, *12*, 768–772. [CrossRef] [PubMed]
4. Gillison, M.L.; Koch, W.M.; Capone, R.B.; Spafford, M.; Westra, W.H.; Wu, L.; Zahurak, M.L.; Daniel, R.W.; Viglione, M.; Symer, D.E.; et al. Evidence for a causal association between human papillomavirus and a subset of head and neck cancers. *J. Natl. Cancer Inst.* **2000**, *92*, 709–720. [CrossRef] [PubMed]
5. Frazer, I.H. Interaction of human papillomaviruses with the host immune system: A well evolved relationship. *Virology* **2009**, *384*, 410–414. [CrossRef] [PubMed]
6. Boldogh, I.; Albrecht, T.; Porter, D.D. Persistent viral infections. In *Medical Microbiology*, 4th ed.; Baron, S., Ed.; University of Texas Medical Branch at Galveston: Galveston, TX, USA, 1996.
7. Haukioja, A.; Asunta, M.; Söderling, E.; Syrjänen, S. Persistent oral human papillomavirus infection is associated with smoking and elevated salivary immunoglobulin g concentration. *J. Clin. Virol.* **2014**, *61*, 101–106. [CrossRef] [PubMed]
8. Rositch, A.F.; Koshiol, J.; Hudgens, M.; Razzaghi, H.; Backes, D.M.; Pimenta, J.M.; Franco, E.L.; Poole, C.; Smith, J.S. Patterns of persistent genital human papillomavirus infection among women worldwide: A literature review and meta-analysis. *Int. J. Cancer* **2013**, *133*, 1271–1285. [CrossRef] [PubMed]
9. Oh, H.Y.; Seo, S.-S.; Kim, M.K.; Lee, D.O.; Chung, Y.K.; Lim, M.C.; Kim, J.-Y.; Lee, C.W.; Park, S.-Y. Synergistic effect of viral load and alcohol consumption on the risk of persistent high-risk human papillomavirus infection. *PLoS ONE* **2014**, *9*, e104374. [CrossRef] [PubMed]
10. Oh, H.Y.; Kim, M.K.; Seo, S.; Lee, D.O.; Chung, Y.K.; Lim, M.C.; Kim, J.; Lee, C.W.; Park, S. Alcohol consumption and persistent infection of high-risk human papillomavirus. *Epidemiol. Infect.* **2015**, *143*, 1442–1450. [CrossRef] [PubMed]
11. Gunnell, A.S.; Tran, T.N.; Torrång, A.; Dickman, P.W.; Sparén, P.; Palmgren, J.; Ylitalo, N. Synergy between cigarette smoking and human papillomavirus type 16 in cervical cancer in situ development. *Cancer Epidemiol. Biomark. Prev.* **2006**, *15*, 2141–2147. [CrossRef] [PubMed]
12. Xi, L.F.; Koutsky, L.A.; Castle, P.E.; Edelstein, Z.R.; Meyers, C.; Ho, J.; Schiffman, M. Relationship between cigarette smoking and human papilloma virus types 16 and 18 DNA load. *Cancer Epidemiol. Biomark. Prev.* **2009**, *18*, 3490–3496. [CrossRef] [PubMed]
13. Peng, S.; Trimble, C.; Wu, L.; Pardoll, D.; Roden, R.; Hung, C.-F.; Wu, T.C. HLA-DQB1*02-restricted HPV-16 E7 peptide-specific CD4+ T-cell immune responses correlate with regression of HPV-16-associated high-grade squamous intraepithelial lesions. *Clin. Cancer Res.* **2007**, *13*, 2479–2487. [CrossRef] [PubMed]
14. Wank, R.; Thomssen, C. High risk of squamous cell carcinoma of the cervix for women with HLA-DQw3. *Nature* **1991**, *352*, 723–725. [CrossRef] [PubMed]
15. Zoodsma, M.; Nolte, I.M.; Schipper, M.; Oosterom, E.; van der Steege, G.; de Vries, E.G.E.; Te Meerman, G.J.; van der Zee, A.G.J. Analysis of the entire hla region in susceptibility for cervical cancer: A comprehensive study. *J. Med. Genet.* **2005**, *42*, e49. [CrossRef] [PubMed]
16. Bernal-Silva, S.; Granados, J.; Gorodezky, C.; Aláez, C.; Flores-Aguilar, H.; Cerda-Flores, R.M.; Guerrero-González, G.; Valdez-Chapa, L.D.; Morales-Casas, J.; González-Guerrero, J.F.; et al. HLA-DRB1 class II antigen level alleles are associated with persistent HPV infection in mexican women; a pilot study. *Infect. Agent Cancer* **2013**, *8*, 31. [CrossRef] [PubMed]
17. Liaw, K.L.; Hildesheim, A.; Burk, R.D.; Gravitt, P.; Wacholder, S.; Manos, M.M.; Scott, D.R.; Sherman, M.E.; Kurman, R.J.; Glass, A.G.; et al. A prospective study of human papillomavirus (HPV) type 16 DNA detection by polymerase chain reaction and its association with acquisition and persistence of other HPV types. *J. Infect. Dis.* **2001**, *183*, 8–15. [CrossRef] [PubMed]

18. Rousseau, M.C.; Pereira, J.S.; Prado, J.C.; Villa, L.L.; Rohan, T.E.; Franco, E.L. Cervical coinfection with human papillomavirus (HPV) types as a predictor of acquisition and persistence of HPV infection. *J. Infect. Dis.* **2001**, *184*, 1508–1517. [CrossRef] [PubMed]

19. Thomas, K.K.; Hughes, J.P.; Kuypers, J.M.; Kiviat, N.B.; Lee, S.K.; Adam, D.E.; Koutsky, L.A. Concurrent and sequential acquisition of different genital human papillomavirus types. *J. Infect. Dis.* **2000**, *182*, 1097–1102. [CrossRef] [PubMed]

20. Zhang, L.; Liao, H.; Yang, B.; Geffre, C.P.; Zhang, A.; Zhou, A.; Cao, H.; Wang, J.; Zhang, Z.; Zheng, W. Variants of human papillomavirus type 16 predispose toward persistent infection. *Int. J. Clin. Exp. Pathol.* **2015**, *8*, 8453–8459. [PubMed]

21. La Torre, G.; de Waure, C.; Chiaradia, G.; Mannocci, A.; Ricciardi, W. HPV vaccine efficacy in preventing persistent cervical HPV infection: A systematic review and meta-analysis. *Vaccine* **2007**, *25*, 8352–8358. [CrossRef] [PubMed]

22. Clifford, G.M.; Smith, J.S.; Aguado, T.; Franceschi, S. Comparison of HPV type distribution in high-grade cervical lesions and cervical cancer: A meta-analysis. *Br. J. Cancer* **2003**, *89*, 101–105. [CrossRef] [PubMed]

23. Zur Hausen, H. Papillomaviruses in the causation of human cancers—A brief historical account. *Virology* **2009**, *384*, 260–265. [CrossRef] [PubMed]

24. Lizano, M.; Berumen, J.; García-Carrancá, A. HPV-related carcinogenesis: Basic concepts, viral types and variants. *Arch. Med. Res.* **2009**, *40*, 428–434. [CrossRef] [PubMed]

25. Katki, H.A.; Cheung, L.C.; Fetterman, B.; Castle, P.E.; Sundaram, R. A joint model of persistent human papillomavirus infection and cervical cancer risk: Implications for cervical cancer screening. *J. R Stat. Soc. Ser. A Stat. Soc.* **2015**, *178*, 903–923. [CrossRef] [PubMed]

26. Lowy, D.R.; Schiller, J.T. Reducing HPV-associated cancer globally. *Cancer Prev. Res.* **2012**, *5*, 18–23. [CrossRef] [PubMed]

27. Gillison, M.L.; D'Souza, G.; Westra, W.; Sugar, E.; Xiao, W.; Begum, S.; Viscidi, R. Distinct risk factor profiles for human papillomavirus type 16-positive and human papillomavirus type 16-negative head and neck cancers. *J. Natl. Cancer Inst.* **2008**, *100*, 407–420. [CrossRef] [PubMed]

28. Weinberger, P.M.; Yu, Z.; Haffty, B.G.; Kowalski, D.; Harigopal, M.; Brandsma, J.; Sasaki, C.; Joe, J.; Camp, R.L.; Rimm, D.L.; et al. Molecular classification identifies a subset of human papillomavirus—Associated oropharyngeal cancers with favorable prognosis. *J. Clin. Oncol.* **2006**, *24*, 736–747. [CrossRef] [PubMed]

29. Newell, G.R.; Krementz, E.T.; Roberts, J.D. Excess occurrence of cancer of the oral cavity, lung, and bladder following cancer of the cervix. *Cancer* **1975**, *36*, 2155–2158. [CrossRef] [PubMed]

30. Ciesielska, U.; Nowińska, K.; Podhorska-Okołów, M.; Dziegiel, P. The role of human papillomavirus in the malignant transformation of cervix epithelial cells and the importance of vaccination against this virus. *Adv. Clin. Exp. Med.* **2012**, *21*, 235–244. [PubMed]

31. D'Souza, G.; Kreimer, A.R.; Viscidi, R.; Pawlita, M.; Fakhry, C.; Koch, W.M.; Westra, W.H.; Gillison, M.L. Case-control study of human papillomavirus and oropharyngeal cancer. *N. Engl. J. Med.* **2007**, *356*, 1944–1956. [CrossRef] [PubMed]

32. Akagi, K.; Li, J.; Broutian, T.R.; Padilla-Nash, H.; Xiao, W.; Jiang, B.; Rocco, J.W.; Teknos, T.N.; Kumar, B.; Wangsa, D.; et al. Genome-wide analysis of HPV integration in human cancers reveals recurrent, focal genomic instability. *Genome Res.* **2014**, *24*, 185–199. [CrossRef] [PubMed]

33. Cullen, A.P.; Reid, R.; Campion, M.; Lörincz, A.T. Analysis of the physical state of different human papillomavirus DNAs in intraepithelial and invasive cervical neoplasm. *J. Virol.* **1991**, *65*, 606–612. [PubMed]

34. Schneider-Maunoury, S.; Croissant, O.; Orth, G. Integration of human papillomavirus type 16 DNA sequences: A possible early event in the progression of genital tumors. *J. Virol.* **1987**, *61*, 3295–3298. [PubMed]

35. Parfenov, M.; Pedamallu, C.S.; Gehlenborg, N.; Freeman, S.S.; Danilova, L.; Bristow, C.A.; Lee, S.; Hadjipanayis, A.G.; Ivanova, E.V.; Wilkerson, M.D.; et al. Characterization of HPV and host genome interactions in primary head and neck cancers. *Proc. Natl. Acad. Sci. USA* **2014**, *111*, 15544–15549. [CrossRef] [PubMed]

36. Steger, G.; Corbach, S. Dose-dependent regulation of the early promoter of human papillomavirus type 18 by the viral E2 protein. *J. Virol.* **1997**, *71*, 50–58. [PubMed]

37. Dong, G.; Broker, T.R.; Chow, L.T. Human papillomavirus type 11 E2 proteins repress the homologous E6 promoter by interfering with the binding of host transcription factors to adjacent elements. *J. Virol.* **1994**, *68*, 1115–1127. [PubMed]

38. Dyson, N.; Howley, P.M.; Munger, K.; Harlow, E. The human papilloma virus-16 E7 oncoprotein is able to bind to the retinoblastoma gene product. *Science* **1989**, *243*, 934–937. [CrossRef] [PubMed]

39. Huibregtse, J.M.; Scheffner, M.; Howley, P.M. A cellular protein mediates association of p53 with the E6 oncoprotein of human papillomavirus types 16 or 18. *EMBO J.* **1991**, *10*, 4129–4135. [PubMed]

40. Jeon, S.; Lambert, P.F. Integration of human papillomavirus type 16 DNA into the human genome leads to increased stability of E6 and E7 mRNAs: Implications for cervical carcinogenesis. *Proc. Natl. Acad. Sci. USA* **1995**, *92*, 1654–1658. [CrossRef] [PubMed]

41. Bastien, N.; McBride, A.A. Interaction of the papillomavirus E2 protein with mitotic chromosomes. *Virology* **2000**, *270*, 124–134. [CrossRef] [PubMed]

42. Groves, I.J.; Coleman, N. Pathogenesis of human papillomavirus-associated mucosal disease. *J. Pathol.* **2015**, *235*, 527–538. [CrossRef] [PubMed]

43. Skiadopoulos, M.H.; McBride, A.A. Bovine papillomavirus type 1 genomes and the E2 transactivator protein are closely associated with mitotic chromatin. *J. Virol.* **1998**, *72*, 2079–2088. [PubMed]

44. Ballestas, M.E.; Chatis, P.A.; Kaye, K.M. Efficient persistence of extrachromosomal KSHV DNA mediated by latency-associated nuclear antigen. *Science* **1999**, *284*, 641–644. [CrossRef] [PubMed]

45. Kapoor, P.; Lavoie, B.D.; Frappier, L. EBP2 plays a key role in Epstein–Barr virus mitotic segregation and is regulated by aurora family kinases. *Mol. Cell. Biol.* **2005**, *25*, 4934–4945. [CrossRef] [PubMed]

46. Shire, K.; Ceccarelli, D.F.; Avolio-Hunter, T.M.; Frappier, L. EBP2, a human protein that interacts with sequences of the Epstein–Barr virus nuclear antigen 1 important for plasmid maintenance. *J. Virol.* **1999**, *73*, 2587–2595. [PubMed]

47. Oliveira, J.G.; Colf, L.A.; McBride, A.A. Variations in the association of papillomavirus E2 proteins with mitotic chromosomes. *Proc. Natl. Acad. Sci. USA* **2006**, *103*, 1047–1052. [CrossRef] [PubMed]

48. McBride, A.A. The papillomavirus E2 proteins. *Virology* **2013**, *445*, 57–79. [CrossRef] [PubMed]

49. Kurg, R. *The Role of E2 Proteins in Papillomavirus DNA Replication*; InTech: Rijeka, Croatia, 2011.

50. McPhillips, M.G.; Ozato, K.; McBride, A.A. Interaction of bovine papillomavirus E2 protein with Brd4 stabilizes its association with chromatin. *J. Virol.* **2005**, *79*, 8920–8932. [CrossRef] [PubMed]

51. Van Tine, B.A.; Dao, L.D.; Wu, S.-Y.; Sonbuchner, T.M.; Lin, B.Y.; Zou, N.; Chiang, C.-M.; Broker, T.R.; Chow, L.T. Human papillomavirus (HPV) origin-binding protein associates with mitotic spindles to enable viral DNA partitioning. *Proc. Natl. Acad. Sci. USA* **2004**, *101*, 4030–4035. [CrossRef] [PubMed]

52. You, J.; Croyle, J.L.; Nishimura, A.; Ozato, K.; Howley, P.M. Interaction of the bovine papillomavirus E2 protein with Brd4 tethers the viral DNA to host mitotic chromosomes. *Cell* **2004**, *117*, 349–360. [CrossRef]

53. Houzelstein, D.; Bullock, S.L.; Lynch, D.E.; Grigorieva, E.F.; Wilson, V.A.; Beddington, R.S.P. Growth and early postimplantation defects in mice deficient for the bromodomain-containing protein Brd4. *Mol. Cell. Biol.* **2002**, *22*, 3794–3802. [CrossRef] [PubMed]

54. McBride, A.A.; McPhillips, M.G.; Oliveira, J.G. Brd4: Tethering, segregation and beyond. *Trends Microbiol.* **2004**, *12*, 527–529. [CrossRef] [PubMed]

55. Abbate, E.A.; Voitenleitner, C.; Botchan, M.R. Structure of the papillomavirus DNA-tethering complex E2:Brd4 and a peptide that ablates HPV chromosomal association. *Mol. Cell* **2006**, *24*, 877–889. [CrossRef] [PubMed]

56. Gauson, E.J.; Wang, X.; Dornan, E.S.; Herzyk, P.; Bristol, M.; Morgan, I.M. Failure to interact with Brd4 alters the ability of HPV16 E2 to regulate host genome expression and cellular movement. *Virus Res.* **2016**, *211*, 1–8. [CrossRef] [PubMed]

57. Helfer, C.M.; Wang, R.; You, J. Analysis of the papillomavirus E2 and bromodomain protein Brd4 interaction using bimolecular fluorescence complementation. *PLoS ONE* **2013**, *8*, e77994. [CrossRef] [PubMed]

58. McPhillips, M.G.; Oliveira, J.G.; Spindler, J.E.; Mitra, R.; McBride, A.A. Brd4 is required for E2-mediated transcriptional activation but not genome partitioning of all papillomaviruses. *J. Virol.* **2006**, *80*, 9530–9543. [CrossRef] [PubMed]

59. Jang, M.K.; Kwon, D.; McBride, A.A. Papillomavirus E2 proteins and the host Brd4 protein associate with transcriptionally active cellular chromatin. *J. Virol.* **2009**, *83*, 2592–2600. [CrossRef] [PubMed]

60. Schweiger, M.-R.; You, J.; Howley, P.M. Bromodomain protein 4 mediates the papillomavirus E2 transcriptional activation function. *J. Virol.* **2006**, *80*, 4276–4285. [CrossRef] [PubMed]

61. Wang, X.; Helfer, C.M.; Pancholi, N.; Bradner, J.E.; You, J. Recruitment of Brd4 to the human papillomavirus type 16 DNA replication complex is essential for replication of viral DNA. *J. Virol.* **2013**, *87*, 3871–3884. [CrossRef] [PubMed]

62. Wu, S.Y.; Nin, D.S.; Lee, A.Y.; Simanski, S.; Kodadek, T.; Chiang, C.M. Brd4 phosphorylation regulates HPV E2-mediated viral transcription, origin replication, and cellular MMP-9 expression. *Cell Rep.* **2016**, *16*, 1733–1748. [CrossRef] [PubMed]

63. Sakakibara, N.; Chen, D.; Jang, M.K.; Kang, D.W.; Luecke, H.F.; Wu, S.-Y.; Chiang, C.-M.; McBride, A.A. Brd4 is displaced from HPV replication factories as they expand and amplify viral DNA. *PLoS Pathog.* **2013**, *9*, e1003777. [CrossRef] [PubMed]

64. Donaldson, M.M.; Mackintosh, L.J.; Bodily, J.M.; Dornan, E.S.; Laimins, L.A.; Morgan, I.M. An interaction between human papillomavirus 16 E2 and TopBP1 is required for optimum viral DNA replication and episomal genome establishment. *J. Virol.* **2012**, *86*, 12806–12815. [CrossRef] [PubMed]

65. Parish, J.L.; Bean, A.M.; Park, R.B.; Androphy, E.J. ChlR1 is required for loading papillomavirus E2 onto mitotic chromosomes and viral genome maintenance. *Mol. Cell* **2006**, *24*, 867–876. [CrossRef] [PubMed]

66. Donaldson, M.M.; Boner, W.; Morgan, I.M. TopBP1 regulates human papillomavirus type 16 E2 interaction with chromatin. *J. Virol.* **2007**, *81*, 4338–4342. [CrossRef] [PubMed]

67. Bang, S.W.; Ko, M.J.; Kang, S.; Kim, G.S.; Kang, D.; Lee, J.; Hwang, D.S. Human TopBP1 localization to the mitotic centrosome mediates mitotic progression. *Exp. Cell Res.* **2011**, *317*, 994–1004. [CrossRef] [PubMed]

68. Hirota, Y.; Lahti, J.M. Characterization of the enzymatic activity of hChlR1, a novel human DNA helicase. *Nucleic Acids Res.* **2000**, *28*, 917–924. [CrossRef] [PubMed]

69. Christensen, N.D.; Budgeon, L.R. Vaccines and immunization against human papillomavirus. *Curr. Probl. Dermatol.* **2014**, *45*, 252–264. [PubMed]

70. Sankaranarayanan, R. HPV vaccination: The most pragmatic cervical cancer primary prevention strategy. *Int. J. Gynecol. Obstet.* **2015**, *131*, S33–S35. [CrossRef] [PubMed]

71. Pils, S.; Joura, E.A. From the monovalent to the nine-valent HPV vaccine. *Clin. Microbiol. Infect.* **2015**, *21*, 827–833. [CrossRef] [PubMed]

72. Mollers, M.; King, A.J.; Knol, M.J.; Scherpenisse, M.; Meijer, C.J.L.M.; van der Klis, F.R.M.; de Melker, H.E. Effectiveness of human papillomavirus vaccine against incident and persistent infections among young girls: Results from a longitudinal dutch cohort study. *Vaccine* **2015**, *33*, 2678–2683. [CrossRef] [PubMed]

73. Stern, P.L.; van der Burg, S.H.; Hampson, I.N.; Broker, T.; Fiander, A.; Lacey, C.J.; Kitchener, H.C.; Einstein, M.H. Therapy of human papillomavirus-related disease. *Vaccine* **2012**, *30*, F71–F82. [CrossRef] [PubMed]

74. Kim, T.J.; Jin, H.-T.; Hur, S.-Y.; Yang, H.G.; Seo, Y.B.; Hong, S.R.; Lee, C.-W.; Kim, S.; Woo, J.-W.; Park, K.S.; et al. Clearance of persistent HPV infection and cervical lesion by therapeutic DNA vaccine in CIN3 patients. *Nat. Commun.* **2014**, *5*, 5317. [CrossRef] [PubMed]

75. Van der Burg, S.H.; Arens, R.; Melief, C.J.M. Immunotherapy for persistent viral infections and associated disease. *Trends Immunol.* **2011**, *32*, 97–103. [CrossRef] [PubMed]

76. Hus, I.; Gonet-Sebastianka, J.; Surdacka, A.; Bojarska-Junak, A.; Roliński, J. Analysis of peripheral blood immune cells after prophylactic immunization with HPV-16/18 ASO4-adjuvanted vaccine. *Postep. Hig. Med. Doświadczalnej* **2015**, *69*, 543–548. [CrossRef] [PubMed]

77. Lee, S.-J.; Yang, A.; Wu, T.C.; Hung, C.-F. Immunotherapy for human papillomavirus-associated disease and cervical cancer: Review of clinical and translational research. *J. Gynecol. Oncol.* **2016**, *27*, e51. [CrossRef] [PubMed]

78. Zhou, Z.-X.; Li, D.; Guan, S.-S.; Zhao, C.; Li, Z.-L.; Zeng, Y. Immunotherapeutic effects of dendritic cells pulsed with a coden-optimized HPV 16 E6 and E7 fusion gene in vivo and in vitro. *Asian Pac. J. Cancer Prev.* **2015**, *16*, 3843–3847. [CrossRef] [PubMed]

79. Mizuuchi, M.; Hirohashi, Y.; Torigoe, T.; Kuroda, T.; Yasuda, K.; Shimizu, Y.; Saito, T.; Sato, N. Novel oligomannose liposome-DNA complex DNA vaccination efficiently evokes anti-HPV E6 and E7 CTL responses. *Exp. Mol. Pathol.* **2012**, *92*, 185–190. [CrossRef] [PubMed]

80. Van de Wall, S.; Nijman, H.W.; Daemen, T. HPV-specific immunotherapy: Key role for immunomodulators. *Anticancer Agents Med. Chem.* **2014**, *14*, 265–279. [CrossRef] [PubMed]

81. Jindra, C.; Huber, B.; Shafti-Keramat, S.; Wolschek, M.; Ferko, B.; Muster, T.; Brandt, S.; Kirnbauer, R. Attenuated recombinant influenza a virus expressing HPV16 E6 and E7 as a novel therapeutic vaccine approach. *PLoS ONE* **2015**, *10*, e0138722.

82. Rosales, R.; López-Contreras, M.; Rosales, C.; Magallanes-Molina, J.-R.; Gonzalez-Vergara, R.; Arroyo-Cazarez, J.M.; Ricardez-Arenas, A.; del Follo-Valencia, A.; Padilla-Arriaga, S.; Guerrero, M.V.; et al. Regression of human papillomavirus intraepithelial lesions is induced by MVA E2 therapeutic vaccine. *Hum. Gene Ther.* **2014**, *25*, 1035–1049. [CrossRef] [PubMed]

83. Corona Gutierrez, C.M.; Tinoco, A.; Navarro, T.; Contreras, M.L.; Cortes, R.R.; Calzado, P.; Reyes, L.; Posternak, R.; Morosoli, G.; Verde, M.L.; et al. Therapeutic vaccination with MVA E2 can eliminate precancerous lesions (CIN 1, CIN 2, and CIN 3) associated with infection by oncogenic human papillomavirus. *Hum. Gene Ther.* **2004**, *15*, 421–431. [CrossRef] [PubMed]

84. García-Hernández, E.; González-Sánchez, J.L.; Andrade-Manzano, A.; Contreras, M.L.; Padilla, S.; Guzmán, C.C.; Jiménez, R.; Reyes, L.; Morosoli, G.; Verde, M.L.; et al. Regression of papilloma high-grade lesions (CIN 2 and CIN 3) is stimulated by therapeutic vaccination with MVA E2 recombinant vaccine. *Cancer Gene Ther.* **2006**, *13*, 592–597. [CrossRef] [PubMed]

85. Adams, M.; Navabi, H.; Jasani, B.; Man, S.; Fiander, A.; Evans, A.S.; Donninger, C.; Mason, M. Dendritic cell (DC) based therapy for cervical cancer: Use of DC pulsed with tumour lysate and matured with a novel synthetic clinically non-toxic double stranded RNA analogue poly [I]:Poly [$C_{12}U$] (Ampligen®). *Vaccine* **2003**, *21*, 787–790. [CrossRef]

86. Santin, A.D.; Bellone, S.; Palmieri, M.; Zanolini, A.; Ravaggi, A.; Siegel, E.R.; Roman, J.J.; Pecorelli, S.; Cannon, M.J. Human papillomavirus type 16 and 18 E7-pulsed dendritic cell vaccination of stage IB or IIA cervical cancer patients: A phase I escalating-dose trial. *J. Virol.* **2008**, *82*, 1968–1979. [CrossRef] [PubMed]

87. Nieto, K.; Gissmann, L.; Schädlich, L. Human papillomavirus-specific immune therapy: Failure and hope. *Antivir. Ther.* **2010**, *15*, 951–957. [CrossRef] [PubMed]

88. Kim, J.H.; Bae, S.N.; Lee, C.W.; Song, M.J.; Lee, S.J.; Yoon, J.H.; Lee, K.H.; Hur, S.Y.; Park, T.C.; Park, J.S. A pilot study to investigate the treatment of cervical human papillomavirus infection with zinc-citrate compound (cizar®). *Gynecol. Oncol.* **2011**, *122*, 303–306. [CrossRef] [PubMed]

89. Carlos de Freitas, A.; da Conceicao Gomes Leitao, M.; Coimbra, E.C. Prospects of molecularly-targeted therapies for cervical cancer treatment. *Curr. Drug Targets* **2015**, *16*, 77–91. [CrossRef]

90. Eyckerman, S.; Titeca, K.; van Quickelberghe, E.; Cloots, E.; Verhee, A.; Samyn, N.; de Ceuninck, L.; Timmerman, E.; de Sutter, D.; Lievens, S.; et al. Trapping mammalian protein complexes in viral particles. *Nat. Commun.* **2016**, *7*, 11416. [CrossRef] [PubMed]

91. Lemmens, I.; Lievens, S.; Tavernier, J. Mappit, a mammalian two-hybrid method for in-cell detection of protein-protein interactions. *Methods Mol. Biol.* **2015**, *1278*, 447–455. [PubMed]

92. Yan, J.; Li, Q.; Lievens, S.; Tavernier, J.; You, J. Abrogation of the Brd4-positive transcription elongation factor B complex by papillomavirus E2 protein contributes to viral oncogene repression. *J. Virol.* **2010**, *84*, 76–87. [CrossRef] [PubMed]

Review

Immunopathogenesis of HPV-Associated Cancers and Prospects for Immunotherapy

Sigrun Smola

Institute of Virology, Saarland University Medical Center, 66421 Homburg/Saar, Germany; sigrun.smola@uks.eu; Tel.: +49-6841-16-23931

Received: 25 August 2017; Accepted: 8 September 2017; Published: 12 September 2017

Abstract: Human papillomavirus (HPV) infection is a causative factor for various cancers of the anogenital region and oropharynx, and is supposed to play an important cofactor role for skin carcinogenesis. Evasion from immunosurveillance favors viral persistence. However, there is evidence that the mere presence of oncogenic HPV is not sufficient for malignant progression and that additional tumor-promoting steps are required. Recent studies have demonstrated that HPV-transformed cells actively promote chronic stromal inflammation and conspire with cells in the local microenvironment to promote carcinogenesis. This review highlights the complex interplay between HPV-infected cells and the local immune microenvironment during oncogenic HPV infection, persistence, and malignant progression, and discusses new prospects for diagnosis and immunotherapy of HPV-associated cancers.

Keywords: human papillomavirus; cervical cancer; skin cancer; epidermodysplasia verruciformis; immune evasion; chronic inflammation; IL-6; JAK-STAT3; immunotherapy; immunoscore

1. Introduction

Approximately 15–20% of all cancers are caused by infectious agents [1] and around 5% by human papillomaviruses (HPVs) [2,3]. The causal relationship between HPV infection and cervical cancer, which harbors HPV in up to 99.7% of cases [4], was highlighted by Harald zur Hausen, who was awarded a Nobel Prize in 2008. In addition to cervical cancer, a significant number of oropharyngeal, penile, anal, vaginal and vulvar cancers are induced by mucosal HPVs [5–7] and cutaneous HPVs have been implicated as cofactors in skin cancer development [8,9].

Invasive cancer is not an immediate consequence of HPV infection. HPV-induced carcinogenesis takes years or decades to occur, and there is increasing evidence that additional tumor-promoting steps are required [10]. It is widely accepted that effective immune control is required to prevent persistent HPV infection. Recent studies indicate, however, that chronic inflammation and misled immune responses in the local immune microenvironment play a critical role during the progression of precancerous lesions to invasive cancer [11–13]. Thus, the unidirectional view of an immune system that primarily serves to attack and eliminate HPV-infected and neoplastic cells needs to be revised [14].

While screening programs have greatly reduced the burden of cervical cancer in developed countries, current diagnostic tests cannot discriminate between lesions that will progress to invasiveness and those that do not. This results in an overtreatment of high-grade lesions that are detected during screening [15]. A better understanding of the immunological mechanisms contributing to HPV-associated cancer development would likely propel not only more accurate diagnosis of progressing precancerous lesions but also novel immunotherapeutic approaches for HPV-driven cancers [16].

This review focuses on the current understanding of the complex interplay between HPV-infected cells and the local immune microenvironment during HPV infection and HPV-associated carcinogenesis, and discusses novel prospects for diagnosis and immunotherapy.

2. Human Papillomaviruses in Mucosal versus Skin Carcinogenesis and Immune Control

Human papillomaviruses are non-enveloped double-stranded (ds) DNA-viruses that are transmitted sexually or by smear infection [17]. More than 200 different HPV types that are contained within 5 different genera have been characterized [18,19]. HPV-induced pathologies vary from benign warts and low- and high-grade neoplasia to malignant cancer and depend on respective HPV types as well as anatomical sites of infection [20]. While almost all cervical cancers are HPV-associated, 64–91% of vaginal, 40–50% of vulvar, 88–94% of anal and 40–50% of penile cancers are HPV-positive [3]. Notably, prevalences of HPV-driven oropharyngeal cancers display larger geographical variations. The highest prevalence is observed in developed countries, with HPV-positivity rates ranging from 35% up to more than 70% in some regions, and numbers of oropharyngeal cancers have significantly increased during the last decades [3,5]. 12–15 mucosal HPV types, all belonging to genus α, have been identified as so-called high-risk HPV (HR-HPV) types [21]. Numerous studies on the HPV life cycle and on the biology of the mucosal HPV oncogenes E6 and E7 have greatly improved our understanding of HPV-induced transformation far beyond the mere inactivation of the tumor suppressor proteins p53 and retinoblastoma (for a review see [22]).

Genus β-HPVs have been implicated in ultraviolet (UV) light-induced non-melanoma skin cancer in patients suffering from the inherited disease Epidermodysplasia verruciformis [23] and a cofactor role for skin carcinogenesis in the normal population has been discussed (for reviews see [8,9,24]). Recent data point to an early role of genus-β HPV in skin carcinogenesis. It has been shown that β-HPV type 8 infection expands the stem cell compartment in Epidermodysplasia verruciformis patients by suppressing the stemness-repressing microRNA-203 [25], an initial key step in skin carcinogenesis. Mechanistically, the E6 protein, the major oncoprotein of HPV8 [26], targets the transcription factor CCAAT/enhancer binding protein (C/EBP)α that serves as a tumor suppressor of UV-induced carcinogenesis. HPV8 E6 thereby prevents microRNA-203 expression, leading to potent up-regulation of the epithelial stemness-maintenance factor ΔNp63 (NH2-terminally deleted p63). ΔNp63 in turn promotes proliferation and inhibits differentiation of keratinocytes [25]. This is in contrast to the E6 protein encoded by mucosal HR-HPV, which suppresses miR-203 via interference with p53 [27], while the E7 protein suppresses a protein kinase C-dependent pathway [28]. Proliferation of hair follicle stem cells was also observed in transgenic mice expressing the HPV8 early region under the control of the keratin 14 (K14)-promoter [29]. Moreover, genus β-HPVs were shown to suppress UV-induced DNA damage repair; they interfere with Notch-signaling that further contributes to ΔNp63 up-regulation and have anti-apoptotic properties in vitro. Different β-HPVs have oncogenic potential in transgenic mice, particularly in synergism with UV-light exposure [26,30–35].

In mucosal as well as cutaneous HPV-associated carcinogenesis it is well accepted that the immune system has an important surveillance function. In immunocompetent individuals, up to 90% of anogenital HPV infections are cleared within two years [36–38] and this is thought to be due to innate immunity as well as adaptive CD8$^+$ T cell-mediated responses directed against viral early proteins [39,40]. Conversely, patients with impaired adaptive immunity, such as transplant recipients or HIV-patients, show higher prevalences of HPV infection and HPV-related diseases, further underlining the importance of immunosurveillance in HPV-associated carcinogenesis [8,41–46].

HPV infections that escape immune control can persist and a certain proportion progresses to cancer. Recent studies have shed light on the immune system as a double-edged sword in HPV-associated carcinogenesis and evidence is increasing that the role of the immune system changes in a stage-dependent manner. At earlier stages, anti-viral immunity predominates and the virus has adopted strategies to counteract immunosurveillance in order to establish persistence in the epithelium. However, at later stages of the disease, HPV-transformed cells reprogram the local immune microenvironment and rather initiate chronic stromal inflammation, which then serves to promote progression of precursor lesions to invasive cancer (Figure 1).

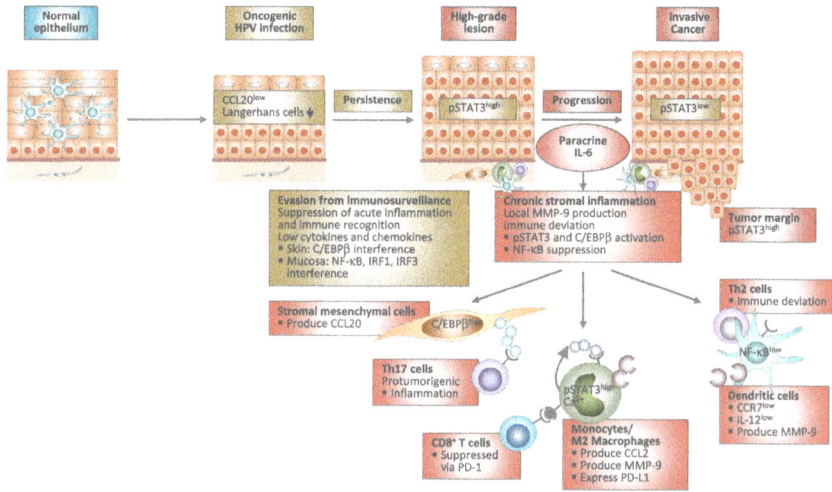

Figure 1. Proposed model of human papillomavirus-induced carcinogenesis. Stage-specific interplay between virally infected keratinocytes and the local immune microenvironment. At early stages, HPV-infected cells suppress acute inflammation in the epithelium and immune recognition. This allows escape from immunosurveillance and viral persistence. During progression to invasive cancer HPV-transformed cells initiate chronic stromal inflammation and immune deviation orchestrated by paracrine IL-6. The IL-6/STAT3 and IL-6/C/EBPβ pathways lead to chemokine induction in stromal mesenchymal and infiltrating immune cells. As a consequence, myelomonocytic cells expressing protumorigenic MMP-9 and Th17 cells are recruited further promoting inflammation. Myelomonocytic cells differentiate into functionally impaired dendritic cells or M2 macrophages expressing PD-L1 that inhibit cytotoxic T cell responses. IL-6 suppresses NF-κB activity in stromal dendritic cells, which are unable to migrate in response to lymph node homing chemokines due to low CCR7 chemokine receptor expression. Instead, they are immobilized within the tumor stroma and produce MMP-9 locally. IL-12 is expressed only at low levels shifting T helper cell responses from Th1 to Th2. Stromal inflammation and immune deviation facilitate progression to invasiveness. HPV: human papillomavirus; IL: interleukin; STAT3: signal transducer and activator of transcription 3; C/EBP: CCAAT/enhancer binding protein; MMP: matrix-metalloproteinase; Th: T helper; PD-L1: programmed death-ligand 1; NF: nuclear factor; CCR: C-C chemokine receptor; CCL: C-C chemokine ligand; IRF: interferon regulatory factor.

3. Immune Escape Paves the Way for HPV Persistence

To maintain a first line of defense against infections agents, skin and mucosal surfaces are equipped with efficient immune sentinels and immune effector mechanisms [47,48]. Keratinocytes, the HPV host cells, form stratified epithelia constituting a physical and immunological barrier against pathogens. They are armed with pathogen recognition receptors, host intrinsic restriction factors and an arsenal of inflammatory cytokines and chemokines orchestrating local immune responses [49]. While the epidermal compartment harbors Langerhans cells and distinct subsets of antigen-presenting cells (APCs) [50], most innate immune cell types including myeloid, dendritic as well as innate lymphoid cells and adaptive resident lymphocytes are located in the dermis [51].

3.1. Passive Mechanisms of Immune Escape

For productive infection HPV depends on the keratinocyte differentiation program. After having entered the proliferating basal keratinocytes, HPV gene expression is low and vegetative replication dramatically increases only in the more differentiated layers of the epithelium that are bound to desquamate shortly. The minor levels of protein expression in the lower epithelial layers, the well-directed non-cytolytic genome amplification restricted to the differentiated layers, and the lack of a viremic phase in the viral life cycle are thought to passively help the virus escape immune recognition [52,53]. Thus, HPV has come to an arrangement with the hostile microenvironment and avoids alerting the immune system.

3.2. Suppression of Cell-Autonomous Immunity and Acute Inflammation in Keratinocytes

Although HPVs encode only a limited number of regulatory genes, they engage various active strategies to counteract immune recognition and cell-autonomous immune responses at different levels [54]. Mucosal as well as cutaneous HPV were shown to suppress recognition by the pattern recognition receptor toll-like receptor 9 [55,56]. Mucosal HPV also specifically inhibits interferon (IFN) expression, IFN signaling and downstream responses [57–60]. The HPV E6 oncoprotein directly targets IFN regulatory factor 3 (IRF3) via direct interaction, while E7 interferes with the anti-viral and pro-apoptotic factor IRF1 [61–64]. Notably, HPV8 has even adapted to an IFN regulatory factor, IRF7, which is activated by UV-light in skin [65]. IRF7 is expressed in suprabasal keratinocytes and it has been demonstrated that it increases HPV8 late promoter activity [66]. In contrast, IRF3-activators, such as dsRNA or RNA bearing 5' phosphates, efficiently repress HPV8 promoter activity and HPV8 E6 does not counteract the suppressive activity of IRF3 as expected from mucosal HPV-encoded E6 proteins [66]. The IRF3-induced state of cell-autonomous immunity against cutaneous β-HPV in keratinocytes was shown to prevail over IRF7 activity. Thus, IRF3 remains an Achilles heel of this cutaneous virus [66] opening the possibility to use IRF3-activating compounds for anti-viral immunotherapy against β-HPV infection.

In contrast to RNA- and other dsDNA-viruses like herpes-simplex or vaccinia viruses that activate the inflammasome in keratinocytes [67], mucosal HPVs rather dampen acute inflammatory responses in the epithelium. Keratinocytes harboring episomal HPV and cervical cancer cells harboring integrated HPV genomes display only low cytokine and chemokine expression in vitro [68–70] and in vivo [12,71] as a direct consequence of HPV oncoprotein-mediated suppression [71–74]. Mechanistically, it was shown that mucosal HPV oncoproteins target the p300/CBP-associated factor/nuclear factor (NF)-κB pathway [70,74–76] and abrogate post-translational processing and secretion of the key inflammatory cytokine interleukin (IL)-1β [77].

In contrast to IL-1β, IL-1α, an intracellularly stored alarmin, is detectable throughout cervical carcinogenesis [78]. IL-1α appears not to be affected by mucosal HPV and this can be exploited for novel immunotherapy strategies. Stimulation of cervical cancer cells with the dsRNA analog PolyIC mimicking dsRNA-virus infection leads to efficient IL-1α release via induction of necroptosis and this was shown to potentiate dendritic cell activation [79]. Notably, PolyIC-induced necroptosis, IL-1α release, and dendritic cell activation are completely dependent on the expression of receptor-interacting protein kinase 3 (RIPK3) in the HPV-transformed cells. In different cervical cancer patients, RIPK3 is expressed at individual levels in the neoplastic cells in situ, which may critically influence their response to dsRNA treatment. Thus, pre-therapeutic RIPK3 expression levels could be used as a novel biomarker to predict the response to immunotherapy with dsRNA or dsRNA-analogs [79,80].

Notably, oncolytic viruses are currently under investigation for cancer therapy and it has become clear that immune activation is an important part of their anti-tumor activity [81]. Whether the immunostimulatory activity of dsRNA oncolytic viruses requires RIPK3 expression in the target cells similar to PolyIC is currently unknown and will be interesting to study.

3.3. Suppression of the Recruitment of Professional APC

Professional antigen-presenting cells connect innate and adaptive immunity. APCs migrate to secondary lymphatic tissues where they encounter specific T cells, which subsequently become activated and are redirected to the sites of infection. Local factors released from epithelial cells critically influence recruitment, differentiation and activation of APCs. Although the role of murine Langerhans cells has been controversially discussed, human Langerhans cells have been shown to prime and cross-prime naive CD8[+] T cells [82], which are known to be critical for the immune control of HPV infection.

Evidence is emerging that HPV infection actively interferes with human Langerhans cell homeostasis in the epidermal compartment. Both cutaneous and mucosal HPV-infected epithelia harbor only low numbers of Langerhans cells [71,83] and the responsiveness of in vitro-generated human Langerhans cells to virus-like particles appears to be restricted [84]. Intriguingly, the expression of chemokines attracting APCs to the epithelium, such as C-C chemokine ligand (CCL)20 and CCL2, were found to be particularly low in HPV-infected epithelia in vivo and in vitro-studies demonstrated that this results from HPV oncoprotein-mediated suppression [11,12,71–74]. While mucosal HPV oncoproteins target the Langerhans attracting chemokine CCL20 by interfering with the NF-κB pathway [74], cutaneous β-HPV employ a different strategy to suppress CCL20 [71]. In skin, stress signals like UV-light can lead to the depletion of Langerhans cells from the epidermis [85]. Subsequent epithelial up-regulation of the chemokine CCL20 leads to C-C chemokine receptor 6 (CCR6)-dependent repopulation of the skin with CD1a[+] Langerhans cell precursors [86]. In normal human keratinocytes, the differentiation-associated transcription factor C/EBPβ has been identified as a novel regulator of CCL20 expression and in human skin. Both C/EBPβ and CCL20 are expressed in the uppermost nucleated epithelial layers. In HPV8-infected skin, however, CCL20 is almost lacking. It has been shown that the HPV8 E7 oncoprotein directly interacts with C/EBPβ in keratinocytes and interferes with its binding to chromatin within the CCL20 promoter region. This suppresses CCL20 expression. As a consequence, Langerhans cell migration is inhibited preventing repopulation of the epithelium with these important APCs [71].

Thus, HPV infection actively suppresses cell-autonomous viral recognition and acute inflammatory signaling in the host keratinocyte as well as recruitment of epithelial APCs. The low levels of inflammatory cytokines produced by HPV-infected cells may further contribute to the lack of APC activation, eventually allowing the virus to escape from local immunosurveillance and to persist in the epithelium.

4. Immunopathogenesis of Transforming HPV Infection during Progression to Invasive Cancer

4.1. Chronic Stromal Inflammation during Progression to Cancer

There is ample evidence that oncogenic HPV infection in the human cervix and skin starts with expansion of the epithelial stem cell compartment [25,87], which may provide a particularly vulnerable and immune privileged milieu [88–90]. Moreover, HPV blocks acute NF-κB- and C/EBPβ-dependent inflammatory signaling in host keratinocytes as outlined above. In persistent low-grade lesions inflammatory cells are barely detectable.

However, with increasing dysplasia a dramatic increment of stromal infiltration with immune cells is noted in cervical patient biopsies [11–13,83,91]. From other cancers it has become clear that chronic inflammation can fuel immune deviation [14,92–94] and the selection pressure set by the local microenvironment can greatly impact the outcome of the neoplastic process [95]. Since HPV oncoproteins suppress acute inflammatory responses [71–74,76] and HPV-positive cancer cells produce only low chemokine levels [68,69,96], the mechanisms underlying immune cell recruitment remained unclear for a long time.

4.2. Paracrine IL-6 Instructs Myelomonocytic Cells to Create a Pro-Tumorigenic and Immunosuppressive Microenvironment in Cervical Carcinogenesis

A clue came from the observation that HPV-transformed cells potently up-regulate chemoattractants in the tumor stroma [11,12]. In monocytes they induce CCL2 production in the nanogram range [11]. This is supposed to attract further myelomonocytic cells and to sustain the inflammatory microenvironment via a CCR2-dependent autocrine amplification loop. Strikingly, CCL2 also leads to a tremendous production of the matrix-metalloproteinase (MMP)-9 in monocytes via intracellular Ca^{2+}-signaling. MMP-9 has been detected in monocytes starting to infiltrate cervical high-grade lesions at the switch to malignancy [11]. Local production of MMP-9 is a particularly interesting consequence of CCL2/CCR2 stimulation. It can promote vasculogenesis and trigger the angiogenic switch during carcinogenesis required for tumor growth [97]. Transgenic mouse models have provided evidence that myeloid cell-derived MMP-9 expression promotes HPV-driven carcinogenesis, and blockage of MMP-9 strongly impairs HPV oncogene-driven carcinogenesis in mice [98,99]. Importantly, in cervical cancer patients high MMP expression correlates with a poor prognosis [100].

Neutralization experiments revealed that HPV-transformed cells induce CCL2 and subsequent MMP-9 expression in monocytes via a combination of IL-6 and macrophage colony-stimulating factor (M-CSF). They activate the janus kinase/signal transducer and activator of transcription 3 (JAK/STAT3) signaling pathway in monocytes and various JAK/STAT3-inhibitors are able to interfere with this pro-tumorigenic response [11].

During productive infection HPV suppresses IL-6 similar to other cytokines [70]. However, both "switch factors" IL-6 and M-CSF, that are necessary for the pro-tumorigenic response in monocytes, are highly up-regulated during later stages of human cervical carcinogenesis in situ [69,101]. Clinically most relevant, IL-6 expression is associated with a negative prognosis for cervical cancer patients [102] further highlighting its pivotal role in linking chronic inflammation and progression to invasive cancer [103].

Stromal myelomonocytic cells can either differentiate into dendritic cells, APCs destined to mount adaptive immune responses, or into macrophages, tissue-resident phagocytes. For the initiation of adaptive immunity, dendritic cells have to mature as indicated by surface expression of CD83, to up-regulate major histocompatibility complex class I and II required for antigen presentation as well as co-stimulatory molecules such as CD80 and CD86 required for T-cell activation, and finally to produce cytokines that polarize T helper 1 (Th1) cells required for efficient CD8$^+$ cytotoxic T cell responses. Under normal conditions, the migration receptor CCR7 becomes expressed on their surface during maturation, ensuring their responsiveness to lymph node homing chemokines [104,105]. A second migration factor, MMP-9, is needed to allow migration through the extracellular matrix [106,107]. In cervical cancer patients, mature CD83$^+$ dendritic cells are present in the tumor stroma. However, they were found to be largely devoid of CCR7 expression [13]. As the underlying mechanism, it was shown that cervical cancer cells actively interfere with NF-κB activation in CD83$^+$ phenotypically mature dendritic cells. As a consequence, expression of the chemokine receptor CCR7 is suppressed in the dendritic cells and their migration towards lymph node homing chemokine is blocked [13]. This may lead to an impaired antigen transport to secondary lymphoid tissues by stromal dendritic cells in cervical cancer patients. Moreover, in high-grade lesions and invasive cancers only low IL-12p40 expression levels required to mount Th1 responses and a shift from Th1 to Th2 responses are observed [108,109].

In contrast to CCR7, MMP-9 is potently up-regulated in immature and mature dendritic cells. Notably, both CCR7 and MMP-9 up-regulation are mediated by IL-6 from the cervical cancer cells [13,110]. Thus, cervical cancer-derived IL-6 immobilizes dendritic cells in the tumor stroma via CCR7 suppression facilitating local MMP-9 production.

In cervical cancer stroma M2-polarized macrophages accumulate with low IFN-γ production and a low capacity to stimulate T cell proliferation [111]. M2 macrophages are also supposed to have a negative impact on cervical cancer therapy, such as therapy with immunoglobulin G (IgG) antibodies directed against epidermal growth factor receptor (EGFR) [112]. Notably, anti EGFR-specific IgA antibodies may overcome this obstacle, since they can engage tumor-associated myeloid cells for tumor cell killing [113]. This may provide a clear advantage for treatment of M2-infiltrated tumors and therefore IgA antibodies might represent a novel future category of antibodies for targeted tumor therapy.

Besides neoplastic cells, M2-polarized macrophages were also found to express the programmed death-ligand 1 (PD-L1) [114]. There is clear evidence that cytotoxic T-lymphocyte-associated protein (CTLA)-4/CD28 and PD-1/PD-L1 interactions between T cells and other cells exert suppressive signals (immune checkpoints) limiting T cell function and maintaining self-tolerance. This has led to regulatory approval and successful implementation of blocking antibodies targeting CTLA-4, PD-1, and PD-L1 for the treatment of different malignancies. PD-1 was detected on most infiltrating $CD8^+$ T cells in cervical cancer [115] suggesting that M2 macrophages might contribute to suppression of cytotoxic T cell responses. Preliminary studies with the PD-1 blocker pembrolizumab, however, showed only low response rates in cervical cancer patients with advanced disease. Currently, predictive biomarkers allowing more precise patient selection are still lacking and it is expected that efficacy might increase when these blockers are applied in combination with other immunotherapies, such as therapeutic vaccines [16,116]. Notably, it has been shown that M2 macrophage differentiation is driven by cervical cancer-derived IL-6 together with prostaglandin E2 [111].

From these studies evidence is increasing that paracrine IL-6 as well as subsequent STAT3 activation and NF-κB suppression are central for reprogramming myelomonocytic cells creating a pro-tumorigenic and immunosuppressive microenvironment in cervical carcinogenesis. As a consequence, phenotypically mature but functionally impaired dendritic cells and macrophages are actively retained in the tumor stroma and this may (1) prevent the initiation of anti-tumoral adaptive Th1 immune responses; (2) suppress cytotoxic T cell activity; and (3) contribute to an aberrant local expression of MMP-9 that promotes tumor growth and vasculogenesis. This suggests that the IL-6/JAK/STAT3 signaling pathway might be an interesting target to revert immune deviation in cervical cancer. In fact, the IL-6/JAK/STAT3 signaling pathway is "druggable" at various levels [117] and clinical trials for different cancer entities are ongoing (see clinicaltrials.gov).

4.3. Paracrine IL-6 Induces CCL20 Chemokine Expression in Stromal Mesenchymal Cells to Support Th17 Recruitment

In human biopsies of cervical carcinogenesis, enhanced infiltration of Th17 cells is observed with increasing stages of disease and this correlates with up-regulation of CCL20 expression in the stromal mesenchymal compartment [12]. It is well known that Th17 recruitment is mediated by CCL20 in a CCR6-dependent manner [118]. Th17 infiltration starts in precursor lesions, and in invasive cancers a high Th17/Treg ratio is observed [119,120]. Th17 cells are a particular subset of the $CD4^+$ T cell lineage that can exert either regulatory or inflammatory functions [121]. In various different cancers Th17 cells promote tumor growth, angiogenesis and also the recruitment of further inflammatory immune cells [122–124]. Mechanisms underlying stromal CCL20 expression remained unclear until cervical cancer explant cultures revealed that cancer-associated fibroblasts produce enormous amounts of CCL20 [12]. These fibroblasts display an activated phenotype characterized by enhanced C/EBPβ expression. C/EBPβ is also known as the NF-IL6 transcription factor and is inducible by pro-inflammatory cytokines like IL-6 [125]. Recently, C/EBPβ has been identified as a novel transcriptional regulator of CCL20 [71]. Thorough analysis demonstrated that cervical cancer-derived IL-6 drives C/EBPβ-mediated CCL20 induction in cancer-associated fibroblasts and CCL20/CCR6-dependent Th17 recruitment [12]. This further substantiates the key role of paracrine IL-6 in cervical carcinogenesis for the initiation and maintenance of chronic inflammation and tumor progression.

These studies provided evidence that paracrine IL-6 shapes a proinflammatory and immunosuppressive microenvironment in cervical carcinogenesis via activation of two pathways in stromal immune and stromal mesenchymal cells, the JAK/STAT3 and the C/EBPβ signaling pathway. In HPV-driven carcinogenesis, IL-6-induced JAK/STAT3- and C/EBPβ-driven stromal inflammation may play a key role in promoting tumor progression. At the same time paracrine IL-6 can limit NF-κB-dependent anti-viral and anti-tumor immune responses in APCs, further highlighting IL-6 as an attractive target for adjuvant immunotherapy of cervical cancer.

4.4. Regulation and Consequences of Epithelial JAK/STAT3 Pathway Activation in HPV-Induced Carcinogenesis

In HPV-driven carcinogenesis STAT3 is not only activated in the tumor microenvironment. Also, epithelial cells in cervical high-grade lesions display strong pTyr705-STAT3 activation by far exceeding activation levels in normal exocervical epithelium or low-grade lesions [11,126]. Studies with transgenic mice expressing the HPV8 early region under the K14-promoter have provided evidence that epithelial STAT3 activation is necessary for HPV8-driven skin tumorigenesis [127] and this may also apply to human HPV-associated carcinogenesis at other body sites.

Notably, in invasive cervical cancers STAT3 activation is often retained at the tumor margin adjacent to the stroma, while in other parts of the tumor STAT3 activation declines compared to cervical high-grade lesions [126]. This suggests a paracrine mode of STAT3 activation at the tumor invasive margin. The overall decline of STAT3 activity in cervical cancer cells can be explained by a decline of the IL-6 binding receptor chain gp80 in cervical cancer cells [69]. Correspondingly, non-malignant HPV-transformed cells strongly respond to IL-6 alone. In cultured cervical cancer cells, however, efficient STAT3 activation can only be elicited by IL-6 in the presence of soluble gp80 (sgp80) [69,126,128].

Strikingly, when the IL-6/STAT3-signaling pathway is activated in cervical cancer cells, they are more efficiently killed by chemotherapeutic drugs, such as cisplatin or etoposide [126]. This was unexpected, since in various other human malignancies STAT3 activation promotes tumor growth and resistance to chemotherapy [129]. As the underlying mechanism, dramatic up-regulation of the pro-apoptotic factor IRF1 has been identified [126]. Obviously, IL-6/STAT3-induced IRF1 activity prevails over the HPV oncogene-mediated inhibition of IRF1 [61–63]. Of most clinical relevance, epithelial IRF1 expression in pre-therapeutic biopsies of cervical cancers significantly correlates with the patient's response to chemo- or radiochemotherapy [126]. This strongly indicates that pre-therapeutic IRF1 expression could be of high value as a novel biomarker to predict response to chemo- or radiochemotherapy in patients [126].

In summary, targeting the IL-6/JAK/STAT3-pathway emerges as a promising way to improve the immune microenvironment in cervical carcinogenesis. However, recent data suggest that these compounds can also interfere with STAT3-induced IRF1 expression in HPV-positive cancer cells, thereby increasing their resistance to chemo- or radiochemotherapy. Therefore, the timing of IL-6/JAK/STAT3-inhibitor administration during therapy appears to be of critical importance. They should not be applied prior to but rather after chemo- or radiochemotherapy of HPV-driven cancers.

5. Conclusions and Prospects for Diagnosis and Immunotherapy

Over recent years it has become evident that the local microenvironment and particularly the immune system plays a pivotal role in carcinogenesis [93,94]. In various cancer types it has been shown that immune cells have a significant prognostic impact, and in colon cancer an "immunoscore" quantifying in situ immune cell infiltrates seems to be superior to the TNM classification [130,131]. Moreover, immune checkpoint inhibitors interfering with PD-1 or CTLA-4 pathways were recently shown to improve therapy response rates in various cancers including melanoma, non-small cell lung cancer, head and neck cancer, renal cell carcinoma and Hodgkin's lymphoma, and numerous clinical trials are ongoing [132]. This has also changed the virologist's view on cancer progression from the strict focus on viral oncogenes necessary for carcinogenesis to a more complex view of the interaction between tumor viruses, the immune system and the tumor microenvironment in promoting cancer progression.

In HPV-induced carcinogenesis, there is now ample evidence for a stage-specific interplay between virally-infected keratinocytes and the local immune microenvironment that can determine the course of disease (Figure 1). This knowledge will pave the way for (1) novel diagnostic tools including immunoscores that allow the discrimination of non-progressing and progressing precursor lesions; (2) novel biomarkers that improve the prediction of therapy response, such as IRF1, which indicates response to chemo- or radiochemotherapy; and (3) novel immunotherapeutic approaches beyond checkpoint inhibitors that target critical stage-specific mechanisms and pathways in HPV-driven carcinogenesis. These include IRF3-activating compounds such as dsRNA to combat oncogenic β-HPV infection, dsRNA-based immunotherapies for patients with high intratumoral RIPK3 levels, IgA-based antibody therapies that engage myeloid cells for tumor cell killing, and last but not least IL-6/JAK/STAT3-pathway blocking regimens for cervical cancer patients after treatment with chemo- or radiochemotherapy.

Acknowledgments: The author wishes to thank all colleagues at the Institute of Virology, Saarland University Medical Center, and all collaborators for helpful discussions, and would like to recognize all research studies that could not be considered in this review due to space limitations. The author is also grateful to the Deutsche Krebshilfe (grant No. 109752), the Saarland Staatskanzlei (grant No. WT/2 LFFP 14/15) and the Gemeinsamer Bundesausschuss (grant No. 01VSF16050) for funding.

Conflicts of Interest: The author declares no conflict of interest.

References

1. Zur Hausen, H. Viruses in human cancers. *Curr. Sci.* **2001**, *81*, 523–527.
2. Parkin, D.M. The global health burden of infection-associated cancers in the year 2002. *Int. J. Cancer* **2006**, *118*, 3030–3044. [CrossRef] [PubMed]
3. Bosch, F.X.; Broker, T.R.; Forman, D.; Moscicki, A.B.; Gillison, M.L.; Doorbar, J.; Stern, P.L.; Stanley, M.; Arbyn, M.; Poljak, M.; et al. Comprehensive control of human papillomavirus infections and related diseases. *Vaccine* **2013**, *31*, 1–31. [CrossRef] [PubMed]
4. Walboomers, J.M.; Jacobs, M.V.; Manos, M.M.; Bosch, F.X.; Kummer, J.A.; Shah, K.V.; Snijders, P.J.; Peto, J.; Meijer, C.J.; Munoz, N. Human papillomavirus is a necessary cause of invasive cervical cancer worldwide. *J. Pathol.* **1999**, *189*, 12–19. [CrossRef]
5. Chaturvedi, A.K.; Engels, E.A.; Pfeiffer, R.M.; Hernandez, B.Y.; Xiao, W.; Kim, E.; Jiang, B.; Goodman, M.T.; Sibug-Saber, M.; Cozen, W.; et al. Human papillomavirus and rising oropharyngeal cancer incidence in the United States. *J. Clin. Oncol.* **2011**, *29*, 4294–4301. [CrossRef] [PubMed]
6. Baldur-Felskov, B.; Hannibal, C.G.; Munk, C.; Kjaer, S.K. Increased incidence of penile cancer and high-grade penile intraepithelial neoplasia in Denmark 1978–2008: A nationwide population-based study. *Cancer Causes Control* **2012**, *23*, 273–280. [CrossRef] [PubMed]
7. Buttmann-Schweiger, N.; Klug, S.J.; Luyten, A.; Holleczek, B.; Heitz, F.; du Bois, A.; Kraywinkel, K. Incidence patterns and temporal trends of invasive nonmelanotic vulvar tumors in Germany 1999–2011. A population-based cancer registry analysis. *PLoS ONE* **2015**, *10*, e0128073. [CrossRef] [PubMed]
8. Smola, S. Human papillomaviruses and skin cancer. *Adv. Exp. Med. Biol.* **2014**, *810*, 192–207. [PubMed]
9. Howley, P.M.; Pfister, H.J. β genus papillomaviruses and skin cancer. *Virology* **2015**, *479–480*, 290–296. [CrossRef] [PubMed]
10. Munoz, N.; Castellsague, X.; de Gonzalez, A.B.; Gissmann, L. Chapter 1: HPV in the etiology of human cancer. *Vaccine* **2006**, *24*, 1–10. [CrossRef] [PubMed]
11. Schroer, N.; Pahne, J.; Walch, B.; Wickenhauser, C.; Smola, S. Molecular pathobiology of human cervical high-grade lesions: Paracrine STAT3 activation in tumor-instructed myeloid cells drives local MMP-9 expression. *Cancer Res.* **2011**, *71*, 87–97. [CrossRef] [PubMed]
12. Walch-Ruckheim, B.; Mavrova, R.; Henning, M.; Vicinus, B.; Kim, Y.J.; Bohle, R.M.; Juhasz-Boss, I.; Solomayer, E.F.; Smola, S. Stromal fibroblasts induce CCL20 through IL6/C/EBPβ to support the recruitment of Th17 Cells during cervical cancer progression. *Cancer Res.* **2015**, *75*, 5248–5259. [CrossRef] [PubMed]
13. Pahne-Zeppenfeld, J.; Schroer, N.; Walch-Ruckheim, B.; Oldak, M.; Gorter, A.; Hegde, S.; Smola, S. Cervical cancer cell-derived interleukin-6 impairs CCR7-dependent migration of MMP-9-expressing dendritic cells. *Int. J. Cancer* **2014**, *134*, 2061–2073. [CrossRef] [PubMed]
14. Shalapour, S.; Karin, M. Immunity, inflammation, and cancer: An eternal fight between good and evil. *J. Clin. Investig.* **2015**, *125*, 3347–3355. [CrossRef] [PubMed]
15. Schiffman, M.; Doorbar, J.; Wentzensen, N.; de Sanjose, S.; Fakhry, C.; Monk, B.J.; Stanley, M.A.; Franceschi, S. Carcinogenic human papillomavirus infection. *Nat. Rev. Dis. Prim.* **2016**, *2*, 16086. [CrossRef] [PubMed]
16. Smola, S.; Trimble, C.; Stern, P.L. Human papillomavirus-driven immune deviation: Challenge and novel opportunity for immunotherapy. *Ther. Adv. Vaccines* **2017**, *5*, 69–82. [CrossRef] [PubMed]

17. Bosch, F.X.; Burchell, A.N.; Schiffman, M.; Giuliano, A.R.; de Sanjose, S.; Bruni, L.; Tortolero-Luna, G.; Kjaer, S.K.; Munoz, N. Epidemiology and natural history of human papillomavirus infections and type-specific implications in cervical neoplasia. *Vaccine* **2008**, *26*, 1–16. [CrossRef] [PubMed]

18. Papillomavirus Episteme. Available online: https://pave.niaid.nih.gov (accessed on 3 June 2017).

19. Bernard, H.U.; Burk, R.D.; Chen, Z.; van Doorslaer, K.; Hausen, H.; de Villiers, E.M. Classification of papillomaviruses (PVs) based on 189 PV types and proposal of taxonomic amendments. *Virology* **2010**, *401*, 70–79. [CrossRef] [PubMed]

20. Egawa, N.; Egawa, K.; Griffin, H.; Doorbar, J. Human papillomaviruses; epithelial tropisms, and the development of neoplasia. *Viruses* **2015**, *7*, 3863–3890. [CrossRef] [PubMed]

21. Bouvard, V.; Baan, R.; Straif, K.; Grosse, Y.; Secretan, B.; El Ghissassi, F.; Benbrahim-Tallaa, L.; Guha, N.; Freeman, C.; Galichet, L.; et al. A review of human carcinogens—Part B: Biological agents. *Lancet Oncol.* **2009**, *10*, 321–322. [CrossRef]

22. Moody, C.A.; Laimins, L.A. Human papillomavirus oncoproteins: Pathways to transformation. *Nat. Rev. Cancer* **2010**, *10*, 550–560. [CrossRef] [PubMed]

23. Orth, G.; Jablonska, S.; Favre, M.; Croissant, O.; Jarzabek-Chorzelska, M.; Rzesa, G. Characterization of two types of human papillomaviruses in lesions of epidermodysplasia verruciformis. *Proc. Natl. Acad. Sci. USA* **1978**, *75*, 1537–1541. [CrossRef] [PubMed]

24. Tommasino, M. The biology of β human papillomaviruses. *Virus Res.* **2017**, *231*, 128–138. [CrossRef] [PubMed]

25. Marthaler, A.M.; Podgorska, M.; Feld, P.; Fingerle, A.; Knerr-Rupp, K.; Grasser, F.; Smola, H.; Roemer, K.; Ebert, E.; Kim, Y.J.; et al. Identification of C/EBPα as a novel target of the HPV8 E6 protein regulating miR-203 in human keratinocytes. *PLoS Pathog.* **2017**, *13*, e1006406. [CrossRef] [PubMed]

26. Marcuzzi, G.P.; Hufbauer, M.; Kasper, H.U.; Weissenborn, S.J.; Smola, S.; Pfister, H. Spontaneous tumour development in human papillomavirus type 8 E6 transgenic mice and rapid induction by UV-light exposure and wounding. *J. Gen. Virol.* **2009**, *90*, 2855–2864. [CrossRef] [PubMed]

27. McKenna, D.J.; McDade, S.S.; Patel, D.; McCance, D.J. MicroRNA 203 expression in keratinocytes is dependent on regulation of p53 levels by E6. *J. Virol.* **2010**, *84*, 10644–10652. [CrossRef] [PubMed]

28. Melar-New, M.; Laimins, L.A. Human papillomaviruses modulate expression of microRNA 203 upon epithelial differentiation to control levels of p63 proteins. *J. Virol.* **2010**, *84*, 5212–5221. [CrossRef] [PubMed]

29. Lanfredini, S.; Olivero, C.; Borgogna, C.; Calati, F.; Powell, K.; Davies, K.J.; De Andrea, M.; Harries, S.; Tang, H.K.C.; Pfister, H.; et al. HPV8 Field Cancerization in a Transgenic Mouse Model Is due to Lrig1+ Keratinocyte Stem Cell Expansion. *J. Investig. Dermatol.* **2017**. [CrossRef] [PubMed]

30. Meyers, J.M.; Spangle, J.M.; Munger, K. The human papillomavirus type 8 E6 protein interferes with NOTCH activation during keratinocyte differentiation. *J. Virol.* **2013**, *87*, 4762–4767. [CrossRef] [PubMed]

31. Schaper, I.D.; Marcuzzi, G.P.; Weissenborn, S.J.; Kasper, H.U.; Dries, V.; Smyth, N.; Fuchs, P.; Pfister, H. Development of skin tumors in mice transgenic for early genes of human papillomavirus type 8. *Cancer Res.* **2005**, *65*, 1394–1400. [CrossRef] [PubMed]

32. Underbrink, M.P.; Howie, H.L.; Bedard, K.M.; Koop, J.I.; Galloway, D.A. E6 proteins from multiple human βpapillomavirus types degrade Bak and protect keratinocytes from apoptosis after UVB irradiation. *J. Virol.* **2008**, *82*, 10408–10417. [CrossRef] [PubMed]

33. Wallace, N.A.; Galloway, D.A. Manipulation of cellular DNA damage repair machinery facilitates propagation of human papillomaviruses. *Semin. Cancer Biol.* **2014**, *26*, 30–42. [CrossRef] [PubMed]

34. Tan, M.J.; White, E.A.; Sowa, M.E.; Harper, J.W.; Aster, J.C.; Howley, P.M. Cutaneous β-human papillomavirus E6 proteins bind mastermind-like coactivators and repress NOTCH signaling. *Proc. Natl. Acad. Sci. USA* **2012**, *109*, E1473–E1480. [CrossRef] [PubMed]

35. Viarisio, D.; Mueller-Decker, K.; Kloz, U.; Aengeneyndt, B.; Kopp-Schneider, A.; Grone, H.J.; Gheit, T.; Flechtenmacher, C.; Gissmann, L.; Tommasino, M. E6 and E7 from β HPV38 cooperate with ultraviolet light in the development of actinic keratosis-like lesions and squamous cell carcinoma in mice. *PLoS Pathog.* **2011**, *7*, e1002125. [CrossRef] [PubMed]

36. Evander, M.; Edlund, K.; Gustafsson, A.; Jonsson, M.; Karlsson, R.; Rylander, E.; Wadell, G. Human papillomavirus infection is transient in young women: A population-based cohort study. *J. Infect. Dis.* **1995**, *171*, 1026–1030. [CrossRef] [PubMed]

37. Ho, G.Y.; Bierman, R.; Beardsley, L.; Chang, C.J.; Burk, R.D. Natural history of cervicovaginal papillomavirus infection in young women. *N. Engl. J. Med.* **1998**, *338*, 423–428. [CrossRef] [PubMed]
38. Moscicki, A.B.; Shiboski, S.; Broering, J.; Powell, K.; Clayton, L.; Jay, N.; Darragh, T.M.; Brescia, R.; Kanowitz, S.; Miller, S.B.; et al. The natural history of human papillomavirus infection as measured by repeated DNA testing in adolescent and young women. *J. Pediatr.* **1998**, *132*, 277–284. [CrossRef]
39. De Jong, A.; van der Burg, S.H.; Kwappenberg, K.M.; van der Hulst, J.M.; Franken, K.L.; Geluk, A.; van Meijgaarden, K.E.; Drijfhout, J.W.; Kenter, G.; Vermeij, P.; et al. Frequent detection of human papillomavirus 16 E2-specific T-helper immunity in healthy subjects. *Cancer Res.* **2002**, *62*, 472–479. [PubMed]
40. Woo, Y.L.; van den Hende, M.; Sterling, J.C.; Coleman, N.; Crawford, R.A.; Kwappenberg, K.M.; Stanley, M.A.; van der Burg, S.H. A prospective study on the natural course of low-grade squamous intraepithelial lesions and the presence of HPV16 E2-, E6- and E7-specific T-cell responses. *Int. J. Cancer* **2010**, *126*, 133–141. [CrossRef] [PubMed]
41. Wieland, U.; Kreuter, A.; Pfister, H. Human papillomavirus and immunosuppression. *Curr. Probl. Dermatol.* **2014**, *45*, 154–165. [PubMed]
42. Proby, C.M.; Harwood, C.A.; Neale, R.E.; Green, A.C.; Euvrard, S.; Naldi, L.; Tessari, G.; Feltkamp, M.C.; de Koning, M.N.; Quint, W.G.; et al. A case-control study of βpapillomavirus infection and cutaneous squamous cell carcinoma in organ transplant recipients. *Am. J. Transplant.* **2011**, *11*, 1498–1508. [CrossRef] [PubMed]
43. Wang, C.J.; Sparano, J.; Palefsky, J.M. Human immunodeficiency virus/AIDS, human papillomavirus, and anal cancer. *Surg. Oncol. Clin. N. Am.* **2017**, *26*, 17–31. [CrossRef] [PubMed]
44. Brickman, C.; Palefsky, J.M. Human papillomavirus in the HIV-infected host: Epidemiology and pathogenesis in the antiretroviral era. *Curr. HIV/AIDS Rep.* **2015**, *12*, 6–15. [CrossRef] [PubMed]
45. Ellerbrock, T.V.; Chiasson, M.A.; Bush, T.J.; Sun, X.W.; Sawo, D.; Brudney, K.; Wright, T.C., Jr. Incidence of cervical squamous intraepithelial lesions in HIV-infected women. *JAMA* **2000**, *283*, 1031–1037. [CrossRef] [PubMed]
46. Halpert, R.; Fruchter, R.G.; Sedlis, A.; Butt, K.; Boyce, J.G.; Sillman, F.H. Human papillomavirus and lower genital neoplasia in renal transplant patients. *Obstet. Gynecol.* **1986**, *68*, 251–258. [PubMed]
47. Kupper, T.S.; Fuhlbrigge, R.C. Immune surveillance in the skin: Mechanisms and clinical consequences. *Nat. Rev. Immunol.* **2004**, *4*, 211–222. [CrossRef] [PubMed]
48. Pasparakis, M.; Haase, I.; Nestle, F.O. Mechanisms regulating skin immunity and inflammation. *Nat. Rev. Immunol.* **2014**, *14*, 289–301. [CrossRef] [PubMed]
49. Kawamura, T.; Ogawa, Y.; Aoki, R.; Shimada, S. Innate and intrinsic antiviral immunity in skin. *J. Dermatol. Sci.* **2014**, *75*, 159–166. [CrossRef] [PubMed]
50. Chopin, M.; Nutt, S.L. Establishing and maintaining the Langerhans cell network. *Semin. Cell Dev. Biol.* **2015**, *41*, 23–29. [CrossRef] [PubMed]
51. Belkaid, Y.; Tamoutounour, S. The influence of skin microorganisms on cutaneous immunity. *Nat. Rev. Immunol.* **2016**, *16*, 353–366. [CrossRef] [PubMed]
52. Einstein, M.H.; Schiller, J.T.; Viscidi, R.P.; Strickler, H.D.; Coursaget, P.; Tan, T.; Halsey, N.; Jenkins, D. Clinician's guide to human papillomavirus immunology: Knowns and unknowns. *Lancet Infect. Dis.* **2009**, *9*, 347–356. [CrossRef]
53. Stanley, M.A. Epithelial cell responses to infection with human papillomavirus. *Clin. Microbiol. Rev.* **2012**, *25*, 215–222. [CrossRef] [PubMed]
54. Smola-Hess, S.; Pfister, H.J. Immune evasion in genital papillomavirus infection and cervical cancer: Role of cytokines and chemokines. In *Papillomavirus Research: From Natural History to Vaccines and Beyond*; Campo, S., Ed.; Caister Academic Press: Norfolk, UK, 2006; Chapter 20; pp. 321–339.
55. Hasan, U.A.; Bates, E.; Takeshita, F.; Biliato, A.; Accardi, R.; Bouvard, V.; Mansour, M.; Vincent, I.; Gissmann, L.; Iftner, T.; et al. TLR9 expression and function is abolished by the cervical cancer-associated human papillomavirus type 16. *J. Immunol.* **2007**, *178*, 3186–3197. [CrossRef] [PubMed]
56. Pacini, L.; Ceraolo, M.G.; Venuti, A.; Melita, G.; Hasan, U.A.; Accardi, R.; Tommasino, M. UV radiation activates toll-like receptor 9 expression in primary human keratinocytes, an event inhibited by human papillomavirus type 38 E6 and E7 oncoproteins. *J. Virol.* **2017**. [CrossRef] [PubMed]

57. Nees, M.; Geoghegan, J.M.; Hyman, T.; Frank, S.; Miller, L.; Woodworth, C.D. Papillomavirus type 16 oncogenes downregulate expression of interferon-responsive genes and upregulate proliferation-associated and NF-κB-responsive genes in cervical keratinocytes. *J. Virol.* **2001**, *75*, 4283–4296. [CrossRef] [PubMed]

58. Chang, Y.E.; Laimins, L.A. Interferon-inducible genes are major targets of human papillomavirus type 31: Insights from microarray analysis. *Dis. Markers* **2001**, *17*, 139–142. [CrossRef] [PubMed]

59. Barnard, P.; McMillan, N.A. The human papillomavirus E7 oncoprotein abrogates signaling mediated by interferon-α. *Virology* **1999**, *259*, 305–313. [CrossRef] [PubMed]

60. Barnard, P.; Payne, E.; McMillan, N.A. The human papillomavirus E7 protein is able to inhibit the antiviral and anti-growth functions of interferon-α. *Virology* **2000**, *277*, 411–419. [CrossRef] [PubMed]

61. Zhou, F.; Chen, J.; Zhao, K.N. Human papillomavirus 16-encoded E7 protein inhibits IFN-γ-mediated MHC class I antigen presentation and CTL-induced lysis by blocking IRF-1 expression in mouse keratinocytes. *J. Gen. Virol.* **2013**, *94*, 2504–2514. [CrossRef] [PubMed]

62. Um, S.J.; Rhyu, J.W.; Kim, E.J.; Jeon, K.C.; Hwang, E.S.; Park, J.S. Abrogation of IRF-1 response by high-risk HPV E7 protein in vivo. *Cancer Lett.* **2002**, *179*, 205–212. [CrossRef]

63. Park, J.S.; Kim, E.J.; Kwon, H.J.; Hwang, E.S.; Namkoong, S.E.; Um, S.J. Inactivation of interferon regulatory factor-1 tumor suppressor protein by HPV E7 oncoprotein. Implication for the E7-mediated immune evasion mechanism in cervical carcinogenesis. *J. Biol. Chem.* **2000**, *275*, 6764–6769. [CrossRef] [PubMed]

64. Ronco, L.V.; Karpova, A.Y.; Vidal, M.; Howley, P.M. Human papillomavirus 16 E6 oncoprotein binds to interferon regulatory factor-3 and inhibits its transcriptional activity. *Genes Dev.* **1998**, *12*, 2061–2072. [CrossRef] [PubMed]

65. Kim, T.K.; Kim, T.; Kim, T.Y.; Lee, W.G.; Yim, J. Chemotherapeutic DNA-damaging drugs activate interferon regulatory factor-7 by the mitogen-activated protein kinase kinase-4-cJun NH2-terminal kinase pathway. *Cancer Res.* **2000**, *60*, 1153–1156. [PubMed]

66. Oldak, M.; Tolzmann, L.; Wnorowski, A.; Podgorska, M.J.; Silling, S.; Lin, R.; Hiscott, J.; Muller, C.S.; Vogt, T.; Smola, H.; et al. Differential regulation of human papillomavirus type 8 by interferon regulatory factors 3 and 7. *J. Virol.* **2011**, *85*, 178–188. [CrossRef] [PubMed]

67. Strittmatter, G.E.; Sand, J.; Sauter, M.; Seyffert, M.; Steigerwald, R.; Fraefel, C.; Smola, S.; French, L.E.; Beer, H.D. IFN-γ Primes Keratinocytes for HSV-1-Induced Inflammasome Activation. *J. Investig. Dermatol.* **2016**, *136*, 610–620. [CrossRef] [PubMed]

68. Altenburg, A.; Baldus, S.E.; Smola, H.; Pfister, H.; Hess, S. CD40 ligand-CD40 interaction induces chemokines in cervical carcinoma cells in synergism with IFN-γ. *J. Immunol.* **1999**, *162*, 4140–4147. [PubMed]

69. Hess, S.; Smola, H.; Sandaradura de Silva, U.; Hadaschik, D.; Kube, D.; Baldus, S.E.; Flucke, U.; Pfister, H. Loss of IL-6 receptor expression in cervical carcinoma cells inhibits autocrine IL-6 stimulation: Abrogation of constitutive monocyte chemoattractant protein-1 production. *J. Immunol.* **2000**, *165*, 1939–1948. [CrossRef] [PubMed]

70. Karim, R.; Meyers, C.; Backendorf, C.; Ludigs, K.; Offringa, R.; van Ommen, G.J.; Melief, C.J.; van der Burg, S.H.; Boer, J.M. Human papillomavirus deregulates the response of a cellular network comprising of chemotactic and proinflammatory genes. *PLoS ONE* **2011**, *6*, e17848. [CrossRef] [PubMed]

71. Sperling, T.; Oldak, M.; Walch-Ruckheim, B.; Wickenhauser, C.; Doorbar, J.; Pfister, H.; Malejczyk, M.; Majewski, S.; Keates, A.C.; Smola, S. Human papillomavirus type 8 interferes with a novel C/EBPβ-mediated mechanism of keratinocyte CCL20 chemokine expression and Langerhans cell migration. *PLoS Pathog.* **2012**, *8*, e1002833. [CrossRef] [PubMed]

72. Kleine-Lowinski, K.; Gillitzer, R.; Kuhne-Heid, R.; Rosl, F. Monocyte-chemo-attractant-protein-1 (MCP-1)-gene expression in cervical intra-epithelial neoplasias and cervical carcinomas. *Int. J. Cancer* **1999**, *82*, 6–11. [CrossRef]

73. Kleine-Lowinski, K.; Rheinwald, J.G.; Fichorova, R.N.; Anderson, D.J.; Basile, J.; Munger, K.; Daly, C.M.; Rosl, F.; Rollins, B.J. Selective suppression of monocyte chemoattractant protein-1 expression by human papillomavirus E6 and E7 oncoproteins in human cervical epithelial and epidermal cells. *Int. J. Cancer* **2003**, *107*, 407–415. [CrossRef] [PubMed]

74. Guess, J.C.; McCance, D.J. Decreased migration of Langerhans precursor-like cells in response to human keratinocytes expressing human papillomavirus type 16 E6/E7 is related to reduced macrophage inflammatory protein-3α production. *J. Virol.* **2005**, *79*, 14852–14862. [CrossRef] [PubMed]

75. Huang, S.M.; McCance, D.J. Down regulation of the interleukin-8 promoter by human papillomavirus type 16 E6 and E7 through effects on CREB binding protein/p300 and P/CAF. *J. Virol.* **2002**, *76*, 8710–8721. [CrossRef] [PubMed]

76. Karim, R.; Tummers, B.; Meyers, C.; Biryukov, J.L.; Alam, S.; Backendorf, C.; Jha, V.; Offringa, R.; van Ommen, G.J.; Melief, C.J.; et al. Human papillomavirus (HPV) upregulates the cellular deubiquitinase UCHL1 to suppress the keratinocyte's innate immune response. *PLoS Pathog.* **2013**, *9*, e1003384. [CrossRef] [PubMed]

77. Niebler, M.; Qian, X.; Hofler, D.; Kogosov, V.; Kaewprag, J.; Kaufmann, A.M.; Ly, R.; Bohmer, G.; Zawatzky, R.; Rosl, F.; et al. Post-translational control of IL-1β via the human papillomavirus type 16 E6 oncoprotein: A novel mechanism of innate immune escape mediated by the E3-ubiquitin ligase E6-AP and p53. *PLoS Pathog.* **2013**, *9*, e1003536. [CrossRef] [PubMed]

78. Iglesias, M.; Yen, K.; Gaiotti, D.; Hildesheim, A.; Stoler, M.H.; Woodworth, C.D. Human papillomavirus type 16 E7 protein sensitizes cervical keratinocytes to apoptosis and release of interleukin-1α. *Oncogene* **1998**, *17*, 1195–1205. [CrossRef] [PubMed]

79. Schmidt, S.V.; Seibert, S.; Walch-Ruckheim, B.; Vicinus, B.; Kamionka, E.M.; Pahne-Zeppenfeld, J.; Solomayer, E.F.; Kim, Y.J.; Bohle, R.M.; Smola, S. RIPK3 expression in cervical cancer cells is required for PolyIC-induced necroptosis, IL-1α release, and efficient paracrine dendritic cell activation. *Oncotarget* **2015**, *6*, 8635–8647. [CrossRef] [PubMed]

80. Smola, S. RIPK3—A predictive marker for personalized immunotherapy? *Oncoimmunology* **2016**, *5*, e1075695. [CrossRef] [PubMed]

81. De Munck, J.; Binks, A.; McNeish, I.A.; Aerts, J.L. Oncolytic virus-induced cell death and immunity: A match made in heaven? *J. Leukoc. Biol.* **2017**, *102*, 631–643. [CrossRef] [PubMed]

82. Klechevsky, E.; Morita, R.; Liu, M.; Cao, Y.; Coquery, S.; Thompson-Snipes, L.; Briere, F.; Chaussabel, D.; Zurawski, G.; Palucka, A.K.; et al. Functional specializations of human epidermal Langerhans cells and CD14+ dermal dendritic cells. *Immunity* **2008**, *29*, 497–510. [CrossRef] [PubMed]

83. Al-Saleh, W.; Delvenne, P.; Arrese, J.E.; Nikkels, A.F.; Pierard, G.E.; Boniver, J. Inverse modulation of intraepithelial Langerhans' cells and stromal macrophage/dendrocyte populations in human papillomavirus-associated squamous intraepithelial lesions of the cervix. *Virchows Arch.* **1995**, *427*, 41–48. [CrossRef] [PubMed]

84. Fausch, S.C.; Da Silva, D.M.; Rudolf, M.P.; Kast, W.M. Human papillomavirus virus-like particles do not activate Langerhans cells: A possible immune escape mechanism used by human papillomaviruses. *J. Immunol.* **2002**, *169*, 3242–3249. [CrossRef] [PubMed]

85. Seite, S.; Zucchi, H.; Moyal, D.; Tison, S.; Compan, D.; Christiaens, F.; Gueniche, A.; Fourtanier, A. Alterations in human epidermal Langerhans cells by ultraviolet radiation: Quantitative and morphological study. *Br. J. Dermatol.* **2003**, *148*, 291–299. [CrossRef] [PubMed]

86. Charbonnier, A.S.; Kohrgruber, N.; Kriehuber, E.; Stingl, G.; Rot, A.; Maurer, D. Macrophage inflammatory protein 3α is involved in the constitutive trafficking of epidermal langerhans cells. *J. Exp. Med.* **1999**, *190*, 1755–1768. [CrossRef] [PubMed]

87. Herfs, M.; Yamamoto, Y.; Laury, A.; Wang, X.; Nucci, M.R.; McLaughlin-Drubin, M.E.; Munger, K.; Feldman, S.; McKeon, F.D.; Xian, W.; et al. A discrete population of squamocolumnar junction cells implicated in the pathogenesis of cervical cancer. *Proc. Natl. Acad. Sci. USA* **2012**, *109*, 10516–10521. [CrossRef] [PubMed]

88. Christoph, T.; Muller-Rover, S.; Audring, H.; Tobin, D.J.; Hermes, B.; Cotsarelis, G.; Ruckert, R.; Paus, R. The human hair follicle immune system: Cellular composition and immune privilege. *Br. J. Dermatol.* **2000**, *142*, 862–873. [CrossRef] [PubMed]

89. Delvenne, P.; Hubert, P.; Jacobs, N. Epithelial metaplasia: An inadequate environment for antitumour immunity? *Trends Immunol.* **2004**, *25*, 169–173. [CrossRef] [PubMed]

90. Hubert, P.; Herman, L.; Roncarati, P.; Maillard, C.; Renoux, V.; Demoulin, S.; Erpicum, C.; Foidart, J.M.; Boniver, J.; Noel, A.; et al. Altered α-defensin 5 expression in cervical squamocolumnar junction: Implication in the formation of a viral/tumour-permissive microenvironment. *J. Pathol.* **2014**, *234*, 464–477. [CrossRef] [PubMed]

91. Mazibrada, J.; Ritta, M.; Mondini, M.; De Andrea, M.; Azzimonti, B.; Borgogna, C.; Ciotti, M.; Orlando, A.; Surico, N.; Chiusa, L.; et al. Interaction between inflammation and angiogenesis during different stages of cervical carcinogenesis. *Gynecol. Oncol.* **2007**, *108*, 112–120. [CrossRef] [PubMed]
92. Coussens, L.M.; Werb, Z. Inflammation and cancer. *Nature* **2002**, *420*, 860–867. [CrossRef] [PubMed]
93. Hanahan, D.; Weinberg, R.A. The hallmarks of cancer. *Cell* **2000**, *100*, 57–70. [CrossRef]
94. Hanahan, D.; Weinberg, R.A. Hallmarks of cancer: The next generation. *Cell* **2011**, *144*, 646–674. [CrossRef] [PubMed]
95. DeGregori, J. Connecting cancer to its causes requires incorporation of effects on tissue microenvironments. *Cancer Res.* **2017**. [CrossRef] [PubMed]
96. Rösl, F.; Lengert, M.; Albrecht, J.; Kleine, K.; Zawatzky, R.; Schraven, B.; zur Hausen, H. Differential regulation of the JE gene encoding the monocyte chemoattractant protein (MCP-1) in cervical carcinoma cells and derived hybrids. *J. Virol.* **1994**, *68*, 2142–2150. [PubMed]
97. Bergers, G.; Brekken, R.; McMahon, G.; Vu, T.H.; Itoh, T.; Tamaki, K.; Tanzawa, K.; Thorpe, P.; Itohara, S.; Werb, Z.; et al. Matrix metalloproteinase-9 triggers the angiogenic switch during carcinogenesis. *Nat. Cell Biol.* **2000**, *2*, 737–744. [PubMed]
98. Coussens, L.M.; Tinkle, C.L.; Hanahan, D.; Werb, Z. MMP-9 supplied by bone marrow-derived cells contributes to skin carcinogenesis. *Cell* **2000**, *103*, 481–490. [CrossRef]
99. Giraudo, E.; Inoue, M.; Hanahan, D. An amino-bisphosphonate targets MMP-9-expressing macrophages and angiogenesis to impair cervical carcinogenesis. *J. Clin. Investig.* **2004**, *114*, 623–633. [CrossRef] [PubMed]
100. Sheu, B.C.; Lien, H.C.; Ho, H.N.; Lin, H.H.; Chow, S.N.; Huang, S.C.; Hsu, S.M. Increased expression and activation of gelatinolytic matrix metalloproteinases is associated with the progression and recurrence of human cervical cancer. *Cancer Res.* **2003**, *63*, 6537–6542. [PubMed]
101. Kirma, N.; Hammes, L.S.; Liu, Y.G.; Nair, H.B.; Valente, P.T.; Kumar, S.; Flowers, L.C.; Tekmal, R.R. Elevated expression of the oncogene c-fms and its ligand, the macrophage colony-stimulating factor-1, in cervical cancer and the role of transforming growth factor-β1 in inducing c-fms expression. *Cancer Res.* **2007**, *67*, 1918–1926. [CrossRef] [PubMed]
102. Srivani, R.; Nagarajan, B. A prognostic insight on in vivo expression of interleukin-6 in uterine cervical cancer. *Int. J. Gynecol. Cancer* **2003**, *13*, 331–339. [CrossRef] [PubMed]
103. Taniguchi, K.; Karin, M. IL-6 and related cytokines as the critical lynchpins between inflammation and cancer. *Semin. Immunol.* **2014**, *26*, 54–74. [CrossRef] [PubMed]
104. Sallusto, F.; Schaerli, P.; Loetscher, P.; Schaniel, C.; Lenig, D.; Mackay, C.R.; Qin, S.; Lanzavecchia, A. Rapid and coordinated switch in chemokine receptor expression during dendritic cell maturation. *Eur. J. Immunol.* **1998**, *28*, 2760–2769. [CrossRef]
105. Forster, R.; Schubel, A.; Breitfeld, D.; Kremmer, E.; Renner-Muller, I.; Wolf, E.; Lipp, M. CCR7 coordinates the primary immune response by establishing functional microenvironments in secondary lymphoid organs. *Cell* **1999**, *99*, 23–33. [CrossRef]
106. Ratzinger, G.; Stoitzner, P.; Ebner, S.; Lutz, M.B.; Layton, G.T.; Rainer, C.; Senior, R.M.; Shipley, J.M.; Fritsch, P.; Schuler, G.; et al. Matrix metalloproteinases 9 and 2 are necessary for the migration of Langerhans cells and dermal dendritic cells from human and murine skin. *J. Immunol.* **2002**, *168*, 4361–4371. [CrossRef] [PubMed]
107. Yen, J.H.; Khayrullina, T.; Ganea, D. PGE2-induced metalloproteinase-9 is essential for dendritic cell migration. *Blood* **2008**, *111*, 260–270. [CrossRef] [PubMed]
108. Giannini, S.L.; Al-Saleh, W.; Piron, H.; Jacobs, N.; Doyen, J.; Boniver, J.; Delvenne, P. Cytokine expression in squamous intraepithelial lesions of the uterine cervix: Implications for the generation of local immunosuppression. *Clin. Exp. Immunol.* **1998**, *113*, 183–189. [CrossRef] [PubMed]
109. Sheu, B.C.; Lin, R.H.; Lien, H.C.; Ho, H.N.; Hsu, S.M.; Huang, S.C. Predominant Th2/Tc2 polarity of tumor-infiltrating lymphocytes in human cervical cancer. *J. Immunol.* **2001**, *167*, 2972–2978. [CrossRef] [PubMed]
110. Hegde, S.; Pahne, J.; Smola-Hess, S. Novel immunosuppressive properties of interleukin-6 in dendritic cells: Inhibition of NF-κB binding activity and CCR7 expression. *Faseb J.* **2004**, *18*, 1439–1441. [CrossRef] [PubMed]
111. Heusinkveld, M.; de Vos van Steenwijk, P.J.; Goedemans, R.; Ramwadhdoebe, T.H.; Gorter, A.; Welters, M.J.; van Hall, T.; van der Burg, S.H. M2 macrophages induced by prostaglandin E2 and IL-6 from cervical carcinoma are switched to activated M1 macrophages by CD4+ Th1 cells. *J. Immunol.* **2011**, *187*, 1157–1165. [CrossRef] [PubMed]

112. Pander, J.; Heusinkveld, M.; van der Straaten, T.; Jordanova, E.S.; Baak-Pablo, R.; Gelderblom, H.; Morreau, H.; van der Burg, S.H.; Guchelaar, H.J.; van Hall, T. Activation of tumor-promoting type 2 macrophages by EGFR-targeting antibody cetuximab. *Clin. Cancer Res.* **2011**, *17*, 5668–5673. [CrossRef] [PubMed]

113. Lohse, S.; Meyer, S.; Meulenbroek, L.A.; Jansen, J.H.; Nederend, M.; Kretschmer, A.; Klausz, K.; Moginger, U.; Derer, S.; Rosner, T.; et al. An Anti-EGFR IgA That Displays Improved Pharmacokinetics and Myeloid Effector Cell Engagement In Vivo. *Cancer Res.* **2016**, *76*, 403–417. [CrossRef] [PubMed]

114. Heeren, A.M.; Punt, S.; Bleeker, M.C.; Gaarenstroom, K.N.; van der Velden, J.; Kenter, G.G.; de Gruijl, T.D.; Jordanova, E.S. Prognostic effect of different PD-L1 expression patterns in squamous cell carcinoma and adenocarcinoma of the cervix. *Mod. Pathol.* **2016**, *29*, 753–763. [CrossRef] [PubMed]

115. Karim, R.; Jordanova, E.S.; Piersma, S.J.; Kenter, G.G.; Chen, L.; Boer, J.M.; Melief, C.J.; van der Burg, S.H. Tumor-expressed B7-H1 and B7-DC in relation to PD-1+ T-cell infiltration and survival of patients with cervical carcinoma. *Clin. Cancer Res.* **2009**, *15*, 6341–6347. [CrossRef] [PubMed]

116. Borcoman, E.; Le Tourneau, C. Pembrolizumab in cervical cancer: Latest evidence and clinical usefulness. *Ther. Adv. Med. Oncol.* **2017**, *9*, 431–439. [CrossRef] [PubMed]

117. Sansone, P.; Bromberg, J. Targeting the interleukin-6/Jak/stat pathway in human malignancies. *J. Clin. Oncol.* **2012**, *30*, 1005–1014. [CrossRef] [PubMed]

118. Singh, S.P.; Zhang, H.H.; Foley, J.F.; Hedrick, M.N.; Farber, J.M. Human T cells that are able to produce IL-17 express the chemokine receptor CCR6. *J. Immunol.* **2008**, *180*, 214–221. [CrossRef] [PubMed]

119. Hou, F.; Li, Z.; Ma, D.; Zhang, W.; Zhang, Y.; Zhang, T.; Kong, B.; Cui, B. Distribution of Th17 cells and Foxp3-expressing T cells in tumor-infiltrating lymphocytes in patients with uterine cervical cancer. *Clin. Chim. Acta* **2012**, *413*, 1848–1854. [CrossRef] [PubMed]

120. Zhang, Y.; Ma, D.; Zhang, Y.; Tian, Y.; Wang, X.; Qiao, Y.; Cui, B. The imbalance of Th17/Treg in patients with uterine cervical cancer. *Clin. Chim. Acta* **2011**, *412*, 894–900. [CrossRef] [PubMed]

121. Bailey, S.R.; Nelson, M.H.; Himes, R.A.; Li, Z.; Mehrotra, S.; Paulos, C.M. Th17 cells in cancer: The ultimate identity crisis. *Front. Immunol.* **2014**, *5*, 276. [CrossRef] [PubMed]

122. Zhang, J.P.; Yan, J.; Xu, J.; Pang, X.H.; Chen, M.S.; Li, L.; Wu, C.; Li, S.P.; Zheng, L. Increased intratumoral IL-17-producing cells correlate with poor survival in hepatocellular carcinoma patients. *J. Hepatol.* **2009**, *50*, 980–989. [CrossRef] [PubMed]

123. Chang, S.H. Tumorigenic Th17 cells in oncogenic Kras-driven and inflammation-accelerated lung cancer. *Oncoimmunology* **2015**, *4*, e955704. [CrossRef] [PubMed]

124. Numasaki, M.; Fukushi, J.; Ono, M.; Narula, S.K.; Zavodny, P.J.; Kudo, T.; Robbins, P.D.; Tahara, H.; Lotze, M.T. Interleukin-17 promotes angiogenesis and tumor growth. *Blood* **2003**, *101*, 2620–2627. [CrossRef] [PubMed]

125. Akira, S. IL-6-regulated transcription factors. *Int. J. Biochem. Cell Biol.* **1997**, *29*, 1401–1418. [CrossRef]

126. Walch-Ruckheim, B.; Pahne-Zeppenfeld, J.; Fischbach, J.; Wickenhauser, C.; Horn, L.C.; Tharun, L.; Buttner, R.; Mallmann, P.; Stern, P.; Kim, Y.J.; et al. STAT3/IRF1 Pathway Activation Sensitizes Cervical Cancer Cells to Chemotherapeutic Drugs. *Cancer Res.* **2016**, *76*, 3872–3883. [CrossRef] [PubMed]

127. De Andrea, M.; Ritta, M.; Landini, M.M.; Borgogna, C.; Mondini, M.; Kern, F.; Ehrenreiter, K.; Baccarini, M.; Marcuzzi, G.P.; Smola, S.; et al. Keratinocyte-specific stat3 heterozygosity impairs development of skin tumors in human papillomavirus 8 transgenic mice. *Cancer Res.* **2010**, *70*, 7938–7948. [CrossRef] [PubMed]

128. Smola-Hess, S.; de Silva, U.S.; Hadaschik, D.; Pfister, H.J. Soluble interleukin-6 receptor activates the human papillomavirus type 18 long control region in SW756 cervical carcinoma cells in a STAT3-dependent manner. *J. Gen. Virol.* **2001**, *82*, 2335–2339. [CrossRef] [PubMed]

129. Tan, F.H.; Putoczki, T.L.; Stylli, S.S.; Luwor, R.B. The role of STAT3 signaling in mediating tumor resistance to cancer therapy. *Curr. Drug Targets* **2014**, *15*, 1341–1353. [CrossRef] [PubMed]

130. Galon, J.; Pages, F.; Marincola, F.M.; Angell, H.K.; Thurin, M.; Lugli, A.; Zlobec, I.; Berger, A.; Bifulco, C.; Botti, G.; et al. Cancer classification using the immunoscore: A worldwide task force. *J. Transl. Med.* **2012**, *10*, 205. [CrossRef] [PubMed]

131. Galon, J.; Mlecnik, B.; Bindea, G.; Angell, H.K.; Berger, A.; Lagorce, C.; Lugli, A.; Zlobec, I.; Hartmann, A.; Bifulco, C.; et al. Towards the introduction of the "Immunoscore" in the classification of malignant tumours. *J. Pathol.* **2014**, *232*, 199–209. [CrossRef] [PubMed]
132. Kavecansky, J.; Pavlick, A.C. Beyond checkpoint inhibitors: The next generation of immunotherapy in oncology. *Am. J. Hematol./Oncol.* **2017**, *13*, 9–20.

MDPI AG
St. Alban-Anlage 66
4052 Basel
Switzerland
Tel. +41 61 683 77 34
Fax +41 61 302 89 18
www.mdpi.com

Viruses Editorial Office
E-mail: viruses@mdpi.com
www.mdpi.com/journal/viruses